EARTH AND SPACE Science
Teacher Edition

purposeful design®
publications

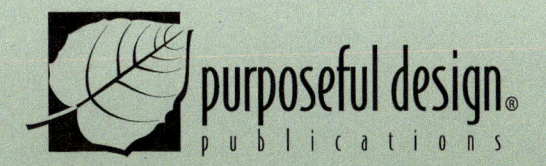

Colorado Springs, Colorado

Development Team

Vice President for Purposeful Design Publications | Steven Babbitt

Directors for Textbook Development | Don Hulin
Lisa Wood

Textbook Training and Development Coordinator | Cindi Banse

Editorial Team | Merrilee Berndt
Lindsey Duncan
Janice Giles
Julie Holmquist
Macki Jones
Adelle Moxness
Jessica Reid
Ian Work

Design Team | Claire Coleman
Steve Learned
Christian Massey

Cover Design | Mike Riester

Art Illustrations | Aline Heiser

Purposeful Design Publications is grateful to Christian Schools International for the contributions they made to the original content of the Purposeful Design Earth and Space Science course.

EARTH AND SPACE Science

Teacher Edition

Purposeful Design

© 2017 by ACSI/Purposeful Design Publications
All rights reserved.

Printed in the United States of America
22 21 20 19 18 17   1 2 3 4 5 6 7

Earth and Space Science – Teacher Edition
Purposeful Design Science series
ISBN 978-1-58331-544-6, Catalog #20082

No portion of this book may be reproduced, stored in a retrieval system, or transmitted, in any form or by any means—mechanical, photocopying, recording, or otherwise—without prior written permission of ACSI/Purposeful Design Publications.

Purposeful Design Publications is the publishing division of the Association of Christian Schools International (ACSI) and is committed to the ministry of Christian school education, to enable Christian educators and schools worldwide to effectively prepare students for life. As the publisher of textbooks, trade books, and other educational resources within ACSI, Purposeful Design Publications strives to produce biblically sound materials that reflect Christian scholarship and stewardship and that address the identified needs of Christian schools around the world.

References to books, computer software, and other ancillary resources in this series are not endorsements by ACSI. These materials were selected to provide teachers with additional resources appropriate to the concepts being taught and to promote student understanding and enjoyment.

Unless otherwise noted, all Scripture quotations are taken from THE HOLY BIBLE, NEW INTERNATIONAL VERSION®, NIV® Copyright © 1973, 1978, 1984, 2011 by Biblica, Inc.® Used by permission. All rights reserved worldwide.

Purposeful Design Publications
A Division of ACSI
PO Box 65130 • Colorado Springs, CO • 80962-5130
Customer Service Department: 800/367-0798 • Website: www.purposefuldesign.com

Table of Contents

Chapter 1: Introduction to Earth Science
1.1	The Study of Earth Science	4
1.2	Age of the Earth	7
1.3	Maps	11

Chapter 2: Minerals
2.1	Introduction to Minerals	18
2.2	Mineral Groups	21
2.3	General Mineral Properties	24
2.4	Special Mineral Properties	28

Chapter 3: Rocks
3.1	Igneous Rock	33
3.2	Sedimentary Rock	37
3.3	Metamorphic Rock	42
3.4	The Rock Cycle	45

Chapter 4: Structure of the Earth
4.1	Core	50
4.2	Mantle	54
4.3	Crust	57

Geology

UNIT 1

UNIT 2

Geologic Changes

Chapter 1: *Weathering and Erosion*

1.1	Mechanical Weathering	64
1.2	Chemical Weathering	67
1.3	Mass Wasting	71
1.4	Wind and Wave Erosion	75
1.5	River Erosion	80
1.6	Glacial Erosion	84

Chapter 2: *Soil*

2.1	Soil Composition	91
2.2	Soil Profile	95
2.3	Soil Types	98
2.4	Soil and Rock Layers	102
2.5	Fossils	106

Chapter 1: *Crust Movement*

1.1	Continental Drift	116
1.2	Plate Boundaries	121
1.3	Rock Stress and Deformation	124
1.4	Mountains	127

Chapter 2: *Earthquakes*

2.1	Earthquake Causes	131
2.2	Earthquake Zones	133
2.3	Faults	136
2.4	Seismic Waves	139
2.5	Prediction and Effects	143

Chapter 3: *Volcanoes*

3.1	Volcano Formation	150
3.2	Volcano Zones	154
3.3	Volcanic Material	157
3.4	Types and Eruptions	161
3.5	Prediction and Effects	164

The Dynamic Earth

UNIT 3

UNIT 4

Water and Water Systems

Chapter 1: *Water*

1.1	Water Properties	172
1.2	The Water Cycle	176
1.3	Glaciers	179
1.4	Groundwater	182
1.5	Surface Water	186
1.6	Rivers	190
1.7	Ponds and Lakes	194
1.8	Dams and Reservoirs	199
1.9	Eutrophication	203

Chapter 2: *Oceans*

2.1	Ocean Water Properties	210
2.2	Ocean Motion	214
2.3	Tides	218
2.4	Ocean Floors	222
2.5	Ocean Exploration	225
2.6	Ocean Resources	228

Chapter 1: *Atmosphere*

1.1	Composition of the Atmosphere	236
1.2	Layers of the Atmosphere	240
1.3	Magnetosphere	243
1.4	Ozone Layer	246

Chapter 2: *Weather*

2.1	Heat Transfer	253
2.2	Air Pressure	257
2.3	Local Winds	261
2.4	Global Winds	265
2.5	Clouds	268
2.6	Precipitation	271
2.7	Air Masses and Fronts	274
2.8	Storms	278
2.9	Forecasting Weather	282
2.10	Controlling Weather	285

Chapter 3: *Climate*

3.1	Climate Factors	290
3.2	Climate Zones	294
3.3	Climate Changes	299

Meteorology

UNIT 5

UNIT 6

The Environment

Chapter 1: *Natural Resources*

1.1	Renewable and Nonrenewable Resources	310
1.2	Air, Water, and Trees	312
1.3	Fossil Fuels	316
1.4	Rocks and Minerals	320
1.5	Land	325
1.6	Soil	328
1.7	Alternative Energy Sources	331

Chapter 2: *Pollution Solutions*

2.1	Land Pollution	339
2.2	Air Pollution	343
2.3	Water Pollution	348
2.4	Stewardship	352

Chapter 1: Solar System

1.1	History of Astronomy	360
1.2	Calendars	364
1.3	Optical Telescopes	368
1.4	Nonoptical Telescopes	372
1.5	Solar System	375
1.6	Sun	379

Chapter 2: Planets

2.1	Planetary Motion	387
2.2	Inner Planets	390
2.3	Outer Planets	394
2.4	Smaller Orbiting Bodies	399

Chapter 3: Sun, Earth, and Moon

3.1	Structure of the Moon	406
3.2	Phases of the Moon	410
3.3	Eclipses	413
3.4	Seasons	416

Astronomy

UNIT 7

UNIT 8

The Great Expanse

Chapter 1: *Stars and the Universe*

1.1	Stars	424
1.2	Life Cycle of Stars	428
1.3	Star Systems	431
1.4	Creation of the Universe	435

Chapter 2: *Space Exploration*

2.1	Rockets	443
2.2	Space Probes	447
2.3	Satellites	450
2.4	Working in Space	454

Preface

We are pleased that you have chosen Purposeful Design Earth and Space Science for your students. This yearlong course builds on the solid foundation of our Purposeful Design Science series for elementary students. And like that series, it is purposefully designed for Christian school educators to provide excellence in science instruction for Christian school students worldwide. We seek to honor God as the Creator by training a generation of students to better understand and enthusiastically embrace both God and His creation.

Purposeful Design Earth and Space Science is the result of a collaborative effort on the part of Christian educators across all educational levels, from elementary teachers to college professors. Their expertise and experience—along with a staff of dedicated and gifted editors and graphic designers—have culminated in the highly interactive, educationally engaging, and academically challenging Earth and Space Science course.

We extend our appreciation to the numerous classroom teachers and content consultants across the country who contributed valuably to this course. In particular, we are deeply grateful for the work of the textbook development team at Christian Schools International who published the earlier editions of this course. Our efforts build on their excellent work.

And like any instructional program or textbook, we acknowledge that instructional effectiveness lies in the hands of skilled and gifted teachers, to incite a love for learning and particularly, a love for science, as they educate a generation of children and future scientists with a God-centered perspective of creation, responsibility, and stewardship. To God be the glory!

Steven Babbitt
Vice President for Purposeful Design Publications

Understanding Purposeful Design Earth and Space Science

Overview

The name Purposeful Design relates to the creative work of God. Every chapter in this book weaves together the wonders of the created world and a biblical worldview. Young people have a natural, God-given inquisitiveness and curiosity about the world and its workings. Purposeful Design Earth and Space Science capitalizes on that curiosity as it engages students in investigating, observing, and thinking about the world around them.

The following *Standards and Content* section explains how this textbook fulfills the National Science Education Standards as it integrates a biblical worldview. The *Components* section describes all the student and teacher materials. The *Instructional Approach*, *Preparing to Teach a Chapter*, and *Assessment* sections show how these materials help teachers lead students in effective learning experiences. The *Additional Features* section highlights some distinctiveness of this textbook.

Standards and Content

Purposeful Design Earth and Space Science instructional materials are designed to help students master appropriate science content and skills within the framework of a biblical worldview.

Science Standards

The scope and sequence of Purposeful Design Earth and Space Science reflect the requirements set forth by the National Science Education Standards (NSES) for Grades 5–8 and the parts of Grades 9–12 that help students develop sophistication in their abilities and understanding of scientific inquiry. Benchmarks for science literacy from the American Association for the Advancement of Science, science education standards from selected states, and portions of the Next Generation Science Standards (NGSS) were also integrated where they best fit. Purposeful Design Publications has not aligned its product to any one standard but has instead collected the most comprehensive and challenging standards available from which to build the series. In addition to meeting science standards, Purposeful Design Science is committed to thoroughly integrating a biblical worldview into all instructional materials. Additional worldview content is provided for teachers, and all instructional materials take a biblical stance toward the natural world and the scientific investigation of that world.

Science Content

The activity-based, hands-on approach found in this textbook meets and frequently exceeds the NSES content standards listed below. Lessons elucidate applications of technology, societal issues, scientific careers, and key events and individuals in the history of science. When appropriate, the faith of scientists is emphasized so students can see that studying God's creation is a natural activity for Christians.
- Unifying Concepts and Processes in Science
- Science as Inquiry
- Earth and Space Science
- Science and Technology
- Science in Personal and Social Perspectives
- History and Nature of Science

The other two NSES categories of Physical Science and Life Science are addressed in the related Purposeful Design Science titles of this series.

Scientific Inquiry

The natural learning process featured in this textbook parallels the best in age-appropriate scientific inquiry. The process begins by asking a question or having students ask a question about something in the physical world that is familiar to them. They are encouraged to predict an outcome. They participate in an investigation to try out their prediction, which not only checks the accuracy of their guess, but also leads them to a more complete understanding of the problem and suggests further questions to be investigated. During this inquiry process, students acquire new information, gain proficiency in research, work collaboratively with other students, think critically, use scientific tools, take measurements, and solve problems—all valuable skills for scientific study. Students then analyze the results and make conclusions. Often, they share the results with other classmates and compare data to further evaluate results.

Biblical Worldview

Biblical worldview sidebars are provided in each chapter of the Teacher Edition. These commentaries help the teacher approach lessons in the chapter from a biblical worldview. They not only strengthen the teacher's grasp of biblical worldview issues but also prepare the teacher to communicate a biblical perspective as a natural part of science instruction. It is assumed that every teacher has access to a Bible. From time to time, teachers will be given specific Bible verses in support of the worldview presented. The goal is to help students develop as whole, unified persons rather than keeping faith and science in separate compartments. God is Lord of all—including science. Therefore, as students center their lives in Him, they can pursue scientific investigation eagerly, confident that whatever they discover will only enhance their sense of awe and stimulate them to worship the Creator to an even greater degree.

A biblical worldview is reinforced throughout these Purposeful Design Earth and Space Science materials. For example, in addition to speaking of God as the Creator, every science concept is approached from the point of view of biblical truth. In plants, animals, and human beings there is evidence that living things have been designed for certain functions. Stable natural laws and predictable cycles of nature point to an orderly Creator. The ideal conditions for life on the earth—conditions that must fall within a narrow range in order for life to exist—point to God's special design and care for this planet and for those He has created in His image (imago Dei). As students study the world around them, whether with unaided eyes, through a microscope, through a telescope, or through mathematical calculations, "the hand of the Lord" that has created everything is evident (Job 12:9).

Teaching Schedule

If your school schedule does not allow for science instruction several times a week throughout the year, this text contains more lessons than you will need. For maximum student benefit, teach complete chapters, implementing all or as many Lab Manual exercises and worksheets as possible. Spend as much or as little time in each lesson as desired. Choose to extend some lessons over multiple days, but it is recommended that all the lessons within a given chapter be covered. Be sure to include at least one chapter from each unit.

Components

All the components of Purposeful Design Earth and Space Science combine to create effective instructional experiences for both teachers and students. Even if science is not their favorite subject, students and teachers can successfully explore the wonders of science from a Christian perspective.

Student Edition
- Objectives listed at the beginning of each chapter prepare students to know what they are about to learn. Teachers and students should read these aloud and then connect each lesson back to the key ideas to reinforce the material presented.
- Engaging pages stimulate students' interest. The *FYI, History, Career, Biography, Challenge, Bible Connection*, and *Try This* sections give opportunities for students to expand their knowledge while integrating the use of technology and scientific inquiry.
- Age-appropriate text challenges students to expand their vocabulary and reading comprehension skills while colorful and realistic illustrations and photographs capture students' curiosity.
- Science vocabulary words are boldfaced in the student text.

Lab Manual

These hands-on activities and worksheets reinforce the science instruction and enrich the students' learning. They provide students practical application of the content covered in the lessons as well as opportunities to further explore the lessons' topics.

Teacher Edition

- Chapter summary pages provide background information for the major concepts covered in the chapter and offer a supplemental materials list. A *Looking Ahead* sidebar supplies an alert for materials that take additional time to collect or for speakers who need to be contacted.
- Each lesson contains a lesson objective sidebar, lists of materials, discussions points for teaching the lesson, independent and classroom activity suggestions, a variety of lesson review questions, and helpful sidebars.
- Each chapter begins with a motivating activity or lively discussion to engage students and stimulate their interest in the chapter topic. Labs and *Try This* activities are referenced in the text for students to complete as they follow instructions on their Lab Manual pages or those in their student text. These activities provide students the hands-on experience of scientific inquiry.
- Student pages appear in readable reduced size for the teacher to reference.
- Blackline masters, transparency masters, and chapter tests supply instructional support for the teacher and enhance student learning and assessment. Most blackline masters are strictly directions for the teacher and do not need to be printed and distributed to students. These downloadable supplemental materials are found on **https://www.acsi.org/textbooks/qrs/midsch**. For blackline masters that require student responses, the answer keys are located on the same web link and all worksheet answers can be found there as well.
- Answer Keys for the labs are located at the end of each chapter.

Instructional Approach

Purposeful Design Earth and Space Science provides students with multiple opportunities to develop their science skills. The framework of instruction is based on connecting students' prior knowledge to new information, constructing a larger science foundation of knowledge by presenting scientific data and terminology, investigating the topic with hands-on activities, extending the chapter topic by applying the information learned, and assessing knowledge gained. The balance among these instructional methods creates a science series that provides a well-rounded science education.

Introduction

Lessons begin instruction by connecting students' prior experiences with major scientific concepts in a meaningful way. Students are encouraged to reflect upon their personal experiences with the subject as they reveal what they already know. These discussions and activities make evident to the teacher any student misconceptions and learning gaps.

Discussion

Students acquire more understanding as the teacher presents specific content in this section and provides opportunities for critical thinking, debate, apologetics, and logical processing. Students' comprehension of the topic expands as students gain new information and proper scientific terminology. They are expected to have read the lesson's text before the teacher enters the Discussion section. The topics can assist students later in the inquiry and activity process.

Activities

The activities include labs, *Try This* activities found in both the Teacher Edition and the Student Edition, worksheets, additional science experiments and demonstrations, as well as literature, biblical integration, field trips, guest speakers, and cross-curricular activities that relate to the lesson concepts. Teachers choose the activities that best align with the school's science curriculum, class time, and resources. Some activities can be assigned as individual or group projects.

Students know, think, and explore science in many different ways but always through the lens of the Bible. Sometimes they work on their own; sometimes they work cooperatively with other students. At times, the teacher directs specific activities; at other times, students design their own activities or experiments. Students spend time reading and writing; they also spend time interacting with others. Students learn that there is not just one way to investigate science and that science is not limited to science class. Science affects all aspects of life and is useful in many ways. The skills they learn will help them in whatever career paths they take in the future, science-based or otherwise.

Preparing to Teach a Chapter
Pages at the beginning of each chapter provide a description of the chapter's content, materials to obtain, and activities to prepare for. Biblical integration is provided on the pages as well.

Chapter Summary
Each chapter begins with a section that gives the teacher helpful information to be used to better plan for lessons. The teacher is given a brief synopsis about the value and purpose of the chapter content.

Background
This section provides the teacher with a brief description of lesson content. The information provided here is usually more in depth to use as a resource and to offer further explanation should questions arise.

Sidebars
Provided in the Chapter Summary are three sidebars—*Looking Ahead*, *Supplemental Materials*, and *Worldview*. *Looking Ahead* gives teachers an advance notice of special or unusual supply needs to allow for ample time to acquire the materials. All supplemental instructional materials necessary for teaching the chapter, such as Lab Manual pages, blackline masters, transparency masters, and tests, are listed in the *Supplemental Materials* sidebar. The teacher can see at a glance the teaching resources for the chapter. The *Worldview* sidebar provides a connection of science content with biblical perspective.

Assessment
Teachers can assess student progress informally as they observe their participation and involvement during activities and experiments. Additionally, tests that include a variety of formats—multiple choice, true or false, short answer, long essay, and matching—are provided as summative assessments and can be downloaded from https://www.acsi.org/textbooks/qrs/midsch. For more versatility, teachers also have the option to select questions from the editable test bank to create their own tests.

Class discussions also provide teachers with valuable information about what students are learning. The teacher text provides key questions to use as formative assessment that can be asked to monitor student understanding at various points in the lesson.

Additional Features
Several lessons suggest possible outings, field trips, or alternative options. Teachers should check the *Looking Ahead* and *Preparation* sidebars well in advance in order to allow ample time to schedule events, recruit parental involvement, and acquire materials.

Science Equipment and Materials
Many of the experiments, demonstrations, and activities can be performed by using common household items or items easily found in nature. However, others will require basic scientific equipment and specific materials. Scientific equipment may include items such as hand lenses, thermometers, metersticks, spring scales, eyedroppers, microscope slides, and triple beam balances. Specific materials are those that are important to the chapter and integral to the directed instruction. There are several places where specialized materials—while not required to teach the lesson—can greatly enhance the learning experience for both teacher and student.

Websites
Many universities and government websites provide activities and research resources for both teachers and students. Teachers should preview these sites for accuracy and age-appropriateness. Additionally, teachers are encouraged to explore the science website links found on the *Textbook Support* tab on the Purposeful Design website.

Cross-Curricular Links
Science is not just limited to "science time." Here are a few examples of how science can be integrated into other subject areas.
- Fine Arts: Some lessons give students the opportunity to draw pictures, play games, role-play, or perform skits and plays.
- Health and Physical Education: Students gain helpful information about healthy diets, sleep patterns, human development, and exercise as they study about the human body.
- Language Arts: Many activities involve written and verbal communication skills. Several of the report options for chapter projects provide opportunities for cross-curricular instruction.
- Mathematics: Computational skills, measurement, logic, and mathematical reasoning are used. Students will see that mathematical skills are vital to scientific inquiry.
- Social Studies: Students can study a wide range of geographic locations, expand their awareness of other people groups and cultures, and discuss events and individuals in the history of science and technology.
- Technology: Suggestions are included for using computers, the Internet, and audiovisual technology in the classroom.

Preparing a Lesson

1 The *Objectives* sidebar clearly indicates the focus of each lesson.

2 Vocabulary words and definitions that students will encounter in the student text are listed in alphabetical order. A glossary containing all of the vocabulary words is located in the back of both the Teacher and Student Editions.

3 Materials for the lesson are listed in this sidebar. Supplemental materials such as blackline masters, transparency masters, and worksheets are also listed.

4 The *Preparation* sidebar identifies materials that need to be obtained or activities that should be performed in advance. The icon denotes the component of the lesson that requires preparation. *Alternatives* and *Safety* are other sidebars that may appear and for which icons are used. These sidebars provide ways to expedite or simplify activities and offer safety reminders.

5 The *Try This* sidebar lists materials and additional information that students need to complete the *Try This* activities in the Student Edition. Note that it too has an icon.

6 *Introduction* contains activities or discussions that will elicit students' interest in the topic.

7 *Discussion* provides topics and questions for critical thinking, debate, apologetics, logical processing, and formative assessment. Students should have read their text by this point.

8 *Activities* including labs, additional experiments, worksheets, research, role-play, literature, biblical integration, field trips, guest speakers, and cross-curricular activities are listed here for the teacher to choose from.

4.2.4 Ocean Floors
Oceans

1 OBJECTIVES

Students will be able to
- model ocean floor features.
- relate floor features in the Atlantic Ocean and the Pacific Ocean.

2 VOCABULARY

- **abyssal plain** a large, nearly flat region beyond the continental margin
- **continental margin** the part of the earth's surface beneath the ocean that is made of continental crust
- **continental rise** the base of the continental slope
- **continental shelf** a broad, relatively shallow underwater terrace that slopes outward from the shoreline
- **continental slope** the steepest part of the continental incline located at the edge of the continental shelf
- **seamount** an underwater volcanic mountain that rises at least 1,000 m above the abyssal plain

3 MATERIALS

- Model-building materials (B)
- WS 4.2.4A Oceans vs. Land

4 PREPARATION

- Obtain materials for *Lab 4.2.4A Predicting Ocean Floor Topography*. Prepare the mystery boxes before class. Shoe boxes with lids work well, and ocean floor models can be made from clay, building blocks, or plaster. Poke several small holes large enough for a probe to be inserted in a straight line along the length of the lid about 2 cm apart. Create a different ocean floor model in each box.

5 TRY THIS

Ocean Floor Features
No additional materials are needed.

Ocean Vacation
No additional materials are needed.

222

Introduction
Have students complete **WS 4.2.4A Oceans vs. Land**. Guide students to answer as many questions as possible without receiving assistance from classmates. Have students place a star next to answers that reference oceans. Ask students if they were able to answer more land questions or more ocean questions. Discuss why there may be such a lack of knowledge regarding oceans. (**Possible answers: It is easier to observe landforms; many people do not spend as much time in the water compared to the amount of time spent on land; and it can sometimes be more dangerous to explore in the ocean than on land.**)

Discussion
- Discuss the importance of studying the ocean floor. Lead students to discuss how studying the ocean can give people a better appreciation of God's creation. Point out that studying organisms that can survive without sunlight provides scientists with an understanding of the different processes some organisms utilize to thrive. Ask how a study of the ocean floor can provide insight into how people can better care for the planet. (**Answers will vary.**)

- Ask the following questions:
  1. Considering what you know about the ocean, what regions of the ocean floor do you think have been explored the most? (**the continental margin that contains the shelf, slope, and rise**)
  2. What regions of the ocean floor have been the least explored? (**abyssal plains and trenches**)
  3. Where do you think Earth's tallest mountains are located? (**on the ocean floor**)
  4. Using what you know about living things, why do you think there are fewer organisms living at the bottom of the ocean compared to the upper regions of the ocean? (**The lack of sunlight in that region means plants cannot use photosynthesis. Plants provide food for many living things in the ocean, so there is a lack of nutritional sources at the bottom of the ocean.**)

Activities

Lab 4.2.4A Predicting Ocean Floor Topography
- mystery boxes, 1 per group
- metric rulers, 1 per group
- probes, 1 per group

A. Assign *Try This* activities for students to complete. Discuss their findings and drawings.

B. Divide the class into groups of three to create large models of individual ocean floors. Models may be built with modeling clay, papier-mâché, or any suitable material. Features should include continental margins, continental shelves, oceanic ridges, continental slopes, continental rises, trenches, abyssal plains, and seamounts. Remind students that the heights and depths of some of the features should vary significantly. Have students label the features and compare characteristics among the other ocean models.

C. Assign student pairs to compose 10 questions about the basic features of the ocean. Instruct students to conduct a survey to assess general knowledge of ocean features by posing their questions to five people of different ages outside of the classroom. As a group, discuss the results with facts.

D. Direct student groups to use their survey results to make an ocean education booklet, poster, or computer presentation to educate the public about basic ocean features.

E. Challenge students to research how many scientists believe plate tectonics formed the distinctive features of the ocean floor. Have students present their findings and discuss any differing opinions or theories.

BLMs, TMs, and tests are available to download at https://www.acsi.org/textbooks/qrs/midsch.

9 Lesson Review

1. What three regions comprise the continental margin? (**continental shelf, continental slope, and continental rise**)
2. Where does the transition from continental crust to oceanic crust occur? (**the continental rise, which is at the base of the continental slope**)
3. What are oceanic ridges? (**underwater mountain chains that form where tectonic plates pull apart**)
4. How do many scientists believe oceanic ridges are formed? (**Many scientists believe that as tectonic plates under the ocean pull apart, magma from Earth's mantle moves up through the crust. Over time, this hardened magma forms the ridges.**)
5. Describe the continental margin in the Atlantic Ocean. (**The continental shelves are wide, and the continental slope drops gradually to a vast underwater plain.**)
6. Where are a majority of Earth's major ocean trenches located? (**the Pacific Ocean**) Why? (**The Pacific Ocean has areas undergoing subduction, which is the process of an oceanic plate being pushed under a continental plate. Many ocean trenches were formed by subduction**).

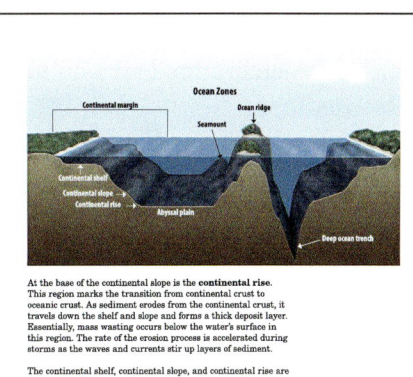

NOTES 11

---

9 *Lesson Review* offers the teacher specific questions to assess students' understanding of lesson content.

10 Readable reductions of each student textbook page allow the teacher to see the student text. Following the student pages are the answer keys for the labs.

11 Space is given for the teacher to write notes about the lesson.

The Scientific Method

Background
The word *science* comes from the Latin word *scire*, which means "to know." A scientist is someone who does science—a person who wants to understand the natural world and all of its complexities. Even after scientific experiments, many scientific hypotheses are disputed within the scientific community; all are subject to revisions and changes as new data and experimental evidence are collected. Theories and laws, however, are supported by a substantial number of experiments and observations. Ideally, scientists are open to constructive criticism about their hypotheses. When scientists communicate in scientific journals or at conventions, they work toward finding new evidence and creating solutions to resolve conflicting interpretations and viewpoints.

The scientific method is an orderly, systematic approach to solving a problem or answering a question. Although different scientific endeavors require different approaches and steps, the following terminology and steps are often used. Students need to become familiar with the vocabulary of science as well as realize that scientists understand that the scientific method is not a rigid set of steps that is always followed. Various orders are possible, and sometimes steps are eliminated or repeated.

Vocabulary
control—the sample in an experiment in which the variables are kept at a base level
hypothesis—a prediction of what you think will happen and which can be tested to see if it is true
inference—an educated guess based on observation
observation—something noticed through the senses
scientific law—a generalization based on observations that describe the ways an object behaves under specific conditions
scientific method—the series of steps that scientists follow when they investigate problems or try to answer questions
theory—an explanation of the scientific laws
variable—a changeable factor that could influence an experiment's outcome

Scientific Method
1. **Identify or define the problem**.

2. **Make a hypothesis**. Not all predictions are hypotheses; they must be measurable. For example, "If the rate of fermentation is related to temperature, then increasing the temperature will increase gas production" is a hypothesis. On the other hand, "If yeast is heated, then more gas will be produced" is not a hypothesis because it offers no proposition to test. It does not show a relationship or suggest variables. A hypothesis may also be a question ("Does temperature affect the fermentation of yeast?") or a conditional statement ("Temperature may affect the fermentation of yeast").

3. **Experiment, controlling the variables**. Scientists repeat and refine the experiment based on the findings. They will not allow a hypothesis to remain untested; it must be tested and shown valid.

4. **Make observations and record the results**. Most scientists accurately record their observations and measurements in a journal or on a computer.

5. **Make inferences and conclusions**. A conclusion is a statement about what a scientist has learned and whether or not the scientist's hypothesis was supported. Scientists often learn as much from an incorrect hypothesis as from a correct one. Students (who often want their experiments to show that their "guess" was correct) need to understand that sometimes scientists arrive at completely unexpected conclusions and that an incorrect hypothesis is never considered a failure. For example, sometimes an unexpected conclusion raises new questions that will lead to another

experiment and conclusion. The purpose of science is not to prove what is already known; the purpose is to increase the understanding of God's world!

6. **Apply the findings**. Scientists add their conclusions to their understanding of the natural and technical universe. To do so with wisdom and stewardship benefits all creation.

Ways to Explore the Scientific Method

Thinking like scientists primarily means being curious about why things happen the way they do. Ways to help students discover how they and others can think like scientists include the following:
- Interviewing a scientist or medical professional to ask how he or she uses steps of the scientific method and experimental procedures on the job
- Researching scientific theories that have been maintained during the centuries but were eventually shown to be inaccurate through experimentation
- Reading biographies about scientists through the ages

Measurements

Background

The metric system is a universal system of measurements that serves as a standard for scientific research throughout the world and in many countries for everyday use. Students need to become more familiar with the metric system and have practice estimating measurements. Throughout this course, the metric system is used predominantly.

Since 1899, international conferences have been held to standardize the metric system. In 1960, the 11th International Conference of Weights and Measures substantially changed the system, renaming it the International System of Units (abbreviated SI).

Using the SI system of measurement has several advantages. It is based on standards that have been recognized by the international science community for use in the sciences and commerce, making the system universally adopted. All of the unit conversions used in the metric system are based on the number 10, and only decimal multiples of the basic units of length, volume, and mass are employed. The metric system is also used by all of the major countries of the world except the United States.

The instrument frequently used to determine the mass of a small object is the triple-beam balance, which is normally capable of measuring masses to the hundredth of a gram. The volume of substances can be determined by using a variety of different methods, depending upon the phase and shape of the substance being measured. Graduated cylinders are used to accurately measure the volume of liquids to the nearest milliliter. Graduated cylinders can also be used to measure granular solids, but this method is inaccurate for substances with large granules. The volume of regularly shaped objects with cubic or rectangular box shapes can be calculated using the formula $V = l \times w \times h$. (Volume equals length times width times height). Water displacement is used to find the volume of irregularly shaped objects that will not dissolve in water. A certain amount of water is poured into a graduated cylinder, and the object is dropped into the cylinder. The volume that the water rises is the volume of the object.

Although students in the United States are more familiar with the Fahrenheit scale, the Celsius (Centigrade) scale is much more common around the world. Scientists use the Celsius scale or the Kelvin scale, which is an absolute scale. Absolute zero was given the value 0 Kelvin; thus the Kelvin scale has no negative numbers. Absolute zero corresponds to $-273.15°C$ and $-459.7°F$. Celsius temperatures can be converted to the Kelvin scale by using the following formula: $K = C + 273.15$. Kelvin is an absolute scale and thus has no degrees. The Fahrenheit scale assigns a temperature of 32° for the freezing point of water and 212° for the boiling point of water. The Celsius scale standardized the temperature scale by assigning a value of 0° for the freezing point of water and 100° for the boiling point of water. The size of the divisions used in the Fahrenheit and Celsius scales are not equal. The formula used to convert Celsius to Fahrenheit is $F = 9/5C + 32$. The formula for converting Fahrenheit to Celsius is $C = 5/9(F - 32)$.

One way to record these measurements is on graphs. A graph is a pictorial representation of statistical data or of relationships between variables. While graphs can serve a predictive function because they show general tendencies in the quantitative behavior of data, as approximations they are sometimes inaccurate and misleading. Most graphs include two axes. The horizontal axis represents the independent variables, and the vertical axis represents the dependent variables. In line graphs (the most common type), the horizontal axis often represents time. Bar graphs can be used to depict the relationship between two nontemporal numerical values. Such information can also be expressed in a circular graph (pie graph) that illustrates the part-to-whole relationship. The size of each sector is directly proportional to the percentage of the whole it represents; this graph is often used to emphasize proportions.

Scientific measurements require specific words to be used in descriptions so the language of science is understood worldwide. Students need to have a working knowledge of the following terms:

Vocabulary

astronomical unit—the average distance between Earth and the sun (about 149,600,000 km)
Celsius scale—a temperature scale in which 0° represents the freezing point of water and 100° represents the boiling point of water
density—the mass per unit of volume of a substance
gram—the standard unit for mass in the metric system
light-year—the distance light travels in a vacuum in one year, approximately 9.46×10^{12} km
liter—the standard unit for the volume of liquid in the metric system
meter—the standard unit for length in the metric system
metric system—a universal system of measurement based on the number 10 used by scientists around the world

Ways to Explore Measurements

Taking scientific measurements provides necessary data for analysis and the resulting conclusions. Ways to help students discover how they and others can measure like scientists include the following:
- Demonstrating how to accurately use a meterstick, a triple beam balance, and a graduated cylinder
- Calculating volume by using the formula $l \times w \times h$ (length times width times height)
- Having students describe why is it important to have standards for measurements
- Determining how many Celsius degrees there are between freezing and boiling
- Measuring a variety of items and comparing the differences
- Labeling three graduated cylinders *1*, *2*, and *3* and having students measure different amounts of water into each one, and then labeling three small objects *1*, *2*, and *3* and having students determine the volume of these objects
- Having students research the history of the French Revolution to discover the reasons why the revolutionaries decided to change their system of measurement
- Guiding students to create metric conversion problems for a partner to complete

Laboratory Safety

Background

Science safety should be one of the first topics of discussion in your classroom, and it is important to review safety rules with students before conducting the first science lab. Reinforcing appropriate laboratory behavior and procedures is always important, especially when students go on to pursue the sciences in high school and college.

The natural immaturity of this age group requires that specific guidelines be established. The most important rule is "Do not engage in horseplay." When this rule is strictly enforced and consequences dealt out immediately for infractions, earth and space science activities become controlled, enjoyable learning experiences for both the teacher and the students.

Each lab has specific safety needs, and students must focus on the teacher's instructions. Allow students to take only their notebooks, student text, and a pen or pencil to the laboratory area. Require that they leave all other items at their desks. Keep safety requirements to a minimum for each activity. When students are presented with too many rules at once, they will tend to lose focus. Write on the board one or two safety rules for each session, and read and review the procedures to be certain that everyone understands the instructions.

Safety Rules

Some standard safety rules include the following:
- Direct students to wash their hands before and after any experiment or hands-on activity. Have students wear protective gear appropriate to the activity. Some activities require goggles or gloves. Sanitize safety goggles after each use.
- Instruct students to report any spills or breakage to the teacher immediately.
- Store equipment and materials in a safe location and make sure they are returned to their designated places after use. Always test and check materials prior to use in the classroom.
- Model safety for the students.
- Direct students to not eat or handle supplies unless instructed to do so.
- Be aware of allergies that students may have to science activity materials and take necessary precaution.
- Carry only one jar at a time.
- Never lean on aquarium glass.
- Never prop a meterstick over your shoulder.
- Carry microscopes with two hands.
- Never eat or drink during a science activity.
- Remove loose or bulky clothing to reduce the chance of spreading fire or knocking over equipment.
- Keep equipment (especially breakable or expensive items such as glassware or microscopes) away from the edge of work surfaces.
- Wear safety goggles, especially for activities involving potentially harmful chemicals, glassware, or heat. (When the teacher is wearing goggles, everyone must wear goggles.)
- Use plastic equipment if possible. However, if glassware is used and does break, back away from the accident and do not touch the shattered glass. Designate someone to retrieve a dustpan and broom while the teacher remains at the scene to ensure that no one else tries to pick up any glass.
- When possible, use hot plates instead of Bunsen burners. Avoiding Bunsen burners will greatly reduce the chance of a fire in the classroom or the chances of a student being accidentally burned. Try to reserve the use of Bunsen burners for teacher-led classroom demonstrations.
- Know basic chemical safety. Chemicals purchased from a chemical supply warehouse are labeled with three hazards—health, flammability, and reactivity—rated on a scale of 0 to 4. A 0 rating indicates that the substance is not a threat for this hazard, and a 4 means it is a substantial threat.
- Be aware of two common safety symbols—flames (signifying flammability) and a test tube being poured on a hand (signifying a corrosive substance). These symbols are primarily for storage

purposes, but they also warn the user of these two dangers. Different chemicals have different disposal guidelines. General guidelines include storing chemical wastes in properly labeled chemical waste jars, storing containers in accordance with the hazards associated with their contents, and not combining incompatible materials.

Safety Procedures
Add to the safety of the classroom and laboratory by incorporating the following procedures:
- Be a good example. Follow the safety rules at all times.
- Inspect the lab area often to look for possible hazards.
- Inspect equipment regularly.
- Correct or report hazards immediately.
- Closely supervise all student activities. Know first-aid procedures, and have a first-aid kit readily available.
- Equip the lab with a fire bucket, fire blanket, fire extinguisher, and safety shower if possible.
- Post emergency numbers next to the classroom phone.

Because earth and space science is often best studied outside, establish a few guidelines for outside activities that will create a safe, enjoyable learning experience. When conducting outdoor activities, review specific guidelines for your particular school area. Have students dress appropriately for the activity and weather conditions, and remember to incorporate all applicable safety procedures.

Ways to Explore Laboratory Safety
Ways to help students discover how they and others can practice safe laboratory and science habits include the following:
- Inviting students to recount experiences in which carelessness or recklessness caused injury. (They do not have to limit their examples to science labs.)
- Relating to students incidents of lab accidents that happened during your middle school, high school, and college lab sessions or those that occurred while you were teaching. Describe the consequences of these accidents. Ask how these accidents could have been prevented.
- Showing students examples of bottled chemicals, and reviewing with them the hazard ratings and the two basic storage symbols.
- Demonstrating how to use the fire bucket, fire blanket, fire extinguisher, eye wash, disposal containers, and the first-aid kit.
- As a class, writing a safety contract for students to follow throughout the year. Students who feel "ownership" of the rules are more likely to follow them. Have students and their parents sign the contract.
- Having student groups role-play proper and improper safety procedures. Have other student groups record which procedures were proper and which were improper.
- Arranging for the class to tour a chemical or research laboratory so students can observe professional laboratory standards used in the scientific world.

UNIT 1

Chapter 1: *Introduction to Earth Science*
Chapter 2: *Minerals*
Chapter 3: *Rocks*
Chapter 4: *The Structure of the Earth*

Key Ideas

Unifying Concepts and Processes
- Systems, order, and organization
- Evidence, models, and explanation
- Change, constancy, and measurement
- Form and function

Science as Inquiry
- Abilities necessary to do scientific inquiry

Earth and Space Science
- Structure of the earth system
- Earth's history
- Origin and evolution of the earth system
- Origin and evolution of the universe

Science and Technology
- Understandings about science and technology

Science in Personal and Social Perspectives
- Science and technology in society
- Natural resources

History and Nature of Science
- Science as a human endeavor
- Nature of science
- History of science
- Nature of scientific knowledge

Geology

Vocabulary

asthenosphere
astronomy
carbon-14 dating
clastic rock
cleavage
contact metamorphism
continental crust
contour line
convection current
core
crust
density
dynamic metamorphism
earth science
environmental science
extrusive rock
felsic rock
fluorescence
foliated structure
fracture
geology
hardness
igneous rock
inner core
intrusive rock
latitude
lava
lithification
lithosphere
lodestone
longitude
luster
mafic rock
magma
magnetosphere
mantle
metamorphic rock
metamorphism
meteorology
mineral
mineralogy
Moho
naturalism
nonfoliated structure
nonsilicate mineral
oceanic crust
oceanography
outer core
phosphorescence
prime meridian
projection
P wave
radioactivity
regional metamorphism
rock cycle
sedimentary rock
sediments
seismic wave
shale
silicate mineral
streak
S wave
theism
varves

SCRIPTURE

Lord, our Lord, how majestic is Your name in all the earth! You have set Your glory in the heavens.

Psalm 8:1

1.1.0 Introduction to Earth Science

LOOKING AHEAD

- For **Lesson 1.1.1**, invite someone to speak about how earth science intersects with a career.
- For **Lesson 1.1.3**, invite someone to share with the class how maps are used in a career. Obtain a variety of geological maps. Invite a surveyor to demonstrate surveying methods and tools. Plan a hike in hilly or mountainous terrain.

SUPPLEMENTAL MATERIALS

- TM 1.1.1A–D Principles of Earth Science
- TM 1.1.2A Venn Diagram: Young Earth and Old Earth
- TM 1.1.3A Map Projections
- TM 1.1.3B–D Types of Maps
- TM 1.1.3E World Time Zones Map

- Lab 1.1.2A Half-Lives
- Lab 1.1.3A Topographic Map

- WS 1.1.3A Traveling Across Time Zones
- WS 1.1.3B World Time Zones Map
- WS 1.1.3C Map Selection
- WS 1.1.3D Tracking Down the Treasures

- Chapter 1.1 Test

BLMs, TMs, and tests are available to download. See Understanding Purposeful Design Earth and Space Science at the front of this book for the web address.

Chapter 1.1 Summary

Earth science is the study of the physical earth and the universe of which it is a part. As such, it can be subdivided into geology, oceanography, meteorology, astronomy, and environmental science. These branches respectively take an in-depth look at Earth's structure and landforms, freshwater and saltwater systems, weather, celestial bodies, and humanity's impact on all of this. Christians acknowledge the God of the Bible as the creator of all things, yet they differ in their theories regarding how He created. Some believe that God created the universe quickly over the course of six literal 24-hour days, but others believe that God created things slowly over the course of millennia. Both views point to evidence in both nature and the Bible to support their claims. Throughout history, people have drawn maps to study Earth and the universe.

Maps are useful tools for depicting and communicating about nearly every feature of the earth and universe. People create maps for specific purposes, such as travel, understanding the movement of tectonic plates, locating natural resources, predicting weather, or depicting the movement of celestial bodies. Types of maps can be distinguished by their features, which are relevant to their intended purposes.

Background

Lesson 1.1.1 – The Study of Earth Science

Earth science is divided into geology, oceanography, meteorology, astronomy, and environmental science. Geology comprises many fields including hydrogeology (the study of how the earth and water interact), hydrology (the study of freshwater's properties and how water moves across the earth and atmosphere), geochemistry (the study of the chemical composition and changes in solid matter on Earth and other celestial bodies), geophysics (the study of the physical forces at work in the earth), and structural geology (the study of how the rock of the earth's crust changes over time because of the movement of the earth). Oceanography includes geological oceanography (the study of ocean floor geology) and biological oceanography (the study of ocean plants and animals). Subsystems within meteorology include microscale (the study of atmospheric phenomena that is limited to less than a few kilometers in scale and less than a day in duration), mesoscale (the study of convection and its relation to storms), synoptic scale (the study of pressure systems and their resulting weather conditions), and global scale (the study of global winds and the heat and moisture associated with them). Several of astronomy's fields include astrophysics (the study of how physics and chemistry help explain the formation and life of cosmic bodies), cosmology (the study of the universe as a whole), and planetary science (the study of planets and other cosmic bodies that are not stars). Environmental science explores humanity's impact on the earth, including how resources are obtained and used, and it interacts with every other branch of earth science.

Lesson 1.1.2 – Age of the Earth

Many scientists are Christians who love God and want to honor Him in their profession. Most scientists currently believe that the earth is 4.6 billion years old. Although some Christians believe that this contradicts Scripture, many other Christians accept this scientific teaching and support their view with Scripture. Both views are supported by reputable Christian scientists. The two major branches of Creationism are old earth, which accepts the scientific stance that Earth is billions of years old, and young earth, which dates Earth as 6,000–12,000 years old.

Scientists use various dating methods to determine the age of an object, a natural phenomenon, or a series of events. Dating methods are classified as relative (calculation of the age of an object or event relative to some other object or event) or absolute (calculation of the actual age of an object using methods that do not depend on environmental conditions). The process of relative dating involves comparing a rock, artifact, or fossil to another object whose age can be surmised. Relative dating methods include the study of geologic strata, accompanying fossils, magnetic orientation, chemical changes, and seasonal variations. Stratigraphic dating assumes that an object cannot be older than the materials composing it and that the lower the layer is, the older it is.

Most relative dating methods are based on naturally occurring annual cycles. For example, streams flowing into some still lakes deposit layers, or varves, of summer silt and winter clay throughout the year. Those layers laid down during the fall and winter are darkened with dead vegetation. Counting each pair of varves can determine the age of the deposit. Scientists consider the varved-clay method is accurate in deposits up to 12,000 years old.

The primary absolute dating methods use radioactive decay. The most reliable absolute dating methods ascertain the radioactivity of certain metals that compose rock. The rate of radioactive decay of these particular isotopes is known and constant, so the age of a rock can be calculated by studying the relative proportions of the remaining radioactive material and its decay product. The half-life of these elements is unaffected by physical and chemical changes. Scientists use carbon-14 dating to determine the age of organic matter up to 50,000 years old; it is useless for inorganic matter. Carbon-14 dating relies on the fact that all organic matter contains carbon, which decays into nitrogen 14. Carbon 14 converts to carbon dioxide in the atmosphere, and plants and organisms ingest it. When the plant or animal dies, it stops acquiring carbon, and the carbon 14 begins to decay. Carbon dating applies only to organisms that obtained their carbon from the air, so it cannot be used for aquatic creatures since their carbon might, for example, have originated from dissolved carbonate rock. Uranium also decays to a stable isotope—lead 206—at a known rate. No natural process alters this decay rate.

Lesson 1.1.3 – Maps
Earth and space scientists are aided greatly by accurate maps. As a sphere, Earth cannot be accurately represented on a flat surface without distortions of distance, direction, size, or shape of areas. All map projections, however, show true location. There are three basic types of projections: azimuthal, conical, and cylindrical. These basic map types can be modified to more accurately suit various purposes or perspectives. For example, in 1569 Gerardus Mercator published a cylindrical projection that is excellent for navigation because relative directions between any two features are accurate; however, the size of areas becomes increasingly distorted in areas farther from the equator.

Geographers use a wide variety of maps. A topographical map shows the location and shape of features on land. A bathymetric map depicts the topography of the ocean floor. Tectonic maps show the boundaries between the earth's plates and often include fault lines and the axes of synclines and anticlines; they are useful for determining the location of earthquakes and volcanoes. A bedrock geologic map shows the distribution of different rock types in a particular region. Different types of rock may be represented by different colors to indicate their type and approximate age. Any of these maps may include a cross-section diagram, which uses lines to represent a constant value of change in rock type or soil density. Paleogeographic maps depict what a region may have looked like long ago, showing how topography or the structure of the crust has changed through plate tectonics or continental drift.

Maps are also essential tools in the field of meteorology. Meteorologists use weather maps to aid in their forecasts. Surface maps show weather conditions in a particular region through the use of weather symbols. Climate maps help meteorologists keep track of a region's long-term climatic conditions.

Astronomers use star charts to study the night sky. Star charts show stars' positions relative to each other, the season, and an observer's location on the earth. Early stargazers saw shapes of people, animals, or objects in the stars. Modern astronomers use many of these shapes to define areas in the night sky, which are helpful in labeling other heavenly bodies as they are discovered.

WORLDVIEW

- Science and faith are inseparable. Despite popular opinion, understanding science and having faith in the God of the Bible are not mutually exclusive. The Bible tells us that God created all things for His own glory and for people to use (Psalm 8). Because people are part of God's creation, our purpose is to glorify Him, including through the way we use the world around us. God wants us to carefully invest the resources He gave us. Matthew 25:14–30 describes a rich man who entrusted three servants with sums of money. The first two servants used that money to create more wealth for their master, but the last did nothing to enhance his master's wealth, merely preserving what he was given. The master praised and rewarded the profitable servants, but he severely punished the one who did nothing with what he was given. In much the same way, God has entrusted us with great riches: the creation, our intellect, and our abilities. He wants us to mirror His own creativity by using these resources to honor Him. Although faith leads us to a right use of the earth, science is a means through which we can act on that faith. The orderly process of science better enables us to understand the intricacies of what God created and to use that creation to glorify the Creator. Through science we develop technologies that enable us to produce beautiful works of art, build comfortable homes, predict the weather, solve complex environmental problems, and explore distant galaxies. When we strive to please our Master by using what He has entrusted us with, it results in others seeing God's glory more clearly.

1.1.1 The Study of Earth Science
Introduction to Earth Science

OBJECTIVES

Students will be able to
- explain the purpose of earth science.
- explain the difference between science and technology.
- identify the major branches of earth science.
- describe a practical application of earth science.

VOCABULARY

- **astronomy** the study of physical things beyond the earth's atmosphere
- **earth science** the study of the earth and the universe around it
- **environmental science** the study of the relationship between organisms and the environment
- **geology** the study of the solid earth
- **meteorology** the study of the atmosphere
- **oceanography** the study of the earth's oceans

MATERIALS

- Items created or derived from minerals (*Introduction*)
- Newspapers and newsmagazines (*Discussion*)
- TM 1.1.1A–D Principles of Earth Science

PREPARATION

- Invite a person to speak to the class about how his or her occupation is influenced by one or more branches of earth science. Suggested occupations include: college or university professor; meteorologist; construction contractor; city water or gas engineer; or someone working at a water treatment plant, concrete mixing plant, recycling center, or garbage treatment plant. (*A*)

Introduction

Show a variety of items created or derived from minerals, such as table salt, a glass beaker, or a paper clip. Have students speculate regarding the origin of each item. Challenge students to keep working backward to the most elemental stage for each item. For example, students may say that a paper clip came from a factory that obtained its metal from a mine; students may guess that the metal used in the paper clip is steel.

Discussion

- Display **TM 1.1.1A–D Principles of Earth Science**. Discuss the purpose of science and particularly earth science. Guide students to consider both theological and practical purposes.

- Discuss the difference between science and technology.

- Have students relate what job or career they would like to have when they are adults or what jobs their parents have. Also have them discuss their hobbies. Discuss how science enhances each one. Even nonscientific jobs benefit from science. For example, writers use computers, which depend on silicon chips; artists use clay and paints made from minerals.

- Discuss the fact that science is a tool that can be used wisely or unwisely. Have students find published examples about how science has been used and discuss whether the usage was wise or unwise. Encourage students to support moral evaluations with Scripture. Provide newspapers or newsmagazines, or have students conduct Internet research.

- The three basic methods of investigation used by scientists are experimental, observational, and historical. Challenge students to consider what methods are used most by earth and space scientists. (**observational and historical**) Why would historical investigation be important to a geologist or an astronomer? (**Many geologic and astrological events happened before they could be directly observed by a scientist.**) Why is experimental investigation more difficult in earth and space science? (**Possible answers: Study subjects won't fit in a lab or are too far away.**) Explain that many of the activities in this course of earth and space science will use observational and historical investigation methods.

- Discuss how environmental science and good stewardship can be applied to daily life.

Activities

A. Present the invited speaker to explain how his or her occupation is influenced by one or more branches of earth science.

B. Have students create a computer presentation or poster showing some of the tools used in an earth science field of their choice. For example, surveyors use transits and levels. Students should label each tool and provide a brief explanation for how unusual tools are used.

C. Direct students to create a computer presentation, poster, or formal report describing a recent advancement in technology that was developed in connection with any of the branches of earth science. The final product should show a picture of the technology, explain the history and purpose of its development, and explain how it is being used.

D. Challenge students to research an occupation associated with a branch of earth science of their choice. Topics should include what tasks people in that occupation perform, the recommended education, average or median salary, and opportunities for employment.

Lesson Review

1. Why is earth science important? (**Answers will vary but should include that studying any part of creation helps people learn more about God's creative powers.**)
2. What is the difference between science and technology? (**Science is the systematic study of creation using methods based on observation and experimentation. Technology is the application of science.**)
3. Give an example of technology that might arise from each of the major branches of earth science. (**Possible answers:** *geology*—glass, medicines, location of oil and natural gas deposits or ores and minerals, exploration of the ocean floor; *oceanography*—systems to predict climate events or clean up oil spills; *meteorology*—mapping and forecasting systems; *astronomy*—devices powered by solar energy, improved diabetes treatment, and global positioning systems; *environmental science*—alternative energy sources)
4. List two ways environmental scientists may work with other scientists to promote good stewardship of the earth. (**Possible answers: Environmental scientists may help geologists find better methods to obtain natural resources without causing permanent damage to the land; they may work with oceanographers to develop a plan for maintaining a healthy environment for the ocean's creatures; they may work with meteorologists to draft laws governing technology that affects the earth's atmosphere.**)

NOTES

1.1.1 The Study of Earth Science

OBJECTIVES
- Explain the purpose of earth science.
- Explain the difference between science and technology.
- Identify the major branches of earth science.
- Describe a practical application of earth science.

VOCABULARY
- **astronomy** the study of physical things beyond the earth's atmosphere
- **earth science** the study of the earth and the universe around it.
- **environmental science** the study of the relationship between organisms and the environment
- **geology** the study of the solid earth
- **meteorology** the study of the atmosphere
- **oceanography** the study of the earth's oceans

Earth is a marvelous planet. Romans 1:20 and Psalm 19 explain that God uses His creation to reveal truth about Himself. Science is the orderly study of creation using observations and experiments. Science helps people understand the truth about God's world and God Himself.

Closely related to science is technology. Technology is the practical application of science. For example, science reveals the properties of minerals and ocean tides. Technology puts this knowledge to use. Minerals can be used for many things, such as ink or concrete. People have converted the power of tides into electricity. Technology is neither good nor bad. God gave people the responsibility to develop technology that glorifies Him. God is glorified when people enhance the safety, beauty, diversity, and usefulness of the world around them.

Science has a long history. The Chinese kept written records of earthquakes as early as 780 BC. The ancient Greeks catalogued rocks and minerals. Nearly 2,000 years ago the Babylonians mapped the positions of stars and planets. Between 300 and 400 AD, the Maya measured the movements of the moon, sun, and planets. Their observations helped them create accurate calendars to aid in farming. The telescope and microscope were invented in the 17th century. These extended the study of the visible world to include both enormous planets and tiny particles and organisms. Over time, studies such as these gave birth to the field of earth science.

Science helps people understand God's amazing creation. From the tiniest cell to the largest galaxy, God's role as Creator is displayed through science.

Earth science is the study of the earth and the universe. You are affected by earth science and technology every day. For example, you may use weather forecasts to plan what you will wear the next day. The concrete in the roads and buildings you see is a careful mixture of various rocks and minerals. When you travel in an automobile, train, or airplane, you are likely using fossil fuels. Your home's lights use electricity generated by the sun, wind, water, or fossil fuels. When you work at a computer, you are benefiting from silicon, a mineral used in computer chips. Studying earth science can help you better appreciate the resources that enable the technology you use every day.

Mayan calendar

Because earth science covers a wide range of subjects, it is divided into separate branches. These branches are geology, oceanography, meteorology, astronomy, and environmental science.

The first branch of earth science, **geology**, is the study of the solid earth. Geology is divided into more specific fields of study, some of which include mineralogy, paleontology, seismology, and volcanology.

The study of rocks and minerals is called *mineralogy*. Mineralogists seek to understand the earth's structure and processes. This study has led to a variety of technologies, such as glass, soft drinks, and medicines.

Paleontology, the study of artifacts and fossils, provides scientists with a glimpse into what the world was like earlier in its history. The data gathered by paleontologists is even used to help locate deposits of oil and natural gas, which some scientists say are linked to the remains of specific organisms.

Silicon is a mineral used to make computer chips.

Seismology is the study of earthquakes as well as Earth's layers and composition. Using technology, seismologists work to predict earthquakes and to assess earthquake risk in different regions. Their studies of the earth's layers have led to technologies that help locate petroleum and minerals and explore features of the ocean floor.

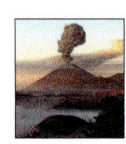

Volcanology is the study of volcanoes and the materials produced by volcanoes. Volcanologists examine the causes and effects of volcanoes. They also try to predict when the next eruption might occur. The technology used by volcanologists is also used to find the locations of ores.

Oceanography, the second branch of earth science, is the study of Earth's oceans, from the surface to the deepest ocean trenches. Oceanography, like geology, has many different specialized fields of study.

Physical oceanography is the study of waves and ocean currents. Physical oceanographers might develop technologies that help predict climate events such as El Niño that depend on ocean currents.

Chemical oceanography is the study of chemicals in the oceans, including those natural to the ocean, such as salt, as well

Chemical oceanographers study the chemical makeup of the earth's oceans.

CAREER

Geologists

Geologists use their knowledge of science to help people stay safe, to care for creation, and to use responsibly the resources God has placed in the earth.

Some geologists create maps that help people study the earth or even just find their way around. Others study earthquakes or volcanoes—they keep track of the tremors in the earth or work with engineers to help design buildings that will withstand the shakes of an earthquake. Some geologists help design waste disposal sites so the garbage people throw away each day does not pollute the water or air. Still other geologists spend weeks at a time on boats, studying the ocean floor. God created a wide and wonderful world, and there are many methods of studying it—inside and outside!

---

as those that enter the ocean from outside sources. Chemical oceanographers might devise technologies to clean up an oil spill.

The third branch of earth science is **meteorology**, which is the study of the atmosphere. Meteorology is usually associated with weather forecasting, but it investigates climates too. Meteorologists do more than help people plan what to wear each day. Their predictions of severe weather have given people in affected areas the opportunity to find shelter before a storm occurs. Their studies of climate patterns have led to mapping and forecasting technologies that help people prepare for floods, harsh winters, and droughts.

Meteorologists do more than predict the weather. They also study climate patterns to help people prepare for floods, severe winters, and droughts.

Astronomy is the fourth branch of earth science; it is the study of physical things beyond the earth's atmosphere. These objects include other planets, stars, asteroids, and comets. Because most objects in space are too far away to study directly, astronomers rely on technology such as various kinds of telescopes and space probes. Astronomy gives a glimpse into the wonder of God's amazing handiwork beyond Earth. Studies linked to astronomy have led to better understandings of Earth's many systems and new technologies such as improved diabetes treatment and global positioning systems (GPS).

"Look up at the sky and count the stars—if indeed you can count them." Genesis 15:5

The final branch of earth science is environmental science. **Environmental science** is the study of humans' relationship to their environment. Therefore, it interacts with many other sciences. Environmental scientists may help geologists find methods to obtain natural resources without causing permanent damage to the land. They may work with oceanographers to develop a plan for maintaining a healthy environment for the ocean's creatures. Environmental scientists may work with meteorologists to draft laws governing technology that affects the earth's atmosphere. God wants people to be good stewards of the earth He gave them. Environmental scientists work to preserve Earth's resources for use by all the planet's organisms.

LESSON REVIEW

1. Why is earth science important?
2. What is the difference between science and technology?
3. Give an example of technology that might arise from each of the major branches of earth science.
4. List two ways environmental scientists may work with other scientists to promote good stewardship of the earth.

Environmental scientists look for the best ways to maintain Earth's resources for the sake of all the organisms that use them.

Introduction to Earth Science

1.1.2 Age of the Earth

Introduction
Share that many scientists are Christians who love God and want to honor Him in their profession. Science and Christianity are not mutually exclusive. Lead students to realize that scientists—regardless if they are Christian or not—have numerous viewpoints on the origin of the universe. Likewise, not all Christians—regardless if they are scientists or not—agree on what scientists conclude. Sometimes the differences in views come from varying interpretations of Scripture. Display **TM 1.1.2A Venn Diagram: Young Earth and Old Earth** to present a comparison of young-earth and old-earth beliefs that Christians have about Creation. Discuss both views. Emphasize that although Christians may express opposing outlooks, they share the common belief that God is the Creator and Sustainer. Read **Colossians 1:15–17**.

Discussion
- Lead a discussion on the two basic positions on the origin of the universe.
  Theism: the belief that the universe was created purposefully by a supernatural being. Christians are biblical theists who recognize the Creator as the God of the Bible.
  Naturalism: the belief that matter and energy are all that exist and that the universe was the result of a series of undirected natural processes.

- Discuss the historical relationship between science and faith. Historically, scientists integrated their faith with scientific study until the Enlightenment and the rise of atheistic thought. Christians have led the way in scientific research because actively pursuing the study of science was and is a way to worship and honor God. Scientists have varying views on faith and scientific findings. Can you be a scientist and be a Christian? (**Yes.**) Many scientists are Christians, but their views on Creation or the age of Earth may vary because Christians do not all agree on these beliefs. Do all scientists who are not Christians agree on how and when Earth was created? (**No. Scientists who are not believers have varied viewpoints because they don't consider God as Creator.**)

- Read **1 Corinthians 13:12**. How does the verse apply to comprehending Earth's creation? (**It underscores that even Christ followers don't know everything. Only God understands fully about Creation. Therefore, we should treat one another with love until we are with Him.**)

- Discuss the need for believers to have unity in essentials of the faith. Is agreement on the age of the earth an essential for salvation? (**No.**) Remind students of what unites believers in Christ.

- Ask the following questions:
  1. Why might some people find the belief in a young Earth an obstacle for believing in Christianity? (**Young-earth Christians believe Earth is thousands of years old. Most scientists believe Earth is billions of years old, so Christianity and science seem to contradict each other.**)
  2. Why might young-earth Christians have difficulty accepting what old-earth Christians believe? (**Young-earth Christians believe in a literal interpretation of Genesis 1.**)
  3. What are varves? How are they used to date rocks? (**Varves are layers of light and dark bands of rock. One varve represents a year's sediment deposits. By counting varves, geologists can estimate the age of rocks.**)

Activities

Lab 1.1.2A Half-Lives
- 1 cm × 8 cm strips of paper per student
- scissors, 1 per student

OBJECTIVES
Students will be able to
- compare naturalism and theism.
- describe various views of the time of Creation.
- explain carbon-14 dating.

VOCABULARY
- **carbon-14 dating** the method used to determine the age of items of organic origin by measuring the radioactivity of their carbon 14 content
- **naturalism** the belief that matter and energy are all that exist and that undirected natural processes formed the universe
- **theism** the belief that the universe was created purposefully by a supernatural being
- **varves** the light and dark layers of sediments deposited in a yearly cycle

MATERIALS
- TM 1.1.2A Venn Diagram: Young Earth and Old Earth

PREPARATION
- Obtain materials for *Lab 1.1.2A Half-Lives*.

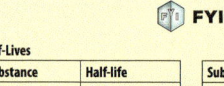

NOTES

A. Consistent with your school's policy, arrange for a class presentation on young-earth and old-earth beliefs. Assign student groups to research either young-earth or old-earth beliefs and to prepare an outline of points and counterpoints, a chart of strengths and weaknesses in viewpoints, or a historical look at the views presented by scientists and the Church. Encourage them to include scientific data and Scripture references. After all is prepared, encourage groups to share their findings. Remind students to keep the unifying factor of God as Creator and Sustainer in mind when they speak with one another. Finish the presentation by asking the questions listed at the end of this activity.

Young Earth: The Book of Genesis outlines a genealogy from the creation of the heavens and earth to Abraham. If no gaps exist, this genealogy cannot be interpreted to extend more than several thousand years. According to this method, the possible range for the age of the earth is about 6,000–12,000 years. The word *day* in Genesis may mean a literal 24-hour day. Assuming that day the word *day* meant a longer period of time at the time of Creation may alter the intent of the original passage. Old-earth interpretations of the Scripture, such as the day-age theory, progressive creation, and the gap theory, explain the meaning of the Genesis account in the Bible differently. Dinosaurs must have been taken on the ark. Dinosaurs are mentioned in Job 40 and 41 as *behemoth* and *leviathan*. The flood distributed dinosaur bones across the earth. Carbon dating has proven unreliable in the past, reporting in some instances that currently living creatures had died thousands of years ago. Rapid sedimentation after the volcanic eruption of Mount Saint Helens showed that some geological processes are rapid rather than very slow. This example of rapid sedimentation demonstrates the possible geologic response to the total devastation that a global flood would produce.

Old Earth: Psalm 90:4 states that to God, 1,000 years is like a day and that God is infinite and humanity is finite. Some theologians interpret Genesis 1 as a logical ordering of the days, not necessarily a chronological order. Sedimentary layers formed by the erosion of hard rocks such as granite have been accumulating for millions of years. Violent movements of the earth's crust occasionally cause nonconformities in sedimentary layers, but most of these layers form a distinct geologic record. Scientists have calculated average sediment deposits based on data collected over a long period of time. An average of 30 cm of sedimentary rock is deposited over a period of 1,000 years. Although floods can dramatically alter this rate, the evidence for most layers points to a gradual accumulation. Fossils have been used to date oil and natural gas deposits. These deposits are formed by intense pressure and chemical reactions on animal and plant materials over millions of years. The rates of decay of isotopes like carbon 14 and uranium 238 are known and are reliable tools to date organic samples and rocks.

1. How can those of differing opinions best pursue truth together?
2. When Christians research science topics, what godly characteristics, such as peace, humility, truth, love, faith, and perseverance, are most useful in their pursuits? Why?
3. What have you learned about not needing to choose between Christian faith and science?

B. Challenge student groups to research the following topics from the chapter and to present their information to the class in a computer presentation or a group oral report: young earth, old earth, special creation, theistic evolution, radiometric dating, carbon-14 dating, uniformitarianism, and appearance of age theory.

C. Have students construct a Venn diagram comparing special creation and theistic evolution. Discuss their findings as a class.

NOTES

---

the age of items of organic origin by measuring the radioactivity of their carbon-14 content. The half-life of carbon 14 is the amount of time it takes for half of the carbon-14 atoms in a sample to decay into nitrogen-14 atoms, which is 5,730 years. If half of the carbon-14 atoms in a bone have decayed into nitrogen-14 atoms, scientists calculate that the bone is 5,730 years old. This method has become more accurate since it was first used in the 1940s as it now takes into account differing levels of carbon dioxide in the atmosphere over time. Carbon-14 dating can only be used to date organic materials, such as bones, shells, wooden objects, or papyrus scrolls, because all organic materials contain carbon. This method cannot be used to date minerals, which do not contain the remains of living things. By using carbon-14 dating, scientists can date materials up to 50,000 years old. Carbon 14 is one of more than 40 radioactive elements scientists use to date materials. Scientists use other radioactive elements with longer half-lives, such as uranium, to date rocks in the earth's crust. Tests using the radiometric dating method indicate the earth is 4.6 billion years old.

Varves

Scientists also use indirect methods to determine when certain rocks were formed. For example, some scientists have estimated it took sedimentary rock 3 million years to form and others have estimated it took 2.4 billion years, depending on what average rate of sedimentation they used. Because of the wide range between estimates, many scientists have reached a general conclusion that sedimentary layers formed by the erosion of hard rocks like granite have been accumulating for millions of years. Other factors, such as soil conditions and site location, could affect such estimates of a rock's age.

Varves, the light and dark layers of sediment deposited yearly in some lakes, are also used by geologists to estimate the age of rocks. Each year, the light layer of coarse rock in a varve collects during the summer, and the finer, dark particles are laid down during winter. Therefore, one varve represents one year. Just as a tree's age can be determined by counting its rings, geologists can estimate how long it took for the rock layers to be formed by counting a sedimentary rock's varves.

According to current scientific data, Earth may be as old as 4.6 billion years old. Not all people accept this scientific view on the age of the earth, however. Some people believe Earth is

FYI
Scientific Data
Why is scientific data generally considered trustworthy? Scientists repeat their experiments and studies over and over to make sure they get the same results. Sometimes they use multiple methods to check their work. For example, when dating an artifact or a rock, scientists double-check their results by using several different radioactive elements. They also repeat other scientists' experiments to verify data is correct. By using different independent methods and comparing their results, scientists are able to assure that their data is reliable.

BIBLE CONNECTION
God as Creator
Does Scripture say anything about microevolution or macroevolution? Read Genesis 1:24–25, 2:7. Could these Scriptures be used as support for different views on the origin of creatures? In considerations like these, it is wise to remember what unites Christians is their belief in God as Creator and Sustainer. As it says in 1 Samuel 2:8b, "For the foundations of the earth are the Lord's; on them He has set the world."

actually very young. Some Christians believe Genesis 1 implies God made everything in the universe in six literal 24-hour days. They believe that Earth as it was created was very different than it is today and that Earth's geologic features, such as rock layers and fossils, were caused by major events like the Flood. By using the genealogies and historical events in Scripture as well as some other scientific data to establish a time line, young-earth supporters believe that the planet is 6,000–12,000 years old.

Other Christians believe that the Earth is very old. Many old-earth supporters believe God made the stars and planets, including Earth, about 12–15 billion years ago, perhaps even through what is called *the big bang*. Some believe the big bang was the creation of energy and matter, described as *light* in Genesis 1:3, on the first day of Creation. Some old-earth believers think the unfolding evolutionary process resulted in the creation of life-forms. Others believe God used evolution, but intervened at times to suddenly create living things. In general, old-earth supporters use scientific discoveries from such fields as biology, astronomy, and geology as evidence that the earth is billions of years old and that the geologic features of the planet are the result of slow changes and natural processes.

Christians also have different views on how God created. One view is called *special creation*, which states that God miraculously created every basic living thing. Any variations evident in living things today, such as the different types of dogs, came about through natural processes. The producing of small changes in organisms is called *microevolution*. In microevolution, new varieties of animals within one species can emerge, but entirely new species of animals cannot be produced from one species.

FYI
Appearance of Age
As geological discoveries indicating that Earth was very old were made in the 18th century, many Christians wondered which to believe—the Bible or science. In 1857, British biologist and preacher Phillip Gosse introduced a new theory—the appearance of age theory. The appearance of age theory states that the Earth is only a few thousand years old, but God created it to appear old. Some Christians believe this today and use this theory to address scientific data from radiometric dating. Other believers disregard this theory because it implies the Creator was deceptive, which conflicts with verses such as Proverbs 14:5 and Leviticus 19:11 that underscore God's condemnation of deception.

NOTES

D. Encourage students to write a personal psalm of praise to God for His creation of the earth.

Lesson Review

1. Compare and contrast naturalism and theism. (**Naturalism says that matter and energy are all that exist and that the universe began as the result of the series of undirected natural processes. Naturalism does not support the idea of any supernatural force or significance to life. All life is random, and the existence of life in its various forms occurred by accident through natural processes. Theism says that a supernatural being, whom Christians recognize as the God of the Bible, personally and purposefully created the universe. The design of creation was intentional rather than accidental.**)

2. Some Christians believe Earth is old and some believe it is young. Describe the old-earth and young-earth views of when Creation happened. (**Old-earth Christians believe God made the stars and planets about 12–15 billion years ago, possibly through the big bang. The earth was formed about 4.6 billion years ago, and Earth's geologic features are the result of slow changes and natural processes that God established. Young-earth Christians believe God made everything in the universe in six 24-hour days, as described in Genesis 1. Earth was very different than it is today, and Earth's geologic features such as rock layers and fossils were caused by catastrophic events like the Flood.**)

3. How do scientists use carbon-14 dating to determine a date for an artifact? (**Carbon 14 decays at a known rate. By analyzing how much carbon 14 is present in a sample from an artifact, scientists can determine its approximate age.**)

4. How can varves help scientists estimate the age of a rock? (**Each year a light layer and a dark layer form in a lake's sediment. By counting the varves, scientists can estimate a rock's age.**)

Advocates of special creation believe God established limits on the amount of natural change that could occur in organisms. For example, there may be new varieties of dogs in the future, but they will still all be dogs and not new creatures.

Another view of how God created is called *theistic evolution*. According to this view, God used the process of evolution rather than sudden works of power to create major groups of living things. He then used natural processes to produce whole new species of creatures. In this view, God did not set limits on how much an organism can change. Therefore, given time, organisms can develop into different species of creatures. This production of entirely new creatures is called *macroevolution*. Theistic evolutionists believe that these changes happen as part of God's plan and that He uses evolution as a tool to create the creatures He wants. Christians are biblical theists who recognize the Creator as the God of the Bible.

No one knows precisely how Earth was created. The most important thing to know is that God is the one who decreed the world into existence. Part of loving God with all your mind includes learning more about His creation. As finite creatures, people will probably never be able to prove the exact time frame of Creation. However, as scientists study the work of an orderly God, they will continue to make amazing discoveries that may adjust what is considered scientific fact today. Such discoveries will continue to reveal the Creator's infinite power to all people.

FYI
Radiometric Dating and Young Earth
All scientific research begins with an assumption about what is fact. Young-earth scientists often disagree with some of the assumptions made in radiometric dating. For example, scientists cannot prove that conditions eons ago were the same as they are now or that a rock that contains a radioactive element was not once contaminated by rainfall or even lava. It is also assumed that radioactive elements like carbon 14 have always decayed at the same rate, an assumption that many young-earth scientists dispute.

LESSON REVIEW
1. Compare and contrast naturalism and theism.
2. Some Christians believe Earth is old and some believe it is young. Describe the old-earth and young-earth views of when Creation happened.
3. How do scientists use carbon-14 dating to determine a date for an artifact?
4. How can varves help scientists estimate the age of a rock?

Name: _____ Date: _____

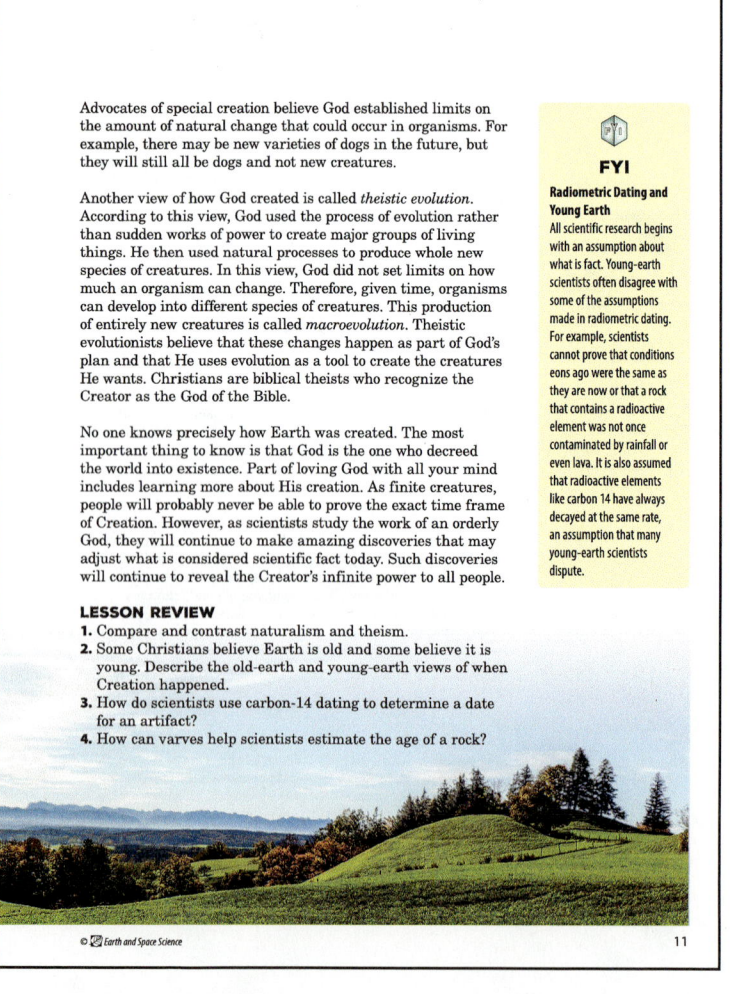

Lab 1.1.2A Half-Lives

QUESTION: What do half-lives teach about Creation?

HYPOTHESIS: Answers will vary.

EXPERIMENT:

You will need:	• 1 cm × 8 cm strip of paper	• scissors

Steps:
1. Create a tally sheet to record the number of cuts made in Steps 2 and 3.
2. The strip of paper represents the carbon 14 in a sample. Cut the paper in half. Discard half of the strip. Record this cut on the tally sheet.
3. Cut the remaining half in half, and discard half of the strip. Record this cut on the tally sheet.
4. Repeat Step 3 until it is no longer possible to cut the strip in half.

ANALYZE AND CONCLUDE:

1. How many times were you able to cut the paper in half until it was too small to cut? Answers will vary but should be around six or seven.

2. What length of time is represented by each cut? 5,730 years.

3. What length of time is represented by the total amount of cuts? Number of cuts times 5,730 years.

4. Could you use the half-life of carbon 14 to prove that dinosaurs died millions of years ago? No. No carbon 14 would remain in a fossil that was millions of years old.

5. Can carbon-14 dating be used to determine the age of the earth? Why? No. Only once-living materials can be analyzed with carbon-14 dating; the rocks and minerals from which the Earth was created are inorganic substances.

6. Uranium 238 has a half-life of 4.5 billion years. Explain why uranium is used instead of carbon 14 to date things older than 50,000 years old. It takes much longer for uranium to decay; therefore, compared with carbon 14, after the same number of half-lives many more years have passed.

Introduction to Earth Science

1.1.3 Maps

Introduction
Point out that in order to study the earth and space, scientists need accurate depictions of those areas. Display a globe. Have students identify the following features:
- North Pole
- Arctic Circle
- South Pole
- Antarctic Circle
- Equator
- Tropic of Cancer
- Tropic of Capricorn
- Prime meridian
- International date line
- Continents
- Country in which they live

Direct students' attention to the world map in their Student Editions. Have students point out differences in the way features appear on the globe and on the map. Ask what might account for these differences. (**The features on a curved surface become distorted when they are displayed on a flat surface.**)

Discussion
- Lead a discussion on how a map might encourage a false understanding of the world. Display **TM 1.1.3A Map Projections**. Have students indicate where each map may be inaccurate depending on what type of map it is. (*azimuthal*: **most distorted farthest away from center of map**; *conical*: **most distorted farther away from where the cone touches the globe**; *cylindrical*: **most distorted near the poles**) Challenge students to consider the possible political implications of geographic misconceptions.

- Display **TM 1.1.3B–D Types of Maps**. Discuss the similarities and differences of the pictured maps. What do these maps depict? (*topographical map*: **heights and shapes of land**; *tectonic map*: **positions of plates**; *weather map*: **weather conditions**) Discuss how scientists use these maps.

- Display **TM 1.1.3E World Time Zones Map**. Point out that time zones are not consistently drawn along lines of longitude. Explain that the colors on the map reflect the times on the clocks along the bottom and top of the map. Discuss why world time zones were created. Guide students to consider the implications of each village, town, or city using its own local time. Point out that time zones in countries such as China and India were established by political decisions.

Activities

Lab 1.1.3A Topographic Map
- modeling clay
- metric rulers, 1 per group
- tubs, 1 per group
- graph paper

A. Complete *Try This: Topographic Map* with students.

B. Display TM 1.1.3E for reference. Divide the class into pairs and have each pair complete **WS 1.1.3A Traveling Across Time Zones** using **WS 1.1.3B World Time Zones Map**. Encourage students to color relevant time zones on WS 1.1.3B for clarity.

C. Assign **WS 1.1.3C Map Selection** for students to complete.

D. Have students complete **WS 1.1.3D Tracking Down the Treasures**, which will also serve as an introduction to minerals.

OBJECTIVES
Students will be able to
- identify lines of longitude and latitude and demonstrate their use in connection with determining location and time.
- explain the challenges in creating an accurate map.
- create a landscape model and craft a topographic map of the features of that model.
- infer the purpose of a map based on its features.

VOCABULARY
- **contour line** a line on a map that joins points of equal elevation
- **latitude** the distance in degrees north or south of the equator
- **longitude** the distance in degrees east or west of the prime meridian
- **prime meridian** an imaginary line that divides the earth into the Western Hemisphere and the Eastern Hemisphere
- **projection** a system of lines drawn on a flat surface to represent curves

MATERIALS
- Globe (*Introduction*)
- Earth science maps (*I*)
- Metersticks, metric tape measures, balls of string (*J*)
- Topographic maps (*J*)
- TM 1.1.3A Map Projections
- TM 1.1.3B–D Types of Maps
- TM 1.1.3E World Time Zones Map
- WS 1.1.3A Traveling Across Time Zones
- WS 1.1.3B World Time Zones Map
- WS 1.1.3C Map Selection
- WS 1.1.3D Tracking Down the Treasures

PREPARATION
- Obtain materials for *Lab 1.1.3A Topographic Map*.
- Obtain materials for *Try This: Topographic Map*.
(*continued*)

PREPARATION

(continued from the previous page)
- Invite a person to share the types and functions of maps used in his or her career. Suggested careers include college or university professor, meteorologist, construction contractor, or a city water or gas engineer. *(F)*
- Invite a surveyor to demonstrate surveying methods and tools. *(G)*
- Arrange a class hike. *(H)*
- Gather a variety of geologic maps. *(I)*

TRY THIS

Topographic Map
- *graph paper*
- *modeling clay*

Modeling clay can be made by mixing 750 mL of flour with 750 mL of salt and adding 300 mL of water.

E. Display TM 1.1.3E. Have students identify the lines of longitude, the time zone in which they live, and the line of longitude used as their time zone's point of reference. Direct students to identify the approximate locations of major cities within their country or around the world and to determine the current time in each city. Share the meaning of some longitudes and latitudes being assigned negative values on different kinds of maps.

F. Introduce the invited speaker to talk about the use of maps in a career in earth or space science.

G. Present the surveyor who will demonstrate surveying methods and tools to the class.

H. Take students hiking in hilly or mountainous terrain. Have them use topographic maps of the region to chart their elevation at different points along the hike.

I. Show students a variety of different maps used by earth scientists, including various geologic maps, weather maps, climate maps, celestial globes, bathymetric, and physiographic maps. Guide students to consider what each map might be used for and how each map might be read.

J. Have student groups each create a scale map of the school grounds or a city block. Each group will need a meterstick, metric tape measure, and a ball of string. Long distances can be paced off, or students can roll out string and then measure the length of string they used. Maps should include all significant features, such as buildings, fields, and bodies of water. Each map should include a map key, a compass rose, the cardinal directions, and a scale of the group's choosing (for example, 1 cm might represent 10 m).

1.1.3 Maps

OBJECTIVES
- Identify lines of longitude and latitude and demonstrate their use in connection with determining location and time.
- Explain the challenges in creating an accurate map.
- Create a landscape model and craft a topographic map of the features of that model.
- Infer the purpose of a map based on its features.

VOCABULARY
- **contour line** a line on a map that joins points of equal elevation
- **latitude** the distance in degrees north or south of the equator
- **longitude** the distance in degrees east or west of the prime meridian
- **prime meridian** an imaginary line that divides the earth into the Western Hemisphere and the Eastern Hemisphere
- **projection** a system of lines drawn on a flat surface to represent curves

Maps are very important to earth scientists. Geologists use maps to search for oil deposits. Oceanographers use maps to travel across and down into the oceans. Meteorologists use maps when they travel to the South Pole to study the ozone hole over Antarctica. Astronomers use maps to recall the placement of each star. Environmental scientists use maps to keep track of the changes in regions of land.

The earth is most accurately represented by a globe. A globe helps people visualize the earth as a sphere. On the top of the globe is the North Pole, the northernmost point on Earth; at the bottom is the South Pole, the southernmost point on Earth.

The equator is an imaginary line that divides the earth into the Northern Hemisphere and the Southern Hemisphere. Parallels are imaginary lines around the earth that are parallel to the equator. Parallels show lines of **latitude**, the distance in degrees north or south of the equator. The equator is at 0° latitude.

Meridians are imaginary lines on Earth's surface that pass through the North Pole and the South Pole. The **prime meridian** is an imaginary line that divides the earth into the Western Hemisphere and the Eastern Hemisphere. Meridians show lines of **longitude**, the distance in degrees east or west of the prime meridian. The prime meridian is at 0° longitude.

The lines of latitude and longitude form an imaginary grid that is used to pinpoint a location on Earth. To provide greater accuracy, each degree is divided into 60 minutes [symbolized by a single prime mark(')], and each minute is divided into 60 seconds [symbolized by double prime marks (")]. For example, Sao Paulo, Brazil, is approximately located at 23°33'1" S 46°37'59" W, and Lima, Peru, is approximately located at 12°2'47" S 77°2'34" W. These numbers show that Lima is northwest of Sao Paulo.

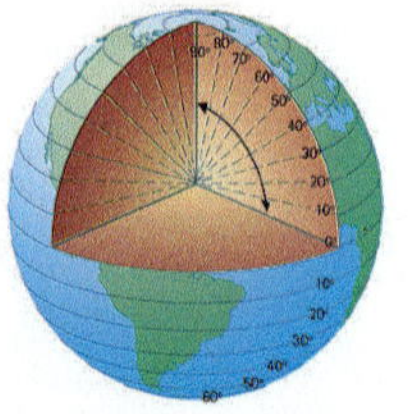

When you look at a globe or a map, you will notice that the hemispheres are divided into additional parts. The Tropic of Cancer (from the equator to 23°27' N) is roughly the northernmost boundary of the tropics. The tropic of Capricorn (from the equator to 23°27' S) is the southernmost boundary of the tropics. The Arctic Circle begins at 66°33' N, and the Antarctic Circle begins at 66°33' S.

Along with helping to determine location, lines of longitude are also used to help determine time. To compensate for the earth's rotation, time zones were established to make sure that daylight hours match daytime and night hours match nighttime. The world has 24 time zones. The prime meridian serves as hour 0, and lines of longitude divisible by 15 serve as the approximate midpoints for the different time zones. The international date line runs approximately along 180° longitude and indicates where one date ends and the next begins. Traveling east from the prime meridian, an hour is added to the time in each time zone until the international date line. Traveling west, hours are subtracted until the international date line is reached. Time zones are not uniform. Each nation may adjust the lines within its borders to accommodate its population or other needs.

FYI

International Date Line
The international date line, which is on the opposite side of the world from the prime meridian, is at 180° longitude. The international date line separates one calendar day from the next. If you travel east around the world and adjust your calendar watch whenever you enter a new time zone, by the time you get back to your starting point, you would be a day behind everyone else. If you travel west, by the time you get back to your starting point, you would be a day ahead of everyone else.

K. Have student pairs use a website or map to identify the latitude and longitude of various cities. Then have groups exchange their lists of coordinates and identify each other's cities.

L. Make a weather map to chart the daily weather conditions for a week. Have a different student group chart each day's weather using the appropriate symbols.

Lesson Review

1. How are lines of latitude related to the equator? (**Lines of latitude show the distance in degrees north or south of the equator.**)
2. What are the two functions of lines of longitude? (**Lines of longitude work with lines of latitude to determine locations on a globe or map, and they serve as approximate boundaries for the world time zones.**)
3. Why is making an accurate map challenging? (**it isn't possible to show a sphere's features on a flat surface.**)
4. What are the three basic types of projection maps, and what are the strengths of each? (***azimuthal projection*: accurate size and shape near the map's center as well as correct direction from the map's center to any other area on the map; *conical projection*: accurate shapes or sizes of all areas as well as accurate distances and directions close to where the cone touches globe; *cylindrical projection*: accurate shapes of all areas as well as accurate size and distance close to where the cylinder touches the globe**)
5. What kind of map would a seismologist use to study earthquake patterns? (**tectonic**)
6. If you were skiing and wanted to avoid steep hills, which of the maps discussed in this lesson would you use? (**topographical**)

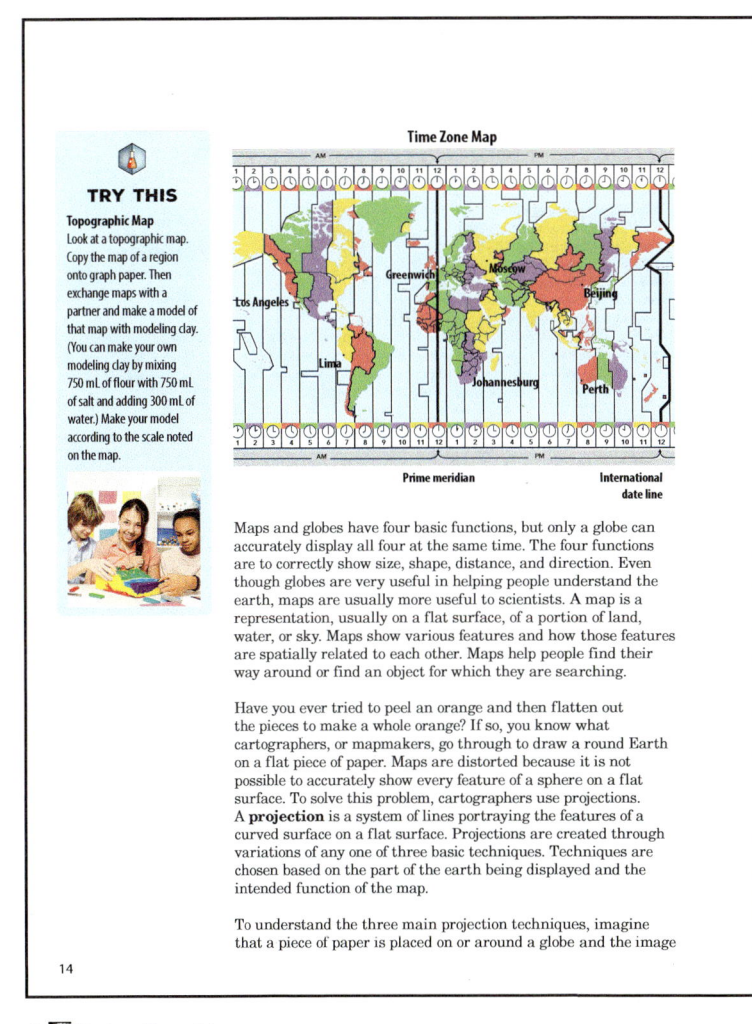

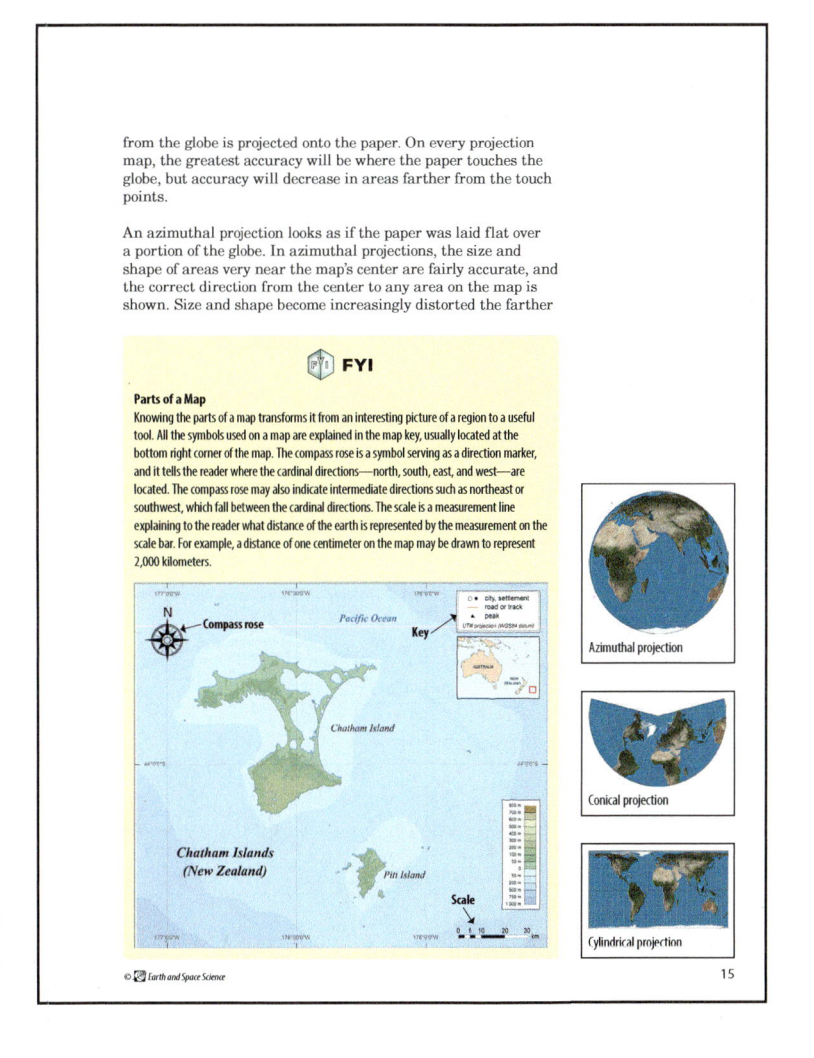

NOTES

TRY THIS
Topographic Map
Look at a topographic map. Copy the map of a region onto graph paper. Then exchange maps with a partner and make a model of that map with modeling clay. (You can make your own modeling clay by mixing 750 mL of flour with 750 mL of salt and adding 300 mL of water.) Make your model according to the scale noted on the map.

Time Zone Map

Maps and globes have four basic functions, but only a globe can accurately display all four at the same time. The four functions are to correctly show size, shape, distance, and direction. Even though globes are very useful in helping people understand the earth, maps are usually more useful to scientists. A map is a representation, usually on a flat surface, of a portion of land, water, or sky. Maps show various features and how those features are spatially related to each other. Maps help people find their way around or find an object for which they are searching.

Have you ever tried to peel an orange and then flatten out the pieces to make a whole orange? If so, you know what cartographers, or mapmakers, go through to draw a round Earth on a flat piece of paper. Maps are distorted because it is not possible to accurately show every feature of a sphere on a flat surface. To solve this problem, cartographers use projections. A **projection** is a system of lines portraying the features of a curved surface on a flat surface. Projections are created through variations of any one of three basic techniques. Techniques are chosen based on the part of the earth being displayed and the intended function of the map.

To understand the three main projection techniques, imagine that a piece of paper is placed on or around a globe and the image from the globe is projected onto the paper. On every projection map, the greatest accuracy will be where the paper touches the globe, but accuracy will decrease in areas farther from the touch points.

An azimuthal projection looks as if the paper was laid flat over a portion of the globe. In azimuthal projections, the size and shape of areas very near the map's center are fairly accurate, and the correct direction from the center to any area on the map is shown. Size and shape become increasingly distorted the farther

FYI
Parts of a Map
Knowing the parts of a map transforms it from an interesting picture of a region to a useful tool. All the symbols used on a map are explained in the map key, usually located at the bottom right corner of the map. The compass rose is a symbol serving as a direction marker, and it tells the reader where the cardinal directions—north, south, east, and west—are located. The compass rose may also indicate intermediate directions such as northeast or southwest, which fall between the cardinal directions. The scale is a measurement line explaining to the reader what distance of the earth is represented by the measurement on the scale bar. For example, a distance of one centimeter on the map may be drawn to represent 2,000 kilometers.

an area is from the map's center. The accuracy of distances depends on how the projection was made. These maps are often used to show the earth's poles.

A conical projection looks as if the paper was rolled into a cone and then placed over the globe. Depending on how the projection was made, either the shapes or sizes of areas will be most accurate. Directions and distances between areas are fairly accurate, but they become distorted in areas farther from where the cone touches the globe. These maps are often used to show areas between the equator and a pole.

A cylindrical projection looks as if the paper was rolled into a cylinder or tube and then placed over the globe. Cylindrical projections usually show accurate shapes of areas. The size of areas and the distance between them become increasingly distorted the farther the area is from where the paper touches the globe. The accuracy of direction depends on how the projection was made. These maps are often used to show the whole world or areas along the equator.

Each branch of earth science has its own types of maps and charts. A geologic map is a map that gives geological information such as rock type, age of the rocks, and the ways that rock has changed. Geologic maps may also show the locations of past volcanic eruptions and earthquakes. Geologic maps are helpful in pointing out dangerous conditions such as potential landslides. They also show information that is helpful for the construction of wells and sewage systems—for example, they may show the level of the water table.

Geologic maps can be of several different types. A tectonic plate map is a map that shows the positions of the earth's plates and may also show the locations of faults, earthquakes, and volcanoes. A bedrock formation map is a map that plots different rock types from a particular time period.

Another type of geologic map is a paleogeographic map, which shows what a region is believed to have looked like long ago. A paleogeographic map shows ancient seas, mountain ranges, and any other important geological features. Natural history museums often use paleogeographic maps to bring the past to life by showing what plants and animals may have lived in a region at a particular time. For example, part of a paleogeographic map might show swamp lands where dinosaurs flourished. Although these maps are interesting, nobody can prove that their details are accurate.

Some geologic maps include a cross-section diagram that shows what rock formations at different depths would look like if an imaginary trench were dug straight down from the surface of land. Using a cross-section diagram is like looking at an ant farm behind a pane of glass. When you look into an ant farm, you can see the tunnels dug by the ants and their activities below the surface of the soil; a cross-section diagram shows you how the different layers of rock formations appear. Sometimes you can see a real cross-section of rock layers along a highway road cut.

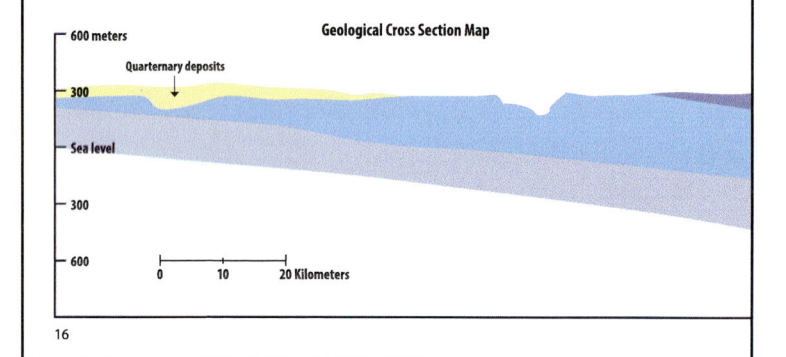

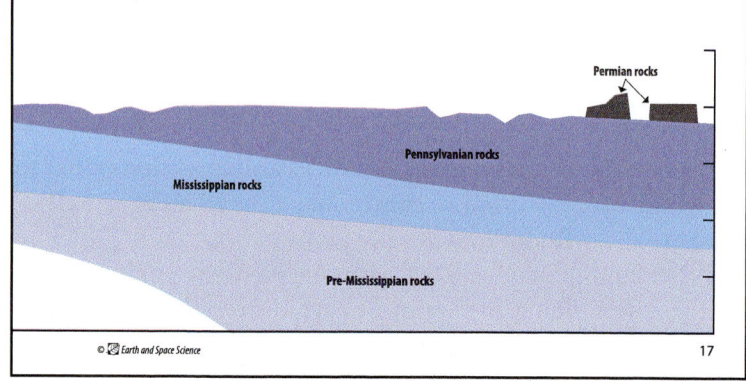

Topographic maps, also called *contour maps*, are very useful to geologists. A topographic map is a map that shows the different heights and shapes of the land using contour lines. A **contour line** is a line on a map that joins points of equal elevation. Contour lines are usually drawn roughly parallel to one another. They form circles or ovals at the tops of hills or in depressions in the land. The spacing of the lines represents the slope of the land. You can tell how steep the terrain is by looking at the distance between each contour line. Contour lines drawn close together show a steep incline; lines drawn farther apart show a gentler slope. The contour lines are usually marked with numbers to signify an elevation in feet above sea level. Topographic maps are helpful in identifying hills on hiking trails or for builders who need to know where the ground is uneven. Topographical maps can also show underwater features. These maps are called *bathymetric maps*.

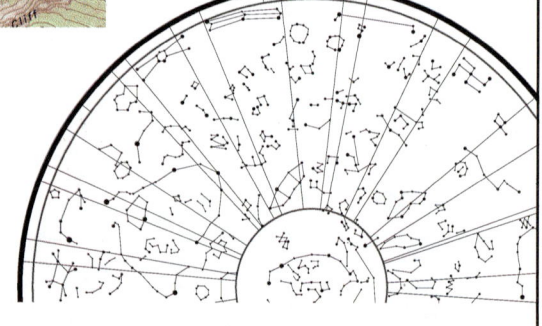

Meteorologists also use maps. When you watch the weather report, you see a weather map, a map that uses symbols to track cloud cover, wind, and other weather conditions. To keep track of the long-term climactic conditions of a region, meteorologists use a climate map, which summarizes an area's annual rainfall, snowfall, and temperatures.

Astronomers use maps called *star charts*. Astronomers usually have their star charts bound together into a star atlas, which is a collection of maps or charts on a related subject. Astronomers also use a sphere called *a celestial globe* to help them locate celestial objects. Earth is pictured at the center of this sphere, and a hollow plastic shell around the Earth represents the night sky. Celestial bodies are imprinted on this shell. The sphere can rotate to show the movement of stars in the night sky.

HISTORY

Star Charts
Early stargazers saw shapes of people, animals, and objects among the stars. Today, astronomers still recognize these shapes and call them *constellations*. The International Astronomical Union recognizes 88 official constellations. All parts of the sky have been assigned to one constellation or another. Star charts show the borders of the constellations, and they may also connect the brighter stars with lines to show the figure that gives the constellation its name. However, stars within the same constellation may be connected in different ways by different cultures. The constellation with the largest area is Hydra. The smallest constellation is Crux, which helps sailors in the tropics and the Southern Hemisphere navigate their boats at night and reminds Christians of the cross of Jesus Christ.

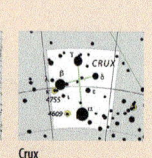

LESSON REVIEW
1. How are lines of latitude related to the equator?
2. What are the two functions of lines of longitude?
3. Why is making an accurate map challenging?
4. What are the three basic types of projection maps, and what are the strengths of each?
5. What kind of map would a seismologist use to study earthquake patterns?
6. If you were skiing and wanted to avoid steep hills, which of the maps discussed in this lesson would you use?

Name: _____   Date: _____

Lab 1.1.3A Topographic Map

QUESTION: What can you learn from a topographic map?

HYPOTHESIS: Answers will vary.

EXPERIMENT:

You will need:	• metric ruler	• graph paper
• modeling clay	• tub	

Steps:
1. Make a model landscape of a mountain and hills of varying slopes with the modeling clay. Make the landscape at least 8 cm high at its highest point.
2. Carefully press the landscape into an empty tub.
3. Hold a metric ruler upright in the tub next to the landscape.
4. Pour water into the tub until the water reaches a depth of 1 cm. Hold the landscape in place and carefully trace around the water level on the clay with a sharp pencil.
5. Repeat Step 4, raising the level of the water 1 cm at a time, tracing the water level each time. Do this until water covers the landscape.
6. Carefully remove the water from the tub. Be careful not to rub off the marks. Trace over the marks again so they show up better.
7. Look straight down at the top of the landscape. On graph paper, sketch the rings as seen from above.
8. Compare the sketch to the landscape.

ANALYZE AND CONCLUDE:

1. How does the number of rings relate to the height of the mountain and hills?

   The more rings, the higher the hill or mountain.

2. How does the distance between rings relate to the slope of the mountain and hills?

   The closer the rings are to each other, the steeper the slope.

1.2.0 Minerals

LOOKING AHEAD

- For **Lesson 1.2.1**, invite a jeweler to speak to the class about distinguishing real gems from synthetic ones. Invite a professor to talk about the difference between rocks and minerals and how minerals help form rocks. Plan a field trip to the geology section of a natural history museum.
- For **Lesson 1.2.2**, grow a sodium borate crystal creation using the method described in *Introduction*. Invite a geologist from a nearby college or university to display samples of several common minerals and to explain their chemical and physical properties, where they can be found, and how they were formed.
- For **Lesson 1.2.3**, plan a field trip to a mineral-rich location where samples may be collected for classification.

SUPPLEMENTAL MATERIALS

- TM 1.2.1A Rocks and Minerals
- TM 1.2.1B Rocks and Minerals Key
- TM 1.2.2A Periodic Table of Elements
- TM 1.2.3A Mineral Luster Chart

- Lab 1.2.3A Mineral Identification

- WS 1.2.2A Precious Jewels
- WS 1.2.2B Periodic Table of Elements
- WS 1.2.2C Chemical Classification
- WS 1.2.4A Special Mineral Properties

- Chapter 1.2 Test

BLMs, TMs, and tests are available to download. See Understanding Purposeful Design Earth and Space Science at the front of this book for the web address.

Chapter 1.2 Summary

Minerals are an important part of creation. They are not only the foundation of the physical earth but also have practical uses ranging from electrical wires to building materials to coins. They are also an essential part of the human body structure and diet. In this chapter students will learn about the properties of minerals, how to identify minerals, and recognize their uses. Students' knowledge of minerals will prepare them to study rocks, the structure of the earth, and the processes that occur on the earth's surface.

Background

Lesson 1.2.1 – Introduction to Minerals

Scientists have not yet identified all of the minerals in God's creation. Around 4,000 different types of minerals are known, and new ones are constantly discovered. Every mineral type contains defined amounts of specific atoms in its structure. As such, minerals are chemical compounds that can be represented by a chemical formula. Some minerals (such as gold, silver, and copper) are native elements, but most are chemical compounds. Most minerals created from elemental compounds have structures that contain an exact proportion of certain atoms. Minute amounts of other atoms mixed among these constituent atoms are known as *impurities*. Impurities can slightly alter the color and other physical properties of minerals. Minerals form by the precipitation from ions in solution, by the cooling and hardening of magma, by the condensation of gases or gaseous action on rock, and by metamorphism. Minerals in rocks may be replaced by other minerals through the action of water or gases. Minerals combine with each other to make rocks, which differ from minerals in that rocks contain grains from multiple minerals and do not have a uniform crystalline structure.

Lesson 1.2.2 – Mineral Groups

Minerals can be classified into several major groups, but because silicate minerals form the largest of these, it is convenient to first group minerals as silicates and nonsilicates. The crystalline structure of all silicates has four oxygen atoms bonded in a pyramid shape to one silicon atom in their center, which is called *a silicon-oxygen tetrahedron*. The wide variety of silicate minerals is partly attributable to the different combinations of the tetrahedrons, including chains and sheets, among others. Silicate minerals include the different varieties of feldspar, quartz, mica, and talc. Of these, the feldspar category is the most common. One property of quartz is piezoelectricity, the ability to vibrate constantly at a precise frequency when an electrical current passes through it; this makes quartz useful in clocks as an oscillator. Mica is particularly distinguishable by its cleavage, which produces thin, shiny, elastic sheets. Rocks composed mostly of talc are commonly called *soapstone* because of talc's soapy, greasy texture.

Nonsilicate minerals include native elements, carbonates, halides, oxides, sulfates, and sulfides. Native elements are comprised of approximately 19 minerals that occur naturally. These do not have chemical formulas other than their own symbols. They are subdivided into metals, semimetals, and nonmetals. The carbonate class of minerals contains at least 80 varieties of which there are three groups that are most commonly found: calcite, dolomite, and aragonite. Minerals in the calcite or dolomite classes are the main components of limestone and marble; minerals in the aragonite class can be found in the shells of shellfish. Halide minerals are composed of a halogen and at least one metal. Halite (sodium chloride) and sylvite (potassium chloride) are among the commonly known and occurring halides. Although there are many other halides, the others are rare and are formed only under precise conditions and in specific locations. For example, laurionite is formed as a result of saltwater interacting with lead slag or from the oxidation of lead ore; as such, it can only form at or near lead mines or where lead slag is dumped. Oxide minerals are distinguishable from other minerals containing oxygen because oxides have a much more simple chemical formula involving oxygen and one or two other metals. In oxides, the negatively charged oxygen atoms are grouped closely together with positive ions, which are usually metal, filling in any open spaces. In contrast, the oxygen in other minerals is covalently bonded to one or more other elements to create a recognizable subgroup of atoms within the mineral. Sulfate minerals are oxygen-sulfur ions bonded with metal ions. Although there are a great many sulfate minerals, the most common are barite, gypsum, and anhydrite. Sulfides are minerals created from sulfur and one or more metals. They are conceptually similar to oxides in that sulfides are the result of sulfur bonding directly to another element, just as oxides are the product of oxygen bonding directly to another element. However, the bonds in sulfides and oxides are dissimilar. Many ore metals, such as lead, copper, and zinc, occur as sulfides.

Lesson 1.2.3 – General Mineral Properties

Certain properties are helpful in mineral identification, including color, luster, hardness, streak, cleavage, fracture, and density. Successfully identifying minerals may require examining two or more properties, although this is not always the case. For example, various native elements with untarnished metallic luster can be identified primarily by color. Although color is a mineral's most obvious property, it is rarely enough to make an accurate identification because trace amounts of chemical impurities can greatly affect a mineral's color. A somewhat more distinctive property related to color is a mineral's streak. A mineral's streak can vary from its observed color. Luster refers to the way that light reflects off a mineral's surface. Categories of luster include metallic, adamantine, pearly, resinous, silky, glassy, waxy, dull, and earthy. A mineral's hardness refers to its resistance to scratching, which is attributable to the strength of the bonds between its constituent atoms. A mineral's relative hardness is measured by the Mohs hardness scale, which German mineralogist Friedrich Mohs developed in 1812. Some minerals tend to break or split along one or more sets of parallel, flat surfaces called *cleavage planes*, which are surfaces of weak chemical bonds. For example, muscovite, which has a sheet structure, has strong chemical bonds within the sheets of silicon-oxygen tetrahedra and weak bonds between those sheets; therefore, muscovite crystals cleave well into thin sheets. Fractures are rough, irregular breaks throughout mineral crystals. If the bonds between a mineral's constituent bonds are strong in all directions, it will fracture rather than cleave. A mineral's density is its mass divided by its volume. Its specific gravity is the ratio of the mineral's density to the density of water. A mineral's density reflects its chemical composition.

Lesson 1.2.4 – Special Mineral Properties

Some minerals have properties in addition to those normally used for identification. A few minerals dissolve in water, and these often have a characteristic taste. Since a water-soluble mineral may be poisonous, taste should not be used as an identifying property. Some minerals have a distinct odor if they are struck or have been recently dug up. In sulfide minerals, the odor is the result of oxygen reacting with the sulfur in the mineral. Another unique property of some minerals is that they release oxygen and fizz when they come in contact with hydrochloric acid. Other minerals are flexible or elastic. Certain minerals react to a magnetic field. The presence of iron accounts for the magnetic properties of minerals in virtually all naturally occurring specimens. Most magnetic minerals are attracted to magnetic fields, but a few, such as bismuth, are repelled by magnetic fields. Luminescence is a mineral's ability to give off light under certain conditions. Fluorescence is a type of luminescence that causes a substance to glow in the visible spectrum when exposed to ultraviolet light. The city of Franklin in New Jersey, United States, calls itself "the fluorescent mineral capital of the world" because many of its minerals are fluorescent. Two minerals commonly found there include willemite, a white or greenish-yellow mineral that fluoresces green, and calcite, a green or white mineral (among other colors) that fluoresces red. Phosphorescence is a type of luminescence exhibited in a few fluorescent minerals in which the mineral glows even after an ultraviolet light source has been removed. Transparent minerals have the ability to refract light. A mineral's texture can sometimes identify it. Texture should not be the only characteristic used to identify a mineral because different people may describe the same texture in different ways. Radioactivity is a property of minerals that contain radioactive elements; thorianite and uraninite are both radioactive minerals.

WORLDVIEW

- Minerals encompass both beauty and strength. Gold and silver have been used as currency and for the creation of highly valued art. Iron and bronze have long been used for tools and weapons because of their malleability and durability. All four of these minerals are mentioned in Daniel's prophecy about world kingdoms being replaced by Christ's kingdom (Daniel 2:27–45). Nebuchadnezzar, the king of Babylon, had a dream in which he saw a statue with a head of gold, a chest and arms of silver, a belly and thighs of bronze, legs of iron, and feet that were part iron and part clay. As Nebuchadnezzar watched, a huge stone was mysteriously cut from a rock and rolled down a hill. It obliterated the statue and grew into an enormous mountain. Daniel explained that each portion of the statue represented a different kingdom, which would become known for either its great splendor or its great might. However, a final kingdom would come, surpassing them all. The rock, representing God's kingdom, will one day bring an end to all human kingdoms and will itself reign supreme in splendor and might. Revelation 18 echoes this prophecy, saying "Babylon," a representation of all that is ungodly, will be thoroughly destroyed. In its place will be a perfect, new world, symbolized by a huge city made from gold and gemstones (Revelation 21). As you lead your students through the study of minerals, marvel not only at their beauty and myriad uses, but also that those who love God will one day experience His loving beauty and strength in a way that far surpasses the glimpse of those qualities that minerals give us now.

1.2.1 Introduction to Minerals

Minerals

OBJECTIVES

Students will be able to
- explain the basic characteristics of a mineral.
- distinguish between minerals and nonminerals.

VOCABULARY

- **mineral** a naturally occurring, inorganic solid with a definite chemical composition and a crystalline structure
- **mineralogy** the study of minerals

MATERIALS

- Samples of rocks and minerals (*Introduction*)
- Bible concordance (*Discussion*)
- TM 1.2.1A Rocks and Minerals
- TM 1.2.1B Rocks and Minerals Key

PREPARATION

- Label rock and mineral samples with numbers or letters and create a corresponding key for identification. (*Introduction*)
- Use a concordance to look up references to minerals in the Bible. (*Discussion*)
- Invite a jeweler to speak to the class about distinguishing real gems from synthetic ones. (*A*)
- Invite a professor to talk about the difference between rocks and minerals and how minerals help form rocks. (*B*)
- Plan a field trip to the geology section of a natural history museum. (*C*)

Introduction

Display a variety of rocks and mineral samples or **TM 1.2.1A Rocks and Minerals**. Direct students to examine the samples and write down whether they think each sample is a rock or a mineral, along with the reasons for their conclusions. Display the key or **TM 1.2.1B Rocks and Minerals Key** and have students compare it against their conclusions. As a class, discuss students' findings and the qualities they used to distinguish between rocks and minerals.

Discussion

- Guide a discussion regarding the use of minerals in the Bible. Have students use the concordance to look up various minerals in the Bible and then identify how these minerals were used in the tabernacle, temple, and the description of New Jerusalem, as well as their use to describe characteristics of people and God. Challenge students to deduce the minerals' use in each application and what they are meant to reveal about God.

- Ask the following questions:
  1. Why are some metals classified as minerals but others are not? (**Naturally occurring metals, such as gold and silver, are minerals, but manufactured alloys like bronze are not found in nature and cannot be considered minerals.**)
  2. Both salt and bone are natural solid substances found in nature. Why is one a mineral and the other is not? (**Salt meets all the criteria for a mineral because it is a naturally occurring, inorganic, crystalline solid, and has a definite chemical composition; bone is not a mineral because it is an organic substance.**)
  3. Are fossil fuels minerals? Why? (**No. They are organic substances, and oil and natural gas are not solids.**)

Activities

A. Present the invited jeweler to speak to the class about distinguishing real gems from those that are synthetic.

B. Introduce the professor to talk about the difference between rocks and minerals and how minerals help form rocks.

C. Take students on a field trip to the geology section of a natural history museum. Many museums offer lectures, videos, and displays of the location, mining, and uses of different minerals.

D. Have students research the effects that efforts to obtain minerals have had historically. Possible topics include the importance of salt in the Roman Empire and China, the European attempts to obtain gold and silver in North and South America, the creation of bronze, and the discovery and use of iron.

E. Direct students to research how fossils are formed and to use their research to explain whether they think fossils are minerals.

F. Encourage students to research and create presentations about mineral discoveries. Presentations should include how the mineral was discovered, an image of the mineral's structure, the chemical formula for the mineral, and any current uses of the mineral.

G. Challenge students to research the synthesis of rubies, emeralds, and diamonds.

H. As a class, research the major sites of mineral deposits around the world. Create a computer presentation or a bulletin board showing the continents and the locations of these mineral sites. Use different symbols and colors to represent different types of minerals.

Lesson Review

1. What is the study of minerals called? (**mineralogy**)
2. Name five minerals and their uses. (**Answers will vary.**)
3. What four basic characteristics do all minerals share? (**naturally occurring, inorganic, having a uniform crystalline solid, and having a definite chemical composition**)
4. Explain whether a rock is a mineral. Why? (**Rocks are not minerals because they do not have a uniform crystalline structure and because they often include at least microscopic traces of multiple minerals or organic substances, which prevents them from having a definite chemical composition.**)

NOTES

1.2.1 Introduction to Minerals

OBJECTIVES
- Explain the basic characteristics of a mineral.
- Distinguish between minerals and nonminerals.

VOCABULARY
- **mineral** a naturally occurring, inorganic solid with a definite chemical composition and a crystalline structure
- **mineralogy** the study of minerals

Throughout history, minerals have been among people's most prized material possessions. Hundreds of years ago North African caravans braved thousands of kilometers of harsh desert to trade salt, which at the time was nearly as valuable as gold. Europeans came to the Americas to search for gold and other precious minerals and waged war to take and keep them. For centuries, people have traveled great distances, fought battles, and taken great risks to get metals and gems.

Minerals are still valuable. Diamonds and gold are some of the most expensive items sold anywhere. Other minerals are valued for their usefulness. Iron is used in construction, aluminum is used in foil or airplane parts, salt seasons food, and uranium can help treat cancer.

A diamond, a lump of iron ore, and a grain of salt have different properties and are made of different elements. However, they have one thing in common: they are all minerals. A **mineral** is a naturally occurring, inorganic solid that has a definite chemical composition and a characteristic crystalline structure.

The study of minerals is called **mineralogy**. Around 4,000 minerals are already known, and scientists are discovering more all the time. To determine whether a substance is a mineral, scientists ask four basic questions. If the answer to all four questions is yes, then they classify the substance as a mineral.

Is the substance natural? All minerals occur naturally; they are formed by the processes of nature, not by humans. Scientists can produce some substances—such as ruby, quartz, and zirconia (which looks like a diamond)—in a laboratory. These substances look a lot like their natural counterparts and they have similar or identical chemical compositions, but they are not minerals because they were never a natural part of the earth's crust. Substances such as steel and bronze are alloys, which are combinations of two or more metals or a metal and another material. Alloys are not considered minerals because they are industrially manufactured.

Is the substance inorganic? An inorganic substance is a substance that is not made up of living things or the remains of living things. Obviously, a worm living in the soil is not a mineral, but neither is coal. Like a worm, coal is an organic substance. It is made of the remains of ancient plants and animals. A few organic substances are classified as minerals because they share all the other properties of minerals. One such substance is aragonite, which is produced by oysters to create their shells and pearls.

Is the substance a uniform crystalline solid (crystal)? All substances on Earth normally exist as solids, liquids, or gases. Minerals always exist in solid form. Although mercury is inorganic and has a uniform composition, it is not a mineral because it is liquid in its natural state. A uniform crystalline solid is a solid that cannot be physically broken down into other components and whose particles are arranged in a regular,

Diamonds, iron ore, and salt are all naturally occurring minerals.

Coal is formed by the remains of plants and animals and is not considered a mineral.

Removing a pearl from a pearl oyster

BIBLE CONNECTION

Salty Facts 
- Marco Polo reported seeing salt coins in Cathay (North China).
- *To salt away* means "to store or lay aside for later need." Before refrigeration was common, meat was salted to preserve it for later.
- Salt was once traded ounce for ounce for gold.
- Rome's major highway was the Via Salacia (salt road). Soldiers used this highway to carry salt up from the Tiber River, where barges brought salt from the salt pans of Ostia, the port city for Rome.
- Roman soldiers were once paid in salt—the word *salary* comes from the Latin word *sal*, which means "salt."
- When Jesus said, "You are the salt of the earth," (Matthew 5:13), He was not only telling people to be "flavorful" Christians and a preservative, He was also telling them that they were valuable.

Mercury is not a mineral because it is liquid in its natural state.

repeating, three-dimensional pattern. The crystalline structure is obvious in some minerals, such as quartz. In other minerals, it can be seen only with the help of a microscope. The different crystalline structures of minerals account for differences in their properties.

Does the substance have a definite chemical composition? Chemical composition is the relative abundance of the different types of atoms in a substance. All minerals have a chemical composition. Scientists use a formula, such as H_2O for water, to represent a substance's chemical composition. Minerals can be elements or compounds. Elements consist of only one kind of atom; compounds consist of multiple kinds of atoms. Gold is a mineral made of only one element; it is made of only gold atoms. The formula for gold is Au. Fluorite, another mineral, is a compound made of calcium and fluoride ions arranged in a specific pattern. The formula for fluorite is CaF_2. Rocks, like minerals, are naturally occurring, inorganic, and solid. However, they are not minerals because they do not have a uniform crystalline structure. In addition, they often include at least microscopic traces of multiple minerals or organic substances, which prevents them from having a definite chemical composition.

LESSON REVIEW
1. What is the study of minerals called?
2. Name five minerals and their uses.
3. What four basic characteristics do all minerals share?
4. Explain whether a rock is a mineral. Why?

The mineral fluorite is colorless when pure, but is colored by impurities.

Minerals

1.2.2 Mineral Groups

Introduction 👋
Display a crystal creation grown using sodium borate (borax). Inform students that the creation was made using a solution of the mineral sodium borate and water and that the creation displays the mineral's crystalline structure. Guide students in growing their own sodium borate crystals by completing the following steps:
- Use 1–2 chenille stems to create a simple design that can be suspended inside a large, wide-mouthed glass jar or beaker without touching the bottom or the sides of the container.
- Tie one end of a thin string to the chenille stem creation and the other end to a pencil or rod placed over the mouth of the container; make sure the chenille stem does not touch the bottom or sides of the container. Remove the chenille stem from the container.
- Dissolve 6 tbsp of sodium borate in 2 cups of boiling water and pour the solution into the glass jar or beaker.
- Suspend the chenille stem creation in the sodium borate solution.
- Allow the chenille stem to remain suspended in the solution until the next day.

Discussion
- Have students review the names and chemical characteristics for the major groups of minerals. Ask students what patterns they observe. (**Answers may vary but should include that all groups ending with *ate* start with an ion of oxygen plus one other constant element, which bonds to a variety of other elements; all groups ending with *ide* start with a single element that bonds to various other elements.**)

- Ask the following questions:
  1. What are the two most common elements in the earth's crust? (**silicon and oxygen**)
  2. Which mineral classification contains minerals that are each composed of a single element? (**native elements**)

Activities 👋
A. Read **Revelation 21:15–21** and have students complete **WS 1.2.2A Precious Jewels**.

B. Assign **WS 1.2.2B Periodic Table of Elements** for students to complete. Display **TM 1.2.2A Periodic Table of Elements** and have students check their work.

C. Have students complete **WS 1.2.2C Chemical Classification**.

D. Direct students to use colored clay and toothpicks to build a molecular model for each of the mineral types listed in the Student Edition. Students should use a different color of clay for each element and prepare a key showing what element each color represents. Alternatively, students may construct electron dot diagrams or drawings of the atomic structures.

E. Introduce the invited geologist to speak to the class about minerals.

F. Divide the class into groups and have each group research a different classification of minerals not described in their Student Editions. Direct each group to prepare a presentation that includes the mineral class's main element or ion; other types of elements or ions that bond with the main element or ion; the name, description, and chemical formula of a specific mineral in the class; at least one place where the specific mineral can be found or mined; and at least one way in which people can use the specific mineral. Possible mineral classes include sulfosalts, nitrates, borates, and phosphates; hydroxides may be included if they are distinguished from oxides.

G. Challenge students to research another simple method of growing crystals and to try the experiment at home. Have students write a report detailing the materials and methods used in

---

OBJECTIVES
Students will be able to
- identify the primary elements in each of several main mineral groups.
- distinguish mineral groups from each other using their chemical composition and structure.

VOCABULARY
- **nonsilicate mineral** a mineral composed of elements or bonded groups of elements other than bonded silicon and oxygen
- **silicate mineral** a mineral formed by bonded silicon and oxygen atoms

MATERIALS
- Large, wide-mouthed glass jars or beakers, chenille stems, string, pencils or rods long enough to span the containers' openings, sodium borate, boiling water (*Introduction*)
- Clay of various colors, toothpicks (*D*)
- TM 1.2.2A Periodic Table of Elements
- WS 1.2.2A Precious Jewels
- WS 1.2.2B Periodic Table of Elements
- WS 1.2.2C Chemical Classification

👋 PREPARATION
- Grow a sodium borate crystal creation using the method described in *Introduction*. (*Introduction*)
- Invite a geologist from a nearby college or university to display samples of several common minerals and explain their chemical and physical properties, where they can be found, and how they were formed. (*E*)

NOTES

the experiment. Students should also keep an observation log describing the crystals after the experiment has run for one hour, two hours, and overnight; students should note the length of time between the observation after two hours and the observation the next morning. Have students share their crystals and reports with the class.

Lesson Review

1. What is the main difference between the elemental components of feldspar and quartz? (**Feldspar's primary elements are aluminum, silicon, and oxygen; quartz is made only from silicon and oxygen.**)
2. How are oxides distinguished from other minerals containing oxygen? (**In minerals other than oxides, oxygen is bonded to one or more other elements and acts as a unit with those elements to bond to a third element. In oxides, however, the oxygen is a unit all by itself.**)
3. Why is quartz classified as a silicate instead of an oxide? (**The structure of quartz's atoms is closer to other silicates than to oxides.**)
4. How are sulfates distinguished from sulfides? (**In sulfates, sulfur and oxygen act as a unit to bond with other elements. In sulfides, sulfur acts as its own unit when bonding to other elements.**)
5. Which mineral classifications have a single element to which other elements are bonded? For each class, what is the primary element to which other elements are added? (*oxides*: **oxygen**; *sulfides*: **sulfur**; *halides*: **a halogen—fluorine, chlorine, bromine, or iodine**)
6. Which mineral classifications have a bonded core group of elements to which other elements may bond? What is the core group of elements for each class? (*silicates*: **silicon and oxygen**; *carbonates*: **carbon and oxygen**; *sulfates*: **sulfur and oxygen**)

Mineral Groups 1.2.2

God created an orderly world; people can study parts of it more efficiently by grouping similar things together. Just as plants and animals can be organized into related groups according to their common features, minerals can also be classified. Minerals are classified according to their chemical compositions and structures.

Minerals can first be divided into silicates and nonsilicates. Silicate minerals are the most plentiful of all the mineral classes. All **silicate minerals** have a silicon atom bonded with four oxygen atoms (SiO$_4$) as a foundation. These minerals make up more than 90% of the earth's crust. (Oxygen makes up 46.6% of the crust by weight, and silicon makes up 27.7%.) The basic combination of silicon and oxygen can bond with other elements (such as iron, aluminum, and potassium) to form different silicate minerals. Silicate minerals can be further subdivided into smaller groups, including feldspar, quartz, mica, and talc.

Feldspar minerals are the most common silicate mineral. They make up more than half of the crust and are a part of all types of rock. They all have a base of silicon, oxygen, and aluminum. Different feldspar minerals are distinguished by their other elements called *additives*. The most common feldspar additives include potassium (KAlSi$_3$O$_8$), sodium (NaAlSi$_3$O$_8$), and calcium (CaAl$_2$Si$_2$O$_8$). The combination of additives gives feldspar a wide range of colors. For example, the range of gray to pink grains visible in granite rock are a type of feldspar. Because feldspar minerals have such a beautiful range of colors, they are often used as decorative stone on buildings.

OBJECTIVES
- Identify the primary elements in each of several main mineral groups.
- Distinguish mineral groups from each other using their chemical composition and structure.

VOCABULARY
- **nonsilicate mineral** a mineral composed of elements or bonded groups of elements other than bonded silicon and oxygen
- **silicate mineral** a mineral formed by bonded silicon and oxygen atoms

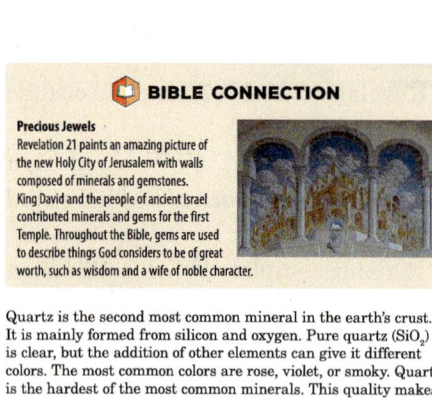

Feldspar minerals are frequently used as decorative stone on buildings because they occur in many beautiful colors.

BIBLE CONNECTION

Precious Jewels
Revelation 21 paints an amazing picture of the new Holy City of Jerusalem with walls composed of minerals and gemstones. King David and the people of ancient Israel contributed minerals and gems for the first Temple. Throughout the Bible, gems are used to describe things God considers to be of great worth, such as wisdom and a wife of noble character.

Quartz is the second most common mineral in the earth's crust. It is mainly formed from silicon and oxygen. Pure quartz (SiO$_2$) is clear, but the addition of other elements can give it different colors. The most common colors are rose, violet, or smoky. Quartz is the hardest of the most common minerals. This quality makes it ideal for sandpaper and industrial cleaners. Quartz is the most common mineral in sand; the clear grains are quartz.

Mica is a soft, often shiny mineral. These minerals form extremely thin, shiny, flexible sheets called *books*, which can easily be separated with a knife. They primarily contain silicon, oxygen, hydrogen, aluminum, and potassium along with a variety of other elements including magnesium, iron, sodium, and lithium. A common form of mica is called *muscovite*. Mica can be found in granite along with feldspar and quartz. Mica is used in insulation, wallpaper, paint, and tile.

Talc is the softest known mineral. It primarily contains silicon, oxygen, hydrogen, magnesium, and usually impurities such as

Muscovite mica

iron, nickel, or aluminum. Talcum powder is made of talc; other uses of talc include roofing materials, paint, and paper.

Less than 10% of the earth's crust is made of nonsilicate minerals. **Nonsilicate minerals** are composed of elements or bonded groups of elements other than bonded silicon and oxygen. Categories of nonsilicate minerals include carbonates, oxides, sulfates, sulfides, halides, and native elements.

Marble rock made of calcium magnesium carbonate

Carbonates are minerals that contain a carbon atom bonded to three oxygen atoms (CO_3); this combination then bonds with a metal. Carbonates such as calcite ($CaCO_3$) are the main components of limestone and marble. Other carbonates help form the shells of many shellfish such as oysters. Carbonate minerals are used in cement and building stones. Carbonates are easy to identify because they fizz and dissolve when they come in contact with acid.

Oxides are minerals in which oxygen is combined with one or two metals. The simple chemical formulas of oxides distinguish

FYI

What a Gem 
Gemstones are minerals whose beauty, durability, and rarity make them valuable. A gem's value depends on its scarcity, color, clarity, cut, and carat weight. Gemstones are usually polished and cut into different shapes so that they shine and sparkle. Translucent and opaque gemstones are usually cut into oval shapes. Transparent gemstones, such as diamonds, are faceted so they reflect light in a dazzling display of sparkles. After being cut and polished, gems are typically set in precious metals, such as gold or silver, and made into jewelry.

Jewelers identify gems according to their chemical and physical properties. Each gem is made of specific elements and has a chemical formula. However, some gems have the same chemical formulas as minerals that are not gems. A common example is that of diamond (a gem) and graphite (a mineral used in pencils). Both minerals are made of silicon and oxygen (SiO_2). However, diamonds are formed under much greater temperature and pressure than graphite, which causes them to become much harder. Jewelers, therefore, also use various physical properties to help them identify gems.

---

them from other mineral categories that include oxygen, such as silicates. In other mineral classes, oxygen bonds to one or more other elements and the bonded group acts as a unit. In oxides, however, the oxygen is a unit all by itself. For example, naturally occurring ice ($H_2 + O$) is considered an oxide; oxygen acts on its own and bonds with hydrogen. In contrast, the oxygen in sodium feldspar is part of the aluminum silicate group that combines with sodium (Na + $AlSi_3O_8$). Quartz is an exception to this principle. Although quartz has a simple formula containing only silicon and oxygen (SiO_2), it is classified as a silicate because the structure of its atoms is closer to other silicates than to oxides. Oxides are often mined for their metals, such as hematite (Fe_2O_3) for iron and cassiterite (SnO_2) for tin.

The oxide hematite (Fe_2O_3) is often mined for iron.

Sulfate minerals are primarily composed of a sulfur atom bonded with four oxygen atoms (SO_4). The bonded oxygen and sulfur then attaches to a metal element. One of the most common sulfates is gypsum. Gypsum has a wide variety of uses, including plaster of paris, fertilizer, paper, and jewelry.

Pyrite, also known as *fool's gold*, is a sulfide.

The sulfide mineral class is distinguished from other categories containing sulfur in much the same way that oxides are separated from other minerals containing oxygen. Just as oxides are distinguished by oxygen bonding directly to metals, sulfides are characterized by sulfur bonding directly to metals. Pyrite, also known as *fool's gold*, is a sulfide in which iron bonds to sulfur (Fe + S_2). Similar to oxide minerals, sulfide minerals are mined for their metals, such as iron, copper, or lead.

Halide minerals contain combinations of a halogen (fluorine, chlorine, bromine, or iodine) with one or more metals. Halite (NaCl), which is called *rock salt* when found in rock form, is often used for melting ice on roads or as table salt. Other halide uses include fertilizers and the making of steel and aluminum.

Native elements are minerals that are made of only one element. At least 19 minerals are native elements. They are divided into three groups: metals, semimetals, and nonmetals. Metals conduct electricity and heat, and they can be bent without breaking.

Commonly known metal native elements include gold, silver, platinum, iron, zinc, tin, mercury, lead, and copper. Semimetals are poor conductors of electricity and heat, and they break easily when bent. The more common semimetal native elements include arsenic and bismuth. Nonmetals do not conduct electricity or heat. Nonmetal native elements include carbon, graphite, and sulfur.

Natural copper

LESSON REVIEW
1. What is the main difference between the elemental components of feldspar and quartz?
2. How are oxides distinguished from other minerals containing oxygen?
3. Why is quartz classified as a silicate instead of an oxide?
4. How are sulfates distinguished from sulfides?
5. Which mineral classifications have a single element to which other elements are bonded? For each class, what is the primary element to which other elements are added?
6. Which mineral classifications have a bonded core group of elements to which other elements may bond? What is the core group of elements for each class?

Halite is used for melting ice on roads or as table salt.

1.2.3 General Mineral Properties

Minerals

OBJECTIVES

Students will be able to
- describe the general properties of minerals.
- classify minerals using mineral properties.

VOCABULARY

- **cleavage** a mineral's tendency to split along definite, flat surfaces
- **density** the mass per unit of volume of a substance
- **fracture** a mineral's tendency to break along irregular lines
- **hardness** a mineral's resistance to being scratched
- **luster** the way a mineral's surface reflects light
- **streak** the color of the powder left by a mineral when it is rubbed against a hard, rough surface

MATERIALS

- TM 1.2.3A Mineral Luster Chart

PREPARATION

- Obtain materials for *Lab 1.2.3A Mineral Identification*. The ACSI Science Equipment Kit contains a mineral test kit that includes mineral samples, a copper coin, and a nail; a rock collection; files; glass plates; streak plates; and graduated cylinders.
- Plan a field trip to a mineral-rich location where samples may be collected for classification. (A)

Introduction

Distribute minerals to student groups. Direct students to devise a classification method to distinguish the different minerals and identify their types. Display various minerals and have students identify them using the classification systems they have created. Lead a discussion regarding the benefits and difficulties of the students' systems.

Discussion

- Discuss which properties are most helpful in distinguishing between minerals.

- Discuss some people's infatuation with crystals and gems as healing therapy. This practice is a common New Age and Hindu tradition. Guide students to understand that although some minerals are essential for health, trust in gems and crystals is putting one's faith in created things instead of in the Creator.

- Ask the following questions:
  1. What challenges do geologists face when trying to identify a mineral by using its color? Have students provide examples to support their answers. (**Many minerals have the same color, such as hornblende, malachite, olivine, and sphalerite, which are all green; some minerals can be found in more than one color, such as corundum that can be red or blue and quartz that can be clear or purple.**)
  2. Are all cleavages or fractures the same? Have students provide examples to support their answers. (**A mineral may cleave in more than one direction, and it may cleave in various degrees of perfection. For example, halite breaks into small cubes, topaz breaks along a single plane parallel to its base, and gypsum breaks perfectly on only one plane but breaks moderately well on two other planes. Minerals can fracture instead of cleave. Fracturing can occur in various ways. For example, pyromorphite has an irregular fracture, chrysotile serpentine has a splintery or fibrous fracture, and quartz has a conchoidal fracture.**)

Activities

Lab 1.2.3A Mineral Identification

- mineral samples
- mineral test kits, 1 per group
- steel files, 1 per group
- glass plates, 1 per group
- streak plates, 1 per group
- triple beam balances, 1 per group
- graduated cylinders of water, 1 per group

If a mineral test kit is not available, obtain a copper coin and a steel nail. Demonstrate the scratch test. Place the streak plate on the tabletop and firmly hold it in place. Use a single determined motion to cut a scratch mark on the streak plate with the sample. Do not gently rub back and forth. Gently wipe away powder to see if a distinct groove or cut in the surface appears.

Display **TM 1.2.3A Mineral Luster Chart** as a reference for students.

Option: Extend the lab by having students test the crystals they grew for Lesson 1.2.2 using the same procedures to identify and describe their mineral's color, luster, streak, hardness, mass, volume, and density.

A. Take students on a field trip to a mineral-rich location where samples may be collected. Have students use the various physical properties of minerals to identify each sample and then research the chemical composition of each mineral represented in the samples.

B. Encourage students to begin or to add to an existing classroom mineral collection. Direct them to collect mineral samples from near their homes and to use rock and mineral guidebooks to identify each sample.

C. Direct students to research whether there are any patterns in minerals' physical properties that point to the presence of particular elements in the mineral, the effects of particular forces that formed the mineral, or the type of atomic bonds in the mineral.

D. Challenge students to design an experiment to determine how different types of atomic bonds affect a mineral's hardness, cleavage, or fracture.

Lesson Review

1. List and define seven physical properties of minerals. (***color***: a mineral's natural color; ***luster***: the way a mineral's surface reflects light; ***hardness***: a mineral's resistance to being scratched; ***streak***: the color of the powder left by a mineral when it is rubbed against a hard, rough surface; ***cleavage***: a mineral's tendency to split along sets of parallel, flat surfaces; ***fracture***: a mineral's tendency to break along irregular lines; ***density***: the mass per unit of volume of a substance)
2. Identify one mineral that has a waxy luster. (**Possible answers: turquoise, sphalerite**)
3. What determines a mineral's hardness? (**the strength of the bonds between the ions or atoms of a mineral and the internal arrangement of the mineral's atoms**)
4. What is the formula for density? (**D = m/V**)
5. Which mineral is often black, has a metallic luster, produces a reddish streak, has a density of approximately 4.9 g/cm³, and has a hardness of about 5.5? (**hematite**)

NOTES

1.2.3 General Mineral Properties

OBJECTIVES
- Describe the general properties of minerals.
- Classify minerals using mineral properties.

VOCABULARY
- **cleavage** a mineral's tendency to split along definite, flat surfaces
- **density** the mass per unit of volume of a substance
- **fracture** a mineral's tendency to break along irregular lines
- **hardness** a mineral's resistance to being scratched
- **luster** the way a mineral's surface reflects light
- **streak** the color of the powder left by a mineral when it is rubbed against a hard, rough surface

Some minerals have identical or nearly identical chemical formulas. Not only this, but some minerals look so much alike that just looking at them is not enough to distinguish between them. Fortunately, minerals possess general physical properties that enable people to distinguish between them. These properties are based not only on the elements that compose each mineral but also on the way that the atoms in the mineral are bonded together. These physical properties include color, luster, hardness, streak, cleavage, fracture, and density. One property is usually not enough to identify a mineral; two minerals might have the same color or hardness, for example. Identifying a mineral requires at least two physical properties.

Many minerals are colored. However, color alone cannot identify a mineral because many minerals are the same color. For example, hornblende, magnetite, and sphalerite can all be black. Even the color of one mineral can vary, depending on the additives or impurities present. Corundum, for example, is a clear mineral made of oxygen and aluminum, but the addition of tiny amounts of chromium results in a ruby, which has a deep red color. Blue sapphire is corundum that contains traces of iron and titanium that give it a blue color. Quartz is often clear, but the addition of manganese makes it a purple amethyst.

Luster is another property of minerals. Luster describes the way that a mineral's surface reflects light. Minerals that contain metals reflect a lot of light, so they are said to have a metallic (shiny) luster. Other minerals, which do not reflect as much light, have a nonmetallic luster. Nonmetallic luster has several categories, including glassy, waxy, earthy, dull, and resinous. Most quartz, for example, has a glassy luster. Turquoise has a waxy luster, as if it were covered with wax. Some minerals, such as kaolinite, have an earthy luster. Diamonds have a luster that sparkles, called *adamantine*.

Rock showing black hornblende with a dull luster

Black sphalerite with a resinous luster

Red corundum (ruby) with a glassy luster

Blue corundum (sapphire) with a glassy luster

28

FYI

Field Hardness Scale		Mohs Hardness Scale	
Hardness	Common Tests	Hardness	Mineral
1	Easily scratched with a fingernail.	1	Talc
2	Scratched with a fingernail.	2	Gypsum
3	Scratched with a penny.	3	Calcite
4	Easily scratched with a knife.	4	Fluorite
5	Scratched with a knife; barely scratches glass.	5	Apatite
6	Scratched with a steel file; easily scratches glass.	6	Feldspar
7	Scratches a steel file and glass.	7	Quartz
		8	Topaz
		9	Corundum
		10	Diamond

Hardness, a mineral's resistance to being scratched, is another way of identifying a mineral sample. Not all minerals are equally hard. A mineral's hardness depends on the strength of the bonds between the ions or atoms and the internal arrangement of the mineral's atoms. Harder minerals have stronger elemental bonds. Diamond, the strongest mineral, has carbon atoms that are each strongly bonded to four other carbon atoms. Graphite, a relatively soft mineral, has sheets of thin carbon atoms weakly bonded together. Although graphite and diamond are both composed entirely of carbon, the way the carbon atoms are arranged and bonded produces minerals of different hardness.

Turquoise has a waxy luster.

Kaolinite has an earthy luster.

Diamond (left, showing adamantine luster) and graphite (above) are both pure carbon, but diamond is significantly harder.

29

FYI

Mineral Properties Chart

Mineral	Color**	Luster	Hardness*	Streak	Density*
calcite	many colors	glassy	3.0	white	2.7
feldspar	white, pink, gray, brown	glassy	6.0	white	2.6
fluorite	many colors	glassy	4.0	white	3.2
galena	lead-gray	metallic	2.5	lead-gray	7.4–7.6
halite	colorless, white, yellow, red, blue	glassy	2.5	white	2.2
hematite	gray, black, red	metallic	5.5–6.5	red to reddish-brown	4.9–5.3
hornblende	green to black	glassy, pearly, dull	5.0–6.0	green-gray to gray	3.0–3.5
magnetite	black	metallic	5.5–6.0	black	5.2
malachite	green	glassy, silky	3.5–4.0	green	3.6–4.0
muscovite	white, gray, colorless	glassy, silky, pearly	2.0–2.5	white	2.8–3.0
olivine	olive green	glassy	6.5–7.0	white, gray	3.2–3.4
quartz	colorless, white	glassy	7.0	white	2.7
sphalerite	yellow, brown, black, red, green, white, colorless	waxy, resinous	3.0–4.0	white, light brown	4.0

*All values are g/cm³ rounded to two significant digits. **Most common colors are listed; other colors may exist.

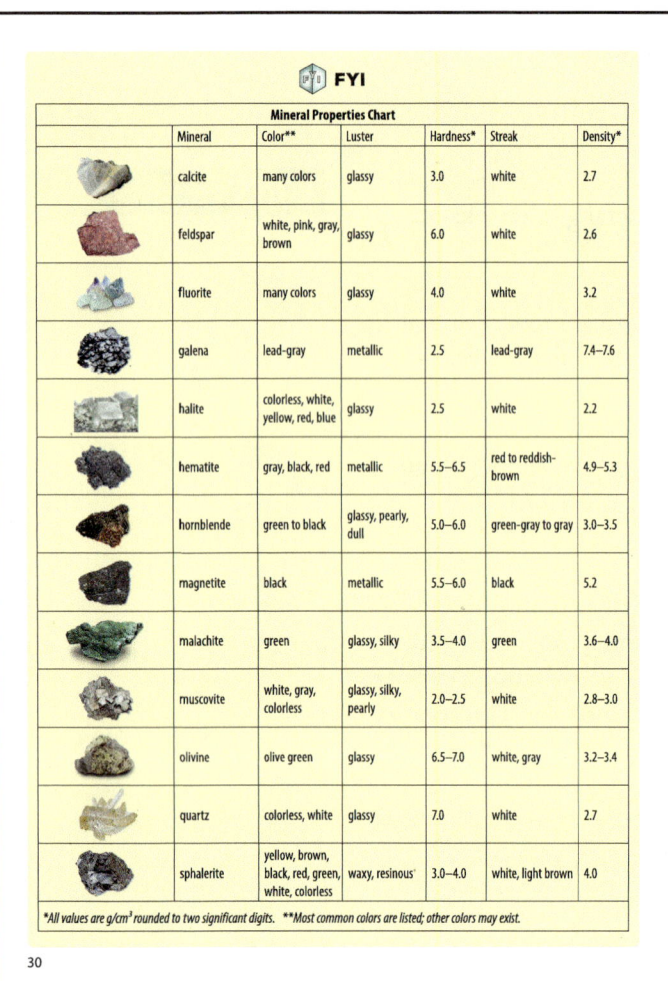

A commonly used scale of hardness is the Mohs hardness scale, which is a list of ten minerals that represent different degrees of hardness. However, a field hardness scale, which uses materials with known Mohs scale hardness ratings, can be used when the minerals from the Mohs scale are not available. To determine the hardness of an unknown mineral, scratch the mineral with each item on the field hardness scale in turn, starting with the least hard. The test stops once the mineral sample has been scratched.

The Mohs scale starts with talc (the softest known mineral) and ends with diamond (the hardest known mineral). Minerals with large numbers on the Mohs and field scales can scratch those with smaller numbers; any mineral can scratch talc. The difference in hardness between any two consecutive minerals on the Mohs scale is not the same. For example, there is only about a 25% difference in hardness between calcite and fluorite, but there is a difference of more than 300% between corundum and diamond.

A mineral with a hardness rating of 6 or less can also be identified by its streak. **Streak** is the color of the powder left by a mineral when it is rubbed against a hard, rough surface, such as a streak plate, which is a ceramic tile with a dull, unglazed surface. A streak plate has a hardness of about 7. A mineral's streak may be a different color than the mineral itself. For example, pyrite is yellow in color, but it has a green-black to brown-black streak. Hematite, which may vary in color from red to brown to black, always produces a rust-red streak. Not all minerals leave colorful streaks. Fluorite, for example, can appear in many different colors, but it always leaves a white streak, just like calcite.

Another identifying property of minerals is **cleavage**, a mineral's tendency to split along definite, flat surfaces. Cleavage occurs along surfaces parallel to the plane of the crystal where the bonds between the atoms of the crystal are weak. Different minerals have different cleavages. For example, halite breaks into small cubes. The cleavage of topaz occurs along a single plane parallel to its base. A mineral that breaks easily and cleanly in one or more directions is said to have *perfect cleavage*.

Streak plates with pyrite (left) and rhodochrosite (right) and their streak colors

Labradorite showing perfect cleavage and pearly luster

Gypsum showing perfect cleavage and silky luster

BIOGRAPHY

Friedrich Mohs

Born the son of a merchant in 1773, German scientist Friedrich Mohs is best known for his contribution to minerology. Mohs studied mathematics, physics, chemistry, and mechanics. After graduating from the University of Halle, Mohs worked in mining. In 1802, Mohs went to Vienna to classify an important private mineral collection. His system of mineral classification stressed the physical properties of minerals, which conflicted with the accepted classification by chemical composition. Mohs went on to become a professor of mineralogy and developed the Mohs scale of hardness that is still used today. He spent much of his later career in Vienna, where he organized the imperial mineral collection and oversaw mining for the government. Mohs died in 1839 during a trip to Italy to inspect volcanic activity.

For example, calcite breaks perfectly along three planes. Some minerals break perfectly in one direction yet not so well in others. Gypsum and labradorite, for example, break perfectly on only one plane, but each cleaves with lesser success along at least one other plane.

Many minerals fracture instead of cleave. Cleavage results in a smooth surface, but **fracture** is the term used to describe a mineral's tendency to break along irregular lines. Various minerals fracture in different ways. An uneven fracture is a break along a flat surface with an irregular pattern; pyromorphite has an uneven fracture. A mineral with a splintery or fibrous fracture breaks into pieces that look like wood; chrysotile serpentine fractures this way. A conchoidal or shell-like fracture of a mineral is smooth and curved, like the inside of a clam shell or broken glass; quartz has a conchoidal fracture.

Finally, density is also useful in identifying minerals. **Density** is the mass per unit of volume of a substance. A mineral's density can be determined by dividing its mass in grams by its volume in cubic centimeters. The formula is $D = m/V$. One quick way to judge mineral density is by heft, which is how heavy samples of equal sizes feel. If you pick up two minerals that are the same size and notice that one is heavier than the other, the heavier mineral is the more dense. For example, a piece of galena feels heavier than a piece of muscovite the same size, which means that the galena is more dense than the muscovite.

Pyromorphite displays an irregular fracture.

Chrysotile serpentine (top left) displays a fibrous fracture.

Rutile quartz (top right) has a conchoidal fracture.

Galena (bottom left, showing a metallic luster) has a higher density than muscovite (bottom right, showing both pearly and glassy luster).

LESSON REVIEW
1. List and define seven physical properties of minerals.
2. Identify one mineral that has a waxy luster.
3. What determines a mineral's hardness?
4. What is the formula for density?
5. Which mineral is often black, has a metallic luster, produces a reddish streak, has a density of approximately 4.9 g/cm³, and has a hardness of about 5.5?

Name: _____ Date: _____

Lab 1.2.3A Mineral Identification

QUESTION: What minerals are these?

HYPOTHESIS: Answers will vary.

EXPERIMENT:

You will need:	• steel file	• triple beam balance
• mineral samples	• glass plate	• graduated cylinder of water
• mineral test kit	• streak plate	

Steps:
1. Describe the specimen's color.
2. Refer to the luster chart and examples in the Student Edition to help you describe the specimen's luster.
3. Scrape the specimen firmly across the unglazed side of the streak plate. You may need to do this a couple times. If you cannot see anything, feel the streak plate for any residue. If there is residue, the streak is white or clear. If there is no residue, there is no streak.
4. Firmly wipe your fingers back and forth across the specimen. If it leaves a residue on your fingers, the hardness is a 1; go to Step 10. If it does not leave a residue, go to Step 5.
5. Try to scratch the specimen with your fingernail. If it leaves a scratch, the hardness is 2; go to Step 10. If there is no scratch, go to Step 6.
6. Use the copper coin to try to scratch the specimen. If it leaves a scratch, it has a hardness of 3; go to Step 10. If there is no scratch, go to Step 7.
7. Try to scratch the specimen with the steel nail. If it leaves a scratch, the hardness is 4–5; go to Step 10. If there is no scratch, go to Step 8.
8. Try to scratch the specimen with the steel file. If it leaves a scratch, the hardness is 6; go to Step 10. If there is no scratch, go to Step 9.
9. Use the specimen to try to scratch the piece of glass. If it scratches the glass, it has a hardness of 8. If it does not scratch the glass, the hardness is a 9–10.
10. Use the triple beam balance to find the specimen's mass.
11. Fill the beaker approximately half full with water and record the volume to the nearest milliliter. Carefully place the specimen in the beaker so as to avoid splashing. Record the volume to the nearest milliliter. Subtract the volume without the specimen from the volume with the specimen and record the difference on the data chart in cubic centimeters. Remember, 1 mL = 1 cm³.
12. Determine the specimen's density by dividing its mass by its volume.
13. Compare your findings to the information on the Mineral Properties Chart in the Student Edition, or to other sources found online or provided by the teacher. Identify the mineral on the data chart.

Lab 1.2.3A Mineral Identification

Mineral Data Chart

	Specimen A	Specimen B	Specimen C	Specimen D
Color				
Luster				
Streak				
Hardness				
Mass (g)		Answers will vary.		
Volume (cm³)				
Density (m/V)				
Identity				

ANALYZE AND CONCLUDE:

1. Of the minerals you tested, which had the greatest mass? Answers will vary.

2. Of the minerals you tested, which had the greatest volume? Answers will vary.

3. Of the minerals you tested, which had the greatest density? Answers will vary.

4. If volume remains constant but mass increases, what will happen to density? Density will increase.

5. If mass remains constant but volume increases, what will happen to density? Density will decrease.

6. Which mineral is the hardest? Answers will vary.

7. Which minerals' colors approximately matched their streaks? Answers will vary.

8. Why are so many tests necessary when identifying minerals? Many minerals have similar properties. The difference becomes apparent only when multiple properties are considered together.

1.2.4 Special Mineral Properties

Minerals

OBJECTIVES

Students will be able to
- describe several special mineral properties.
- assess the usefulness of special mineral properties in producing a definitive identification of a mineral sample.

VOCABULARY

- **fluorescence** the ability of a mineral to glow and change color under ultraviolet light
- **lodestone** a piece of magnetite that naturally acts as a magnet
- **phosphorescence** the ability of some fluorescent minerals to continue to glow after an ultraviolet light is no longer focused on them
- **radioactivity** the ability of an element to give off nuclear radiation as a result of a change in the atom's nucleus

MATERIALS

- Minerals that display special properties such as magnetite, calcite, graphite, talc, and sphalerite (obtain at least one sample that fluoresces), metric tape measure, iron nail, eyedropper or pipette, weak solution of hydrochloric acid, box at least 20 cm square with a peephole, black light, laser pointer (*Introduction*)
- Wintergreen-flavored hard candies (*A*)
- WS 1.2.4A Special Mineral Properties

PREPARATION

- Establish one station for each mineral listed in *Materials*. Provide each station with the materials listed. (*Introduction*)

Introduction

Explain that students will perform various tests on unidentified mineral samples to determine the samples' identity. Direct students' attention to the resources they will use to identify the minerals. Divide the class into groups so that each group has a station. Distribute sufficient copies of **WS 1.2.4A Special Mineral Properties** so each student has a copy for each mineral sample being identified.

Discussion

- Guide students to discuss the results of the activity in *Introduction*. Point out any discrepancies in results or observations and guide students to determine methods for reducing such discrepancies.

- Lead a discussion regarding which of the special mineral properties listed in the Student Edition would be the best and worst suited for helping to identify a mineral.

- Have students consider why different samples of the same mineral might refract light differently.

Activities

A. Demonstrate triboluminescence by having students crunch wintergreen-flavored hard candies in the dark. Triboluminescence is the emission of light caused by friction or breakage. Encourage students to experiment with minerals that exhibit triboluminescence, such as quartz, sphalerite, fluorite, calcite, or muscovite, and to check other minerals for this quality.

B. Assign students to choose a special mineral property and to invent a new use for it.

C. Assign students to research the effect lodestones had on open-sea navigation among the great sailing civilizations of history such as the Phoenicians, Greeks, Vikings, Portuguese, or Spanish.

D. Have students research locations rich in fluorescent minerals.

E. Challenge students to research how minerals react with other substances. Have them design and perform an experiment to confirm their findings. Students should consider both the reason why reactions occur as well as what the reactions produce, such as a gas (fizz), heat, or light.

F. Direct students to investigate why some minerals have an odor and the circumstances under which that odor is released.

G. Encourage students to research what enables a mineral to stretch and bend. Students should distinguish among flexibility, elasticity, ductility, and malleability.

H. Have students research what makes minerals fluorescent and phosphorescent. Students should distinguish between fluorescence and phosphorescence and explain why a mineral can be fluorescent but not phosphorescent.

I. Direct students to design an experiment to determine what crystal structures are best for creating a clear double refraction of an image.

J. Challenge students to research why certain minerals refract light better than other minerals. Students should consider both the chemical composition of the minerals as well as their crystalline structures.

Lesson Review

1. Why is taste not a good property to use when determining the identity of an unknown mineral? (**Some minerals are toxic if ingested.**)
2. What element causes some minerals to be magnetic? (**iron**)
3. Explain the difference between fluorescence and phosphorescence in minerals. (**Fluorescent minerals respond to ultraviolet light by visibly glowing in darkness. Phosphorescent minerals continue glowing for a short time after an ultraviolet light is no longer focused on them.**)
4. Why is fluorescence an unreliable property for determining a mineral's identity? (**Minerals of the same species that are from different places may fluoresce different colors or not at all.**)
5. What is refraction? (**the bending of light rays as they pass through two substances of different densities**)
6. What causes nuclear radiation? (**Nuclear radiation occurs when an atom's nucleus decays, emitting energy in the form of alpha or beta particles, gamma rays, or X-rays.**)

NOTES

1.2.4 Special Mineral Properties

OBJECTIVES
- Describe several special mineral properties.
- Assess the usefulness of special mineral properties in producing a definitive identification of a mineral sample.

VOCABULARY
- **fluorescence** the ability of a mineral to glow and change color under ultraviolet light
- **lodestone** a piece of magnetite that naturally acts as a magnet
- **phosphorescence** the ability of some fluorescent minerals to continue to glow after an ultraviolet light is no longer focused on them
- **radioactivity** the ability of an element to give off nuclear radiation as a result of a change in the atom's nucleus

All minerals have color, luster, streak, hardness, cleavage or fracture, and density. Some minerals have other special identifying properties as well.

Taste is one such property. Only minerals that dissolve in water have taste. Some taste salty or bitter; others do not have a taste that is easily described. Halite, common table salt (NaCl), and sylvite (KCl) are alike in most of their physical properties, but they have distinctly different tastes. Some minerals are toxic, so it is unsafe to routinely taste minerals.

A few minerals have a distinctive smell. Usually the odor is strongest in a mineral that has been struck or recently dug up. Sulfides such as pyrite have a rotten egg smell. Arsenopyrite smells like garlic. Clay minerals have an "earthy" smell.

Some minerals react with other substances. For example, calcite will fizz if weak hydrochloric acid is dropped on it. This test helps identify limestone and marble rock because they are composed mainly of calcite.

Flexibility is a property of some minerals. The chlorites, for example, are so flexible that thin layers can be bent without breaking. Biotite, a type of mica, is both flexible and elastic.

Some minerals contain enough iron to make them magnetic. *Ferromagnetism* describes a mineral's strong attraction to magnetic fields. In contrast, *paramagnetism* describes a mineral's weak, indistinct, attraction to magnetic fields. *Diamagnetism* is the property of being repelled from magnetic fields; bismuth, a metal, is an example of a diamagnetic mineral.

Magnetite is a naturally occurring mineral that is naturally magnetic. A **lodestone** is a piece of magnetite that is strongly

Halite (left) and sylvite (middle) are minerals that can be safely identified through taste. Sulfur (right) has a distinctive smell.

Biotite mica (left) has thin, flexible layers. Lodestone (middle) and bismuth (right) both have a magnetic field.

magnetic. Early sailors created compasses using lodestones, or needles magnetized by lodestones, to help them navigate. When allowed to turn freely, one end of the lodestone or magnetized needle always points to the magnetic North Pole.

Fluorescence is the ability of a mineral to glow and change color under ultraviolet light. Fluorescent minerals contain particles known as *activators*, which respond to ultraviolet light by visibly glowing in darkness. Fluorescence is an unreliable method for mineral identification because two different samples of the same type of mineral from the same general area may fluoresce in different colors or not at all. Once a mineral has been identified, however, its fluorescent color can sometimes be used to identify its place of origin. **Phosphorescence** is the ability of a few fluorescent minerals to continue glowing for a short time after an ultraviolet light is no longer focused on them. For example, calcite is sometimes phosphorescent. Like fluorescence, phosphorescence is rarely used to identify a mineral.

Some minerals can be identified by the way light interacts with them. Refraction is the bending of a light wave caused by a change in the wave's speed as it passes from one medium to another. Light rays refract, or bend, when they pass through two substances of different densities, such as moving through air and a transparent crystal. The ability of a mineral to produce two images is called *double refraction*. Double refraction occurs when a transparent substance divides light rays into two parts, creating a

Wernerite in daylight

The same wernerite sample under ultraviolet light

double image of any object viewed through it. Calcite crystals have this property.

Another special property of some minerals is texture. Some metallic minerals have a jagged texture. Other minerals, such as talc and graphite, have a greasy or oily feel. Copper feels rough. Because diamonds absorb heat so well, they feel cold at room temperature; similarly, graphite, which is a good conductor of both electricity and heat, also feels cold at room temperature. Fibrous minerals, such as chrysotile (a type of asbestos), have a distinct, silky feel. Texture alone is not a reliable characteristic for distinguishing minerals because different people may use different descriptions to explain how something feels.

A few minerals are radioactive. The ability of an element to give off nuclear radiation as a result of a change in the atom's nucleus is called **radioactivity**. Nuclear radiation occurs when an atom's nucleus decays, emitting energy in the form of alpha or beta particles, gamma rays, or X-rays. A radioactive mineral is unstable; its elements continually break down, very slowly destroying the mineral's crystal structure. Radioactive minerals can be detected with an instrument called *a Geiger counter*. Radioactive elements include uranium and radium. The mineral uraninite is the chief source of uranium.

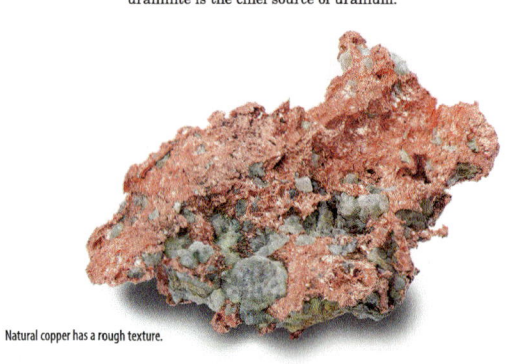

Iceland Spar calcite

Natural copper has a rough texture.

CAREER

Gemologist 
Gemology is the science of gems, both natural and artificial gemstone materials. It is a rewarding and challenging career with many job opportunities. People who choose gemology as a career use their science and math skills to work in the field or in a lab. Gemologists may do a variety of things, such as traveling to find gems, identifying gems' physical properties, giving gems a grade value to indicate their worth, and turning gems into beautiful jewelry or art. In order to become a certified gemologist, people must take classes to learn how gems are created, how to treat and care for gems, how to tell if gems are natural or synthetic, and how to collect or sell gems. There is so much to know about gems that some gemologists specialize in one type of gem! Many gemologists join organizations, such as the International Gem Society, to share their knowledge and learn from others in their field. Gemology would be an interesting career to consider.

When you work at a computer, you are benefiting from silicon, a mineral used in computer chips.

LESSON REVIEW
1. Why is taste not a good property to use when determining the identity of an unknown mineral?
2. What element causes some minerals to be magnetic?
3. Explain the difference between fluorescence and phosphorescence in minerals.
4. Why is fluorescence an unreliable property for determining a mineral's identity?
5. What is refraction?
6. What causes nuclear radiation?

Zircon gives off nuclear radiation.

1.3.0 Rocks

Chapter 1.3 Summary
Rocks have always played an important role in society, from being utilized as building materials to the use of fossil fuels to weapons. They have a wide diversity of shapes, textures, and colors. Rocks are classified according to how they were formed. There are three main types of rock: igneous, sedimentary, and metamorphic. Igneous rock is created from cooled and hardened magma. Sedimentary rock is formed from the compression and cementation of mineral particles, rock fragments, shells, leaves, bones, and other remains of once-living things. Metamorphic rock forms when the structure and constitution of existing rocks change because of heat, pressure, or chemical reactions. Each of these three types of rock is interrelated according to the rock cycle. They are continually changing and transforming into other rock types.

Background

Lesson 1.3.1 – Igneous Rock
Rock inside of the earth is slightly flexible because of the extreme heat. Flexible rocks that reach the earth's surface melt because of the lower pressure. The melted rock is called *magma*. Igneous rock forms from cooled and hardened magma.

Igneous rock is classified by mineral content, texture, and location of the cooling process. The classification of an igneous rock as determined by mineral content is divided into two categories: felsic and mafic. A felsic igneous rock is typically light-colored, lightweight, and has a high concentration of silicon (Si), aluminum (Al), sodium (Na), and potassium (K). A mafic igneous rock is typically dark-colored, heavy, and has a high concentration of iron (Fe), magnesium (Mg), and calcium (Ca).

There are four basic igneous rock textures: glassy, fine-grained, coarse-grained, and porphyritic. Each of these textures varies in crystal size, shape, arrangement, and distribution of minerals. Scientists often have to utilize high-powered microscopes to observe crystals or even the lack of crystals.

If magma cools beneath the earth's surface, an intrusive igneous rock forms. If magma cools on the earth's surface (lava), an extrusive igneous rock forms. Intrusive igneous rock can form large, mineral crystals with coarse grains or fine-grained mineral crystals depending on how quickly they cool.

Lesson 1.3.2 – Sedimentary Rock
The most common type of rock on the earth's surface is sedimentary rock. Sedimentary rock forms in two steps: compaction and cementation, a process called *lithification*. As sedimentary rock builds up, layers are created called *strata*. These layers often contain fossils and give scientists clues as to how the rocks were formed.

Sedimentary rock is classified according to how it is formed. The three main groups of sedimentary rock are clastic, chemical, and carbonate (organic). Clastic sedimentary rock is made of separate rock particles and fragments that were eroded from an older rock. This group of rock is further classified by the sediment size from which it forms. Chemical sedimentary rock is composed of minerals that were once dissolved in water. Carbonate sedimentary rock is composed of organic materials from decaying organisms. There are also biochemical rocks that are created by a combination of chemical and organic processes.

Scientists study sedimentary rock formation in an attempt to determine historic events (for example, the Flood) and to determine the evolution of mountain systems, continental blocks, and ocean basins. Sedimentary rock formation has also been used by some scientists to explain the progression of organism complexity according to the theory of evolution in the plant and animal kingdoms.

LOOKING AHEAD

- For **Lesson 1.3.2** *Try This: Shifting Sediments*, prepare a sample two or three days before this lesson. If possible, plan a field trip to an active sand and gravel mine, caves that contain stalactites and stalagmites, or an offshore oil platform. Also, invite a sedimentologist to come in for a class discussion.
- For **Lesson 1.3.3**, schedule a guided tour of a location with visible foliation or an art museum with marble artwork.
- For **Lesson 1.3.4**, arrange for a tour of the local area to view buildings made of stone or that have stone facings.

SUPPLEMENTAL MATERIALS

- TM 1.2.2A Periodic Table of Elements
- TM 1.3.1A Extrusive Igneous Rock
- TM 1.3.1B Intrusive Igneous Rock
- TM 1.3.2A Submarine Sandwich
- TM 1.3.2B Clastic Sedimentary Rock
- TM 1.3.2C Chemical Sedimentary Rock
- TM 1.3.2D Fossils
- TM 1.3.4A The Rock Cycle

- Lab 1.3.1A Formation of Igneous Rock
- Lab 1.3.1B Crystal Size of Igneous Rock
- Lab 1.3.2A Take a Closer Look
- Lab 1.3.2B Formation of Sandstone
- Lab 1.3.3A Gneiss Foliation
- Lab 1.3.4A How Rocks Change

- WS 1.3.1A Igneous Rock Scavenger Hunt
- WS 1.3.1B Identification of Igneous Rock
- WS 1.3.2A Sedimentary Rock Scavenger Hunt
- WS 1.3.3A Metamorphic Rock Classification
- WS 1.3.3B Metamorphic Rock Scavenger Hunt
- WS 1.3.4A The Rock Cycle

- Chapter 1.3 Test

BLMs, TMs, and tests are available to download. See Understanding Purposeful Design Earth and Space Science at the front of this book for the web address.

WORLDVIEW

- God created a world that is constantly changing and being renewed. He designed systems or cycles to ensure that all natural processes work together properly. Two examples of systems designed by God are the water cycle, which allows water to continuously circulate in the earth-atmosphere system and the carbon cycle, which ensures carbon is circulated in various forms throughout nature. These two examples are essential for sustaining life. Another important system designed by God is the rock cycle. Igneous rock is continuously weathering and eroding to form sedimentary rock. Metamorphic rock melts to form magma that may cool and harden to become igneous rock. The rock cycle can be compared to the changes that individuals undergo the moment they accept Christ as their personal Savior. We are reminded that "if anyone is in Christ, the new creation has come: the old has gone, the new is here!" (2 Corinthians 5:17). The individual's speech, actions, perspectives, and treatment of others has been changed or metamorphosed from the old self to the new self with the goal to become more like Christ. The water and carbon cycles may be necessary for the sustainment of life on Earth, but our faith in Christ is essential for everlasting life. For some, the changes may not always be easy or comfortable. However, as written in Psalm 19:14, our faith is put on the promise that Jesus is our Rock and our Redeemer. He will sustain His followers and equip them with the resources and talents that are needed to prosper and flourish.

The economic significance of sedimentary rock is crucial. God has entrusted people with the earth and all of its resources including oil, natural gas, coal, phosphates, salt deposits, groundwater, and other natural resources. As good stewards, individuals must protect these gifts and find ways to invest in these resources wisely.

Lesson 1.3.3 – Metamorphic Rock

Metamorphic rock is formed through a process called *metamorphism*. During the process, the texture, mineral composition, or chemical composition of existing rocks is changed because of exposure to elevated temperature and pressure. A completely new rock is formed, but it still remains solid. The metamorphic rock is generally coarser, denser, and less porous than the parent rock.

There are three main types of metamorphism: dynamic, contact, and regional. Dynamic metamorphism is produced by mechanical forces with direct pressure as the primary cause. When the heat of magma comes in contact with existing rocks, contact metamorphism occurs. Regional metamorphism occurs when large regions of the earth's crust are affected by high temperatures and high pressures. However, there are a few less common types of metamorphism. Hydrothermal metamorphism is defined as changes that occur in rocks near the earth's surface where there is intense hot water activity. Examples of locations where hydrothermal metamorphism occurs are Yellowstone National Park, the Salton Sea in California, and Wairakei in New Zealand. Other types of metamorphism include cataclistic, burial, and shock metamorphism. By studying how a rock metamorphoses, scientists are able to interpret the conditions inside the earth's crust.

New metamorphic rock will develop into one of two basic structures: foliated or nonfoliated. Metamorphic rock with a foliated structure has visible layers or bands aligned in planes. In contrast, metamorphic rock with a nonfoliated structure has no visible layers or bands. Foliation is one of the single most identifying characteristics of metamorphic rock.

Metamorphic rock is very versatile. It uses include building materials, cosmetic products, and fireproofing and insulating materials.

Lesson 1.3.4 – The Rock Cycle

There are many examples of systems that operate in cycles throughout creation. Rocks are no exception. James Hutton (1726–1797), the founder of modern geoscience, authored the concept of the rock cycle, which relates igneous, sedimentary, and metamorphic rock. Hutton noticed that some rocks have straight layers, and others have layers that are tilted. He observed that some rocks are weathered and others are not. These observations, combined with Hutton's research, led to the conclusion that rocks undergo constant change over time. The earth serves as a rock recycler. In accordance with the law of conservation of mass, rock is neither created nor destroyed but is redistributed and transformed into different kinds of rock.

Petrology, the study of rocks and their origins, has pinpointed the interrelationships of the rock cycle. Igneous rocks are formed when molten rock material solidifies either at the earth's surface or below it. The minerals of igneous rocks are destabilized by uplift and exposure of rocks at the earth's surface. The minerals break down into smaller grains that are blown or carried as sediments. Compaction and cementation lithify the sediments into sedimentary rocks. Physical changes, such as pressure and temperature, and chemical changes transform igneous and sedimentary rocks into metamorphic rocks. Rock from any of the three categories may melt and crystallize into igneous rocks, although some specific types of rock will not melt. It is important to note that rocks do not always complete the entire rock cycle, nor do they always complete the cycle in a uniform order. For example, during the metamorphosis of granite to gneiss, the igneous rock changes directly to a metamorphic rock. It skips the sedimentary stage completely.

Rocks

1.3.1 Igneous Rock

Introduction

Share with students that the properties of rock depend largely on how it is formed. Some rock is made of small mineral grains that lock together, like pieces of a puzzle. Others are made of grains of sand tightly held together or solidified lava that once flowed from a volcano. By examining rocks closely, it is sometimes possible to tell what they are made of. Present samples of the three main types of rock (igneous, sedimentary, and metamorphic). Ask students to point out how they are different and how they are alike.

Select an igneous rock and place it next to an unlit candle. Light the candle and have students observe what happens to the wax as it melts close to the burning flame and starts to solidify as it travels down the candlestick. Relate this process to the formation of igneous rock. Introduce the fact that the Latin word for igneous is *ignis* which means "fire." Igneous rock is formed when molten magma cools and solidifies. (Note: A cross section of a volcano could also be used at this time to indicate the location of the magma inside and lava located outside the volcano.)

Discussion

- Discuss the formation of new land or islands from current volcanoes, both above and below ground. One example would be Hawaii. Guide students to consider how much of the earth was created by this method. Make sure students understand that this newly created land may not be viable yet, but it will be some day.

- Display **TM 1.2.2A Periodic Table of Elements**. Discuss how the names *mafic* and *felsic* were created. The term *mafic* comes from *Ma* (magnesium) and *Fe* (iron). Mafic rock, such as basalt, has higher amounts of iron and magnesium than felsic rock. The term *felsic* comes from *feldspar*, which contains K (potassium), Na (sodium), Ca (calcium), and *silica*. Felsic rock includes granite.

- Discuss the formation of coarse-grained and fine-grained rocks. Pass around examples of each (coarse-grained rocks include gabbro, syenite, diorite, peridotite, pyroxenite, and granite; fine-grained rocks include: basalt, trachyte, dacite, and andesite) and discuss regions where these two types of rocks can be found.

- Explain that it may be difficult sometimes for scientists to classify rocks from their composition, texture, or color. For example, monzonite looks like granite and has a similar composition, but it can only be definitively identified in a laboratory.

- Discuss the formation of intrusive and extrusive rock. If possible, have students examine samples of each or display **TM 1.3.1A Extrusive Igneous Rock** and **TM 1.3.1B Intrusive Igneous Rock**. Extrusive igneous rock includes volcanic tuff and volcanic breccia; intrusive igneous rock includes obsidian, basalt, and felsites. Point out that many extrusive rocks have intrusive components.

- Have students start to brainstorm about the relationships between different types of igneous rock. For example, basalt is mafic, fine-grained, and intrusive. Discuss the relationships students find.

Activities

Lab 1.3.1A Formation of Igneous Rock

- candy molds, 1 tray per group
- nonstick cooking spray
- hot plates, 1 per group
- pure maple syrup
- pans, 1 per group
- candy thermometers, 1 per group
- hand lenses, 1 per group

OBJECTIVES

Students will be able to
- define and identify igneous rock.
- describe how the cooling rate of magma affects the texture of igneous rock.
- classify igneous rock according to mineral composition and texture.
- simulate the formation of igneous rock in order to understand how formation affects the crystal size.

VOCABULARY

- **extrusive rock** an igneous rock formed when lava cools on the earth's surface
- **felsic rock** a light-colored, lightweight igneous rock that is rich in silicon, aluminum, sodium, and potassium
- **igneous rock** a rock formed from cooled and hardened magma
- **intrusive rock** an igneous rock formed when magma cools beneath the earth's surface
- **lava** the magma that has reached the earth's surface
- **mafic rock** the dark-colored, heavy igneous rock that is rich in iron, magnesium, and calcium
- **magma** the melted rock beneath the earth's surface

MATERIALS

- Samples of igneous, metamorphic, and sedimentary rock; taper candle; match (*Introduction*)
- Igneous rock samples, several per group (*Discussion, B*)
- Hand lenses, rocks and minerals field guides (*A, B*)
- Jar of water, pumice, igneous rocks (*C*)
- TM 1.2.2A Periodic Table of Elements
- TM 1.3.1A Extrusive Igneous Rock
- TM 1.3.1B Intrusive Igneous Rock
- WS 1.3.1A Igneous Rock Scavenger Hunt
- WS 1.3.1B Identification of Igneous Rock

© Earth and Space Science

PREPARATION

- Obtain materials for *Lab 1.3.1A Formation of Igneous Rock* and *Lab 1.3.1B Crystal Size of Igneous Rock*.
- Have students collect igneous rocks for class. (A)

Lab 1.3.1B Crystal Size of Igneous Rock

- microscope slides, 2 per group
- eyedroppers or pipettes, 1 per group
- concentrated iodine solution
- microscopes, 1 per group
- stopwatches, 1 per group
- matches, 1 per group

A. Conduct an igneous rock scavenger hunt. Have students bring in igneous rock samples they find near their home or school to complete **WS 1.3.1A Igneous Rock Scavenger Hunt**. Distribute hand lenses and field guides. If students bring in rocks that are not igneous rock, have them save them for similar activities on metamorphic and sedimentary rock in Lessons 1.3.2 and 1.3.3. Challenge students to determine the conditions under which the igneous rocks were formed, depending on their observations.

B. Divide the class into groups. Give each group several igneous rock samples, hand lenses, and a rocks and minerals field guide. Provide students with a copy of **WS 1.3.1B Identification of Igneous Rock** to complete.

C. Have a student volunteer slowly place several igneous rocks, one at a time, into a jar of water. Include pumice. Have students speculate about why pumice floats.

1.3.1 Igneous Rock

OBJECTIVES
- Define and identify igneous rock.
- Describe how the cooling rate of magma affects the texture of igneous rock.
- Classify igneous rock according to mineral composition and texture.
- Simulate the formation of igneous rock in order to understand how formation affects the crystal size.

VOCABULARY
- **extrusive rock** an igneous rock formed when lava cools on the earth's surface
- **felsic rock** a light-colored, lightweight igneous rock that is rich in silicon, aluminum, sodium, and potassium
- **igneous rock** a rock formed from cooled and hardened magma
- **intrusive rock** an igneous rock formed when magma cools beneath the earth's surface
- **lava** the magma that has reached the earth's surface
- **mafic rock** the dark-colored, heavy igneous rock that is rich in iron, magnesium, and calcium
- **magma** the melted rock beneath the earth's surface

Rocks have always played an important role in society. Throughout the Bible, there are countless examples of God's people harvesting and fashioning rocks to create useful items and structures that would better serve their needs. Some examples include the creation of great cities using the rocks that were quarried and the use of rocks such as flint to fashion weapons and tools. Today, individuals still build cities from rocks—usually in combination with cement, which is made from a type of rock known as *limestone*. A rock is a hard substance composed of one or more minerals. Many of the substances used in the modern world are found in the earth's crust. Fossil fuels derived from coal provide energy. Ores provide useful metals. The silicon used to make computer chips is taken from rocks.

Not only do rocks play an important role in society, they also show God's creativity and love of variety. He fashioned rocks to have a wide array of shapes, textures, and colors. Some rock was formed when magma from deep inside the earth cooled and hardened. Other rock formed gradually over many years as grains of sand and various sediments were compressed together and eventually turned to stone. Still other rock was formed when extreme heat and pressure deep inside the earth caused some rocks to change into completely different rocks.

Rocks are all around even though they may not be obvious at first glance. As Moses mentioned in Deuteronomy 33:19, there are "treasures hidden in the sand." Rock is classified in an orderly way according to how it was formed. The three main types of rock are igneous, sedimentary, and metamorphic. Look around at the many products and materials at home and school. How many of these things can be traced to substances obtained from rock? Imagine trying to melt a rock. How much heat would

Cliffs on Svalbard, a Norwegian archipelago

BIOGRAPHY

Florence Bascom
Florence Bascom (1862–1945) was an American educator and geological survey scientist. She received the first PhD awarded to a woman at Johns Hopkins University, Baltimore, and she was the first woman hired by the United States Geological Survey. Florence founded the department of geology at Bryn Mawr College and is best known for her innovative use of petrography, which is the description and systematic classification of rocks.

be required to do that, and what kind of result would you expect? The inside of the earth is so hot that some of the rocks found there are slightly flexible. Below the earth's crust, pockets of flexible rocks rise toward areas of lower pressure that are close to the surface. The reduction of pressure allows the rocks to melt. These rocks are referred to as *magma*. **Magma** is melted rock beneath the earth's surface. Magma, which is very hot, is composed not only of melted rock but also water vapor, carbon dioxide, and rock crystals. **Lava** is magma that has reached the earth's surface. **Igneous rock** are rocks formed from cooled and hardened magma. The name *igneous* comes from the Latin word *ignis*, which means "fire."

One way igneous rocks are classified is by their mineral content. **Felsic rock** is light-colored, lightweight igneous rock that is rich in silicon, aluminum, sodium, and potassium. These rocks form from minerals that are acidic and have low melting points (600°C–750°C). Rhyolite and granite are examples of felsic rock. **Mafic rock** is dark-colored, heavy igneous rock that is rich in iron, magnesium, and calcium. These rocks form from minerals that are alkaline (basic) and have higher melting points (1,000°C–1,200°C). Gabbro and basalt are examples of mafic rock. Some rocks are classified somewhere in between felsic and mafic rock depending on their composition. These rocks are called *intermediate rock*.

Although igneous rock forms from magma, it does not all have the same texture. Therefore, igneous rock is also classified by texture—the size, shape, arrangement, and distribution of the minerals that make up the rock. Igneous rock has four basic textures: fine-grained, coarse-grained, glassy, or porphyritic. Fine-grained igneous rock, such as basalt, has interlocking mineral crystals that can be seen only under a microscope. Coarse-grained igneous rock, such as granite, has interlocking mineral crystals of roughly the same size that can be seen

Lava cooling to form igneous rock

Rhyolite is an example of a felsic rock.

D. Have students research where different types of igneous rocks are found in the world. For example, basalt is found on the ocean floor. Challenge students to determine the most prevalent type of igneous rock in the region.

E. Guide students in creating graphic organizers to help categorize the information studies in this lesson.

Lesson Review
1. What is igneous rock? (**Igneous rock is rock formed from cooled and hardened magma.**)
2. How does the cooling rate of magma affect the texture of igneous rock? (**The texture of igneous rock is determined by how slowly or quickly the magma forming the rock cools. Magma that cools quickly produces fine-grained crystals, and magma that cools slowly produces coarse-grained crystals. The crystal size and type determines the texture of the igneous rock.**)
3. What are the mineral contents and textures used to classify igneous rock? (*mineral content*: **felsic, mafic;** *textures*: **fine-grained, coarse-grained, glassy, porphyritic**)
4. What is the difference between an intrusive igneous rock and an extrusive igneous rock? (**Intrusive igneous rock forms when magma beneath the earth's surface cools. Extrusive igneous rock forms when lava cools on the earth's surface.**)
5. Compare the formation of a large-grained igneous rock with that of a fine-grained igneous rock. (**The large-grained igneous rock probably cooled slowly underground. The fine-grained igneous rock cooled more quickly, probably aboveground.**)

NOTES

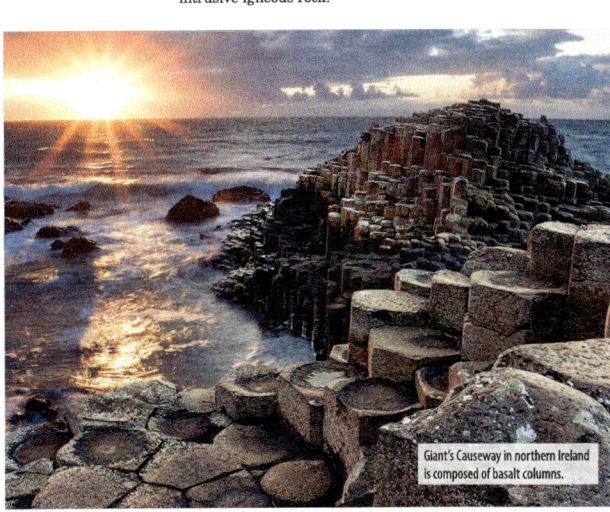

Lab 1.3.1A Formation of Igneous Rock

QUESTION: How do the crystals in igneous rock form on the basis of their cooling pattern?

HYPOTHESIS: Answers will vary.

EXPERIMENT:

You will need:	• hot plate	• thermometer
• candy molds	• pure maple syrup	• hand lens
• nonstick cooking spray	• pan	

Steps:
1. Spray the candy molds with nonstick cooking spray.
2. Using the hot plate, heat pure maple syrup in a pan to the "hard-crack" stage (about 150°C). Use the thermometer to find the correct temperature.
3. Before the syrup starts to crystallize, quickly pour some syrup into a few molds.
4. Do not return pan to the hot plate. When the remaining syrup begins to crystallize, quickly pour some of it into several other molds.
5. Return syrup to hot plate and heat for a moment longer. Quickly pour the rest of the syrup into the remaining molds.
6. Allow the candy to cool.

ANALYZE AND CONCLUDE:

1. With the hand lens, examine the candy that you poured before the syrup began to crystallize. Record your observations. This candy is clear, brittle, hard, smooth, and shiny.

2. Examine the candy that you poured as the syrup began to crystallize. Record your observations. This candy has large, even crystals.

3. Examine the candy that was reheated and poured after the syrup began to crystallize. Record your observations. This candy has large, coarse crystals.

4. Explain how the formation of crystals in maple syrup compares to the formation of crystals in igneous rock. Include the duration of the cooling process in your discussion. If magma cools quickly, fine-grained crystals are formed. As with maple syrup, if magma cools slowly, large coarse-grained crystals are formed.

Lab 1.3.1B Crystal Size of Igneous Rock

QUESTION: What size crystals form in igneous rock when magma cools slowly? Quickly?

HYPOTHESIS: Answers will vary.

EXPERIMENT:

You will need:	• concentrated iodine solution	• match
• 2 microscope slides	• microscope	
• eyedropper or pipette	• clock or stopwatch	

Steps:
1. Place one drop of iodine on each microscope slide.
2. Using a stopwatch or clock, record the time it takes for iodine crystals to form at room temperature.

   Crystal formation time at room temperature: _____

3. Place the slide under the microscope and sketch your observations as the crystals form.

   Drawings will vary.

4. Use the match to gently heat the underside of the second microscope slide. The heat will cause the iodine crystals to form rapidly. Record the time it takes for the iodine crystals to form with heat applied.

   Crystal formation time with added heat: _____

5. Place the slide under the microscope and sketch your observations of the crystals that have formed.

   Drawings will vary.

Lab 1.3.1B Crystal Size of Igneous Rock

ANALYZE AND CONCLUDE:

1. What represents magma in this laboratory experiment? The Iodine solution represents magma in this laboratory experiment.

2. Which microscope slide best represents extrusive rock formation? Slide 1, which was at room temperature, represents extrusive rock formation.

3. Which microscope slide best represents intrusive rock formation? Slide 2, which had additional heat, represents intrusive rock formation.

4. Why do you think that magma inside the earth's crust cools slower than magma on the earth's surface? Magma inside the earth's crust cools slower than magma on the earth's surface because it is closer to the very hot core. There is additional pressure, which limits the cooling process.

Rocks

1.3.2 Sedimentary Rock

Introduction
Present students with a large, layered submarine sandwich (to be enjoyed after class, if desired) or display **TM 1.3.2A Submarine Sandwich**. Explain to students that there are many layers, with varying thicknesses, that make up this sandwich (examples: meat layer, tomato layer, lettuce layer). Compare the sandwich to a sedimentary rock, which is also made up of many layers of varying thicknesses. Display a sample of sedimentary rock. Share with students that although sedimentary rock accounts for only 5% of all rocks in the earth's crust, it covers 75% of the land surface in stacked layers that average 2.26 km thick, with a range of 0.0–12.9 km.

Have students begin to consider which layers were created first and what happens to these layers as more and more layers begin to settle above over time. Challenge students to further consider whether they think layers can shift, which would cause the lower layers to transpose and be on top of the original upper layers. If they think that could happen, what would cause such a layer shift to occur? (**Answers will vary.**) What kind of problems would scientists face if this occurred? (**Answers will vary but should include that it would no longer be obvious which layers were the oldest, so dating such rock would be difficult.**)

Complete *Try This: Shifting Sediments* with students to visualize the process of stratification. Display a sample made two or three days ago and discuss results. Large pebbles will settle to the bottom first, followed by the gravel, and in descending size order, the sand, the mud, and clay. Explain that this process is the first step in lithification, called *compaction*. If time permits, have students measure the layer of one size of sediment every 10 seconds and record their findings in a data table they create. Once the layers have settled, have students measure the layer of each size group. From these two sets of data, have students create two graphs. One graph will show the thickness of a layer changing with time and one graph will show the different layer depths for size groups. Explain to students that results can vary depending on several factors including the amount of time the jar contents were shaken, the materials used, the water, the waiting time between measurements, and the measuring of the layers (human error). Also, explain that in nature, the results could vary depending on erosion, weather, and other factors.

Discussion
- Display **TM 1.3.2B Clastic Sedimentary Rock** and **TM 1.3.2C Chemical Sedimentary Rock**. Discuss how clastic, chemical, and carbonate (organic) sedimentary rock are commonly formed.

- Have students in small groups determine why the locations of mouths of rivers and along beaches may explain why conglomerate rocks have rounded edges.

- Discuss with students that chemical reactions occurring in seawater may result in the precipitation of minute mineral crystals that settle to the ocean floor and form chemical sediment. Evaporation in shallow ocean basins, for example, can produce a sequence of evaporate sediments, including gypsum and rock salt.

- Discuss the formation of fossils in sedimentary rock. Display **TM 1.3.2D Fossils** or a set of fossils for students to observe, preferably an animal and a plant remnant and an impression.
  1. For fossils to form, sediments, such as mud or sand, must immediately cover the deceased animal or plant to prevent other animals from ingesting the remains or aerobic bacteria from decomposing the organism.
  2. Beneath the sediments, anaerobic bacteria slowly break down the soft tissues, leaving only the harder parts of the organism, such as bones, teeth, claws, shells, or scales.
  3. Plants have strong cells made of cellulose, which enable them to be fossilized in their entirety.

OBJECTIVES
Students will be able to
- explain the process of compaction and cementation.
- describe how the three types of sedimentary rocks are formed.
- list and describe the major types of clastic sedimentary rock.
- identify sedimentary rock and simulate its formation.

VOCABULARY
- **clastic rock** a sedimentary rock made of rock particles and fragments deposited by water, wind, or ice
- **lithification** the process that transforms layers of rock fragments into sedimentary rock
- **sedimentary rock** a rock formed from sediments that have been compacted and cemented together
- **sediments** particles of minerals, rock fragments, shells, leaves, and the remains of once-living things
- **shale** a clastic rock composed of silt- and clay-sized particles in flat layers

MATERIALS
- Submarine sandwich, sedimentary rock sample (*Introduction*)
- Fossil set (*Discussion*)
- Hand lenses, rocks and minerals field guides (*A*)
- Sedimentary rock samples, including sandstone (*B*)
- Sedimentary rock samples, including some found in water and some found on land; grit; rock tumbler (*C*)
- TM 1.3.2A Submarine Sandwich
- TM 1.3.2B Clastic Sedimentary Rock
- TM 1.3.2C Chemical Sedimentary Rock
- TM 1.3.2D Fossils
- WS 1.3.2A Sedimentary Rock Scavenger Hunt

PREPARATION

- Obtain materials for *Lab 1.3.2A Take a Closer Look* and *Lab 1.3.2B Formation of Sandstone*.
- Obtain materials for *Try This: Shifting Sediments*. Prepare a sample two to three days before presenting lesson.
- Have students bring samples of sedimentary rock to class. (A)
- Schedule a field trip to an active sand and gravel mine or caves that contain stalactites and stalagmites. (D)
- Invite a sedimentologist to speak to the class. (E)

TRY THIS

Shifting Sediments
- clear jars with lids, 1 per student
- samples of sand, gravel, mud, large pebbles, clay
- Epsom salt, 50 mL per student

4. Water moves through the sediments as this slow decay process takes place. Minerals in the water collect in the decaying cells of the organism. In time, the organism is replaced by these minerals collecting in the cells and preserving the outline of the hard structures.

- Ask the following question:
  1. What are some important natural resources that can be found in sedimentary rock? (**Possible answers: oil, natural gas, coal, phosphates, salt deposits, groundwater**)

Activities

Lab 1.3.2A Take a Closer Look

- sedimentary rock samples
- hand lenses, 1 per group
- rulers, 1 per group
- eyedroppers or pipettes, 1 per group
- paper towels
- vinegar or dilute hydrochloric acid (0.05 M)

Rock samples for each group should include both conglomerate and breccia rock.

Lab 1.3.2B Formation of Sandstone

- 150 mL of sand
- 125 mL of plaster of paris
- large paper cups, 1 per student
- spoons, 1 per student
- 250 mL beakers, 1 per student
- 125 mL of water
- small shells or bones, 1 per student

1.3.2 Sedimentary Rock

OBJECTIVES
- Explain the process of compaction and cementation.
- Describe how the three types of sedimentary rock are formed.
- List and describe the major types of clastic sedimentary rock.
- Identify sedimentary rock and simulate its formation.

VOCABULARY
- **clastic rock** a sedimentary rock made of rock particles and fragments deposited by water, wind, or ice
- **lithification** the process that transforms layers of rock fragments into sedimentary rock
- **sedimentary rock** a rock formed from sediments that have been compacted and cemented together
- **sediments** particles of minerals, rock fragments, shells, leaves, and the remains of once-living things
- **shale** a clastic rock composed of silt- and clay-sized particles in flat layers

Unlike the earth's crust, which is dominated by igneous and metamorphic rock, the most common rock visible on the earth's surface is sedimentary rock. After igneous rock develops from cooled magma, it is often exposed to the atmosphere and hydrosphere. This exposure to wind and water causes fragments of the rock to loosen and break away into tiny particles called *sediments*. **Sediments** are particles of minerals, rock fragments, shells, leaves, and the remains of once-living things. Different sediments have different textures—from very coarse to very fine. Some examples of sediments include gravel, sand, silt, and mud.

Sediments form in a variety of ways. Water from rain or runoff soaks into pores of rocks. When the water freezes, the pores crack open and portions of the rock fall away, creating sediments. Acids dissolved in rainwater also break down rocks into tiny fragments. Another way that sediments form is when rocks are exposed to the heat of the sun, which causes rock molecules to expand and contract. This molecular movement can create cracks in rocks until at some point, fragments and particles completely separate from the original rock. Water, wind, or ice carry these fragments, along with organic sediments, away and deposit them. When sediments collect, sedimentary rock can form.

Sedimentary rock is rock formed from sediments that have been compacted and cemented together. The process that transforms layers of rock fragments into sedimentary rock is called **lithification**, which means "to turn into stone." Lithification occurs by two processes: compaction and cementation.

Many deposits of small pieces of earth collect on top of each other to form layers called *strata*. The weight of the upper layers

TRY THIS

Shifting Sediments
Fill a clear jar with sand, gravel, mud, large pebbles, and clay. Add 50 mL of Epsom salt. Add water until there is only about 5 cm of space left at the top. Place the lid on the jar and shake for several seconds. When all is thoroughly mixed, place the jar on a flat surface and allow it to sit undisturbed overnight. The next day, observe how the layers have settled. Carefully pour the water out and let the layers dry completely. The Epsom salt (magnesium and sulfur) acts as glue to hold the "rock" together.

Stratification of sedimentary rock

puts tremendous pressure on the bottom layers, compressing the sediments together until the bottom layers slowly turn into hard rock. During this compaction, as the sediments are pushed together, inner pore spaces become smaller and some of the water is squeezed out. The remaining water surrounding the sediments can contain dissolved minerals, which later recrystallize as new minerals in the pore spaces.

Cementation occurs following compaction and recrystallization. In cementation, the crystals interlock and connect the sediment grains. This process essentially glues the sediments together. The resulting strata layers range in varying degrees of thickness and color, which helps to easily distinguish the incorporated sedimentary rock. This visible stratification, or layering, of sedimentary rock gives geologists clues about how rocks formed. Most sedimentary layers are deposited in a nearly horizontal position. However, there are times when scientists examine a rock layer that is folded or tilted. Scientists then assume this folding or tilting is a result of a disturbance in the earth's crust.

Sedimentary rock is classified into three main groups: clastic, chemical, and carbonate, or organic. Geologists classify sedimentary rock depending on how it is formed. The most common sedimentary rock, **clastic rock** (from the Greek word *klastos*, meaning "broken into pieces"), is made of separate rock particles and fragments

Folded rock layers

A. Conduct a sedimentary rock scavenger hunt. Have students bring in sedimentary rock samples they find near their home or school and complete **WS 1.3.2A Sedimentary Rock Scavenger Hunt**. Distribute hand lenses and field guides. If students bring in rocks that are not sedimentary rock, have them save them for a similar activity on metamorphic rock in Lesson 1.3.3. Challenge students to hypothesize the conditions under which the sedimentary rock samples formed.

B. Have students examine samples of a wide variety of sedimentary rock, including sandstone. Challenge students to formulate ideas as to how these rocks may have formed, depending on their observations.

C. Present students with examples of rocks found in or near the water and samples of rocks found away from the water. Have students observe the differences in texture and determine if the rocks are conglomerate or breccia. Utilizing a rock tumbler, show students the changes that rocks undergo from the erosive action of water and grit.

D. Plan a field trip to visit an active sand and gravel mine or caves that contain stalactites and stalagmites.

E. Introduce the invited sedimentologist to the class. Encourage the class to ask questions.

Lesson Review

1. How does compaction and cementation form layers of rock? (**During compaction, powerful pressure pushes the sediments together, squeezing air and water out of the spaces between the**

NOTES

---

that were eroded from an older rock. These fragments come together by wind, water, or ice to form a new rock by compaction and cementation.

Clastic sedimentary rock is further classified by the sediment size from which it forms. A conglomerate rock is a clastic rock composed of rounded, gravel-sized rock fragments usually larger than 2 mm in diameter. Conglomerate rock forms where sediments are deposited, such as at the mouths of rivers and along beaches. Individual rock fragments can be seen in conglomerate rock. These rock fragments are usually cemented together by tiny mineral particles that form what is called a *clastic matrix*.

Breccia is a type of clastic rock composed of sharp-cornered, angular fragments larger than 2 mm in diameter that are cemented together with carbonate, silica, or silt material. Breccia often forms at the base of a steep cliff where rockslides have occurred.

Sandstone is a clastic rock composed of rounded, sand-sized grains usually between 0.063 mm and 2 mm in diameter. This clastic rock is the second most common sedimentary rock. It comprises 10%–20% of the sedimentary rock in the earth's crust. Sandstone has many pores through which water can easily move. One of the most common minerals in sandstone is quartz, which can comprise 90% of the rock.

Shale is a clastic rock that forms in flat layers composed of silt- and clay-sized grains smaller than 0.004 mm in diameter. These layers are brittle and can be easily broken apart into flat pieces. Many of the particles are so small that they are barely visible without a microscope. Some geologists chew the sediments to estimate the size of their particles. (Silt is gritty, and clay is smooth.) Shale is the most abundant sedimentary rock, accounting for roughly 70% of sedimentary rock. These clastic rocks are often found with layers of sandstone or limestone. They typically form in environments where mud, silts, and other sediments were deposited by gentle transporting water currents. These sediments are then compacted in areas such as the ocean floor, basins of shallow seas, and river floodplains.

Conglomerate rock

The White Hoodoos near Wahweap Creek in Utah's Grand Staircase-Escalante National Monument are sandstone.

Chemical sedimentary rock is divided into two groups: allochemicals and orthochemicals. Examples of allochemicals include some limestone and chert. Examples of orthochemicals include bedded deposits of halite, gypsum, anhydrite, and banded iron formations. These sedimentary rocks do not form from separate rock pieces. They are composed of minerals that were once dissolved in water. Their structure is made up of interlocking crystals that result in a small, fine grain. As the water evaporates, the minerals that are left behind build up into rock masses. For example, some chemical sedimentary rock forms when dissolved salts in a body of water are deposited and the water evaporates away. Rock salt and gypsum are two examples of chemical sedimentary rock.

Carbonate sedimentary rock is composed of organic materials from decaying organisms. Coal, for example, is an organic sedimentary rock made of carbon from ancient plant remains. Other carbonate sedimentary rock is composed of the skeletal remains of marine creatures. Given that the skeletons are mineral and not technically organic, they are sometimes termed *biochemical*. An example of a biochemical rock is limestone, which forms from the mineral calcite. Limestone deposits often develop from the shells of clams, plankton, and other aquatic

BIBLE CONNECTION

The Importance of Salt
Salt is mentioned in at least 36 places in the Bible. It was crucial in ancient cultures as a seasoning, preservative, disinfectant, component of ceremonial offerings, or as a unit of exchange. Salt was a necessity of life, both literally and metaphorically. In Matthew 5, Jesus uses salt as a metaphor, suggesting the children of God should preserve themselves from impurities just as salt preserves food.

FYI

Too Much Salt?
The Dead Sea in Israel is highly concentrated with various salts that allow individuals to easily float. On the western shore near Ein Gedi, one can find pebbles cemented with halite.

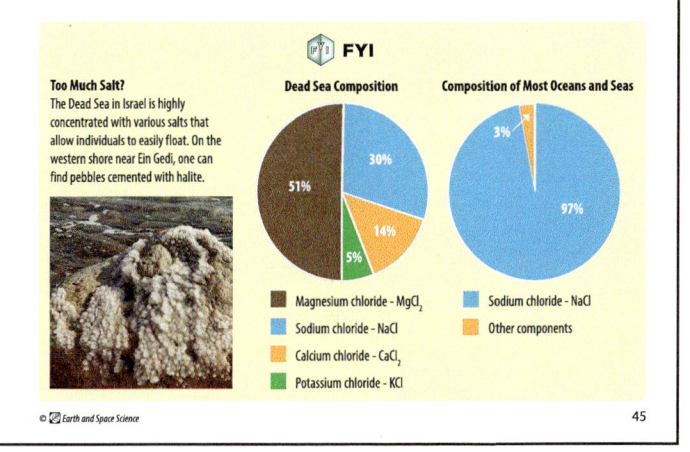

Dead Sea Composition
- 51%
- 30%
- 14%
- 5%

Composition of Most Oceans and Seas
- 97%
- 3%

- Magnesium chloride - MgCl$_2$
- Sodium chloride - NaCl
- Calcium chloride - CaCl$_2$
- Potassium chloride - KCl
- Sodium chloride - NaCl
- Other components

NOTES

fragments. During cementation, dissolved minerals are carried through sediments by water. These minerals are left in the spaces between the sediments, which acts as a glue to hold the sediments together.)

2. What are the three types of sedimentary rock? (**clastic, chemical, and carbonate**)
3. How are the three types of sedimentary rock formed? (***clastic***: **formed by separate rock particles and fragments that are compacted and cemented,** ***chemical***: **formed by water-soluble minerals left behind when water evaporates,** ***carbonate***: **formed by organic materials from decaying organisms**)
4. Would it be more likely to find fossils in an igneous rock or a sedimentary rock? Why? (**It would be more likely to find a fossil in sedimentary rock because igneous rocks form from melted rock and the fossils would probably not survive the heat. Sedimentary rocks form from particles that are laid down. This often occurs around once-living things, forming fossils.**)
5. Describe the major types of clastic sedimentary rock. (***conglomerate***: **rounded, gravel-sized rock fragments cemented by mineral particles;** ***breccia***: **sharp-cornered angular fragments cemented by carbonate, silica, or silt;** ***sandstone***: **rounded, sand-sized grains, porous;** ***shale***: **silt- and clay-sized grains, flat layers, brittle**)
6. Assume that the volume of a layer of mud will decrease by 40% during deposition and compaction. If the original sediment layer is 25 cm thick, what will be the thickness of the shale layer after compaction? (**If the mud layer loses 40% of its thickness during compaction, the final thickness will be 60% of 25 cm: 25 cm × 0.60 = 15 cm.**)

CAREER

Sedimentologist
Sedimentologists are a specific group of geologists who study sedimentary rock and how it forms. They study the origin and deposition of sediments and their conversion to sedimentary rock. Sedimentologists look at sediment cores to determine how the process of sedimentation produces rocks that are in the process of forming at the bottom of oceans, deltas, and lakes. They are often involved in searching for and finding oil, natural gas, and economically important minerals.

creatures. Chalk is a fine-grained limestone made of microscopic shells, fragments of larger shells, and calcite. It is soft enough to write with because chalk particles are tiny and rather loosely packed.

Some limestone is created entirely by chemical processes instead of organic processes. For example, as rainwater lands on the earth, it has the opportunity to enter caves through the cracks in rocks. The rainwater will then pass through organic material and incorporate carbon dioxide gas along the way, creating carbonic acid. This weak acid passes through joints and cracks in limestone. The mineral calcite is dissolved from the limestone rock. This process is what forms a cave. When the water that holds the dissolved rock is exposed to the air in the cave, it releases the carbon dioxide gas, much like a fizzy drink does when it is first opened. As the carbon dioxide is released, calcite is redeposited on cave walls, ceilings, and floors. This redeposited mineral will build up after countless water drops complete the chemical process, eventually forming a stalactite. If the water that drops to the floor of the cave still contains some dissolved calcite, it can deposit more dissolved calcite there, forming a stalagmite.

By analyzing and interpreting the sedimentary rock record, scientists attempt to date and document many of the significant events that have occurred in Earth's history. This record provides information on ancient geography. A map of the distribution of sediments that formed in shallow oceans bordering rising mountains or in deep, subsiding ocean trenches will indicate past relationships between seas and landmasses. An accurate interpretation allows scientists to form conclusions about the evolution of mountain systems, continental blocks, and ocean basins. Some scientists also attempt to draw conclusions about the origin and evolution of the atmosphere and hydrosphere. Other scientists examine the sedimentary rock record containing fossils of once-living creatures in an attempt to document the theory of evolutionary advancement from simple to complex organisms in the plant and animal kingdoms.

It is also important to understand the economic significance of sedimentary rock. For example, sedimentary rock essentially contains the world's entire supply of oil and natural gas, coal, phosphates, salt deposits, groundwater, and other natural resources. As good stewards, humanity is responsible for taking care of and preserving the many gifts God has entrusted to people.

A west Texas oil pumpjack

LESSON REVIEW
1. How does compaction and cementation form layers of rock?
2. What are the three types of sedimentary rock?
3. How are the three types of sedimentary rock formed?
4. Would it be more likely to find fossils in an igneous rock or a sedimentary rock? Why?
5. Describe the major types of clastic sedimentary rock.
6. Assume that the volume of a layer of mud will decrease by 40% during deposition and compaction. If the original sediment layer is 25 cm thick, what will be the thickness of the shale layer after compaction?

Mount Sodom salt cave near the Dead Sea in Israel

Name: _____ Date: _____

Lab 1.3.2A Take a Closer Look

QUESTION: How is sedimentary rock identified or classified?

HYPOTHESIS: Answers will vary.

EXPERIMENT:

You will need:	· ruler	· vinegar or dilute hydrochloric acid (0.05 M)
· sedimentary rock samples	· eyedropper or pipette	
· hand lens	· paper towels	

Steps:
Complete the charts by answering the questions below.
1. On Charts 1 and 2, list the sedimentary rock samples that have been provided.
2. Examine the grain size for each clastic sedimentary rock using a hand lens. Identify the grains that you believe make up your clastic sedimentary rock samples (round, angular, flat; gravel size [>2 mm], sand size [0.063 mm–2 mm], silt and clay size [not visible–<0.063 mm]).
3. Which samples show the best evidence of stratification (layering)?
4. Place 1 drop of water on each specimen. Which are porous? (If the water begins to seep into the rock, it is porous.)
5. Dip each of the samples in water. Which ones exhibit an earthy smell?
6. Dry the samples with paper towels. Then test to see whether the sample is a carbonate sedimentary rock or has calcium carbonate for its matrix by placing a drop of weak acid on the rock and seeing whether it will fizz. (Note: If there is a sandstone, shale, or conglomerate that fizzes, it is probably because there is the presence of a calcium carbonate cement. This reaction does not mean that the rock is a limestone; it means that the cementing agent is made of calcium carbonate.) Rinse off the samples.
7. Examine the conglomerate and breccia rock samples. What color are the matrices? Note the wide range of fragment size in each. Are the fragments interlocking or separate in each?

Lab 1.3.2A Take a Closer Look

Chart 1

Sample	Grain Size and Shape	Best Stratification	Porous?	Earthy Smell?	Reacts with Weak Acid?
		Answers will vary.			

Chart 2

Sample	Classification (Clastic, Chemical, or Carbonate?)	Matrix Color? (May Be N/A)	Fragments: Interlocking or Separate? (I, S or n/a)
		Answers will vary.	

ANALYZE AND CONCLUDE:

1. What do geologists look for in sedimentary rocks in order to identify them? They look for grain size, mode of formation, and mineral content.
2. What does grain size tell geologists? Grain size tells geologists the rock's original location, the distance sediments traveled, and the conditions in which sediments traveled.
3. What does the acid test indicate? The acid test indicates whether sedimentary rock is limestone or if the rock's matrix contains calcium carbonate.
4. What are the two steps that sediments undergo to become sedimentary rock? compaction and cementation

---

Name: _____ Date: _____

Lab 1.3.2B Formation of Sandstone

QUESTION: How does heat, pressure, and time affect the formation of sandstone?

HYPOTHESIS: Answers will vary.

EXPERIMENT:

You will need:	· large paper cup	· 125 mL of water
· 150 mL of sand	· spoon	· small shell or bone
· 125 mL of plaster of paris	· 250 mL beaker	

Steps:
1. Place sand and plaster of paris in a paper cup and mix thoroughly with a spoon.
2. Pour 100 mL of water into the cup and stir the mixture for several minutes. If the mixture is too thick to stir, add a little more water. If too much water is added and the mixture becomes soupy, use the tip of a pencil or a pin to punch a very small hole near the bottom of the cup to drain the excess water.
3. Place a small shell or bone in the center of the mixture. (This is intended to simulate a fossil when the sandstone is split in half.)
4. Let the mixture dry overnight in an area where it will not be disturbed.
5. The next day, carefully peel away the paper cup to reveal a pillar of sandstone.
6. Cut the sandstone in half to reveal the "fossil."

ANALYZE AND CONCLUDE:

1. How did the "rock" form? The rock formed when the plaster of paris cemented the sand grains together.
2. Under what conditions might the rock have formed faster? The rock may have formed faster with increased heat and pressure.
3. Compare the time it took the mixture to set as sedimentary rock with the natural formation of sandstone. The classroom rock formed in 24 hours, unlike sedimentary rock in nature, which forms over many years.

1.3.3 Metamorphic Rock

Rocks

OBJECTIVES

Students will be able to
- identify metamorphic rock and discuss how it forms.
- identify the agents of change in the process of metamorphism.
- discuss features and examples of two categories of metamorphic rock.
- compare and contrast the different types of metamorphism.

VOCABULARY

- **contact metamorphism** metamorphism that occurs when the heat of magma comes in contact with existing rocks
- **dynamic metamorphism** metamorphism that is produced by mechanical force
- **foliated structure** a rock structure with visible layers or bands aligned in planes
- **metamorphic rock** a rock formed when the structure and mineral composition of existing rocks change because of heat, pressure, or chemical reactions
- **metamorphism** the process of change in the structure and mineral composition of a rock
- **nonfoliated structure** a rock structure with no visible layers or bands
- **regional metamorphism** metamorphism that occurs when large regions of the earth's crust are affected by high temperatures and pressures

MATERIALS

- Metamorphic rock samples (*Introduction, A*)
- Hand lenses, rocks and minerals field guides (*A*)
- WS 1.3.3A Metamorphic Rock Classification
- WS 1.3.3B Metamorphic Rock Scavenger Hunt

Introduction

Ask students if they have ever seen a rock change. Share with students that rocks do change. However, this change is not usually visible to human eyes because the change is very slow or occurs beneath the earth's surface in the presence of heat and extreme pressure. Present samples of common metamorphic rock including schist, gneiss, quartzite, slate, and marble. Be sure to point out the rocks that exhibit foliation and those that do not. Have students make observations and speculate what causes the differences in structure.

Relate geologic metamorphism to other types of metamorphism—insect metamorphosis, frog metamorphosis, and the metamorphosis of a Christian heart. Have students compare and contrast these processes to the creation of metamorphic rock.

Discussion

- Ask the following questions:
  1. How do the layers in foliated metamorphic rock look? (**Possible answers: The layers or bands are aligned in planes; they look like the pages of a book.**)
  2. What kinds of conditions change rocks? (**heat, pressure, and chemical processes**)
  3. What types of rocks are changed into metamorphic rock? (**any type of rock: igneous, sedimentary, or other metamorphic rock**)
  4. Where does the heat that changes rocks originate? (**The heat originates from molten magma.**)
  5. What are the two categories of metamorphic rock? (**foliated and nonfoliated**) How do these categories differ? (**Metamorphic rock with a foliated structure has visible layers. Metamorphic rock with a nonfoliated structure does not have visible layers.**)

Activities

> ### Lab 1.3.3A Gneiss Foliation
> - gneiss samples
> - modeling clay
> - colored sequins
> - 25 cm string, 1 per group
> - wooden blocks, 2 per group
>
> Demonstrate the proper technique to use to cut the ball of clay in half with the string.

A. Conduct a metamorphic rock scavenger hunt. Divide the class into groups. Distribute hand lenses and field guides. Have students bring in metamorphic rock samples they find near their home or school and complete **WS 1.3.3A Metamorphic Rock Classification** and **WS 1.3.3B Metamorphic Rock Scavenger Hunt**. Challenge students to hypothesize the conditions under which the sedimentary rock samples formed.

B. Plan a guided geologic tour of a location with visible foliation or an art museum with marble artwork.

C. Have students create a concept map for metamorphic rock using the following points:
  - Metamorphic rock in the earth's crust is changed by three major processes: dynamic metamorphism, contact metamorphism, and regional metamorphism. Dynamic metamorphism is produced by mechanical forces. Contact metamorphism occurs when magma and rock come in direct contact. Regional metamorphism results from the heat and pressure of tectonic activity.
  - Direct pressure is the primary cause of dynamic metamorphism.
  - Only rocks that are near or touching the magma are changed during contact metamorphism.
  - Large regions of the earth's crust are affected by high temperatures and pressures with regional metamorphism. Most metamorphic rock is formed through regional metamorphism.

Lesson Review

1. What is metamorphic rock? (**Metamorphic rock is rock formed when the structure and mineral composition of existing rocks change because of heat, pressure, or chemical reactions.**)
2. How does heat change rocks? (**Extreme heat speeds up chemical reactions in minerals, breaking down chemical bonds and forming new ones. This process forms new compounds. The increased heat and the change of the water content inside minerals work together to soften rocks, forcing them to "flow" under high pressure.**)
3. How does pressure change rocks? (**Extreme pressure applied from opposite directions causes mineral layers to develop perpendicular to the pressure direction.**)
4. How are foliated and nonfoliated structures different? Name examples of each. (**Metamorphic rock with a foliated structure has visible layers or bands aligned in planes. Slate, schist, and gneiss are three examples of foliated metamorphic rock. Metamorphic rock with a nonfoliated structure has no visible layers or bands. Marble and quartzite are two examples of metamorphic rock with a nonfoliated structure.**)
5. Compare and contrast the three types of metamorphism. (**Dynamic metamorphism is produced by mechanical forces with direct pressure as the primary cause. Contact metamorphism occurs when the heat of magma comes in direct contact with existing rocks. Regional metamorphism occurs when large regions of the earth's crust are affected by high temperatures and pressure. All types of metamorphism can change igneous, sedimentary, or other metamorphic rock.**)

PREPARATION

- Obtain materials for *Lab 1.3.3A Gneiss Foliation*.
- Have students bring samples of metamorphic rock to class. (*A*)
- Schedule a guided geologic tour of a location with visible foliation or an art museum with marble artwork. (*B*)

1.3.3 Metamorphic Rock

OBJECTIVES
- Identify metamorphic rock and discuss how it forms.
- Identify the agents of change in the process of metamorphism.
- Discuss features and examples of two categories of metamorphic rock.
- Compare and contrast the different types of metamorphism.

VOCABULARY
- **contact metamorphism** metamorphism that occurs when the heat of magma comes in contact with existing rocks
- **dynamic metamorphism** metamorphism that is produced by mechanical force
- **foliated structure** a rock structure with visible layers or bands aligned in planes
- **metamorphic rock** a rock formed when the structure and mineral composition of existing rocks change because of heat, pressure, or chemical reactions
- **metamorphism** the process of change in the structure and mineral composition of a rock
- (continued on next page)

Rocks seem permanent and indestructible. Perhaps there is a boulder near your house or school that you see every day, and it never seems to change. We often use the word *rock* to describe something that is unchanging. However, God created a dynamic world where even rocks can change over time.

If you were a scientist and you decided to measure the pressure and temperature of the layers of the earth, you would find that both the pressure and temperature increase the closer you get to the center of the earth. At some point, the temperature and pressure reach high enough levels for rocks to melt and become magma. However, before reaching this point, there is a region where temperature and pressure are at high levels, but not so high that rocks will melt. When rocks are exposed to these conditions, their texture, mineral composition, or chemical composition can be affected. If one of these changes occurs, a new rock is formed. These new rocks have undergone **metamorphism**, which is the process of change in the structure and mineral composition of a rock.

The products of metamorphism are **metamorphic rock**, rocks formed when the structure and mineral composition of existing rocks change because of heat, pressure, or chemical reactions. During metamorphism, a rock changes form, but it remains solid. Metamorphic rock can form from any type of rock. The chemical and physical properties of the new rock are usually very different from those of the old rock.

Rocks within the earth's crust can be changed by three main types of metamorphism: contact, regional, or dynamic. **Dynamic metamorphism** is produced by mechanical forces. Direct pressure is the primary cause of this type of metamorphism. One example of dynamic metamorphism is the mineralogical changes

Metamorphic Rock

Gneiss—foliated structure

Pink quartzite—nonfoliated structure

that occur along the flat surface between two pieces of land that have shifted during an earthquake. **Contact metamorphism** occurs when the heat of magma, such as an igneous intrusion, comes in contact with existing rocks. This process produces a local effect, only changing the rocks that are near or touching the magma. In contrast, **regional metamorphism** occurs when large regions of the earth's crust are affected by high temperatures and high pressures. Regional metamorphism changes minerals and rock types and is often accompanied by folding and rock layer deformation in the area.

Regional metamorphism forms most metamorphic rock. It is not uncommon to find rocks formed through contact metamorphism near regional metamorphic rock. This connection becomes evident when observing any volcanic activity that accompanies plate movements. As the magma that is produced by the volcano comes in contact with rocks and changes them through contact metamorphism, the movement of tectonic plates can also affect a large region of the earth's crust, resulting in regional metamorphism. By studying how a rock metamorphoses, scientists can interpret the conditions inside the earth's crust.

VOCABULARY
- (continued from previous page)
- **nonfoliated structure** a rock structure with no visible layers or bands
- **regional metamorphism** metamorphism that occurs when large regions of the earth's crust are affected by high temperatures and pressures

Contact and Regional Metamorphism

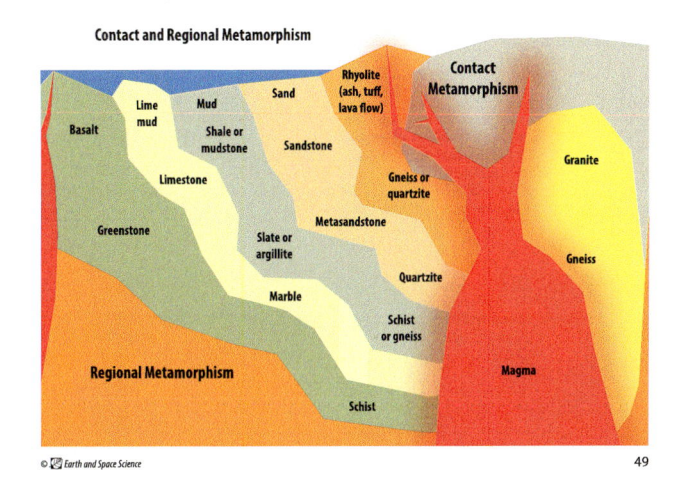

There are different grades of metamorphism. Lower temperatures and lower pressures produce a low grade of metamorphism. Higher temperatures and higher pressures produce a high grade of metamorphism. There is also an intermediate grade of metamorphism with varying temperatures and pressures.

Whether metamorphic rocks are formed by dynamic, contact, or regional metamorphism, the result is that the mineral crystals in the original rocks are converted and rearranged into new minerals that are stable under new temperature and pressure conditions. These new rocks will develop one of two basic structures—foliated or nonfoliated. Metamorphic rock with a **foliated structure** has visible layers or bands aligned in planes, similar to the pages in a book. Slate, schist, and gneiss are three common foliated metamorphic rocks.

How does foliation occur? Foliation is the product of pressure that has been applied from opposite directions. The foliation develops perpendicular to the pressure direction. By analyzing metamorphic structures and mineral composition, geologists are able to identify metamorphic rock.

Metamorphic rock with a **nonfoliated structure** has no visible layers or bands. Quartzite and marble are two examples of metamorphic rock with a nonfoliated structure. Quartzite is metamorphosed from quartz-rich sandstone. Marble is formed from limestone and dolostone. This metamorphic rock is valued as a building and monument stone because of its durability and ability to transmit light. Marble is usually white, but it may be almost any color from white to black.

The unique physical properties and symmetries of one class of metamorphic minerals result in the formation of large single crystals. These minerals are called *porphyroblasts*. It is interesting that other surrounding crystals may still remain small. A popular example of a porphyroblast is garnet.

Garnet is a porphyroblast.

Michelangelo's statue of Moses was chiseled from marble.

Many of the commercial products that are used throughout the world are the result of metamorphism of igneous and sedimentary rock. Three such products include talc and asbestos, which are minerals, and coal. The extreme softness of talc has been utilized in cosmetic powder products, in lubricants, and in textured paints. Asbestos easily separates into long, flexible fibers and has the useful property of being resistant to the effects of heat and fire. It has been used in fireproofing and insulating materials. Asbestos was also widely used in construction materials until in the 1970s it was discovered that asbestos poses a serious health risk. Regulatory agencies in the United States and abroad began placing tight restrictions on the use and exposure of asbestos. The metamorphism of coal, a sedimentary rock, may produce graphite, which is the main ingredient of the lead in pencils. Take time to observe the various uses of metamorphic rock in your area.

LESSON REVIEW
1. What is metamorphic rock?
2. How does heat change rocks?
3. How does pressure change rocks?
4. How are foliated and nonfoliated structures different? Name examples of each.
5. Compare and contrast the three types of metamorphism.

Coal, a sedimentary rock, may sometimes metamorphose to produce graphite, the main ingredient of "lead" in pencils.

Shale is a sedimentary rock that can change into at least four different metamorphic rocks.

Increased pressure → / Increased temperature →
Shale
Slate
Phyllite
Schist
Gneiss

Name: _____ Date: _____

Lab 1.3.3A Gneiss Foliation

QUESTION: How do a rolling pin and clay represent the formation of metamorphic rock? What conditions will cause an igneous rock to change into a metamorphic rock?

HYPOTHESIS: Answers will vary.

EXPERIMENT:

You will need:	• modeling clay	• 25 cm string
• gneiss sample	• colored sequins	• 2 wooden blocks

Steps:
1. Sketch the gneiss sample below. Make sure the sketch includes the mineral grain arrangement.

Drawings will vary.

2. Pour the sequins onto the work area and roll the ball of clay over the sequins. Once a large number of sequins stick to the outside of the ball, knead the ball until the sequins are thoroughly mixed throughout the clay.
3. Form the clay into a ball.
4. Cut the ball in half with the string and draw a picture of your observations of how the sequins are arranged below.

Drawings will vary.

Lab 1.3.3A Gneiss Foliation

5. Form the clay into a ball again and place it on one end of one of the wooden blocks.
6. With the second block, slowly smear the clay ball across the surface of the first block and draw your observations below.

Drawings will vary.

ANALYZE AND CONCLUDE:

1. What changed the arrangement of the sequins? The pressure applied by the block rearranged the sequins, which are now perpendicular to the pressure direction.

2. What does the arrangement of the sequins tell you about the rock's history? Answers will vary but should include that the amount of pressure or the direction of pressure can be determined.

3. What features did the gneiss sample and the smeared clay containing sequins have in common? Answers will vary but should include mention of distinct mineral/sequin layer.

Rocks

1.3.4 The Rock Cycle

Introduction
Present students with examples of various cycles in creation. Some examples include the water cycle, the nitrogen cycle, the carbon cycle, the food chain cycles, and the earth's cycle around the sun. Explain to students that God designed these cycles to ensure that life on Earth would be renewable and sustainable. Share with students that rocks also travel through a cycle.

Discussion
- Display **TM 1.3.4A The Rock Cycle**. Compare and contrast the rock cycle with other cycles in creation.

- Have students discuss different ways rocks are formed or destroyed. Some examples are as follows:
  1. Magma cools to form igneous rock.
  2. Weathering agents—rain, snow, sleet, and hail—pound rocks, wash away tiny fragments of rocks, and erode rocks into sediments.
  3. The heat of the sun cracks open stones. Freezing and thawing of water in rocks widens cracks and causes large rocks to fracture into smaller fragments.
  4. The pounding surf polishes jagged rocks into smooth stones and eventually turns them into small pebbles and sand.
  5. Organisms wear away minerals and rocks. For example, the acidic excretions of lichen dissolve small depressions in boulders to open up spaces for fungus to grow.
  6. Sediments are carried by wind, water, and glaciers for great distances where they are deposited into beds. The sediment layers build up and exert tremendous pressure on the layers below. Through the processes of compaction and cementation, sediments are converted into sedimentary rock.
  7. When igneous or sedimentary rock is exposed to extreme pressure, heat, or chemical processes beneath the earth's surface, they can be transformed into metamorphic rock.

- Ask the following questions:
  1. Do all rocks travel through all steps of the rock cycle in a specific order? (**No. Igneous rock can change directly into metamorphic rock without first changing into a sedimentary rock.**)
  2. What processes ensure that an igneous rock will become a sedimentary rock? (**weathering; erosion; and lithification, or compaction and cementation**)
  3. What conditions must exist for a sedimentary rock to be transformed into a metamorphic rock? (**extreme heat and pressure**)

Activities

Lab 1.3.4A How Rocks Change
• samples of gneiss, granite, limestone, marble, quartzite, sandstone, shale, and slate • hand lenses, 1 per group • triple beam balances or electronic scales, 1 per group • 100 mL graduated cylinders or beakers (large enough to hold rock samples), 1 per group

A. Have students complete **WS 1.3.4A The Rock Cycle**.

B. Set up a rock display with the rocks collected from the student scavenger hunts. Explain to students that the rocks now represent sports stars. The rocks will need to be given a name and a trading card. Direct students to select a rock and to use an index card to create a trading card that will include the following: a picture of the rock, the type of rock, the minerals contained in the rock, how the rock formed, where the rock can be found, what the rock is used for, how to recognize the rock, a slogan, and the rock's statistics. Following this activity, collect and shuffle

OBJECTIVES
Students will be able to
- demonstrate an understanding of the process that forms the basic substances involved in the rock cycle.
- label a diagram of the rock cycle.
- use the rock cycle to explain how rocks are classified.

VOCABULARY
- **rock cycle** the process by which one rock type changes into another

MATERIALS
- Index cards, rock photos (B)
- TM 1.3.4A The Rock Cycle
- WS 1.3.4A The Rock Cycle

PREPARATION
- Obtain materials for *Lab 1.3.4A How Rocks Change*.
- Take photos or gather photos from the Internet of the rocks collected on the scavenger hunts. (B)
- Arrange for a tour of the local area to view buildings made of stone or that have stone facings. (C)

NOTES

the trading cards, distribute the cards, and have students try to match the trading card with the appropriate rock.

C. Take a tour of the local area and point out particular structures and buildings that are made of stone or have stone facings. Have students form a hypothetical company with one or two other classmates. The directive for this company is to formulate a proposal to build a building or building facing using igneous, sedimentary, or metamorphic rock. Students will need to choose the building material and support their proposal with evidence.

D. Direct students to create a graphic organizer to compare and contrast types of rocks.

E. Have students create a computer presentation or a poster of the rock cycle that shows the primary substances and processes involved in the cycle. Assign the task of finding pictures or samples of different types of igneous, sedimentary, and metamorphic rock along with sediments to include in the presentation or affix to the poster.

F. Direct students to research information about the type of rock used to build structures in the local area as well as any type of rock that occurs naturally in the local area. Have students create a promotional brochure that describes a tour focused on local geology.

Lesson Review

1. To what type of rock can all rock be traced? (**Igneous rock.**) Why? (**The cooling of magma is the start of the cycle. Magma always cools to form igneous rock.**)

1.3.4 The Rock Cycle

OBJECTIVES
- Demonstrate an understanding of the process that forms the basic substances involved in the rock cycle.
- Label a diagram of the rock cycle.
- Use the rock cycle to explain how rocks are classified.

VOCABULARY
- **rock cycle** the process by which one rock type changes into another

Earlier in the chapter, you learned rocks are classified according to how they are formed. Igneous rock forms when hot magma cools. Sedimentary rock forms from sediments that have been compacted and cemented together. Metamorphic rock is formed deep in the earth's crust where extreme pressure, temperature, and chemical reactions change them into different rocks. Any of these types of rocks can be changed into the other two types. The process by which one rock type changes into another is called the **rock cycle**.

The rock cycle begins with the cooling of magma to form igneous rock. Once a mass of igneous rock is exposed to the earth's surface, it begins to break down into smaller fragments. Rain, snow, sleet, and hail pound the rocks and wash away tiny fragments of the rock. The sun's heat cracks rocks open, and the freezing of water and the thawing of ice widens the cracks and breaks large rocks into smaller fragments. Pounding surf polishes jagged rocks first into smooth stones and then into small pebbles and sand. Even organisms can wear away rocks. The acidic excretions of lichen, for example, dissolve small depressions in boulders to open up spaces for fungus to grow. Over thousands of years, water and atmospheric forces erode igneous rock into small rock particles called *sediments*.

If rocks are continuously weathering, why are sediments not piled up everywhere? Wind, rivers, streams, and even glaciers carry sediments for hundreds of kilometers and deposit them in beds. These sediment layers build up over time, and the top layers put tremendous pressure on the bottom layers. The processes of compaction and cementation turn the sediments into sedimentary rocks.

Sedimentary rock exposed to extreme pressure, heat, or chemical processes beneath the earth's surface can be transformed into metamorphic rock. The types of metamorphic rock formed depends on the amount of heat and pressure to which the rocks are exposed. If the heat and pressure become more intense, the metamorphic rock melts and reforms into magma. This magma can then cool and form igneous rock, starting the rock cycle over again. Most of the rocks in the earth's crust have probably passed through the rock cycle many times.

Large rocks breaking apart through the process of freezing water and thawing ice

Rocks do not always complete the entire rock cycle. For example, igneous rock never exposed to the weather at the earth's surface will not be eroded into sediments. Likewise, not all rocks complete the cycle in a uniform order. Entire steps may be omitted. For example, igneous rock exposed to heat and pressure may change directly into metamorphic rock without first converting to sedimentary rock. Both igneous and sedimentary rock may melt into magma without first becoming metamorphic rock. The rock cycle is continually changing and remaking rocks.

Igneous intrusion

LESSON REVIEW
1. To what type of rock can all rock be traced? Why?
2. Explain the basic processes in the rock cycle.
3. What are the five basic substances involved in the rock cycle?
4. Use the rock cycle to explain how rocks are classified.

The Rock Cycle

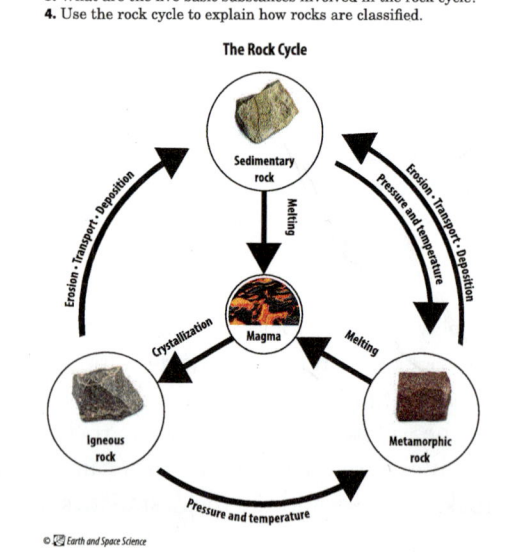

2. Explain the basic processes in the rock cycle. (**Magma cools and hardens into igneous rock. Igneous rock erodes into sediments, which are compacted and cemented into sedimentary rock. Sedimentary rock is exposed to pressure, heat, or chemical processes allowing it to transform into metamorphic rock. Intense heat and pressure melt metamorphic rock back into magma.**)
3. What are the five basic substances involved in the rock cycle? (**The five basic substances of the rock cycle are magma, igneous rock, sediments, sedimentary rock, and metamorphic rock.**)
4. Use the rock cycle to explain how rocks are classified. (**Rocks are classified by how they are formed. The rock cycle is the process by which one rock type changes into another, so the rock cycle is how each classification of rock is produced. For example, magma cools and becomes igneous rock. Weathering breaks down igneous rock into sediment, which collects and is compacted into sedimentary rock. And sedimentary rock can be transformed into metamorphic rock through exposure to pressure, heat, or chemical processes.**)

NOTES

Name: _____ Date: _____

Lab 1.3.4A How Rocks Change

QUESTION: How do the characteristics of sedimentary and igneous rock compare to metamorphic rock?

HYPOTHESIS: Answers will vary.

EXPERIMENT:

You will need:	• sample of gneiss, granite, limestone, marble, quartzite, sandstone, shale, and slate	• 100 mL graduated cylinder or beaker
• hand lens		
• triple beam balance or electronic scale		

Steps:
1. Examine each rock sample using the hand lens and record your observations in the table.
2. Determine a method to measure the mass and volume of a rock sample.
3. Calculate the density of each rock sample and record it in the table. Remember: Density = mass ÷ volume

Sample	Rock Type (I, S, or M)	Physical Characteristics	Mass (g)	Volume (mL)	Density (g/mL)
Gneiss	M				
Granite	I				
Limestone	S		Answers will vary.		
Marble	M				
Quartzite	M				
Sandstone	S				
Shale	S				
Slate	M				

Lab 1.3.4A How Rocks Change

ANALYZE AND CONCLUDE:

1. Describe how sandstone and quartzite are similar and different. Sandstone is a clastic sedimentary rock made of mostly sand-sized minerals or rock grains. Quartzite is a hard, nonfoliated metamorphic rock that was originally sandstone. Sandstone is converted into quartzite through heat and pressure. Sandstone can have visble bands, but quartzite does not. In sandstone, the cement holding the grains together is softer than the grains themselves, so sandstone breaks around the grains. In contrast, quartzite is very hard throughout, so it will break through the grains.

2. Describe how the grain size of sandstone changes during metamorphism. The grain size of sandstone increases with increasing temperature and pressure during metamorphism.

3. Describe the textual differences observed between shale and slate. Shale is softer than slate and has a claylike texture. Slate has a foliated texture.

4. Compare your calculated densities to those calculated by other students. Infer why the values may differ. Answers will vary but may include inaccurate measurements.

5. Evaluate the changes in density between shale and slate, sandstone and quartzite, limestone and marble, and granite and gneiss. Does density always change? Explain. Density usually changes because water is either lost or gained during metamorphism.

1.4.0 The Structure of the Earth

LOOKING AHEAD

- For **Lesson 1.4.1**, obtain a copy of *Journey to the Center of the Earth* by Jules Verne.
- Obtain samples of basalt and granite that are small enough to fit inside a graduated cylinder for **Lesson 1.4.3**.

SUPPLEMENTAL MATERIALS

- TM 1.4.1A Cross Section of the Earth
- TM 1.4.1B Kola Peninsula
- TM 1.4.1C Earth's Magnetic Field
- TM 1.4.2A Moho Map

- Lab 1.4.1A Making an Electromagnet
- Lab 1.4.2A Making Waves
- Lab 1.4.2B Asthenosphere

- WS 1.4.1A Model of the Earth
- WS 1.4.1B Seismic Waves
- WS 1.4.3A The Oceanic Crust
- WS 1.4.3B The World's Highest Mountains
- WS 1.4.3C Diagram of Earth's Structure

- Chapter 1.4 Test
- Unit 1 Test

BLMs, TMs, and tests are available to download. See Understanding Purposeful Design Earth and Space Science at the front of this book for the web address.

Chapter 1.4 Summary

Over 2,000 years ago, the ancient Greek Eratosthenes calculated the earth's circumference; but not until the beginning of the 20th century did scientists determine that Earth is made up of three main layers: the crust, the mantle, and the core. Modern scientists glean their knowledge of the earth's interior through several methods. These include the analysis of variations of the earth's gravitational field, comparisons with meteorites, interpretations of rock nodules brought to the surface by volcanic eruptions, and seismology—the study of seismic waves passing through the earth. Scientists have concluded that the earth consists of layers of varying size, composition, state, and density. The earth is much denser near its core than at the crust.

In this chapter, students will explore the structure and composition of the interior of the earth as geologists understand it today. Each lesson covers a different layer of the earth, starting with the core and ending with the earth's crust. The chapter discusses the data known about each layer, revealing to students that God uses many parts woven together to complete His design and to maintain His creation.

Background

Lesson 1.4.1 – Core

The science of seismology has allowed scientists to theorize about the composition of the earth, even the core. Seismologists measure the velocities at which seismic waves travel through the earth. Primary seismic waves, known as *P waves*, travel faster through the earth. Secondary seismic waves, known as *S waves*, travel slower and only through solids.

By studying how P and S waves travel, scientists have learned that the earth's core, believed to be composed primarily of iron and nickel, has two parts: the inner core and the outer core. The inner core is considered to be solid despite its high temperature because of the extremely high pressure exerted on it. Its density is approximately 13 g/cm^3, and it is hotter than the liquid outer core, which has a density range of 9.9–12.2 g/cm^3. S waves, which cannot travel through liquids, are reflected back to the surface of the earth when they strike the outer core, indicating that this outer core is molten.

Geophysicists believe that the outer core is the source of the earth's magnetic field. The dynamo theory suggests that heat is converted to mechanical energy (currents in the liquid core), which is then converted to electromagnetic energy—the magnetic field. The magnetic field has tremendous implications on the exploration of the earth, as it makes the use of a compass possible. The magnetosphere, which extends 60,000 km into space, also shields the earth from much of the solar wind. When the solar wind reaches earth's magnetic field, it is deflected, much like water around a ship's bow.

The earth's core is hot for several reasons. Some of the core's heat is residual; it came from the initial formation of the earth. The core is the hottest layer because heat is lost from the earth's surface. Heat is generated by friction as denser core material sinks to the center of the planet. Also, heat is generated by decaying radioactive elements.

Lesson 1.4.2 – Mantle

The mantle fills the space between the earth's outer core and its crust. The entire mantle constitutes about 83% of the earth by volume. It is too deep beneath the earth's surface to directly observe, but seismology and gravity studies have helped scientists map its boundaries, and volcanoes eject what may be samples of the upper mantle. Rare intrusive rocks known as *kimberlite* also contain possible fragments of the upper mantle. The mantle is relatively flexible, so it flows instead of fracturing.

The mantle is often divided into three parts: the upper mantle, the lower mantle, and a transition zone. Along with the earth's crust, part of the solid upper mantle makes up the lithosphere to a depth of about 100 km. Within the lithosphere is the Moho, the boundary that separates the upper mantle and the crust. Studies of P and S waves suggest that the rocks below the Moho are less rigid and slightly more dense than crustal rock. Although the top layer of the mantle is very hot, it is solid because the immense pressure of the lithosphere on the mantle rock raises its melting temperature.

The asthenosphere lies in the upper part of the mantle. It is a zone of low seismic velocity and rigidity just below the lithosphere—from about 100 km to 250 km. The plasticity of the asthenosphere makes it crucial to plate tectonics. Here the rocks come close to melting. According to the theory of plate tectonics, the upper part of the mantle released heat energy that broke the earth's crust into vast plates that slide around on the plastic zone, developing stresses along the plate margins that cause the earth's crust to fold and fault. The asthenosphere and lithosphere, as well as inner and outer core, are divisions of the earth classified according to physical properties, whereas the terms *crust*, *mantle*, and *core* are classified according to composition.

The transition zone of the mantle is the main source for all basaltic magmas. This layer is located in the range of 410–660 km. The transition zone is made of the same minerals as the lower mantle, but the minerals are in a less-dense phase. The temperatures here are 1,500°C at the top ranging to 1,700°C toward the bottom.

The lower mantle is held solid by high pressure. It contains 72.9% of the mantle and is composed mostly of iron and magnesium silicates, along with iron, calcium, and aluminum. The lower mantle extends from a depth of 660–2,700 km, making it roughly 2,000 km thick. The lower mantle is 1,700°C–3,600°C.

Lesson 1.4.3 – Crust

The crust beneath the continents is divided into two rock layers. The continental layer consists chiefly of silicon, oxygen, aluminum, calcium, sodium, and potassium. This rock layer is sometimes called *the sial*. The sial, in turn, rests on a layer consisting mainly of silicon, oxygen, iron, and magnesium. This layer of rock is sometimes called *the sima*. The sima extends under the oceans and forms the ocean basins. No sial is found under the oceans.

The crust is relatively light. Compared with the other layers, it is also cold, which makes it brittle. Brittleness accounts for the faulting of the earth's crust. Folding of the earth's crust implies plasticity; folded rocks were once buried, pressurized, and warmed to make them plastic. About 80% of the earth's crust was formed by igneous activity.

Oceanic crust underlies the two-thirds of the earth's surface that is covered by oceans. Its thickness is about 6 km, and its composition is uniform. New crust is constantly generated by sea floor spreading at mid-ocean ridges (places where basalt magma comes up from the upper mantle and cools). Older crust is recycled into the mantle at subduction zones. As a result, the oceanic crust is much younger than the continental crust. The density of the oceanic crust is about 3 g/cm^3.

The oldest parts of the continental crust are the shields, which serve as the stable interior to the continents. Most of the remaining crust consists of orogens, the roots of mountain belts formed at different times; most of these are covered by younger sedimentary rock. The average thickness of the continental crust is about 40 km, but beneath parts of the Andes and the Himalayas the crust can be 100 km thick. The density of the continental crust is about 2.7 g/cm^3.

> **WORLDVIEW**
>
> - Unlike the random, undirected acts of atheistic evolution or the actions of sophomoric Greek gods who cause storms and pestilence at every emotional whim, God designed all aspects of the earth with intentionality and purpose. God asks Job, "Where were you when I laid the earth's foundation? Tell me, if you understand. Who marked off its dimensions? Surely you know! Who stretched a measuring line across it? On what were its footings set, or who laid its cornerstone—while the morning stars sang together and all the angels shouted for joy?" (Job 38:4–7) God made it clear to Job that He gave thought to all parts of the earth, from the atmosphere in the heavens to the depths, or foundation, of the earth. In this chapter, the structure of the inner earth is studied. Scientists debate how the motion in the outer core began, and the scientific method falls short in determining the origin of the motion. However, it is clear that God set the core in motion to produce the magnetosphere, an important protection for human life. Even the inner workings of terra firma were designed with a plan and a purpose. In any endeavor to understand the planet God made for His people, we are to join the angels and shout for joy at the mighty works of His hands.

1.4.1 Core

The Structure of the Earth

OBJECTIVES

Students will be able to
- explain how seismic waves are used to determine the features of the core.
- describe the two parts of the earth's core.
- explain how the core generates a magnetic field.
- describe the benefits of the magnetic field around the earth.

VOCABULARY

- **core** the central portion of the earth
- **inner core** the solid center of the earth
- **magnetosphere** the area around the earth that is affected by the earth's magnetic field
- **outer core** the liquid layer of the earth's core that surrounds the inner core
- **P wave** the fastest seismic wave, which travels through solids, liquids, and gases
- **seismic wave** a wave of energy that travels through the earth
- **S wave** the seismic wave that travels only through solids

MATERIALS

- *Journey to the Center of the Earth* by Jules Verne, apples, knife (*Introduction*)
- Extendable spring toy (*Discussion*)
- Four colors of modeling clay; triple beam balances, 1 per group; fishing line (*B*)
- Compasses, magnets (*D*)
- Layer cake, clear straws, graph paper, colored pencils, metric rulers, knife (*E*)
- TM 1.4.1A Cross Section of the Earth
- TM 1.4.1B Kola Peninsula
- TM 1.4.1C Earth's Magnetic Field
- WS 1.4.1A Model of the Earth
- WS 1.4.1B Seismic Waves

Introduction

Read excerpts from Jules Verne's *Journey to the Center of the Earth*, focusing on descriptions of the center of the earth. Ask students to repeat back some of the descriptions. Next, ask students what they believe the center of the earth is like and why. (**Answers will vary.**)

Cut an apple in half with a knife. Tell students that the apple core represents the earth's core. The flesh of the apple represents the mantle, and the skin represents the crust. The deeper layers of the apple can be studied because the apple can be cut in half. Take a second apple and do not cut it. Ask students how the inside of the apple can be studied without cutting it in half. (**Answers will vary.**)

Discussion

- Display **TM 1.4.1A Cross Section of the Earth**. Discuss the layers of the core. Refer to *FYI: Core Knowledge* in the Student Edition to discuss the percentages of each section.

- Display **TM 1.4.1B Kola Peninsula**. Draw students' attention to the location geologists chose to drill. Refer to map skills learned in Chapter 1. Determine latitude and longitude of the Kola Peninsula. Have students generate ideas as to why the scientists chose to drill in this location. (**Answers will vary.**)

- Demonstrate P and S waves. Stretch an extendable spring toy, push one end quickly toward the other, and then pull it back to its original position. The resulting wave pattern is a model of how P waves travel through rock. The spring can also be used to illustrate S waves; instead of pushing the end forward, move the end side-to-side. Ask for ideas about how the spring would behave differently if it were immersed in different substances, such as water or gelatin. (**Answers will vary.**)

- Discuss how much knowledge about the inner earth has been obtained through the indirect evidence gleaned from studying seismic waves. Detail how scientists have used these waves to make predictions concerning the composition and state of inner earth materials. Explain how scientists use P and S waves to gather data about the earth's inner core. P waves carry energy from a movement in the earth through the interior of the earth. They pass through both the outer and inner core, but they bend. The degree to which P waves bend provides information about the composition of the core. S waves are stopped by the dense liquid of the outer core.

- Discuss the meaning of the word *core*: the central, innermost, or most essential part of anything. The awe expressed for God's creation is often limited to the parts that can be seen. Have students discuss how even the deepest parts of the earth are an integral part of creation and are part of the wide variety of means that God uses to maintain His world.

Activities

Lab 1.4.1A Making an Electromagnet

- large iron nails, at least 7.5 cm long, 1 per group
- paper clips, 1 per group
- insulated copper wire, 60 cm per group (24 gauge is best; not thicker than 18 gauge)
- electrical or cellophane tape
- 1.5 volt batteries (any cell size, AA, C, D), 1 per group

Strip the ends of the wires to expose the copper.

A. Complete *Try This* activities with students. Use **TM 1.4.1C Earth's Magnetic Field** to verify student answers for *Try This: A Magnetic Core*.

B. Guide students to build a model of the earth, using **WS 1.4.1A Model of the Earth**. Students will need to look through the three lessons of this chapter to find all the necessary data for this activity.

C. Assign **WS 1.4.1B Seismic Waves**.

D. Encourage students to explore how a compass works. Have each group examine a compass and note which way it points as they move around the room. Then instruct students to hold a strong magnet near the compass and record what happens. Ask students to infer how the earth's magnetic field plays a part in the workings of the compass. (**Answers will vary.**)

E. Simulate how scientists see beneath the earth's surface. Provide student groups with a clear plastic straw, graph paper, colored pencils, and a metric ruler. Have them lay out a coordinate system for straw drill holes. Direct students to use their straws to drill into the prepared layer cake. Have groups create a cross section on graph paper. Then cut into the cake and compare it to the cross section. Explain that in a similar way, drilling into the earth helps geologists understand the earth's composition.

F. Challenge students or student groups to create a method to study the center of the earth. Give students some ideas to get them started: going down into caves, digging, drilling, studying materials like igneous rock that came up from underground, studying surface rocks, or sending X-rays through the earth.

PREPARATION

- Obtain materials for *Lab 1.4.1A Making an Electromagnet*.
- Obtain materials for *Try This* activities.
- Bake a cake with at least three layers of different colors and thicknesses. Assemble and frost the cake. (*E*)

TRY THIS

Pressure Practice
- large marshmallows, 2 per group
- books, several per group

A Magnetic Core
- foam cups, 1 per group
- cardboard, 11" × 17", 1 per group
- iron filings
- bar magnets, 1 per group

1.4.1 Core

OBJECTIVES
- Explain how seismic waves are used to determine the features of the core.
- Describe the two parts of the earth's core.
- Explain how the core generates a magnetic field.
- Describe the benefits of the magnetic field around the earth.

VOCABULARY
- **core** the central portion of the earth
- **inner core** the solid center of the earth
- **magnetosphere** the area around the earth that is affected by the earth's magnetic field
- **outer core** the liquid layer of the earth's core that surrounds the inner core
- **P wave** the fastest seismic wave, which travels through solids, liquids, and gases
- **S wave** the seismic wave that travels only through solids
- **seismic wave** a wave of energy that travels through the earth

God carefully planned the creation of the earth, even its deepest parts. He designed the deep layers to sustain the rest of His earthly creation. In Psalm 95:4, the psalmist says that the depths of the earth are in the Creator's hand.

Four hundred years ago, Galileo used a telescope to view the planets, which are millions of kilometers away, but geologists can only explore short distances into the earth. The world's deepest mine is only 3.9 km deep, and no drill has bored more than 12 km into the earth. Earth scientists have never come close to reaching the center of the earth.

Throughout history, people could only guess what the inside of the earth was like. In 1864, Jules Verne wrote an imaginative description of the earth's layers in his book *Journey to the Center of the Earth*. Verne described vast forests of giant mushrooms and great seas crawling with terrifying creatures. Since that time geologists have learned that the center of the earth is very different from what Jules Verne imagined.

During the 1970s, Russian scientists began drilling in the Kola Peninsula. After 20 years, they were able to penetrate the crust to 12 km. Although this may appear to be a great feat, the earth

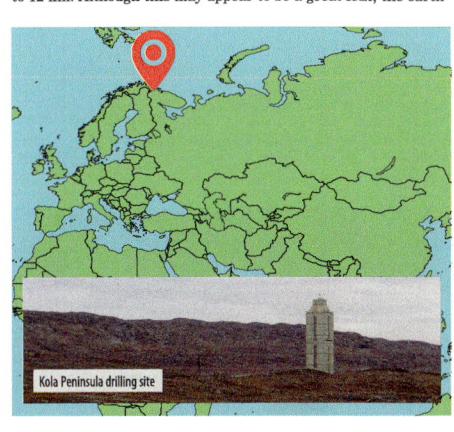

Kola Peninsula drilling site

The Crust, Mantle, and Core of the Earth

goes much deeper and can be divided into three major sections: the core, the mantle and the crust.

The **core** is the central portion of the earth. The diameter of the core is slightly larger than the diameter of the planet Mars, which is 6,787 km. The inside of this core is nearly as hot as the surface of the sun. Also, Earth's core is not a perfect sphere. Scientists have found evidence that the core has huge peaks that poke up into the mantle and that parts of the mantle carve deep valleys into the core.

Even though geologists have not drilled directly to the core, they have gathered information about the core by measuring the speed and behavior of **seismic waves** as they travel through the earth. Seismic waves are generated by a movement in the earth. There are two types of seismic waves: surface waves and body waves. **P waves** and **S waves** are body waves that carry energy from a movement in the earth through the interior of the earth. P and S waves behave very differently when they travel through a solid, liquid, or gas. P waves (primary seismic waves) are the fastest seismic waves. They travel through solids, liquids, and gases. S waves (secondary seismic waves) travel only through solids. The velocity of both types of waves is affected by the density and composition of the materials they go through. Using

TRY THIS
Pressure Practice
The inner core is solid and not liquid because the pressure of the earth at this depth counteracts the extreme temperature. To see how this could happen, take two large, soft marshmallows; leave the first one alone and place a book on top of the second one. Remove the book and compare the marshmallows. Now, place two books on top of the second marshmallow. Compare. Continue to add weight until you have a marshmallow that is completely solid. How much pressure did it take to solidify the marshmallow? How are the marshmallows like the earth's core? How are they different?

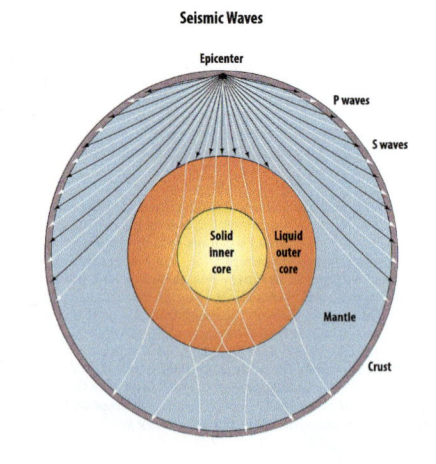

NOTES

G. Have students research different careers in geology, including paleontologist, stratigrapher, geomorphologist, mineralogist, petrologist, sedimentologist, structural geologist, geophysicist, surveyor, or instrumentation technician. Suggest students use websites for local geological organizations to gather information.

Lesson Review

1. How are seismic waves used to determine the composition and features of the core? (**Scientists can measure the behavior of P and S waves. Each type of wave moves differently and travels through different substances. Monitoring the behavior helps scientists know what substances the waves are traveling through.**)
2. What are the basic characteristics of the two parts of the earth's core? (**The outer core is liquid. It is probably iron and sulfur. It is responsible for generating a magnetic field around the earth. The inner core is solid, and it is probably an iron-nickel alloy. The core becomes solid at this point because the extreme pressure at this depth overcomes the intense heat.**)
3. How does the core generate a magnetic field? (**The constant motion of the liquid iron in the outer core produces electric currents, which generate a magnetic field.**)
4. What benefits result from the magnetic field around the earth? (**The magnetosphere protects life on the earth from dangerous solar winds and it also makes the use of a compass possible.**)

Name: _____ Date: _____

Lab 1.4.1A Making an Electromagnet

QUESTION: Will an electric current in motion create a magnetic field?

HYPOTHESIS: Answers will vary.

EXPERIMENT:

You will need:	· insulated copper wire, 60 cm long with each end stripped, exposing the copper	· electrical or cellophane tape
· large iron nail, at least 7.5 cm long		· 1.5 volt battery (AA, C or D cell)
· paper clip		

Steps:

1. Touch the nail to the paper clip. Is there a magnetic attraction? No.
2. Tape each end of the wire to each end of the battery. Be sure the copper is making contact with the battery.
3. Tape the nail to the wire so the nail is parallel to the wire and the tip is still exposed.
4. Place the paper clip near the tip of the nail. Is there a magnetic attraction? No.
5. Remove the wire from the battery and the nail from the wire.
6. Tightly coil the wire around the nail. Be sure to leave about 3 cm at each end of the coil.
7. Tape an end of the wire to each end of the battery. Be sure the copper is making contact with the battery.
8. Place the paper clip near the tip of the nail. Is there a magnetic attraction?
   Yes, there is an attraction.

ANALYZE AND CONCLUDE:

1. Was there a magnetic attraction between the paper clip and the nail in Step 1?
   No.
2. Was there a magnetic attraction in Step 4? No.
3. Was there a magnetic attraction in Step 8? Yes.
4. What caused the magnetic field in Step 8? the motion of the electric current through the coiled wire
5. How is this experiment like the core of the earth? The coil of the wire around the magnet is like the motion of the liquid iron in the earth's outer core. The nail is magnetized just like the earth is magnetized or has a magnetic field around it.

1.4.2 Mantle

The Structure of the Earth

OBJECTIVES

Students will be able to
- describe the separate sections of the mantle, including the asthenosphere.
- explain what convection currents are and how they are created.
- compare and contrast direct observations with indirect observations.

VOCABULARY

- **asthenosphere** the layer of the upper mantle composed of low-density rock material that is semiplastic
- **convection current** the circular movement of heated materials to a cooler area and cooled materials to a warmer area
- **lithosphere** the outermost, rigid layer of the earth composed of the stiff upper layer of the mantle and the crust
- **mantle** the portion of the earth's interior extending from the outer core to the bottom of the crust
- **Moho** the boundary between the mantle and the crust

MATERIALS

- Four colors of modeling clay; triple beam balances, 1 per group; fishing line (A)
- TM 1.4.1A Cross Section of the Earth
- TM 1.4.2A Moho Map
- WS 1.4.1A Model of the Earth

PREPARATION

- Obtain materials for *Lab 1.4.2A Making Waves* and *Lab 1.4.2B Asthenosphere*.

Introduction

Ask students what a mantle is. (**a shelf above a fireplace, a shelf on the wall**) Read a dictionary definition to students. Ask again what a mantle is and give a possible example. (**A mantle can cover or protect something.**) Convey to students that the mantle of the earth is a covering for the core.

Remind students about the apple demonstration from the previous lesson. Remind students that the flesh of the apple represents the mantle. The mantle makes up the majority of the earth's mass.

Discussion

- Review the layers of the earth using **TM 1.4.1A Cross Section of the Earth**. Discuss the layers of the mantle.

- Display **TM 1.4.2A Moho Map**. Ask students to look at the map and tell what they think the map is showing. (**Possible answers: how deep the Moho is, where the Moho is deep and where it is shallow, the Moho inside the entire planet**) Ask students to identify the depth of the Moho at different locations and to hypothesize about where scientists might choose to drill. (**Answers will vary.**)

- Discuss the discovery of the Moho. Ask students how the Moho was discovered. (**Possible answers: by studying seismic waves, by looking at indirect evidence, by watching what happens during an earthquake**)

- Discuss scientists' desire to actually see the mantle. Scientists have not yet been able to drill to the mantle because the technology does not exist to get there. A major obstacle is the drill. Ask the following questions:
  1. What does a drill do? (**It makes holes.**)
  2. How does it work? (**It spins really fast. It has a point on the end to dig into something. It has to move down into the substance being drilled.**)
  3. What should it be made of or what properties does it need to have? (**mineral qualities such as hardness or tensile strength**)
  4. Why do you think geologists have not been able to get to the mantle? (**Answers will vary.**)

Activities

Lab 1.4.2A Making Waves

- long hallway, preferably a tile floor, 1 per group
- items that make sound as they move across the floor

Lab 1.4.2B Asthenosphere

- beakers, 2 per group
- cornstarch
- disposable aluminum pans, pie pans, or baking pans, 1 per group
- spoons, 1 per group
- gloves, 1 pair per student

A. Continue building the model of the earth using **WS 1.4.1A Model of the Earth** if students have not completed it.

B. Direct students to research basic features of mantle minerals olivine, pyroxene, and feldspar and to determine if these minerals can be found in the local area.

C. Challenge students to research the formation of kimberlite, a rock composed of olivine, pyroxene, dark mica, calcite, and other minerals. It is formed by the crystallization of molten mantle materials that intruded into the crust. Kimberlite may contain diamonds produced by the extreme temperatures of the upper mantle.

NOTES

Lesson Review

1. How are the three layers of the mantle alike? How are they different? (**The three layers are all made of the same materials and are very hot. They are different in temperature and pressure. The lower mantle has the most pressure and highest temperature. The transition zone lies between the lower and upper mantle, so its rock is less flexible than that in the lower mantle. The upper mantle is the coolest and has the least pressure. The upper mantle above the asthenosphere is solid and part of the lithosphere. It includes the asthenosphere.**)
2. What is the asthenosphere? (**The asthenosphere is part of the upper mantle. It is a semisolid layer near the earth's crust composed of low-density rock material that is semiplastic. It flows very easily and slowly.**)
3. What are convection currents and what causes them? (**Convection currents are circular movements of a substance from a warmer area to a cooler area and a cooler area to a warmer area. This is caused by the extreme differences in temperatures and the expansion and condensation of the substance.**)
4. Compare and contrast direct observations with indirect observations. (**Direct observations are made by using any of the five senses. Usually people see the object they are studying. Indirect observations are made by studying data that results from the behavior of the object being studied.**)

1.4.2 Mantle

OBJECTIVES
- Describe the separate sections of the mantle, including the asthenosphere.
- Explain what convection currents are and how they are created.
- Compare and contrast direct observations with indirect observations.

VOCABULARY
- **asthenosphere** the layer of the upper mantle composed of low-density rock material that is semiplastic
- **convection current** the circular movement of heated materials to a cooler area and cooled materials to a warmer area
- **lithosphere** the outermost, rigid layer of the earth composed of the stiff upper layer of the mantle and the crust
- **mantle** the portion of the earth's interior extending from the outer core to the bottom of the crust
- **Moho** the boundary between the mantle and the crust

The **mantle** is the portion of the earth's interior that extends from the outer core to the bottom of the crust. It accounts for most of the earth's volume and mass. Earth's mantle is made of very hot rock. You would not be able to live if the earth's crust lay directly on top of the hot core. In fact, the land beneath your feet would melt! The mantle protects you from the core's intense heat. In later chapters, you will learn how the mantle's forces can give rise to continental drift, earthquakes, and volcanic eruptions.

The rock in the mantle has the property of plasticity, or flexibility. Plasticity means that the rock can flow or change shape. The rock flows at different rates, depending on its composition and temperature. Just as temperature differences in the atmosphere cause air to move and produce wind, the heat differences in the mantle's magma cause convection currents. A **convection current** is the circular movement of heated materials to a cooler area and cooled materials to a warmer area. As hotter fluid near the core moves up toward the crust, it expands, and the cooler rock near the surface sinks toward the core as it condenses. This cycle happens continuously.

The mantle is very complex, so scientists have divided it into separate sections to make it easier to study and to discuss. There is a lower mantle, a transition zone, and an upper mantle. All three parts of the mantle are formed with the same minerals rich in the elements iron, magnesium, silicon, and oxygen. But the density and temperature of the mantle sections change. The

Convection Currents in the Mantle

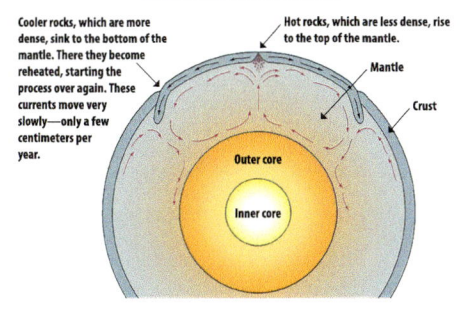

Cooler rocks, which are more dense, sink to the bottom of the mantle. There they become reheated, starting the process over again. These currents move very slowly—only a few centimeters per year.

Hot rocks, which are less dense, rise to the top of the mantle.

Mantle
Crust
Outer core
Inner core

FYI

Mantle Measurement
- Thickness: 2,900 km
- Volume: 83% of Earth's volume
- Mass: 67% of Earth's mass
- Temperature: 1,000°C–3,700°C
- Density: 3.4–5.6 g/cm³

greater the depth in the mantle, the greater the density and temperature become.

The lower mantle is hot, flexible, and under great pressure. In the transition zone, the rock becomes less flexible. The upper mantle is solid rock except for the **asthenosphere**, a semisolid layer near the earth's crust. This layer of the upper mantle is composed of low-density rock material that is semiplastic, like putty. This region flows very easily and slowly—at about the speed that your fingernails grow. The **lithosphere**, the outermost, rigid layer of the earth, is the stiff upper layer of the mantle and the crust. It rides on the asthenosphere.

Geologists have spent decades attempting to make direct observations of the earth by drilling deep into the crust to reach the mantle. Scientists use many types of observations, but the preferred method is direct observation, where the object being studied can actually be seen and perceived with all five senses. Until the mantle is actually reached, scientists must rely on indirect observations of the mantle.

In October of 1909, Croatian scientist Andrija Mohorovicic recorded seismic waves at several different stations during an earthquake. From these indirect observations, Mohorovicic concluded that the unexpected changes in the velocity of the seismic waves marked a boundary between the mantle and the crust. This boundary was named the *Mohorovicic discontinuity*, or the **Moho** for short. Later studies using more sophisticated instruments confirmed his findings.

LESSON REVIEW
1. How are the three layers of the mantle alike? How are they different?
2. What is the asthenosphere?
3. What are convection currents and what causes them?
4. Compare and contrast direct observations with indirect observations.

HISTORY

Drilling Continues
Between 2002 and 2011, several holes were drilled in the eastern Pacific. Scientists believed they reached rock just above the Moho, but they were unable to drill deeper. The SloMo project began drilling through the thinner crust in the Indian Ocean in an attempt to reach the mantle and possibly the Moho. Several setbacks slowed the project: the team got started late because supplies were not ready, a drill bit was broken, and a crew member had to be taken to shore for medical reasons. A week later, the crew and team of the ship, *JOIDES Revolution*, returned to the site to restart drilling. By January of 2016, drilling had reached a depth of 710 m, almost 610 m shy of the goal. However, the next phase of the project was planned and researchers believe they would achieve their goal of reaching the mantle in five years.

Ophiolite rocks are thought to form in the upper mantle.

Lab 1.4.2A Making Waves

QUESTION: Can indirect observations be used to draw correct conclusions?
HYPOTHESIS: Answers will vary.

EXPERIMENT:

You will need:	
• long hallway, preferably with a tile floor	• items that make sound as they move across the floor

Steps:
1. Divide the class into two groups—*Wavemakers* and *Listeners*.
2. Wavemakers should create a plan for making several different types of sound waves. For example, in Trial 1, a person could slowly bounce a ball toward the back of a listener. For Trial 2, a person could quickly run in a zigzag pattern toward a listener. Make plans for at least five different trials. Wavemakers should record the plan in the data table provided.
3. Listeners should set up chairs at one end of a hallway facing away from the other end.
4. Listeners should sit in the chairs. Each Listener should use a lab sheet to record his or her indirect observations in the data table provided.
5. As they are sitting in the chairs, Listeners should listen to the sounds and record the indirect observations. Each Listener should hypothesize about the characteristics of the sound.
6. Repeat Step 6 until the Wavemakers have finished making all the different sounds.
7. Listeners should now ask the Wavemakers what items actually made the sounds. Record actual items in data table.

Wavemakers' Data

Trial	Speed (slow, fast)	Direction (toward, away, back and forth)	Substance (basketball, metal chair)
		Answers will vary.	

Lab 1.4.2A Making Waves

Listeners' Data

Trial	Speed (slow, fast)	Direction (toward, away, back and forth)	Substance (basketball, metal chair)	Actual (fill this in when all trials are complete
		Answers will vary.		

ANALYZE AND CONCLUDE:

1. Which items were guessed correctly? Answers will vary.
2. Which items were guessed incorrectly? Answers will vary.
3. Why do you think some guesses were wrong? Possible answers: I couldn't actually see what was happening. The sound moved too fast or didn't last long enough. The sound was too much like something else I've heard.
4. How is this lab similar to scientists using seismic waves? Scientists can't actually see what the seismic waves are traveling through and we couldn't actually see what was making the sound. We had to use indirect observations to determine what was making the sound. Scientists use indirect observations to determine what is inside the earth.
5. How is this lab different? We were listening to sound waves. Scientists are recording movement of seismic waves. We could find out the actual source of the sound. Scientists cannot yet observe the actual motion inside the earth.
6. Do you think indirect observations are a good tool for scientists to use? Answers will vary.

Lab 1.4.2B Asthenosphere

QUESTION: What causes the asthenosphere to behave like both a solid and a liquid?
HYPOTHESIS: Answers will vary.

EXPERIMENT:

You will need:		
• 2 beakers	• disposable aluminum pan, pie pan, or baking pan	• gloves
• cornstarch	• spoons	

Steps:
1. Measure 240 mL of cornstarch and put it into the pan.
2. Add 150 mL of water to the cornstarch and stir. Stirring will be difficult at first but keep stirring until it is a white liquid. Describe the qualities of the mixture.
   Answers will vary. The mixture is mostly solid, but sometimes it's a liquid.
3. Add another 120 mL of cornstarch to the mixture and stir. As you stir, the mixture should become more solid and more difficult to stir.
   Describe the qualities of the mixture. It's becoming more solid when I'm stirring. When I stop, it returns to liquid.
4. As you try to stir, the mixture should become very solid. If it does not, add some more cornstarch. If it does not liquefy when you stop stirring, add some water.
5. Wearing a glove, pick up some of the mixture with your hand. Describe what happens. When I'm grabbing the mixture, it's almost solid. When I open my hand, the mixture runs through my fingers like a liquid.
6. Squeeze the mixture into a ball, then relax your hand. Describe what happens.
   When I squeeze, it's totally solid. As soon as I let go, it returns to liquid.
7. Clean up everything. Wash the mixture down the sink with hot water. Continue running hot water after the mixture is flushed down the drain.

ANALYZE AND CONCLUDE:

1. Would you classify the cornstarch mixture as a solid or a liquid? Explain.
   It is both. Sometimes it's liquid and sometimes it's solid.

Lab 1.4.2B Asthenosphere

2. When does the mixture behave like a solid and when does it behave like a liquid? The mixture becomes more solid when pressure is applied to it. Without pressure, the mixture is a liquid.
3. How is the mixture like the asthenosphere? This mixture is like the asthenosphere in that sometimes it is a solid and sometimes it is a liquid.
4. What causes the asthenosphere to behave like a solid? What causes it to behave like a liquid? The asthenosphere becomes solid when more pressure is applied to it. It is liquid when the heat counteracts the pressure.

The Structure of the Earth

1.4.3 Crust

Introduction
Emphasize the thinness of the crust when compared to Earth's interior. Discuss the fact that this thin layer is home to all life and displays most of the beauty of creation people see every day. Use the following illustration to give students an idea of the thinness and variations of the earth's crust: *If you jogged for 45 minutes at a speed of 8 kph, you would run the thickness of the oceanic crust. If you drove a car for 30 minutes at 80 kph, you would travel the thickness of the continental crust.* As a class, solve for the distances and then determine comparable distances away from the school. (Note: As another option, have students calculate how long it would take to jog or drive through the other layers of the earth.)

Discussion
- Review the layers of the earth using **TM 1.4.1A Cross Section of the Earth**. Discuss the crust, its composition, and its variety of surface features. Ask students for types of landforms and what they look like. (**Possible answers:** *mountains—rocky, jagged, covered with pine trees; rolling hills; wide plains; deserts—sandy, lots of cactus*)

- Discuss the specific regions of the continental crust: shields, fold belts, and sedimentary basins. Ask students what makes these similar and what makes them different. (**Possible answers: They are similar in that they are all part of the earth's crust; they have different shapes, locations, compositions, and formations.**)

- Lead a discussion on the elements or minerals that are found in the earth's crust. Inform students that the elements silicon, oxygen, aluminum, calcium, sodium, potassium, iron, and magnesium are all found in the crust. Ask what any of these elements is used for. (**Answers will vary.**) Why are such usable elements in the earth's crust where they are easily accessible? (**God created many features of the earth for people's benefit. God placed the minerals in the crust so people could use them to meet their needs.**)

- Discuss the layers of the oceanic crust. Point out the similar composition of the layers and the difference of formation depending on the depth. Ask students what may account for the difference in the formations. (**Possible answers: temperature differences, volcanic activity, lava flow, and currents**)

Activities
A. Complete *Try This: "Eureka"* with students.

B. Finish building the model of the earth using **WS 1.4.1A Model of the Earth** if students have not completed it.

C. Have students complete **WS 1.4.3A The Oceanic Crust**.

D. Direct students to research the highest mountain on each of the continents using **WS 1.4.3B The World's Highest Mountains**.

E. Assign students to use **WS 1.4.3C Diagram of Earth's Structure** to complete a diagram of the structure of the earth.

F. Have students make a chart comparing the features, compositions, and depths of the continental crust and the oceanic crust. When students have completed the chart, direct them to write a paragraph describing how the two are similar and how they are different.

G. Challenge students to write their own "journey to the center of the earth" story.

OBJECTIVES
Students will be able to
- describe the features of the earth's crust.
- compare and contrast the continental crust with the oceanic crust.

VOCABULARY
- **continental crust** the crust on which the continents rest
- **crust** the thin, hard outer layer of the earth
- **oceanic crust** the crust beneath the oceans

MATERIALS
- Four colors of modeling clay; triple beam balances, 1 per group; fishing line (*B*)
- TM 1.4.1A Cross Section of the Earth
- WS 1.4.1A Model of the Earth
- WS 1.4.3A The Oceanic Crust
- WS 1.4.3B The World's Highest Mountains
- WS 1.4.3C Diagram of Earth's Structure

PREPARATION
- Obtain samples of basalt and granite, small enough to fit inside a graduated cylinder for *Try This: "Eureka."*

TRY THIS
"Eureka"
- *graduated cylinders*
- *water*
- *basalt samples*
- *granite samples*
- *triple beam balance*

NOTES

H. Encourage students to make a wall mural or salt map showing the types of continental crust and the layers of the oceanic crust.

I. As a class, research the major types of microorganisms whose skeletons compose deep-sea sediments. Stress that this thin veneer lies on top of the true basaltic ocean crust.

Lesson Review

1. Briefly describe the structure of the earth. (**There are three layers to the earth: the core, the mantle, and the crust. The core is solid and very hot, the mantle is molten liquid, and the crust is the hard outer layer of the earth.**)
2. Describe the features of the earth's crust. (**It is the only layer of the earth that can actually be observed. The crust is the thinnest layer of the earth. Its depth ranges from about 6–40 km. It is made of rock and is covered with soil, sediment, or water in many places. It is much cooler than the other layers.**)
3. Even though the crust is thinner than the other layers, scientists have been unable to observe below it. Why? (**The crust is too thick and dense for drills to go through.**)
4. Compare the continental and oceanic crusts. (**The continental crust is made of mostly granite but has other types of minerals and rock as well. It has regions classified by composition and formation, such as shields, fold belts, and sedimentary basins. The oceanic crust is thinner and denser than the continental crust. Basalt is the primary rock in the oceanic crust. The oceanic crust is composed of layers of sediments and dikes made of basalt and gabbros. These vary according to the movement of lava in the different layers—pillow lava, sheet flows, and sheeted dikes.**)

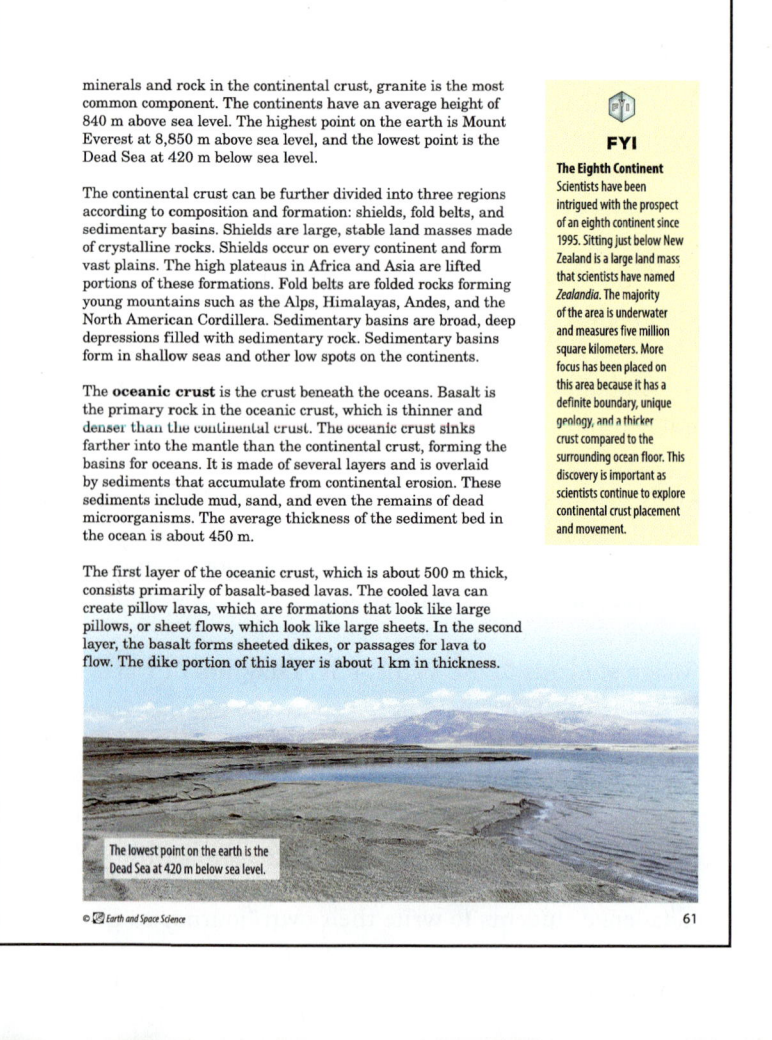

1.4.3 Crust

OBJECTIVES
- Describe the features of the earth's crust.
- Compare and contrast the continental crust with the oceanic crust.

VOCABULARY
- **continental crust** the crust on which the continents rest
- **crust** the thin, hard outer layer of the earth
- **oceanic crust** the crust beneath the oceans

FYI
Crust Counts
- Thickness: 6 km (oceanic), 40 km (continental)
- Volume: <1%
- Mass: <1%
- Temperature: 0°C–400°C
- Density: 3 g/cm³ (oceanic), 2.7 g/cm³ (continental)

The crust of the earth is the only layer that can actually be observed. The **crust** is the thin, hard outer layer of the earth. It is home to all life on Earth, and it has a great variety of features. A view from an airplane gives you the opportunity to see much of the varied scenery over the crust's different regions. Some regions are thick with trees. Other regions have farmland, deserts, or mountains. Flying over a river system gives a good outline of the design of watersheds, and flying over the ocean can make one feel very small indeed. Consider the vastness of the earth's continents. The earth's surface is truly a beautiful mosaic.

All of these wonderful features lie on the crust. This thin layer of rock is covered with soil, sediment, or water in many places. The crust is a very complex and varied layer, partly because of the action of the mantle that lies underneath. The hot mantle beneath the crust constructs and destroys the ocean floor, builds and splits continents, and rearranges land masses into towering mountain chains.

The crust is by far Earth's thinnest layer. The depth of the crust ranges from about 6–40 km. The crust is too thick in most places for drills to reach the mantle, however. The density of the crust varies depending on the individual rocks found in different locations. The temperature of the crust at Earth's surface is cool, but it is very hot near the mantle. Without this temperature difference, most life on Earth would cook.

Scientists divide the crust into two categories: continental and oceanic. **Continental crust** is the crust on which the continents rest. The continental crust has a more complex rock structure than the oceanic crust. Although there are many types of minerals and rock in the continental crust, granite is the most common component. The continents have an average height of 840 m above sea level. The highest point on the earth is Mount Everest at 8,850 m above sea level, and the lowest point is the Dead Sea at 420 m below sea level.

The continental crust can be further divided into three regions according to composition and formation: shields, fold belts, and sedimentary basins. Shields are large, stable land masses made of crystalline rocks. Shields occur on every continent and form vast plains. The high plateaus in Africa and Asia are lifted portions of these formations. Fold belts are folded rocks forming young mountains such as the Alps, Himalayas, Andes, and the North American Cordillera. Sedimentary basins are broad, deep depressions filled with sedimentary rock. Sedimentary basins form in shallow seas and other low spots on the continents.

The **oceanic crust** is the crust beneath the oceans. Basalt is the primary rock in the oceanic crust, which is thinner and denser than the continental crust. The oceanic crust sinks farther into the mantle than the continental crust, forming the basins for oceans. It is made of several layers and is overlaid by sediments that accumulate from continental erosion. These sediments include mud, sand, and even the remains of dead microorganisms. The average thickness of the sediment bed in the ocean is about 450 m.

The first layer of the oceanic crust, which is about 500 m thick, consists primarily of basalt-based lavas. The cooled lava can create pillow lavas, which are formations that look like large pillows, or sheet flows, which look like large sheets. In the second layer, the basalt forms sheeted dikes, or passages for lava to flow. The dike portion of this layer is about 1 km in thickness.

FYI
The Eighth Continent
Scientists have been intrigued with the prospect of an eighth continent since 1995. Sitting just below New Zealand is a large land mass that scientists have named *Zealandia*. The majority of the area is underwater and measures five million square kilometers. More focus has been placed on this area because it has a definite boundary, unique geology, and a thicker crust compared to the surrounding ocean floor. This discovery is important as scientists continue to explore continental crust placement and movement.

Highest point on the earth: Mount Everest at 8,850 m in the Himalayan mountain range in both Nepal and Tibet

The lowest point on the earth is the Dead Sea at 420 m below sea level.

TRY THIS

"Eureka"

The Ancient Greek scientist Archimedes was once asked to determine whether someone had stolen the king's crown and replaced it with a fake. Archimedes knew he could not melt the crown to see whether it was pure gold, but he had to find its true composition. History tells us that Archimedes came up with the solution as he was bathing. He saw the bath water rise as he got in the tub. "Eureka!" he exclaimed as he ran down the street. Archimedes realized that a solid placed in water will displace its own volume of water. Use Archimedes' principle to determine the volume and ultimately the density of the main components of the continental crust and the oceanic crust. Measure the volume of a sample of basalt and a sample of granite using a graduated cylinder and water. Determine the mass of each by using a scale. Calculate the density of each rock sample using the formula: density = mass/volume. Record the results and observations about the properties of each rock. How are the samples similar? How are they different? How do you think these differences affect the earth's crust?

The third layer of the oceanic crust is composed of dikes made of gabbro, which is basalt rock with coarse mineral grains. This portion is about 4.5 km thick.

Most people think of the ocean floor as a place without interesting features, probably because people cannot see the ocean floor. But the oceanic crust is a work of art. Its features look much like the features of the continental crust. Undersea ridges form extended mountain ranges—in fact, the mid-ocean ridge in the Atlantic Ocean is perhaps the earth's single most dramatic feature. Individual volcanic mountains called *seamounts* rise from the ocean floor. The tallest mountain in the world begins on the ocean floor. The base of Mauna Kea is almost 6,000 m below sea level and the summit is about 4,000 m above. This makes the full height of Mauna Kea about 10,000 m!

The ocean floor has a flat region that extends hundreds of kilometers. Deep, steep valleys and ravines plunge into the oceanic crust. Submarine canyons in the ocean floor were cut by river water or shallow ocean currents flowing out to sea. The oceanic crust, like the continental crust, reveals God's creativity and love of variety.

Reynisdrangar, basalt sea stacks, in Iceland

LESSON REVIEW
1. Briefly describe the structure of the earth.
2. Describe the features of the earth's crust.
3. Even though the crust is thinner than the other layers, scientists have been unable to observe below it. Why?
4. Compare the continental and oceanic crusts.

UNIT 2

Chapter 1: *Weathering and Erosion*
Chapter 2: *Soil*

Key Ideas

Unifying Concepts and Processes
- Systems, order, and organization
- Evidence, models, and explanation
- Change, constancy, and measurement
- Evolution and equilibrium
- Form and function

Science as Inquiry
- Abilities necessary to do scientific inquiry

Earth and Space Science
- Structure of the earth system
- Earth's history
- Energy in the earth system
- Geochemical cycles
- Origin and evolution of the earth system

Science and Technology
- Abilities of technological design
- Understandings about science and technology

Science in Personal and Social Perspectives
- Populations, resources, and environments
- Natural hazards
- Risks and benefits
- Science and technology in society
- Population growth
- Natural resources
- Environmental quality
- Natural and human induced hazards
- Science and technology in local, national, and global challenges

History and Nature of Science
- Science as a human endeavor
- Nature of science
- History of science
- Nature of scientific knowledge

Geologic Changes

Vocabulary

abrasion
bedrock
carbonation
carbonization
channel
chemical weathering
deflation hollow
desiccation
drumlin
earthflow
erratic
fault
floodplain
fossil
geologic column
glacial drift
glacier
gully
horizon
humus
hydrolysis
ice wedging
index fossil
intrusion
landslide
law of superposition
mass wasting
mechanical weathering
moraine
mudflow
oxidation
petrifaction
pore space
regolith
rock pedestal
runoff
soil creep
soil profile
subsoil
till
topography
topsoil
trace fossil
unconformity

SCRIPTURE

Before the mountains were born or You brought forth the whole world, from everlasting to everlasting You are God.
Psalm 90:2

2.1.0 Weathering and Erosion

LOOKING AHEAD

- For **Lesson 2.1.1**, contact a teacher in another climate and arrange for a collaborative activity comparing the effects of weathering in different climates.
- For **Lesson 2.1.2**, arrange for a field trip to an old cemetery to survey the weathering rates of different types of stone of various ages.
- For **Lesson 2.1.3**, plan a field trip to visit the site of a mass wasting event.
- In **Lesson 2.1.4**, invite an insurance agent to explain how erosion factors into homeowner's insurance. Invite a representative from an engineering firm or a contractor to explain ways in which erosion is prevented or remedied. Plan a field trip to a beach that has breakwaters.
- For **Lesson 2.1.5**, plan a field trip to a local river, stream, or ocean shoreline to search for examples of erosion.

SUPPLEMENTAL MATERIALS

- TM 2.1.2A Weathering

- Lab 2.1.1A Ice Wedging
- Lab 2.1.2A Chemical Weathering
- Lab 2.1.3A Landslide
- Lab 2.1.4A Deflation
- Lab 2.1.4B Wave Erosion
- Lab 2.1.5A Water Erosion and Time
- Lab 2.1.5B Water Erosion and Force
- Lab 2.1.6A Glacial Erosion

- WS 2.1.1A Weathering Observations
- WS 2.1.2A Sculpture Preservation
- WS 2.1.4A The Art of Wind and Water
- WS 2.1.6A Scrape, Rattle, and Roll

- Chapter 2.1 Test

BLMs, TMs, and tests are available to download. See Understanding Purposeful Design Earth and Space Science at the front of this book for the web address.

Chapter 2.1 Summary

Some weathering and erosion forces are imperceptible; others are huge and powerful, but all are part of God's orderly plan. Whether small or large, forces of weathering and erosion shape the earth. These forces break rocks into soil, carve canyons, form lakes, direct streams, and sculpt mountains. Weathering and erosion are natural processes, but human activity affects the rate at which they occur. People have a responsibility to carefully consider the effect their actions will have on the creation around them.

Background

Lesson 2.1.1 – Mechanical Weathering

Mechanical weathering, also called *physical weathering*, is the process by which rock at or near the earth's surface is physically disintegrated. Agents of mechanical weathering include temperature extremes, ice, plant roots, and surface water. Climate is a major factor in weathering. Wide temperature fluctuations increase the effects of exfoliation and ice wedging. The freeze-thaw weathering process is common in mountains and in glacial environments. Water in a crack of a rock increases in volume by about 9% as it turns to ice, which exerts pressure great enough to enlarge the crack. This ice wedging occurs where the temperature regularly fluctuates below and above the freezing point of water. Frost is another agent of the freeze-thaw cycle. Other mechanical forces that work to enlarge cracks in rocks include the growth of minerals introduced through water solution, the release of pressure because of excavation or natural uplift, and the intrusion of plant roots. Roots secrete chemicals to extract nutrients, which also contributes to chemical weathering. Other agents of mechanical weathering include animals that burrow into soft rocks and human activities such as mining, farming, construction, and excavation. Sometimes factors combine to affect the rate of weathering. The mineral composition of a rock is also a factor in weathering. For example, sedimentary rock weathers faster than igneous rock. Other factors include rock joints, grain size, and topography.

Lesson 2.1.2 – Chemical Weathering

Chemical weathering changes the chemical structure of a material, forming a new material. Water, oxygen, or plants usually cause chemical weathering. Water can chemically alter a mineral through hydrolysis, which is the breaking down of a substance by a chemical reaction with water. A second way water chemically alters minerals is through hydration, in which water is absorbed into the crystal structure of a mineral and alters it. Water can also combine with gases in the air or the soil to produce acids, which dissolve minerals in rocks. When water combines with carbon dioxide it creates carbonic acid, which particularly dissolves certain minerals in limestone. Water can also combine with nitrogen oxides and sulfur dioxide to form nitric and sulfuric acids, respectively. Although all of these acids can form naturally, nitric and sulfuric acids are most commonly the by-products of burning fossil fuels. Water-related chemical weathering is prevalent in regions that experience plenty of precipitation, particularly if there is abundant vegetation to help retain the moisture. A second agent of chemical weathering is oxygen. Oxidation is the combination of oxygen with iron-rich minerals. Corrosion is the atmospheric oxidation of metals. The rusting of iron is the most common form of corrosion. A third common form of chemical weathering is biological. Biological weathering occurs when plants' roots grow into fissures in rocks and produce weak acids, which help break down the rocks.

Lesson 2.1.3 – Mass Wasting

Mass wasting is the downhill movement of loose rocks and soil, and it can be divided into several categories. When gravity causes a rapid, downhill movement of a large mass of rock and soil along a distinct plane, it is called *a landslide*. Sometimes water helps lubricate this process; at other times gravity works alone. Rock slides are landslides of large masses of rock with little or no water involved. Mudflows are streams of mud that pour down canyons or mountains during heavy rainstorms, especially in areas with little vegetation to protect hillsides from erosion. The mud and debris funnels through natural channels until it spreads out at the base of a slope. In contrast, when water-saturated soil or clay on a slope slides slowly downhill, it is called *an earthflow*. Earthflows are primarily distinguished from mudflows in that they have less water content and are not necessarily confined to a channel. Soil creep is another slow form of mass wasting in which topsoil almost imperceptibly moves down a slope over an extended period of time. Solifluction is the movement of a top layer of regolith over an impermeable layer, such as permafrost or clay.

Lesson 2.1.4 – Wind and Wave Erosion
Wind is a powerful force of erosion. Abrasion occurs when wind drives small particles against objects, which wears them smooth or knocks previously weathered portions loose. Wind deflates areas with loose soil or sand, carrying the loose material away to leave desert pavement or deflation hollows. Large deflation hollows can extend below the water table, creating oases. The wind eventually deposits its load of sand and soil, which can form loess hills from soil or dunes from sand.

Waves are another powerful force of erosion. The force of their impact, combined with abrasion from the sediments they carry, can wear away rock to form caves, arches, and stacks in the sea. Beaches protect inland features from waves' erosive powers. Beach sand migrates with the seasons as the amount of incoming material and the intensity of the waves change. People's actions can either promote or prevent erosion to some extent. Stripping land of vegetation or impeding the natural path of waves can increase the effects of wind and wave erosion. However, people can lessen the effect of wind erosion by constructing various forms of windbreaks, and they can protect property from wave erosion by building further inland.

Lesson 2.1.5 – River Erosion
Rivers and their tributaries are the primary movers of soil and sediments. Signs of river erosion include cracks in a riverbank, grass clumps in a river, overhangs of a riverbed, exposed tree roots or trees leaning into a river, a collapsed riverbank, and brown water. As a river moves materials along its bed, the sides of the channel and the riverbed are changed according to its topography. On steep slopes, gravity pulls water swiftly downward, which primarily results in erosion along the channel bed. Water can deposit its load of sediment at several points: on gradual slopes, along the inside curves of meanders, during transitions from a steep to a shallow slope, or on reaching a still body of water. Lowland rivers can attain a greater velocity than highland streams because their many tributaries give them a greater volume of water. Rivers simultaneously erode and deposit materials when they flood. During flooding, a river's volume and rate of flow increase, which increase its force and enable it to erode a greater amount of material. Flooding can be controlled by woodlands, marshes, dams, and levees.

Lesson 2.1.6 – Glacial Erosion
Glaciers are responsible for much of Earth's topography. They shape landforms by removing, transporting, and depositing materials that vary in size from tiny particles to enormous boulders. The movement of glaciers is not always obvious. Glaciers move as a solid rather than as a liquid, as indicated by the formation of crevasses (cracks in the surface caused by tension on the brittle ice). Friction slows the sides and bottom of a glacier, but the center and surface are able to move more rapidly. The flow rate of a glacier is also affected by its volume, the slope of the ground, the slope of the glacier's top, the amount of debris that it carries, and the temperature. The factors behind glacier movement are complicated; not all of the factors affect each glacier at the same time. These factors include that a glacier may melt under pressure and then refreeze (this may push the glacier along the path of least resistance), that a glacier's ice layers may slide, or that a glacier's granules may be rearranged when a glacier melts under pressure.

WORLDVIEW

- Many of the earth's features seem to be timeless, barring any cataclysmic activity. Mountains, hills, valleys, and riverbeds seem to appear the same to us as they did to the generations before us. Yet, when we look closely, all of these seemingly permanent features have actually changed. Rivers alter their courses as banks fall in or as sediment deposits grow. Mountain slopes form new crevices and wear away. Valleys deepen or widen. If we watch long enough and carefully enough, we will see even the most stable features of our planet wear away. God, however, is a true constant. Psalm 102:25–28 reminds us that God is more enduring than the earth He created. God's essence and will are the same today as when He created the universe, and they will still be the same when this world is replaced by the new heaven and new earth. Psalm 102 also gives hope that although this world will one day perish, God will provide a permanent place for His children in His presence. The future of those who love God is every bit as certain as God's changelessness because He decreed it. As you lead your students through this chapter on weathering and erosion, encourage them to place their faith in the one true constant—their Creator and Redeemer, God.

2.1.1 Mechanical Weathering

Weathering and Erosion

OBJECTIVES

Students will be able to
- identify and explain several processes of mechanical weathering.
- assess the effects that climate, rock type, exposed surface area, and topography have on mechanical weathering.
- apply concepts of mechanical weathering to technological designs.

VOCABULARY

- **ice wedging** the mechanical weathering process in which water in the cracks of rocks freezes and expands, widening the cracks
- **mechanical weathering** the breaking down of rocks by physical processes
- **topography** the surface features of a place or region

MATERIALS

- Sealable plastic bottle, ziplock bag (*Introduction*)
- WS 2.1.1A Weathering Observations

PREPARATION

- Fill a plastic bottle of water to the brim, seal it, place it in a large ziplock bag, and freeze it. The bottle should crack. (*Introduction*)
- Obtain materials for *Lab 2.1.1A Ice Wedging*.
- Obtain materials for *Try This: Bean Power* and *Try This: Moving Mountains*.
- Contact a teacher in another climate to arrange for a collaborative activity in which students compare the effects of weathering in different climates. (*C*)

TRY THIS

Bean Power
- lima bean seeds, 4–8 per group
- plaster of paris
- small paper cups, 2 per group

The plaster in the cup with the bean (*continued*)

Introduction

Display a frozen bottle of water that has expanded and burst. Ask students why the bottle broke. (**The expanding ice broke the bottle.**) Lead students in a discussion regarding how roads and rocks might be affected in a manner similar to the frozen water bottle.

Ask students why sidewalks are divided into sections with spaces or cracks. (**The spaces or cracks allow the sections of sidewalk to expand and contract without cracking in undesired places.**) Ask why one section of sidewalk sometimes heaves upward. (**Possible answers: The rocks underneath it were forced upward because of ground frost; the tree roots pushed it up.**) How can these problems be addressed? (**Answers will vary.**)

Discussion

- Guide a discussion regarding the evidence of mechanical weathering in the region.

- Challenge students to consider how the earth would be different if God designed rock to weather much faster or slower.

- Ask the following questions:
  1. What is the difference between surface area and topography? (**Surface area is the outermost layer of something. Topography is the surface features of a place or region.**)
  2. How are surface area and topography connected to the mechanical weathering rate? (**The greater the exposed surface area a place has, the greater the potential for mechanical weathering. Topography can also affect the weathering rate. For example, areas with high elevation weather at an increased rate because of temperature extremes.**)

Activities

> **Lab 2.1.1A Ice Wedging**
> - plastic drinking straws, 1 per student
> - modeling clay
> - ziplock bags, 1 per student
> - permanent markers, 1 per group
>
> This lab requires access to a freezer.

A. Complete *Try This* activities with students.

B. Distribute **WS 2.1.1A Weathering Observations** and have students record their observations over the next several days. As a class, discuss the ramifications of these observations.

C. Arrange for a collaborative activity with a class in another climate. Assign students to compare weathering, erosion, and any related issues in the two regions. Related issues may include protection of monuments; maintenance or repair of roads, structures, or landscapes; and safety precautions when hiking, climbing, or swimming.

D. Have students design an experiment that tests the effects of mechanical weathering. For example, they may choose to moisten different-sized balls of modeling clay and expose them to cycles of freezing and thawing or to expose the same moistened ball to several freeze-thaw cycles. Each experiment should include a control.

E. Challenge students to design experiments that will compare the following factors in weathering in the local area: type of material being weathered, exposed surface area, and topography. If practicable, have students complete the experiments.

64

F. Direct students to research the materials best suited for building a bridge in their region. Students should select a site for the bridge, such as over a road, over water, or between mountain sides or hills. Guide them to consider the physical forces that will likely act on the bridge, and have them determine what can be done to minimize the effects of mechanical weathering on the bridge's materials and its foundation.

Lesson Review

1. What is mechanical weathering? (**Mechanical weathering is the breaking down of rocks by physical processes.**)
2. How can water cause mechanical weathering? (**Water can freeze and expand in cracks, which eventually breaks apart rocks.**)
3. Explain how climate affects weathering processes. (**In cold, wet climates, the frequent freeze-thaw cycles encourage ice wedging. In dry climates, rapid changes between hot and cold temperatures cause rapid expansion and contraction of rocks, which causes rocks to split.**)
4. How do the following factors work together to affect the rate at which rock weathers: type of rock, exposed surface area, and topography? (**The exposed surfaces of a soft rock will weather faster than exposed surfaces of a hard rock. Greater exposed surface area enables more weathering. There are greater temperature extremes at high elevations, which encourage weathering. Weathered material falls away from rocks at higher elevations because of the effects of gravity, which expose more surface area for new weathering.**)
5. What kinds of mechanical weathering might an architect need to consider when designing a house to be built on a forested mountain cliff? (**Possible answers: temperature extremes, tree roots, and burrowing animals**)

TRY THIS

(continued from previous page) seeds should crack as the seeds germinate in about a day because of the heat and moisture created by the chemical reaction of the plaster powder and water.

Moving Mountains
Possible experiments include filling a plastic container with water, covering it with a plate and a heavy weight, and freezing the container; or alternately heating a rock with a Bunsen burner and dunking it in a beaker of cold water. Be sure to stress proper lab safety if students are permitted to heat and cool rocks.

2.1.1 Mechanical Weathering

OBJECTIVES
- Identify and explain several processes of mechanical weathering.
- Assess the effects that climate, rock type, exposed surface area, and topography have on mechanical weathering.
- Apply concepts of mechanical weathering to technological designs.

VOCABULARY
- **ice wedging** the mechanical weathering process in which water in the cracks of rocks freezes and expands, widening the cracks
- **mechanical weathering** the breaking down of rocks by physical processes
- **topography** the surface features of a place or region

Some of Earth's features appear timeless. Mountains, cliffs, canyons, and valleys may look as though they have not changed for centuries. However, they are actually changing all the time. God uses tools such as wind, water, and chemical action to continuously shape the earth's surface. An example of how rocks change over time may be seen in stone monuments such as those used to mark graves or to commemorate important people, places, or events. New monuments have crisp inscriptions and sharp corners, but the lettering on old stone monuments is worn and smooth, and some may even be unreadable. This change is the result of weathering and erosion.

Weathering is the breaking down of rocks into smaller pieces by mechanical or chemical processes. Weathering works together with erosion to produce visible changes in the earth's surface. Erosion is the removal and transport of material by wind, water, or ice. Weathered material stays in one place, but eroded material moves.

Mechanical weathering, also known as *physical weathering*, is the breaking down of rocks by physical processes. Most of the weathering on monuments is mechanical weathering. The engraving is worn away, but the chemical makeup of the stone has not changed. Most mechanical weathering is caused by temperature extremes and the activities of living things, which is known as *biological weathering*. How fast weathering happens, known as its *rate*, is affected by the climate, the type of rock, the amount of exposed surface area, and the topography.

New inscription

Weathered inscription

Climate is probably the greatest factor in weathering. Ice wedging is common in climates that receive frequent precipitation and experience rapid temperature extremes. **Ice wedging** is the mechanical weathering process in which water freezes and expands in the cracks of rocks, widening the cracks. This process is common in mountainous regions during the spring and fall because temperatures can be warm during the day but drop below freezing at night. During the day, rainwater flows into cracks in the rocks and seeps into the rocks' pores. At night, the water freezes and expands. The pressure of the expanding ice is so great that it wedges the cracks apart. Through repeated thawing and freezing, the ice is wedged deeper into the rocks, widening the cracks. Over time ice wedging shatters the rock surfaces into angular fragments and blocks.

Hot, dry climates are free from ice wedging, but they still experience the mechanical weathering that comes from temperature differences. Even though deserts are often blistering hot during the day, temperatures can drop below freezing at night because the air has very little water vapor to hold in heat. The rocks expand in the heat and contract in the cold. This continuing cycle weakens natural cracks in the rocks, causing them to eventually split apart. Often, these splits separate surface layers from the layers underneath. When this process occurs on massive rock layers of granite or other hard rocks, the rocks weather into curved slabs in a process called *exfoliation*. Exfoliation is similar to peeling the layers off an onion.

Living things also contribute to mechanical weathering. Trees sometimes sink their roots into rocks. The roots follow cracks and crevices, splitting the rocks apart as the trees grow. Burrowing animals digging in the soil expose rocks, and some mollusks can

TRY THIS

Bean Power
Plant 2–4 lima bean seeds in about 2 cm of prepared plaster of paris in a small paper cup. Prepare a control cup of plaster of paris without bean seeds. Observe the cups for several days at the same time each day and sketch what you see. Create a hypothesis to explain the results.

Rock broken by repeated freezing and thawing

even bore holes into rocks. This exposes new rock surfaces to the forces of weathering.

The type of rock, the rock's amount of exposed surface area, and topography affect how quickly a rock weathers in any climate. Certain rocks are softer than others. In general, sedimentary rocks are easily weathered, especially by mechanical processes. The particles in shale and sandstone are relatively loose and easily broken up into fragments of clay and sand. Igneous and metamorphic rocks are quite hard, so they weather more slowly than sedimentary rocks. Even after all of the sedimentary rock of a region has weathered or eroded away, igneous and metamorphic rocks often appear unchanged. The mineral in igneous and metamorphic rocks that is most resistant to weathering is quartz. Quartz often remains as tiny grains of sand after other minerals in a rock have eroded away.

The amount of rock that is exposed to the elements also determines the rate of weathering. The greater the exposed surface area, the greater the potential for weathering. Small rocks weather faster than large rocks, just as sugar grains dissolve faster than sugar cubes in hot tea. Surface area is increased when cracks form or widen and when loose particles or soil erodes away.

Topography, the surface features of a place or region, also affects the rate of weathering. High elevation encourages weathering because of its greater temperature extremes. Also, gravity pulls weathered rock fragments from steep slopes, and the water from heavy rain helps gravity by washing away particles more rapidly. As particles are swept from the slopes of mountains, new rock is constantly exposed, speeding up the weathering processes.

Weathering and erosion affect structures people build as well as natural features. Architects and engineers have to take weathering factors into consideration as they choose building sites, select building materials, and design structures. They may need to reinforce a foundation, seal building stones, or improve drainage in order to prevent the erosion of a building site or a structure.

LESSON REVIEW
1. What is mechanical weathering?
2. How can water cause mechanical weathering?
3. Explain how climate affects weathering processes.
4. How do the following factors work together to affect the rate at which rock weathers: type of rock, exposed surface area, and topography?
5. What kinds of mechanical weathering might an architect need to consider when designing a house to be built on a forested mountain cliff?

TRY THIS
Moving Mountains
Design an experiment to determine whether temperature changes can split rocks.

HISTORY
Measuring Weathering
Research on tombstone weathering has contributed to the general knowledge of weathering factors. In the late 1880s Sir Archibald Geikie of Edinburgh, Scotland, calculated the weathering rate for marble tombstones to be 8.5 mm per century. In the 1960s, professor E. M. Winkler measured a vein in a marble marker in Indiana and calculated the weathering rate for the marble surface to be 1.5 mm in 43 years. If all factors remained constant, Professor Winkler's calculations show marble weathering in that region to be 3.49 mm per century. Both of these studies were done in humid climates. A measurement taken in a dry climate showed a weathering rate of only 1 mm per century. The monuments on the Nile River in Egypt (a very arid climate) testify to the resistance of the granite there to weathering. In 1916, the American geologist D.C. Barton estimated a weathering rate of 1–2 mm every millennium for granite in dry climates. The greatest rate he calculated was only 0.5 mm per century.

Name: _____ Date: _____

Lab 2.1.1A Ice Wedging

QUESTION: How can water break up rocks?
HYPOTHESIS: Answers will vary.

EXPERIMENT:

You will need:	• modeling clay	• permanent marker
• plastic drinking straw	• ziplock bag	

Steps:
1. Seal one end of a drinking straw with clay. Do not allow the clay to extend past the end of the straw.
2. Fill the straw with water.
3. Seal the other end of the drinking straw with clay. Do not allow the clay to extend past the end of the straw.
4. Place the straw inside the ziplock bag.
5. Use a permanent marker to write your name on the bag.
6. Place the straw in a freezer overnight.
7. Predict what will happen to the clay when the water in the straw freezes. Record your prediction.
8. Observe the straw the next day and record your observations. Answers will vary but should include that the expanding ice pushed the clay out of the straw.

ANALYZE AND CONCLUDE:
1. What happened to the clay? Why did this happen? The frozen water pushed the clay plugs out of the straw; a column of ice extended past the end of the straw. This happened because water expands as it freezes.
2. What do you think would happen if the water in the cracks of a rock froze? The rock would crack or break.

Weathering and Erosion

2.1.2 Chemical Weathering

Introduction
Show the class a piece of limestone. Have students suggest easy ways to wear away the surface of the stone. (**Answers will vary.**) With an eyedropper, place several drops of dilute sulfuric acid on the limestone to show an example of chemical weathering. Explain how acidic precipitation can chemically weather buildings and statues. Natural, unpolluted rainwater is mildly acidic because of carbon dioxide (CO_2) in the atmosphere. Water vapor in clouds can be made even more acidic if water dissolves the gaseous pollutants nitrogen oxide (NO_2) and sulfur dioxide (SO_2), which first oxidizes to form sulfur trioxide (SO_3) before reacting with water.

$NO_2 + H_2O \longrightarrow HNO_3$ (nitric acid) $SO_3 + H_2O \longrightarrow H_2SO_4$ (sulfuric acid)

Discussion
- Guide a discussion regarding the evidence of chemical weathering in the region.

- Lead the class in discussing how an iron cannon could be protected from oxidation. Have students consider all of the cannon's exposed surfaces and the effects normal use and the environment might have on the various methods of preservation identified in the Student Edition.

- Display **TM 2.1.2A Weathering** and have students explain whether each picture shows mechanical weathering, chemical weathering, or both. Ask students to hypothesize what factors likely contributed to the weathering.
  1. Mechanical and chemical weathering: *cracked wall*—expansion and contraction because of heat, ice wedging, or settling of the house because of frost upheaval; *flaking stucco*—ice wedging or temperature fluctuation
  2. Mechanical weathering: *sidewalk cracks*—pressure from tree roots
  3. Mechanical and chemical weathering: *small cracks and surface chips*—acid precipitation or hydrolysis; *large cracks*—ice wedging
  4. Chemical weathering: *cave*—acid precipitation; *stalactites and stalagmites*—calcium deposits as water evaporated from calcium-laden water
  5. Mechanical and chemical weathering: *eroded rock*—water, ice wedging; *rusty bridge*—oxidation. Plant roots are likely mechanically and chemically widening cracks in the rock.

Activities

Lab 2.1.2A Chemical Weathering
- steel wool, 2 pieces per group
- permanent markers, 1 per group
- ziplock bags, 2 per group
- paper towels, 2 per group

Consider having students use wet steel wool to test their hypotheses for retarding chemical weathering.

A. Complete *Try This: Speed Up, Slow Down* with students.

B. Have students complete **WS 2.1.2A Sculpture Preservation**.

C. Direct students to research the effects of weathering on rocks commonly used for tombstones, such as marble, granite, sandstone, and slate. Take students on a field trip to an old cemetery to survey the weathering effects of different types of stone of various ages. Have students compare their observations to their research and write hypotheses to explain any seeming discrepancies. Emphasize to students the proper, respectful demeanor for such a place.

OBJECTIVES
Students will be able to
- identify and explain several agents of chemical weathering.
- contrast the types of chemical weathering that involve water.
- apply concepts of chemical weathering to technological designs.

VOCABULARY
- **carbonation** the process in which carbon dioxide from the atmosphere or soil dissolves in water to form carbonic acid
- **chemical weathering** the breaking down of rocks by chemical processes
- **hydrolysis** the breaking down of a substance by a chemical reaction with water
- **oxidation** a chemical change in which a substance combines with oxygen

MATERIALS
- Limestone, eyedropper or pipette, dilute sulfuric acid (*Introduction*)
- TM 2.1.2A Weathering
- WS 2.1.2A Sculpture Preservation

PREPARATION
- Obtain materials for *Lab 2.1.2A Chemical Weathering*.
- Obtain materials for *Try This: Speed Up, Slow Down*.
- Arrange for a field trip to an old cemetery to survey the weathering rates of different types of stone of various ages. Before the trip, identify markers with unique weathering patterns, markers made of the same rock but with different ages, and markers of the same age but made from different rocks. (*C*)
- Assign students to bring in water samples, other than tap water, for acidity testing. (*D*)

TRY THIS

Speed Up, Slow Down
- pH test kits, 1 per group
- vinegar or lemon juice
- chalk

For example, students may choose to create water samples of various acidity using vinegar or lemon juice, test the pH of each sample, immerse a piece of chalk in each sample, and time how long it takes for each piece to dissolve. Alternatively, students could measure the chalk's mass prior to immersion, immerse chalk in each solution for the same amount of time, allow the chalk to dry completely, and compare the chalk's mass before and after immersion.

D. Have students bring in water samples from a variety of sources and measure its acidity. Possible sources include well, ponds, streams, or rainwater. Lead a discussion regarding the possible reasons for the various acidity levels.

E. Assign students to design an experiment that compares the rates of mechanical and chemical weathering in different climates. For example, students could soak clay balls in water and compare the weathering of one that is frozen and thawed repeatedly with one that remains frozen. Alternatively, students could compare the effects of a few drops of vinegar on chalk or steel wool placed in a freezer with similarly treated steel wool that is exposed to a humidifier.

F. Challenge students to research the impact of tourism on the rates of mechanical and chemical weathering on the ancient Nabataean and Roman architecture in the city of Petra, Jordan. Students should include recommended solutions for restoring and preserving the architecture.

G. Direct each student to choose three of the six major land biomes and for each chosen biome to select a stone for a monument that will withstand the common mechanical and chemical weathering factors found in that biome. Students should consider whether the monument will be in a city or in the country and the particular weathering factors that might be found there.

H. Challenge students to suggest ways that power companies, car manufacturers, and other industries can reduce sulfur dioxide and nitrous oxide emissions but still provide the services on which people depend. Direct students to note possible problems or challenges that the industries might face when trying to implement the suggested changes.

2.1.2 Chemical Weathering

OBJECTIVES
- Identify and explain several agents of chemical weathering.
- Contrast the types of chemical weathering that involve water.
- Apply concepts of chemical weathering to technological designs.

VOCABULARY
- **carbonation** the process in which carbon dioxide from the atmosphere or soil dissolves in water to form carbonic acid
- **chemical weathering** the breaking down of rocks by chemical processes
- **hydrolysis** the breaking down of a substance by a chemical reaction with water
- **oxidation** a chemical change in which a substance combines with oxygen

Chemical weathering is the breaking down of rocks by chemical processes. Chemical reactions occur between the minerals in rocks and either water, carbon dioxide, oxygen, or acids. These reactions either break down or change the chemical structure of some of the minerals. The rate of chemical weathering is affected by many of the same factors as the rate of mechanical weathering. Certain rocks contain minerals that are more easily affected by chemical weathering. Rocks that have been broken up into fragments by mechanical weathering are more susceptible to chemical weathering because the smaller rock fragments have more exposed surface area. Rocks in warm climates are more susceptible to chemical weathering because high temperatures speed up chemical reactions.

Water is one of the primary agents of chemical weathering. Many people think of water as pure—not as something strong enough to be part of a chemical reaction. However, water weathers minerals in rocks in three main ways: hydrolysis, hydration, and carbonation. **Hydrolysis**, the breaking down of a substance by a chemical reaction with water, dissolves minerals, which then drain into lower layers of soil and rock. Rocks that are weathered by hydrolysis have rough surfaces and are often pitted or grooved. Feldspar weathers into kaolinite, a clay mineral, through hydrolysis.

Some minerals can change by absorbing water in a process called *hydration*. In these cases, the mineral's chemical formula changes very little, but the crystalline structure is significantly changed. For example, anhydrite can absorb water and become gypsum.

Water can combine with other substances to form acids, which weather rocks. **Carbonation** is the process in which carbon dioxide from the atmosphere or soil dissolves in water to form carbonic acid. The interaction between carbonic acid and the calcite in limestone and marble produces calcium bicarbonate, which dissolves easily in water and washes away. Over time, acidic water can carve caves out of rock. The dissolved calcium bicarbonate can be deposited from slowly dripping water to form stalactites (mineral forms hanging from the ceiling of a cavern) and stalagmites (mineral forms rising from the floor of a cavern).

In addition to carbon dioxide, water in the atmosphere can form sulfuric and nitric acids when it combines with the gases sulfur dioxide and nitrogen oxide. These gases are commonly released into the atmosphere as a result of burning fossils fuels, such as oil or coal. Molecules of nitric and sulfuric acids increase the acidity of normal precipitation (rain, fog, snow, sleet, or dew), which is commonly called *acid rain*.

Oxygen is a component in another major chemical weathering process. **Oxidation** is a chemical change in which a substance combines with oxygen. For example, iron combines with oxygen to form rust. Iron oxide produced by oxidation provides the red, rusty color of soil in the southeastern United States, southeastern China, and the tropics of South America and Africa. Protection against rust is crucial in the industrial world. Preventative

Rock weathered by hydrolysis

Slowly dripping water deposits dissolved calcium bicarbonate to form stalactites.

HISTORY
Climate and Weathering
A good example of a climate's influence on the rate of weathering is Cleopatra's Needle, an Egyptian granite obelisk. The obelisk stood exposed to the elements in the hot, dry climate of Egypt for 3,000 years. During this time its surface scarcely changed. In 1881 the obelisk was set up in Central Park in New York City. After a century of exposure to pollution and acid rain, Cleopatra's Needle has been weathered severely. Chemical and mechanical weathering processes in New York produced far more damage in 100 years than the desert climate of Egypt did in 3,000 years.

I. Have students research to find the type of stone that is best for a countertop. Emphasize that the stone should be low maintenance and resistant to acids and scratching.

J. Encourage students to design graphic organizers to help them remember the processes of hydrolysis, carbonation, and oxidation. Have them include an example of each.

Lesson Review

1. What are three agents of chemical weathering? (**Possible answers: water, carbon dioxide, oxygen, and acids**)
2. Explain the differences among the three types of chemical weathering caused by water. (**Water can react chemically with a mineral through hydrolysis, which changes the mineral's chemical formula and forms a new mineral. Water can be absorbed by a mineral and change its crystalline structure, which results in a new mineral. Water can combine with a gas to form an acid, which dissolves minerals in rocks.**)
3. What is oxidation? (**a chemical change in which a substance combines with oxygen**)
4. How can plants contribute to chemical weathering? (**Plant roots secrete a weak acid that dissolves rock surfaces, creating small depressions and forming cracks in the rock.**)
5. Compare the rate of chemical weathering in moist, hot climates with that in dry, cold climates. (**Chemical weathering occurs faster in moist, hot climates because water is a major agent in chemical weathering and because hot temperatures encourage chemical reactions.**)
6. What agents of chemical weathering should be considered when building a bridge over a river? (**Possible answers: hydrolysis, hydration, carbonation, acid rain, oxidation, and biological agents**)

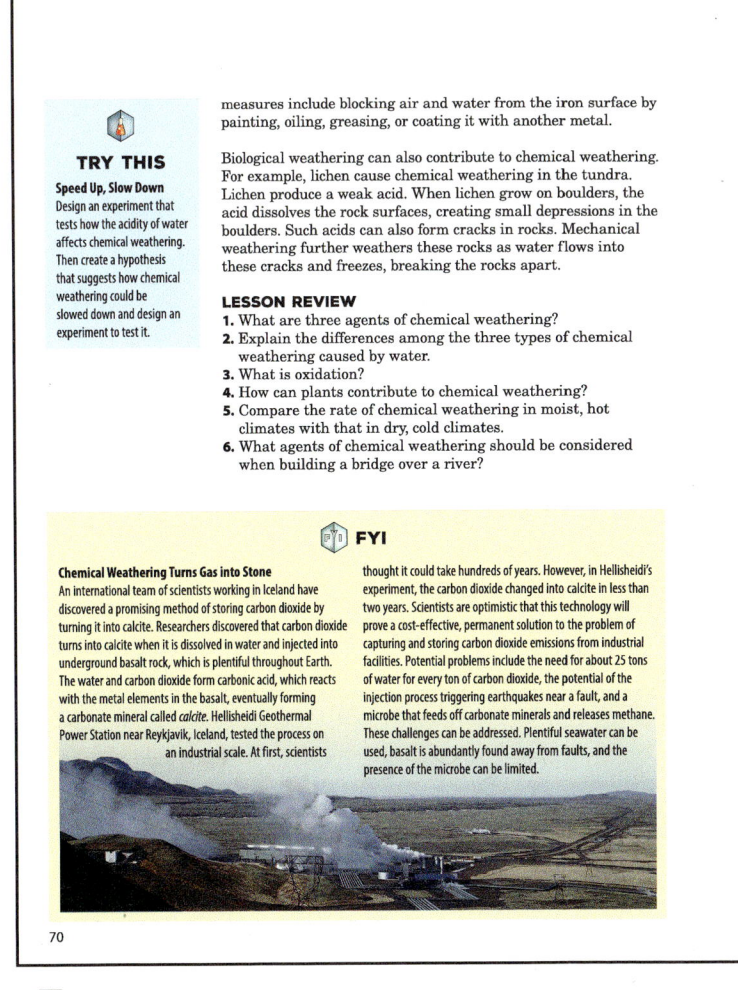

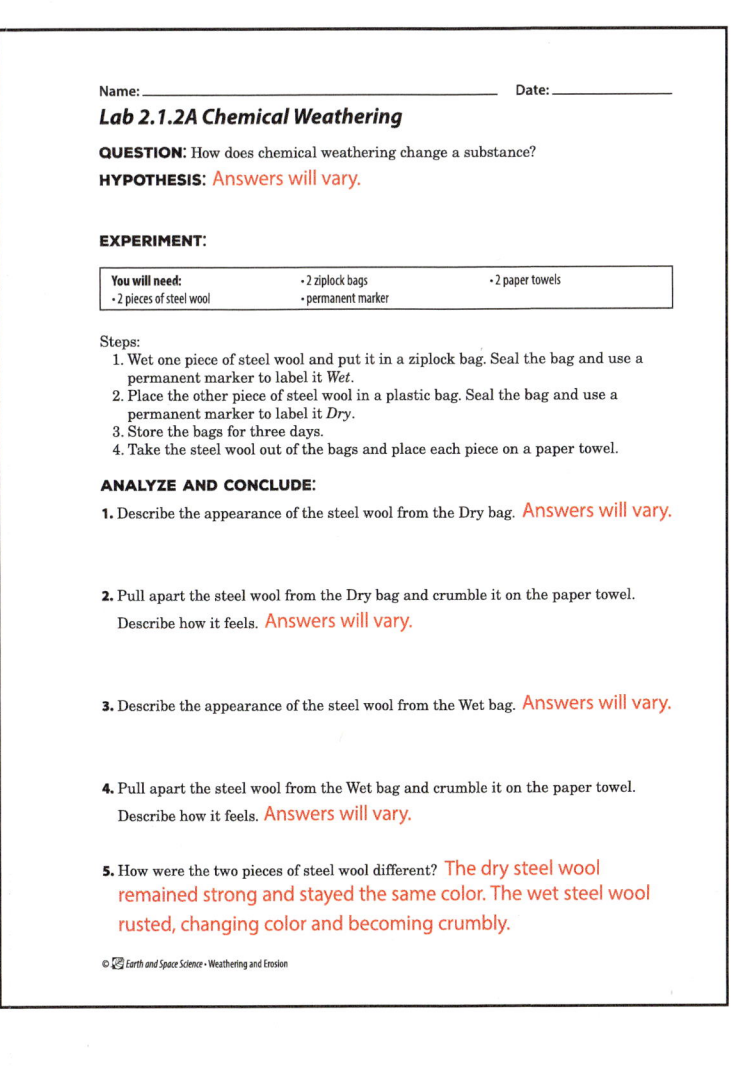

NOTES

Lab 2.1.2A Chemical Weathering

6. How could this kind of weathering happen to a rock? **If a rock contains iron, it could rust if it is exposed to moisture for extended periods of time.**

7. What kind of climates might experience this type of weathering? **moist climates**

8. Suggest a way to slow down or stop chemical weathering on the wet steel wool. **Possible answers: Freeze the wet steel wool or coat it with petroleum jelly before moistening it.**

Weathering and Erosion

2.1.3 Mass Wasting

Introduction ✋
Use the following demonstration to show the class the eroding power of rain on a steep slope. Clip a large sheet of white paper to a piece of cardboard and place it flat on a table with the cardboard on the bottom. Use an eyedropper to drop several drops of colored water on the flat paper, holding the eyedropper approximately 1 m above the paper. Have the class observe the splash patterns made by the drops. Incline the paper and cardboard by putting a wood block under one end. Repeat the process using a different color of water and holding the eyedropper at the same height. Continue increasing the slope of the paper and cardboard by adding more blocks, and drop different colors of water with each new incline. Observe the differences in the splash patterns. As the slope of the board is increased, the splash patterns of the drops should lengthen downhill. Emphasize to students that when rain falls on the slope of a hill or mountain, the drops of rain throw small particles a greater distance than drops hitting flat ground do. Trickles and streams of water will also form more easily on sloped terrain.

Discussion
- If there is nearby evidence of mass wasting, guide students to consider what may have caused it and whether there are any measures that can be taken to mitigate future mass wasting in the same location. If there is no nearby evidence of mass wasting, guide students to consider what factors make such an event unlikely. Lead a discussion regarding where there might be a potential for local mass wasting and what events might prompt it to occur.

- Ask the following questions:
  1. How is a mudflow different from a landslide? (**Mudflows have a higher water content than landslides, which causes them to behave like rivers. Mudflows follow natural channels until they reach the base of a slope, but landslides move along a distinct plane.**)
  2. What is common among causes of mass wasting events? (**They all loosen the earth.**)
  3. Should building in areas prone to mass wasting be prohibited? Why? (**Answers will vary but should include balancing public safety and providing for the needs of a community.**)

Activities ✋

Lab 2.1.3A Landslide
- beakers, 1 per group
- sand
- large plastic tubs, 1–2 per group
- protractors, 1 per group
- 15 mL measuring spoons, 1 per group
- timers, 1 per group

Prepare an example slope ahead of time, noting the final sand-to-water proportion so students can create their slopes with the same moisture content. Demonstrate making the slope and measuring its steepness. (Note: Each group will conduct only one of the experiments listed in the lab.) Students may experiment with two of the variables at the same time to determine if multiple variables produce a larger landslide in less time. For an extension, students can repeat the experiment with different slope materials, such as clay or gravel.

A. Direct students to use clay and a protractor to construct a 45° slope against one side of each of four large, clear plastic tubs. In each tub, have students add 2" of sand over the clay and place an identical brick at the top of each slope. Direct students to sketch the hills and predict what will happen if they increase the hill's water content in each of the following ways: a sudden downpour at the hill's crest, a sudden downpour at the hill's base, a gentle sprinkle at the hill's crest, and a gentle sprinkle at the hill's base. Provide a different watering can for each tub and note the volume of water in the can. Caution students to mimic rainfall, not a waterfall. Have one student note the time elapsed between beginning to pour the water and the landslide. Direct students to stop pouring water once the landslide begins, measure the volume of water

---

OBJECTIVES
Students will be able to
- distinguish among various forms of mass wasting.
- determine how various factors contribute to mass wasting events.
- infer reasons why people might want to build in locations prone to mass wasting.
- identify methods of reducing the risk of mass wasting.

VOCABULARY
- **earthflow** the movement of wet soil down a slope
- **landslide** the rapid downhill movement of a large amount of rock and soil
- **mass wasting** the downhill movement of rock and soil caused by gravity
- **mudflow** the rapid downhill movement of a large mass of mud and debris
- **regolith** a loose layer of rock and soil
- **soil creep** the extremely slow downhill slide of soil

MATERIALS
- White paper, cardboard, beakers of colored water, eyedropper, wood blocks (*Introduction*)
- Clay, protractors, plastic tubs, sand or gravel, bricks, watering cans (*A*)

⚙ PREPARATION
- Prepare four beakers, each with a different color of water. (*Introduction*)
- Obtain materials for *Lab 2.1.3A Landslide*.
- Plan a field trip to visit the site of a mass wasting event. (*B*)

NOTES

left in the can, and then subtract that number from the original volume to find out how much water was poured on the hill. Direct students to sketch each landslide, carefully noting where the slide began and how the material moved. Have students make a chart showing the volume of water used on each hill, the manner in which the water was poured, and where on the hill the water was poured. Lead a discussion regarding how rainfall and soil content affect the likelihood of landslides. Consider repeating the experiment using bricks of different sizes, a stack of two or more bricks, and bricks covering the slope. As another option, consider changing the composition of the hill or leaving a tub with a dry hill and brick somewhere in the classroom to see how long it will take the brick to slide without being disturbed.

B. Take students on a field trip to study a local area that has experienced mass wasting. Have students record the visible effects of the mass wasting event. Assign students to write a report regarding the contributing factors of the mass wasting event and what could be done to reduce the likelihood of a future mass wasting event in the same area. Soil creep is the most common mass wasting event, but the effects of landslides, mudflows, or rock falls in recent years may still be readily observable.

C. Direct students to create flip books or diagrams of a landslide, rock slide, rock fall, slump, mudflow, soil creep, earthflow, and solifluction.

D. Have student groups conduct research about large mass wasting events. Student reports should include the following information: how it started, how humans may have contributed to it, its effects, and what was or can be done to reduce future occurrences in the same area.

Mass Wasting 2.1.3

Weathering and erosion work together to slowly smooth broken or carved rock, create caves, or split boulders. These forces erode such tiny fragments and take such a long time to occur that they are difficult to notice. There is a form of erosion, however, that is hard to miss! **Mass wasting** is the downhill movement of rocks and soil caused by gravity. Mass wasting may affect only a few square meters, or it may involve an entire mountainside. **Regolith** is a loose layer of rock and soil that moves during mass wasting. Regolith soaked with water flows down a mountainside, but dry regolith tumbles or slides. Some of the regolith typically accumulates at the bottom of the slope in a bulge called *a toe*. The size and speed of the movement depends on the steepness of the slope, the type of rock, the presence of vegetation, and the amount of moisture in the ground.

You may have heard of a "landslide victory" such as when a candidate wins an election by a huge margin over an opponent. In nature, landslides are similarly dramatic mass wasting events. A **landslide** is a rapid, downhill movement of a large amount of rock and soil that separates from the bedrock beneath it. Landslides happen when the weight of materials on a slope can no longer resist the gravitational pull to slide downhill. If a landslide occurs as a large block of soil and rock that slides down a concave slope, it is called *a slump*. Often, the curved slope causes the sliding mass to tilt backward against the slope. The effect is similar to what happens when a large spoonful of pudding is slowly tilted upward; the pudding moves along the spoon's curved surface, and some spills over the end. Slumping

OBJECTIVES
- Distinguish among various forms of mass wasting.
- Determine how various factors contribute to mass wasting events.
- Infer reasons why people might want to build in locations prone to mass wasting.
- Identify methods of reducing the risk of mass wasting.

VOCABULARY
- **earthflow** the movement of wet soil down a slope
- **landslide** the rapid downhill movement of a large amount of rock and soil
- **mass wasting** the downhill movement of rocks and soil caused by gravity
- **mudflow** the rapid downhill movement of a large mass of mud and debris
- **regolith** a loose layer of rock and soil
- **soil creep** the extremely slow downhill slide of soil

typically occurs along steep slopes. A slump block may measure up to 5 km long and 150 m thick. Landslides can be triggered by earthquakes, volcanic eruptions, spring thaws, or heavy rains.

On May 12, 2008, a tragic earthquake struck China's Sichuan province, about 80 km west-northwest from the city of Chengdu. The earthquake caused massive damage and loss of life, not only from the earthquake itself but also from the landslides it caused. In some places, landslides dammed rivers, creating new lakes, and in other places they blocked roads and destroyed property.

Landslides that primarily involve rocks are called *rockslides*. These typically start on steep slopes when huge chunks of rock come loose from the bedrock and slide downhill. Similarly, rockfalls are common on steep cliffs or hills where individual rocks and boulders occasionally break free. Rockfalls are not true landslides because they are an abrupt movement of rocks that

E. Assign students to build a landform in a plastic container and to choose a type of mass wasting to replicate. Have students draw the landform before the mass wasting, during each stage of the experiment, and after the mass wasting occurred. Direct students to record the materials used to create the landform, the angles of all slopes on the landform, the forces used to induce the mass wasting event, the manner in which forces were applied, the strength or amount of force applied, and the time frame in which the forces were applied. Have students present their projects and results to the class.

F. Direct students to design an experiment that tests how vegetation affects the likelihood of a mass wasting event under various conditions.

Lesson Review
1. What is a landslide called that slides along a curved slope? (**slump**)
2. How is an earthflow different from soil creep? (**An earthflow is faster than soil creep.**)
3. What factors might trigger a mass wasting event? (**gravity, heavy rainfall, earthquakes, volcanic eruptions, spring thaws, and construction**)
4. Describe a scenario in which people might build in a location prone to mass wasting. (**Possible answers: to have a house with a beautiful view, to construct apartments to solve overcrowding, to build a factory to provide jobs, to build a highway or a road**)
5. What are some methods engineers can use to reduce the risk of mass wasting? (**grading properly, building retaining walls, providing proper drainage, covering loose rocks with steel mesh, planting vegetation, and building rock sheds**)

NOTES

Name: _____ Date: _____

Lab 2.1.3A Landslide

QUESTION: How does steepness, moisture content, or an earthquake affect the stability of a slope?

HYPOTHESIS: Answers will vary.

EXPERIMENT:

You will need:	• 1–2 large plastic tubs	• timer
• beaker	• protractor	
• sand	• 15 mL measuring spoon	

Steps:
1. Use a beaker to measure sand and water into a large plastic tub according to the proportion given by your teacher. Thoroughly mix the sand and water together.
2. Create a hill with a 50° slope against one of the narrow ends of the tub. Verify the angle of the slope with a protractor.
3. On a separate piece of paper create a data table to record your observations.
4. Sketch your hill on a separate piece of paper.
5. Follow the steps given for your assigned experiment.

Steepness:
1. Prepare a second batch of sand and water exactly like the first one.
2. Use the second batch of sand to gradually increase the steepness of the slope by 5°–10° at a time.
3. Record the slope's steepness and describe your observations after each addition.
4. Sketch the hill, including where the landslide started and how the sand moved.

Moisture Content:
1. Use a 15 mL measuring spoon to gently pour water on the top of the hill. Be sure to pour the water the same way each time.
2. Record the amount of water added to the hill. Describe how you added the water and your observations of the hill after each water addition.
3. After a landslide occurs, tally the total amount of water added to the hill.
4. Sketch the hill, including where the landslide started and how the sand moved.

Earthquake:
1. Gently shake the tub from side to side to simulate an earthquake. Count the number of shakes needed to cause a landslide (one shake equals one side-to-side movement) and time how long the hill shakes before a landslide occurs.
2. Record the number of shakes and the time taken for a landslide to occur.
3. Calculate the shaking frequency. Divide the number of shakes by the number of seconds and label the answer as *shakes per second*. Record your answer.
4. Sketch the hill, including where the landslide started and how the sand moved.

Lab 2.1.3A Landslide

ANALYZE AND CONCLUDE:

1. How did the results of your experiment differ from similar experiments? Why might this be the case? Answers will vary.

2. Compare your results with groups doing the other two experiments. Which method triggered a landslide most quickly? Answers will vary.
   Which method caused the greatest amount of material to slide?
   Answers will vary.

3. Compare the results to your hypothesis and revise it if needed.
   Answers will vary.

4. Use what you learned to suggest methods of reducing the chances of a landslide.
   Possible answers: reduce the steepness of slopes, prepare drainage channels, remove loose soil, plant vegetation on slopes, anchor rocks together, or build retaining walls

Weathering and Erosion

2.1.4 Wind and Wave Erosion

Introduction
To demonstrate wind erosion, fill a large, clear plastic tub with sand. Have students wear goggles. Select volunteers to use a hair dryer to blow air over the sand at different speeds. Have students experiment with keeping the hair dryer in one place for 1–2 minutes and with moving it from side to side. Place a block in the middle of the tub and later scrape a long trough in the sand and have students observe the effect of the block and trough on the sand as air is blown from behind them.

Discussion
- Refer to the wind experiment in *Introduction* while leading a discussion about the effect various landforms and wind intensity have on wind erosion.

- Discuss how waves shape beaches and create sandbars. Guide students to consider how waves affect cliffs and what happens to the sediment created from cliff erosion. Direct students to refer to the wave experiment in *Introduction* to support and illustrate their conclusions.

- Have students respond to each of the following statements with one of the words in parentheses and explain their answers. Encourage respectful discussion.
  1. Natural environments should (always/often/sometimes/never) be preserved at all costs.
  2. The government should (always/often/sometimes/never) enforce codes that tell people where they may build on a beach and what the structures must be like.
  3. If a community suffers from a storm, other citizens must (always/often/sometimes/never) provide shelter and food for the affected people until those people can move or rebuild.
  4. The government (always/often/sometimes/never) has the right to tell people how to farm or to limit what people may grow on their private property.

Activities

Lab 2.1.4A Deflation
- safety goggles, 1 per student
- ice cube trays, 1 per group
- wood blocks, 1 per group
- metric rulers, 1 per group
- varied-grain sand, 300 mL per group
- hair dryers, 1 per group

Lab 2.1.4B Wave Erosion
- 500 mL beakers, 1 per group
- sand, 1 L per group
- paint roller trays, 1 per group
- timers, 1 per group
- metric rulers, 1 per group
- capped, empty water bottles, 1 per group

The paint roller tray should hold at least 3 L. Use 500 mL water bottles for the best results.

Option: Observe the effects of waves on a shore with a headland by piling 0.5 L of aquarium gravel in the center of the shoreline and extending into the water. Repeat the original steps. Compare the results to the lab completed without a headland. For the best comparison, trace the contours of the first beach on the tray and use the outline to sculpt the second beach.

A. Have students complete *Try This: Sea Stacks*.

B. Assign students to complete **WS 2.1.4A The Art of Wind and Water**.

C. Have students shake pebbles in plastic containers to illustrate abrasion. Place 10 dolomite pebbles and 10 hard pebbles in each of three 1 L containers with lids. Shake one container 100 times, one 1,000 times, and one not at all. Have students record their observations.

OBJECTIVES
Students will be able to
- distinguish between wind erosion through abrasion and deflation.
- measure the process and effects of erosion caused by waves.
- model the relationship between erosion and the formation of features.
- summarize ways in which people accelerate or reduce the effects of wind and wave erosion.

VOCABULARY
- **abrasion** the wearing down of rock surfaces by other rocks or sand particles
- **deflation hollow** a soil depression scooped out by the wind
- **rock pedestal** a mushroom-shaped rock formed by the erosion of the rock's base

MATERIALS
- Tubs, sand, goggles, hair dryer with multiple speeds, block (*Introduction*)
- 30 dolomite pebbles, 30 hard pebbles, three 1 L containers with lids (*C*)
- WS 2.1.4A The Art of Wind and Water

PREPARATION
- Obtain materials for *Lab 2.1.4A Deflation* and *Lab 2.1.4B Wave Erosion*.
- Obtain materials for *Try This: Sea Stacks*.
- Invite an insurance agent to explain how erosion factors into homeowner's insurance. (*D*)
- Invite an engineer or a contractor to explain ways in which erosion is prevented or remedied. (*E*)
- Plan a field trip to a beach that has breakwaters. (*F*)

TRY THIS

Sea Stacks
- sand
- plates, 1 per group
- coins, several per group
- watering cans with water, 1 per group

NOTES

D. Introduce the insurance agent to explain how erosion factors into homeowner's insurance.

E. Present the engineer or contractor to explain ways in which erosion is prevented or remedied.

F. Take students on a field trip to a beach that has breakwaters, which prevent sand from moving along the coast and eroding away the beach. Have students observe evidence of sideways coastal movement and how the walls prevent excess beach erosion.

G. Assign students to build a model of a beach, simulating wind, waves, and precipitation. Direct students to use the model to design a beach protection plan.

H. Have students create models of farming methods that are used to reduce wind erosion.

I. Challenge students to research the effects of a hurricane on a populated beach. Students should identify any protective measures that were in place before the storm, assess how well those measures worked, and explain any subsequent plans enacted to prevent future damage. Direct students to create charts or graphs comparing the rates of erosion during hurricanes as opposed to normal conditions for their particular beach.

Lesson Review

1. Concerning wind erosion, what is the difference between abrasion and deflation? (**Abrasion wears down rock surfaces using other rocks or sand particles, and deflation is the removal of loose soil or sand from the earth's surface.**)

Wind and Wave Erosion 2.1.4

Have you ever sat on a beach when a strong wind was whipping across the sand? The strong wind pelts tiny grains of sand against your skin. The tiny grains can bite into your skin until it is raw. Wind is the most influential sculptor of dry landscapes. Its force can move large amounts of soil and sand over great distances. When these blowing particles rub against surfaces, those surfaces weather and erode.

The way in which wind moves particles depends on their size. Tiny dust particles can be suspended in wind and moved great distances. Wind can briefly lift grains of sand before they fall and bounce along the ground; this movement is called *saltation*. Bouncing particles and wind can also push or roll larger grains that are too heavy for the wind to lift in a movement called *creep*.

Deflation is a form of erosion in which wind carries away sand and soil particles. Wind deflates dry land more quickly than wet land because moisture weighs down soil and glues the grains together, making it harder for the wind to pick them up. Major deflation events come in the form of dust storms. Under the right conditions, strong winds can lift sand and soil over 1 km into the air and carry the dust 40–80 km. Some of these storms can stretch up to 160 km. Deflation can also occur in smaller events.

Deflation commonly produces two distinct landforms. *Desert pavement* is the hard, packed ground left after all the loose soil or sand has been washed or blown away. **Deflation hollows** are soil depressions scooped out by the wind. These can range in area from about 1 m² to several thousand square kilometers. Egypt's Qattara Depression, the world's largest deflation hollow, is about 18,100 km².

Abrasion is the wearing down of rock surfaces by other rocks or sand particles. Abrasion smooths or polishes the surfaces of rocks that face the wind. It can also help erode rocks into interesting shapes. For example, abrasion can help form **rock pedestals**, which are mushroom-shaped rocks. Blowing sand primarily chips away weathered material at the rocks' bases because the sand is too heavy to reach the top of the rocks.

Another effect of wind erosion is wind deposition. When wind takes soil or sand from one place, it deposits the particles someplace else to form either loess or dunes. Loess is a porous deposit of fine silt that is usually rich in minerals. These qualities make loess excellent for farming. The most well-known sand deposits are dunes, mounds of sand deposited by the wind. Dunes form in places where the wind is strong and the soil is dry and unprotected by vegetation, such as deserts or beaches. Wind carries sand until it comes against a barrier that traps or slows the grains, enabling them to collect in a pile. Barriers can include rocks, stands of grass, bushes, or even soft ground. The pile of sand grows and becomes sloped on the side facing the wind. Eventually, the sheltered side becomes too steep, and the top of the pile collapses away from the wind. This process of growing and collapsing causes the dune to move in the direction of the wind.

If you have ever tried to build a sand castle near waves or splashed water against a hill of dirt, then you know that water is also a force of erosion. One type of water erosion is wave action. The force of the striking waves can cause weathered rock to crumble. The sand and small rocks carried by waves create an abrasive force

OBJECTIVES
- Distinguish between wind erosion through abrasion and deflation.
- Measure the process and effects of erosion caused by waves.
- Model the relationship between erosion and the formation of features.
- Summarize ways in which people accelerate or reduce the effects of wind and wave erosion.

VOCABULARY
- **abrasion** the wearing down of rock surfaces by other rocks or sand particles
- **deflation hollow** a soil depression scooped out by the wind
- **rock pedestal** a mushroom-shaped rock formed by the erosion of the rock's base

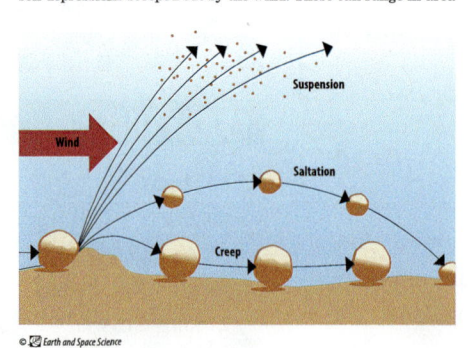

TRY THIS
Sea Stacks
Pour sand on a plate, making a mound in the center. Place several coins on the mound. Use a watering can to sprinkle water on the mound. Describe the results. Compare the results to ocean wave erosion.

2. Wind deposition forms what two land features? **(loess and dunes)**
3. How do waves erode cliffs? **(through the force of their impact and through abrasion caused by the sand and rocks suspended in the waves)**
4. What often happens to sand eroded from a beach during the winter? **(It is deposited offshore in sandbars.)**
5. Summarize the primary means by which people can accelerate or reduce the effects of wind and wave erosion. (**Possible answers:** *acceleration of wind erosion*—cutting forests, excavating for construction, tilling fields; *reduction of wind erosion*—cutting trees selectively and replanting, sprinkling water on construction sites, laying gravel on machine paths, constructing or planting windbreaks, using contour plowing, planting cover crops; *acceleration of wave erosion*—building walls or structures that create different wave patterns; *reduction of wave erosion*—bringing in sand to maintain beach areas, building fences and planting vegetation to encourage dune formation)

NOTES

HISTORY

The Great Dust Bowl 
During the Great Depression, the combination of poor farming methods, drought, and strong winds caused terrible erosion and dust storms throughout the United States' southern Great Plains. The primary states affected were Colorado, Kansas, Oklahoma, Texas, and New Mexico. This event was called *the Great Dust Bowl*. The Dust Bowl started in 1931 as drought killed wheat crops. These crops had replaced most of the natural grassland. Without the crops or the grass, the wind began to erode the deeply plowed land. The ensuing dust storms could be enormous. The worst ones reduced visibility to nothing, and a few even reached the east coast of the US. As the Dust Bowl continued, the federal government, under President Franklin Roosevelt, implemented several disaster relief programs. One such program was the Soil Conservation Service (SCS), which was established in April 1935. The SCS taught farmers conservation techniques such as strip cropping, contour plowing, and planting cover crops. In March 1937, the Shelterbelt Project began. This project encouraged farmers to plant native trees as windbreaks. These and other conservation measures reduced the wind erosion by 65%. Rain returned to the area in 1939, and the land began to recover.

similar to windblown sand. Waves can also carry and deposit sediment to both build up and erode beaches.

Waves battering a rock shoreline create steep cliffs called *sea cliffs*. The waves carve out the bottom of the cliff until the overhang drops into the sea. The cliff gradually wears back and becomes steeper. Soft rocks such as those on the coast of Dover, England, weather quickly, but cliffs of hard rock such as granite show little wear even after centuries of being pounded by waves.

Sea cliffs do not usually weather evenly. Waves often cut deep into the weakened rock at the cliff's base to form *sea caves*. If the cave forms in a promontory—a high piece of ground that juts out into the water—the waves will continue eroding the rock until the center of the cave is completely eroded and a *sea arch* is formed. Sea arches exposed to constant wave action may weaken until their middles collapse, leaving isolated columns of rock called *sea*

Sea cave, sea arch, and sea stack at Cliffs Porte d'Aval in Étretat, France

---

Sandbar

Contour farming

stacks. Eventually even the sea stacks undergo weathering until they no longer rise above the water's surface.

Pieces of rock that are broken apart by waves continuously grind against the shore and each other until they are reduced to small pebbles or sand. A beach forms when waves deposit sand or pebbles along an ocean shore or lakefront. Erosion changes beaches throughout the seasons. The large waves of winter storms carry enormous amounts of sand away from the beach, depositing it offshore. The deposits often form long, underwater ridges called *sandbars*. Sandbars may become exposed during low tides. In the summer, currents and waves return sand from the sandbars back to the shore, widening the beach once again.

Beach materials can be found in a variety of sizes and colors. The size of materials is largely determined by the force and frequency of the waves. Strong, frequent wave action has the ability to bring in pebbles, and it tends to carry lighter sand away. The color of beach sand depends on the materials in the sand. White sand is typically quartz or crushed shells. Black sand comes from basalt, a volcanic rock, or from heavy metals such as magnetite. Pink

FYI

A Room with a View 
People are drawn to beautiful areas and frequently build houses on cliffsides and beaches so they can enjoy the beauty of creation. Unfortunately, such areas are prone to erosion and dramatic mass wasting events. In some parts of the world, available land is scarce, so apartment complexes, factories, and roads have even been built on hilly areas that are at risk for landslides or mudflows. Such construction can have serious consequences. For example, in India, heavy machinery that was used to build hydroelectric dams in the Himalayas contributed to huge mudslides in the region. Engineers and city planners work to balance the wants and needs of the population with the very real risks of building on potentially unstable ground.

sand generally comes from corals. Rare green sand typically comes from olivine, a mineral that erodes out of basalt. These materials can come from a variety of sources. Obviously, coral and marine shell fragments come from the sea. Minerals, however, may be deposited from rivers that carried the sand for many kilometers, or they may come directly from coastal rocks.

A jetty protects the coastline from erosion.

Erosion is a natural process, but human activity can accelerate or hinder it. People may encourage wind erosion by clear cutting large swaths of forest, by excavating or bulldozing sites in preparation for construction, or by tilling fields in preparation for planting. Various methods can help lessen human impact. Selective cutting and replanting in forests not only maintains healthy plants and trees, but also provides strong roots to hold the soil. Sprinkling water over construction sites helps prevent wind erosion, and laying gravel over machine paths and roadbeds helps keep soil in place. Farmers can plant windbreaks and plow along the contours in their fields to slow the wind and catch particles driven by the wind. By planting cover crops in dormant fields, farmers help return nutrients to the soil and limit wind erosion.

People can affect beach erosion too. In some places people build walls to either deflect waves and currents or encourage sand deposits. Care must be taken when building these because they can create different wave and current patterns, increasing erosion in other areas. Another way to maintain the size of a beach is to bring in sand from elsewhere. People can help protect structures from the effects of storms or sea surges by encouraging the formation of beach dunes by building fences or planting vegetation to hold the sand.

LESSON REVIEW
1. Concerning wind erosion, what is the difference between abrasion and deflation?
2. Wind deposition forms what two land features?
3. How do waves erode cliffs?
4. What often happens to sand eroded from a beach during the winter?
5. Summarize the primary means by which people can accelerate or reduce the effects of wind and wave erosion.

Erosion-control fence at a beach

Name: _____ Date: _____

Lab 2.1.4A Deflation

QUESTION: How do grains erode?
HYPOTHESIS: Answers will vary.

EXPERIMENT:

You will need:		
• safety goggles	• wood block	• hair dryer
• ice cube tray	• metric ruler	
	• 300 mL of varied-grain sand	

Steps:
1. Put on safety goggles.
2. Lay the ice cube tray on the lab bench and place the wood block next to the tray at one end.
3. Measure the distance from the center of each section of the tray to the wood block and record these in a data table on a separate sheet of paper.
4. Form a small pile of sand on the wood block and record its height. Answers will vary.
5. Hold the hair dryer close to the sand pile and record the distance between them. Answers will vary.
6. Sketch the setup. Label all distances and the height of the sand pile.

Drawings will vary.

7. Blow sand into the ice cube tray for 1 minute. Hold the hair dryer the same distance away from the sand pile at all times.
8. Use the metric ruler to measure or estimate the size of the sand grains that jumped into each section of the tray.
9. Create another data table to show the number and approximate size of the particles you found in the different sections of the ice cube tray.

Lab 2.1.4A Deflation

10. Sketch the setup again.

Drawings will vary.

ANALYZE AND CONCLUDE:
1. Explain the relationship between the size of the sand grains and the distance they traveled. The smallest grains travel the farthest.

2. Describe how the shape of the sand pile changed. Answers will vary.

3. How might the sand pile look if the force of the blowing air was greater or if the air blew for a longer time? Answers will vary but should include that the sand pile would become smaller.

Name: _____ Date: _____

Lab 2.1.4B Wave Erosion

QUESTION: How do waves affect a sandy shoreline over time?
HYPOTHESIS: Answers will vary.

EXPERIMENT:

You will need:		
• 500 mL beaker	• paint roller tray	• capped, empty water bottle
• sand	• timer	
	• metric ruler	

Steps:
1. Use 1 L of sand to create a beach in the shallow end of the paint tray. The beach should slope toward the deep end of the tray so water can cover part of it.
2. Pour 1.5 L of water into the deep end of the tray. Try not to create waves.
3. Let the sand and water sit for 5 minutes.
4. On a separate piece of paper, draw a picture of the beach.
5. Measure the width of the beach between the tub and the shoreline. If the shoreline is irregular, then take several measurements. Mark all measurements on your drawing.
6. Gently place a capped water bottle in the tray's deep end so it floats parallel to the shoreline.
7. Bob the bottle up and down for 1 minute to create waves. Try to keep the bobbing force and frequency consistent and count the number of times you bobbed the bottle.
8. Repeat Steps 3–5. Label the drawing *1 Minute*.
9. Below your drawing, describe what you observed as the bottled was bobbed. Include the approximate force of the waves (gentle, moderate, hard) and the number of times you bobbed the bottle.
10. Repeat Steps 3–5 and 9, but this time bob the bottle for 2 minutes. Try to keep the bobbing force and frequency the same as that used for the 1-minute trial. Label the drawing as *2 Minutes*.

ANALYZE AND CONCLUDE:
1. Compare the amount of erosion on the beach between the 1-minute trial and the 2-minute trial. Answers will vary but should include that more erosion occurred after 2 minutes than after 1 minute.

Lab 2.1.4B Wave Erosion

2. What happened to the sand that eroded from the shoreline? The sand was deposited offshore, and some of it accumulated into a sandbar.

3. How could the lab be modified to make the waves more consistent?
   Answers will vary.

4. Explain whether the lab would be more realistic with identical waves or with varying waves. The lab would be more realistic with waves that vary somewhat because ocean waves are not perfectly consistent.

5. Review your hypothesis. If necessary, write a more accurate answer.
   Answers will vary.

2.1.5 River Erosion

Weathering and Erosion

OBJECTIVES

Students will be able to
- determine factors that affect river erosion.
- evaluate the benefits and hazards of living on a floodplain.
- explain how erosion and deposition can change the landscape.

VOCABULARY

- **channel** the path that a stream follows
- **floodplain** a flat area along a river formed by sediments deposited when a river overflows
- **gully** a narrow ditch cut in the earth by runoff
- **runoff** water from precipitation that flows over the land

MATERIALS

- Stream table, basin, sand, aquarium gravel, flat rock (*Introduction*)

PREPARATION

- Set up a stream table on an angle and arrange a basin to catch overflow drainage. (Note: The lid of a long, plastic container will provide a reasonable stream table substitute.) Spread 3 cm of fine sand mixed with some aquarium gravel over the bottom of the table. Place a flat rock on the sand at the upper end of the table. Scoop out some sand at the lower end to form a lake. Create a 1 cm deep channel from the rock to the lake. (*Introduction*)
- Obtain materials for *Lab 2.1.5A Water Erosion and Time* and *Lab 2.1.5B Water Erosion and Force*.
- Obtain materials for *Try This: A Changing View*.
- Plan a field trip to a local river, stream, or ocean shoreline to identify examples of erosion. (*B*)

Introduction

Inform students that the Mississippi River in the United States carries about 395,533 metric tons of material daily to the Gulf of Mexico. This action has formed seven deltas. Demonstrate the formation of a delta with the prepared stream table by gently pouring water over the rock. Have students observe the delta that forms at the mouth of the stream. Vary the amount of water flowing down the river to demonstrate the different amounts of sand that are deposited. Forceful streams will deposit pebbles as well as sand. The width and course of the river may change with the amount of water. If the course changes, a new delta is formed.

Discussion

- Lead students in a discussion concerning the possible benefits and detriments of building along a river in a floodplain. Guide students to consider the impact of natural and human flood control.

- Ask students the following questions:
  1. Why are the beds and banks of steep mountain streams shaped differently from lowland streams? (**In steep mountain streams, gravity swiftly pulls water downward; this creates streams with steep banks and narrow beds. Gravity does not have such a strong effect on the gentle slopes of lowland streams; this creates more gradually sloping banks and broad beds.**)
  2. How can erosion and deposition change a river's channel? (**Erosion and deposition change the path and depth of a river's channel. Swiftly flowing water causes erosion, which deepens or widens channels. Sediment builds up where water slows down. This buildup makes the channel more shallow or narrow. Sediment is deposited along the inside bends of a river because the water is forced to slow down, but the outer parts of the bends are eroded by the swiftly flowing water.**)
  3. What natural features and human-made structures help control floods? (*natural*: **forests and wetlands**; *human-made*: **dams and levees**)
  4. Why do levees have to be built up from time to time? (**After a river floods, the slowly receding water allows sediment to build up along the riverbed. Levees have to be built taller to keep up with the rising river.**)
  5. How are alluvial fans and deltas similar and different? (**Alluvial fans and deltas are both the result of a river suddenly slowing down and depositing much of its load. Alluvial fans form on land, usually at the base of a mountain; deltas form in water, usually where a river empties into a lake or an ocean.**)

Activities

Lab 2.1.5A Water Erosion and Time

- baking trays, 1 per group
- sand
- fine screen pieces, 1 per group
- bricks, 1 per group
- protractors, 1 per group
- hoses with faucet connections, 1 per group
- 500 mL beakers, 1 per group
- timers, 1 per group
- hose clamps, 1 per group

The baking tray should have a rim about 2–4 cm high. The screen should be large enough to cover the sink drain. Demonstrate how to use a protractor to measure the angle of slope.

Option 1: Have students experiment with different flow rates. Guide them to reform the sand after each trial. Emphasize that the angle of the slope and the amount of time the water runs should be constant among all trials.

Option 2: Lead students to experiment with different angles of slope. Have them reform the sand after each trial. Remind them to keep the rate and time of water flow constant for all trials.

Lab 2.1.5B Water Erosion and Force

- shoe boxes, 1 per group
- scissors, 1 per group
- scales, 1 per group
- gravel
- sand
- plastic tubs, 1 per group
- blocks, 1 per group
- hoses, 1 per group
- metric rulers, 1 per group
- fine screen pieces, 1 per group

Emphasize to students that the first trial should use slow moving water.

Option: Have students strain the water from the plastic tub into a beaker and record the volume of runoff after each trial. After measuring the runoff volume of the slow flow, direct students to reform the slope, then suddenly dump the runoff onto the slope; measure, describe, and draw the results. Repeat the process with the runoff of the fast moving water, but this time slowly pour the runoff onto the reformed slope. Challenge students to consider whether speed or volume has greater erosive powers.

TRY THIS
A Changing View
- sand
- small rock fragments
- chalk
- fine silt
- large plastic tubs, 1 per group

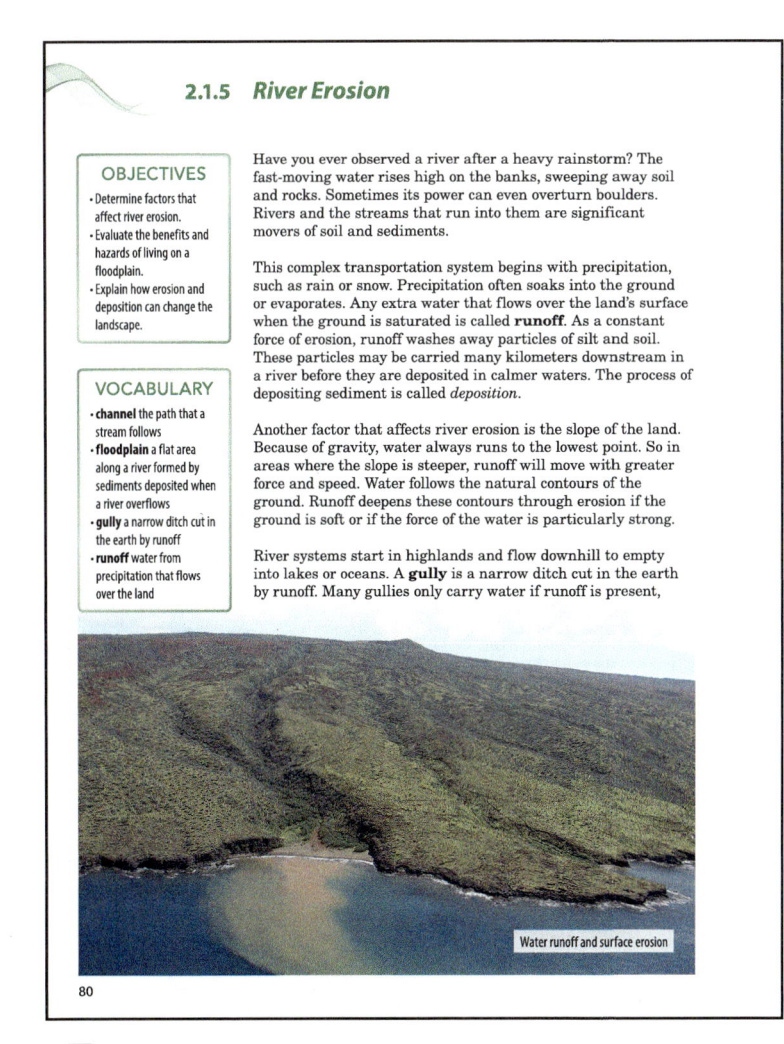

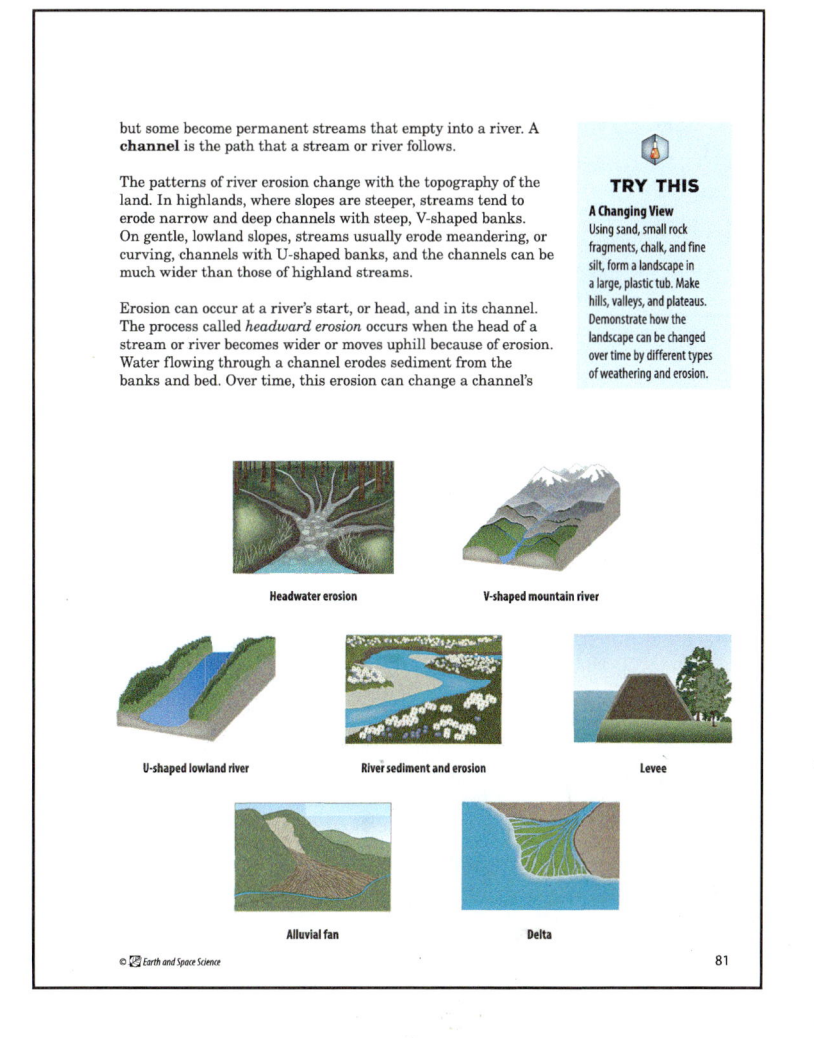

A. Complete *Try This: A Changing View* with students.

B. Take students on a field trip to a local river, stream, or ocean shoreline to identify examples of erosion. Have them determine the causes of the erosion patterns they find. Challenge students to determine whether the erosion is harmful or helpful to the local ecosystem. Guide them to predict what the location will look like in the future.

2.1.5 River Erosion

OBJECTIVES
- Determine factors that affect river erosion.
- Evaluate the benefits and hazards of living on a floodplain.
- Explain how erosion and deposition can change the landscape.

VOCABULARY
- **channel** the path that a stream follows
- **floodplain** a flat area along a river formed by sediments deposited when a river overflows
- **gully** a narrow ditch cut in the earth by runoff
- **runoff** water from precipitation that flows over the land

Have you ever observed a river after a heavy rainstorm? The fast-moving water rises high on the banks, sweeping away soil and rocks. Sometimes its power can even overturn boulders. Rivers and the streams that run into them are significant movers of soil and sediments.

This complex transportation system begins with precipitation, such as rain or snow. Precipitation often soaks into the ground or evaporates. Any extra water that flows over the land's surface when the ground is saturated is called **runoff**. As a constant force of erosion, runoff washes away particles of silt and soil. These particles may be carried many kilometers downstream in a river before they are deposited in calmer waters. The process of depositing sediment is called *deposition*.

Another factor that affects river erosion is the slope of the land. Because of gravity, water always runs to the lowest point. So in areas where the slope is steeper, runoff will move with greater force and speed. Water follows the natural contours of the ground. Runoff deepens these contours through erosion if the ground is soft or if the force of the water is particularly strong.

River systems start in highlands and flow downhill to empty into lakes or oceans. A **gully** is a narrow ditch cut in the earth by runoff. Many gullies only carry water if runoff is present, but some become permanent streams that empty into a river. A **channel** is the path that a stream or river follows.

The patterns of river erosion change with the topography of the land. In highlands, where slopes are steeper, streams tend to erode narrow and deep channels with steep, V-shaped banks. On gentle, lowland slopes, streams usually erode meandering, or curving, channels with U-shaped banks, and the channels can be much wider than those of highland streams.

Erosion can occur at a river's start, or head, and in its channel. The process called *headward erosion* occurs when the head of a stream or river becomes wider or moves uphill because of erosion. Water flowing through a channel erodes sediment from the banks and bed. Over time, this erosion can change a channel's

TRY THIS
A Changing View
Using sand, small rock fragments, chalk, and fine silt, form a landscape in a large, plastic tub. Make hills, valleys, and plateaus. Demonstrate how the landscape can be changed over time by different types of weathering and erosion.

Water runoff and surface erosion

Headwater erosion

V-shaped mountain river

U-shaped lowland river

River sediment and erosion

Levee

Alluvial fan

Delta

NOTES

C. Challenge students to choose a river system and trace its erosion patterns. Have them research portions of the river that are particularly prone to erosion or flooding and what factors affect these processes. Direct students to research building codes in these places that apply to building houses near riverbanks. Have students determine whether codes require a minimum distance from the water and whether the codes were written for flooding, erosion, or plumbing systems.

D. Assign students to research a river of historical significance that floods seasonally, such as the Nile or Amazon. Have students research the effect the flooding has had on local populations over time including farming, industry, transportation, and living conditions.

Lesson Review

1. What factors affect river erosion? (**the topography of the land and the amount of runoff**)
2. How does erosion change a river's head? (**A river's head may become wider or move uphill due to erosion.**)
3. Why does the inside of a river bend gain material while the outer side of the bend loses material? (**Water slows down on the inside of a river bend, which enables deposition. Water speeds up along the outer part of a bend, which encourages erosion.**)
4. How are the processes of erosion and deposition connected? (**The material that rivers erode from one place is eventually deposited someplace else.**)
5. What benefits and hazards are associated with living in a floodplain? (**The land in a floodplain is flat, and the soil is very rich in minerals. The flat land, rich soil, and access to water make floodplains good for farming. Floodplains are prone to frequent flooding.**)
6. What landforms can river deposits create? (**floodplains, alluvial fans, and deltas**)

---

depth and shape. All the sediment carried by a stream or river is called its *load*. When quickly flowing water slows down, it deposits all or part of its load, which can change a channel's depth. In a meandering river, water flows slowly along the inside of the bends but flows quickly along the outer curves in a motion similar to that of a merry-go-round. A person on the edge of the merry-go-round travels faster than a person in the center. In a river, the difference in speed deposits sediment along the inside of a bend, but erodes the bend's outer curve. Over time, the shape of the river's channel is changed.

A river can also deposit its load after flooding. The increased volume of flood waters carries sediment over a river's banks, and the sediment is left behind as the flood waters slowly recede. After many years, these sediments build up what are called *floodplains* such as those found along the Nile River in Egypt, the Mississippi River in the United States, and the Ganges River in India. A **floodplain** is a flat area along a river formed by sediments deposited when a river overflows. Ever since ancient times, people have settled in floodplains to farm the soil, which is rich in minerals and soil deposits left by the floods. The Egyptians, for example, settled on the floodplains of the Nile.

A levee breach

Namibia floodplain

Floodplains can be dangerous because they are naturally prone to flooding. Two elements of natural flood control are forests and wetlands. For example, wetlands absorb some of the floodwater, and they slow the speed of the water as it drains back to the channel. Human-made methods of controlling floods include dams and levees. Dams are structures built on a river to regulate its flow. Levees are earthwork constructions built to stop water from flowing in a specific direction. Levees must be maintained to keep them functioning properly. As floodwaters drain, the sediments are deposited along the levee and in the riverbed, which is the ground at the bottom of a river. These deposits build up the riverbed over time. The levee must then be made higher as the riverbed rises, or the river must be dredged to remove sediments.

Rivers slow down when their channels become wider. This slowing happens suddenly when a mountain stream enters a broad plain or when a river reaches a large body of water, such as a lake or an ocean. This sudden slowing deposits much of the river's load in one location. On land, this kind of deposit is called *an alluvial fan*. When a river empties into another body of water, the sediment deposit is called *a delta*. Deltas tend to form triangular shapes, and the term *delta* comes from the name of the triangular Greek letter that looks like this: Δ. Alluvial fans and deltas form in stages. Deposited sediments first split a river into channels of shallow water called *distributaries*. The distributaries change course as sediment builds. The changed course takes sediment to new locations and widens the alluvial fan or delta.

Alluvial fan

River delta

LESSON REVIEW
1. What factors affect river erosion?
2. How does erosion change a river's head?
3. Why does the inside of a river bend gain material while the outer side of the bend loses material?
4. How are the processes of erosion and deposition connected?
5. What benefits and hazards are associated with living in a floodplain?
6. What landforms can river deposits create?

Name: _____ Date: _____

Lab 2.1.5A Water Erosion and Time

QUESTION: How is gully formation affected by the amount of time a slope is exposed to runoff?

HYPOTHESIS: Answers will vary.

EXPERIMENT:

You will need:	• brick	• timer
• baking tray with raised edges	• protractor	• hose clamp
• sand	• hose with faucet connection	
• fine screen	• 500 mL beaker	

Steps:
1. Pack a baking tray with damp sand.
2. Place the screen over a sink drain. Place one end of the tray over the sink and place the other end on a brick.
3. Use a protractor to determine the angle of elevation for the surface of the sand.
4. On a separate piece of paper, draw and label the setup.
5. Attach the hose to the faucet and place the other end in the beaker.
6. Turn on the water so only a trickle comes out of the hose.
7. Time how long it takes to fill the 500 mL beaker.
8. Clamp the hose.
9. Calculate the rate of flow as liters per second (L/s) and record it on your drawing.
10. Move the hose to the top of the elevated pan and let the water run for 1 minute. Note how much time has passed when erosion begins.
11. Clamp the hose but do not turn off the faucet. Do not move the hose.
12. Measure, draw, and describe the erosion.
13. Allow the water to run for 1 more minute. Note how much time has passed when erosion begins.
14. Turn off the faucet.
15. Measure, draw, and describe the erosion.

ANALYZE AND CONCLUDE:
1. Compare the amount of erosion after 1 minute to the amount of erosion after 2 minutes. Answers will vary. The erosion from the fast-moving water should be greater than the erosion from the slow-moving water.
2. In the first trial, how soon were signs of erosion visible after the water started flowing over the sand? Answers will vary.
3. In the second trial, how soon were additional signs of erosion visible after the water started flowing over the sand? Answers will vary but should include that extra erosion began soon after the water began flowing.

Name: _____ Date: _____

Lab 2.1.5B Water Erosion and Force

QUESTION: How does the speed of water affect erosion on a slope?

HYPOTHESIS: Answers will vary.

EXPERIMENT:

You will need:	• gravel	• hose
• shoe box	• sand	• metric ruler
• scissors	• plastic tub	• fine screen
• scale	• block	

Steps:
1. Remove the lid from a long shoe box.
2. Hold one narrow end of the shoe box facing you with the open side turned up and make a mark in the center of the end. Cut from the end's two top corners to the mark and remove the triangular piece.
3. Weigh the empty shoe box and record its mass in the data table provided.
4. Lightly pack the box with a mixture of gravel and damp sand up to the bottom of the cutout. Make the surface as flat as possible.
5. Weigh the full shoe box and record its mass in the table.
6. Subtract the mass of the empty shoe box from the mass of the full shoe box and record the mass of the material in the table.
7. Weigh the plastic tub and record its mass in the table. Place the tub in a sink.
8. Set the open edge of the shoe box on the edge of the sink so the runoff will flow into the tub.
9. Use a block to elevate the closed end of the shoe box.
10. Attach a hose to the faucet and place the other end of the hose on the upper end of the shoe box.
11. Turn on a slow stream of water and let it flow through the shoe box for 2 minutes.
12. Turn off the water.
13. Measure, draw, and describe the erosion in the shoe box.
14. Drain the contents of the tub through a screen then pour the contents of the screen back into the tub.
15. Weigh the tub and eroded material and record its mass in the table.
16. Subtract the mass of the empty tub from the mass of the full tub and record the mass of the eroded material in the table.
17. Repeat Steps 4–16, but this time use a fast stream of water.

Lab 2.1.5B Water Erosion and Force

	Slow Water	Fast Water
Mass of Empty Box		
Mass of Full Box		
Mass of Materials	Answers will vary.	
Mass of Empty Tub		
Mass of Full Tub		
Mass of Eroded Materials		

ANALYZE AND CONCLUDE:
1. How much more eroded material resulted from the fast water than the slow water? Answers will vary but should include that the faster water eroded more material than the slower water.
2. Compare the dimensions of the gullies formed by the slow water and the fast water. Answers will vary but should include that the faster water carved larger gullies than the slower water.
3. What differences did you observe in the erosion patterns formed by the slow and the fast water? Answers will vary but should include that the faster water caused more erosion than the slower water.
4. Compare your data table with other groups' findings. What might account for the differences you see? Answers will vary but should include the rate of water flow, the elevation of the boxes, the size of the outflow openings in the shoeboxes, how firmly the sand was packed down, and the proportions of gravel to sand.

2.1.6 Glacial Erosion

Weathering and Erosion

OBJECTIVES

Students will be able to
- determine how glaciers created various landforms.
- explain what clues glacial landforms provide regarding the glaciers that formed them.
- compare ice age theories.

VOCABULARY

- **drumlin** a long, tear-shaped mound of till
- **erratic** a piece of till that is not native to the place where it was deposited
- **glacial drift** the general term for any sediment deposited by a glacier
- **glacier** a large mass of moving ice that forms on land and remains from year to year
- **moraine** an accumulated deposit of till
- **till** unsorted rocks and sediments left behind when a glacier melts

MATERIALS

- Card stock, large wood block, polished board, prepared ice cubes, clay, trays (*Introduction*)
- WS 2.1.6A Scrape, Rattle, and Roll

PREPARATION

- Cut several 20 cm × 20 cm squares of card stock. Fill all compartments in an ice cube tray approximately half full of fine gravel; fill the rest of each compartment with water and freeze. (*Introduction*)
- Obtain materials for *Lab 2.1.6A Glacial Erosion*. Prepare 1 ice cup for each group by filling each paper cup with 20 g sand and 50 mL water, stirring, and freezing until solid.
- Obtain materials for *Try This: Kettle Lake*.

Introduction

Use card stock and a wood block to demonstrate how glaciers move. Stack the cut squares of card stock on a polished board or a smooth, inverted tray. Place a large wood block on the stack of card stock squares. Slowly lift one end of the board and have the class observe the block gradually slip down the slope as the cards slide over each other. The block represents a glacier, and the card stock squares represent sheets of compressed snow and ice crystals beneath the glacier. These sheets of ice slide over each other easily, just as the squares in the demonstration slide over each other. The weight of a glacier will cause the ice sheets to slide down a slope.

Guide students to use the prepared ice cubes and clay to demonstrate some effects of glacial erosion. Have each student spread clay on a tray in an even layer, approximately 2–5 mm thick, 10 cm wide, and 30 cm long. Distribute the prepared ice cubes to students and have them firmly press and move the gravel side of the ice cube along the length of the clay. Direct students to allow what remains of their cubes to melt on their trays. Have students describe and draw conclusions about the effects of dragging their ice cubes through clay as well as what happens when their leftover ice cubes melt.

Discussion

- Discuss what the *Introduction* activities suggest about how glaciers affect the land over which they move. Guide students to consider why the activities might yield somewhat different results from actual glacial erosion.

- Ask students the following question:
1. How do old-earth scientists and young-earth scientists agree and disagree regarding ice ages and glacial erosion? (*agree*: **there was at least one ice age, something happened that reduced the sun's ability to warm Earth, and glacial erosion shaped a variety of Earth's features;** *disagree*: **the length and number of ice ages, when one or more ice ages occurred, what reduced the sun's ability to warm Earth, how quickly glaciers shaped various Earth features**)

Activities

Lab 2.1.6A Glacial Erosion

- 90 mL paper cups, 1 per group
- 20 g sand per group
- 50 mL water per group
- trays with sides, 1 per group
- sand
- blocks, several per group
- protractors, 1 per group
- metric rulers, 1 per group
- timer
- thermometers, 1 per group
- graph paper

Prepare the ice cups ahead of time. Keep the ice cups frozen until students are ready to use them. Students may need several attempts to raise their trays to an angle that does not cause the sand to spill out. Make sure that each group elevates its tray to a different angle. Reset the timer every 5 minutes to help all students make their measurements at the same time.

Option 1: Have students use clear ice and note what happens to the ice as it moves downhill.

Option 2: Vary the amount of sand in the ice cups but place all the trays on the same angle.

A. Complete *Try This: Kettle Lake* with students.

B. Assign students to complete **WS 2.1.6A Scrape, Rattle, and Roll**.

C. Direct students to find a picture or photograph of an eroded area or feature in a nonscientific source and write a paragraph about the forces of erosion that shaped this feature.

D. Encourage students to research the different landforms created by glacial erosion and deposition. Have students create a computer presentation that illustrates and defines each landform.

E. Challenge students to research and present different theories regarding the presence, duration, and effects of ice ages. Have students explain whether the ice age theories fit their understanding of creation as detailed in the Bible.

F. Have students research the Little Ice Age (1300–1850) in regard to when and why it occurred; what it reveals about the formation, movement, and effects of glaciers; or how information gleaned from it influences current climate change theories.

> **TRY THIS**
>
> **Kettle Lake**
> - *paper cups, 1 per group*
> - *trays, 1 per group*
> - *sand*

Lesson Review
1. What features do valley glaciers create through erosion? (**cirques, tarns, arêtes, and horns**)
2. How did continental glaciers most likely form lakes? (**Large lakes were likely formed through erosion, but smaller lakes could have been created either by erosion or deposition.**)
3. Which glacial landform indicates where the edges of a glacier stopped? (**moraine**)
4. What information about glaciers might drumlins provide? (**the direction a glacier traveled**)
5. What information about glaciers might erratics provide? (**an area from which a glacier traveled**)
6. Explain two ice age theories. (**Possible answers: Ice ages were caused by changes in the earth's orbit and tilt; ice ages were caused by the release of superheated water from under the crust that raised the ocean temperature, which increased both evaporation and snow; an ice age was caused by smoke from volcanic eruptions that occurred during the Flood, which blocked sunlight and caused temperatures to drop.**)

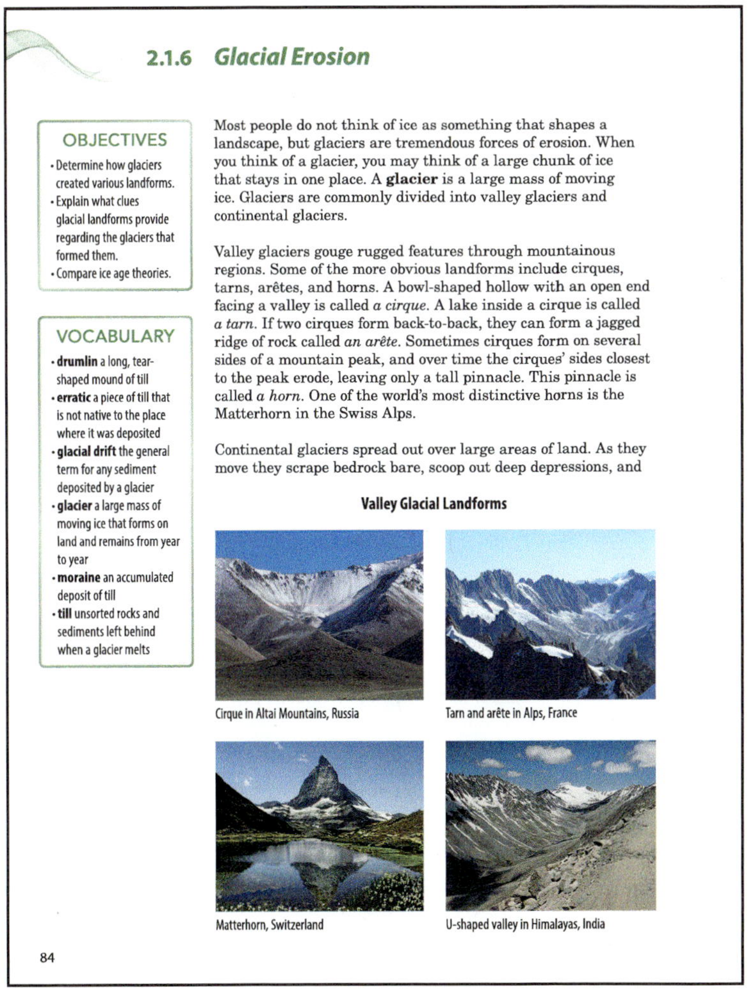

2.1.6 *Glacial Erosion*

OBJECTIVES
- Determine how glaciers created various landforms.
- Explain what clues glacial landforms provide regarding the glaciers that formed them.
- Compare ice age theories.

VOCABULARY
- **drumlin** a long, tear-shaped mound of till
- **erratic** a piece of till that is not native to the place where it was deposited
- **glacial drift** the general term for any sediment deposited by a glacier
- **glacier** a large mass of moving ice that forms on land and remains from year to year
- **moraine** an accumulated deposit of till
- **till** unsorted rocks and sediments left behind when a glacier melts

Most people do not think of ice as something that shapes a landscape, but glaciers are tremendous forces of erosion. When you think of a glacier, you may think of a large chunk of ice that stays in one place. A **glacier** is a large mass of moving ice. Glaciers are commonly divided into valley glaciers and continental glaciers.

Valley glaciers gouge rugged features through mountainous regions. Some of the more obvious landforms include cirques, tarns, arêtes, and horns. A bowl-shaped hollow with an open end facing a valley is called *a cirque*. A lake inside a cirque is called *a tarn*. If two cirques form back-to-back, they can form a jagged ridge of rock called *an arête*. Sometimes cirques form on several sides of a mountain peak, and over time the cirques' sides closest to the peak erode, leaving only a tall pinnacle. This pinnacle is called *a horn*. One of the world's most distinctive horns is the Matterhorn in the Swiss Alps.

Continental glaciers spread out over large areas of land. As they move they scrape bedrock bare, scoop out deep depressions, and carve U-shaped valleys. Some scientists theorize that retreating continental glaciers carved deep depressions in some places. Deep, narrow, U-shaped valleys that are partially submerged and open into the sea are called *fjords*. Fjords have a distinctive shape, which is why some scientists believe that they were at least partially eroded by glaciers. Fjords are common in Norway, Alaska, and other places near the Arctic Circle, but they can also be found in the far southern end of Chile.

Significant signs of continental glacial erosion are also found farther inland. Some geologists believe glacial erosion may be responsible for carving out lake beds. For example, some scientists theorize that the Great Lakes in Canada and the United States were formed by glacial erosion. Smaller round lakes called *kettle lakes* may have resulted from glacial erosion. Some kettle lakes may even have been formed by glacial deposition. If sediment covered a block of ice that was separated from a retreating glacier, the sediment would have fallen in on itself as the ice melted and created a water-filled pit.

> **TRY THIS**
>
> **Kettle Lake**
> Kettle lakes are formed when till covers a block of ice that became separated from a retreating glacier. As the ice slowly melts, the till falls in on itself, creating a pit that retains the melted ice. Make your own kettle lake. Freeze water in a small paper cup. Peel the paper off the ice cube and place it on a tray. Cover the ice cube with sand. Observe what happens as the ice melts.

Valley Glacial Landforms

Cirque in Altai Mountains, Russia
Tarn and arête in Alps, France
Matterhorn, Switzerland
U-shaped valley in Himalayas, India

Continental Glacial Landforms

Tarn in Cascades, United States
Moraine in Rocky Mountains, Canada
Kettle lake in Isunngua, Greenland
Drumlin in Andechs, Bäckerbichl, Germany
Erratic in Alberta, Canada
Fjord in Norway

FYI
Icy Sculptors
Glaciers have shaped much of the landscape of the Northern Hemisphere, Antarctica, and many mountain ranges around the world. During ice ages, large sheets of ice moved south from the North Pole and north from the South Pole. The friction as they rubbed across the land slowed their journey across the continents, and the glaciers left behind dramatic patterns.

Glaciers deposit materials as well as erode them. As glaciers melt, they deposit all of the materials that they scooped up during their journeys. Glacial deposits can be identified by their unique shapes and composition. **Glacial drift** is the general term for any sediment deposited by a glacier; it is categorized into stratified drift and till. Stratified drift sediment was deposited by melting glacier water. The particles of stratified drift are about the same size and weight because they are sorted by streams running under a glacier or by lakes or streams at a glacier's edge.

Till is unsorted rocks and sediments left behind by a glacier. Till forms distinctive geologic features. A **moraine** is an accumulated deposit of till. Moraines commonly occur as large ridges or mounds. Moraines can form on the ground or on the glacier itself. They are often found on the sides of valley glaciers, at the place where two valley glaciers meet, or at the leading edges of both valley and continental glaciers. Moraines can even indicate where the boundaries of a glacier once existed. Another type of moraine, called *a ground moraine*, can yield fertile soil. Ground moraines are a widespread covering of till left behind by retreating glaciers.

A **drumlin** is a long, tear-shaped mound of till. How they form is unknown, but some scientists believe they were formed by deposits from the underside of glaciers as the glaciers advanced. Drumlins taper to a point in the direction the depositing glacier was traveling.

An **erratic** is a piece of till that is not native to the place where it was deposited. Erratics can be as small as a pebble or as large as a boulder. One erratic block in Alberta, Canada, weighs 16,500,000 kg or 16,500 metric tons! Erratics with distinctive minerals indicate areas from which a glacier traveled.

Most old-earth scientists believe that a majority of the erosion and deposition caused by glaciers happened during several long periods of time when sheets of ice covered large areas of Earth's surface. Such periods are called *ice ages*. Scientists disagree regarding the number and duration of ice ages that the earth has experienced. Some believe that there were more than four ice ages and that the most recent major ice age ended about 11,000 years ago. Most young-earth scientists believe that there was only one major ice age, which began shortly after the Flood.

The cause of an ice age remains a mystery. According to one theory, regular changes in Earth's orbit and in the tilt of its axis sparked the beginning of an ice age. These changes are thought to affect how much the sun warms the earth. If Earth's average temperature drops by even a few degrees Celsius, then glaciers could spread widely. This theory is called *the Milankovitch Theory*; it is named after Milutin Milankovitch, a Yugoslavian scientist who first proposed it in the 1920s.

Another theory suggests that during the Flood the average ocean temperature rose because superheated water vented from beneath the crust. The warmer ocean water increased evaporation, which in turn caused more snow to fall during winter. Many young-earth scientists believe the Flood was accompanied by major volcanic eruptions. The smoke of these eruptions blocked energy from the sun and caused average temperatures to drop, which enabled glaciers to grow.

Smoke from volcanic eruptions may have contributed to an ice age.

LESSON REVIEW
1. What features do valley glaciers create through erosion?
2. How did continental glaciers most likely form lakes?
3. Which glacial landform indicates where the edges of a glacier stopped?
4. What information about glaciers might drumlins provide?
5. What information about glaciers might erratics provide?
6. Explain two ice age theories.

FYI
Interpreting Erosion 
Many old-earth scientists believe the theory of uniformitarianism helps people determine Earth's age. They believe that weathering and erosion rates have always been the same. Using these rates as constants, they work backward to calculate when Earth's features began forming. Uniformitarians conclude that different parts of Earth's topography must have developed over thousands or millions of years. In contrast, many young-earth scientists believe that cataclysmic events sped up erosion to shape Earth's features in a much shorter period of time. They base their theory on the observable effects of massive landslides, volcanoes, earthquakes, and powerful storms. Proponents of this view believe that many of Earth's features were shaped by the Flood.

Name: _____ Date: _____

Lab 2.1.6A Glacial Erosion

QUESTION: How is a glacier's erosion pattern affected by its environment?

HYPOTHESIS: Answers will vary.

EXPERIMENT:

You will need:		
• ice cup containing 20 g sand and 50 mL water	• sand	• timer
	• blocks	• thermometer
• tray	• protractor	• graph paper
	• metric ruler	

Steps:
1. Pack the tray with dry sand and use the blocks to elevate one end of the tray.
2. Use a protractor to measure the tray's angle of elevation and record it. _____
3. Get an ice cup from your teacher. Tear the paper cup off the ice.
4. Measure the height of the ice cube.
5. Place the ice cube at the top of the tray, touching the inside of the elevated edge.
6. Place a thermometer on the sand near the ice cube's anticipated path.
7. Every 5 minutes, measure the temperature as well as the ice cube's height and its distance from the tray's elevated end (measure from the inside edge of the tray). Record the measurements in the table.

	0 min	5 min	10 min	15 min	20 min	25 min	30 min	35 min	40 min
Temp.									
Ice Cube Height				Answers will vary.					
Distance Ice Cube Traveled	0.0 cm								

8. On a piece of graph paper, graph the tray's temperature in relation to time.
9. Graph the ice cube's height in relation to time.
10. Graph the distance the ice cube traveled in relation to time.
11. After 40 minutes, sketch what the tray of ice and sand looks like.

Drawings will vary.

Lab 2.1.6A Glacial Erosion

ANALYZE AND CONCLUDE:
1. Describe the environment your tray is in. For example, is the tray in direct sunlight or in shade? What is the temperature of the classroom? Is there any air movement near the tray? Are there any heat sources near the tray, such as people, animals, or running machinery? Did the environment change at all over the course of the experiment? Answers will vary.

2. Describe the path the ice cube made as it traveled down the tray. Answers will vary.

3. Compare your data to the data obtained by other groups. What might account for variances among the data collected by different groups? Answers will vary but should include the different angles of slope and any other variations in the environment.

2.2.0 Soil

Chapter 2.2 Summary
Soil is the biologically active porous medium that forms in the top layer of the earth's crust. This amazing mixture is crucial to the sustenance of life on Earth. Soil stores water and nutrients, filters and breaks down wastes, and participates in the recycling of carbon and nitrogen. The structure of soil contains a soil profile made up of six layers called *horizons*. These horizons differ in texture, color, organic material content, thickness, and other properties. Many different soil types are found around the world. The differences are determined by the regional climate, the slope of the land, the parent material (basalt, limestone, granite, or shale), the type of vegetation present, and the amount of time that the parent rock has been exposed to weathering and erosion processes.

Soil and sediment layers hold a great deal of information about the history of Earth. Scientists have studied the geologic column for years to find clues to what happened in Earth's past. They made a few assumptions during their study. The first assumption is the law of superposition, which states that the layers of strata that are found lower in the geologic column are older than the layers located higher in the column. Sometimes geologists come across a break in the geologic record. These breaks are called *unconformities*. There are three main types of unconformities: disconformities, nonconformities, and angular unconformities.

When attempting to describe Earth's past, the presence of fossils in rock strata has proved to be very useful. Many paleontologists assume that the age of a fossil is the same age as the strata where the fossil was found. Some fossils are easily recognized, abundant, and widely distributed geographically. These fossils are called *index fossils*. Fossils are formed in a variety of ways including petrifaction, mold and cast formation, freezing, carbonization, and desiccation. Some fossils may also form if the organism becomes entrapped in tar or resins.

Background

Lesson 2.2.1 – Soil Composition
The organic matter in soil consists of decomposed plant and animal material and living plant roots. Soil is second only to the oceans as the largest carbon repository on the planet because of its large quantity of organic matter. Soil's natural tendency to store carbon is essential for mitigating and adapting to climate change as well as improving flood and drought resilience.

Soil is Earth's largest natural water filter. Water that flows onto soil is purified by natural filtering processes, which makes runoff far less toxic when it eventually reaches its destination. Soil, working in conjunction with microorganisms, absorbs nutrients like calcium, magnesium, and potassium and removes them from the water supply. The microbes in soil decompose organic pollutants and make nutrients available for plants.

Many of the antibiotics used in medicine come from soil. Some examples include streptomycin and cyclosporine, a medication used to prevent transplant organ rejection.

The inorganic part of soil may include various sizes and shapes of rocks and minerals. Sand is the largest inorganic soil particle, followed by silt and clay. Gravel and stone, which are larger fragments, are not considered part of soil. Coarser soil has a lower capacity to retain organic plant nutrients, gases, and water, which are essential for plants. Soils with higher clay contents tend to retain these substances and are usually better suited for agriculture. The arrangement of particles, known as *soil structure*, affects the soil's pore space and density, which determines its capacity to hold air and water.

Factors such as vegetation type, climate, parent rock material, topography, and geologic age determine the nature of soil. Acidic soils occur in humid regions because alkaline minerals are leached downward. Alkaline soils occur in dry regions because alkaline salts remain concentrated

---

LOOKING AHEAD

- For **Lesson 2.2.1**, five weeks before the lesson, create humus farms. Obtain hair or fur, acorns, feathers, and vermicompost supplies.
- For **Lesson 2.2.2**, obtain rock samples, an analysis of minerals and soils kit; moisture meters; soil pH meters; fertilizer meters; stainless steel soil thermometers; soil test kits; humus test kits; and soil sampling tubes.
- For **Lesson 2.2.3**, prepare red cabbage juice prior to the activity; let it stand for one hour. Obtain soil test kits, soil pH meters, soil biology and chemistry experiment kits, soil organism study kits, soil sampling tubes, and chemical composition of soil kits. Arrange a visit to a local farm.
- For **Lesson 2.2.4**, create a geologic column prior to presenting the lesson.
- For **Lesson 2.2.5**, obtain fossil samples including trace fossils. Arrange a field trip to a natural history museum or national park. Invite a paleontologist to speak. Arrange a fossil hunting field trip.

SUPPLEMENTAL MATERIALS

- TM 2.2.2A Soil Horizons
- TM 2.2.2B Regional Soil Profiles
- TM 2.2.3A Common Soil Types
- TM 2.2.4A Unconformities
- TM 2.2.4B The Great Unconformity
- TM 2.2.5A Trace Fossils

- Lab 2.2.1A Pore Space
- Lab 2.2.1B Capillarity Investigation
- Lab 2.2.3A Soil Fertility
- Lab 2.2.3B Water-Holding Capacity

- WS 2.2.2A The Layered Look
- WS 2.2.4A Law of Superposition

- Chapter 2.2 Test
- Unit 2 Test

BLMs, TMs, and tests are available to download. See Understanding Purposeful Design Earth and Space Science at the front of this book for the web address.

> **WORLDVIEW**
>
> - Soil is the foundation of physical life on Earth. The primary role of soil is to provide the environment and resources that foster the health and growth of the plant kingdom, bacteria, protists, and other small organisms. If the soil in a variety of ecosystems was destroyed, the effect would be catastrophic. Life on Earth would not exist. The food chain would completely unravel, and the supply of oxygen needed to sustain life would be severely depleted. The planet would be subject to more prevalent drought and flooding; it would not be able to adapt to climate changes. However, life on Earth would not exist if God had not first created life. People are not just physical bodies that eat and breathe. God created individuals with a spirit that also needs to be fed in order that it may grow and thrive. The foundation of a spiritual life is Christ, as Paul reminds us in 1 Corinthians 3:11: "For no one can lay any foundation other than the one already laid, which is Jesus Christ." Without Christ, a person's life is dramatically affected. Our relationships with others, how we view the world, and how we view what it means to be made in the image of God are all transformed and have greater meaning when Christ is at the center of our lives.

near the surface. Geologically young soils resemble their parent material more than older soils that have been altered over time by climate and vegetation.

Lesson 2.2.2 – Soil Profile

Undisturbed soil tends to form horizons that are roughly parallel to the land surface. Soil classification is determined by the distinctive horizons of the soil profile. The O horizon (organic) is the thin surface layer of loose organic debris and humus. The A horizon (topsoil), formed at or near the surface layer, contains rich topsoil and some humus. Between the A and B horizons is the E horizon (eluviated), which exhibits a significant loss of clay and humus and appears bleached. It contains little or no organic matter. The B horizon (subsoil) contains inorganic compounds formed by the decomposition of organic material, a process known as *mineralization*. The material accumulates in the B horizon by the downward leaching action of water. The lowest soil layer, the C horizon (substratum) represents the weathered mineral parent material. Below the five soil layers is a layer of consolidated bedrock known as *the R horizon*.

A vertical cross section of the soil exposes obvious layers. Each layer differs from the other, physically or chemically. The differences arise from many factors, such as parent material, slope of the land, vegetation, weathering, time, and climate. Factors that influence the environment also affect the soil, including additions such as fallen leaves, windblown dust, or chemicals. Other factors include the loss of minerals through deep leaching or the loss of surface area through erosion, translocations by leaching of minerals or evaporating water, and transformations, such as decay, weathering, and chemical reactions. These processes occur in different ways at varying depths and within horizons.

Lesson 2.2.3 – Soil Types

Russian scientist Vasily Dokuchaev began the modern classification of soils in the 1880s. Europe, Canada, and the United States began to classify soils only a few years later. There are many ways to classify soils, and several systems are used worldwide. All systems recognize soil as a product of the environment, but they differ in groupings and horizon definitions. The current taxonomy is determined by soil shape and form. Taxonomy helps to derive facts related to crops, formation, management, and the environment.

Climate is the major factor that affects soil type. Light, temperature, and water content influence the rate of decomposition of organic matter, which is crucial to the formation of fertile soil. Soil holds the moisture and heat required for microorganisms to thrive and to perform the decomposition process that changes organic materials into humus. Organic matter in soil that is too wet, dry, or cold will decompose slowly. The sun's energy warms the soil and promotes evaporation, which affects moisture content.

The pH value of soil is a measure of its acidity or alkalinity. Soil pH affects the quality of plant growth because it relates not only to nutrient availability but also to the concentration of materials that are toxic to plants in the soil. For example, in highly acidic soils, aluminum and manganese are often prevalent, which are toxic to plants. At low pH values, calcium, phosphorus, and magnesium are less available to plants. At pH values of 6.5 and above, phosphorus and most of the micronutrients become less available. Different plants thrive in different pH ranges. Various factors affect the pH value of soil. Fertilizers that contain ammonium or urea increase soil acidity, as does the decomposition of organic matter.

Lesson 2.2.4 – Soil and Rock Layers

Scientists use several techniques to determine the ages of rocks. Before more radiometric dating methods became available, geologists used relative dating techniques. They divided the geologic record into time units depending on the correlation of rock formations. Early European geologists built on these time units and incorporated interpretations of the fossil record to devise the Geologic Time Scale. This scale includes the Cenozoic, Mesozoic, Paleozoic, and Precambrian eras. These time units are still used today.

In the 17th century, Danish geologist Nicolas Steno formulated the law of original horizontality, one of the fundamental laws of stratigraphy. According to this law, sediments are usually deposited in horizontal layers. Exceptions to this rule occur in sand dunes and deltas. James Hutton was baffled when he visited the Scottish coast in the 1780s and noticed that the surface sedimentary rocks were horizontal and the lower layers of rock tilted at a high angle. This arrangement of layers was a great example of angular unconformity. Hutton concluded that such an unconformity would take a very long time to form and that the unconformity represented a gap in the record of Earth's history. The actual amount of time represented by the gap could not be determined, which led Hutton to conclude that the earth was older than a few thousand years. His interpretation of angular unconformity was a watershed in the history of geology.

Today many scientists use radiometric dating to determine the age of certain rocks. They measure the amounts of certain radioactive isotopes in the rocks in relation to the amount of corresponding decay isotopes. However, many young-earth scientists question the reliability of this method because the results can be inconsistent. Researchers continue to study geologic columns and to develop new methods of determining the age of the earth.

Lesson 2.2.5 – Fossils

A fossil is the remains or impression of an organism preserved by natural conditions. Fossils may be preserved in sedimentary rock, ice, tar pits, resins, or other materials. Fossils were known to the ancient Greeks, but scientists did not study them until the 18th century. Until around 1800, fossils were not recognized as the remains of living things of the past. During the 19th century, the study of fossils made rapid progress thanks to several key individuals and the discussion of the concept of Earth as ancient in the scientific community.

In order to form a fossil, an organism must be quickly buried in material that inhibits weathering and blocks oxygen and bacteria from inducing decay. Sediment-covered shells and bones leave reproductions of external and internal structures in tar pits, bogs, volcanic ash, and quicksand. Some specimens are encased in rock and dissolved by water, which leaves a natural mold. Scientists can use these molds to make casts. Natural casts can form if subsurface water fills a mold with a mineral material. Molds of thin objects like leaves are known as *imprints*.

Seldom are an organism's soft tissues preserved, although impressions of dinosaur skin, animal footprints, and even excrement have been found. In rare cases, entire preserved animals have been discovered. For example, in May 2013, scientists in Siberia found a nearly complete woolly mammoth carcass with its skin and flesh intact preserved in ice.

Some organisms are preserved through petrifaction, a process in which organic material is infiltrated or replaced with minerals. Petrifaction can occur in several ways. The minerals from groundwater may seep into porous materials such as bone, wood, and shells, making the material more compact and resistant to disintegration. The minerals may completely replace the organic matter so the original structure is maintained. Imprints of plant tissues—the carbon residue of the organism—are sometimes found in shales.

Desiccation is a rare type of fossilization that happens when the hard and soft portions of the organism dry out. Synthetic desiccation, also known as *mummification*, involved bodies of people or animals being packed and covered with natron, a salty drying agent, and left to dry for 40 to 50 days. By that time, all of the body's liquid has been absorbed, leaving only the hair, skin,

NOTES

NOTES

and bones. In ancient Egypt, the dried body cavity was stuffed with resin, sawdust, or linen and shaped to restore the deceased's form and features. In contrast, natural mummies are the result of environmental conditions such as freezing soil, ice, and dryness. Mummified remains are usually found in desert areas where the climate further inhibits decay.

Scientists attempt to date strata layers by using well-preserved fossils. For example, if one type of fossil is found in a particular strata layer, scientists assume that if they find the same fossil in another region on Earth, the two sets of fossils must be the same age. Some groups of fossils that have been frequently observed in strata layers throughout the world are used for this purpose. These fossils are called *index fossils*. An example of an index fossil is a trilobite. Trilobite fossils are easy to recognize, plentiful, and found throughout much of the world. Scientists use index fossils because they are believed to have only existed for a relatively short period of geologic time. Some scientists disagree with trying to assign ages to strata in this manner because strata layers are used to date fossils and fossils are used to date strata layers, which could be regarded as circular reasoning. Scientists grapple with other challenges, such as the extent of human impact on the fossil record, the existence of polystrate fossils that extend through multiple layers of strata, and the lack of transition fossils.

Soil

2.2.1 Soil Composition

Introduction
Fill a clear basin with water and explain to students that this represents the surface of the earth. Ask students to indicate if they think this is a good representation of the surface of the earth and why. Have students form groups to determine various ways that soil affects Earth's structure, agriculture, ecosystems, and their personal lives. Build on students' ideas with specific examples locally and globally.

Present the humus farm. Have students compare the height of the debris with the original height of the organic matter. Direct students to observe the varieties of mold and other fungi that are growing in the various jars. Explain that the role of the bacteria and fungi is to break down the organic matter into nutrient-rich humus. Divide the class into pairs and have students use microscopes to explore different portions of the humus in various stages of formation.

Discussion
- Ask students to name examples of living things that reside in soil. How do these organisms receive air? (**Answers will vary.**) To demonstrate that air is trapped in soil, fill a jar half full of soil and slowly pour water over it. Air bubbles will rise through the water.

- Discuss why the soil on Earth is different from the soil found on the moon or on Mars. (**Only soil on Earth has an organic component.**) Why are living things important for the formation of rich soil? (**Living things radically change the nature of soil, converting it from a layer of tiny rock fragments into a rich ecosystem teeming with life and able to sustain a vast array of plant species.**)

- Ask the following questions:
  1. Why are water and air important for the formation of rich soil? (**Once water and air are present in soil, chemical reactions occur to allow living things such as bacteria, fungi and plant roots to function and thrive.**)
  2. Describe the effect of plants on soil formation. (**Plant roots bind up soil particles to prevent them from eroding. Plants' leaves shelter the soil from too much water. Roots draw minerals up to the soil surface. Plants also decay to form organic food sources for burrowing insects, worms, and larger animals.**)
  3. Why are bacteria and fungi important for the formation of rich soil? (**Bacteria and fungi thrive on dead plant materials and animal droppings. They help form the ingredients that comprise the dark, nutrient-rich organic material called *humus*.**)

Activities

Lab 2.2.1A Pore Space
- 250 mL beakers, 2 per group
- dry topsoil samples, 20 g per group
- spoons, 1 per group

Lab 2.2.1B Capillarity Investigation
- graduated cylinders, 1 per group
- small dishes, 1 per group
- small peat pots, 1 per group
- metric rulers, 1 per group
- stopwatches, 1 per group

(continued)

OBJECTIVES
Students will be able to
- distinguish among the three basic types of soil particles.
- identify the inorganic and organic components of soil and compare their functions.
- summarize how soil forms.

VOCABULARY
- **bedrock** the layer of solid rock beneath the soil
- **humus** the nutrient-rich, organic material in soil
- **pore space** the amount of space between soil particles

MATERIALS
- Clear basin, humus farms, microscopes (*Introduction*)
- Wide-mouthed jar, soil (*Discussion*)
- Samples of gravel, sandy soil, silty soil, clay soil, rocks and minerals field guide (*A*)
- Beaker, hot plate, watchglass (*B*)
- Grass clippings, hair or fur, leaves, acorns, feathers (*C*)
- Vermicompost supplies (*D*)

PREPARATION
- Five weeks before teaching this lesson, label 2 wide-mouthed jars *A* and *B* and the date they were created. Place 5 cm of soil in each jar. Water the soil in the jars to form a soggy mixture. Cover the soil in Jar A with 2 cm of grass clippings. Cover the soil in Jar B with 3 cm of leaf litter. Mark the level of the contents on the outside of each jar. Place both jars in direct sunlight. Water both jars every other day. Using the same procedure, add two additional jars to the humus farm each week to demonstrate the progression of soil formation. (*Introduction*)
- Obtain materials for *Lab 2.2.1A Pore Space* and *Lab 2.2.1B Capillarity Investigation*.

NOTES

(continued from previous page)

Have one large wastebasket available and an extra supply of dry peat pots. Determine the capillarity prior to the lab. Small disposable pie plates can be used for water dishes. If peat pots are unavailable, clay pots may be used. (Time will vary.) Peat pots may be dried and reused.

Share with students that peat is a surface deposit consisting of decayed plant materials. It forms a very spongy soil, displaying obvious capillary action. The capillary action of soil allows for water to be drawn toward the surface of soil. Plant roots can then absorb the water, increasing the growth and health of the plant.

A. Divide the class into groups and provide students with samples of gravel, sandy soil, silty soil, and clay soil. Before class, study the rock particles to determine each type of soil using a rocks and minerals field guide. Encourage students to discover basic properties about each soil type, including particle size, color, and composition.

B. To demonstrate that soil holds water, place a sample of soil in a small beaker and then gently heat the beaker on a hot plate. Place a watchglass over the beaker. Water will condense on the underside of the watchglass and at the top of the beaker. The condensation is water that was trapped in the soil.

C. Show students organic substances that can be found in soil: grass clippings, hair or fur, leaves, acorns, and feathers. Explain that organic substances, microorganisms, decayed animals, and

---

2.2.1 Soil Composition

OBJECTIVES
- Distinguish among the three basic types of soil particles.
- Identify the inorganic and organic components of soil and compare their functions.
- Summarize how soil forms.

VOCABULARY
- **bedrock** the layer of solid rock beneath the soil
- **humus** the nutrient-rich, organic material in soil
- **pore space** the amount of space between soil particles

At first, you might not think of soil as important, but imagine a world without it. Life on Earth would not be sustainable. Healthy soil, along with water, air, and sunlight, is essential to the biosphere. Scientists have found that the world's soil is one of the largest reservoirs of biodiversity. It contains almost one-third of life on this planet including worms, insects, small vertebrates, fungi, and microorganisms. These organisms work with soil to provide nutrients to plants. When water flows through soil, it is purified by a natural filtering process. Soil absorbs nutrients like calcium, magnesium, and potassium, which prevents these elements from entering the water supply. At the same time, microorganisms found in the soil decompose the organic pollutants in the water and supply the nutrients absorbed by the soil to plants. This complex process also makes runoff far less toxic when it eventually reaches its destination.

Soil is a mixture of organic materials, rock particles, water, and gases. Organic materials are materials that are obtained from living or once-living things. Microorganisms, decaying plant and animal material, and living plant roots are organic materials. These organisms and materials turn lifeless fragments of rock into soil that is rich in nutrients. Without this mixture, plants could not grow, the food supply would be depleted, and most living things could not survive. Soil is the backbone of every land ecosystem on Earth. Current research indicates that unlike the moon or the other terrestrial planets where the top layer of crust is nothing more than tiny rock fragments, Earth's soil contains a mixture of nonliving and living things.

Soil teeming with life

The main ingredient of soil is small rock particles. These particles form the mineral content of soil. Soil can take thousands of years to form as most rocks break up into particles of different sizes. However, not all rocks disintegrate. Deep beneath the earth's surface lies a layer of solid rock that is not exposed to weathering processes unless it has been brought to the surface. This layer of solid rock beneath the soil is called **bedrock**. Bedrock is like a bowl that holds Earth's soil and groundwater. The loose layer of rock and soil above bedrock is regolith. The rocks near the bottom of the regolith are larger than the ones near the top because they undergo little weathering and erosion. The top layer of regolith is constantly exposed to weathering and erosion. Eventually these processes grind the rock particles into pieces that are small enough to become part of the soil.

Undisturbed Soil Components
- Organic matter
- Air
- Rock particles
- Water

Three basic types of soil are determined by the size of their rock particles. In a soil sample, the amount of each soil type depends on the "parent" rock that was weathered. For example, when the mineral feldspar goes through chemical weathering, small grains that contain aluminum and water are created. These grains form clay, which has a particle diameter of less than 0.002 mm. Individual grains of clay are so small that you cannot see them with the unaided eye or feel them between your fingers. When clay is squeezed, it can be easily molded and retains its shape. Clay becomes hard and loses its plasticity when heated. There are two categories of clay: residual clay and sedimentary clay. Residual clay is found where it formed. Most residual clay

🟧 BIBLE CONNECTION

The Potter and the Clay

In Isaiah 64:8, God's people are likened to potter's clay. Without any assistance, clay looks like a globule that lacks any form or purpose. As soon as the potter begins to work the clay, it starts to take shape. God is the potter. If people are ready, God will mold and shape their lives into something brand new that is better than they could ever have imagined.

God never makes mistakes. Clay on Earth will crack and break when exposed to heat if it was not formed properly. When God is allowed to be the potter, people can withstand the trials and pressures of this world. They will withstand the heat of evil with God by their side.

© Earth and Space Science

88   89

excrement all play an important part in soil formation. Point out that these things add nutrients to the soil that make it possible for plants to grow and for other life to thrive on Earth.

D. Start a school vermicompost pile. Have students research how to build and care for a worm farm. Once the worms have created vermicompost, use the worm compost to fertilize the school garden or potted plants.

E. Have students research one type of soil classification system and write an essay including the individuals or group of individuals that created the classification system, the soil groups, the characteristics used to determine group placement, and whether the system is still used today.

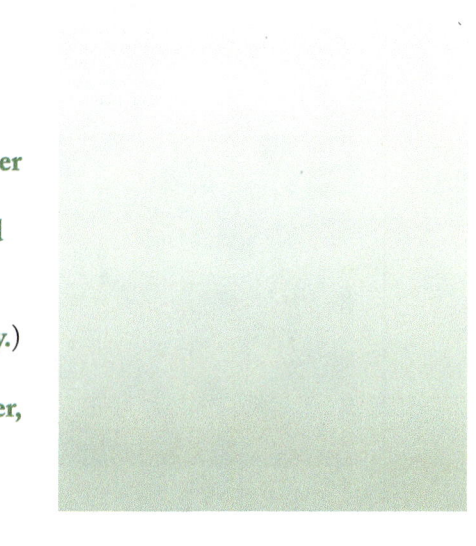

NOTES

Lesson Review
1. Name and describe the three basic types of soil particles. (**clay, particle diameter of less than 0.002 mm; silt, particle diameters between 0.002 mm and 0.05 mm; and sand, particle diameter between 0.05 mm and 2.0 mm**)
2. Name three things that comprise organic material in soil. (**microorganisms, decaying plant and animal material, and living plant roots**)
3. What inorganic components are found in soil? (**rock particles, water, and gases**)
4. How does pore space in clay compare to pore space in silt? (**Silt has larger pore spaces than clay.**)
5. How does soil form? (**Soil is formed when rocks break down into small particles through the processes of weathering and erosion and those rock particles mix with organic materials, water, and gases.**)

---

is formed by surface weathering. In contrast, sedimentary clay, or transported clay, was transported from its place of origin by erosion.

Silt particles are the next largest soil particles with particle diameters between 0.002 mm and 0.05 mm. These particles are not easily seen, and they feel like flour. Silt, which is composed primarily of quartz particles, is a product of the weathering and decomposition of preexisting rock. Silt is made from a variety of rocks. It is often carried by rivers and deposited on riverbanks. Hardened silt forms the sedimentary rock siltstone, which is deposited in thin layers. Siltstone is hard and flat, and it breaks into nearly rectangular slabs.

Sand particles are the largest soil particles. The rounded or angular particles of sand have a diameter between 0.05 mm and 2.0 mm. Most sand grains are large enough to be seen without aid, and their small crystals are easily felt. Sand is the product of the weathering and decomposition of igneous, sedimentary, or metamorphic rock. Most sand is made up of silica, usually quartz. However, sand grains may also be comprised of other

Magnified Size Comparison of Sand, Silt, and Clay Particles

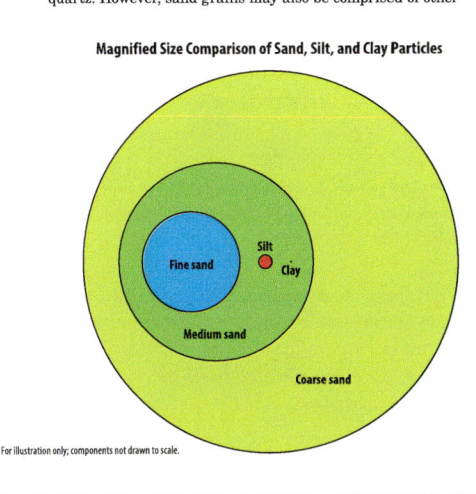

For illustration only; components not drawn to scale.

Sand, Silt, and Clay Particles

Sand	Silt	Clay
0.05 mm–2 mm	0.002 mm–0.05 mm	<0.002 mm

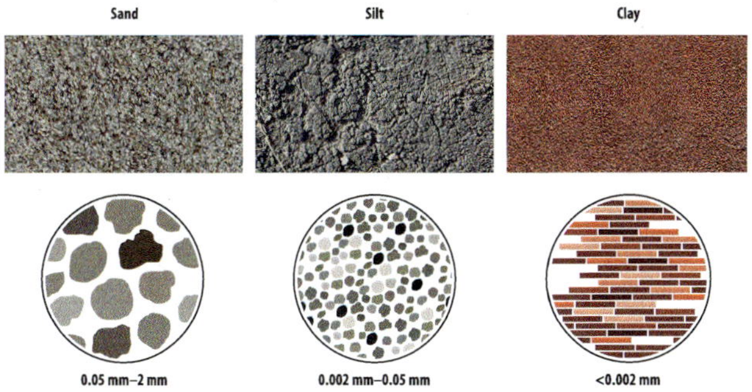

minerals. Sand may even be organic in origin. For example, organic sand can come from coral or shell. Sand deposits are formed by wind, running water, waves, and glaciers. Generally, sand makes poor soil. However, some sand in soil is helpful because it permits the free movement of air, improves drainage, and offers less resistance to roots.

Rock fragments that are larger than sand are not considered soil. These fragments are called *gravel*. Gravel consists of rock particles that are usually round with a particle diameter between 2.0 mm and 75.0 mm. Many kinds of rock comprise gravel, but quartz is the most common component. Gravel deposits are a product of the weathering of rocks and erosion by waves and running water. Gravel is useful when forming roads and concrete. In areas lacking natural gravel deposits, gravel is produced by quarrying and crushing durable rocks, such as sandstone, limestone, or basalt.

The mineral content of soil is generally made up of different proportions of clay, silt, and sand. Once the rock is broken into different-sized particles, air and water move in to fill the gaps between the larger particles. The amount of space between soil particles is called **pore space**. Different kinds of soil have different pore spaces. Fine soil particles, such as clay, have small pore spaces. Larger particles, such as silt or sand, have larger pore spaces. These spaces hold water and air. When soil is moist, the spaces hold the water that plants need to grow. During a drought, the spaces are almost entirely filled with air.

The organic material, rock particles, water, and gases found in soil are transformed by living things to create a rich ecosystem. Plant roots bind soil particles together to prevent erosion. Plant leaves shelter the soil from absorbing too much water. Roots draw minerals up to the surface. When plants decay, they form food sources for burrowing insects, worms, and larger animals. Bacteria and fungi thrive on dead plant materials and animal droppings. These ingredients comprise the dark, nutrient-rich, organic material called **humus**. Some scientists claim that humus makes soil more fertile. Others state that humus helps prevent disease in plants. Humus contains many useful nutrients such as nitrogen, which is a key nutrient for plants. Farmers depend on nitrogen and other nutrients found in humus for healthy crops. Although humus can be found naturally in many forests, it can also be created in a process known as *composting*.

LESSON REVIEW
1. Name and describe the three basic types of soil particles.
2. Name three things that comprise organic material in soil.
3. What inorganic components are found in soil?
4. How does pore space in clay compare to pore space in silt?
5. How does soil form?

Nutrient-rich humus can be found in many forests.

92

Name: _____ Date: _____

Lab 2.2.1A Pore Space

QUESTION: What determines the amount of pore space in a soil sample?
HYPOTHESIS: Answers will vary.

EXPERIMENT:

You will need:	• 50 g dry topsoil
• two 250 mL beakers	• spoon

Steps:
1. Pour 150 mL of water into one of the beakers.
2. Place 50 grams of soil into a second beaker.
3. Carefully pour the water into the beaker of soil.
4. Stir gently until bubbles cease to rise to the surface of the water.
5. Measure and record the volume of the water and soil mixture.
   - Volume of water: Answers will vary.
   - Volume of dry soil: Answers will vary.
   - Calculated volume of mixture: Answers will vary.
   - Actual volume of mixture: Answers will vary.
   - Difference between calculated and actual volume of mixture: Answers will vary.
6. Calculate the percentage of pore space filled with air in the soil sample.

   % of pore space = (difference in Step 5 ÷ actual volume of mixture) × 100 = Answers will vary.

ANALYZE AND CONCLUDE:
1. Why do some gardeners think soil that contains a large amount of pore space is better for plant growth than soil with a small amount of pore space? A large amount of pore space in soil allows water to flow freely to reach plant roots and allows gases to be easily exchanged.

2. Design a controlled experiment to determine the amount of water in soil samples collected around the school.
Answers will vary.

© Earth and Space Science • Soil

Name: _____ Date: _____

Lab 2.2.1B Capillarity Investigation

QUESTION: How does the capillarity in peat pots relate to capillary action of soil?
HYPOTHESIS: Answers will vary.

EXPERIMENT:

You will need:	• small dish	• metric ruler
• graduated cylinder	• small peat pot	• stopwatch or watch with second hand

Steps:
1. Measure 75 mL of water and pour it into a shallow dish or pan.
2. Place the peat pot upside down in the water.
3. As the water begins to be absorbed by the peat, measure and record the height of the absorbed water at 1-minute intervals in the data table below.

Time (minutes)	Height of Water (cm)
0	0.0 cm
1	
2	
3	
4	
5	
6	
7	
8	
9	
10	

4. After 10 minutes, remove the pot from the dish or pan and measure how much water is left. Calculate the amount of water that was absorbed by the peat pot.
   - Amount of water in the dish at 0 minutes: _____
   - Amount of water in the dish at 10 minutes: _____
   - Total amount of water absorbed by the peat pot: _____

ANALYZE AND CONCLUDE:
1. Calculate the average rate (cm/min) of capillary action for 10 minutes.
   (Rate = distance ÷ time)
   - _____ (Rate) = _____ (distance) ÷ _____ (time)

© Earth and Space Science • Soil

Lab 2.2.1B Capillarity Investigation

2. Draw a line graph illustrating time (*x*-axis) and height of water (*y*-axis).

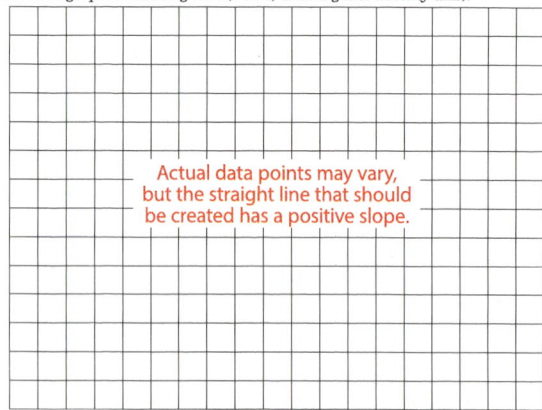

Actual data points may vary, but the straight line that should be created has a positive slope.

3. Compare how time relates to the height of water as illustrated in the table.
Answers will vary.

4. What type of soil is most similar to peat pots? What type of soil is the most different from peat pots? Topsoil or humus is most similar to peat pots. Clay is least like peat pots.

5. How would knowing about the absorbent abilities of peat be helpful to a farmer or a gardener? The rate of capillarity for peat and topsoil is affected by the amount of pore space in the soil. If the rate of capillarity is small, there is not enough pore space in the soil and the plants on a farm or in a garden would not thrive.

© Earth and Space Science • Soil

Soil

2.2.2 Soil Profile

Introduction
Ask students how soil forms. (**Soil forms when rocks break down into small particles through weathering and erosion and those rock particles mix with organic materials, water, and gases.**) Using tongs, heat a piece of glass for a few seconds over a Bunsen burner flame and then submerge the piece of glass into a metal pan of cold water. The sudden cooling will cause the glass to crack and fracture. Heat a granite chip over the flame and plunge it into the cold water. (Note: Do not heat a limestone rock.) Continue to heat and cool the granite chip until it fractures. Explain that one way soil is formed is when rocks are broken into fragments by temperature changes. Ask students where frequent and extreme temperature changes occur in the soil. (**at the surface**)

Crush samples of soft rocks such as shale, limestone, sandstone, or slate by wrapping them in a towel and using a hammer or rolling pin to pound them into fine fragments of soil. Ask students which soil layer would experience the most weathering and erosion processes. (**the top layer**)

Display a soil profile so students can visualize the different horizon layers. If a profile is not available, present **TM 2.2.2A Soil Horizons**. Explain to students that most soil is created from the weathering of sedimentary rock.

Discussion
- Ask students if the characteristics of soil at the surface level are the same as soil located several meters below the surface. Why? (**No. The soil at the surface layer is different from the soil deeper. Each layer is different because natural processes change the soil in each layer.**) What might affect these layers? (**Answers will vary but should include climate, slope, vegetation, erosion, and natural processes such as rain.**) Students may know that topsoil is the most fertile layer. Guide them to discuss possible reasons for this. Discuss how the activities of living things help form rich topsoil.

- Display **TM 2.2.2B Regional Soil Profiles** to show varying thicknesses of soil horizons found in different climates. Encourage students to consider the following:
  1. Why do all soil profiles not have an O layer?
  2. Why is the A layer for grassland soil so thick compared to most other A layers in various climates?
  3. How does the plant root structure compare between different climates? What does this indicate about the soil profiles?

- Ask the following questions:
  1. Explain how the O horizon affects the A horizon. (**The O horizon contributes the organic material to the A horizon.**)
  2. What factors determine the texture and composition of soil? (**The parent rock determines the texture and composition of soil. The type of soil depends on the topography of the land, the climate of the region, the amount of time the soil has been forming, and the type of vegetation that grows in the region.**)

Activities
A. Have students complete *Try This: Topsoil Versus Subsoil*. Collect the top few centimeters of good garden soil. In addition, collect subsoil from about 50 cm to 65 cm below ground level. All of the other variables (amount of sunlight, amount of soil, number and type of seeds, location of seeds, and amount of water) should be constant. One possible experiment would be to have students monitor the growth of the plants in both trays for two weeks and measure how quickly each tray produces a sprout, count the number of sprouts in each tray on different dates, measure the height of the tallest plants on specific dates, and observe the health of the plants.

OBJECTIVES
Students will be able to
- model a soil profile and compare the properties of the different horizons.
- explain how the activities of living things help form rich topsoil.

VOCABULARY
- **horizon** a layer in a soil profile
- **soil profile** a cross section of soil layers and bedrock in a particular region
- **subsoil** soil that is rich in minerals that have drained from the topsoil
- **topsoil** rich soil formed from mineral fragments, air, water, and organic materials

MATERIALS
- Tongs, piece of glass, Bunsen burner, metal pan, granite chip, rock samples, towel, hammer or rolling pin, soil profile (*Introduction*)
- Moisture meters, soil pH meters, fertilizer meters, stainless steel soil thermometers, soil test kits, humus test kits, and soil sampling tubes (*C*)
- Analysis of minerals and soils kit (*D*)
- TM 2.2.2A Soil Horizons
- TM 2.2.2B Regional Soil Profiles
- WS 2.2.2A The Layered Look

PREPARATION
- Gather rock samples. (*Introduction*)
- Obtain materials for *Try This: Topsoil Versus Subsoil* and *Try This: The Layered Look*.

TRY THIS
Topsoil Versus Subsoil
- garden soil
- soil about 50–65 cm below ground level
- metric rulers, 1 per group
- planting trays or flowerpots, 2 per group
- seeds, 8 per group

© *Earth and Space Science*

> **TRY THIS**
>
> **The Layered Look**
> - shovels, 1–2 per group
> - soil markers, 6 per group
> - metric rulers, 1 per group
> - pH meters, 1 per group
> - soil thermometers, 1 per group

B. Complete *Try This: The Layered Look* using **WS 2.2.2A The Layered Look**.

C. Direct student groups to create an experiment comparing and contrasting various soil profiles or horizon layers using the following equipment: moisture meters, soil pH meters, fertilizer meters, stainless steel soil thermometers, soil test kits, humus test kits, or soil sampling tubes.

D. Have students utilize an analysis of minerals and soils kit to act as crime scene investigators attempting to identify or eliminate a suspect. Students will explore forensic science by analyzing soil, glass, dust, and metal traces using chemical and physical techniques. Analysis of minerals and soils kits are available from local or online distributors.

E. Assign students to research the Great Dust Bowl, paying special attention to how the farmers prevented humus formation and the effect this had on the topsoil. Have students write a firsthand account of this event in the role of a rancher, farmer, federal government official, President Franklin D. Roosevelt, a member of the National Guard, or a doctor treating Dust Bowl survivors. Have students propose ways they would have revitalized the area after the Great Dust Bowl.

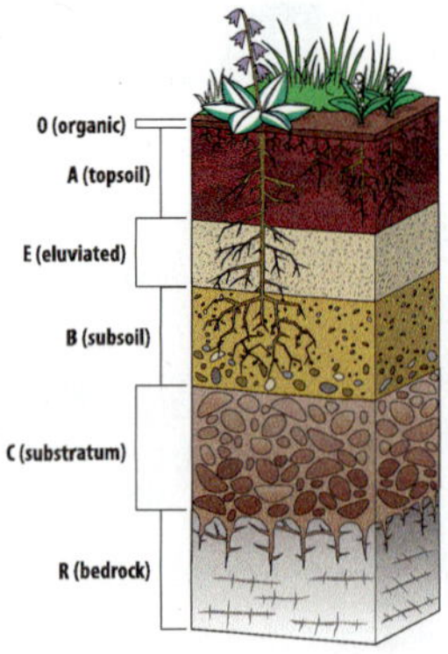

Lesson Review

1. What is topsoil? How do living things help form topsoil? (**Topsoil is rich soil formed from mineral fragments, air, water, and organic materials. Large burrowing animals create wide tunnels through which water, gases, and organic materials can travel. Microorganisms feed on dead animal and plant materials to form humus, and they are a food source for many types**

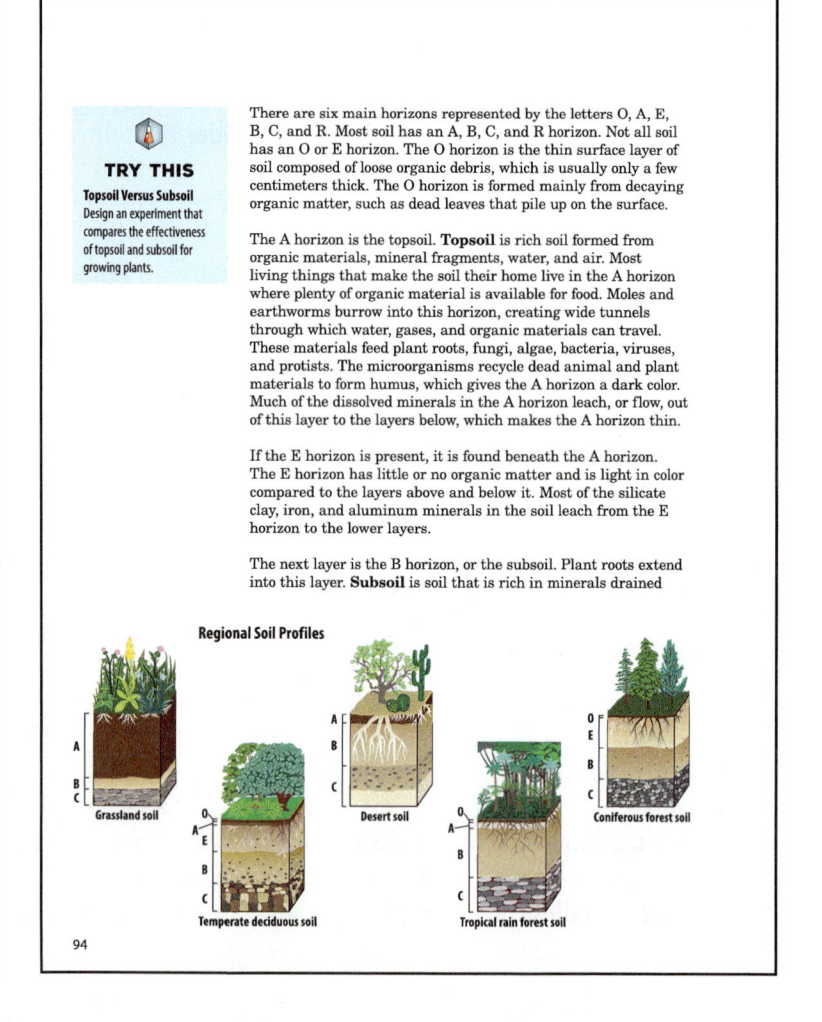

of larger animals. Arthropods also create tunnels for the rapid transport of water and other ingredients. The droppings and decaying bodies of animals and decaying plant materials encourage more microscopic activity, which creates more humus.)

2. What is a soil profile? (**A soil profile is a cross section of soil layers and bedrock in a particular region.**)

3. Name and describe the different horizons. (**The A horizon is topsoil, the rich soil formed from mineral fragments, air, water, and organic materials. Most of the living things that inhabit soil live in the A horizon where plenty of organic material is available for food. Much of the dissolved minerals in the A horizon leach out of this layer. The E horizon is one of the upper layers of soil, but it doesn't have much organic material. It is lighter in color than the other soil layers. The B horizon is the subsoil, the layer of soil beneath the A horizon. It is rich in minerals that have leached out of the topsoil. The B horizon is less fertile than the A horizon and is paler in color than the topsoil. The C horizon, the bottom layer of soil, is composed of partially weathered pieces of bedrock. The R horizon is consolidated bedrock.**)

4. Explain why A horizons often are darker than B or C horizons. (**The A horizon contains organic material that eventually leads to the formation of humus, which is dark in color.**)

NOTES

2.2.3 Soil Types

Soil

OBJECTIVES

Students will be able to
- analyze the effects of climate on soil type.
- explain how a soil's fertility is connected to its formation.
- describe factors that affect the development of soil.

MATERIALS

- Seed packets (*Discussion*)
- Soil samples, glass beakers, red cabbage juice (*B*)
- Samples of different soils (*C*)
- Soil test kits, soil pH meters, soil biology and chemistry experiment kits, soil organism study kits, soil sampling tubes, chemical composition of soil kits (*D*)
- TM 2.2.3A Common Soil Types

PREPARATION

- Obtain materials for *Lab 2.2.3A Soil Fertility* and *Lab 2.2.3B Water-Holding Capacity*.
- Obtain materials for *Try This: Map the Soil* and *Try This: Under the Weather*.
- Prepare red cabbage juice in advance by heating water and adding half of a red cabbage, chopped, to 1 L of water. Let the mixture stand for about an hour. Use a strainer to sift out the cabbage pieces. (*B*)
- Select an area where students can collect soil samples. (*E*)
- Arrange a visit to a local farm. (*F*)

TRY THIS

Map the Soil
- soil maps, 1 per group

Under the Weather
- jars, 1 per student
- soil and vegetable matter
- plastic wrap
- markers

Introduction

Ask students if they have lived outside of the country where they were born. (**Answers will vary.**) Have student volunteers describe the climate and type of vegetation they observed in those areas. Compare and contrast the type of vegetation mentioned to local vegetation. Why are there similarities and differences? (**Different climates have different vegetation.**)

Discussion

- Ask students if too much water ever poses a problem for plants and why. (**Answers will vary.**) How does excess water affect soil acidity? Discuss how gardeners use this information when planting. Read information from various seed packets. Have students name other factors gardeners must consider. Why is soil depth important? (**Possible answers: Soil layers differ in nutrients and water; some plants have different water needs.**)

- Guide students to name plants that can grow in one region of the world but not in another and then provide explanations for this. Remind students to consider solar radiation, temperature, humidity, amount and frequency of precipitation, atmospheric pressure, and wind.

Activities

Lab 2.2.3A Soil Fertility

- planting trays or flowerpots, 3 per group
- rich topsoil samples, 1 per group
- sandy soil samples, 1 per group
- clay soil samples, 1 per group
- bean or alfalfa seeds, 12 per group
- metric rulers, 1 per group

Lab 2.2.3B Water-Holding Capacity

- funnels, 3 per group
- ring stands, 3 per group
- 250 mL beakers, 3 per group
- wax marking pencils, 1 per group
- filter paper
- potting soil samples, 30 g per group
- dry sand samples, 30 g per group
- dry silt samples, 30 g per group
- scale
- 100 mL graduated cylinders, 1 per group
- stopwatches, 1 per group

A. Have students complete *Try This* activities. To extend *Try This: Under the Weather*, challenge students to design their own optimal decomposer using any combination of variables uncovered during their observations and discussion. Emphasize that they should justify their choice of conditions and predict how each factor will contribute to decomposition.

B. Direct students to test soil acidity by doing the following:
   1. Provide student groups with several different soil samples in glass beakers and a beaker full of cabbage water. Have them add 200 mL to each soil sample and observe the color change.
   2. Discuss how red cabbage water serves as an indicator for soil acidity. An acidic soil is indicated by very red water. A basic or alkaline soil is indicated by very blue water. Another prepared indicator may also be used to determine a more accurate pH value for each soil sample.
   3. Display **TM 2.2.3A Common Soil Types** and use the following information to discuss the pH of several common types of soil and how the climate contributed to the soil type. Have students use online resources to determine what climates are related to these soils.
      - Podzols: This acidic soil has an ashy composition and is commonly found in cool, damp, coniferous forests in the Northern Hemisphere.
      - Desert soils: Desert soil contains no humus. It is salty and highly alkaline.

- Grumosols (also known as *vertisols*): This water-logged soil is gray or blue in color and highly acidic. Found in warm, subhumid or semiarid climates.
- Chernozems: This black, humus-rich soil is found in the American prairies and Russian steppes.
- Ferrosols: Iron oxide compounds turn this acidic soil deep red or yellow in color. This deep soil is found in tropical rain forests.
- Gelisol: This soil contains permafrost and is found in very cold climates.

C. Distribute samples of many different kinds of soil to student groups. Direct them to develop a classification system for these soils and then to describe how their system uses factors that affect the development of the soil. Have students answer the following questions:
  1. What determined your system? (**Possible answers: particle size, appearance**) Which soils would you group together according to your system? (**Answers will vary.**)
  2. Some soils are not easily tested in the classroom. What are other ways these soils could be classified? (**Possible answers: chemical composition, agricultural use**)

D. Encourage students to experiment when examining diverse soil types using various soil tests and materials. Some suggested tests and materials include the following: soil test kits (to measure pH, phosphorus, potassium, and nitrogen levels), soil pH meters, soil biology and chemistry experiment kits (to identify a wide variety of soil microorganisms and calculate soil fertility), soil organism study kits, soil sampling tubes, and chemical composition of soil kits that test for the presence of calcium, carbonates, magnesium, phosphates, sulfates, ammonium ions, chlorides,

NOTES

2.2.3 Soil Types

OBJECTIVES
- Analyze the effects of climate on soil type.
- Explain how a soil's fertility is connected to its formation.
- describe factors that affect the development of soil.

TRY THIS
Map the Soil
Locate your home on a local soil map. What type of soil is most common in your region? What do you think affects this?

Perhaps you have hiked through forests with rich, black soil or slipped on the red clay of a stream bank. Maybe you have noticed scrubby plants growing in sandy soil near a beach or planted vegetables in dark brown garden soil. In all of these situations, you have experienced different varieties of soil.

Scientists have several ways to classify soil. In fact, over the centuries, scientists have created and used a variety of soil classification systems. The earliest systems grouped soils by color or by how suitable they were for the production of certain types of crops. For example, soils were grouped as rice soils, wheat soils, or vineyard soils. Another system classified soils into sandy soils, clay soils, and loam, a fertile soil that is rich in humus. Over the years, many classification systems have been developed depending on the origin of the soil or its properties. However, there is not one system that has been designated as the best system for soil classification.

Characteristics of soil differ from place to place. The main factor that determines the type of soil found in a region is the climate. Climate factors, such as solar radiation, temperature, humidity, precipitation, atmospheric pressure, and wind, all play important roles in the development of different soil types. Specifically, the climate helps determine the acidity, fertility, and depth of soil. Climate also indicates the type of weathering processes that will occur to form the soil.

To measure the acidity of the soil in a particular region, scientists use the pH scale. This scale ranges from 0 to 14 with 7 being neutral. An acidic substance has a pH value of zero to less than seven. A basic, or alkaline, substance has a pH value greater than 7. Scientists have found that the soil in areas of

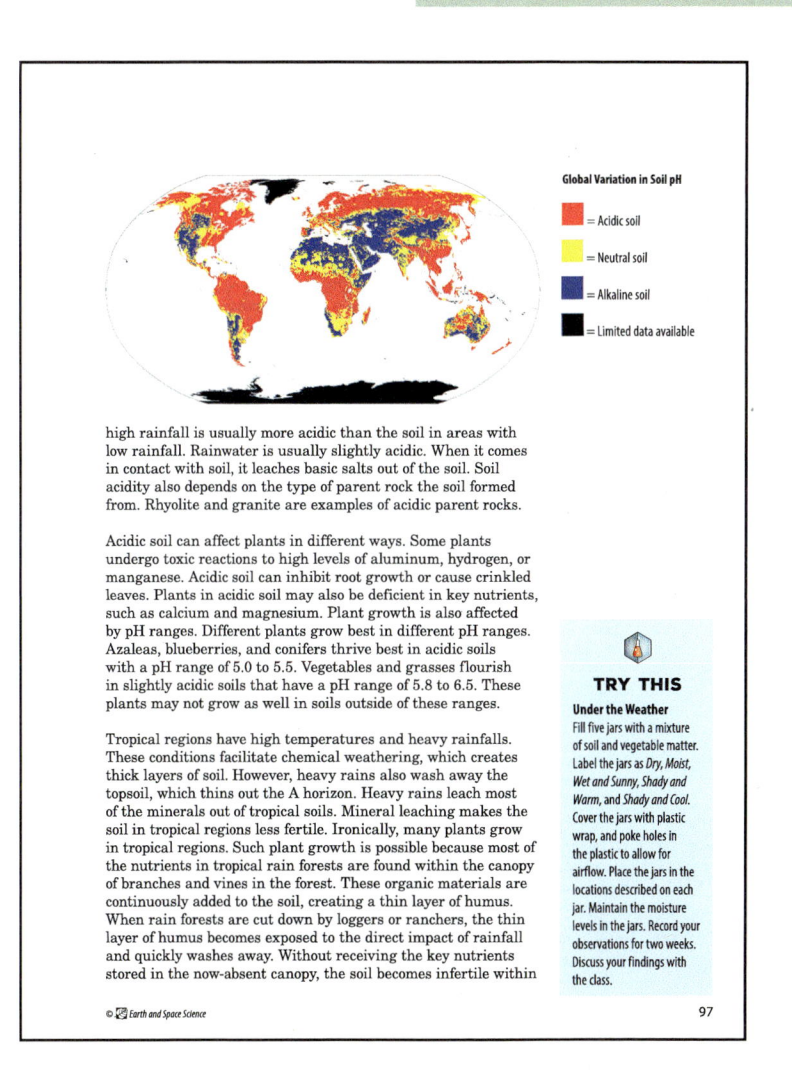

pH Scale

Global Variation in Soil pH
- = Acidic soil
- = Neutral soil
- = Alkaline soil
- = Limited data available

high rainfall is usually more acidic than the soil in areas with low rainfall. Rainwater is usually slightly acidic. When it comes in contact with soil, it leaches basic salts out of the soil. Soil acidity also depends on the type of parent rock the soil formed from. Rhyolite and granite are examples of acidic parent rocks.

Acidic soil can affect plants in different ways. Some plants undergo toxic reactions to high levels of aluminum, hydrogen, or manganese. Acidic soil can inhibit root growth or cause crinkled leaves. Plants in acidic soil may also be deficient in key nutrients, such as calcium and magnesium. Plant growth is also affected by pH ranges. Different plants grow best in different pH ranges. Azaleas, blueberries, and conifers thrive best in acidic soils with a pH range of 5.0 to 5.5. Vegetables and grasses flourish in slightly acidic soils that have a pH range of 5.8 to 6.5. These plants may not grow as well in soils outside of these ranges.

Tropical regions have high temperatures and heavy rainfalls. These conditions facilitate chemical weathering, which creates thick layers of soil. However, heavy rains also wash away the topsoil, which thins out the A horizon. Heavy rains leach most of the minerals out of tropical soils. Mineral leaching makes the soil in tropical regions less fertile. Ironically, many plants grow in tropical regions. Such plant growth is possible because most of the nutrients in tropical rain forests are found within the canopy of branches and vines in the forest. These organic materials are continuously added to the soil, creating a thin layer of humus. When rain forests are cut down by loggers or ranchers, the thin layer of humus becomes exposed to the direct impact of rainfall and quickly washes away. Without receiving the key nutrients stored in the now-absent canopy, the soil becomes infertile within

TRY THIS
Under the Weather
Fill five jars with a mixture of soil and vegetable matter. Label the jars as *Dry*, *Moist*, *Wet and Sunny*, *Shady and Warm*, and *Shady and Cool*. Cover the jars with plastic wrap, and poke holes in the plastic to allow for airflow. Place the jars in the locations described on each jar. Maintain the moisture levels in the jars. Record your observations for two weeks. Discuss your findings with the class.

NOTES

iron compounds, nitrates, and potassium. Challenge students to use such tests to determine the possible climate where a soil is found.

E. Have students classify the local soil samples they collected and then record where each soil sample was found. Guide them to describe the factors that led to the development of the soil.

F. Take students to a local farm. Have the farmer discuss the importance of soil and the contributing factors of soil development.

Lesson Review

1. How does climate affect soil type? (**The climate of a region determines the type of weathering processes forming the soil, and these processes determine the soil's composition.**)
2. Describe the soil of deserts and how the climate of deserts forms this soil. (**Desert soil is composed mainly of regolith. It is thin and infertile. The low rainfall slows the chemical weathering processes. Any water that falls on the desert evaporates quickly, leaving a whitish crust of salts.**)
3. Why is the soil on mountain slopes infertile and the soil in valleys fertile? (**Rainwater washes topsoil off of the slopes. Any topsoil that clings to slopes is usually too thin to support much plant life. Without organic matter added by vegetation, thick humus cannot form. This combination of factors makes the soil on mountainsides rocky, thin, and infertile. Soil in lowland valleys retains water, and rich organic materials form a thick layer of humus.**)
4. List four factors that affect soil development. (**climate, parent material, type of vegetation, and time**)

---

two or three growing seasons. Because the land is no longer viable, more forests are cut down and burned.

Bom Futura tin mine in Brazil

The low rainfall of deserts slows the chemical weathering process that produces soil. Desert soil is composed mainly of regolith. Deserts are immature, weakly developed environments with mostly alkaline soil. Any water that falls on the desert evaporates quickly and leaves behind a white crust of salts. The high salt content of desert soil, along with the absence of rapid chemical weathering, water, organic materials, and vegetation, makes desert soil thin and infertile. Sometimes the soil's salt level can become so high that no plants, even desert plants, can grow there.

Arctic climates also slow down chemical weathering. Most of the weathering in the Arctic is mechanical weathering, which does not produce soil very efficiently. Similar to desert soil, the soil in Arctic regions is thin, composed mainly of rock fragments that are highly acidic and do not have suitable drainage systems. In addition, located above the mineral horizon in the Arctic is a layer of variable organic matter that has not decomposed. Depending on the season, the tundra soils of Arctic regions are often either frozen or waterlogged. The upper layer of tundra soil is rich in peat and looks like bluish mud.

FYI

Biological Soil Crusts
In arid regions throughout the world, crusts of soil particles are forming that are bound together by organic materials. Because vegetative production is very sparse in these areas, highly specialized communities of cyanobacteria, mosses, and lichens congregate in the open spaces. Biological soil crusts can be found throughout the United States, Antarctica, Australia, and Israel. Crust thicknesses can reach up to 10 cm. Usually, the crust is darker than the surrounding soil partly because of the density of organisms and the dark color of the enclosed organisms. In these dry climates, biological soil crusts help to stabilize the soil, prevent erosion, fix atmospheric nitrogen, and contribute nutrients to plants.

Temperate regions have moderate temperatures. These areas receive consistent amounts of rain. Both mechanical and chemical weathering processes in these regions produce forest soil that is rich in humus several meters deep. Soil in regions receiving more than 65 cm of rainfall annually have fertile soils made of clay, quartz, and iron compounds. Regions that receive less than 65 cm of rain a year have soil that is high in calcium carbonate, a compound that makes soil more alkaline and very fertile. The light rainfall in temperate regions helps to prevent essential minerals from being leached out of the topsoil.

Soil type is not only affected by climate. It is also influenced by the shape or slope of the land. For example, rainwater removes topsoil as it flows rapidly down steep slopes. The runoff produces a thick layer of soil at the bottom of a slope in comparison to the top or incline portions of the slope. Scientists researching soils in Manitoba, Canada, found that the flat portions of land have twice the thickness of topsoil compared to land with only a 10° slope. Any topsoil that clings to slopes is generally too thin and dry to support much plant life. Thick humus cannot form without the organic matter added by vegetation, so the soil on mountain sides is rocky, thin, and infertile. This soil quality makes growing crops on mountainsides difficult. In lowland valleys, soil retains water and rich organic materials that allow for the formation of a thick layer of humus. The soil in valleys is fertile and produces good farmland.

Three other factors that affect soil formation are parent material, type of vegetation, and time. Basalt minerals weather to form clay soils. Limestone, granite, and shale disintegrate to form sandy soil. Silty soil is composed of minerals such as quartz and fine organic particles. The local vegetation in an area tends to produce humus acids that are powerful erosion agents. Once the soil is formed, plants act as stabilizers for the soil profiles. The longer the parent rock has been exposed to weathering the less likely it is that soil will resemble the parent rock. However, if weathering occurred for a short period of time, the parent rock will determine the soil characteristics.

LESSON REVIEW
1. How does climate affect soil type?
2. Describe the soil of deserts and how the climate of deserts forms this soil.
3. Why is the soil on mountain slopes infertile and the soil in valleys fertile?
4. List four factors that affect soil development.

BIBLE CONNECTION

The Parable of the Sower
In Mark 4, Jesus taught a parable comparing the fertility of soil with the condition of people's hearts. Some people's hearts are like a worn path. Seed dropped there does not take root because the birds eat it. These people hear the Word, but Satan quickly steals it away. Other people's hearts are like rocky places, where seed springs up, but dies because it has no deep roots. These people's joy on hearing the Word lasts only a short time, especially when trouble comes. The third group of people in Jesus' parable have hearts like fields filled with thorns. Seed planted there is choked out, just as people's worries keep the Word from bearing fruit. But other people's hearts are like good soil. When the Word is planted there, it is heard, accepted, and produces a bountiful crop. Do not let the Word land in rocky or thorny places. Read the Word daily so your heart can produce a bountiful crop.

Name: _____ Date: _____

Lab 2.2.3A Soil Fertility

QUESTION: Will rich topsoil, sandy soil, and clay soil each yield the same fertility result?

HYPOTHESIS: Answers will vary.

EXPERIMENT:

You will need:	· sandy soil sample	· metric ruler
· 3 planting trays or flowerpots	· clay soil sample	
· rich topsoil sample	· 12 bean or alfalfa seeds	

Steps:
1. Put each type of soil into its own tray or flowerpot and label the containers.
2. Plant 4 seeds in each tray or flowerpot.
3. Place the trays or pots next to each other on a windowsill or in an area that receives plenty of sunlight. Water each soil sample every other day with the same amount of water. Use enough water to moisten the soil. Rotate the trays or pots daily.
4. Observe the trays for two weeks. Record the date that each seed sprouts.

ANALYZE AND CONCLUDE:
1. After two weeks, describe the characteristics of each type of soil, including color, particle size, moisture, and compactness. Record the number of seeds that sprouted in each soil sample.

	Color	Particle Size	Moisture	Compactness	Number of Seeds
Topsoil					
Sandy Soil		Answers will vary.			
Clay Soil					

2. Measure and record the height of each sprout along with the appearance of each sprout.

	Topsoil	Sandy Soil	Clay Soil
Day 1			
Day 2		Answers will vary.	
Day 3			

Lab 2.2.3A Soil Fertility

	Topsoil	Sandy Soil	Clay Soil
Day 4			
Day 5			
Day 6			
Day 7			
Day 8			
Day 9		Answers will vary.	
Day 10			
Day 11			
Day 12			
Day 13			
Day 14			

3. Which soil was the most fertile? topsoil
4. What characteristics of this type of soil made it more fertile? Pore space, presence of organic matter, and size of particles made topsoil the most fertile.
5. How would knowing about soil fertility affect a farmer's decisions on what crops to plant? Different plants grow better in soils that have different amounts of water and organic matter. Some plants grow better in sandy soils. Examples include rhododendron, rosemary, and achillea. Other plants grow better in clay soils like junipers and honeysuckle.

Name: _____ Date: _____

Lab 2.2.3B Water-Holding Capacity

QUESTION: Will soil, sand, or silt hold more water?

HYPOTHESIS: Answers will vary.

EXPERIMENT:

You will need:	· wax marking pencil	· dry silt sample
· 3 funnels	· filter paper	· scale
· 3 ring stands	· potting soil sample	· 100 mL graduated cylinder
· three 250 mL beakers	· dry sand sample	· stopwatch

Steps:
1. Place the funnels in the ring stands. Place a beaker under each funnel. Use the marking pencil to label each beaker as *Soil*, *Sand*, and *Silt*.
2. Form 3 cones from filter paper. Insert them into each of the funnels. Moisten the top of the filter paper so it will adhere to the funnel.
3. Use the scale to measure out 30 g of soil, sand, and silt. Keep the materials separate. Add the materials to each appropriately labeled funnel.
4. Slowly add 100 mL of water to each funnel. Do not allow the water to rise above the top of the filter paper. Record the time the water is added to each funnel. After 10 minutes has elapsed, remove the beaker. Replace the beaker with another receptacle if water is still dripping from the funnel.
5. Measure the quantity of water collected in each beaker. Record your measurements and observations below.

	Initial Time	Final Time	Amount of Water Collected
Soil			
Sand			
Silt			

ANALYZE AND CONCLUDE:
1. Which soil held the most water? Why? Answers will vary.

2. Would a combination of the soils be the best environment for plants? Why?
A combination of soils would make a great environment for plants because the sand would provide a good drainage system, the soil would provide the appropriate nutrients that are required for optimal growth, and the silt would hold water.

2.2.4 Soil and Rock Layers

Soil

OBJECTIVES

Students will be able to
- summarize how geologists apply the law of superposition.
- explain how the principle of crosscutting relationships is applied to dating rock layers.
- compare and contrast the three basic types of unconformities.

VOCABULARY

- **fault** a fracture in the earth's crust along which rocks move
- **geologic column** the order of rock layers
- **intrusion** a large mass of igneous rock forced between or through layers of existing rock
- **law of superposition** a law that states that layers found lower in the sedimentary rock formation are older than layers found closer to the top of the formation
- **unconformity** the eroded surface that lies between two groups of strata

MATERIALS

- 10-gallon aquarium; large, flat rocks; string; medium-sized pebbles; colored, coarse aquarium gravel; colored sand; dried leaf or bone; scissors; fine silt; soil (*Introduction*)
- Cross section of a tree trunk (*Discussion*)
- Jar containing rocks, pebbles, and sand (*Discussion*)
- Clear basin and sand (*C*)
- Pictures or illustrations of rock deformations (*D*)
- TM 2.2.4A Unconformities
- TM 2.2.4B The Great Unconformity
- WS 2.2.4A Law of Superposition

PREPARATION

- Prepare a jar with three layers using rocks, pebbles, and sand. (*Discussion*)
- Obtain pictures or illustrations of rock deformations. (*D*)

Introduction

Display **TM 2.2.4A Unconformities** for reference. Demonstrate how a geologic column forms by doing the following:
- Place the aquarium on a table where students can easily observe all four sides.
- Lay large, flat rocks on the bottom of the aquarium to represent the bedrock.
- Securely tie individual strings around three or four flat rocks of different sizes. Place the flat rocks on the bedrock near the glass and drape the strings over the side of the aquarium.
- Lay down a layer of medium-sized pebbles.
- Pour a 1–2" layer of colored, coarse aquarium gravel over the pebbles.
- Place two more different colored layers of gravel over the first layer. Spread out each layer unevenly over the previous layer. The layers should be different thicknesses.
- Pull one of the rocks tied to a string up but not out of the aquarium to represent a sudden Earth movement, such as an earthquake or an intrusion of magma that has cooled to form igneous rock. Moving the rock will create tilted layers of sediment and rock.
- Lay down several layers of colored sand over the gravel. Use a different color of sand for each layer. Coarse sand should be used first, followed by fine-grained sand.
- Embed a dried leaf, bone, or other organic material in one of the sand layers to represent a fossil.
- Lift one or more of the tied rocks up and out of the aquarium to cause a violent Earth movement in the upper layers of the geologic column. Rocks tied with strings may also be placed in the upper layers of the geologic column. Cut the strings and submerge ends to leave no trace of how the geologic record was arranged.
- Cover the geologic column with fine silt, then soil.

Discussion

- Ask a student volunteer to write on the board a list of their immediate family members in order from oldest to youngest without listing their ages. Have classmates determine if there is any way they could determine the exact age of the listed family members from the information provided on the board. Explain to students that the information provided can only help determine the relative age of each family member. For example, the mother is older than the daughter. Share with students that many old-earth geologists use relative age to assign an age to different layers of rock. Some of the assumptions these scientists make are derived from the law of superposition. Ask students what type of information is needed to determine the exact age of the family members. (**birthdates**)

- Display a cross section of a tree trunk. Ask students if they know what the rings represent. (**Answers will vary.**) Have students visually compare the tree rings with the sedimentary strata of the geologic column made in *Introduction*. Ask what the similarities and differences are. (**Answers will vary.**) Point out the *History* account about tree rings used by scientists in the Student Edition to help them understand historical events.

- At the front of the classroom, place the prepared jar that contains layers of rocks, pebbles, and sand. Ask students which layer was laid down first. (**the bottom layer**) Is it possible to determine which layer was laid down first if you did not see the layers being placed in the jar? (**Answers will vary.**) Point out that the materials form horizontal layers.

- Ask the following questions:
  1. How do geologists explain sedimentary rock that is tilted, folded, or vertical? (**It has been altered by a past movement in the earth's crust.**)
  2. What clues do geologists use to determine the age of sedimentary rock that is not horizontally layered? (**ripple marks, particle size**)

Activities

A. Have students complete **WS 2.2.4A Law of Superposition**.

B. Display **TM 2.2.4B The Great Unconformity** and have students identify the two main types of unconformities in the Grand Canyon cross section. (Disconformities are represented by the red lines above and below the seventh rock layer and an angular unconformity is represented by the red lines outlining the slanted rock layer.)

C. Direct students to fill a clear basin three-quarters full with water. Have students add a small amount of sand to the water. Allow time for the sand to settle to the bottom. Direct students to gently rock the basin back and forth without spilling the water. Have students observe the ripples that are produced in the sand to those ripple marks that are observed in sedimentary layers.

D. Set up at least five different stations around the classroom. At each station, display pictures of rock deformations that simulate the principles of crosscutting relationships and inclusions. Have students form small groups to solve the mystery behind each rock deformation. After groups have visited the stations, discuss the groups' theories about what led to each rock deformation.

E. Encourage students to write letters or e-mails to both young-earth and old-earth scientists to ask them about the scientists' views on the formation of the geologic column and any changes or anomalies that have been discovered. Or have students inquire about what methods of dating soil layers some scientists favor and why. Provide an opportunity for students to share the scientists' replies with the class. Remind students that they may receive very different answers,

---

2.2.4 Soil and Rock Layers

OBJECTIVES
- Summarize how geologists apply the law of superposition.
- Explain how the principle of crosscutting relationships is applied to dating rock layers.
- Compare and contrast the three basic types of unconformities.

VOCABULARY
- **fault** a fracture in the earth's crust along which rocks move
- **geologic column** the order of rock layers
- **intrusion** a large mass of igneous rock forced between or through layers of existing rock
- **law of superposition** a law that states that layers found lower in the sedimentary rock formation are older than layers found closer to the top of the formation
- **unconformity** the eroded surface that lies between two groups of strata

Have you ever observed the rings in a cross section of a tree trunk? A tree produces one dark and one light ring in a single growth period, which is usually a year. The rings can provide a great deal of information regarding the environmental conditions during the tree's growth. For example, a very thin ring may indicate a year of drought. Sometimes portions of rings on one side of the tree show that the tree was burned and scarred. A tree's age can be determined by counting the pairs of rings. The rings on the innermost portion of the tree are older than those closer to the outermost portion.

In the same way, geologists study Earth's layers of sediment and soil to learn more about the age of the earth and to discover what may have happened throughout Earth's history. It may seem impossible for geologists to determine the age of layers of sediment when no one was alive to see them form. However, there are some methods that geologists are able to employ.

Sedimentary rock generally forms horizontal layers. The order of the rock layers is called the **geologic column**. Geologists have to make a few assumptions when attempting to determine the age of the geologic column. One common assumption made by old-earth geologists when observing strata is that layers found lower in the formation are older than layers found closer to the top of the formation. This principle is called the **law of superposition**. One problem with this law that young-earth geologists point out is that it assumes that layers of strata are formed one at a time. However, many layers of strata can be deposited at the same time.

Geologic column

HISTORY

The Mysterious Anasazi Migration
Geologists and archaeologists have been working together at the Davis Ranch site near the San Pedro River in Arizona to try to explain the mass migration of the Anasazi people. The Anasazi was a complex tribe of Native Americans who created a diverse community dedicated to native religious beliefs. About 700 years ago, this community migrated thousands of people to another location in the United States. Some scientists believe this migration happened because of a severe drought. Others believe it may have been because warfare was increasing in the region. A third school of thought proposed that the Anasazi people experienced a type of religious movement that they wanted to expand to southern regions. In an effort to study these hypotheses, scientists have examined varying widths of tree rings, the thickness of pollen layers, and ancient artifacts such as pottery. Evidence suggests that there was a dry spell and a cold climate environment during the last quarter of the 13th century when the Anasazi decided to migrate. However, additional evidence revealed that droughts had occurred in this region prior to the migration and the people had not moved during that time. In fact, when the Anasazi's crops began to suffer and they began to feel pressure from immigrating tribes, the people changed their practices to hunting and gathering and built fortified structures to protect against outsiders. The Anasazi seem fairly adaptable, which is why their complete relocation is a mystery. The real reason for the Anasazi migration may never be determined with certainty. However, an in-depth geological study may provide researchers with valuable clues to why the Anasazi moved.

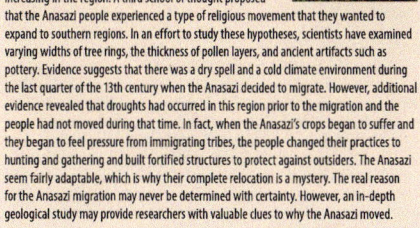

Another problem that geologists may encounter when studying the geologic column is that not all sedimentary rock layers are horizontal. The varying orientation and unevenness of the layers indicate that these layers have been disturbed over the years. Sedimentary layers that do not lie horizontally may have been tilted by movements in the earth's crust. For example, the activity of the many earthquakes that occur over time cause older sedimentary layers to be transposed above younger layers.

Earthquakes and volcanoes not only transpose layers but also introduce faults or intrusions. A **fault** is a fracture in the earth's crust along which rocks move. An **intrusion** occurs when a large mass of igneous rock is forced between or through layers of existing rock. Intrusions form when hot magma is pushed into rock layers from beneath the earth's crust and it cools before reaching the surface. Geologists have determined that faults and intrusions that cut across layers of rock are younger than the layers of rock they cut across, a concept called *the principle of*

NOTES

which illustrates why such topics can cause tension and a lack of understanding between proponents of either view. Ask how a Christian can discuss the earth's age or other divisive topics without conveying contempt for a fellow believer whose view is different. (**Possible answers: by controlling tone of voice or choice of words, by using nonthreatening body language, by keeping the focus on gaining understanding of the other's view rather than winning an argument**)

Lesson Review

1. What is a geologic column? How does the law of superposition apply to the column? (**A geologic column is the order of sedimentary rock layers. Old-earth geologists assume that the strata found lower in the formation is older than layers found closer to the top of the formation according to the law of superposition.**)
2. Using faults and intrusions to date a geologic record can be difficult. Why? (**Over time, the many movements in the earth's crust combine to make a complex mosaic of rock layers, faults, and intrusions.**)
3. Compare and contrast the three basic types of unconformities. (*disconformities*: **occur when there is an eroded portion of rock between two horizontal sedimentary layers, hard to see if the eroded portion is even;** *nonconformities*: **occur when there is an eroded portion of rock between a sedimentary layer and a layer of igneous or metamorphic rock, easy to recognize;** *angular unconformities*: **occur when horizontal layers of rock that are brought to the surface tilt or fold and are exposed to the effects of erosion, frequently occur when mountains form**)
4. What can the layers of the earth teach people about its history? (**Possible answers: Layers were laid down at different times; certain patterns show times of erosion or earth movement.**)

---

crosscutting relationships. The analysis of faults and intrusions that cut across existing rock is very complex. Geologists study and observe these formations very carefully before they attempt to assign dates to different rock layers.

Another phenomenon geologists have studied in nature is related to rock fragments called *inclusions*. As rocks are eroded, small parts of the rock break away. These small fragments may stay in the original rock layer or they may travel and become ingrained in a separate, nearby layer of rock. Geologists have concluded that rock inclusions that have traveled must be older than the new rock layer they are embedded in. This conclusion is called *the principle of inclusions*.

Shifting rock layers and the incorporation of faults and intrusions make it difficult for geologists to apply the law of superposition. In such cases, geologists search for other clues to determine the age of different rock layers. One clue they use is the particle size of each layer. Larger particles form the bottom sedimentary rock layers and small particles are generally located in the upper layers. Ripple marks—small waves formed by wind or water on the surface of sand—can also be used to determine the original order for sedimentary layers. Ripples are often preserved when sand is transformed into sandstone. When ripples form, the crest is located at the top of the ripple and the curved trough is located at the bottom of the ripple. If ripple crests are not facing upward in a disturbed layer of rock, many geologists conclude that they turned since their original formation.

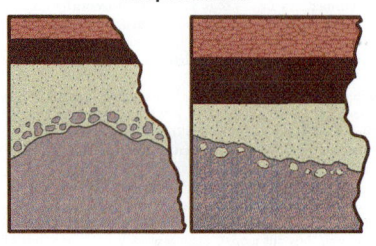

Sometimes geologists come across a break in the geologic record where layers of rock are missing because they have been eroded away. The eroded surface that lies between two groups of strata is called an **unconformity**. Unconformities occur when movements in the earth's crust lift buried rock layers to the surface, exposing the layers to the effects of weathering and erosion. These eroded layers are eventually buried again under newly formed sedimentary layers. The portion of rock that was eroded leaves a gap of missing information.

There are three basic types of unconformities: disconformities, nonconformities, and angular unconformities. A disconformity occurs when there is an eroded portion of rock between two horizontal sedimentary layers. If the eroded portion is uneven, this type of unconformity is easily recognized. However, if the eroded portion is even, a disconformity is hard to discern.

Nonconformities occur when there is an eroded portion of rock between a sedimentary layer and a layer of igneous or metamorphic rock. When an igneous rock that is not layered, such as granite, is lifted to the surface, eroded, and buried by sediments, a nonconformity occurs at the boundary between the new sandstone and the old granite. This boundary indicates

Intrusive rock

Principle of Inclusions

Types of Unconformities

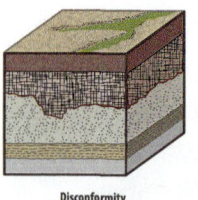

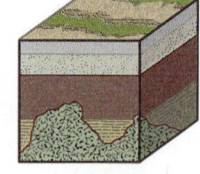

Disconformity | Angular unconformity | Nonconformity

Ripples in the desert sand

Angular unconformity

that the granite eroded for an unknown period of time. Nonconformities are often easy to recognize.

Angular unconformities occur when horizontal layers of rock that are brought to the surface tilt or fold and are exposed to the effects of erosion. The tilted layer of rock is then covered by a new horizontal layer of sedimentary rock. This type of unconformity frequently occurs when mountains are being formed.

Earth scientists are interested in understanding what was happening during some of Earth's most historic events. For decades, scientists have wrestled with trying to determine the age of the earth. Some scientists tried to determine the age of the earth by calculating how long it may have taken for Earth to cool from an original molten state. Others measured the depth of sediments and tried to determine the amount of time that it would take for those sediments to accumulate. Both of these methods have proven unreliable. With advances in technology and more detailed information, scientists are currently using three main methods to determine the age of the earth: radiometric dating, stratigraphic superposition, and the fossil record. The fossil record will be discussed in the next lesson.

The introduction of radiometric dating helped with some of the challenges early old-earth geologists faced as they attempted to determine the age of the earth. However, even though this system is widely used, it is not foolproof. Both young-earth and old-earth scientists will continue to study geologic columns and to develop new methods of determining the age of the earth.

LESSON REVIEW
1. What is a geologic column? How does the law of superposition apply to the column?
2. Using faults and intrusions to date a geologic record can be difficult. Why?
3. Compare and contrast the three basic types of unconformities.
4. What can the layers of the earth teach people about its history?

2.2.5 Fossils

Soil

OBJECTIVES

Students will be able to
- distinguish among different methods by which fossils are formed.
- describe how scientists use fossils to interpret Earth's physical history.

VOCABULARY

- **carbonization** a process of converting organic material into carbon
- **desiccation** a type of fossilization where the organic material becomes dehydrated
- **fossil** the preserved remains or impression of an organism that lived in the past
- **index fossil** a fossil that is useful for dating and correlating the strata in which it is found
- **petrifaction** a process in which the organic portion of an organism is infiltrated or replaced with minerals
- **trace fossil** a fossil of a track, trail, burrow, or other trace of an organism

MATERIALS

- Fossil samples (*Introduction*)
- Trace fossil images (*B*)
- TM 2.2.5A Trace Fossils

PREPARATION

- Invite students who collect fossils to bring in their collections. Obtain fossil samples. (*Introduction*)
- Obtain materials for *Try This: Carbonization in Action*.
- Obtain trace fossil images. (*B*)
- Arrange for a field trip to a local museum or national park. (*C*)
- Invite a paleontologist to discuss current topics in the field. (*D*)
- Arrange a fossil hunting field trip. (*E*)

TRY THIS

Carbonization in Action
- leaves, 4 per pair
(*continued*)

Introduction

Ask students if they can remember what they had for lunch three days ago. What physical evidence would help them remember? (**Possible answers: reading the lunch menu, checking the contents of the trash**) Point out that some challenges may arise when studying the behavior and lifestyle, including their diets, of creatures that lived many years in the past. Explain how paleontologists use various methods in an attempt to solve these challenges.

Read **Psalm 24:1–2** or **Psalm 104:24–30** aloud to establish the association of fossils with God's creation. Remind students that fossils were once living things that lived on Earth and that all parts of creation were made by God.

Have students who brought in their fossil collections share them with the class. Present students with a variety of fossil types, such as plant and animal imprints, insects in amber, fossilized teeth or bones, and petrified wood. Have a discussion with students regarding what they know about fossils and how fossils form. Clarify any inaccuracies or misunderstandings.

Discussion

- Share with students that the Latin word *petram* means "rock." Remind students of Jesus' words to Peter in Matthew 16:18: "And I tell you that you are Peter, and on this rock I will build My church." Ask students what type of fossilization process has this word as its root. (**petrifaction**) What do people mean when they claim to have been petrified? (**To be petrified means to be too frightened to move—as though turned to stone.**) How can wood or any organic material be turned into stone? (**Possible answers: Groundwater carries dissolved minerals that fill the cavities of organic materials; groundwater containing dissolved minerals removes the original organic material and replaces it with the minerals.**)

- Discuss various dilemmas that are produced from the study of the fossil record. For example, how is the relative age of a fossil determined if it spans multiple strata? Have students consider how both a young-earth scientist and an old-earth scientist would determine the fossil's age. Remind students that the focus of the discussion should be to gain an understanding not only what both scientists would conclude but how they would have determined their findings. Ask what role bias might play in these scientists' findings. (**Answers will vary.**)

- Ask students if they think it is ethical for the general public to retain fossils without first presenting their finds to a museum or other public sector. Why? (**Answers will vary.**) Relate this idea to the attempts made throughout history to squelch the sharing of the gospel. How has God prevented this from happening through His people?

- Discuss the possible "rebirth" of woolly mammoths. Should scientists try to clone a mammoth? Why? Does such an undertaking raise ethical questions? (**Answers will vary.**)

- Ask the following questions:
  1. What kind of information can be revealed from a study of the fossil record? (**discovery of unknown species; understanding of geologic events in Earth's history; climate, vegetation, organism interaction during a particular time**)
  2. Why do the hard parts of an organism fossilize much easier than the soft tissues? (**Soft tissues normally decay rapidly.**)
  3. For a fossil to form, what must occur? (**The dead organism must be buried quickly to delay decomposition and to prevent the ravaging of scavengers, and the organism must possess hard parts that are capable of being fossilized.**)

4. In what type of rocks are fossils usually found? Why are they found in this type of rock? (**Fossils are usually found in sedimentary rock because sediments cover the body and eventually preserve the organism in fossil form.**)

Activities

A. Have students complete *Try This: Carbonization in Action*. Discuss how this activity is related to the formation of fossils. Ask which type of fossilization this activity represents. (**carbonization**)

B. Display **TM 2.2.5A Trace Fossils** or a variety of trace fossil images. Have students speculate about what type of organism created each particular type of trace fossil.

C. Take students on a field trip to study fossils. Ask students to consider how the fossils found in this location are used to interpret Earth's physical history. Possible prominent fossil-rich destinations include the La Brea Tar Pits in southern California, the Petrified Forest National Park in Arizona, the Badlands National Park in South Dakota, the Dinosaur Provincial Park in Alberta, the Museum für Naturkunde in Berlin, the Royal Belgian Institute of Natural Science in Brussels, the National Dinosaur Museum in Australia, the Iziko Museum in South Africa, and the Zigong Dinosaur Museum in China.

D. Present the paleontologist to speak to students about the hard work and dedication that this career requires, the interesting finds he or she has made and how these finds have contributed to the fossil record, how such fossils are used by scientists, and the locations he or she has traveled to locate fossils. Ask the paleontologist to bring the tools or equipment that are used in the field.

TRY THIS

(continued from previous page)
- waxed paper, 4 sheets per pair
- heavy textbooks, 6 per pair

Fossils 2.2.5

Have you ever hunted for fossils or seen fossils in a museum? Fossils provide people with a broadened view of God's creation. Without fossils of extinct animals like dinosaurs, you would never know about many of the amazing creatures God created.

The word *fossil* means "dug up." A **fossil** is the preserved remains or impression of an organism that lived in the past. The study of fossils is called *paleontology*. Scientists who study fossils to learn about Earth's history are called *paleontologists*. These individuals use fossils as clues to past events and to help determine the relative age of rock strata, or layers.

It is important to understand that fossil formation is a very rare event. The right conditions must be present to limit destructive physical and biological processes for a plant or animal to be preserved. There are two general conditions that must be met for an organism to be preserved as a fossil. First, the burial of an organism must be rapid to delay its decomposition and to prevent scavenging. In addition, hard components that are able to be fossilized must be present. Without these conditions, the organisms would decay and there would be no evidence of their existence. Significant portions of the fossil record have been completely eliminated because of the erosion, deformation, and metamorphism of the original rock. Many fossils may be hidden under the rock surfaces in areas that are not easily accessible or are located in geographic areas that have not been studied. Given this information, it is easy to understand why only an extremely small percentage of species that once lived on Earth have been preserved and discovered in the rock record.

Fossils are usually found in sedimentary rock layers. Most fossils have been preserved in marine or freshwater environments where there are low oxygen levels, high salinities, or relatively high rates of sediment deposition. Most fossils are the skeletal remains of shells, teeth, or bone. Shells of marine animals are the most common fossils. Major storm events may account for the inclusion of shells into the sedimentary record by burial. Soft tissues and organisms with nonmineralized skeletons are rarely preserved as fossils. The majority of species in marine and freshwater communities have soft bodies. These species are able to be preserved because they typically are quickly covered by many layers of sediment. Interestingly, many preserved soft tissues have been found in recent years throughout the world. Some regions that contain very well-preserved soft tissues are the Burgess Shale in an area of the Canadian Rocky Mountains, the Chengjiang Maotianshan Shales in China, the Mazon Creek Formation near Morris, Illinois, and the Messel Pit in Germany.

Fossils form through a variety of processes depending on the physical and chemical conditions of the environment. Most organisms decay when they die. A lion that dies on the African savanna, for example, is quickly picked apart by vultures and decays in the sun. Such a lion does not form a fossil. Fossils form only when dead organisms are protected from decomposition, scavenging, mechanical destruction, transportation, and chemical dissolution and alteration. An organism that dies and is soon after buried by sand or mud may form a fossil. The soft tissues usually decay quickly, but the harder parts, such as those in a skeleton, gradually undergo chemical change and can be preserved in several different ways.

Some fossils form when dead plants and animals are buried by sediment on the seafloor. The hard and soft portions of their bodies chemically change or dissolve away, leaving no original or altered material. A hollowed-out impression, called *a mold*, is left in its place. Minerals may fill this mold to form a natural cast. If the mold is not filled, an imprint of the organism can be left

OBJECTIVES
- Distinguish among different methods by which fossils are formed.
- Describe how scientists use fossils to interpret Earth's physical history.

VOCABULARY
- **carbonization** a process of converting organic material into carbon
- **desiccation** a type of fossilization where the organic material becomes dehydrated
- **fossil** the preserved remains or impression of an organism that lived in the past
- **index fossil** a fossil that is useful for dating and correlating the strata in which it is found
- **petrifaction** a process in which the organic portion of an organism is infiltrated or replaced with minerals
- **trace fossil** a fossil of a track, trail, burrow, or other trace of an organism

Fossil jewel beetle found in the Messel Pit of Germany

Ammonite fossil

NOTES

E. If your school is located near a place where fossils can be collected, take the class on a fossil hunting trip. Obtain permission to collect fossils at the site.

F. Direct students to write a short essay or create a computer presentation or poster about what fossils can teach people about God's creation that would not otherwise be known. Ideas may include information about fossils that represent creatures that have been extinct for years, information about the behavior of certain animals, and information about what kinds of plants and animals were common to a region at an earlier time.

G. Have students present a process of fossilization in comic strip form.

Lesson Review

1. What two types of fossilization produce fossils that are the most well preserved? (**freezing and desiccation**)
2. What two assumptions do scientists make regarding fossils and rock strata? (**Fossils are the same age as the strata they are located in. Similar fossils located around the world are the same age.**)
3. Why have more fossils not been found? (**Most organisms decay before they have the opportunity to fossilize. Many fossils are located under the surface of the earth in areas that are not easily accessible. A number of fossils have been removed because of weathering and erosion.**)
4. How do scientists use fossils to interpret Earth's physical history? (**Scientists use fossils to study the earth's climate, physical and chemical conditions of environment, and relative age of rock strata.**)

---

in the rock layers. Magma intrusions can also naturally fill a mold, creating an igneous cast of the fossil.

The most common type of fossilization is the very slow process of **petrifaction** in which the organic portion of an organism is infiltrated or replaced with minerals. Eventually, the organism turns into a rock or a rocklike substance. You may have seen examples of this process if you ever visited the Petrified Forest National Park in the United States, which is famous for its petrified conifer trees. Petrifaction works best on the hard remains of an organism. It involves two processes: permineralization and replacement. Permineralization occurs when groundwater carrying dissolved minerals, such as apatite, calcite, and pyrite, infiltrates the pore spaces and cavities of bone, shell, or wood. The minerals precipitate out of the water and are deposited in the hard remains. Much of the original material of the specimen remains following permineralization. Replacement occurs when groundwater containing dissolved minerals removes the original organic material and replaces it with minerals such as silica, calcite, and pyrite. This gradual substitution results in a near-perfect replica of the original organism.

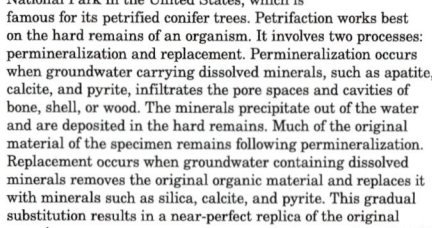

Mold and cast fossil

Petrified wood

Rarely, entire specimens are preserved by freezing soil and ice. Such specimens are one type of natural mummies. The organisms do not decay because the freezing temperatures delay decomposition. The remains of many mammoths have been excavated out of the ice in Alaska and Russia. These specimens were very well-preserved because of the climate. Woolly rhinoceroses have also been preserved in frozen soil.

Some fossilized organisms were preserved by tar. In some locations, tar pits resembled lakes of thick liquid oil oozing onto the earth's surface. Often these tar pits were covered with plant debris or water, making the pits invisible to unfortunate organisms that fell into them. It is not uncommon to find predator fossils in tar as well because when an herbivore became stuck in the tar, predators were often attracted to the site. During the wet season, the tar beds were covered with water rich in sediments, which added to the sedimentary fossil record. One of the most famous fossil sites is the La Brea Tar Pits in southern California. Bones of animals that lived thousands of years ago have been unearthed from these tar pits.

Fossils are also created when organisms become trapped in various types of natural resins. A resin is a viscous liquid produced by plants that can harden into a solid. One of the most common resins is amber, which is hardened tree sap. An entire organism can be preserved with this type of fossilization. Many insects and other small creatures that became trapped in sticky tree sap many years ago were perfectly preserved when the sap hardened into amber.

Sometimes there is only indirect fossil evidence that shows an organism existed in a particular area. Another name for an indirect fossil is a **trace fossil**. Trace fossils include burrows, tracks, trails, and waste products. When paleontologists examine trace fossils, they can glean a great deal of information regarding

Scorpion in amber

FYI

Fossil Formation
Fossils provide a fascinating glimpse of extinct parts of God's creation, such as the dinosaurs. Both old-earth and young-earth scientists study fossils for clues about the age of the earth. Both groups of scientists believe that in order for fossils to form, the animal or plant had to have been buried quickly. Old-earth scientists believe that the sedimentary rock layers where most fossils are found developed over thousands or millions of years, so fossilization likewise takes time. In contrast, young-earth scientists believe the key to fossilization is the presence of the right chemical conditions, such as cementing agents, rather than time.

TRY THIS

Carbonization in Action
Place two leaves side by side between layers of wax paper. Set one heavy textbook on top of the wax paper. Place two more leaves between another set of layers of wax paper. Set five heavy textbooks on top of the second set of leaves. Leave the books undisturbed for two days. Remove the books and separate the wax paper layers. What do you observe? Which set of leaves produced the most liquid residue?

the behavior of an organism, its walking characteristics, and eating habits. Rarely are trace fossils found near the actual animal remains. This may create challenges when examining the fossil record, but paleontologists still value trace fossils because they provide more history regarding the life patterns of various creatures.

Many plants are preserved through another type of fossilization called **carbonization**. This process occurs when leaves, stems, and other plant materials are flattened between two layers of rock. Pressure from the rocks causes liquids and gases inside the plant material to be forced out. The liquids and gases go through a series of chemical reactions that produce a thin carbon residue on one layer of rock. The other layer of rock holds an impression of the plant material. Carbonization is not restricted to plants alone. Fish and jellyfish have also been fossilized by this process.

Desiccation, or mummification, is a unique and rare form of fossilization. This type of fossilization provides specimens that are second only to frozen fossils in quality. In very dry climates, the hard and soft portions of a deceased organism dry out before any method of decay can take place, forming natural mummies. In years past, Egyptians mimicked natural mummification with the use of natron, a drying agent, to preserve their loved ones and many animals. After an elaborate method of preservation was employed, the Egyptians buried their dead in the desert. The majority of these were simply buried in hollows in the sand rather than in tombs. It has been estimated that over 70 million people were mummified in Egypt.

Exploring the fossil record is not only interesting because new species are being uncovered all the time, but it also helps scientists in their attempt to explain and correlate different events in Earth's history. For example, paleontologists would be able to deduce the type of climate and environment in a particular layer if they found certain species of coral fossils that required an environment of warm, shallow water in that layer. Paleontologists have investigated what the climate was like during a particular period of time, how organisms interacted with one another, and what type of vegetation grew in a particular area.

Scientists also attempt to provide relative ages for different strata by studying the fossils that are found throughout the stratigraphic record. Many old-earth scientists assume that the particular layer that a fossil is found in is the same age as the fossil itself. Scientists cannot provide an exact age for the fossil or strata. They can only provide a relative age, meaning this layer and fossil are younger or older than another layer and fossil. Paleontologists also infer that the same fossils found in various locations around the world are roughly the same age. There are some fossils that are easy to recognize, plentiful, and found throughout much of the world. These fossils are referred to as **index fossils**. Another aspect of index fossils is that they are believed to have existed only for a relatively short period of geologic time. Old-earth paleontologists refer to index fossils in their attempt to make definite boundaries in the geologic time scale and to correlate different strata. Many young-earth scientists disagree with the use of index fossils for dating because some index fossils are found in different layers of the geologic column around the world, which indicates that the geologic time scale is incorrect. The ongoing study of fossils will continue to fascinate scientists as it reveals God's creation.

LESSON REVIEW
1. What two types of fossilization produce fossils that are the most well preserved?
2. What two assumptions do scientists make regarding fossils and rock strata?
3. Why have more fossils not been found?
4. How do scientists use fossils to interpret Earth's physical history?

FYI

Baby Lyuba
In 2007, a reindeer herder discovered the extremely well-preserved remains of a female woolly mammoth calf on the banks of a frozen river on the Yamal Peninsula in Siberia. This specimen, named Lyuba, is an excellent example of a naturally formed mummy. Researchers determined that Lyuba had mud from a lake bottom in her trunk and airway by using a CT scan to examine the mummy. They theorize that she fell through a frozen lake, inhaled the mud, and died. The discovery of this one baby mammoth helped scientists learn more about normal mammoth biology, time in the womb, lifestyle habits, climate conditions, and mammoth migration.

CAREER

Paleontologist
Paleontologists are scientists who study the history of life on Earth through the fossil record. Paleontologists may study microscopic fossils; fossil plants, including fossil algae and fungi; pollen and spores; processes of decay and preservation; fossil tracks and trails; or the ecology and climate of the past. It is estimated that more than 99% of all species that have existed on Earth are extinct, so the field of paleontology will be thriving for quite some time. The majority of paleontologists spend most of their time in the field collecting fossils. Some field assignments involve working in areas that are not easily accessible, such as a steep mountaintop. When not working outside in the field, paleontologists might be working at a university, in a museum, for federal or state governments, or for private industry.

UNIT 3

The Dynamic Earth

Chapter 1: *Crust Movement*
Chapter 2: *Earthquakes*
Chapter 3: *Volcanoes*

Key Ideas

Unifying Concepts and Processes
- Systems, order, and organization
- Evidence, models, and explanation
- Change, constancy, and measurement
- Evolution and equilibrium
- Form and function

Science as Inquiry
- Abilities necessary to do scientific inquiry
- Understandings about scientific inquiry

Earth and Space Science
- Structure of the earth system
- Earth's history
- Energy in the earth system
- Origin and evolution of the earth system

Science and Technology
- Abilities of technological design
- Understandings about science and technology

Science in Personal and Social Perspectives
- Populations, resources, and environments
- Natural hazards
- Risks and benefits
- Science and technology in society
- Natural and human induced hazards
- Science and technology in local, national, and global challenges

History and Nature of Science
- Science as a human endeavor
- Nature of science
- History of science
- Nature of scientific knowledge
- Historical perspectives

Vocabulary

aa	footwall	plateau	subduction
aftershock	hanging wall	plate boundary	subduction zone
caldera	hot spot	plug	tectonics
compressional stress	isostasy	pluton	tensional stress
continental drift	liquefaction	pyroclast	tiltmeter
deformation	Love wave	Rayleigh wave	trench
epicenter	magnitude	reverse fault	tsunami
faulting	normal fault	rift	volcanic bomb
fissure	oceanic ridge	seafloor spreading	volcano
focus	orogenesis	seismograph	
folding	pahoehoe	sheering stress	

SCRIPTURE

At that moment the curtain of the temple was torn in two from top to bottom. The earth shook, the rocks split, and the tombs broke open. The bodies of many holy people who had died were raised to life.

Matthew 27:51–52

3.1.0 Crust Movement

LOOKING AHEAD

- For **Lesson 3.1.2**, collect cardboard boxes to be used to make plate boundary models.

SUPPLEMENTAL MATERIALS

- BLM 3.1.1A Continental Drift Reader's Theater

- TM 3.1.1A World Maps
- TM 3.1.2A Plate Map
- TM 3.1.3A Formations
- TM 3.1.4A Types of Mountains

- Lab 3.1.1A Isostasy
- Lab 3.1.3A The Effects of Stress

- WS 3.1.1A Graph the Continents
- WS 3.1.1B It's a Puzzle
- WS 3.1.1C Digging Deeper
- WS 3.1.3A Under Stress
- WS 3.1.4A Mountain Range Scavenger Hunt
- WS 3.1.4B Mountain Perspectives

- Chapter 3.1 Test

BLMs, TMs, and tests are available to download. See Understanding Purposeful Design Earth and Space Science at the front of this book for the web address.

Chapter 3.1 Summary

The theory of plate tectonics combines many features and characteristics of continental drift and seafloor spreading into a coherent model. The understanding of plate tectonics has revolutionized geologists' understanding of continents, ocean basins, and mountains. Plate tectonics is a field of research as volatile as the earth itself; since its conception, it has been changed, adapted, and argued. This topic offers a good opportunity to discuss the constantly changing nature of science and the fact that scientists often disagree.

In 1915, German geologist Alfred Wegener published *Die Enstehung der Kontinente und Ozeane* (*The Origin of the Continents and Oceans*), which proposed that the continents were once part of one large landmass and that they are still changing and moving. In the 1960s, American geologist Henry Hess, who had been working on mapping the ocean floor since the 1940s, developed the theory of seafloor spreading. Anomalies on the ocean floor led him to believe the tectonic plates in the ocean were spreading apart. As a result, fresh magma from the mantle of the earth flowed up through weaknesses in the crust. As the magma hit the cold ocean water at the surface of the earth, it cooled and formed new crust. This theory sparked debate and excitement. Geologists, teachers, and students argued about whether Hess's idea could be true. In 1984, NASA presented pictures and notable evidence that the plates are moving.

Background

Lesson 3.1.1 – Continental Drift

The theory of continental drift developed over nearly a century as scientists studied world maps, examined fossils on different continents, and theorized about the forces that could move continents. In 1898, American Frank Bursey Taylor was the first to propose the theory of continental drift, which states that the relative positions of the continents have changed considerably over time. In 1915, Alfred Wegener put forth the first detailed theory on continental drift when he proposed that the continents were once part of one large landmass and that they are still moving. This theory initially stirred up a great deal of controversy. Some scientists agreed that the continents looked as though they were once one mass, but they could not explain how continents could move. Over the next few decades, the theory was supported by the work of other scientists. In 1921, South African geologist Alexander Du Toit found similar geological and paleontological features on both sides of the Atlantic. He expanded the idea of Pangaea in his book, *Our Wandering Continents,* which was published in 1937. In 1929, Scottish geologist Arthur Holmes suggested that the continents were moved by thermal convection in the mantle, a view that is widely accepted today, although many young-earth scientists disagree with this theory. In the 1950s, the theory of continental drift was revisited when British geophysicists reported that the magnetic pole of each continent had apparently changed position through geologic time. The pole paths for Europe and North America could be made to coincide by bringing the continents together. Scientists reevaluated Wegener's theory, and as a result the theory of plate tectonics was accepted by many in the scientific community. Plate tectonics and seafloor spreading are not accepted as fact by all scientists, however. Some scientists, many of whom are young-earth proponents, contend that these theories cannot be proven because of unanswered questions on how plate movement is energized, uncertain dating of geologic examples used to support the idea of seafloor spreading, and doubts about subduction.

Clarence Dutton, who proposed the theory of isostasy in 1882, believed the asthenosphere supported the earth's crust. He also believed that as the weight of a mountain decreases from erosion, the higher the mountain floats in the asthenosphere. Although the crust is not able to react quickly to changes in the mountain's weight, the weight of the mountains changes so slowly that the asthenosphere is able to keep up. The continental crust floats higher than the oceanic crust because the continental crust is less dense. Accordingly, isostatic adjustments to changes in mass distribution on the earth's surface cause a compensatory uplift of mountains and plateaus as erosion wears them down. The mass of eroded material is then added to and depresses the continental shelves and the ocean floor. The growth and melting of continental ice sheets also cause such adjustments. The theory of isostasy has wide acceptance in the scientific community today.

Lesson 3.1.2 – Plate Boundaries

According to the theory of plate tectonics, the earth's plates do not all float as one; they float in different directions so as to interact with each other at their boundaries. These plate boundaries behave in various ways. The three major types of plate boundary movement are convergent, divergent, and transform. These three boundary interactions can lead to different types of surface action.

When continental crust converges with oceanic crust, the heavy, older oceanic plate is subducted underneath the younger, lighter continental plate. Convergent interaction can result in mountains, earthquakes, volcanoes, folding, and faulting. Continent-to-continent convergent activity results in mountains, earthquakes, folding, faulting, and very thick crust, whereas convergent ocean-to-ocean activity results in earthquakes, volcanoes, and island arcs.

Continent-to-continent divergent interaction can result in rift valleys, earthquakes, and faults. Ocean-to-ocean divergent interaction can cause earthquakes, volcanoes, and rift valleys. According to the prevailing theory, new seafloor spreads in both directions to form a ridge system. The basaltic magma rises from the earth's mantle, hardens into new oceanic crust, and welds to the older crust.

Seafloor spreading is believed to be caused by stresses; in response to the plate separation, the mantle swells up beneath the spreading axis. The oceanic trenches bordering the continents mark regions where the oldest oceanic crust is recycled into the mantle's steep subduction zones, an area where many earthquakes occur. This pull of the deeply plunging lithosphere may be one of the forces that drives plate separation.

Considerable evidence supports the theory of seafloor spreading. Basaltic oceanic crust and its sediments are progressively younger and thinner as they approach an oceanic ridge. Also, the rock on the ocean floor is younger than that on the continents. Older ocean crust has been recycled in ocean trench systems.

Evidence from the study of the fracture zones in sections of the ridge also support this theory, as do magnetic surveys. The North Pole has not always been north. Scientists believe the earth has, in fact, reversed its polarity many times. This back-and-forth rotation of the poles creates a type of zebra-striped pattern that shows the alternating magnetic relationships of the ocean's crust. A time scale has shown 171 magnetic reversals. Magnetic surveys near an oceanic ridge showed elongated patterns of normal and reversed polarity of the ocean floor in bands that parallel the rift and are symmetrically distributed as mirror images on either side. This magnetic record shows the movement of the ocean floors.

Transform interaction results in fault zones and earthquakes. Students will study these in greater depth in Chapter 3.2.

Lesson 3.1.3 – Rock Stress and Deformation

Earth's rocks are subjected to stress when the plates of Earth's thick crust move or when magma pushes up toward the surface. Changes in such giant landmasses result in rock deformation, such as fracturing, tilting, or folding, which changes the appearance of Earth's surface. For example, as two plates push together, compression stress can cause rock masses to fold and form mountain ranges. Scientist believe this is how the Alps were formed. Compression stress can also cause a thrust fault where one wall slides over the other; many scientists believe this is how the Rocky Mountains were formed. Rift valleys are formed from tensional stress when the block of land between two normal

---

WORLDVIEW

- Agree to disagree. In the realm of science, Christians can be divided about many things: the length of a creation day; the age of the earth; how the continents move; or the creation of geological formations. The one area we should be in agreement about, however, is "In the beginning God ..." (Genesis 1:1). The belief in God as Creator unites us as believers. He is the Creator and Sustainer of life. God's Word tells us to have an attitude of humility and gentleness, like that of our Lord and Savior, Christ Jesus. The apostle Paul writes to the Philippians, "Then make my joy complete by being like-minded, having the same love, being one in spirit and of one mind. Do nothing out of selfish ambition or vain conceit. Rather, in humility value others above yourselves" (Philippians 2:2–3). Even though some of the theories discussed in this chapter are not embraced by all Christians, we should be united in purpose. When we disagree about how the earth has changed, instead of trying to persuade others to our "right" way of thinking, we should remember Paul's exhortation in Philippians 2:5 to have the same attitude as Christ Jesus. We should exemplify Christ's heart and glorify God in all our endeavors.

faults slides downward. Shearing stress, which pushes rocks in opposite directions until they break apart, are believed to have formed the San Andreas fault.

Many factors determine whether a rock will fold or fault. One factor is temperature. Rock becomes very hard when it is compressed, but the hotter it is, the more it will soften and fold. Cooler rock is more likely to fracture. Likewise, the greater the pressure on the rock, the more likely it is to fold. Pressure may be produced by the crustal rocks of continents pressing down on the rock underneath.

Rock type also determines how the rock will respond to pressure. Certain elements are brittle; others are more pliable. The process by which the rock formed—such as whether it cooled quickly or slowly—also factors into its reaction to stress. Sudden stress may cause a rock to fracture, whereas gradual stress may cause it to bend.

A fold is a deformed arrangement of stratified sedimentary or igneous rocks. When oceanic crust meets less dense continental crust, the oceanic crust is forced under the continental crust. The continental crust buckles, forming folded mountains. Although stratified rocks are usually deposited in horizontal layers, they often incline. Arches are called *anticlines*; depressions are called *synclines*. A monocline is a steplike structure that slopes in one direction only. Folds on a grand scale, which might run most of the length of a continent, are known as *geosynclines* and *geanticlines*. Geologists can determine the nature of the original fold from the exposed portions: two outcrops dipping toward each other mark a syncline; two outcrops dipping away from each other mark an anticline.

Certain domes are very short anticlines; some basins are synclines. Usually folds form at some distance below the surface, and they may be exposed by erosion. The crests of anticlines, for example, may be eroded until only worn-down stumps remain, and synclines may be eroded with the edges projecting over the surface. The ridge crests of the Appalachian Mountains, for example, are eroded limbs of folds. Porous and permeable rocks of anticlines are often a source of oil and natural gas. Many scientists contend that when the Appalachian Mountains folded, they trapped the organic remains of late Paleozoic tree fern swamps and converted them to anthracite coal.

Geologists come from all over the world to study the curious geologic formations in the Ouachita Mountains of Arkansas and Oklahoma. The Ouachita Mountains are folded mountains. The sedimentary rock was compressed and folded to form these mountains. Brittle rock faulted, allowing large sheets of land to override the land to the north. Over time, water and wind erosion whittled away on these mountains. The middle parts of the mountains are now visible, offering a valuable study of the folding and faulting process.

A break or crack in a rock is called *a fracture*; a fault is a fracture in the earth's crust where the rock on one side moves in relation to the rock on the other side. Evidence of faults can be found either at the surface or underground. The movement of a fault generally has both a vertical and horizontal component, although sometimes the fault itself is vertical.

Faults are most evident in outcrops of sedimentary formations, where they conspicuously offset previously continuous strata. A fault plane may move vertically, horizontally, or obliquely; it may also consist of the rotation of one or both fault blocks. The two classes of faults are the dip-slip, up and down movement, which is further divided into normal and reverse faults, and strike-slip, or lateral. Normal faults are commonly associated with tensional stress. Reverse faults are usually associated with compressional stress. Strike-slip faults are frequently associated with shearing stress. In strike-slip faulting, the two blocks move either to the left or to the right relative to one another.

In normal faulting, the largest, most compressive stress is vertical; the smallest and intermediate stresses are horizontal. The fault makes an angle of less than 45° with the major vertical stress direction. In reverse faulting, the smallest, least compressive stress is vertical; the smallest and intermediate stresses are horizontal. The fault makes an angle of less than 45° with the major horizontal stress direction. In strike-slip faulting, the intermediate stress is vertical; the largest and smallest stresses are horizontal. The fault makes an angle of less than 45° with the major horizontal stress direction.

A fault is often a rich mineral source. Gas and oil companies spend millions of dollars searching for faults that may be oil sources. Silver is also found in faults. Faults can also contribute to earthquakes.

Lesson 3.1.4 – Mountains

Much of the earth is covered with mountains, most of which stretch in vast rows called *ranges* flanked by flat lands called *plateaus*. Related mountain chains and ranges make up mountain systems such as the Tethyan Mountain System, which includes the Alps and Himalayas. Mountain chains are adjacent to plate boundaries or former plate boundaries, where compressional forces crumpled continental margins.

In general, scientists believe stress near the earth's surface results in faulted mountains; deeper stress results in folded mountains. Folded mountains such as the Himalayas and Appalachians form from the collision of two continental plates. In this type of collision, no subduction occurs, and the two plates pile up on each other. Folds are especially evident when contrasting rock layers are present. Folding occurs at depths with high pressures and high temperatures, two factors that cause rocks to become ductile, which causes them to deform before they break. The application of stress must be very slow in order for folding to occur. Folds occur at all different scales, from those visible under the microscope to those that cover many kilometers.

Fault-block mountains usually form at lower temperatures than folded mountains, and the faulting occurs at shallower depths than folding. Faulting is caused by the movement of large crustal blocks along faults formed when tensional forces pull apart the crust. Tension is often the result of part of the crust uplifting; it can also be produced by opposite-flowing convection cells in the mantle.

Plateaus form on the earth's crust when hard, relatively erosion-resistant horizontal rock layers are covered by other horizontal layers of rock. Plateaus are large areas of flat topped rocks high above sea level formed by vertical uplift. Thick, horizontal layers of rock are slowly uplifted by colliding continents. Plateau areas are pushed up gently; they do not fold and crack into mountains, but instead they remain flat. Plateaus are generally found next to mountain ranges. Volcanic plateaus can also form when lava flows harden to strata and pile up in flat, horizontal layers.

NOTES

3.1.1 Continental Drift

Crust Movement

OBJECTIVES

Students will be able to
- experiment with the theory of isostasy.
- explain the theory of continental drift.
- relate how scientific progress depends on the work of past scientists.

VOCABULARY

- **continental drift** the theory that the continents can move apart from each other and have done so in the past
- **isostasy** the equilibrium in the earth's crust maintained by a flow of rock material in the asthenosphere
- **tectonics** the study of the movement and changes in the rocks that make up the earth's crust

MATERIALS

- Magazine pages (*Introduction*)
- BLM 3.1.1A Continental Drift Reader's Theater
- TM 3.1.1A World Maps
- WS 3.1.1A Graph the Continents
- WS 3.1.1B It's a Puzzle
- WS 3.1.1C Digging Deeper

PREPARATION

- Tear magazine pages into pieces. (*Introduction*)
- Obtain materials for *Lab 3.1.1A Isostasy*. Fill petri dishes with colored water and freeze.
- Obtain scissors, glue sticks, and large sheets of construction paper. (*B*)

Introduction

Divide the class into groups. Give each group a magazine page that you have torn into sections. Have students try to assemble the picture and record how they decided where each piece went. Discuss how this is similar to what Alfred Wegener did. He mentally moved the continents together. Once he had the mental picture in place, he studied the fossils, climates, and geology of the continents. He found matching fossils of the same extinct species on adjacent areas of different continents. Wegener also found similar connections in the climate and the geology. He concluded that the continents must have been together at one time for the climate, geology, and fossils to so closely correspond.

Discussion

- Use **TM 3.1.1A World Maps** to compare the changes of the world map over time. Discuss the history of mapmaking and how this technology could affect the study of geography, climate, and geology. Ask students why the first maps are not as accurate as the ones printed currently. (**Current technology wasn't available.**) What evidence did cartographers use to make the maps? (**Answers will vary.**) Why do you think the theory of a single continent was not suggested before world maps were made? (**People couldn't see how the continents resemble puzzle pieces.**)

- Discuss God as the Creator and Sustainer of the earth. God created a dynamic earth that is infinitely complex and beneficial for its inhabitants. Discuss the fact that humans have been mandated to explore, discover, and understand this dynamic planet. Remind students that humans have limited knowledge and wisdom, and they can argue about various interpretations. The succession of historic explanations in this lesson about the earth's dynamics demonstrates the complexity of creation, the limited human capacity to understand, and God's infinite wisdom, creativity, and majesty. Encourage students to watch for examples of each throughout this unit.

- Discuss the theory of Pangaea, one supercontinent. This can be a controversial topic, but it can fit into both a young-earth and an old-earth worldview. Ask the following questions:
  1. How could young-earth creationists explain the continents moving from one whole continent into several smaller continents? (**Young-earth creationists can fit Pangaea into their understanding by using the worldwide Flood. God could have created a single continent in one day of creation and broken it apart during the Flood.**)
  2. How could an old-earth creationist explain it? (**Old-earth creationists could explain that God created one large continent and that the continents are slowly moving. These continents could come apart, given enough time.**)
  3. Has either interpretation been proven right and the other wrong? (**No.**)
  4. What part of the discussion is the most important? (**God created the heavens and the earth.**)
  5. Can we know how He did it? (**No. The Bible doesn't say how God created, just that He did.**)
  6. How does God instruct Christians to handle disagreements? (**Answers will vary but should include with humility, gentleness, and kindness.**)

  Convey to students that in classical terms, an argument is not bickering between sides, but it is a set of reasons given to support an idea in the attempt to persuade others to agree. Ask students in light of the differing views of the age of the earth and its formation, what gives an argument authority. (**Answers will vary.**)

- Have students look at their fingernails. Do they think their nails are moving? (**Answers may vary.**) Can they prove it? (**No.**) Nails grow so slowly that the growth cannot be seen, but eventually the growth is evident because the nails must be trimmed. Convey that the crust of the earth behaves similarly. Although the earth's crust does not appear to be moving, it is.

- Ask the following questions:
  1. What is the theory of isostasy? (**This theory is about the balance of the crust on the asthenosphere. As one part erodes, another part gains mass.**)

2. What evidence did Wegener use to support the theory of continental drift? (**He found evidence of similar fossils on different continent coastlines. He found evidence of climate similarities: coal deposits in the polar regions and glacial features near the equator.**)
3. What evidence was found in the 1950s to support the theory of continental drift? (**Scientists found that the magnetic poles had changed directions on each continent many times.**)
4. What type of evidence, direct or indirect, did NASA supply in the 1980s? Does the evidence support or deny the theory of continental drift? (**NASA supplied direct evidence in actual photos of the continents changing locations. The evidence seems to support the theory.**)

Activities

Lab 3.1.1A Isostasy

- plastic petri dishes, 2 per group
- stackable metric weights
- waterproof clear plastic tubs, 1 per group
- cold water
- metric rulers, 1 per group
- ice disks, 1 per group

Use petri dishes to make ice disks ahead of time. Add food coloring to make them easier to see in the water. Have students use the grams listed on the weights to figure the mass of the stacks.

A. As a class, read through **BLM 3.1.1A Continental Drift Reader's Theater**.

B. Have students work in groups to complete **WS 3.1.1A Graph the Continents**.

NOTES

3.1.1 Continental Drift

OBJECTIVES
- Experiment with the theory of isostasy.
- Explain the theory of continental drift.
- Relate how scientific progress depends on the work of past scientists.

VOCABULARY
- **continental drift** the theory that the continents can move apart from each other and have done so in the past
- **isostasy** the equilibrium in the earth's crust maintained by a flow of rock material in the asthenosphere
- **tectonics** the study of the movement and changes in the rocks that make up the earth's crust

The continental crust consists of seven major continents—large landmasses that are completely surrounded by water, although sometimes a land bridge may link two continents. Europe and Asia have traditionally been considered different continents even though they are part of one landmass called *Eurasia*. Generally, continents are separated by deep oceans, not shallow seas. The Baltic Sea and North Sea, for example, are merely flooded portions of a continent.

If they think about it at all, most people probably assume that Earth's landmasses and oceans have always looked like they do now. However, scientists have made discoveries that indicate the earth has changed and is still changing. It seems the processes God set in place have been at work reshaping the land throughout history. For centuries, scientists have studied the earth, trying to understand these processes.

You know that some scientific data is gathered through direct observation and other data is gathered through indirect observation. Since scientists have no direct observations of how the continents may have moved, they must rely on data gathered through indirect observation. For example, Scottish geologist James Hutton believed that basaltic lava formed the continents and that the continents were part of a renewing cycle. Hutton theorized that God created a recycling world and volcanoes were pathways to a molten underworld. Hutton's theory was called *volcanism*. In 1807, the newly formed Geological Society of London accepted volcanism as the correct explanation for the formation of the continents. In the 1850s, volcanism was nudged a step further when British geologist Charles Lyell wrote *Principles of Geology* and pointed out that geologists are not able to look at the complete picture. Contrary to the young-earth belief that events like the Flood shaped Earth's landmasses, Lyell believed if geologists could see everything, they would discover that catastrophes had nothing to do with the earth's geological features. Lyell agreed with Hutton's theory of uniformitarianism, a belief that natural processes have acted on the earth in the same way and at about the same intensity throughout history.

Then in 1882, American geologist Clarence Dutton presented the idea of **isostasy**, the theory that the continents float on the asthenosphere, much like an iceberg floats in water. Dutton suspected that people can see only the tip of the mountains just as people can see only the tips of icebergs. Although not accepted at the time, most scientists today accept the theory of isostasy. Much like large ships float and move across the water in the oceans, the earth's crust floats and moves on the asthenosphere. Different parts of the crust float at different levels depending on their mass. Think about a heavy ship in the ocean; the heavier the load, the deeper into the water the ship sinks. As the ship is unloaded, it slowly rises in the water. The earth's crust behaves in the same way. The mountains that rise majestically above the

FYI

Volcanic Eruptions in Sumatra
Sumatra is located in the Pacific Ocean's Ring of Fire, and the country's 130 active volcanoes regularly contribute to the volatility of that region. Mount Sinabung has been erupting almost daily since 2013. In the summer of 2016, three of Sumatra's volcanoes, Mounts Sinabung, Rinjani, and Gamalama, all erupted over a few days, spewing ash, causing evacuations, and disrupting travel in and out of the country. The eruptions were preceded by earthquakes.

Mount Sinabung

Continental Drift

Before — After

NOTES

C. Assign **WS 3.1.1B It's a Puzzle**. Students will need scissors, glue sticks, and a large piece of construction paper.

D. Divide the class into groups and challenge students to dig deeper into their understanding of the scientific process. Assign one or more of the following topics to each group. Have each group research the topic using the Student Edition and the Internet and then direct them to discuss their answers before they write them down. Help students differentiate between fact and opinion driven by worldview and see why there are debates about scientific facts. Remind students Christians can agree that God is the Creator and Sustainer of the earth but can have different opinions about creation. Explain that humility, gentleness, and grace-filled hearts should dictate how well Christians can agree to disagree. When all groups are finished, have each group share their findings with the class. Extend this activity by using **WS 3.1.1C Digging Deeper**.

- Charles Lyell said that catastrophes have nothing to do with creating the earth's geological features. Is there any evidence that contradicts this statement? What should a scientist do if evidence contradicts a scientific conclusion? What would you do if you found evidence that contradicted what you believed to be true?
- Clarence Dutton is credited with the theory of isostasy. What did he do to come up with the idea of isostasy? Did he use direct or indirect observations?
- Alfred Wegener strongly believed in the idea of Pangaea, and he believed he had proof of a supercontinent. Are the similar fossils found on two different continents proof that the continents were once connected? Why? Are the glacial features found near the equator proof that all continents were once one continent? Why?

---

ground also sink deep into the asthenosphere because of their greater mass.

Isostasy accounts for the balance between the forces that raise landmasses and the forces that depress them. The thicker continental crust floats higher than the thinner oceanic crust because the rock that forms the continental crust is less dense than the very dense rock of the oceanic crust. Heavy sections of the crust, such as a section of crust that holds a mountain, sink farther into the asthenosphere. As rain, wind, and pollution erode the mountain, it becomes lighter and does not push as heavily into the asthenosphere. The crust slowly rises out of the asthenosphere as the crust erodes.

When accurate maps of the entire world became available, people noticed that the continents looked like pieces of a jigsaw puzzle that had been moved apart from each other. Scientists wondered if that was a coincidence or if the continents had once been connected. In 1911, German meteorologist Alfred Wegener made a discovery that indicated the continents had once been part of a larger landmass. By studying fossils, he determined that certain climate patterns had once been found only on the "matching" coastlines of Africa and South America. Wegener also found the same kinds of fossils—such as the fossil of the seed fern, a plant that is now extinct—on adjoining points on different continents. These findings added to the scientific evidence that the continents had once been joined in a larger landmass.

This map, which was produced in 1852, may be the map that Wegener used.

114

From his findings Wegener developed the theory of **continental drift**, which states that the continents can move apart from each other and have done so in the past. Wegener believed that the continents began as one supercontinent. He named this supercontinent *Pangaea*, which is Greek for "all lands." He named the water surrounding this mass *Panthalassa*, which is Greek for "all seas." He said that these continents had slowly separated into what are now the world's seven continents.

In 1915, Wegener presented his ideas to the scientific community in his book *The Origin of the Continents and Oceans*. The evidence that Wegener cited included the unusual presence of coal deposits in south polar regions, glacial features near the equator, and the jigsaw fit of the opposing Atlantic continental shelves. He hypothesized that the earth's interior must include a plastic, or malleable, layer to allow for vertical adjustments caused by the creation of new mountains and the wearing down of old mountains by erosion. In addition, the earth's rotation horizontally adjusted rock in this plastic layer, causing the continents to drift. Friction along the leading edges of the drifting continents built the mountains.

In the 1920s, South African geologist Alexander Du Toit's finding of similar geological and paleontological features on both sides of the Atlantic reinforced Wegener's theory. His later book, *Our Wandering Continents*, suggested that the supercontinent broke into two masses—the one to the north he called *Laurasia* and the one to the south he called *Gondwanaland*. In 1929, Scottish geologist Arthur Holmes suggested the continents moved by thermal convection in the mantle, a theory that is accepted by many scientists today. Many scientists had cast Wegener's ideas aside because he could not thoroughly explain how or why the continents could move. What kind of force on Earth was strong enough to push or pull something as heavy as a continent? Many old-earth scientists have theorized that the earth's rotation and convection form a self-sustaining dynamo that developed over millions of years and drives continental drift. In contrast, prominent young-earth scientists theorize that the decaying electrical current in the metallic core provides the energy for continental movement and indicates that Earth cannot be millions of years old.

In the 1950s, Canadian geologist Lawrence Morley and British geologists Frederick Vine and Drummond Matthews studied the

Wegener's Pangaea

Du Toit's continents

115

- Scientists found changes in magnetic polarity in each continent in the 1950s. How did scientists find the changes? Are the changes facts? Do these changes prove the continents have moved? Are there any other explanations besides continental drift that could explain the changes in the magnetic poles? (Note: Young-earth creationists attribute these changes to the Flood.)
- NASA has the ability to take photos of the continents from outer space. Do the photos taken by NASA prove that the continents are moving? Why? Are these photos direct or indirect evidence? Do the photos prove that all the continents were one giant supercontinent? Why?

Lesson Review

1. What causes the different crusts to rise and lower? (**When crust has less mass, it floats higher. When crust has more mass, it sinks lower into the asthenosphere.**)
2. Which sink lower into the asthenosphere—mountains or flatlands? Why? (**Mountains sink lower because they have more mass than flatlands.**)
3. What is stated in the theory of continental drift? (**The continents can move apart from each other and have done so in the past.**)
4. Starting with the idea of isostasy and ending with plate tectonics, explain how scientists' understanding of the earth's crust has changed. (**Dutton's idea of isostasy was somewhat accepted. With the exploration of continental drift, scientists started to believe that the earth's crust could be moving, but they didn't know how. More evidence was gathered with the discovery of magnetic pole changes and NASA photographs. As technology improved, scientists were able to gather more information. Decades of study have resulted in the current understanding of plate tectonics.**)

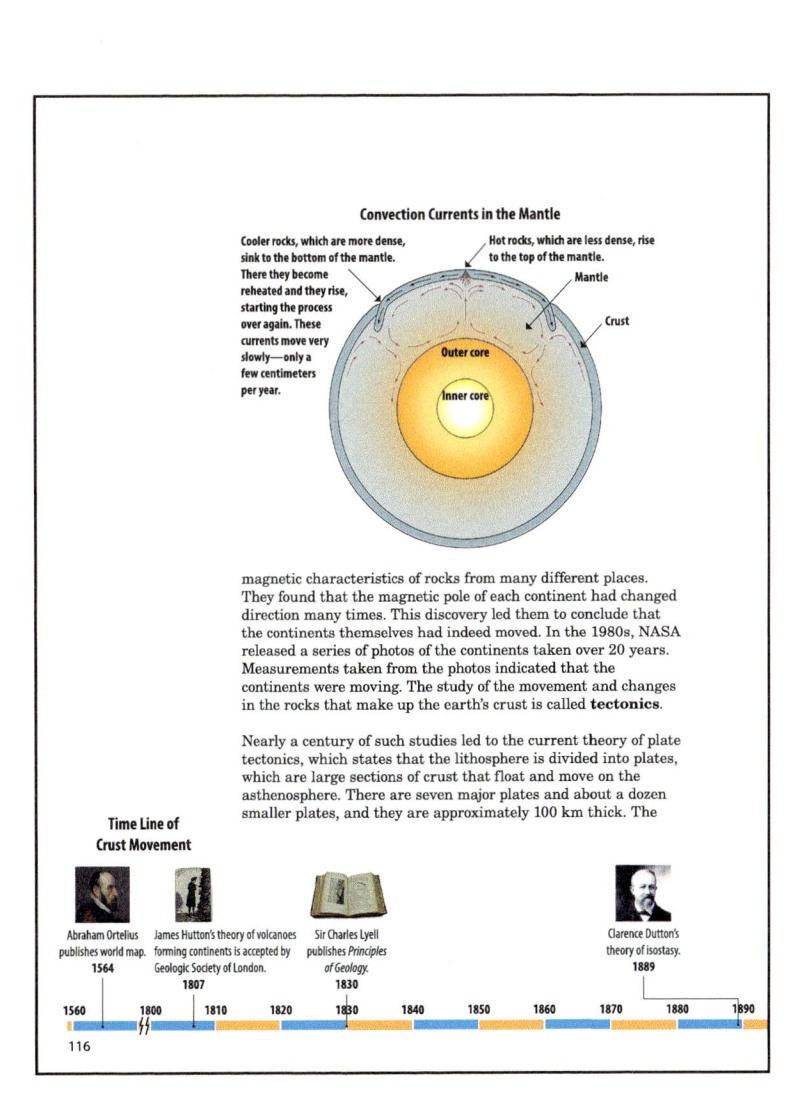

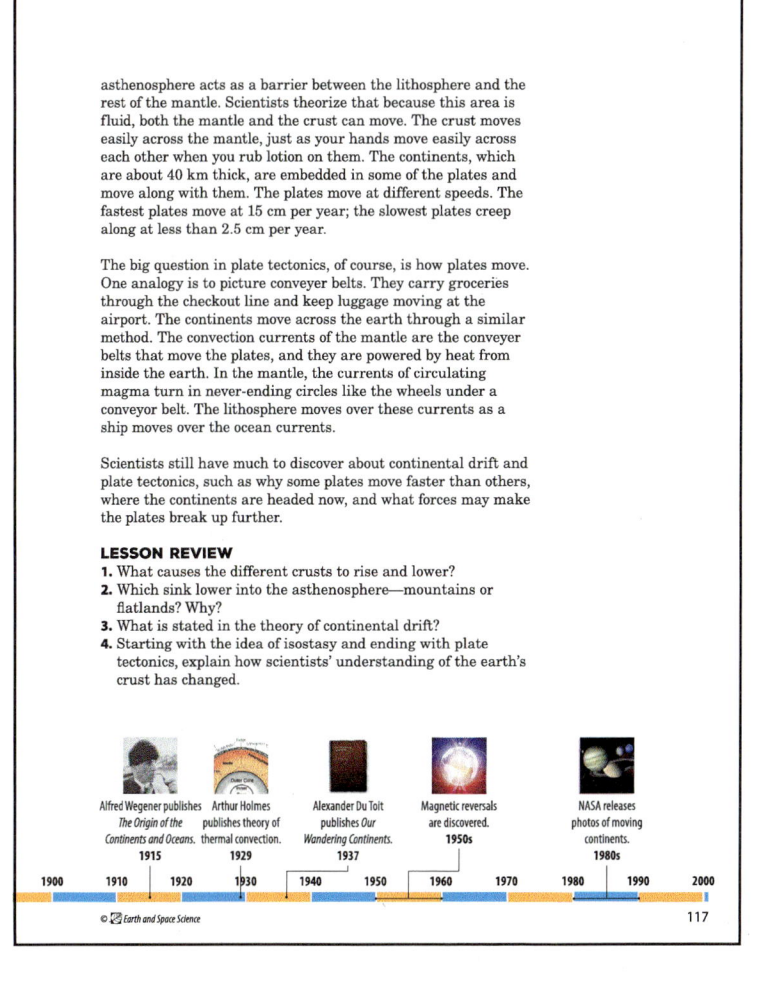

NOTES

Name: _____ Date: _____

Lab 3.1.1A Isostasy

QUESTION: What would cause different parts of the crust to float at different depths in the asthenosphere?

HYPOTHESIS: Answers will vary.

EXPERIMENT:

You will need:	• waterproof clear plastic tub	• ice disk
• 2 plastic petri dishes	• cold water	
• stackable metric weights	• metric ruler	

Steps:
1. Fill the tub about half full of water. The water represents the asthenosphere.
2. Float the empty petri dishes on the water. The dishes represent the continental crust. Make a drawing below to show where the bottom of the dishes are in relation to the surface of the water.
3. Stack the weights into two columns. One stack should be about twice as high as the other. These will need to sit on the floating petri dishes, so do not make them too tall. Record the height and mass of each stack. Answers will vary. Height should be measured in centimeters. Mass should be the total grams listed on the stackable weights.
4. Carefully place a stack of weights in the center of each petri dish. The taller stack represents a mountain and the shorter stack represents a flatland.
5. Look through the side of the tub and draw what you see now, showing where the bottom of each dish is in relation to the surface of the water.
6. Ask your teacher for an ice disk. The disk represents the oceanic crust. Place the disk in the water. Make a drawing to show where the bottom of the disk is in relation to the surface of the water.

Floating Crust

Empty Petri Dishes	Stacked Petri Dishes	Ice Disk
Drawing should show both dishes floating completely on top of the water.	Drawing should show the taller stack sitting lower in the water and the shorter stack sitting higher in the water.	Drawing should show the disk sitting about halfway into the water. It should sit lower than the empty petri dish.

Lab 3.1.1A Isostasy

ANALYZE AND CONCLUDE:
1. What does the empty petri dish represent? continental crust
2. What does the ice disk represent? oceanic crust
3. Which one sits lower in the water? the ice disk
4. How is the ice disk like the oceanic crust? It is more dense than the continental crust because basalt is more dense than granite. So it sits lower in the water.
5. Which dish sat lower in the water? Why? The dish with the taller stack sat lower in the water because it had a greater mass, which correlates to the mass of mountains compared to the mass of flatlands.
6. Explain how this lab demonstrates the theory of isostasy. This lab shows how different parts of the crust float at different levels in the asthenosphere. The thick, mountainous continental crust, represented by the tall stack, sits lower in the asthenosphere than the thinner flatland crust, represented by the short stack. The ice disk is like the oceanic crust in that it is thinner than the continental crust but more dense. Even thought it is thin, it sits lower.

Crust Movement

3.1.2 Plate Boundaries

Introduction
Demonstrate how the motion of convection currents could cause tectonic plate movement. Place a plastic tub on two stacks of books, which should be high enough for a heat lamp (150–200 watt bulb and a socket) to fit underneath. Fill the tub with water and place four shallow aluminum foil boats in the tub. The boats should be touching and forming a square. Place the lamp underneath the tub. Turn on the lamp and place a drop of food coloring in the water where the boats meet. Have students observe for several minutes. Ask them to describe what they see happening. (**Answers will vary but should include that the food coloring, the water, and the boats are all moving.**)

Explain that the motion of the water is similar to the convection currents in the earth's mantle. The crust of the earth is divided into several plates and sits on the asthenosphere. The plates move somewhat like the boats, and as they move, many different things can happen. Ask students to describe how the boats moved. (**Answers will vary.**)

Discussion
- Review convection currents. Water, air, magma, and many other fluids travel in circles when heated, with the heated fluid or air rising to the top. Ask students why. (**Answers will vary.**) Explain that substances expand as they are heated. Expanded, the substance is less dense than it was before and will rise through more dense material. As the heated substance rises, it cools, becoming more dense. The increased density causes the substance to move back toward the bottom where it begins the cycle again.

- Display **TM 3.1.2A Plate Map**. Direct students' attention to the names of the plates. Ask students to identify which plates are under continents, which are under oceans, and if any plates are under both. (**Answers will vary.**) How do you think scientists are able to determine where the plate boundaries lie? (**Answers will vary but should include that the boundaries are along coastlines or mountain ranges, or scientists use indirect observations.**)

- Discuss the difference in interpretation of scientific data between old-earth and young-earth scientists regarding seafloor magnetic reversal.

Activities
A. Complete *Try This: Moving Plates* with students.

B. Divide the class into groups. Have each group research a specific type of plate boundary and find an example of it. Direct each group to make a working model of the boundary. Supply students with various supplies, such as construction paper, scissors, cardboard boxes, and clay. Have groups present their findings and their models to the class. **Option:** Have groups make a computer presentation with an animated model of the plate boundary.

C. Challenge students to discover what can happen when tectonic plates collide. Give each student a hard-boiled egg. Direct them to draw plate boundaries on the eggshell with a permanent marker. Guide them to use a butter knife to crack the shell along the boundaries and to then move the eggshell plates. Have students share what happened to the plates as they moved. (**The plates cracked, crumpled up, or slid under each other.**) Relate how the eggshell plates are similar to tectonic plates moving on the asthenosphere.

D. Direct students to research a new island and report on its location and its formation process.

E. Assign students to research the historical use of magnetometers in the discovery of magnetic seafloor reversal.

OBJECTIVES
Students will be able to
- describe the characteristics of different plate boundaries.
- explain the process of seafloor spreading.
- infer how seafloor spreading and subduction work together to recycle the earth's crust.

VOCABULARY
- **oceanic ridge** a mountain chain that forms on the ocean floor where tectonic plates pull apart
- **plate boundary** the point at which one tectonic plate meets another
- **seafloor spreading** the process by which a new oceanic lithosphere is formed at an oceanic ridge as older materials are pulled away from the ridge
- **subduction** the process of one tectonic plate being pushed under another tectonic plate
- **trench** a deep underwater valley

MATERIALS
- Plastic tub, several books, heat lamp, aluminum foil, food coloring (*Introduction*)
- Various art supplies (*B*)
- Hard-boiled eggs, permanent markers, butter knives (*C*)
- TM 3.1.2A Plate Map

PREPARATION
- Make four shallow boats from aluminum foil. (*Introduction*)
- Obtain materials for *Try This: Moving Plates*.
- Collect various sizes of cardboard boxes. (*B*)
- Obtain eggs and hard boil them. (*C*)

TRY THIS
Moving Plates
- sandwich cookies

Earth and Space Science

NOTES

Lesson Review

1. What are three types of plate boundaries? What is the direction of movement at each boundary? (*convergent*: **plates move toward each other;** *divergent*: **plates move away from each other;** *transform*: **plates move horizontally past each other**)
2. What can result from each type of plate boundary movement? (**Converging movement can form trenches or mountains. Diverging movement can form ocean ridges or mountains. Transforming movement does not create or destroy any crust.**)
3. What do many scientists believe causes the plates to move? (**They believe convection currents in the mantle cause the plates to move.**)
4. Using the terms *subduction* and *seafloor spreading*, explain how the earth's crust recycles. (**Through the process of subduction, crust is lost as the ocean floor is subducted into the mantle. This oceanic crust is destroyed in the process. At another location, the crust is replenished through the process of seafloor spreading. New magma comes up through the oceanic ridge and creates new crust. The older crust is pushed away from the ridge. Eventually, that older crust could be subducted under the continental crust at a convergent boundary. Then it will be recycled back into the mantle.**)
5. What findings did Henry Hess discover that led him to the theory of seafloor spreading? (**He first discovered that the ocean was warmer in some areas than in other areas. He found the warmer crust was thinner and newer than the cooler crust.**)
6. What other evidence exists to support the theory of seafloor spreading? (**The existence of magnetic striping on the ocean floor supports the theory. The magnetic stripes mirror each other on either side of an oceanic ridge. This gives evidence that change is happening on the oceanic ridge and the crust is spreading away from it.**)

3.1.2 Plate Boundaries

OBJECTIVES
- Describe the characteristics of different plate boundaries.
- Explain the process of seafloor spreading.
- Infer how seafloor spreading and subduction work together to recycle the earth's crust.

VOCABULARY
- **oceanic ridge** a mountain chain that forms on the ocean floor where tectonic plates pull apart
- **plate boundary** the point at which one tectonic plate meets another
- **seafloor spreading** the process by which a new oceanic lithosphere is formed at an oceanic ridge as older materials are pulled away from the ridge
- **subduction** the process of one tectonic plate being pushed under another tectonic plate
- **trench** a deep underwater valley

If you were to compare the land with the oceans, would you find more similarities or more differences? In many ways the two are opposites. The land and the ocean seem to have nothing in common. But like all parts of creation, they fit into one interconnected master design.

As scientists have studied continental drift, they have theorized that the continents are connected to a large section of the lithosphere and that the ocean floors are connected in the same manner. They have determined that some of the continents and oceans even share the same piece of the earth's lithosphere.

Most scientists believe the continents and the oceans rest on large, moving plates. Some plates, such as the North American Plate, support mostly land. Other plates, such as the Pacific Plate, support mostly ocean. Most plates support a combination of water and land. Each plate touches several other plates. The point at which one tectonic plate meets another is called a **plate boundary**.

Plates interact in many ways at the plate boundaries. To imagine the different ways that these tightly packed, moving plates interact, clap your hands in slow motion. Plates that are moving toward each other, like your hands, are said to be converging. The boundary between two plates that are moving together is called *a convergent boundary*. For example, the Nazca Plate and the South American Plate are converging at a rate of about 8 cm per year, which is considered an average speed. In comparison, other plates ram into each other at a rate of 5–10 cm per year.

TRY THIS

Moving Plates
The current plate tectonics theory maintains that the earth's crust sits on huge plates of rock. These plates ride on the asthenosphere and move with the convection action of the mantle. Take a sandwich cookie and remove one side. Break that side into two pieces. Put those pieces back on the cookie. Hold the cookie with two hands. Now use your thumbs to move the broken pieces toward each other, then away from each other, and then sliding in opposite directions along the broken fault. How is the cookie like the plates of the earth? Describe which type of boundary is modeled with each movement of the cookie pieces.

Plate Boundaries

Scientists theorize that convergent boundaries occur because of the way Earth's internal convection currents power the movement of plates. According to scientific models, some convection currents are flowing downward. Downward-flowing currents pull together the plates that lie above them. This convergence may either form mountains like the Himalayas or pull one of the plates under the other. The process of an oceanic plate being pushed under a continental plate is called **subduction**. The descending plate, which slides under the other one, can release fluids and gases, which cause the hard mantle to partially melt. According to geologic data, the Andes Mountains of South America are currently undergoing this type of action on the boundary between the Nazca Plate and the South American Plate. On the continental side of the subduction, the Andes Mountains are slowly rising up. On the oceanic side, the oceanic crust is being pushed under the continental crust and a deep underwater valley, or a **trench**, is forming.

In other areas, plates are moving apart from each other, or diverging. Scientists call the boundary between two tectonic plates that are moving away from each other *a divergent boundary*. The North American Plate and the Eurasian Plate are diverging at a rate of 2.5 cm per year. Many scientists believe divergent boundaries occur above upward-flowing convection currents. At the boundary of divergent oceanic plates, magma flows up through the separation and forms new crust. The mountain chains that form on the ocean floor where tectonic plates pull apart are called **oceanic ridges**.

The prevailing theory about how new oceanic crust is formed at an oceanic ridge is called **seafloor spreading**. Henry Hess developed this theory after serving as a U.S. Navy captain during World War II. During his command, he created a map of the ocean's topography. Hess observed that the ocean was warmer in some areas than in others. He analyzed the warmest crust and theorized that the crust was much newer at the center of the ridge than anyplace else. He concluded that new crust was being formed from magma flowing up through faults in the ocean. Because the earth's circumference never changes, he knew that older crust was also being destroyed somewhere else. These counterpoints were at the earth's subduction zones.

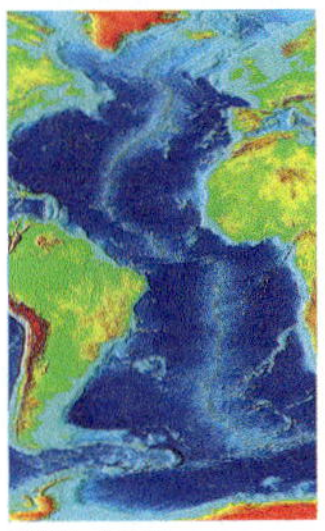

Henry Hess

There is scientific evidence to support Hess's theory that sections of the earth are separating and that new crust is forming through the gap. According to radiometric dating, the crust near the Mid-Atlantic Ridge and other ocean ridges is younger than the crust farther away from the ridges. Old crust is pushed farther away from the ridge by new crust, and all crust is eventually recycled. This crust recycling explains why there is relatively little sediment on the ocean floor.

Magnetic studies offer additional evidence for seafloor spreading. The ocean floor is striped with magnetic fields. Canadian geologist Lawrence Morley and British geologists Frederick Vine and Drummond Matthews hypothesized that the magnetic striping was produced by repeated reversals of the earth's magnetic field. These stripes are different widths and are mirror images of each other across the ridge. The consistent pattern of magnetic reversals on both sides of the ridge suggests the seafloor spreads outward from the ridge. Seafloor spreading also explains how the continents move. German meteorologist Alfred Wegener had suggested that continents simply "plow" through the ocean floor, but this is physically impossible. The continents do not power themselves; they are carried along as the ocean floor spreads from the oceanic ridges.

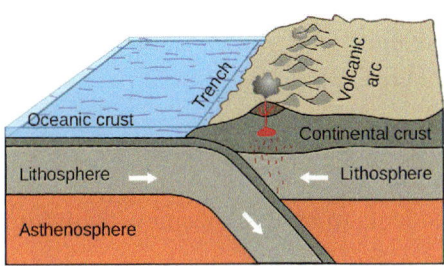

Mid-Atlantic Ridge

The third type of plate boundary is called *a transform boundary*, which is the boundary between two tectonic plates that are sliding past each other horizontally. Crust is neither created nor destroyed at these boundaries. The boundary between the Pacific Plate and the North American Plate is considered a transform boundary. The North American and the Pacific Plates slide past each other at a rate of 4–6 cm per year.

LESSON REVIEW
1. What are the three types of plate boundaries? What is the direction of movement at each boundary?
2. What can result from each type of plate boundary movement?
3. What do many scientists believe causes the plates to move?
4. Using the terms *subduction* and *seafloor spreading*, explain how the earth's crust recycles.
5. What findings did Henry Hess discover that led him to the theory of seafloor spreading?
6. What other evidence exists to support the theory of seafloor spreading?

Subduction

Seafloor Spreading

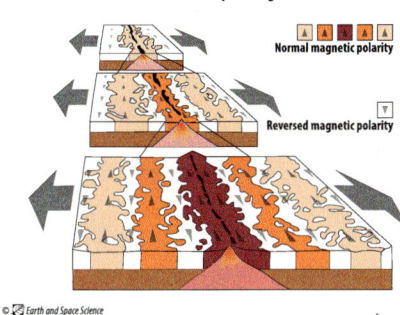

CHALLENGE

Earth's Polarity Reversals
The scientific community agrees that Earth's magnetic field has reversed polarity many times in the past, meaning that the north magnetic pole became the south magnetic pole and then changed back. What scientists do not agree on is how much time it takes for such a reversal to occur. Old-earth scientists contend that testing and dating of sediments indicate such a change takes thousands of years, and based on their models, such reversals happen millions of years apart from one another. Young-earth scientists point to recent findings using radiometric dating on lake sediments that offer evidence that the polarity reversed in less than 100 years in the past, which matches their models. Both interpretations are based on scientific evidence. In light of the differing scientific views, what gives an argument authority?

3.1.3 Rock Stress and Deformation

Crust Movement

OBJECTIVES

Students will be able to
- describe the different types of tectonic stress.
- explain how the different types of stress can affect the earth's crust.
- list the properties of rock that determine how it deforms under stress.

VOCABULARY

- **compressional stress** the stress produced by two tectonic plates coming together
- **deformation** a change in the shape or volume of rocks
- **faulting** the breaking of the earth's crust and the sliding of the blocks of crust along the break
- **folding** the bending of rock layers from stress in the earth's crust
- **footwall** the landmass below a fault
- **hanging wall** the landmass above a fault
- **shearing stress** the stress produced by two tectonic plates sliding past each other horizontally
- **tensional stress** the stress produced by two tectonic plates moving apart

MATERIALS

- Clay, bread dough, cooked noodles, rubber bands, hair ties, empty balloons, cereal, uncooked noodles, eggshells (*Introduction*)
- Uncooked lasagna or spaghetti noodles (*Discussion*)
- Gelatin blocks, balloons, cotton balls, clay, raw noodles, crackers, sponges, paper towels (*B*)
- TM 3.1.3A Formations
- WS 3.1.3A Under Stress

PREPARATION

- Obtain materials for *Lab 3.1.3A The Effects of Stress*.
- Obtain materials for *Try This: Many Faults*.
- Make gelatin blocks. (*B*)

Introduction

Display several ductile items, such as clay, bread dough, or cooked noodles; several elastic items, such as rubber bands, hair ties, or empty balloons; and several brittle items, such as cereal, uncooked noodles, or eggshells. Make three columns on the board, and title them *Ductile*, *Elastic*, and *Brittle*. Have students classify the items. Ask students if some substances could belong under more than one category. (**Yes. For example, bread is elastic before it is cooked but after it is cooked, it can be brittle or maybe ductile.**)

Discussion

- Discuss the changes in physical properties. Distribute an uncooked lasagna or spaghetti noodle to each student and instruct them to bend it. Have them describe how brittle the noodle is. (**It can bend up to a point, but then the building pressure causes the noodle to snap.**) Ask students what could be done to the noodle to make it more ductile. (**It could be soaked in water or oil; it could be cooked in boiling water.**)

- Lead students in a discussion about how the properties of rock and a rock's environment affect a rock's ability to fault or fold when stress is added to it.

- Discuss the three types of stress and what deformations can happen as a result of these stresses.

- Mention local landforms and encourage students to hypothesize how these forms may have been created.

- Display **TM 3.1.3A Formations**. Discuss each type of stress and the resulting landforms.

- Share the "principle of least astonishment" (paraphrased) from old-earth scientist Dr. Ronald Merrill with students: It is easier to believe that a natural process did not accurately record what happened than to believe that there is something fundamentally wrong with conventional scientific wisdom. Ask students to consider what role Christian virtues such as humility might play in scientific research. What virtues would a scientist need? (**Answers will vary but should include patience, humility, kindness, courage, and wisdom.**)

Activities

Lab 3.1.3A The Effects of Stress
- modeling clay, 4 colors per group
- knives, 1 per group

A. Complete *Try This: Many Faults* with students.

B. Have students complete **WS 3.1.3A Under Stress**. Items for students to classify may include gelatin blocks, balloons, cotton balls, clay, raw noodles, crackers, sponges, or paper towels.

C. Assign students to research and then demonstrate the process of "strike and dip." This process is used by geologists to measure the angle of a fault. Evaluate students' class demonstration of the method.

D. Extend students' knowledge of rock stresses and deformations by dividing the class into groups and having each group research one or two of the following terms: *plunge, anticlinorium, synclinorium, axial plane, monocline, isocline, pericline, overfold, recumbent,* and *overturned*. Direct each group to create an illustration or model of each term and to present its findings to the class.

E. Have students collect and label rocks or rock pictures that show signs of folded layers.

F. Challenge students to create a computer presentation about each type of tectonic stress.

Lesson Review

1. Describe three kinds of tectonic stress. (**Tensional stress is when two plates move apart. Compressional stress is when two plates come together. Shearing stress is when two plates slide past each other.**)
2. How does each kind of stress affect the earth's crust? (**Tensional stress can cause valleys to form or deepen, water to flow into the valleys, magma to come up through a gap and form new crust, or a normal fault to form. Compressional stress can cause rocks to decrease in volume and increase in density, or it can cause folded mountains or reverse faults. Shearing stress can cause earthquakes, broken rocks, or a strike-slip fault.**)
3. What is the difference between faulting and folding? (**Faulting occurs when the rock not only deforms but also breaks and moves under stress. Folding occurs when the rock is elastic so it bends and deforms under stress but it doesn't break.**)
4. What properties determine whether a rock will fault or fold under stress? (**These factors are involved in determining whether a rock will fault or fold: composition, breaking point, elasticity, brittleness, heat, and direction and duration of stress.**)

> **TRY THIS**
>
> **Many Faults**
> - 10 cm foam blocks, 1 per group
> - knives, 1 per group

3.1.3 Rock Stress and Deformation

OBJECTIVES
- Describe the different types of tectonic stress.
- Explain how the different types of stress can affect the earth's crust.
- List the properties of rock that determine how it deforms under stress.

VOCABULARY
- **compressional stress** the stress produced by two tectonic plates coming together
- **deformation** a change in the shape or volume of rocks
- **faulting** the breaking of the earth's crust and the sliding of the blocks of crust along the break
- **folding** the bending of rock layers from stress in the earth's crust
- **footwall** the landmass below a fault
- **hanging wall** the landmass above a fault
- **shearing stress** the stress produced by two tectonic plates sliding past each other horizontally
- **tensional stress** the stress produced by two tectonic plates moving apart

Have you ever been stressed? You feel stressed when pressure is being put on you. Maybe you feel a lot of pressure to score a goal at a soccer game, or to play your piano piece without any mistakes at the piano recital, or to get a good grade on your history project. How do you react to stress?

Just like you, the earth experiences pressure. And when the pressure becomes too intense, the earth reacts. In geology, stress is the force that causes Earth's crust to change its shape or volume. How can something solid undergo stress? According to scientists, the earth's crust is not one large section like an orange peel; it is more like the many sections of the outside of a soccer ball. The movement of the tectonic plates causes stress. Imagine what a soccer ball would look like if its sections shifted around. The sections would collide and separate. In the same way, the earth's plates would collide and separate. Most of the time you cannot feel this movement, although if the collisions are strong enough you may feel an earthquake.

According to the principles of plate tectonics, stress from the movement of plates can cause the earth's crust to move up, down, and sideways. Stress can even change the shape and volume of a mass of rock. Scientists have defined three basic types of stress that are caused by plate movement: tensional, compressional, and shearing.

Tensional stress is the stress produced by two tectonic plates moving apart. This kind of stress occurs at divergent boundaries. You see an example of this kind of stress when you grab a slice of pizza and the cheese stretches in an attempt to stay with the rest of the pizza. When two plates move apart, the rock increases in volume and decreases in density, becoming thinner in the middle than on the edges. In this way, tensional stress forms valleys. Iceland and Africa both have valleys formed by tensional stress. The land separates in sections called *grabens* and *horsts*. This makes the valleys look like they are made from large steps.

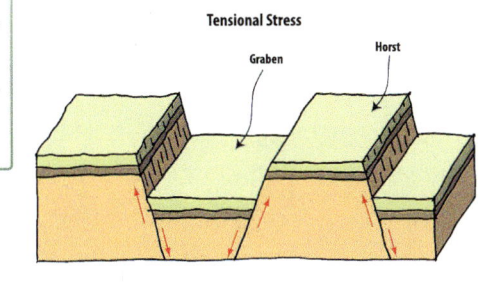

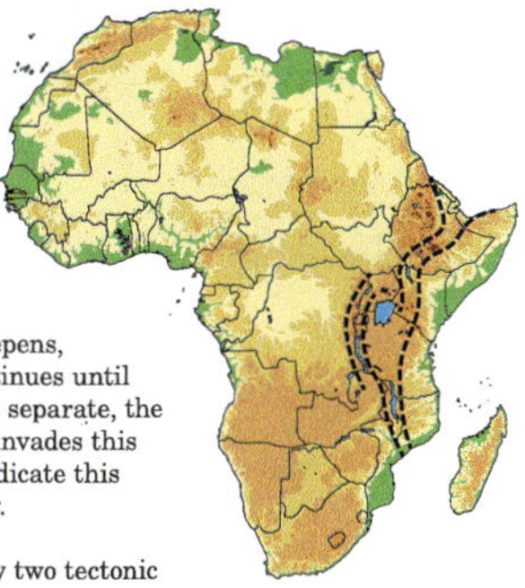

As the plates continue to separate, the valley deepens, and water flows through this low area. This continues until the plates are free from each other. As the plates separate, the earth's crust gaps. Hot magma from the mantle invades this opening, forming new crust. Scientific studies indicate this process happened in the East African Rift Valley.

Compressional stress is the stress produced by two tectonic plates coming together. This kind of stress occurs at convergent boundaries. The particles of the rock squeeze closer together, increasing in density and decreasing in volume. Compressional stress is like a car crash in very slow motion. In fact, the mountains formed by such collisions look like the fronts of cars that have collided with each other. Scientists believe the Alps, for example, were formed by compressional stress.

Shearing stress is the stress produced by two tectonic plates sliding past each other horizontally. This kind of stress occurs at transform boundaries. Rocks that undergo shearing stress are not compressed or stretched—they simply break or bend apart. Obviously, shearing is very destructive to landmasses. Many violent earthquakes strike where shearing occurs. For example, the San Andreas Fault in California experiences a lot of shearing stress because the Pacific Plate is moving north along the state's coastline and the North American Plate is moving south.

Sometimes as the earth's crust undergoes stress, it experiences **deformation**, a change in the shape or volume of rocks. If the crust receives too much stress, it may get bent out of shape permanently. Or, the rock can return to its former shape after the stress has passed, much like gelatin does if you poke your finger into it. Two kinds of deformation are faulting and folding.

Tensional Stress — Graben, Horst

East African Rift Valley

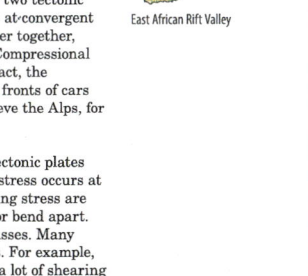

Traveling through the Alps

Faulting is the breaking of the earth's crust and the sliding of the blocks of crust along the break. **Folding** is the bending of rock layers resulting from stress in the earth's crust.

Whether rock faults or folds depends on several things. One factor is the rock's point of deformation, or its breaking point. Robert Hooke, a 17th century British physicist, developed a formula and a measurement guide for specific rock deformation. Hooke stated that the earth's crust can be stretched or compressed depending on its elasticity. For example, chalk is very brittle and does not handle stress well; it snaps if you try to bend it. In the same way, brittle deformation causes a section of the earth to break away from the adjoining section. Granite is more ductile, or bendable, than chalk and can withstand a large amount of stress. A ductile deformation folds and bends like a foam mattress but does not snap apart.

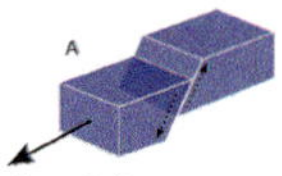

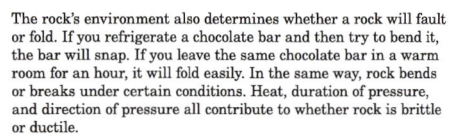

Fault line in Uzbekistan

A — Reverse fault
B — Normal fault 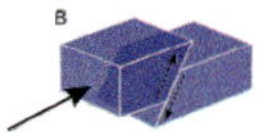
C — Strike-slip fault 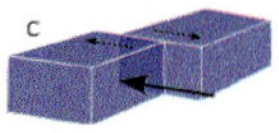

The rock's environment also determines whether a rock will fault or fold. If you refrigerate a chocolate bar and then try to bend it, the bar will snap. If you leave the same chocolate bar in a warm room for an hour, it will fold easily. In the same way, rock bends or breaks under certain conditions. Heat, duration of pressure, and direction of pressure all contribute to whether rock is brittle or ductile.

All faults are not alike. They depend on the kind of stress that is acting on that section of the earth. A normal fault involves downward movement caused by tensional stress. The **hanging wall**, or the landmass above the fault, slides down the **footwall**, the landmass below the fault. The hanging wall is wide at the top and rests on what looks like the feet of the footwall.

Compressional stress causes a reverse fault. In a reverse fault the hanging wall climbs up the footwall. The two walls move toward each other, causing the landmass to compact.

Shearing stress causes a strike-slip fault. In a strike-slip fault, the hanging wall and footwall try to slide past each other. You can demonstrate this using your hands: if you push your hands together very hard and try to slide them past each other, your hands will slide, then stop, then slide, then stop. The earth behaves the same way in a strike-slip fault—it jerks and causes disturbances not only around the fault but farther out as well.

Folded rock, resulting from ductile deformation, decorates the earth with some of the most beautiful mountains in the world, including the Andes. The downward arches of folded rock are called *synclines*. The upward arches of folded rock are called *anticlines*. The series of synclines and anticlines in folded rocks look like large waves frozen in rock.

LESSON REVIEW
1. Describe three kinds of tectonic stress.
2. How does each kind of stress affect the earth's crust?
3. What is the difference between faulting and folding?
4. What properties determine whether a rock will fault or fold under stress?

HISTORY
Hooke's Law
In 1660, Robert Hooke discovered a law of elasticity, finding a direct correlation between the force applied to a solid body and the ability of that body to stretch. This law, which is written as $F = -kx$, helps scientists understand how a rock will respond to different types of stresses—shearing, twisting, and stretching. How much stress can a particular rock take before it breaks or becomes permanently deformed? Hooke's Law can help answer that question.

TRY THIS
Many Faults
Color several horizontal layers on a 10 cm foam block to represent layers of rock. Cut the block at an angle to represent a fault line. Label one piece as the footwall and the other piece as the hanging wall. Now move the walls to represent a normal fault, a reverse fault, and a strike-slip fault. What direction did the hanging wall move in each case? What type of stress is used to create each type of fault?

The Andean fold in Argentina

Name: _____ Date: _____

Lab 3.1.3A The Effects of Stress

QUESTION: How do different types of stress affect rock formations?

HYPOTHESIS: Answers will vary.

EXPERIMENT:

| You will need: | • 4 colors of clay | • knife |

Steps:
1. Shape each color of clay into a long flat rectangle about 3 cm thick. Each color should be the same shape and size.
2. Layer the clay rectangles.
3. Cut the block into four equal sections.
4. Take one block and push in from opposite sides. Record the changes made to Block 1. The block changed shape. It is taller and thinner.
5. Take the second block and slowly pull the block from opposite sides. Record the changes made to Block 2. The block has been stretched. It's longer and thinner.
6. Take the last two blocks and rub one side of one block against a side of the other block. Record the changes made to Blocks 3 and 4. Some of the clay rubbed off onto the other block. The shape of the blocks didn't change much.
7. Cut through each block vertically. Sketch the layers in each block.

Block 1	Block 2	Block 3	Block 4
	Sketches will vary.		

ANALYZE AND CONCLUDE:
1. What type of stress was applied in Step 4? compressional stress
2. Record the type of stress applied in Step 5. tensional stress
3. What type of stress was applied in Step 6? shearing stress

Lab 3.1.3A The Effects of Stress

4. Describe how each type of stress deforms the block. Include the terms *volume*, *density*, and *shape*. Compressional stress decreases volume and increases density. The shape is taller and more narrow. Tensional stress increases volume and decreases density. The shape is much longer or it is wider and flatter. Shearing stress caused some of the clay to transfer to the other block, but didn't change the block shape much.

5. What properties of the clay affected its deformation? The clay is very ductile and flexible. It changes shape quite easily and doesn't break. It's not elastic so it stays deformed. It does not go back to its original shape.

6. What factors could change how the clay deformed? Possible answers: Temperature change would affect how it deforms—colder clay might be more brittle, warmer clay might be more ductile; baking the clay might make it so brittle it wouldn't deform or it might break if under extreme stress.

7. How does what you observed in the clay deformation relate to what happens when rock is under stress? Answers will vary but should include that rock under stress deforms just like the clay did when it was under stress in the lab.

Crust Movement

3.1.4 Mountains

Introduction
Read **Isaiah 40:4**, which states, "Every valley shall be raised up, every mountain and hill made low; the rough ground shall become level, the rugged places a plain." Other verses to consider include Psalm 36:6, 46:2, 90:2; Isaiah 54:10; Matthew 17:20; and 1 Corinthians 13:2. Discuss the fact that mountains, the most dominant feature on the landscape, give such a powerful impression that they are often used as imagery in the Bible.

Discussion
- Using **TM 3.1.4A Types of Mountains**, discuss how the different types of mountains and plateaus are formed. Compare old-earth and young-earth viewpoints on mountain formation.

- Discuss the difficulties of climbing a mountain. Consider the oxygen levels, temperatures, severe weather, and steep inclines. Ask students about any mountains they have visited and what makes the altitude more challenging than being near sea level. (**Answers will vary.**)

Activities
A. Complete *Try This: Mountain Model* with students. Challenge students to make the model to scale. For example, the Cascade Range is 1,100 km long and Mount Rainier is 4,392 m high. If the model is made with a scale of 1 cm = 500 m, the range would extend 2,200 cm, or 22 m, and Mount Rainier would be 8.78 cm high. For practical purposes, you may choose to make only one dimension to scale.

B. Have students complete **WS 3.1.4A Mountain Range Scavenger Hunt**.

C. Display a mountain from TM 3.1.4A for the class. Assign **WS 3.1.4B Mountain Perpsectives**.

D. Challenge students to research Edmund Hillary and Tensing Norgay. Have students make a presentation about the first successful ascent to the summit of Mount Everest using one of the following ideas:
   1. Pretend you are a talk show host and interview Edmund Hillary and Tensing Norgay.
   2. Write a news article about the historic climb.
   3. Create scrapbook pages that one of the men would have made about the climb.

Lesson Review
1. What is orogenesis? (**Orogenesis is the process of forming mountains.**)
2. How is a folded mountain formed? (**Scientists who explain the earth's processes using plate tectonics assert that a folded mountain is formed when the plates squeeze together and push upward.**)
3. How is a fault-block mountain formed? (**Old-earth scientists believe that a fault-block mountain is formed at divergent boundaries where the plates move away from each other. The rocks break apart. Uplifted rocks can reach hundreds of meters in the air.**)
4. What conditions are different in the formation of a folded mountain and a fault-block mountain? (**A fault-block mountain is usually formed where temperatures are cooler than where folded mountains are formed. A cooler rock would be more brittle than a warmer rock, so a cooler rock would be more apt to break than fold.**)
5. What are the two ways a plateau can form? (**Many scientists believe that plateaus can be formed by a gentle collision of plates or they can result from lava flow that builds up on the land's surface.**)
6. How is a plateau different from a mountain? (**A plateau has a large flat top instead of a peak. Many scientists believe that a plateau can be formed by a process similar to that of a folded mountain but the tectonic plates converge so slowly that the rock is not squeezed up into a full mountain.**)

OBJECTIVES
Students will be able to
- compare and contrast the formation of folded mountains with fault-block mountains.
- describe how a plateau is formed and how it is different from a mountain.

VOCABULARY
- **orogenesis** the process of mountain formation
- **plateau** a large area of flat-topped rock high above sea level

MATERIALS
- TM 3.1.4A Types of Mountains
- WS 3.1.4A Mountain Range Scavenger Hunt
- WS 3.1.4B Mountain Perspectives

PREPARATION
- Obtain materials for *Try This: Mountain Model*. Obtain clay or prepare clay by mixing 750 mL of flour, 750 mL of salt, and 300 mL of water per group.

TRY THIS
Mountain Model
- flour
- salt
- food coloring

© *Earth and Space Science*

3.1.4 Mountains

OBJECTIVES
- Compare and contrast the formation of folded mountains with fault-block mountains.
- Describe how a plateau is formed and how it is different from a mountain.

VOCABULARY
- **orogenesis** the process of mountain formation
- **plateau** a large area of flat-topped rock high above sea level

Perhaps you have heard someone reply to the question, "Why do you want to climb a mountain?" with the answer, "Because it's there!" That is the answer that British expeditionist George Mallory gave in 1924 when a journalist asked him why he wanted to climb Mount Everest, the world's highest mountain. It was not until 1953 that New Zealand native Edmund Hillary and his sherpa (local guide), Tenzing Norgay, reached the top of that mountain. They had climbed 8,848 m to the top of the world.

If you were to climb Mount Everest today, you would break that record. Why? Because Mount Everest is still growing. The latest measurement is 8,850 m. Since 1994, global positioning satellites placed on a plateau below the summit have measured the entire mountain range as moving 40–50 mm per year toward China. Scientists believe that as the Indian Plate moves steadily and slowly under the Eurasian Plate, Mount Everest gains height.

Typically, mountains are considered to be steadfast, unchanging parts of creation, but God-designed processes are constantly reshaping the earth with forces almost too great to understand. These forces result in changes over time and across the earth. Some of these changes produce mountains; some of these changes move mountains.

Orogenesis is the process of mountain formation. Old-earth and young-earth scientists disagree about the process. Old-earth

Mount Everest

BIBLE CONNECTION

Mountains 
From the beginning of time, mountains have been an important part of human existence. The first mention of mountains is found in Genesis 7 where the Flood waters are exceeding the height of the mountains. In Exodus, Moses was called up the Sinai wilderness mountain to meet with God and receive the Ten Commandments. Mountains are a symbol of power, majesty, and refuge. God's dwelling place is often referred to as "His holy mountain." King David wrote in Psalm 15:1, "Lord, who may dwell in Your sacred tent? Who may live on Your holy mountain?" Mountains appear to be steadfast and immovable. However, the God who created the mountains also moves and changes them. The author of Job writes, "He moves mountains without their knowing it" (Job 9:5) and "A mountain erodes and crumbles" (Job 14:18). As scientists continue to study the earth, its mountains, and their constantly changing formations, the evidence bears witness to the truth of God's Word.

scientists believe that orogenesis occurs where the earth's plates collide and disrupt the earth's crust. Young-earth scientists assert that the strata laid down by the Flood was then folded, eroded, and uplifted, which resulted in mountains. Different mountains formed in different ways, but mountain formation is always a complicated process. According to old-earth scientists, land features that begin as small faults and folds when plates collide can grow into magnificent mountain ranges. A mountain range is a series of connected mountains. Most mountain ranges are found along the edges of tectonic plates.

Folded mountains are formed when converging rock layers are squeezed together and pushed upward. Folded mountains have zigzag folds or wrinkles where the rock has bent. The folding is especially evident in contrasting layers of rock that have folded. Some of these folds are visible only under a microscope, but others cover many kilometers. The tops of these folded rock formations are worn away by wind, water, and moving rock. Eventually the weaker layers are eroded more deeply, forming valleys. The regions that do not wear away as quickly form mountain ranges. Many scientists believe this is how the Alps and the Appalachian Mountains were formed.

The Appalachian Mountains are folded mountains.

Most scientists who affirm plate tectonics agree that fault-block mountains are formed at divergent boundaries. In regions where the plates move away from each other, large mountains form when rock blocks break or fault. The uplifted blocks can reach hundreds of meters into the air. Faulting usually occurs at lower temperatures than folding. When sedimentary rocks are tilted up through the faulting process, they produce mountains with sharp, jagged peaks. The Teton Range in western Wyoming and the Sierra Nevada mountains in California and Nevada were formed this way. This process is also occurring today in the East African Rift Valley.

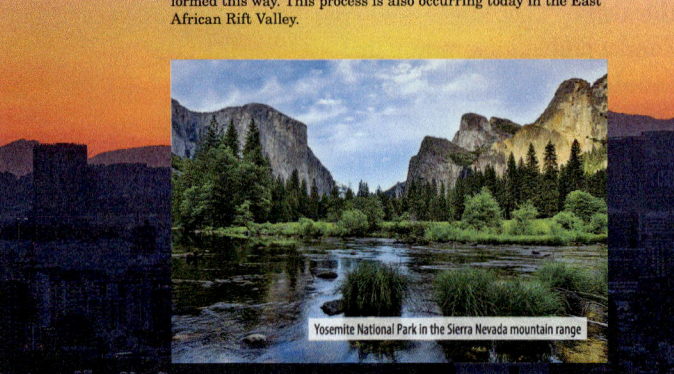

Yosemite National Park in the Sierra Nevada mountain range

Sometimes when volcanoes erupt and spew molten rock onto the earth's surface, mountains form. Mount Fuji in Japan is a volcanic mountain of this kind. Many scientists believe volcanic mountains occur along convergent plate boundaries.

Most mountain ranges are flanked by flat lands called **plateaus**, large areas of flat-topped rock high above sea level. They form on the earth's crust when hard, erosion-resistant rock caps other horizontal layers of rock. Plateaus are formed in much the same way as mountains. A largely accepted concept is that some plateaus are formed from colliding plates that are pushed together too gently to fold or crack into mountains; instead they remain flat. Other plateaus result from lava flows that build up on the land's surface.

Plateau

LESSON REVIEW
1. What is orogenesis?
2. How is a folded mountain formed?
3. How is a fault-block mountain formed?
4. What conditions are different in the formation of a fault-block mountain and a folded mountain?
5. What are the two ways a plateau can form?
6. How is a plateau different from a mountain?

TRY THIS

Mountain Model
Research the dimensions and features of an actual mountain range. Make a model of the range with modeling clay or with clay made from 750 mL of flour, 750 mL of salt, 300 mL of water, and food coloring.

Mount Fuji rises over Tokyo, Japan.

3.2.0 Earthquakes

Chapter 3.2 Summary

Several thousand earthquakes occur on Earth each year, but most are too small for people to notice. Some young-earth scientists believe earthquakes are a result of pressure at fault lines in the earth; others agree with the majority of scientists who believe most earthquakes are caused by the movement of tectonic plate boundaries. According to this theory, when plates catch on each other, potential energy builds up until the plates release, sending seismic waves through the earth and shaking the crust. Earthquakes can be categorized according to plate movement and depth. Typically, deeper earthquakes are not felt as strongly on the surface as shallow earthquakes because the seismic waves' energy has more time to dissipate. Faults can occur along plate boundaries and within plates, but both behave the same way. Faults are generally categorized into strike-slip and dip-slip faults, and dip-slip faults can be subdivided into normal, low-angle normal, reverse, and thrust. If a fault moves slowly and nearly constantly, it is said to be a creep fault. In order of detection, seismic waves include the body P and S waves as well as the surface Love and Rayleigh waves. Each wave has different properties. An earthquake's size can be measured by its felt effects, or intensity, and by its strength, or the amount of energy released, called its *magnitude*. An earthquake's effects partially depend on where the earthquake strikes and what the population's level of preparedness is. Earthquakes occurring in or near the ocean can create tsunamis. Areas with structures not designed to withstand earthquakes or with poor infrastructure experience greater earthquake-related damage. Predicting when earthquakes will strike has not been discovered, but it is possible to identify areas of greater risk.

Background

Lesson 3.2.1 – Earthquake Causes

Many scientists believe the lithosphere, which includes the crust and the upper mantle, has been moving, growing, and distorting since God set the world in motion, or a least since the Flood. According to plate tectonics, the majority of earthquakes originate with compressional or tensional stresses that build up at the margins of Earth's lithospheric plates. Most shallow earthquakes originate with the sudden release of stress along a fault, which causes opposing blocks of rock to move past each other. These movements send waves through and around Earth. Local earthquakes may also be caused by volcanic eruptions, rockfalls, landslides, and even explosions.

Lesson 3.2.2 – Earthquake Zones

Like the rest of creation, earthquakes can be explained by Earth's design. Earthquakes occur frequently all over the world, both along plate edges and along faults. In the 1920s, scientists observed that most earthquakes occurred along narrow zones. In 1954, French seismologist J.P. Rothé published a map that showed the concentration of earthquakes along these zones, which correspond to plate boundaries where the mantle's motion moves the plates. In fact, earthquakes help scientists define plate boundaries. Earthquakes at divergent boundaries are usually shallow, within 30 km of the surface. Earthquakes at transform boundaries also tend to occur at shallow depths. At convergent boundaries, earthquakes can vary in depth. Deep earthquakes erupt in subduction zones where crust that has been pushed deep into the mantle releases water and gas. Interior earthquakes can erupt along faults formed in response to stresses from plates moving against each other, but these account for less than 10% of all earthquakes.

Lesson 3.2.3 – Faults

Earthquakes occur over the area of a fault called *the rupture surface*; the whole fault plane does not usually slip at once. A rupture begins at a point on the fault plane called *the focus*, or *hypocenter*. The epicenter is the point on the surface directly above the focus. Movement along a fault plane may be vertical, horizontal, or oblique. One or both of the fault blocks may also rotate. Evidence of faults can be found both at the fault surface and underground at the fault plane. Surface faulting generally affects a long, narrow zone with a total area that is small compared with the total area affected by the ground shaking. The severity of the damage depends in part on the surface fault displacement. Although faults cannot always be seen, they can be obvious where they conspicuously offset

---

LOOKING AHEAD

- For **Lesson 3.2.1**, invite a local university professor who is knowledgeable about young-earth and old-earth beliefs to give a presentation on earthquakes.
- For **Lesson 3.2.3**, obtain two long ropes. Invite a surveyor to demonstrate how measurements of the earth are made and how these measurements can help identify fault movement.
- For **Lesson 3.2.4**, obtain a large spring to use when demonstrating P waves.
- For **Lesson 3.2.5**, invite an architect or engineer to present ways in which buildings can be built and retrofitted to make them more resistent to earthquakes. Invite a representative from a disaster relief organization to present ways in which students can best help people affected by earthquakes.

SUPPLEMENTAL MATERIALS

- BLM 3.2.3A Fault Tug-of-War

- TM 3.2.2A Collision Forces

- Lab 3.2.2A Earthquake Energy
- Lab 3.2.4A Seismograph

- WS 3.2.2A Earthquake Zone Map
- WS 3.2.2B Earthquake Zones
- WS 3.2.3A Analyze a Strike-Slip Fault
- WS 3.2.4A Measuring Earthquakes Graph
- WS 3.2.4B Measuring Earthquakes
- WS 3.2.5A Earthquake Prediction
- WS 3.2.5B Earthquake Prevention

- Chapter 3.2 Test

BLMs, TMs, and tests are available to download. See Understanding Purposeful Design Earth and Space Science at the front of this book for the web address.

WORLDVIEW

- The Bible uses earthquake imagery in two ways that should comfort Christians. First, earthquakes depict the intensity of God's judgment on His enemies. Christians around the world face persecution, and sometimes it may appear as though God's enemies cannot be stopped. Yet in response to believers' pleas for deliverance, Psalm 18:7 declares, "The earth trembled and quaked, and the foundations of the mountains shook; they trembled because He was angry." God's judgment against the wicked will be so fierce that even the most immoveable of Earth's features, the mountains, will shake with fear. Secondly, God's love for His people is just as great as His wrath against His enemies. God promises His people in Isaiah 54:10, "Though the mountains be shaken and the hills be removed, yet My unfailing love for you will not be shaken nor My covenant of peace be removed." In other words, even if the seemingly impossible happened and mountains shook and hills were uprooted, God's love for His people cannot fail. Psalm 97 ties these two concepts together. It describes the earth trembling at God's judgment on the wicked, yet it affirms the ultimate security that God's people have in Him.

otherwise continuous strata, such as streams, linear lakes, and scarps. Blind thrust faults do not reach the earth's surface; the ground above a blind thrust fault bends instead of breaking, thereby creating rolling hills. Active faults can be still for decades and then move meters in a matter of seconds. Other fault movement is more subtle. Faults may creep 1–10 cm per year. Fault movements are typically measured using GPS.

Lesson 3.2.4 – Seismic Waves

Seismic waves include body waves, which move through Earth's interior, and surface waves, which move on or just under Earth's surface and are detected after the body waves. Primary waves (P waves) are compressional body waves. Secondary waves (S waves) are transverse body waves that cause the earth to vibrate perpendicularly to the direction of their motion. P waves travel through solids, liquids, and gases; S waves travel only through solids. Love waves are surface waves that move parallel to the Earth's surface and perpendicular to the direction of wave propagation. Rayleigh waves are surface waves that move in an elliptical motion, producing both a vertical and horizontal component of motion in the direction of wave propagation.

A material's density and rigidity affect how fast body waves travel through it as well as the degree to which the waves will be reflected or refracted. Measuring body waves' speed and direction, or velocity, helps scientists determine the boundaries between the crust, mantle, and core. For example, the fact that S waves seem to vanish at depths below 2,900 km suggests the presence of liquid around Earth's outer core. Computer imaging using seismic waves creates three-dimensional views of Earth's interior, which helps scientists develop better theories regarding Earth's composition. This technology is called *seismic tomography*.

Lesson 3.2.5 – Prediction and Effects

Predicting and preventing earthquakes has long been an elusive goal of Earth science. Scientists cannot yet predict when earthquakes will occur or how powerful they might be. However, where earthquakes will occur is better understood. When dealing with faults that have been thoroughly studied, scientists can usually identify which portions are more likely to cause an earthquake, and they can estimate the potential earthquake's magnitude. Scientists use the size of the fault segment, the flexibility of the fault's rock, and the amount of accumulated stress in the fault to make these determinations. However, even when scientists know that strain has built up, they cannot tell whether a large earthquake or a series of small earthquakes will strike. Although earthquakes cannot be reliably predicted, the tsunamis that follow them often can be once the earthquake erupts. A rare example of a successful earthquake prediction occurred in 1975 when a 7.3 magnitude earthquake in Haicheng, China, was predicted. Clues of the impending earthquake arose from changes in land elevation, an unexpected drop in groundwater levels, reports of strange animal behavior, and several foreshocks. When the number of foreshocks increased, an evacuation warning was issued. The earthquake struck the day after the evacuation warning. Most earthquakes, however, do not have such obvious warnings. A year after the successful Haicheng prediction, a 7.6 magnitude earthquake struck Tangshan, China, without any warning.

New technology may save lives, however. In Japan, when an earthquake's P waves are detected, the Earthquake Early Warning system announces an imminent earthquake and its estimated magnitude before tremors begin. The system can warn people farther out from an earthquake's epicenter about a minute before tremors occur. There is little warning time possible the closer a person is to the epicenter. The warnings are sent out to TV, radio, businesses, schools, and even by mobile phone.

Although long-range forecasting of earthquakes is not currently possible, the technology to construct buildings and bridges that are more earthquake resistant and to identify faults does exist. Scientists are using technology to build smarter in earthquake-prone regions and to help people avoid the most dangerous areas.

Earthquakes 3.2.1 Earthquake Causes

Introduction
Have students cup their hands, hook their fingers together, and then pull hard. Ask them to share what happens. (**Answers will vary.**) Point out how far their hands fly apart once they overcome the pressure holding them together. Explain that earthquakes behave in a similar way; the more the rock is caught, the more pressure builds up. When the pressure is finally released, the tectonic rock moves as far as it can. Only air pressure is acting on students' hands once they are released, so their hands move a distance away. Ask students what stops tectonic plates. (**Answers will vary.**) Convey that friction between the plates slows them down until they reach new stopping points.

Discussion
- Invite students to share their knowledge of earthquakes, either earthquakes they have personally experienced or seen in the news. Guide students in creating a chart that lists similarities and differences among the various accounts.

- Present young-earth and old-earth viewpoints regarding the factors that cause earthquakes. Read **Hebrews 12:14** and then have students share what they would do in an effort to better understand a position contrary to their own beliefs.

- Lead a discussion regarding why the Bible sometimes uses earthquakes as a sign of God's judgment or displeasure. Consider passages such as Psalm 99:1–3; Isaiah 13:9–13, 29:5–6; Ezekiel 38:17–23; Joel 2:1–11; and Hebrews 12:25–29. (**Possible answer: God's judgment comes after much patience and mercy. It is in response to sin and wickedness and those who act in a way contrary to God and His commands.**)

- Ask the following questions:
  1. In considering the theory of plate tectonics, how would earthquakes be affected if Earth's mantle cooled? (**Convection currents would no longer be produced, so Earth's plates would stop moving. Earthquakes would cease.**)
  2. How would earthquakes be affected if the boundaries between the plates were straight and smooth? (**Possible answers: Since the plates wouldn't catch on each other, they would move slowly and steadily instead of moving in separate jolts; without jagged edges to lock the plates together, the plates would move much more quickly past each other without anything to stop them and cause large earthquakes.**)

Activities
A. Complete *Try This: Causing Shocks* with students and discuss their answers.

B. Introduce the invited professor to give a presentation on earthquakes. Encourage students to ask questions about young-earth and old-earth viewpoints.

C. Have students collect and then describe eyewitness accounts of an earthquake, such as published accounts, diaries, and letters from friends or relatives. Presentations should include the earthquake, the date, the local time, the location, the magnitude, a description of the damage it caused, and an account of how the area recovered from it. Encourage students to include pictures of the area before and after the earthquake.

D. Guide students to postulate reasons for specific earthquakes, their magnitudes, and the resulting damage.

E. Direct students to design a device that would allow scientists to study plate boundaries and movements or "springs of the great deep" (Genesis 7:11). Encourage realistic applications of ideas.

OBJECTIVES
Students will be able to
- summarize the connection between tectonic plate movement and earthquakes.
- infer why earthquakes vary in magnitude.
- indicate the causes of earthquakes.

VOCABULARY
- **magnitude** the strength of an earthquake

MATERIALS
- No additional materials are needed.

PREPARATION
- Obtain materials for *Try This: Causing Shocks*.
- Invite a local university professor who is knowledgeable about young-earth and old-earth beliefs to give a presentation on earthquakes. (*B*)

TRY THIS
Causing Shocks
- paper cups, 1 per group
- stools or chairs, 1 per group

NOTES

F. Assign students to research the effects of earthquakes caused by human activities. Possible sources include fracking, bomb testing, and mining. Then divide the class into groups to discuss the pros and cons (economically, politically, and fiscally) of the activities that led to the earthquakes.

G. Have students make a map showing the general location of both ancient and modern cultures from around the world that have legends, myths, or prominent beliefs about earthquakes.

H. Challenge students to create a shoebox diorama depicting the underground causes of an earthquake and to write a description of each cause depicted. Allow time for students to present their dioramas and descriptions to the class.

Lesson Review

1. What is the connection between tectonic plate movement and earthquakes? (**According to plate tectonic theory, the earth's plates are constantly moving against each other. Since the plates don't have smooth edges, they get caught on each other, which builds up stress on the plates. The stress continues to build until the plates move apart, which triggers an earthquake.**)
2. When do Earth's plates cause earthquakes? (**when the pressure is released that built up after they caught on each other**)
3. What term is used to describe the strength of an earthquake? (**magnitude**)
4. What determines how large an earthquake will be? (**the amount of pressure that is released**)
5. Other than shifting plates, what else can cause earthquakes? (**erupting volcanoes, explosions caused by humans, collapsing buildings, and any other force that is strong enough to produce seismic waves**)

3.2.1 *Earthquake Causes*

OBJECTIVES
- Summarize the connection between tectonic plate movement and earthquakes.
- Infer why earthquakes vary in magnitude.
- Indicate the causes of earthquakes.

VOCABULARY
- **magnitude** the strength of an earthquake

When people think of earthquakes, they typically think of feeling the earth moving. However, according to the plate tectonic theory, the earth's crust is always moving, even though people cannot feel the motion. The thousands of earthquakes that happen each year can be explained from an old-earth point of view as occurring because of Earth's plates moving against each other. Young-earth scientists believe that today's earthquakes are much smaller versions of what occurred when the "springs of the great deep broke forth" as stated in Genesis 7:11.

Many scientists believe earthquakes are caused by the movement of tectonic plates, which are moved at rates varying from 1–10 cm per year by convection currents in the mantle. The plates remain in constant contact with each other as they move away from each other, toward each other, and past each other. Because plates are not perfectly smooth, they tend to catch on each other. The sudden, jarring movement of the plates as they move produces seismic waves, which are waves of energy that travel through the earth. Seismic waves move the ground when they reach the crust's surface, and this motion is called *an earthquake*.

Thousands of earthquakes occur every year, yet the **magnitude**, or strength of an earthquake, varies greatly. Many earthquakes are so small that they do not produce any damage, and most cannot even be felt. However, sometimes the plates get caught for a long time and considerable pressure is built up. The release of this great stress produces great energy as the plates jolt past each other. Depending on the amount of stress released, the

Earthquake damage in Nepal

HISTORY

A Terrific Shock
American author, Samuel Clemens (better known as Mark Twain), experienced his first earthquake while walking along a California street in 1865. He wrote about it in his book, *Roughing It*: "Before I could turn and seek the door, there came a terrific shock; the ground seemed to roll under me in waves, interrupted by a violent joggling up and down, and there was a heavy grinding noise as of brick houses rubbing together. I fell up against the frame house and hurt my elbow. I knew what it was, now.... A third and still severer shock came, and as I reeled about on the pavement trying to keep my footing, I saw a sight! The entire front of a tall four-story brick building in Third Street sprung outward like a door and fell sprawling across the street, raising a dust like a great volume of smoke!"

plates may move between a few millimeters to several meters before their friction temporarily stops them. This movement and the large seismic waves produced can cause significant damage.

The causes of earthquakes are not limited to the interactions between plate boundaries. The stress of moving plates can create faults, or breaks, in the crust. These faults move similarly to the plates and create their own earthquakes. Earthquakes can be caused by any force that is strong enough to produce seismic waves including erupting volcanoes, explosions caused by humans, and collapsing buildings.

LESSON REVIEW
1. What is the connection between tectonic plate movement and earthquakes?
2. When do Earth's plates cause earthquakes?
3. What term is used to describe the strength of an earthquake?
4. What determines how large an earthquake will be?
5. Other than shifting plates, what else can cause earthquakes?

Part of the San Andreas fault zone in California, United States

TRY THIS

Causing Shocks
Place an empty paper cup on the ground. Then jump up and down one time next to the cup. Try jumping at different distances from the cup. Jump off a stool next to the cup, and hop next to the cup. Observe what happens to the cup after each jump or hop. What explains the difference in the way the cup behaves?

Earthquakes

3.2.2 Earthquake Zones

Introduction
Display **TM 3.2.2A Collision Forces**. Direct students' attention to the damaged and undamaged parts of each vehicle. Ask students why only part of each vehicle is damaged. (**Possible answers: The damaged part is the only part that was hit; the part that was hit absorbed all the force.**) How is this similar to the collision of tectonic plates? (**Possible answer: The parts where the plates hit each other are most affected.**) How could the damage of the cars fit a young-earth viewpoint of the reason for earthquakes? (**Answers will vary but should include that the damage comes from pressure and strain at fault lines.**)

Discussion
- Ask students the following questions:
  1. What is a convergent boundary? (**a place where two plates are moving toward each other**)
  2. What is a divergent boundary? (**a place where two plates are moving apart from each other**)
  3. Why do earthquakes happen at plate boundaries? (**most movement happens at boundaries**)
  4. Where do the highest number of earthquakes take place? (**along the Pacific Plate**)
  5. What happens when a plate is subducted? (**It goes underneath another plate.**)
  6. How does an earthquake's force on the surface compare to its source at its focus? (**The force on the surface is less than the force at the focus.**)

Activities

Lab 3.2.2A Earthquake Energy
- plastic drop cloths, 1 per pair
- desks, 1 per pair
- metersticks, 1 per pair
- chalk or masking tape, 1 per pair
- disposable cups, 1 per pair
- sand, 50 g per pair
- rubber bands, 1 per pair
- tape measures, 1 per pair

For each pair, place a desk in an open area with a plastic drop cloth underneath it to help with sand cleanup. Inform students that on their graphs in *Analyze and Conclude Question 1*, they will need to write their own distance labels for how far the cup moved. These will vary among pairs of students because of the variance among the distances the rubber bands will be stretched.

Option 1: Direct students to stay the same distance from the desk but to vary the length of their stretched rubber band. Have students explain the correlation of the experiment to earthquake magnitude, location, and surface effects.

Option 2: Challenge students to produce similar cup movements from two different distances. Guide students to explain the correlation of the experiment to earthquake magnitude, location, and surface effects.

A. Direct students to use **WS 3.2.2A Earthquake Zone Map** to complete **WS 3.2.2B Earthquake Zones**.

B. Assign students to research the magnitude and intensity of earthquakes at their epicenters. Students may research earthquakes near where they live or earthquakes from around the world. Have students make a scatter plot showing magnitude along the horizontal axis and intensity along the vertical axis. Students should create a key that identifies each earthquake's epicenter and depth below the surface. Have students draw conclusions about the relationships among magnitude, depth, and epicenter intensity. Consider challenging students to find intensity measurements in places farther from the epicenter, compare those measurements with the epicenter intensity, and draw conclusions about energy expenditure, basic ground composition, and potential building risks for the general area.

OBJECTIVES
Students will be able to
- identify where earthquakes are most likely to occur.
- describe the typical relationship between an earthquake's zone and its magnitude.
- explain how the depth of an earthquake's focus affects how it is experienced on the surface.

MATERIALS
- TM 3.2.2A Collision Forces
- WS 3.2.2A Earthquake Zone Map
- WS 3.2.2B Earthquake Zones

PREPARATION
- Obtain materials for *Lab 3.2.2A Earthquake Energy*.

NOTES

C. Have students research a particular location along a plate boundary to discover why people live there. Assign students to create an advertising campaign to promote the location. Students should consider natural resources, transportation, historical factors, the frequency of earthquakes, and how people have prepared for earthquakes; include tsunamis if the location is on a coast or island.

D. Assign students to research a noted scientist who contributed to understanding seismic activity. Have students complete the assignment in pairs and present their findings in a mock interview with one student playing the scientist and the other student playing the role of the interviewer.

Lesson Review

1. Where are earthquakes most likely to occur? (**along plate boundaries**)
2. Identify the typical magnitude of earthquakes that occur along each of the three types of plate boundaries. (*divergent*: **low magnitude**; *convergent*: **low to high magnitude**; *transform*: **moderate to high magnitude**)
3. Identify the type of boundary along which earthquakes of each depth category typically erupt. (*shallow-focus*: **divergent and transform boundaries**; *intermediate-focus*: **convergent boundaries**; *deep-focus*: **convergent boundaries**)
4. Using your answers to Questions 2–3, identify the likely magnitude of earthquakes occurring in each depth zone. (*shallow-focus*: **low magnitude**; *intermediate-focus*: **low to high magnitude**; *deep-focus*: **low to high magnitude but usually high magnitude**)
5. Explain how the depth of an earthquake's focus affects how it is experienced on the surface. (**Shallow-focus earthquakes tend to do more damage than deep-focus earthquakes because their energy does not have as much space to diminish before it affects the crust's surface.**)

3.2.2 Earthquake Zones

OBJECTIVES
- Identify where earthquakes are most likely to occur.
- Describe the typical relationship between an earthquake's zone and its magnitude.
- Explain how the depth of an earthquake's focus affects how it is experienced on the surface.

Earthquakes can happen anywhere in the world. However, according to many scientists who support the tectonic plate theory, most earthquakes are caused by tectonic plate movement at the boundaries. When earthquakes are mapped, they appear to form a dot-to-dot outline of the plates. Earthquakes can be divided into zones according to the movement that causes them and their depth of origin.

The type of earthquake that strikes in a certain area depends in part on the relationship of the plates in that area. Plates move across the earth in several different directions. Divergent boundaries occur where tectonic plates are moving away from each other. Earthquake maps show numerous earthquakes along the Mid-Atlantic Ridge, which is the result of the North American and Eurasian Plates slowly separating. The ridge is often studied where it surfaces in Iceland, which straddles the two plates. Earthquakes along divergent boundaries generally have low magnitudes.

Earthquakes range from low to high magnitude at convergent boundaries where the plates are moving toward each other. Along these boundaries, the plates with greater density sink under the plates with less density. If the plates have similar densities, then they smash together to create folds and faults. Regardless of what happens to the plates, the convergent movement produces earthquakes. One such boundary is marked by the Southern Alps in New Zealand. Here, the Australian Plate is moving down and under the Pacific Plate.

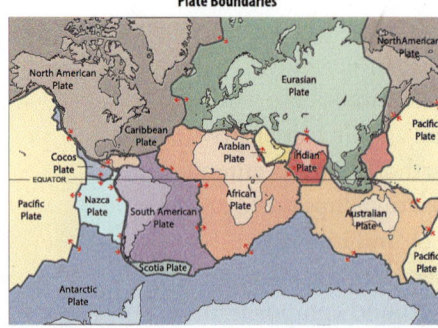

Transform boundaries, where plates are sliding past each other horizontally, often join sections of other types of boundaries. The earthquakes at these places can be highly destructive, partly because their magnitude can be moderate to high. A good example of a transform boundary is in New Zealand, which straddles the boundary between the Australian and Pacific Plates. The Pacific Plate is being forced under the Australian Plate near New Zealand's North Island, but the opposite is happening under South Island. The two convergent boundaries are connected by a transform boundary where the plates slide parallel to each other.

Earthquakes can also be zoned according to the depths at which they occur. An earthquake that occurs within 70 km of the earth's surface is considered a shallow-focus earthquake. Shallow-focus earthquakes frequently occur along divergent and transform

The Mid-Atlantic Ridge passes through Thingvellir National Park, Iceland.

BIBLE CONNECTION
Prison Break
Read Acts 16:16–40, which tells about the earthquake that freed Paul from prison. This earthquake occurred in Philippi, near the boundary between the African and Eurasian Plates. This region is one of the most seismically active areas in the Mediterranean. Statistically, most earthquakes are of low magnitude. Such quakes rarely generate much damage and cause few, if any, fatalities. The earthquake that freed Paul fits this description. It was strong enough to shake the prison's foundation and to skew the doorframes but not of a magnitude strong enough to destroy the prison.

Plate Boundaries

BIOGRAPHY

Inge Lehmann
Inge Lehmann (1888–1993) was a seismologist who determined that the center of Earth is a solid core around which a liquid mantle floats.

Lehmann grew up in Denmark. When she was a teenager, she and her family felt an earthquake tremor. This experience was her first introduction to a field of study in which she became notable.

After many years at the University of Copenhagen in Denmark and the University of Cambridge in England, she earned a master's degree in mathematics. She later studied in Germany, France, Belgium, and the Netherlands, earning a second master's degree. Her second degree was in a branch of mathematics called *geodesy*. This field applies mathematics to the study of Earth's exact shape.

Her work as a seismologist began in 1925, when she helped the director of the Royal Danish Geodesic Institute set up a network of seismic stations in Denmark and Greenland. These stations recorded any Earth tremors. Lehmann eventually became the head of the institute's seismology department, analyzing and recording seismograms coming from those stations.

Her study of P waves helped her determine that Earth's core has both inner and outer layers. Until that point, seismologists had thought that the unusual wave patterns they saw were a result of diffraction. Lehmann theorized that the waves' patterns made more sense if there were a solid inner core within the liquid outer core. Later in her career, she became an expert on Earth's mantle.

Inge Lehmann was blessed with a long, productive career. In 1971 at the age of 83, she was given the American Geophysical Union's highest honor, the William Bowie Medal. She also received honorary doctorates from Columbia University and the University of Copenhagen. Lehmann published her last professional paper when she was 99 years old.

CAREER

Seismologist
A seismologist is a person who studies how energy waves from earthquakes travel through the earth and impact the surface. Seismologists examine these waves for a variety of purposes. They use waves to study large-scale issues including the composition of Earth's layers and the elasticity of different substances located below Earth's surface. On a smaller scale, seismologists use earthquake waves to locate possible oil deposits. Seismologists also examine seismic data to help governments determine where nuclear devices are being tested.

People interested in becoming seismologists should gain a strong background in mathematics, physics, and basic geology. They should also be skilled in working with computers. A bachelor of science degree may be sufficient to work in some oil industry careers related to seismology, but a master of science degree in geophysics is usually preferred. Typically, a doctorate degree is required to work as a researcher in seismology.

boundaries, such as in the mid-Atlantic region and in the middle of continents. Shallow-focus earthquakes are extremely common, but most of them have low magnitudes. Intermediate-focus earthquakes occur 70–300 km below the surface, and anything deeper than that is considered to be a deep-focus earthquake. Most intermediate-focus and all deep-focus earthquakes occur around the Pacific Plate along convergent boundaries. Deep-focus earthquakes typically have the greatest magnitudes.

Amatrice, Italy, after the August 24, 2016, earthquake

Great magnitude does not always equal great destruction. Deep-focus earthquakes tend to exert more initial force, which means their magnitude rating is higher compared to shallow-focus earthquakes. However, shallow-focus earthquakes tend to do more damage because their energy does not have as much distance to diminish before it affects the crust's surface. For example, when two unrelated earthquakes erupted on August 24, 2016, the depth of one earthquake greatly reduced its destructive power. One of the earthquakes erupted about 4–10 km beneath Norcia, Italy, with a magnitude of about 6.0. The same day, another earthquake erupted about 84 km below Chauk, Myanmar, with a magnitude of 6.8. Both earthquakes were damaging, but the shallower Italian earthquake caused much more damage than the deeper one in Myanmar, even though the Italian earthquake had a lower magnitude.

Epicenter of Myanmar August 24, 2016, earthquake

LESSON REVIEW
1. Where are earthquakes most likely to occur?
2. Identify the typical magnitude of earthquakes that occur along each of the three types of plate boundaries.
3. Identify the type of boundary along which earthquakes of each depth category typically erupt.
4. Using your answers to Questions 2–3, identify the likely magnitude of earthquakes occurring in each depth zone.
5. Explain how the depth of an earthquake's focus affects how it is experienced on the surface.

FYI

Differing Views
Scientists are trained to be independent thinkers even within a group. Although some young-earth scientists agree with the theory of plate tectonics and believe that it explains the formation of earthquakes, others do not. Both groups of scientists carefully examine the work of past researchers and their peers and construct intricate models to study the earth, but they arrive at different conclusions. For example, some young-earth researchers who believe plates are real structures and that they continue to move have pointed to earthquakes as proof that subduction zones exist in the lithosphere where rock is deformed and becomes ductile. Other scientists, who also believe the earth is young, refute this idea by saying that the models of plate movement are too simple, that measurements of many large features on the ocean floor do not indicate plate movement, and that the evidence for subduction zones is flawed.

Name: _____ Date: _____

Lab 3.2.2A Earthquake Energy

QUESTION: What happens to energy as it is spent over a distance?

HYPOTHESIS: Answers will vary.

EXPERIMENT:

You will need:		
• desk	• chalk or masking tape	• rubber band
• meterstick	• disposable cup	• tape measure
	• 50 g sand	

Steps:
1. Measure a 2 m distance from the edge of your team's desk. Mark the distance with chalk or masking tape.
2. Fill a large disposable cup with 50 g of sand.
3. Set the cup on the edge of the desk closest to the chalk line.
4. One team partner should place a rubber band over his finger and stretch it back. The other partner should measure and record the length of the stretched rubber band and make sure that the first partner's fingertip is aligned with the edge of the desk and facing the cup.
5. Keeping the rubber band in place, aim at the cup and release the rubber band. Measure and record the distance the cup moved back from the edge of the table. In the same space, note whether the cup fell over.
6. Replace the cup and any sand that spilled from it.
7. Increase the distance from the desk by 1 m.
8. Repeat Steps 1–6, making sure to stretch the rubber band the same length as in the first trial.
9. Repeat the experiment two more times, each time increasing the distance from the desk by 1 m and always stretching the rubber band to the same length as in the first trial.

Trial	Length of Stretched Rubber Band	Distance from Desk	Distance Cup Moved
1.		2 m	
2.	Answers will vary.	3 m	Answers will vary.
3.		4 m	
4.		5 m	

Lab 3.2.2A Earthquake Energy

ANALYZE AND CONCLUDE:
1. Create a line graph of the results. On the x-axis plot the distance from the desk, and on the y-axis plot the distance the cup moved.

Answers will vary but should include a decline in distance moved as the distance from the desk increased.

(y-axis: Distance Cup Moved (cm); x-axis: Distance from Desk (m), 0 to 6)

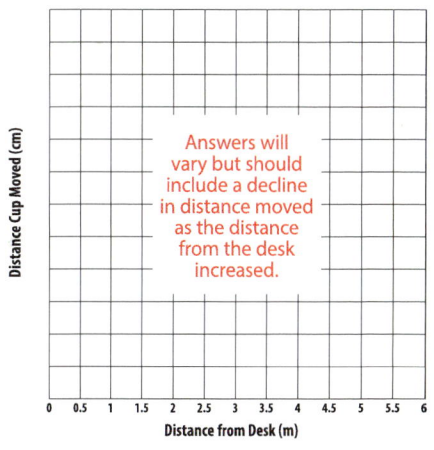

2. What is the relationship between the distance the rubber band traveled and the distance the cup moved? The farther the rubber band traveled, the less the cup moved.

3. What accounts for the relationship between the distance the rubber band traveled and the distance the cup moved? The rubber band expended some of its force as it flew because of friction between it and air molecules.

4. How does this experiment correspond to an earthquake's depth and the amount of its force that is felt on the earth's surface? The deeper an earthquake's focus is located, the less force its waves will have on the surface.

3.2.3 Faults

Earthquakes

OBJECTIVES

Students will be able to
- describe the three basic fault movements.
- distinguish between the fault movement of an earthquake and fault creep.

VOCABULARY

- **normal fault** a fault in which the hanging wall slides down the footwall
- **reverse fault** a fault in which the hanging wall climbs up the footwall

MATERIALS

- 2 ropes (*A*)
- Chairs or desks (*D*)
- BLM 3.2.3A Fault Tug-of-War
- WS 3.2.3A Analyze a Strike-Slip Fault

PREPARATION

- Invite a surveyor to demonstrate how measurements of the earth are made and to explain how these measurements can help identify fault movement. (*C*)

Introduction

Have students snap their fingers. Then have them describe each step of the process. (**We pushed our fingers together and then sideways. While our fingertips were being pressed together, friction kept them from moving. When they pushed sideways hard enough to overcome this friction, our fingers moved suddenly. Energy in the form of sound waves was released.**) Inform students that this process illustrates how an earthquake happens. Scientists, both young-earth and old-earth, who accept the theory of plate tectonics believe that Earth's crust pushes the plate edges together. The friction across the surface of the fault holds the rocks together so they do not immediately slip. Eventually, however, enough stress builds up and causes the rocks to slip suddenly, releasing energy in waves that travel through the earth to cause the shaking that is felt during an earthquake.

Discussion

- Direct students to imagine that they are facing a fault before an earthquake occurs. Pose the following scenarios in which the fault noticeably moves and have students identify where they would end up in relation to their original location.
  1. Students are standing next to a strike-slip fault. (**We would move either to the left or right of our original position.**)
  2. Students are standing on the hanging wall of a normal fault. (**We would move down from our original position.**)
  3. Students are standing on the footwall of a reverse fault. (**We would not move, but it would look like we had gone down because the hanging wall would have moved up.**)

- Guide a discussion regarding whether students would rather live near a normal fault or in an area that experiences fault creep. (**Answers will vary.**) Some students may prefer to live near fault creep and forego the risk of a damaging earthquake; others may prefer to risk an earthquake in order to avoid the constant movement of land.

- Lead students in a discussion of the following scenario: You are a city council member for a town that is near a known fault. Identify challenges your town might face in relation to the fault. Propose possible solutions to those problems. Identify possible limitations that might affect the proposed solutions.

Activities

A. Use **BLM 3.2.3A Fault Tug-of-War** to illustrate how fault movements cause earthquakes.

B. Direct students to complete **WS 3.2.3A Analyze a Strike-Slip Fault**.

C. Introduce the surveyor so he or she can demonstrate how measurements of the earth are made and how these measurements can help identify fault movement.

D. Arrange chairs or desks in two parallel lines so each chair or desk in the line is touching the one next to it. Explain that the desks represent the sides of a strike-slip fault. Direct students to solve each of the following scenarios:
  1. How long will it take until one chair in each row does not have any part of another chair opposite it if both sides move 5 cm per year in different directions? (**Answers will vary on the basis of the dimensions of the chairs.**)
  2. How long will it take until one chair in each row does not have any part of another chair opposite it if one side stays still but the other side moves 5 cm per year? (**Answers will vary.**)
  3. How long will it take until no chairs have any part of another chair opposite them if both sides move 5 cm per year in different directions? (**Answers will vary.**)
  4. How long will it take until no chairs have any part of another chair opposite them if one side stays still but the other side moves 5 cm per year? (**Answers will vary.**)

5. How far offset would each end of the rows be after 20 years if both sides of the fault maintained a steady movement of 5 cm per year for 18 years but for 2 years the rate of movement doubled? (**Each end will be offset by 220 cm.**)
6. How far offset would each end of the rows be after 20 years if one side of the fault did not move but the other side maintained a steady movement of 5 cm per year for 18 years but for 2 years the rate of movement doubled? (**Each end will be offset by 110 cm.**)

E. Have students research a historic earthquake and prepare a first-person narrative as if they had lived through it. Students should consider how people would have described the intensity of the earthquake, the likely side effects of the earthquake, how rescue and rebuilding efforts would have been conducted, the quality of medical care available and its implications for survival of the wounded, whether people would be likely to move away or rebuild in the same location, the reasons people would have given for the earthquake, and their response in light of those reasons.

Lesson Review
1. What term describes a fault in which the rocks move past each other sideways? (**strike-slip**)
2. In which direction does the hanging wall slide in a normal fault? (**down**)
3. In what type of fault does the hanging wall climb up the footwall? (**reverse fault**)
4. What is fault creep? (**Fault creep is the slow, almost continuous movement of rock along a fault.**)
5. How is fault creep different from fault movement in an earthquake? (**Fault creep is slow and steady; the movement of a fault that causes an earthquake is sudden and jarring.**)
6. What kind of fault movement does the Hayward Fault demonstrate? (**fault creep**)

NOTES

3.2.3 Faults

OBJECTIVES
- Describe the three basic fault movements.
- Distinguish between the fault movement of an earthquake and fault creep.

VOCABULARY
- **normal fault** a fault in which the hanging wall slides down the footwall
- **reverse fault** a fault in which the hanging wall climbs up the footwall

You have already learned that a break in the earth's crust along which rocks move is called *a fault*. Your textbook slides easily over your desk, and skates glide effortlessly over ice, but Earth's plates do not move easily past each other. They scrape against each other like giant pieces of sandpaper. The movement can be sideways (horizontal) or up and down (vertical). A fault in which two fault blocks move past each other horizontally is called *a strike-slip fault*. The rock on either side of a strike-slip fault moves parallel to the line formed by the fault, known as *the strike*. As the fault moves, rocks that were once next to each other can be separated by many kilometers. The San Andreas Fault in California is a strike-slip fault that is part of a larger fault zone that extends over 1,200 km. This fault is responsible for the terrible 1906 earthquake in San Francisco, California.

A fault in which two fault blocks move past each other vertically is called *a dip-slip fault*. Dip-slip faults can be of several types. In each type, the rock above the fault surface is called *the hanging wall*, and the rock below the fault surface is called *the footwall*.

A **normal fault** is a fault in which the hanging wall slides down the footwall. Normal faults result from tensional forces that pull rocks apart, and they have steep dips, or slopes. Normal faults can create ridges, called *horsts*, and valleys, called *grabens*. In

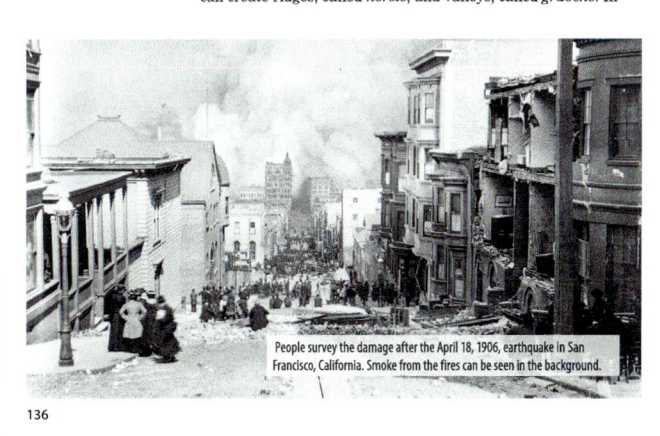

People survey the damage after the April 18, 1906, earthquake in San Francisco, California. Smoke from the fires can be seen in the background.

FYI

Fault Terms
The terms used to describe faults can be traced back to English coal mines. Every so often, the veins coal miners followed would be disrupted by a fault. The miners would follow the fault up or down to find the vein again. If the fault continued in the direction the miners had been traveling originally, it was called *a normal fault*. If the fault backtracked from the miners' original direction, it was called *a reverse fault*. The terms *hanging wall* and *footwall* also come from mining. The hanging wall was the rock hanging above the miners' heads, and the footwall was the rock on which the miners walked.

California, the Sierra Nevada Mountains are a horst and the adjacent Owens Valley is a graben. A normal fault with a very gentle dip is called *a low-angle normal fault*. A good example is the Mai'iu Fault in Papua New Guinea.

A **reverse fault** is a fault in which the hanging wall climbs up the footwall. Reverse faults result from compressional forces that push rocks together. The Sierra Madre fault zone in southern California demonstrates reverse faults. In this system, the San Gabriel Mountains are thought to have been formed as the hanging walls of reverse faults were pushed up their footwalls,

FYI

Fault Facts
- One of the longest faults in the world is the Sunda Megathrust in Southeast Asia at 5,500 km long. The movement of this fault in 2004 caused a tsunami that killed 230,000 people.
- The Main Uralian Fault through the Ural Mountains reaches 15 km below the crust surface.
- The Alpine Fault in New Zealand moves very fast compared to other faults in the world.
- Istanbul, Tokyo, Seattle, San Francisco, and Los Angeles are all built on active faults.
- In October 2016, a new fault was discovered that runs parallel to the San Andreas Fault. Researchers theorize that this new fault may absorb stress from the San Andreas, which could explain why that fault has not had a major earthquake in 100 years.

Owens Valley and Sierra Nevada Mountains in California

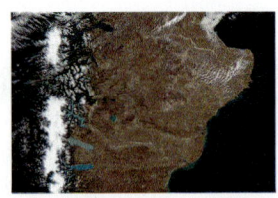

The Andes Mountains mark part of the boundary between the Nazca and South American Plates.

which are the San Fernando and San Gabriel Valleys.

A reverse fault with a slope measuring 45° or less is called *a thrust fault*. Scientists who favor the tectonic plate theory believe the subduction of the Nazca Plate under the South American Plate creates thrust faults above where the subduction occurs. These faults contribute to the height of the Andes Mountains. This process causes earthquakes that vary in both depth and intensity.

In general, the movement of faults jars the earth a little bit at a time. However, in some cases the rock on either side of the fault line moves fairly smoothly. This slow, almost continuous movement of rock along a fault is called *fault creep*. These faults experience fewer noticeable earthquakes than faults in which rock faces catch and temporarily hold each other.

One notable example of fault creep is the Hayward Fault, a strike-slip fault that runs from San Jose, California, up to San Pablo Bay. The effects are especially visible in the town of Hayward, where various features are misaligned because the town has straddled the fault for several decades. Until it was repaired in 2016, a section of curb at the corner of Rose and Prospect Streets served as a measurable example of fault creep. Two parts of the curb slowly moved out of alignment as the fault moved. Photographs taken in the 1970s showed the curb sections overlapping by approximately 2 cm. By 2004, a photograph of the same curb showed that the two sections were separated by at least 2 cm. Such pictures indicate the location and movement of the fault.

A section of offset sidewalk shows the effects of fault creep.

LESSON REVIEW
1. What term describes a fault in which the rocks move past each other sideways?
2. In which direction does the hanging wall slide in a normal fault?
3. In which type of fault does the hanging wall climb up the footwall?
4. What is fault creep?
5. How is fault creep different from fault movement in an earthquake?
6. What kind of fault movement does the Hayward Fault demonstrate?

Earthquakes

3.2.4 Seismic Waves

Introduction
Drop a small rock into a bucket of water. Ask students what they saw. (**Waves moved out in concentric circles from where the rock hit the water.**) Have students speculate whether the water under the surface was affected. Explain that waves are the result of energy moving through a substance.

Place a shallow tray near one edge of a desk and spread a thin layer of sand over it. Use a rubber mallet to gently hit the side of the desk closest to the tray. Have students record what happened to the sand. Smooth out the sand and repeat the experiment with the tray in the center of the desk and on the far side of the desk from the mallet. With each experiment, have students note how the sand is affected. Ask students what relationship they observe between the sand's movement and the proximity of the tray to the mallet. (**The sand moved most when the tray was closest to the mallet.**) Explain that P waves and S waves are energy from earthquakes that affect the ground like the energy from the mallet affected the sand.

Demonstrate P waves by pulling and releasing a spring. Demonstrate S waves by snapping a rope tied to a doorknob or desk up and down and then side to side. Explain that the movement of an S wave is perpendicular to the direction of its force; it does not matter if the movement is vertical or horizontal.

Discussion
- Complete *Try This: Wave Refraction* with students. Ask students whether a seismograph could detect seismic waves that originated from a point on the exact opposite side of Earth from it. (**No. Because Earth is not made from a single substance, waves will bend instead of traveling in a continuous straight line to the opposite side of Earth.**)

- Guide a discussion that considers the purpose of intensity and magnitude scales. Ask students whether this information is important for places where no people live. (**Possible answers: This information might be important to developers who are considering building new communities or to geologists who are interested in learning how Earth is changing or what it may have looked like a long time ago.**)

- Ask the following questions:
  1. What are the two categories of waves? (**body waves and surface waves**)
  2. Which waves are body waves? (**P waves and S waves**)
  3. Do body waves or surface waves cause more damage during an earthquake? (**surface waves**)
  4. Which waves are surface waves? (**Love waves and Rayleigh waves**)

Activities

Lab 3.2.4A Seismograph
- string
- lead weights, 1 per pair
- ring stands, 1 per pair
- metersticks, 1 per pair
- graph paper, 2 pieces per pair
- tape
- felt-tip pens, 1 per pair

Divide the class into pairs. If students' seismographs do not record much movement sitting on a table, direct students to move the seismographs to the floor.

For the best results, have only one student jump at a time. Consider having all pairs move paper through their seismographs as any student is jumping and record the student's name and distance to their seismographs on the appropriate graph. Have pairs compare their results.

OBJECTIVES
Students will be able to
- demonstrate seismic waves and how they are recorded.
- explain the difference between intensity and magnitude.
- compare different scales used to measure intensity and magnitude.

VOCABULARY
- **epicenter** the point on the earth's surface directly above an earthquake's focus
- **focus** the point inside the earth where an earthquake begins
- **Love wave** a fast surface wave that moves in a side-to-side pattern as it travels forward
- **Rayleigh wave** a slower surface wave that moves in an elliptical pattern as it travels forward
- **seismograph** an instrument that measures and records seismic waves

MATERIALS
- Rock, bucket, tray, sand, rubber mallet, large spring, rope (*Introduction*)
- Pan, flashlight (*C*)
- Large sheets of paper, tape (*D*)
- WS 3.2.4A Measuring Earthquakes Graph
- WS 3.2.4B Measuring Earthquakes

PREPARATION
- Obtain materials for *Lab 3.2.4A Seismograph*.
- Obtain materials for *Try This: Wave Refraction* and *Try This: Energy Waves*.

TRY THIS

Wave Refraction
- large coins, 1 per group
- bowls, 1 per group

(**Waves refract as they pass through Earth's layers like the light bent for the coin to be seen.**)

TRY THIS

Energy Waves
- string

(**Answers will vary.**)

A. Complete *Try This: Energy Waves* activities with students.

B. Direct students to complete **WS 3.2.4A Measuring Earthquakes Graph** and **WS 3.2.4B Measuring Earthquakes**.

C. Engage students in the following activity to demonstrate how a seismograph works. Place a pan of water on a table. Dim the lights in the room and shine a flashlight on the water to create a spot of reflected light on the wall or ceiling. Have a student stamp near the water and guide the class to observe the resulting movement of the spot of light. Demonstrate how greater or closer "earthquakes" cause greater shimmering and how earthquakes that are less intense or farther away cause less shimmering. Ask students how this is similar to the way a seismograph works. (**Possible answers: The base of a seismograph also moves with the ground as the pendulum records vibrations so seismic waves can be measured. The size of the seismic waves depends on how far from the seismograph they started and how large the earthquake was.**)

D. Divide the class into pairs or small groups and have each pair create a simple seismograph by taping a large piece of paper to a desk. One partner should try to slowly draw a straight line as the other partner shakes the desk. Ask students how the intensity of the shaking relates to the seismogram. (**The harder the desk shakes, the wider the seismogram squiggles are.**)

E. Assign students to research and compare different earthquake intensity scales, particularly noting when and why each was created or modified and why each is used in a particular region.

Seismic Waves 3.2.4

An earthquake has several parts. An earthquake's **focus**, or hypocenter, is the point inside the earth where an earthquake begins. It is where the built-up pressure in the earth is released. A focus can be at any depth in Earth's crust; some are as deep as 700 km below the surface. The **epicenter** is the point on the surface directly above an earthquake's focus. Ground movement is usually felt most strongly at the epicenter because it is the point on the surface closest to the focus.

When faults move or plates that are caught jar free, a great deal of energy is released. This energy travels in seismic waves. Seismic waves leaving an earthquake's focus flow much the same way as ripples from a pebble thrown into still water. The waves are tallest and strongest when they are closest to the focus. As they move out farther, they become shorter and weaker.

Earthquake waves are divided into two categories known as *body waves* and *surface waves*. Seismic waves that travel underground are body waves. They consist of P (primary) waves and S (secondary) waves. P waves are the faster of the seismic body waves. They can travel through solids, liquids, and gases. As P waves travel, they cause consecutive sections of the earth to push together, then spread back apart like a giant spring. P waves usually cause little to no damage. S waves are slower than P waves, and they travel only through solids. S waves move the earth perpendicular to the direction they are traveling. S waves can cause greater damage than P waves because they are stronger than P waves and because their motion can cause buildings to sway.

Seismic waves that travel on or just under the earth's surface are surface waves. They are detected after the body waves pass. Surface waves cause more damage than body waves because

OBJECTIVES
- Demonstrate seismic waves and how they are recorded.
- Explain the difference between intensity and magnitude.
- Compare different scales used to measure intensity and magnitude.

VOCABULARY
- **epicenter** the point on the earth's surface directly above an earthquake's focus
- **focus** the point inside the earth where an earthquake begins
- **Love wave** a fast surface wave that moves in a side-to-side pattern as it travels forward
- **Rayleigh wave** a slower surface wave that moves in an elliptical pattern as it travels forward
- **seismograph** an instrument that measures and records seismic waves

P Waves
As P waves travel, they alternate between a compressed position and a stretched position.

S Waves
S waves travel back and forth, perpendicular to the direction they are traveling.

Love Waves
Love waves shake Earth's surface from side to side.

Rayleigh Waves
Rayleigh waves move in an elliptical motion as their energy travels forward.

TRY THIS

Wave Refraction
Place a large coin in a bowl. Place the bowl on a table and have a partner move it away from you until the coin just disappears from your sight. Then have your partner pour water into the bowl until you can see the coin again. Trade roles with your partner and repeat the experiment. What does this experiment suggest about the way seismic waves behave when they encounter different substances in the ground?

they have greater force and last longer. The two kinds of surface waves are Love waves and Rayleigh waves. **Love waves** are fast surface waves that move in a side-to-side pattern as they travel forward. Their pattern looks like a snake squiggling across sand. **Rayleigh waves** are slower surface waves that move in an elliptical pattern as their energy travels forward. Rayleigh waves feel like ocean swells; as they pass under a person, the person will feel the ground move up and down as well as back and forth.

Most of the thousands of earthquakes that happen every year are so small that people do not even notice them. Scientists who study earthquakes are known as *seismologists*. They use technology to identify earthquakes that would otherwise go unnoticed. A **seismograph** is an instrument that measures and records seismic waves. Some seismographs are so sensitive that they are placed in remote areas where they will not pick up the vibrations of passing trucks or trains. The two basic parts of a seismograph are a base that moves with the earth and a pendulum that hangs over the base and records its movements. Early seismographs had a pen attached to the pendulum and a roll of continuously turning paper attached to the base. If the ground moved, the pen traced the base's movements on the paper. Now most seismographs have a pendulum wrapped with a wire coil suspended in a magnetic field. The coil's movement creates an electric current, which computers then translate into the seismic waves' movements.

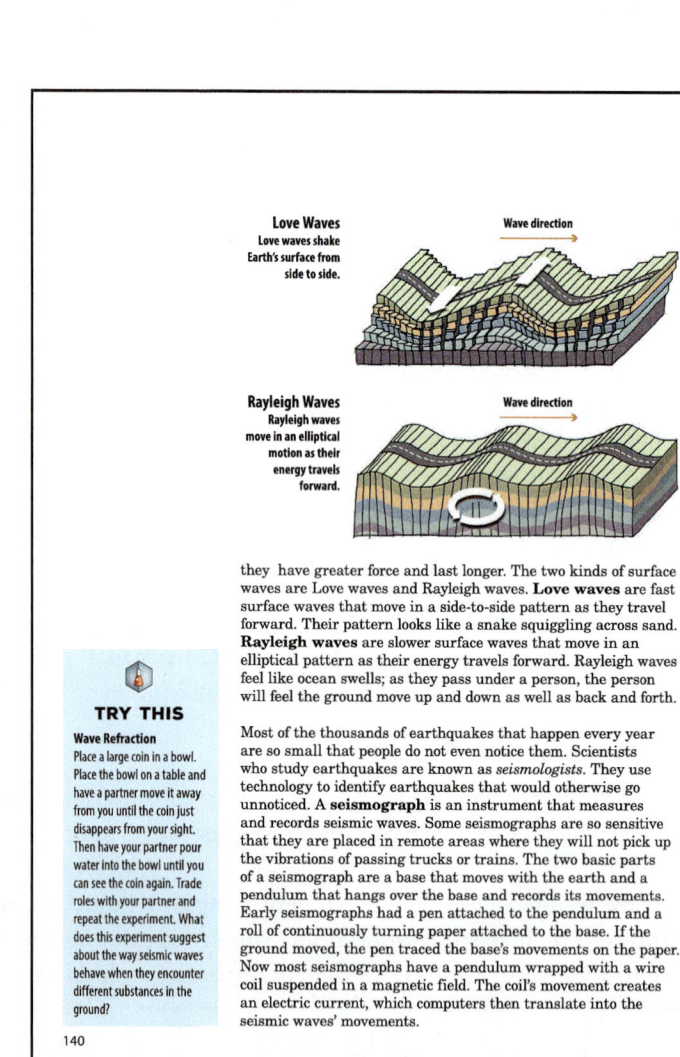

F. Have students prepare presentations illustrating past and present technologies used in earthquake detection, recording, and reporting. Students should include information on the leading geologists or inventors involved and what events or factors led them to the invention of their technology.

G. Challenge students to make a more complex seismograph than that used in Lab 3.2.4A Seismograph.

Lesson Review

1. Which seismic waves travel through the earth, are detected first by seismograph machines, and have a compressional wave pattern? (**P waves**)
2. Which seismic waves that travel through the earth are detected second by seismograph machines and move the earth perpendicular to the direction they are traveling? (**S waves**)
3. Which seismic waves travel on or near the surface and have a side-to-side motion? (**Love waves**)
4. Which seismic waves travel only on or near the surface and move the earth in an elliptical pattern as their energy continues to move forward? (**Rayleigh waves**)
5. What is the difference between intensity and magnitude? (**Intensity is a subjective description of an earthquake's effects on Earth's surface, but magnitude is a measurement of the strength of an earthquake.**)
6. Compare three scales that can be used to measure an earthquake's intensity, and two scales that can be used to measure an earthquake's magnitude. (*intensity*: **Japan Meteorological Agency Seismic Intensity Scale, used in Japan and Taiwan; Modified Mercalli Intensity Scale, used in the United States; European Macroseismic Scale-98, used in Europe and other areas around the world;** *magnitude*: **Richter scale—mathematical formula, uses seismograph information,**

NOTES

Over the years, scientists have developed a variety of scales to chart earthquakes. These scales can measure either the intensity or the magnitude of an earthquake. An earthquake's intensity is a description of its effects on the earth's surface. A single earthquake often produces different intensity measurements depending on the materials in the ground. Tightly compacted materials will shake less than loose materials. Because the force of seismic waves lessens as they travel away from the epicenter, the location where the waves are measured also affects the intensity data.

Different places around the world use different intensity scales. For example, the Japan Meteorological Agency Seismic Intensity Scale, which is used in Japan and Taiwan, describes a level 3 earthquake as, "Felt by most people in buildings," but the term *Level 3* is described somewhat differently in the Modified Mercalli Intensity Scale, used in the United States, and the European Macroseismic Scale-98, used in Europe and other areas around the world. None of the scales can be called right or wrong; they are just different ways of categorizing what is seen or felt during an earthquake.

An earthquake's strength, its magnitude, is determined by measuring the energy released by an earthquake. Magnitude is a more consistent measurement than intensity because it measures force released at a single point instead of force experienced wherever seismic waves travel. However, two different people may arrive at different magnitude measurements depending on various factors, including how their seismographs are adjusted and how far away from the earthquake's focus their seismographs are.

In 1936, Charles F. Richter developed a mathematical formula that used printed information from seismograph machines, called *seismograms*, to calculate an earthquake's magnitude. This formula is called *the Richter scale*. Each level on the scale is 10 times greater than the level below it. For example, level 5 is 10 times greater than level 4, and level 8 is 10,000 times greater than level 4. Most earthquakes rate less than 4 on the Richter scale and cause little or no damage. Usually, only a couple earthquakes a year register over an 8. The Richter scale is not used much anymore because it was designed for earthquakes in California, which differ from those occurring in other parts of the world. It is also less accurate on very large magnitude earthquakes.

Consequently, the moment magnitude scale has become more widely used than the Richter scale. Like the Richter scale, the moment magnitude scale is a mathematical formula that is used to interpret the measurements recorded by a seismograph. The term *moment magnitude* refers to the distance a fault moved and the force used to move it. Unlike the Richter scale, the moment magnitude scale works with earthquakes of many sizes around the world. There are other scales that measure magnitude, including ones that measure local magnitude, surface-wave magnitude, and body-wave magnitude, but they are less reliable for measuring large earthquakes.

TRY THIS
Energy Waves
Tie a piece of string to a desk. Wrap the other end of the string around your finger and hold it to your ear. Pluck the string. What kind of sound do you hear? Move the string away from your ear and pluck it again. What kind of sound do you hear? How do you think this is similar to earthquake waves?

Charles F. Richter

Early seismograph

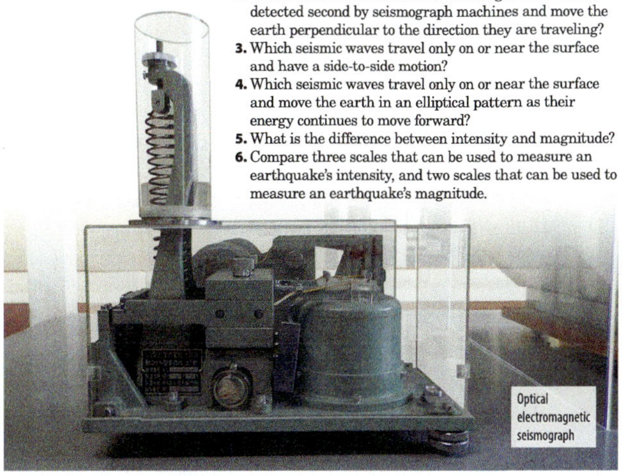

Optical electromagnetic seismograph

LESSON REVIEW
1. Which seismic waves travel through the earth, are detected first by seismograph machines, and have a compressional wave pattern?
2. Which seismic waves that travel through the earth are detected second by seismograph machines and move the earth perpendicular to the direction they are traveling?
3. Which seismic waves travel only on or near the surface and have a side-to-side motion?
4. Which seismic waves travel only on or near the surface and move the earth in an elliptical pattern as their energy continues to move forward?
5. What is the difference between intensity and magnitude?
6. Compare three scales that can be used to measure an earthquake's intensity, and two scales that can be used to measure an earthquake's magnitude.

NOTES

good for California earthquakes, less accurate for large magnitude earthquakes, not used much anymore; moment magnitude scale—mathematical formula, uses seismograph information, good for use around the world, accurate for any size earthquake, widely used.)

Name: _____ Date: _____

Lab 3.2.4A Seismograph

QUESTION: How does a seismograph work?
HYPOTHESIS: Answers will vary.

EXPERIMENT:

You will need:	• ring stand	• tape
• string	• meterstick	• felt-tip pen
• lead weight	• graph paper	

Steps:
1. Find a partner and use string to suspend a lead weight from a ring stand so the weight is approximately 5 cm from the work surface.
2. Place a piece of graph paper underneath the ring stand.
3. Tape the pen to the weight so the point of the pen just touches the paper.
4. Have one partner jump up and down while the other partner slowly and steadily pulls the paper forward.
5. Label the graph paper with the name of the person who jumped. Measure the distance the person was from the seismograph with a meterstick.
6. Replace the graph paper with a fresh piece.
7. Trade jobs with your partner and repeat Steps 4–5.
8. Try jumping in different locations. For each location, label the graph paper with the type of force exerted and the distance the force was from the seismograph.

ANALYZE AND CONCLUDE:

1. Why must the paper be moving and why must it move at a steady rate? The paper must be moving to record the fluctuation of the vibrations over time. The paper must move at a steady rate so the time intervals will be the same.

2. What did you observe regarding the marks on the graph and the approximate force of the jumper? Stronger forces produced longer marks.

3. What did you observe regarding the marks on the graph and the distance between the jumper and the seismograph? Closer forces produced longer marks and farther forces produced shorter marks.

4. Why is more than one seismograph needed to accurately describe an earthquake? Seismographs closer to the focus will record stronger vibrations. Seismographs can also be calibrated differently or they can be over different materials in the ground. Both could affect measurements.

Earthquakes

3.2.5 Prediction and Effects

Introduction
Display the prepared pan of thick gelatin. Have students or groups of students create structures out of marshmallows, toothpicks, and coffee stirrers. Each structure should have at least three levels. Direct students to place their structures in turn on the gelatin's surface, to shake the pan to simulate an earthquake, and then to record their observations.

Discussion
- In reference to the *Introduction* activity, lead a discussion regarding why the longest-lasting structures worked so well, how the structures moved in relation to the shaking surface, and the implications this activity has for designing buildings that will resist earthquakes.

- Emphasize to students that knowledge and wisdom are required for making decisions regarding preparedness for earthquakes or any other force of nature. Ask students what the difference is between knowledge and wisdom. (**Knowledge is information stored in one's memory. Wisdom is the correct application of knowledge.**) Read Bible passages that discuss wisdom, such as **Deuteronomy 4:1–8, 2 Chronicles 1:7–12, Psalm 111:10, Proverbs 8:12–16,** and **James 3:13–18**. Lead a discussion regarding the likelihood that people from different places or cultures—even within one nation—will agree regarding what a wise decision is and why those differences of opinions exist. Have students consider whether all Christians will have a common understanding of wise decisions. Engage students in considering how Christians should handle disagreements; have students refer to passages such as Romans 14:19–15:6, 1 Corinthians 6:1–11, 1 Corinthians 10:23–33, and James 4:1–12. How can knowing how to handle disagreements be helpful when preparing for earthquakes? (**Answers will vary.**)

Activities
A. As a class, complete **WS 3.2.5A Earthquake Prediction** and **WS 3.2.5B Earthquake Prevention**. Divide the class into two or more groups for each scenario and direct each group to develop its own response. Have groups working on the same scenario discuss their proposed responses and create a unified response to share with the class; strong disagreements should be noted and presented as dissenting opinions.

B. Challenge students to work in small groups to write one or more scenarios similar to those in WS 3.2.5A and WS 3.2.5B. Direct student groups to exchange scenarios, to provide a reasoned response, and to share their response with the class. Lead a class discussion of the responses.

C. Present the invited architect or engineer. Have the speaker explain ways that buildings or other structures can be built or retrofitted to make them more resistent to earthquakes. Discuss what modifications the speaker mentioned would reduce earthquake damage.

D. Introduce the invited representative from a disaster relief organization to present ways students can best help people affected by earthquakes. Ask students what causes the most injuries during an earthquake. (**collapsing buildings**) What needs do people have following an earthquake? (**Answers will vary but should include water, food, shelter, and medical care.**) Elicit ideas from students about how the infrastructure in an area can be protected during an earthquake.

E. Direct students to research safety precautions and devise or revise a preparedness plan for people living in places prone to earthquakes.

F. Guide student groups to research recent earthquakes from around the world and to create a presentation that shows their earthquake's location, lists the earthquake's magnitude and local intensity measurements, identifies factors that contributed either to structure preservation or damage, and includes photographs of the damage done by the earthquake.

OBJECTIVES
Students will be able to
- relate various side effects of earthquakes.
- interpret clues that suggest an earthquake may be about to occur.
- summarize ways to reduce damage caused by earthquakes.

VOCABULARY
- **aftershock** a tremor that follows a large earthquake
- **fissure** a tear in the crust caused by the friction of a fault
- **liquefaction** the process by which soil loses strength and acts as a liquid instead of a solid
- **tsunami** a very large ocean wave caused by an underwater earthquake or volcanic eruption

MATERIALS
- Marshmallows, toothpicks, coffee stirrers (*Introduction*)
- WS 3.2.5A Earthquake Prediction
- WS 3.2.5B Earthquake Prevention

PREPARATION
- Make a pan of thick gelatin. (*Introduction*)
- Invite an architect or engineer to present ways in which buildings can be built or retrofitted to make them more resistent to earthquakes. Encourage students to write questions for the speaker to answer during the presentation. (*C*)
- Invite a representative from a disaster relief organization to present ways in which students can best help people affected by earthquakes. (*D*)

NOTES

G. Assign student groups to research folklore concerning earthquakes and to explain how science either refutes or supports the ideas. Consider having each group script a conversation between someone who believes the folklore and a scientist; emphasize speaking truth in love, especially if science disproves the folklore.

Lesson Review

1. List four naturally occurring side effects of earthquakes. (**aftershocks, fissures, liquefaction, and tsunamis**)
2. How can liquefaction change the earth's surface? (**Liquefaction can result in sand boils, landslides, and the tilting or collapse of ground that was once flat.**)
3. What causes a tsunami to become taller than a normal wave? (**When a tsunami enters shallower water, the wave's bottom slows down because of friction. As the wave's underside slows, the wave's length decreases and its height increases.**)
4. List several clues that seismologists consider when identifying the likelihood that an earthquake will occur at a fault. (**seismic gap between earthquakes, new cracks beneath the earth's surface that cause an increase in radon gas emission or the sudden lowering of water table, expansion of crust, and halt of fault creep**) Which clue seems the most helpful? Why? (**Answers will vary.**)
5. Summarize ways in which earthquake damage can be reduced. (**Understanding more about how faults and earthquakes work leads to building and technology innovations. Some examples are the construction of bridges and buildings that can sway, the installation of automatic switches to shut off electricity and gas lines when tremors are detected, and the placement of tsunami warning systems along coastlines.**)

Prediction and Effects 3.2.5

Even if you have never experienced an earthquake, you may have seen or read reports of an earthquake or its aftermath. Smaller earthquakes may do minimal damage, such as knocking things off shelves and bookcases. Larger earthquakes can collapse buildings and bridges and leave people homeless. Even though only a small number of earthquakes cause damage, the ones that do can be devastating. An earthquake's effects do not stop with the initial tremors. The natural side effects of an earthquake can also be damaging.

When a fault releases its built-up pressure, there is typically one large, primary earthquake followed by one or more smaller earthquakes. These smaller earthquakes, or **aftershocks**, are tremors that follow a larger earthquake. They can topple buildings already destabilized by the initial earthquake. An earthquake's epicenter can have aftershocks for weeks as the fault continues to release its stress and readjust its position.

Another way seismic waves affect the earth's surface is through the creation of **fissures**, which are tears in the crust caused by the friction of a fault. Imagine chopping a pillow with the

OBJECTIVES
- Relate various side effects of earthquakes.
- Interpret clues that suggest an earthquake may be about to occur.
- Summarize ways to reduce damage caused by earthquakes.

VOCABULARY
- **aftershock** a tremor that follows a large earthquake
- **fissure** a tear in the crust caused by the friction of a fault
- **liquefaction** the process by which soil loses strength and acts as a liquid instead of a solid
- **tsunami** a very large ocean wave caused by an underwater earthquake or volcanic eruption

CHALLENGE
Earthquake Resistant Construction
Use your knowledge of seismic waves to suggest ways a building could be designed to withstand earthquakes and their side effects. Consider both the structure of the building and the building's electric, water, and gas utilities.

Portable seismometer, Iceland

BIBLE CONNECTION

Fissures
Earthquakes rarely cause large fissures, or tears in the crust. However, there is one record of God using a fissure to punish people who rebelled against Him. Numbers 16 tells the account of Korah, Dathan, Abiram, and their followers. These men refused to accept God's appointment of Moses and Aaron as their leaders. God punished these rebellious men by causing the ground to suddenly open and swallow them before closing over them again. Others who rebelled were burned by fire from heaven or killed by a plague. God's justice declares that the punishment for sin and rebellion is death and destruction (Genesis 2:17, Deuteronomy 28:15—68). God is also merciful. He accepts the death of a substitute in the sinner's place. God showed mercy to the rebellious Israelites by stopping the deadly plague when Aaron made atonement. Hebrews 10 explains that the Old Testament sacrifices pointed to Jesus. Just as Aaron's atonement saved the lives of the Israelites, so also Jesus' atonement saves all who trust in Him.

edge of your hand. The once-smooth pillow now has a large wrinkle in the center, with smaller wrinkles radiating from it. Those smaller wrinkles are like the fissures caused by the readjustment of the earth.

Seismic waves also affect the earth below the surface. Sometimes the soil loses strength and acts like a liquid instead of a solid, which is called **liquefaction**. Liquefaction only occurs in sandy, loosely packed soil that is saturated with water. As seismic waves pass through this soil, their energy increases the water pressure. The increased water pressure breaks the bonds between the soil grains and suspends them in the water. Once enough soil grains are suspended, the soil behaves like a liquid. If the water pressure becomes too great, the water will break through the surface, carrying sandy soil with it. These eruptions are called *sand boils*. Liquefaction can also cause landslides and tilt or collapse ground that was once flat.

Sand boils

Earthquakes can affect oceans too. A **tsunami** is a very large ocean wave that is caused by a sudden movement in the ocean floor. This movement could be an earthquake or a volcanic eruption. Picture what would happen if you were carrying a pan full of water and then stopped very suddenly. The water would keep moving at the rate you were going before you stopped. Earthquakes that strike oceanic crust or occur near the ocean have this effect on the water. When the ocean floor undergoes a sudden powerful shift, the water is forced to move. This movement of water turns into a tsunami.

Tsunamis are very powerful. In the open ocean, tsunamis may have wavelengths of up to several hundred kilometers and can travel at speeds up to 800 kph, yet their wave heights are less than 1 m, so people on ships at sea do not even notice them. But when a tsunami enters shallow water, the bottom of the wave slows down due to friction. As the wave's underside slows, the wave's length decreases and its height increases, sometimes reaching higher than 30 m. You can imagine the destruction of piers, buildings, beaches, and human life when such a wave breaks on shore. On March 11, 2011, a 9.0 magnitude earthquake struck Japan. It generated a tsunami that was felt as far away as the western coasts of North and South America. In Japan, the tsunami's waves reached up to 10 m high and swept up to 10 km inland.

Some of the effects of a tsunami caused by the March 11, 2011, earthquake in Japan.

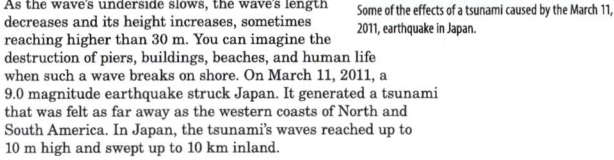

Scientists have not yet discovered a dependable way to predict when earthquakes will strike. They have, however, developed tools and methods that help them forecast the most likely places that earthquakes will happen. Scientists know that any fault can produce an earthquake. The main questions are when the earthquake will occur and how large it will be.

Time is a significant factor in an earthquake's size. The larger the time span between earthquakes in one location, the greater the probability that an earthquake will occur. The time period between earthquakes is called *a seismic gap*.

Another indication that an earthquake may be getting ready to occur is the formation of new cracks beneath the surface. One way these cracks are detected is by the sudden increase of radon gas emissions near faults containing uranium. Radon is created when the element uranium breaks down. Radon slowly makes its way through small cracks in rock and soil until it reaches the surface and dissipates safely. A sudden increase in radon emissions could indicate that old cracks widened or new cracks formed.

The sudden lowering of the water table can indicate new underground cracks as well. New cracks provide new places for underground water to flow, which reduces the amount of water held in aquifers and wells. When an area experiences an abrupt drop in the water table, an earthquake might be getting ready to strike.

Tsunami emergency shelter

Base isolators, similar to these, along with other technology enabled many Japanese buildings to withstand the March 11, 2011, earthquake.

HISTORY

Chilean Preparedness Saves Lives
On September 16, 2015, an 8.3 magnitude earthquake struck near Illapel, Chile. The earthquake created a tsunami with waves that reached up to 4.75 m. Despite the strength of both the earthquake and the tsunami, lives and property were saved because of high levels of preparedness. Chile's strict building codes have ensured that most modern buildings can withstand a major earthquake with minimal damage. Chile has also invested in sea-level monitoring systems and tsunami warning systems, which would enable people to evacuate to high ground before a tsunami hit.

Clues suggesting an impending earthquake can also be found aboveground. For instance, Earth's crust may expand slightly before an earthquake strikes. In addition, if a slow fault creep area stops creeping, pressure may be building up that will be released in an earthquake. Scientists track surface movements with satellites. These satellites include radar and lidar technology. Lidar means "light detection and ranging," and it works similarly to radar, but uses lasers instead of radio waves.

All of these clues are used by scientists to determine the likelihood of an earthquake in a particular location. However, none of the clues happen consistently enough to enable seismologists to predict exactly when an earthquake will strike. Nevertheless, knowing how faults and earthquakes work enables people to design buildings and technology that can save both lives and property. Some of these innovations include constructing bridges and buildings to sway with an earthquake, installing automatic switches to shut off electricity and gas lines when tremors are first detected, and establishing tsunami warning systems along coastlines.

LESSON REVIEW
1. List four naturally occurring side effects of earthquakes.
2. How can liquefaction change the earth's surface?
3. What causes a tsunami to become taller than a normal wave?
4. List several clues that seismologists consider when identifying the likelihood that an earthquake will occur at a fault. Which clue seems the most helpful? Why?
5. Summarize ways in which earthquake damage can be reduced.

FYI
Earthquake Warning
Located in the Pacific Ocean's Ring of Fire, Japan experiences frequent earthquakes. In order to protect people, Japan's Earthquake Early Warning system warns people about an imminent earthquake before tremors begin. When seismologists detect an earthquake's P waves, warnings are sent out to the media, businesses, schools, and even mobile phones before the S waves hit. Although less warning time is possible if a person is close to the epicenter, the system can warn people farther away about a minute before tremors occur.

3.3.0 Volcanoes

LOOKING AHEAD

- For *Lab 3.3.1A How Volcanoes Form* in **Lesson 3.3.1**, collect small cardboard boxes to be used for volcano models.
- For **Lesson 3.3.3**, burn wood, marshmallows, and toast. Obtain samples of several types of ash and melted items. Gather samples of pyroclastic materials, such as pumice, aa, pahoehoe, igneous rock, and volcanic ash.
- Locate sources for both a young-earth and an old-earth creationist viewpoint on volcanoes for **Lesson 3.3.5**.

SUPPLEMENTAL MATERIALS

- TM 3.3.2A Ring of Fire
- TM 3.3.3A Plutons and Pyroclasts
- TM 3.3.3B Pyroclastic Materials
- TM 3.3.4A Volcano Type Visuals
- TM 3.3.4B Volcano Types
- TM 3.3.4C VEI
- TM 3.3.5A Volcano Damage

- Lab 3.3.1A How Volcanoes Form
- Lab 3.3.1B How Magma Moves
- Lab 3.3.3A Viscosity

- WS 3.3.1A Decade Volcanoes
- WS 3.3.1B Decade Volcanoes Map
- WS 3.3.2A Volcano Zone
- WS 3.3.4A Comparing Eruptions
- WS 3.3.4B Artistic Interpretations

- Chapter 3.3 Test
- Unit 3 Test

BLMs, TMs, and tests are available to download. See Understanding Purposeful Design Earth and Space Science at the front of this book for the web address.

Chapter 3.3 Summary

The eruption of volcanoes has occurred throughout recorded history. Volcanoes play an important role in maintaining the planet: they are a cooling system for the earth and a recycling system for atmospheric gases. They also replenish the soil with nutrients. Students will come to appreciate God's plan for volcanoes as they learn about the orderly way in which volcanoes form and the variety of ways they display the forces within inner Earth.

Background

Lesson 3.3.1 – Volcano Formation

In the 13th century, German scientist Albertus Magnus presented the theory that volcanoes erupt because wind blows under the earth's surface. Many other theories followed. A significant breakthrough came when French scientist Nicholas Desmarest recognized that basalt originates from volcanoes. In 1775, Desmarest grouped volcanoes into three major categories: recently erupted volcanoes, older volcanoes that show evidence of long-ago eruption, and volcanoes that are almost totally obscured by the surrounding area. Scientists classify volcanoes similarly today, using the terms *active*, *dormant*, and *extinct*. As volcanology has become its own branch of science, more specific classifications of volcanoes have been added.

In the late 18th century, James Hutton theorized that volcanoes are the natural ever-occurring result of the earth's internal heat, forming where internal materials are able to come to the earth's surface. In 1825, German geologist Leopold von Buch published a physical description of the Canary Islands. His observations of the islands led him to the conclusion that ocean islands are volcanic islands formed by the heating of the interior of the earth.

Today more than 1,500 volcanoes around the world are potentially active; hundreds more are dormant, or inactive. Volcanoes begin with magma from deep within the earth, which flows out of the mantle and onto the earth's surface through weaknesses in the crust. Sometimes this lava flows easily, such as in Hawaii. At other times, the lava bursts violently out of the volcano, as it did from Mount Saint Helens in 1980 and from Mount Pinatubo in 1991.

Magma density and pressure cause volcanoes to erupt. Low-density magma rises to the surface or to a depth determined by the density of the magma and the weight of the rocks above it. As the magma rises, gas dissolved in the magma forms bubbles, which exert pressure. This pressure forces the magma to the surface.

Hot spot volcanoes, such as those in Hawaii, form when hot magma melts holes in the earth's crust and the magma forces the crust upward to form a volcano. New islands form as the continental plate drifts over the hot spot. Rift volcanoes form where continental plates are moving apart, such as at the Mid-Atlantic Ridge. Subduction zone volcanoes, such as Mount Saint Helens and Mount Fuji, form where two continental plates collide. One plate is forced under the other, and the lower plate melts as it is pushed downward. The magma pushes through the resulting weak spots in the earth's crust.

Lesson 3.3.2 – Volcano Zones

Volcano zones relate to plate tectonics. Plate collisions are of three types: oceanic-oceanic, oceanic-continental, or continental-continental. In oceanic-oceanic collisions, one plate can be subducted under the other. If either slab contains water or carbon dioxide, the gases can be released, causing liquid rock to rise through the overlying mantle and forming a new crust. The emerging lava can create underwater volcanoes or even volcanic islands.

In oceanic-continental collisions, the oceanic crust is always the descending slab because it is more dense than the continental crust. The high levels of silica in the granite make the lava very viscous, so these eruptions are more explosive than the island arc eruptions, such as those that formed the Aleutian Islands. The Cascade volcanoes, including Mount Rainer and Mount Saint Helens, are of this type.

Rift volcanoes form when plates separate. Although no rock is forced down to depths where it can melt, something must fill the immense cracks that form where the plates separate. Hot, soft rock from the mantle fills these cracks. In the mantle, this rock is under great pressure, but as the rock rises to fill the gap, the pressure drops faster than the rock can cool. A small amount of basaltic rock melts and flows to the surface. Because this lava contains little or no dissolved gas, these eruptions are not explosive; only a small amount of lava erupts at any given time. These continuous eruptions occur along all of the oceanic ridges, making this type of volcano the most common in the world. They are often overlooked because most happen underwater. This type of volcanism is also found on Iceland and the East African Rift.

Lesson 3.3.3 – Volcanic Material

Many types of volcanic material issue forth from erupting volcanoes. Magma that reaches the earth's surface is called *lava*. Lava pours or shoots through the vents at temperatures that range from about 700°C–1,200°C. Lava has a wide-ranging viscosity. Some lava is quite fluid, like syrup; other lava is so thick it can scarcely flow.

Magma that does not reach the surface of the earth but instead hardens underground forms an intrusion called *a pluton*. Plutons that are horizontal are often confused with buried lava flows. However, surface details reveal the difference: lava flows are weathered, pockmarked with chambers from old gas bubbles, and have a wavy surface; plutons are not weathered and since they have been compressed, they do not have a wavy surface.

Plutons take on many forms including dikes, sills, and batholiths. Batholiths are the largest of the plutons and are made of solid magma. When composed of many layers of magma, they are called *layered complexes*. One of the largest batholiths is located in South Africa. Within this batholith, geologists and miners have discovered minerals such as olivine, pyroxene, plagioclase, platinum, chromium, gold, and other minerals.

Ash is another material issued from volcanoes. It is composed of fragments of rock, minerals, and volcanic glass measuring 0.025–4 mm in diameter. Volcanic ash is created during explosive eruptions by the shattering of solid rocks and the violent separation of magma into tiny pieces. The ash of burned wood or paper is soft, but volcanic ash is hard and does not dissolve in water. Volcanic ash is as abrasive as crushed glass. It is also a corrosive material and a conductor of electricity.

Hot ash flows, which can race across the ground at several hundred kilometers per hour, burn everything in their paths. Inhaling hot ash can asphyxiate both animals and people. Swirling particles of ash can generate lightning, which can, as in the case of Mount Saint Helens, ignite forest fires. Ash can even alter the climate somewhat.

Lesson 3.3.4 – Types and Eruptions

The earth's land has more than 500 active volcanoes. An active volcano is one that has erupted recently or that volcanologists think could erupt soon. Active volcanoes show no signs of weathering. In fact, as soon as weathering occurs, a volcano is declared dormant.

Each eruption of an active volcano is different. Giant eruptions can affect the whole world by sending up enough ash to temporarily change the climate. Small eruptions may affect only a single hillside or valley. Volcanic eruptions are classified according to their explosiveness.

> **WORLDVIEW**
> - The prophet Isaiah wrote of the Lord's favor for His people, how He will comfort those who have suffered a time of devastation and grief. He does not promise to withhold trials, but He will "bestow on them a crown of beauty instead of ashes, the oil of joy instead of mourning, and a garment of praise instead of a spirit of despair." (Isaiah 61:3) Volcanic eruptions can be extremely devastating to a community; natural resources can be wiped out, homes can be destroyed, and entire families can be lost. These events can lead people to turn away from God or even curse Him. But according to Romans 8:28, God brings new life and benefits even after catastrophic events: "And we know that in all things God works for the good of those who love Him, who have been called according to His purpose." Volcanic soil can be rich with new minerals to grow strong new crops, precious metals can be formed by magma intrusions, and even diamonds can be brought up from the depths to be used for industry and fine jewelry. God can bless despite the tragedy, giving us beauty from ashes and a garment of praise instead of a spirit of despair. Natural disasters such as volcanic eruptions can cause great despair and disillusionment, but we know that our hope lies with the Father of heaven and Earth.

Icelandic eruptions, named for those in Iceland, are mild emissions of gas and very hot, thin lava through many long fissures in the volcano's surface. These eruptions do not shatter the earth—they build the earth. The Icelandic eruptions in the Mid-Atlantic Ridge make new crust to replace the crust that is being destroyed on the opposite side of the earth. Despite their mild nature, these volcanoes pose a threat to the people of Iceland because the lava flow can be just as deadly as a more explosive eruption. Volcanoes that are characterized by Icelandic eruptions erupt repeatedly, forming thick plateaus of lava such as the Columbia Plateau in the western United States.

Hawaiian eruptions are named after the Hawaiian Islands. Like Icelandic eruptions, Hawaiian eruptions have many fissures that bring magma to the surface and have beautiful fire fountains of hot, thin, fast-flowing lava. The main difference between these two types of eruptions is that most of the lava in Hawaiian eruptions flows out of the main vent, not along the side.

Strombolian eruptions are short-lived explosive eruptions that shoot very thick, pasty lava into the air along with bursts of steam and gas. This eruption type is named after the Italian volcano Stromboli, which erupts almost continuously. Because these eruptions produce little or no lava, they usually form cinder cones. The explosions rip apart the vent opening. Strombolian eruptions are known for volcanic bombs.

Vulcanian eruptions, named after the island Vulcano off Italy's coast, are blasts of dark ash, steam, and gas. Such eruptions can shoot gas and ash 15 km high.

Peléan eruptions are characterized by fine ash, gas-charged fragments of lava, and superheated steam that travels downhill at great speed. Peléan eruptions are named for the 1902 eruption of Mount Pelée on the island of Martinique in the Caribbean Sea. The eruption and the pyroclastic flow that followed killed nearly 30,000 people. *Nuée ardente*, or a "glowing cloud" of gas and ash, flew down the mountain at over 100 kph. The dense ash cloud hugged the ground as it approached the coast. One of the few survivors rescued was a prisoner—Louis-Auguste Cyparis.

Plinian eruptions are the most powerful eruptions. Plinian volcanoes are characterized by hot ash clouds and deadly pyroclastic flows. They are named after Pliny the Elder of Rome, a victim of the eruption of Mount Vesuvius in 79 AD. His nephew, Pliny the Younger, recorded his observations of the event.

A measure of an eruption's size is the Volcanic Explosivity Index, or VEI. Like the Richter scale, this scale is logarithmic—each unit increase in VEI, from 2 to 3 for example, corresponds to a tenfold increase in the volume of material erupted onto the surface. Descriptions such as *small* or *moderate* eruptions are only loosely tied to the VEI scale.

Lesson 3.3.5 – Prediction and Effects

A goal of geologists is to accurately predict volcanic eruptions, but that goal has been elusive. A forecast is a general statement of the time, place, and the nature and size of a coming eruption according to past events. The time frames given in forecasts may span several years. A prediction is a precise statement of time and place made as a result of monitoring movements of thrust faults or atmospheric gas levels. Although volcanologists continue to strive to make accurate predictions, usually data and technology allow only a forecast. Both forecasts and predictions can be used for planning purposes and public safety policies, however.

Volcanologists often look to past events because the analysis of past eruptions may help them forecast and predict future volcanoes. Deposits from past volcanoes help them better predict the behavior of a future eruption. Mapping out past lava flows, debris flows, or pyroclastic flows helps them better formulate disaster plans. At the same time, they look for warning signs of an eruption.

For example, sulfur dioxide and other gases arise from magma that is rising in the volcano chamber. The movement of magma produces small earthquakes and causes the volcano's slopes to bulge. Scientists use tiltmeters to measure the tilt of the slope and to track changes in the rate of swelling of an active volcano. A bulge caused by rising magma and gases is a sign that the volcano may erupt.

Volcanoes normally have continuing low-level seismic activity, but an increase in activity can also signify a coming eruption. Other signs volcanologists study include the type of local earthquakes and where they start and end. For example, a short-period earthquake is related to the fracturing of brittle rock as the magma forces its way upward, which points to the growth of a magma body near the surface. A long-period earthquake indicates increased gas pressure in a volcano's vents. A harmonic tremor occurs when there is sustained movement of magma below the surface.

Another warning sign is a gas emission from the volcano. Gases escape from magma that is nearing the surface as the magma's pressure decreases. Sulfur dioxide is one of the main components of volcanic gases. In March and April of 1991, magma was moving below the surface of Mount Pinatubo in the Philippines. Small earthquakes were felt and steam emissions were on the rise. For three months leading up to the June 12 eruption, thousands of small earthquakes occurred, and thousands of tons of sulfur dioxide gas were emitted from the volcano. After the gas was released, the gas-poor magma filled the vent and formed a lava dome. But on June 12, Pinatubo blew its top and millions of cubic meters of magma reached the surface in a spectacular eruption.

Despite these signs of impending eruptions, the science of prediction is far from accurate. Currently it is impossible to pinpoint an eruption exactly. Sometimes moving magma cools instead of erupts. Furthermore, monitoring potential eruptions is very expensive, especially given the large number of volcanoes on Earth and the fact that most erupt only every few hundred or thousand years. Still, once a volcano shows the early signs of an eruption, monitoring devices can help scientists track it further and help alert communities to potential disaster.

Scientists forecasted America's greatest volcanic eruption, the May 18, 1980, eruption of Mount Saint Helens. In the 1950s, geologists began an extensive study of the deposits around Mount Saint Helens. By 1975, they predicted that Mount Saint Helens was the volcano in the lower 48 states most likely to erupt by the end of the century. On March 16, 1980, a series of small earthquakes began to wake up the volcano. About a week later, an explosion of steam blew a 75 m crater through the mountain's ice cap. By mid-May, a bulge extended 137 m from the side of the mountain. Scientists forecasted that Mount Saint Helens would erupt soon, but they could not predict when. Evacuation orders issued months before the eruption saved countless lives. Still, 57 people died because they ignored the warnings and because scientists had underestimated the radius of the disaster, which was classified by volcanologists as a Plinian eruption.

However, the eruption of Mount Saint Helens helped advance the field of "disturbance ecology," which is the recovery of ecosystems after disasters. This ecosystem's recovery has been surprisingly quick and serves as a good example of the recycling nature of the creation. Volcanoes not only recycle rock matter and atmospheric gases, but also deposit nutritious ash that aids the return of life to the desolated area. Much of the forest around Mount Saint Helens was still covered by snow, which protected the plants. Seeds blew in from other areas to join the existing plants. Scientists changed their thinking about forestry, realizing that it is not necessary to clean up the land after trees are cleared to allow for regrowth—trees actually grow back faster when the land is "messy." This ecological recovery is a wonderful example of God's methods of restoring and maintaining His creation.

NOTES

3.3.1 Volcano Formation

Volcanoes

OBJECTIVES

Students will be able to
- discuss three methods by which volcanoes can form.
- explain how volcanoes grow.
- infer the benefits of volcanoes.

VOCABULARY

- **hot spot** a place on the earth's surface that is directly above a column of rising magma
- **volcano** a vent in the earth's crust through which magma, steam, ashes, and gases are forced

MATERIALS

- Can of soda (*Introduction*)
- WS 3.3.1A Decade Volcanoes
- WS 3.3.1B Decade Volcanoes Map

PREPARATION

- Obtain materials for *Lab 3.3.1A How Volcanoes Form*.
- Obtain materials for *Lab 3.3.1B How Magma Moves*. Prepare unflavored gelatin in small bowls and chill to set.
- Choose a video about Pompeii. (*B*)

Introduction

Shake a can of soda. Ask students what would happen if you opened the can. (**The drink would shoot out of the opening.**) Why would this happen? (**The pressure inside the can has built up.**) What would happen if I shook the can but did not open it? (**Nothing—the can would contain the beverage.**) Convey that magma deep within the earth is pressurized just like the soda can. When no opening presents itself, the magma stays inside until the pressure inside the earth overcomes the resistance of the crust. In both cases, the pressure arises from heat. When a can of soda is left in a hot car, it will eventually explode from the pressure. Ask what would happen if you shook the can and then poked a hole in it with a pin. (**The soda pop would squirt through the hole.**) That reaction is what happens when magma finds a crack in the earth's crust. Ask what would happen if the pressure inside the can increased greatly but there was no escape route for the beverage. (**Possible answers: the can would contain the pressure, it would explode**) Share that nobody knows what would happen if the pressure and heat inside the earth could not escape, but it is known that volcanoes are an important part of the earth's venting system.

Discussion

- Discuss the three ways volcanoes can form.

- Discuss the activity of the closest volcano. Discuss the benefits and disadvantages of living near a volcano. Ask how the soil around a volcano is different from other soil. (**Answers will vary.**)

- Ask the following questions:
  1. How do most volcanoes form? (**Most volcanoes form through subduction at plate boundaries.**)
  2. Explain the difference between a hot spot volcano and a rift volcano. (**Hot spot volcanoes form when a deep pocket of hot magma rises up to the crust. Rift volcanoes form when divergent plates move apart and magma rises through the gap.**)

Activities

Lab 3.3.1A How Volcanoes Form

- boxes, 1 per group
- newspaper
- plastic tubing
- balloons, 1 per group
- clamps to fit plastic tubing, 1 per group
- tape
- flour
- metric rulers, 1 per group

Lab 3.3.1B How Magma Moves

- unflavored gelatin
- small bowls, 1 per group
- aluminum pans, 1 per group
- hot pads, 2 per group
- large bowls, 1 per group
- bricks or large blocks, 2–4 per group
- large syringes, 1 per group
- chocolate syrup

Be sure the aluminum pans are wider than the diameter of the small bowls.

A. Have students complete **WS 3.3.1A Decade Volcanoes** and **WS 3.3.1B Decade Volcanoes Map**.

B. Show a video about Pompeii to demonstrate that not all volcanoes destroy things with lava. Pompeii was preserved in ash exactly as it was when the eruption happened.

C. Divide the class into four groups. Have each group answer the lesson review questions. Then, assign each group to write an answer on the board. Have groups improve the answer for another group. Finally, have each group grade another group's answer and defend the grade they give.

D. Have students create a flip book about the formation of a volcano.

E. Challenge students to research volcano legends of a specific country or region. Some options include Hawaii, Greece, Japan, and New Zealand.

Lesson Review

1. Describe three ways a volcano can form. (**A volcano can form over a hot spot. Magma expands and fills a magma chamber. Pressure builds up and the magma searches for cracks or weak spots to escape. The pressure of the expanding magma and gases can force the gas and lava through the volcano's vents. Volcanoes can form in subduction zones at plate boundaries. As the subducted plate heats up, the rock melts and gases are released. Pressure can bring the magma and gases to the surface and erupt in a volcano. Volcanoes can form at divergent plate boundaries. As the plates spread apart, magma can rise up through the gap. These rift volcanoes occur mostly below sea level.**)
2. How do volcanoes help scientists study the interior of the earth? (**Volcanoes give scientists direct evidence of what is inside the earth. Volcanoes bring the inside of the earth to the outside where scientists can actually see the contents of the mantle.**)
3. How do volcanoes grow? (**Volcanoes can grow either by intrusion or extrusion. Intrusion is when magma moves up inside the volcano and stays there. Extrusion is when the lava comes out of the volcano and then cools on the outside.**)
4. What important roles do volcanoes play in creation? (**Volcanoes can make the soil fertile. Volcanoes allow the earth to vent pressure that builds up inside the earth. The eruptions bring balance and stability to the inner earth.**)

NOTES

3.3.1 Volcano Formation

OBJECTIVES
- Discuss three methods by which volcanoes can form.
- Explain how volcanoes grow.
- Infer the benefits of volcanoes.

VOCABULARY
- **hot spot** a place on the earth's surface that is directly above a column of rising magma
- **volcano** a vent in the earth's crust through which magma, steam, ashes, and gases are forced

Have you ever wished that you could peek through a window and see the inside of the earth for yourself? God did create windows to the inner earth—volcanoes. A **volcano** is a vent, or opening, in the earth's crust through which magma, steam, ashes, and gases are forced.

An erupting volcano shouts of God's astonishing design; the strength of a volcano shows a glimpse of His power. Volcanoes are an obvious example of God's awesome deeds. They not only put on a marvelous show, but they are also an important part of the framework of creation. God wisely planned for volcanoes to work together with the rest of creation. Still, volcanoes are one of creation's more mysterious and terrifying displays.

Magma beneath the earth's surface moves around in chambers, oozing through cracks in Earth's surface or melting through solid rock to reach Earth's surface. Magma that has reached the earth's surface is called *lava*. Lava reaches the earth's surface through volcanoes. Some volcanoes are little more than oozing cracks in the earth. Others—the kind that usually come to mind when you think about volcanoes—form cone-shaped mountains out of volcanic material. A circular depression called *a crater* often forms around the vents of these volcanoes.

Volcanoes may form in one of three ways. About 5% of volcanoes form over **hot spots**, places on the earth's surface that are

Tavurvur's 2009 eruption in Papua New Guinea

Mount Etna crater in Sicily, Italy

directly above a column of rising magma. Scientists believe that hot spots form when a portion of the deep mantle is heated. This part of the mantle, which is now less dense than the surrounding material, begins to expand and rise like a hot air balloon. The magma gathers in a porous region of rock called a *magma chamber*. The pressure builds from the expanding magma, which searches for cracks and weaknesses through which to escape. As it approaches Earth's surface, a hot spot can push up on Earth's crust. The pressure forces gas, ash, and magma through the volcano's vents.

Lava from the eruption hardens to form layers of rock and builds a volcanic cone as it cools. The Hawaiian Islands were formed by this process. Scientists believe the islands all formed over one hot spot because as the oceanic crust moved, each eruption formed a new island. If the crust had stayed in one place, only one island would have formed. In fact, the summit of the youngest Hawaiian

An island builds up over a hot spot. As the island moves past the hot spot, the island stops growing and begins to erode. Meanwhile, another island is forming.

volcano, Loʻihi, is still 970 m below sea level. Most hot spots lie beneath the ocean. However, two continental volcano systems are associated with a hot spot. One is in Yellowstone National Park in the United States, and the other is in the Cosgrove hot spot track in Australia, which is the world's longest chain of continental volcanoes.

Fresh lava on the Hawaiian Islands

Hot spots are not the only cause of volcanoes. About 80% of volcanoes form at plate boundaries when one tectonic plate is forced under another one. The subducted plate usually melts when the edges reach a depth greater than 100 km. With the increased temperatures, the plate releases water and gases. The melted crust and gases rise to Earth's surface under pressure and may erupt in a volcano.

Finally, a volcano may form at a divergent plate boundary. As the plates move away from each other, magma can rise through the gap. These rift volcanoes occur mostly below sea level and account for the other 15% of the earth's volcanoes.

Volcanoes grow in one of two ways. They can grow by intrusion, when magma moves up inside the cone and remains there without erupting through the vent. Volcanoes can also grow by

Volcano Formation

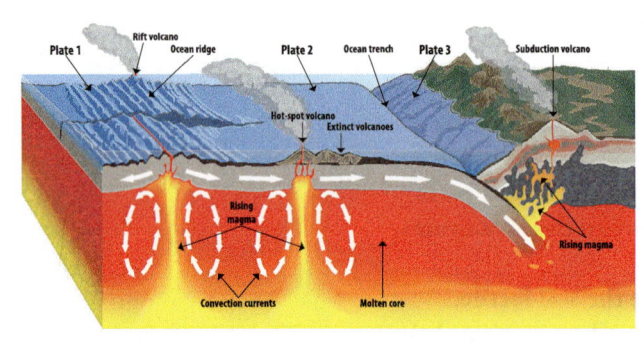

150

FYI

SAR and InSAR images
Satellite data helps volcanologists predict volcanic eruptions. Using SAR (Synthetic Aperture Radar) and Interferometric Synthetic Aperture Radar (InSAR), scientists have been able to measure the deformation of volcanoes prior to eruption. InSAR technology uses two or more SAR images to detect tiny surface changes and movements. The SAR image below shows Teide, a volcano in the Canary Islands. NASA and other space agencies monitor volcanic activity around the world using satellites with InSAR technology. InSAR can precisely map large areas that are difficult to reach on land. Knowing that a volcano is getting ready to erupt gives authorities time to alert citizens and get them to safety before the eruption occurs.

extrusion, when magma and other materials come through the vent and pile up as lava on the outside of the cone.

Although volcanoes are regarded as very destructive, they are actually an important process that God uses to sustain creation. Volcanic soil can be rich in minerals and good for growing grapes, coffee, and tea. Eruptions release gases that are trapped in the ground. Just as the vents of your house carry heat away from the furnace to every room of your house, volcanoes are part of the earth's intricate venting system. Think about the tremendous power and heat from all the volcanoes that have erupted since the time of creation. Imagine what would have happened if this heat and pressure had not been able to escape through volcanoes.

LESSON REVIEW
1. Describe three ways a volcano can form.
2. How do volcanoes help scientists study the interior of the earth?
3. How do volcanoes grow?
4. What important roles do volcanoes play in creation?

151

Name: _____ Date: _____

Lab 3.3.1A How Volcanoes Form

QUESTION: What creates the different features of a volcano?
HYPOTHESIS: Answers will vary.

EXPERIMENT:

You will need:	• plastic tubing	• clamp to fit plastic tubing
• box	• balloon	• flour
• newspaper	• tape	• metric ruler

Steps:
1. Line the box with newspaper. Punch a hole through the middle of the bottom of the box and the newspaper. The hole should be just large enough to allow the plastic tubing through.
2. Pass the tubing through the hole.
3. Tape and seal the deflated balloon on the end of the tubing inside the box.
4. Carefully pile flour on top of the balloon and the surrounding area.
5. Slowly blow through the tubing from the other end to inflate the balloon to approximately 10 cm in diameter. Then clamp the outside end of the tube to hold the air in the balloon.
6. If any part of the balloon is showing through the flour, cover it and mold the flour into the shape of a volcano cone.
7. Illustrate and label your observations.
8. Release the clamp to deflate the balloon.
9. Illustrate and label your observations.

Balloon Volcano Observations

Inflated Balloon Volcano	Deflated Balloon Volcano
Illustrations will vary. The first illustration should show the shape of the volcano and the balloon inflated underneath. The outside should be labeled as the volcano or the cone. The inflated balloon should be labeled as the magma chamber or as magma.	The second illustration should show the volcano with a crater at the top. In addition, the flour (rock) may be shown on the sides, up in the air, or around the bottom. The deflated balloon should be labeled as an empty magma chamber.

152

Lab 3.3.1A How Volcanoes Form

ANALYZE AND CONCLUDE:
1. How did the volcano form? The balloon filled up with air and pushed the flour out to form a volcano.
2. What part of a volcano does the balloon represent? The balloon represents the magma chamber filling up with magma.
3. What formed the crater in the volcano? The crater was formed when the balloon lost all its air.
4. How is this similar to a crater forming in a real volcano? A crater can form in a volcano when it erupts and releases lava and gases just like the crater formed in the flour when the balloon released all its air.
5. What factors might affect the size of a volcano's crater? The amount of magma released or the amount of energy released might affect the size of the crater formed.
6. Design an experiment to test this hypothesis. Experiments will vary. Possible experiments include varying the amount of flour used or the amount of air released from the balloon.

Name: _____ Date: _____

Lab 3.3.1B How Magma Moves

QUESTION: What determines where and how magma will move in a volcano?

HYPOTHESIS: Answers will vary.

EXPERIMENT:

You will need:	• 2 hot pads	• large syringe
• small bowl of set, unflavored gelatin	• large bowl	• chocolate syrup
• aluminum pan	• 2–4 bricks or large blocks	

Steps:
1. Punch 3 or 4 holes in the pan, just large enough for the syringe to go through.
2. Heat about 2 cups of water. Using hot pads, pour the hot water into the large bowl. Dip the small bowl in the hot water to loosen the gelatin.
3. When the gelatin has slightly loosened from the bowl, place the pan on the gelatin bowl. Turn the pan and bowl over together to release the gelatin from the bowl. The set gelatin represents a volcano.
4. Place the pan with the gelatin volcano on top of the bricks. This pan should be raised up enough to get the syringe underneath and through the holes.
5. Fill the syringe with chocolate syrup, which represents the magma.
6. Predict what will happen when you inject the syrup into the gelatin volcano. How will it travel through the gelatin volcano? Will it break through the surface? Where? Answers will vary.

7. Insert the syringe through one of the holes in the pan and into the gelatin volcano. Very slowly inject the syrup.
8. Record your observations. Answers will vary.

9. Repeat Step 7 using a different hole in the pan.
10. Record your observations. Answers will vary.

Lab 3.3.1B How Magma Moves

ANALYZE AND CONCLUDE:
1. What caused the chocolate syrup to move through the gelatin? the pressure from the syringe
2. What causes magma to move through a volcano? the pressure built up from convection currents and gases
3. What do you think determined the path the syrup would take through the gelatin? the direction the syringe was pointing and weak spots in the gelatin
4. What factors affect the path magma takes through a volcano? the amount of magma, the amount of pressure, and where the thinner or weaker spots are in the volcano

3.3.2 Volcano Zones

Volcanoes

OBJECTIVES

Students will be able to
- explain where most volcanoes occur and why.
- compare the volcanoes along the Mid-Atlantic Ridge with the volcanoes along the Ring of Fire.

VOCABULARY

- **rift** a deep crack that forms between two tectonic plates as they separate
- **subduction zone** a place where one tectonic plate is pushed under another tectonic plate

MATERIALS

- Dark-colored tealight or votive candle, 500 mL beaker, sand, hot plate (*Introduction*)
- TM 3.3.2A Ring of Fire
- WS 3.3.2A Volcano Zone

PREPARATION

- Obtain materials for *Try This: Calm Volcanoes*.

TRY THIS

Calm Volcanoes
- metal pans, 1 per group
- mud
- pieces of cardboard, 1 per group
- metric rulers, 1 per group
- scissors, 1 per group

(**The cardboard represents a plate with a rift. The mud represents the magma. The mud oozed up when we pressed on the cardboard. Magma flows up through a rift at the Mid-Atlantic ridge just like the mud moved up through the cut in the cardboard.**)

Introduction

Ask students to explain what a volcano is and where volcanoes can be found. (**Answers will vary but should include on plate boundaries or over hot spots.**) Explain that many volcanoes are below sea level, but can eventually grow to be islands above sea level. Demonstrate what an underwater eruption looks like by placing a dark-colored tealight or votive candle in a 500 mL beaker. Cover the candle completely with sand. Slowly add water until the beaker is almost completely full. Place the beaker on a hot plate and turn the heat to the medium-high heat setting. Wait for the wax to melt and bubble to the top. Ask students to explain what is happening. (**Answers will vary.**) How do you think islands can form from underwater volcanoes? (**Answers will vary but should include that lava from the eruption builds up to form the island.**)

Discussion

- Using **TM 3.3.2A Ring of Fire**, discuss the location of many of the earth's volcanoes. Point out the Ring of Fire. Inform students that nearly 90% of all active volcanoes form in this area. Have students refer to Lesson 3.2.2 and examine the plate boundaries on the map in the Student Edition. Ask how the locations of the volcanoes correspond to the plate boundaries. (**Answers will vary but should include most volcanoes are on plate boundaries.**) Are there more islands in the Pacific Ocean or the Atlantic Ocean? (**Pacific**) Why? (**Answers will vary.**) Explain that most islands are formed by volcanic activity. Most of the islands in the Pacific were formed by volcanic activity along the subduction zones. Point out volcanic islands around the world, such as Japan, the Philippines, the Hawaiian Islands, the Aleutian Islands, Indonesia, and Iceland. Remind students that both oceans experience volcanic activity, but that the activity in the Atlantic is more quiet because it occurs along the Mid-Atlantic Ridge, a divergent plate boundary.

- Discuss possible relationships between earthquakes and volcanic activity.

- Ask the following questions:
  1. What is a subduction zone? (**A subduction zone is a place where one tectonic plate is pushed under another tectonic plate.**)
  2. What happens to the crust when it is subducted? (**It is melted by the heat of the mantle.**)
  3. Why are the volcanoes more violent around the Pacific Ocean than they are in the Atlantic Ocean? (**The Pacific Plate is the site of numerous subduction zones. The collision of the plates adds energy. In the Atlantic Ocean, divergent plates move away from each other, allowing the new magma to flow in.**)
  4. What would happen if there were no cracks, fissures, faults, or plate boundaries in the earth's crust? (**Answers will vary.**)

Activities

A. Complete *Try This: Calm Volcanoes* with students.

B. Have students complete **WS 3.3.2A Volcano Zone**.

C. Direct students to contact volcanologists via the Internet with questions about the Ring of Fire. Have students share their results with the class.

D. Challenge students to investigate the sources and uses of geothermal energy. Assign students to make a computer presentation of their findings.

Lesson Review

1. Where do most volcanoes occur? Why? (**Most volcanoes occur on plate boundaries because that is where magma can escape.**)

2. How are volcanoes formed along the Mid-Atlantic Ridge? (**Volcanoes along the mid-Atlantic ridge are formed along divergent plate boundaries. As the plates separate, magma from the mantle can come up through the rifts between the plates.**)
3. How are volcanoes formed along a subduction zone? (**As the oceanic plates subduct under the continental plates, the collision causes rock to melt and gases to be released. The gases and melted rock then find their way up to the surface.**)
4. Why are volcanoes along a subduction zone more violent than volcanoes on the Mid-Atlantic Ridge? (**The heat generated by collisions in subduction zones melts the mantle that is trapped above the subducted plate, which increases the pressure on the magma. The pressurized magma is then forced to the surface.**)
5. Why are the volcanoes formed from oceanic-oceanic plate collisions the most violent? (**Volcanoes formed by these plate collisions are the most violent because the magma in these volcanoes contains dissolved water and carbon dioxide. When the magma reaches the earth's surface, the gases are released in very explosive, lava-spewing eruptions.**)

NOTES

3.3.2 Volcano Zones

OBJECTIVES
- Explain where most volcanoes occur and why.
- Compare the volcanoes along the Mid-Atlantic Ridge with the volcanoes along the Ring of Fire.

VOCABULARY
- **rift** a deep crack that forms between two tectonic plates as they separate
- **subduction zone** a place where one tectonic plate is pushed under another tectonic plate

No matter where you stand on the earth, you are standing far above the mantle. The earth's entire crust lies above the mantle, which holds the heat that forms volcanoes. For that reason, you might expect to find volcanoes randomly scattered over the earth, but that is not the case.

Just like earthquakes, most volcanoes occur along plate boundaries, where the heat from the mantle can escape. According to the theory of plate tectonics, the plates do not rest quietly on the earth's surface; the edges of the plates are continuously being formed and destroyed.

One large divergent plate boundary between two oceanic plates is the Mid-Atlantic Ridge. One plate moves toward Europe, and the other moves toward North America. As the plates separate, heat, gases, and magma can escape through the **rifts**. Volcanic eruptions are fairly mild along the Mid-Atlantic Ridge where new land is being formed. For example, scientists believe Iceland was formed by magma that escaped and cooled, just like the nearby island of Surtsey. The mild volcanic eruption that formed Surtsey began in 1963. By 1967, the volcano had become the island of Surtsey, with an area of 2.5 km² and an altitude of 170 m.

On the other side of the planet, however, volcanic eruptions are not so quiet. Many subduction zones occur in the Pacific Ocean where the oceanic and continental plates meet. A **subduction zone** is a place where one tectonic plate is thought to push under another tectonic plate. The dense oceanic plate subducts under the continental plate because the basalt is heavier than the granite. The old crust of the earth is then destroyed as the plates grind into each other. Scientists have identified three types of plate collisions: oceanic-oceanic, oceanic-continental, or continental-continental. The heat generated by these collisions melts the mantle trapped above the subducted plate, which increases the pressure on the magma. The presurized magma is forced to the surface, and unlike the mild eruptions on the Mid-Atlantic Ridge, these eruptions roar.

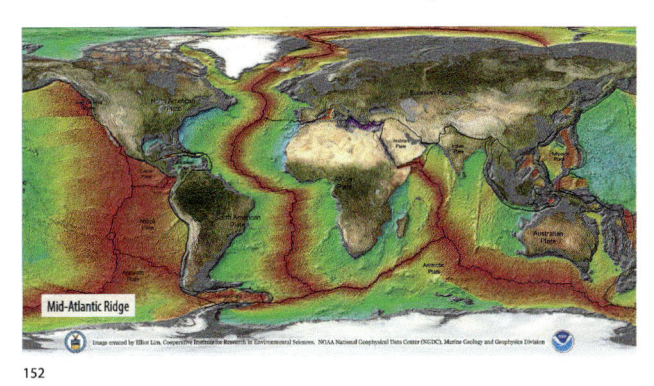

Approximately 90% of all volcanoes form along the subduction zones bordering the Pacific Plate, the Cocos Plate, and the Nazca Plate, which subduct under the continental plates they touch. This rim of the Pacific Ocean is so famous for its volcanoes that it is called *the Ring of Fire*. Scientists noticed and described this fiery ring of volcanic activity even before the plate tectonics theory was developed. Other volcanoes in the Ring of Fire can form through oceanic-oceanic collisions. Because the magma in these volcanoes contains dissolved water and carbon dioxide, when the magma reaches the surface the gases are released in very explosive, lava-spewing eruptions.

In contrast, the rock on both plates of a continental-continental plate collision is too light to sink into the mantle, so instead of one plate subducting under the other, the edges of the plates crumble and fold. The Himalayan Mountains in central Asia are

FYI
Island Laboratory
The island of Surtsey offers a unique natural laboratory that is untainted by human population. Volcanologists, biologists, and botanists have the opportunity to study the genesis of a geological formation, the colonization of species, and the new growth of plant life. Three years after the formation of the island, birds were found nesting on the island, including the fulmar and guillemot. Lichen and mosses have been observed as well. The island has been protected since its formation and authorities only permit scientists onto to the island for research purposes.

TRY THIS

Calm Volcanoes
Volcanoes formed along the Mid-Atlantic Ridge are not very violent. Because the plates are spreading apart instead of colliding, very little energy is added to the magma that escapes through the rifts. Fill a pan with thick mud and cut a piece of cardboard to fit the pan. Cut two slits in the cardboard, about 5 cm wide and at least 12 cm long. Press the cardboard down onto the mud. What does the cardboard represent? What does the mud represent? What did the mud do when you pressed on the cardboard? How is this like a volcano along the Mid-Atlantic Ridge?

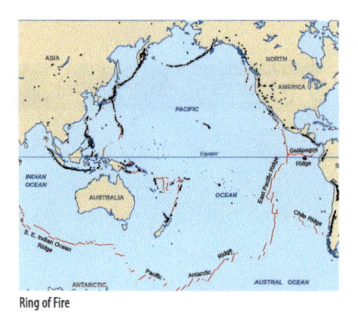

thought to have formed this way when the Indo-Australian Plate crashed into the Eurasian Plate.

Many volcanic islands resulted from plate collisions that produced violent eruptions. Three documented examples are Tambora, Indonesia, in 1815; Krakatoa, Indonesia, in 1883; and Mount Pinatubo, Philippines, in 1991. Volcanic islands are often part of an arc-shaped group. The arc shape is evident in the islands of Japan, the Aleutian Islands extending from the southwest coast of Alaska, and the Timor-Java-Sumatra chain in Indonesia.

Ring of Fire

LESSON REVIEW
1. Where do most volcanoes occur? Why?
2. How are volcanoes formed along the Mid-Atlantic Ridge?
3. How are volcanoes formed along a subduction zone?
4. Why are volcanoes along a subduction zone more violent than volcanoes on the Mid-Atlantic Ridge?
5. Why are volcanoes formed from oceanic-oceanic plate collisions the most violent?

Map of Indonesia

Volcanoes

3.3.3 Volcanic Material

Introduction
Display several different types of ash and melted or burnt items. Items may include wood ash, burnt marshmallow ash, melted plastic, or burnt toast. Ask students to identify what the items were before they were exposed to high heat. (**Answers will vary.**) Why are the results, or products, of combustion all different? (**Answers will vary but should include these were different substances to begin with so they burn differently, they may burn at different rates, or they were exposed to different temperatures.**)

Discussion
- Introduce the concept of viscosity by asking students to name fluids with different flow rates, such as water and corn syrup. Inform students that semifluids, such as lava, also flow at different rates. Viscosity is the resistance of fluids or semifluids to flow. Explain that the viscosity of lava is connected to the amount of silicate it contains. The higher the silicate content in lava, the more viscous, or resistant to flow, it becomes.

- Discuss the different types of pyroclastic particles—pumice, ash, lava blocks, and volcanic bombs. Explain that the size of the particles is used to classify them. Ask if these materials are extrusive or intrusive. (**extrusive**) How do these particles travel? (**They are thrown out from underground by the force of a volcanic eruption.**) Discuss the kinds of damage pyroclasts can cause.

- Display **TM 3.3.3A Plutons and Pyroclasts**. Discuss the visible differences between plutons and pyroclasts. Explain how the different cooling times affect the size of the crystals contained in each. Point out that the surface of a pluton is smoother than pyroclastic rock because it forms and cools inside the earth in an enclosed space.

- Explore the destruction of Sodom and Gomorrah by having students read **Genesis 19:23–29**. Explain that Zoar lies along the Dead Sea rift zone where volcanoes erupted thousands of years ago. Some people believe that the burning sulfur in this passage could have been cinder and ashes from a volcanic eruption. Discuss this idea and the Scriptural account of the cities' destruction.

- Ask the following questions:
  1. What substances are released in a volcanic eruption? (**extrusive elements, such as lava, pumice, volcanic ash, volcanic blocks, and volcanic bombs**)
  2. What makes a volcanic block different from a volcanic bomb? (**A block leaves the volcano as solid rock, and a bomb leaves the volcano as molten rock. They both have the same minimum size of 64 mm.**)
  3. Describe the different types of plutons. (**Sills are long and horizontally shaped. Laccoliths are also horizontal but are larger than sills and have a dome shape. Vertical plutons are called *dikes*. Dikes can vary in thickness. Batholiths can be made up of several plutons. They are the largest plutons and must have a surface area of at least 100 km².**)

Activities

Lab 3.3.3A Viscosity
- corn syrup
- molasses
- isopropyl alcohol
- vegetable oil
- waxed paper
- pipettes, 5 per group
- 20 cm × 30 cm pieces of cardboard, 1 per group
- tape
- metric rulers, 1 per group
- stopwatches, 1 per group

A. Complete *Try This: Plutons* with students. Have students explain how plutons form. Ask students if plutons are intrusive or extrusive rock. (**intrusive**)

OBJECTIVES
Students will be able to
- compare pahoehoe lava to aa lava.
- explain why intrusive rock has larger crystals than extrusive rock.
- classify volcanic materials as plutons or pyroclasts.

VOCABULARY
- **aa** lava that has a rough surface
- **pahoehoe** lava that has a smooth or billowy surface
- **pluton** a body of magma that has hardened underground
- **pyroclast** a solid volcanic material such as ash and rock that has been ejected during an eruption
- **volcanic bomb** a fragment of molten rock that is shot into the air by a volcano

MATERIALS
- Wood ash, melted plastic, burnt marshmallows, burnt toast, other types of ash or melted items (*Introduction*)
- TM 3.3.3A Plutons and Pyroclasts
- TM 3.3.3B Pyroclastic Materials

PREPARATION
- Burn wood, marshmallows, pieces of toast, or other items. Melt plastic or other items. (*Introduction*)
- Obtain materials for *Lab 3.3.3A Viscosity*. Cut cardboard into 20 cm × 30 cm pieces.
- Obtain materials for *Try This: Plutons*.
- Gather samples of pyroclastic materials, such as pumice, aa, pahoehoe, igneous rock, and volcanic ash. (*B*)

TRY THIS
Plutons
- modeling clay
- sculpting tools
- plaster of paris

Encourage students to make different types of plutons.

NOTES

B. Divide the class into groups. Give students samples of pyroclastic materials such as pumice, aa, pahoehoe, volcanic bombs, and ash, or display **TM 3.3.3B Pyroclastic Materials**. Have students describe these materials and use what they know about extrusive crystal formation to hypothesize why the materials have the characteristics they do.

C. Challenge students to research one of the following topics and to create a travel brochure for the specific location or one of their own choosing:
- Dike: The Great Dyke of Zimbabwe
- Sill: Hadrian's Wall, England
- Geyser: Old Faithful in Yellowstone National Park, USA
- Batholith: Cornubian batholith, England

D. Watch a video about a pahoehoe lava flow in Hawaii. Challenge students to research what steps government agencies take to protect people and resources from the flow.

Lesson Review

1. What is the difference between pahoehoe lava and aa lava? (**Pahoehoe is lava that has a smooth surface. It flows easily. The lava on top cools, but the warmer lava underneath keeps flowing under the cooler surface lava. Aa lava has a rough surface and carries rough rock and cooled lava fragments as it flows. Like pahoehoe, the lava under the rough surface keeps moving downhill. Aa lava has a rough layer of rock and cooled lava fragments at the bottom and the top of a flow.**)

Volcanic Material 3.3.3

What is the first thing you think about when you picture a volcano? Most people think of boiling red lava shooting out of a mountain. It is hard to imagine how hot that melted rock is. Lava reaches temperatures of 1,200°C. The main ingredients in lava are silicate and gas. The more silicate in the lava, the less easily it flows. Lava is extrusive; it cools aboveground.

Pahoehoe is lava that has a smooth or billowy surface. This type of lava flows easily. Pahoehoe takes several forms. For example, it can be puffy layers of lava that look like large black clouds sitting on the earth. It can also be ropy-textured lava, which forms when the surface lava hardens but the lava beneath it continues to flow. As pahoehoe makes its way down the sides of the volcano cone, hot lava continues to flow under the cooled, hardened surface lava. After the hot lava has flowed past, the cooled crust forms a lava cave. Within the cave, the cooled cones of lava look like stalactites.

In contrast, **aa** lava has a rough surface. As it flows, it carries rough rock and cooled lava fragments from the volcano. If you walked across aa, its sharp edges would shred the soles of your shoes. The rough surface covers the dense, actively flowing core of lava. As this core of lava moves downhill, the rough fragments float along the surface until they eventually fall down. Some fragments are covered by the moving flow. A layer of rock and cooled lava fragments are found at the top and the bottom of an aa flow. Ancient peoples used aa to make weapons and tools.

Sometimes magma cools and hardens underground. The process is called *intrusive action* because foreign matter,

OBJECTIVES
- Compare pahoehoe lava to aa lava.
- Explain why intrusive rock has larger crystals than extrusive rock.
- Classify volcanic materials as plutons or pyroclasts.

VOCABULARY
- **aa** lava that has a rough surface
- **pahoehoe** lava that has a smooth or billowy surface
- **pluton** a body of magma that has hardened underground
- **pyroclast** a solid volcanic material such as ash and rock that has been ejected during an eruption
- **volcanic bomb** a fragment of molten rock that is shot into the air by a volcano

Pahoehoe on the coastal plain of Kilauea volcano in Hawaii

Pumice is so light it can sit on a rolled $20 bill.

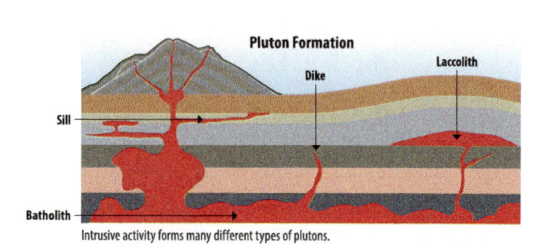

Intrusive activity forms many different types of plutons.

magma, intrudes into the existing rocks. As magma makes its way through the rocks underground, its heat changes the rocks around it. In contrast, lava's heat changes only the rocks underneath the lava. The crystalline textures of rocks differ depending on whether they are extrusive or intrusive. In general, rocks formed from lava, which is extrusive, have small or no crystals. However, the size of crystals in plutonic rock is quite a bit larger. This difference in crystal size occurs because plutonic rock is intrusive, and it takes a long time to cool, providing time for larger crystals to grow.

A body of magma that has hardened underground is called a **pluton**. Many plutons stay buried, but some are uncovered when the surrounding rock is worn away. Plutons form in different shapes. A long, horizontal pluton is called *a sill*. The Romans built part of Hadrian's Wall on a sill in Britain around 120 AD. Laccoliths also intrude horizontally. These plutons form from larger puddles of magma and are characterized by a dome shape. Plutons that run vertically and are usually straight are called *dikes*. Dikes can vary in thickness from a few centimeters to hundreds of meters. The largest intrusion, often made of several plutons, is called *a batholith*. Some batholiths extend to a depth of nearly 15 km and almost reach the base of the earth's crust. By their definition, batholiths must have a surface area of at least 100 km². In North America, the Idaho Batholith is about 40,000 km² and another large batholith forms the entire Sierra Nevada mountain range, including Yosemite National Park.

Volcanic eruptions hurl materials other than lava into the air. **Pyroclasts** are solid extrusive materials such as ash and rocks that are thrown aloft. Ash clouds of finely broken rock fragments and gas pose a serious threat to living things. Volcanic ash is extremely fine but also very sharp. Winds can carry these clouds

Dike in Makhtesh Ramon in Israel

2. Classify the different types of pyroclasts from smallest particle to largest. (**Volcanic ash is made of particles smaller than 4 mm. It is made of finely broken rock fragments. Pumice is a lightweight porous rock. The holes in pumice were caused by trapped gases. Solid rock fragments greater than 64 mm in size are considered volcanic blocks. Volcanic bombs are blobs of molten rock hurled into the air. When the bomb hits the ground it squishes and changes shape. Volcanic bombs are also greater than 64 mm in size. Scientists give the bombs names based on their shape–spindle, pancake, ribbon, or bread-crust bombs.**)
3. What is a pluton? (**A pluton is an intrusive body of magma that has cooled and hardened underground.**)
4. Why are crystals larger in plutonic rock than in other volcanic materials like pyroclasts? (**Plutonic rock is intrusive, so it cools and hardens underground. It takes longer to cool than pyroclasts, which cool aboveground. Because plutonic rock cools more slowly, its crystals have more time to grow.**)

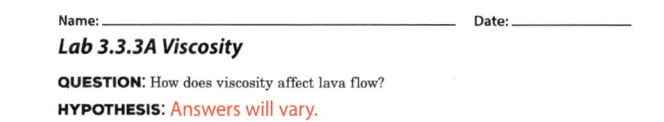

NOTES

Lab 3.3.3A Viscosity

10. Repeat Steps 8–9 for the other four liquids.

Observations

	Corn Syrup	Molasses	Isopropyl Alcohol	Vegetable Oil	Water
Prediction Time					
Actual Time					

Answers will vary.

ANALYZE AND CONCLUDE:

1. In Step 3, which liquid did you predict to be the least viscous? How does your prediction compare with your results? **Answers will vary.**

2. In Step 5, which liquid was the most viscous? **corn syrup or molasses**

3. Was the most viscous liquid in Step 5 the slowest liquid to reach the bottom in Step 10? Why? **Answers will vary.**

4. How do your results in the first trial compare with your results in the second trial?
   Answers will vary.

5. Is the same liquid the least viscous in both cases? **Answers will vary.**
   Is the same liquid the most viscous in both cases? **Answers will vary.**

6. If your results are not the same, explain why you think that is.
   Answers will vary.

7. How does silicate content affect the viscosity of lava? **The greater the silicate content, the more viscous the lava.**

Volcanoes

3.3.4 Types and Eruptions

Introduction
Demonstrate the following types of eruptions using the materials described, or display **TM 3.3.4A Volcano Type Visuals**. Have students write a description of each type. Explain the name of each type of eruption and guide students to label their descriptions accordingly. Ask students which demonstration or photo resembles a volcanic eruption most closely. (**Answers will vary but should include that they all do.**)
Icelandic: Use a baby bottle with several holes punched in the nipple to demonstrate water trickling out as though through a leak.
Hawaiian: Allow water to flow from a squeeze bottle—more than a leak and with a little pressure.
Strombolian: Squirt water from a small squirt toy.
Vulcanian: Squirt water under pressure from a hose or a squirt bottle.
Peléan: Display a picture of water coming from a fire hose.
Plinian: Display a picture of a large waterfall.

Discussion
- Display **TM 3.3.4B Volcano Types** and discuss the different shapes of volcanoes and what types of eruptions created these volcanoes.

- Introduce the Volcanic Explosivity Index (VEI) using **TM 3.3.4C VEI**. Discuss the characteristics in the index and relate these to the types of eruptions. Note that the index includes a classification of Ultra-Plinian and excludes Peléan. Explain that scientists create many different taxonomies and do not always use the same classifications. Ask students if using different systems of classification could be problematic for scientists. (**Yes.**) In what ways could it create problems? (**Scientists using different systems might not understand one another's research because the names are different. Using different systems could create confusion.**)

- Ask the following questions:
  1. What is a volcanic plug? (**It is hardened magma that plugs the volcano's center.**)
  2. How is a caldera different from a crater? (**A caldera is larger than a crater and results from magma erupting or withdrawing from the chamber. A void is left under the overlying rock so it collapses. A crater is smaller and is formed during an eruption when the plug or overlying layer either collapses or becomes part of the pyroclastic flow.**)

Activities
A. Complete *Try This: Volcanic Plug* with students and discuss answers.

B. Have students complete **WS 3.3.4A Comparing Eruptions**.

C. Direct students to complete **WS 3.3.4B Artistic Interpretations**.

D. Assign students to make a computer presentation about or a papier–mâché model of a specific volcano. Have students include facts such as location, activity, lava type, volcanic classification, eruption type, eruption history, pyroclastic flow, and casualties. Schedule time for students to present their projects to the class.

E. Challenge students to find and read eyewitness accounts of a volcanic eruption, such as Pliny the Younger's account of Mount Vesuvius in 79 AD. Some suggestions are Krakatoa in 1883; Parícutin in 1948; Mount Saint Helens in 1980; Mount Pinatubo in 1991; Eyjafjallajökull in 2010, or any other eruption with an eyewitness account. Have students answer the following questions:
  1. When and where did the eruption happen?
  2. Who wrote the eyewitness account?

OBJECTIVES
Students will be able to
- describe three categories of volcanoes.
- relate the type of volcanic eruption to the kind of volcano produced.
- identify the components of a volcanic eruption and how they affect the type of eruption that occurs.

VOCABULARY
- **caldera** a volcanic crater that is greater than 2 km in diameter and is formed by the collapse of surface rock into an empty magma chamber
- **plug** a structure of hardened magma that forms inside a vent

MATERIALS
- Baby bottle, empty squeeze bottle, squirt toy, hose, pictures (*Introduction*)
- TM 3.3.4A Volcano Type Visuals
- TM 3.3.4B Volcano Types
- TM 3.3.4C VEI
- WS 3.3.4A Comparing Eruptions
- WS 3.3.4B Artistic Interpretations

PREPARATION
- Obtain materials for *Try This: Volcanic Plug*.

TRY THIS
Volcanic Plug
- 2 L bottles, 1 per group
- baking soda, 5 mL per group
- vinegar, 120 mL per group
- bathroom tissue, 7.5 cm per group
- corks, 1 per group

This activity should be done outside. If corks are not available, make plugs using potatoes and the top of a 2 L bottle by cutting a 2.5 cm cube from the potato. Press the mouth of the bottle into the center and twist to cut about halfway through the cube. Carefully remove the bottle and take out the potato core. The potato cube is the plug.

NOTES

3. What details about the eruption were given?
4. From the description, what type of eruption do you think it was? Why?
5. What details about people were given?
6. How do you think the eyewitness felt about the event? Why?
7. Do you think the account is truthful? Why?

Encourage students to share their answers to the questions with the class. As a class, discuss how knowing God's Word can help people cope with the destruction of a volcanic eruption.

Lesson Review

1. What are three categories of volcanoes and how are these determined? (**The three categories are active, dormant, and extinct. Active volcanoes have erupted recently or will erupt soon. Dormant volcanoes have no recent history of eruptions and show no signs of eruption. Extinct volcanoes can no longer erupt. Usually, the only thing left is a plug.**)
2. What is a shield volcano? Describe the two types of eruptions that can form shield volcanoes. (**A shield volcano has a gentle slope and is spread out over a large area. It is made mostly of lava flows. Shield volcanoes can be formed by Icelandic or Hawaiian eruptions. These eruptions contain little to almost no gas and the lava is thin. Icelandic eruptions are mild and Hawaiian eruptions are mild to moderate in nature.**)
3. Compare a cinder cone to a composite volcano. Describe their physical features. (***cinder cone***: **small, cone-shaped volcano made of ash and cinders, one vent, rarely more than 300 m high;** ***composite volcano***: **much larger and steeper than cinder cone, alternating layers of pyroclastic materials and lava, sides reinforced with many dikes**) Specify the eruptions associated with each.

3.3.4 Types and Eruptions

OBJECTIVES
- Describe three categories of volcanoes.
- Relate the type of volcanic eruption to the kind of volcano produced.
- Identify the components of a volcanic eruption and how they affect the type of eruption that occurs.

VOCABULARY
- **caldera** a volcanic crater that is greater than 2 km in diameter and is formed by the collapse of surface rock into an empty magma chamber
- **plug** a structure of hardened magma that forms inside a vent

Around 1,500 volcanoes dot the earth. According to historical records, approximately 500 of these volcanoes have erupted. A volcano that has recently erupted or will soon erupt is called *an active volcano*. The sides of an active volcano show little erosion. Some active volcanoes release gases, smoke, and ash, and they heat up surrounding lakes, streams, and rocks. A volcano that is currently erupting is called *an alive volcano*.

Volcanoes can go dormant just as plants go dormant in the winter until warmer temperatures allow them to grow. A dormant volcano is a one that has not erupted for a long time but may erupt again. After a volcano unleashes its built-up pressure, it can sit for years without erupting. Volcanoes that have no recent history of eruptions and that show no signs of eruption are classified as *dormant volcanoes*. But these volcanoes are not dead—they are only sleeping. For example, in 1991, one of these sleeping giants awoke. Mount Pinatubo in the Philippines erupted after 600 years of being dormant.

Hardened magma can form a **plug** in the volcano's center. Weathering and erosion wear down the outside of the volcano, but the plug resists these forces. After the outer material has weathered and eroded away, the volcano's vent remains filled with a column of cooled magma sticking up out of the earth. When only a volcanic plug remains, the volcano is usually considered extinct. An extinct volcano is a volcano that can no longer erupt. However, it is hard to tell which volcanoes are truly extinct. The magma below the volcano might be gathering strength and preparing to erupt again. Scientists are careful about declaring a volcano extinct.

Volcanoes erupt differently depending on the magma composition, the temperature of the magma, and the amount of pressure on the magma. Heat, pressure, gases, lava, and pyroclastic materials are all components of an eruption. Eruptions are classified into six different types according to their explosiveness and kind of volcanic material they emit: Icelandic, Hawaiian, Strombolian, Vulcanian, Peléan, and Plinian. The higher the amount of gas, the greater the violence. The varying action of the eruptions creates different sizes and shapes of volcanoes. As eruptions become more violent, cone shapes become steeper.

Roque de Agando is an exposed volcanic plug on La Gomera in the Canary Islands.

A shield volcano is a gently sloping volcano built almost entirely of lava flows. Some shield volcanoes have several vents. These vents spread the lava over wide areas. For example, Mauna Loa in Hawaii has an area of 5,271 km². Shield volcanoes form from Icelandic and Hawaiian eruptions. An Icelandic eruption has very little gaseous content with thin lava that pours out from several vents. A Hawaiian eruption is a mild to moderate eruption of gases and thin lava through one vent or several fissures. Neither of these eruptions is very explosive. The lava exiting the volcano has low viscosity and has very few gases mixed with it, so the

---

TRY THIS

Volcanic Plug
The existence of a plug in a volcanic vent can create large amounts of pressure inside the volcano. The gases build up pressure because they are contained under the plug. Eventually, so much pressure can build up that the plug is blown out of the vent. Create a simulation of an eruption with a volcanic plug. Pour 120 mL of vinegar into an empty 2 L bottle. Place 5 mL of baking soda on a 7.5 cm strip of bathroom tissue. Spread the baking soda in a line; roll the paper around the baking soda. Keep the baking soda secure by twisting the ends of the tissue. Drop the soda packet into the bottle and quickly place the cork in the opening, barely sealing the bottle. Stand about 1 m away from the bottle and watch what happens. Did the cork come out of the bottle? Did the "magma" erupt? Would you call the eruption gentle or violent? What caused the cork to come out of the bottle? What does the cork represent? What do the baking soda and vinegar produce that is similar to a product of a volcanic eruption?

HISTORY

Mount Vesuvius and the City of Pompeii
Italy's Mount Vesuvius erupted on August 24 in 79 AD. Within two days, 6–7 m of ash and volcanic debris had choked the city. But this devastating ash also shielded the ruins from air and weather. The buried city of Pompeii was discovered in the late 16th century, and systematic excavations began in the early to mid-18th century. Excavators found the city preserved in the ash. When archaeologists uncovered the city, it was so well preserved that they could see the features of the victims' faces. Pliny the Younger witnessed the destruction of the city of Pompeii by Mount Vesuvius. He watched from a safe distance as most of the inhabitants of the city were buried in ash. Pliny wrote in his journal about the flashes of light and the flames. The preserved ruins of Pompeii, along with Pliny's eyewitness account, taught the world a great deal about volcanic activity, Roman culture, and how people lived in the ancient world.

(*cinder cones*: sometimes formed from Strombolian eruptions; *composite volcano*: Vulcanian, Peléan, or Plinian)

4. List the eruptions in order from least violent to most violent. (**Icelandic, Hawaiian, Strombolian, Vulcanian, Peléan, Plinian**)

5. What are the components of an eruption? (**heat, pressure, gases, lava, and pyroclastic materials**) Give three examples of eruptions and how their components affect the eruption. (**Possible answers:** *Icelandic*: very little gaseous content, thin lava, several vents; *Hawaiian*: mild to moderate eruption of gases, thin lava, one vent or several fissures; *Strombolian*: runny lava, steam or gases, pyroclasts, volcanic bombs, single vent; *Vulcanian*: dark ash, thick lava, steam, gases, *Peléan*: fine ash, thin lava, gases, pyroclastic material, superheated steam; *Plinian*: hot ash clouds, pyroclastic flows, thick magma, abundant gases)

NOTES

Page 160

HISTORY

Birth of a Volcano
Parícutin, a volcano in Mexico, is a cinder cone. On February 20, 1943, smoke began to rise from a cornfield 320 km west of Mexico City. Several weeks prior, residents felt tremors and rumblings from the earth, hints of the tumultuous events to come. The newly formed volcano spewed lava and ash from a growing vent. It rose 50 m in one day; within a week it had more than doubled. When it was finished developing, Parícutin was over 350 m tall and had swallowed a town. By 1952, Parícutin was no longer erupting. But during those nine years, this fast and furious volcano excited the world and expanded volcanologists' knowledge of volcanoes.

Shield volcano   Cinder cone   Composite

lava flows easily from a shield volcano, oozing down like syrup over pancakes.

A cinder cone is a small, cone-shaped volcano built of ash and cinders. Cinder cones have a single vent. During an eruption, lava rushes through the vent and bursts up toward the sky. As it cools in the atmosphere, the fragmented lava falls back down and adds to the size of the cone. Each eruption adds more ash to the cone. Although the ash layers are thick, cinder cones are smaller than other volcanoes—rarely more than 300 m high. Cinder cones are sometimes formed from a Strombolian eruption, an explosive eruption of runny lava and steam or gases. Strombolian eruptions also shoot out rocky pyroclasts and volcanic bombs.

Composite volcanoes are the largest group of volcanoes. A composite volcano is a steep-sided volcano composed of lava, ash, and cinders, which are particles of burned material. Sometimes composite volcanoes shoot rock fragments and ash into the sky. At other times, thick lava flows down the steep sides of these volcanoes, so they have alternating layers of pyroclastic material and lava. The lava of composite volcanoes is thick and moves slowly, and it often carries large chunks of pyroclastic materials with it. Sometimes magma does not reach the surface, so it forms dikes within the volcano walls. The dikes add strength and structure to the walls, giving the volcano the support necessary to grow taller. Italy's Mount Vesuvius, Sicily's Mount Stromboli, Indonesia's Krakatoa, the Philippines' Pinatubo, and Japan's Mount Fuji are all composite volcanoes.

Composite cones form from Vulcanian, Peléan, and Plinian eruptions. A Vulcanian eruption is an eruption composed of bursts of dark ash, thick lava, steam, and gases. Vulcanian eruptions can shoot gases and ash 5–10 km into the air. Peléan eruptions are named after Mount Pelée, a volcano in the Caribbean. Peléan eruptions have a spectacular fiery cloud of fine ash, thin lava, gases, pyroclastic material, and superheated

---

Page 161

steam. Weighed down by its heavy load, this cloud does not float over the volcano peak but rather rolls swiftly down the volcano's sides like an avalanche. These clouds are extremely destructive.

A Plinian eruption is an eruption characterized by hot ash clouds, deadly pyroclastic flows, or both. The most violent eruption, Plinian eruptions often start unexpectedly after a long period of quiet volcanic activity. Thick magma and abundant gases explode deep in the volcano, and the gases shoot upward into a high cloud. The volcano's top collapses, forming a large crater.

Sometimes large circular depressions called **calderas** form at the top of a volcano. Exceeding the maximum diameter of a crater at 2 km or more, calderas form when magma is withdrawn or erupts from a magma chamber, leaving a void beneath the surface. The overlying rock collapses into the void, forming a caldera.

LESSON REVIEW
1. What are the three categories of volcanoes and how are these determined?
2. What is a shield volcano? Describe the two types of eruptions that can form shield volcanoes.
3. Compare a cinder cone to a composite volcano. Describe their physical features. Specify the eruptions associated with each.
4. List the eruptions in order from least violent to most violent.
5. What are the components of an eruption? Give three examples of eruptions and how their components affect the eruption.

The Aniakchak Caldera in Alaska is 10 km wide and 762 m deep, formed during a massive volcanic eruption 3,500 years ago.

Strombolian eruption, Italy

3.3.5 Prediction and Effects

Volcanoes

OBJECTIVES

Students will be able to
- describe the changes volcanologists measure to predict volcanic eruptions.
- identify the negative and positive effects of a volcanic eruption.

VOCABULARY

- **tiltmeter** an instrument that uses liquid to register changes in the earth

MATERIALS

- 2 paper cups, drinking straw, modeling clay, large flat pan, water, straight pin (*Discussion*)
- Volcano video (*B*)
- News articles, eyewitness accounts, scientific journal articles on a recently predicted volcanic eruption (*C*)
- Young-earth and old-earth resources on volcanoes. (*D*)
- TM 3.3.5A Volcano Damage

PREPARATION

- Assemble the tiltmeter model by putting one hole just large enough for the end of the straw in the side of each cup. Insert one end of the straw into the holes. Seal around the openings with modeling clay. Fill each cup with water just above the holes. (*Discussion*)
- Gather materials for *Try This* activities.

TRY THIS

Carbon Dioxide
- short candles and holders, 1 per group
- deep bowls, 1 per group
- vinegar, 240 mL per group
- matches
- baking soda, 20–30 g per group

The sides of the bowl must be taller than the lit candle. The holder should be heavy enough to keep the candle from floating. Do not use too much baking soda—the candle will get wet and go out instead of the CO_2 pushing out the oxygen.

Introduction

Divide the class into groups and have students discuss the following questions:
- Do you think that volcanic eruptions can be predicted? If so, how?
- If written records do not exist, how do scientists know the last time a volcano erupted?
- If it was possible, should scientists stop volcanoes? Why?
- How would the earth be different without volcanoes?

Have student groups present their opinions to the rest of the class. (**Answers will vary.**)

Discussion

- Read **Isaiah 61:3** and **Romans 8:28**. Challenge students to think about a difficult time after which God faithfully orchestrated a good result. Christians can trust God to work things for good according to His purpose. A volcanic eruption can be a horrible tragedy, but God can bring beauty from ashes.

- Demonstrate how a tiltmeter works and how it can help forecast eruptions. Set the prepared tiltmeter on the pan. Prop up one end to simulate the tiltmeter sitting on the side of a volcano. Use the straight pin or thumbtack to poke pinholes in each cup to mark the water levels. Then change the tilt of the pan, which represents the volcano, modeling the changes that can take place when magma and gases build up below the surface. Compare the new water levels to the pinholes. Repeat the process, tilting the "volcano" in a different way. Indicate that a tiltmeter is like a level.

- Display **TM 3.3.5A Volcano Damage** and then lead a discussion about why improving technology for predicting volcanoes is important.

- Divide the class into groups to discuss and compare the positive and negative results of volcanic eruptions. Encourage students to suggest ideas beyond the information given in the Student Edition.

Activities

A. Complete *Try This* activities with students and discuss the questions.

B. Show a video about a volcanic eruption and the effect it had on the surrounding area. After the video, discuss how the eruption affected the area. Ask students what the negative effects were for the area around the volcano. (**Answers will vary.**) Were any positive effects mentioned? (**Answers will vary.**) What do you think scientists learned from the eruption? (**Answers will vary.**)

C. Choose a recent volcanic eruption that was anticipated by scientists to analyze as a class. Have students read the provided materials to determine how scientists knew an eruption was imminent and what technology was used to predict the eruption. Ask what preparations, if any, were made by the local community or government before the eruption. (**Answers will vary.**) Under what conditions should the scientific community warn citizens of a possible event before it occurs? (**Answers will vary.**) How should the government respond to such a warning? (**Answers will vary.**) What responsibility do people who live near a volcano have when they are told an eruption may occur? (**Answers will vary.**)

D. Challenge students to research how both young-earth scientists and old-earth scientists determine the age of a volcano. Direct students to interview or write e-mails to both categories of scientists to obtain information. Have students share with the class the information they receive from the scientists they contact.

E. Assign groups of students to research one of the four basic types of volcanologists: physical, geophysical, geodesic, and geochemical. Direct students to write a description of each career, describing it in first person to include field study, lab study, data collected and how it is used, and a leading volcanologist of that type. Have groups present their projects to the class.

F. Encourage students to research technology for monitoring volcanic activity and forecasting eruptions. Assign them to answer the following prompt: How do volcanologists protect themselves when studying active volcanoes?

Lesson Review

1. What changes do volcanologists monitor? (**They monitor any changes in gases, any change in the temperature of the volcano and nearby streams, and any change in the ground size and shape.**)
2. What causes a volcano to bulge? (**Rising magma can cause a volcano to bulge.**)
3. What geologic event can precede a volcanic eruption? (**an earthquake**)
4. What problems can be caused by heavy volcanic ash? (**Volcanic ash can block out the sun, collapse roofs, cause breathing problems for humans and animals, smother crops, pollute rivers, and drop global temperatures.**)
5. What other problems can be caused by volcanic eruptions? (**Pyroclastic flow can tear down trees and destroy buildings.**)
6. What benefits come from volcanic eruptions? (**Possible answers: Volcanic debris makes soil fertile, provides building materials, and can be used in hand soaps and cleaners. Diamonds are brought to the surface and can be used in jewelry and for industrial purposes. Eruptions give scientists the only visible information about the inside of the earth. Volcanoes release trapped atmospheric gases.**)

TRY THIS

Expanding Volcano
- 7" or 9" balloons, 1 per group
- clear 2 L bottles, 1 per group
- vinegar, 230 mL per group
- baking soda, 20 g per group
- bathroom tissue

(**The mixture begins to foam. The balloon inflates. The reaction in the bottle represents the rising magma. The balloon represents the surface of the volcano.**)

3.3.5 Prediction and Effects

OBJECTIVES
- Describe the changes volcanologists measure to predict volcanic eruptions.
- Identify the negative and positive effects of a volcanic eruption.

VOCABULARY
- **tiltmeter** an instrument that uses liquid to register changes in the earth

David A. Johnston

"Vancouver! Vancouver! This is it!" This radio transmission came from volcanologist Dr. David A. Johnston, who was monitoring Mount Saint Helens from a ridge about 8 km north of the mountain in the United States. It was 8:32 AM on May 18, 1980. Scientists knew that Mount Saint Helens was going to erupt; increased earthquake activity, small eruptions of steam and ash, and bulges on the surface warned that magma was rising. Unfortunately, scientists did not know the size of the coming eruption. If they had, Dr. Johnston would not have been on a ridge that close to the mountain. The radio transmission was the last anyone ever heard from him. He and his equipment were never found.

Volcanoes can be frightening, and their effects can be deadly. At times, scientists can predict their eruptions to some degree. Although scientists cannot predict exactly when a volcano will erupt or how large the eruption will be, knowing that a volcano is likely to erupt within a certain time frame can prepare people for when the eruption does happen. In the 1970s, scientists predicted that Mount Saint Helens would erupt before the year 2000. Dr. Johnston and other volcanologists warned that the eruption could be devastating. Careful preparation by local agencies and a public evacuation plan saved many lives.

A change in the type or volume of volcanic gases being released is one sign of a coming eruption. Such an alteration indicates a change in the magma within the volcano. Aircraft are used to fly over volcanic vents and to measure levels of gases like sulfur dioxide and carbon dioxide. Such flights can be risky, so

Car covered in ash after 1991 eruption of Mt. Pinatubo

162

TRY THIS

Carbon Dioxide
Carbon dioxide (CO_2) is a gas that can be produced by volcanoes; it is also in carbonated beverages. In small concentrations, it is pleasant in a drink. But in large concentrations, it can be harmful to humans. If CO_2 levels are too high, people cannot breathe. Set a small candle on a candleholder inside a deep bowl. Pour about 240 mL of vinegar in the bowl to the top of the candleholder. Be careful not to get the candle wet. Light the candle. It should burn brightly. The lit candle represents a person who is breathing normally. Without getting it in the flame, carefully sprinkle 20–30 g of baking soda all around the vinegar. What happens in the vinegar? Wait a few seconds. What happens to the flame? The reaction of the vinegar and the baking soda produces CO_2. How is this similar to a volcano? How is the candle like a person breathing the gases given off by a volcano?

The upward movement of gases under the earth's surface triggers minor movements of the crust. Earthquakes are a sign that a volcano could be waking up. For example, more than 10,000 small earthquakes were recorded in the area before the eruption of Mount Saint Helens. Another clue that a volcano heats up, nearby streams, springs, and rocks also heat up. Steam or smoke

Near Mount Saint Helens, Spirit Lake was still filled with broken trees in 2012.

163

The bulge on Mount Saint Helens grew 2 m a day for three weeks.

Vulcanologist sampling lava on Kilauea, Hawaii

may even escape through Earth's crevices. One way to measure the temperature is through thermal imaging. Scientists mount infrared cameras on helicopters and fly near the volcano. The images provide temperature readings to monitor volcanic activity and indications of cooler areas where scientists can place their instruments.

Volcanologists also employ other technology to measure ground deformation. A **tiltmeter** is an instrument that uses water levels and laser technology to register changes in the earth. Rising magma can cause ground inflation or deflation prior to and during an eruption. The ground may bulge, as was noted before Mount Saint Helen's erupted. Water levels then change with the newly modified angle of the volcano's slope. Scientists measure deformations using information received from electronic distance measuring devices, global positioning satellites (GPS), and radio telescopes. All these instruments provide critical data about the activity inside a volcano.

Although these prediction methods can give people enough warning to leave the area, the effects of the volcanic eruption cannot be predicted. Volcanic ash can block out the sun for days and travel thousands of kilometers. Unlike soft, fluffy ash from burned paper or wood, volcanic ash is hard. It can blanket communities with a "snow" that does not melt and collapse rooftops with accumulations as little as several centimeters. The ash can cause breathing problems for people and animals, smother crops, and pollute rivers. When it mixes with rainwater or water from melted glaciers, it does not dissolve. Instead, it acts like a large mass of wet cement. In addition, pyroclastic flow can tear down trees and destroy buildings.

Volcanoes can change the climate temporarily, although some scientists contend that these changes are long lasting. As volcanic ash and gases from large eruptions reach the upper atmosphere and spread around the globe, they block sunlight, which can drop temperatures around the world. For example, after the Philippines' Mount Pinatubo erupted in 1991, the average global temperature dropped by 0.4°C.

TRY THIS

Expanding Volcano
A volcano can expand or develop a bulge on its side before it actually erupts as a result of magma motion and gas buildup. Take a balloon and stretch it out. Pour 230 mL of vinegar into an empty 2 L bottle. Place 20 g of baking soda into a small piece of bathroom tissue or paper towel and roll into a small package. Drop the baking soda package into the bottle and then quickly stretch the opening of the balloon over the mouth of the bottle. Be sure you have a tight seal. As the baking soda begins to react with the vinegar, what happens in the bottle? What happens to the balloon? Watch the system for about two minutes. What does the reaction in the bottle represent? What does the balloon represent?

Although volcanic eruptions are sometimes devastating, they also provide benefits. Volcanic soil is one of the most fertile types of soil. The lush foliage of the Hawaiian Islands offers proof of this. Rich with calcium, potassium, and magnesium, the volcanic fields are the ideal growing area for many types of plants. Indonesia's agriculture also benefits from its volcanic soil. Volcanic deposits also provide building materials, such as those used to make the concrete shields of nuclear reactors. Pumice, an abrasive pyroclast, is used as an ingredient in many hand soaps and household cleaners. Gems such as diamonds are pushed up to the surface by volcanic activity. Diamonds are popular in fine jewelry, but they are also utilized in industrial applications. The hardness of diamonds makes them ideal for cutting, grinding,

Benchmark installation for GPS surveys on the north flank of Mount Saint Helens, 2012

drilling, and polishing. Volcanic eruptions also provide scientists with the only visible information about the inside of the earth. And along with releasing built-up heat and pressure, volcanoes release atmospheric gases that have been trapped in the ground.

Although volcanoes can be terrifying and destructive, God is faithful to continue supporting the earth and the life on it. Five years after the Mount Saint Helens eruption, established plants were growing again. Within 10 years of the eruption, many animals had returned. Scientists predict that a forest will grow in that area within the next 100 years.

LESSON REVIEW
1. What changes do volcanologists monitor?
2. What causes a volcano to bulge?
3. What geologic event can precede a volcanic eruption?
4. What problems can be caused by heavy volcanic ash?
5. What other problems can be caused by volcanic eruptions?
6. What benefits come from volcanic eruptions?

The rich, volcanic soil in Bali, Indonesia, is perfect for growing rice.

UNIT 4

Water and Water Systems

Chapter 1: *Water*
Chapter 2: *Oceans*

Key Ideas

Unifying Concepts and Processes
- Systems, order, and organization
- Evidence, models, and explanation
- Change, constancy, and measurement
- Evolution and equilibrium
- Form and function

Science as Inquiry
- Abilities necessary to do scientific inquiry
- Understandings about scientific inquiry

Earth and Space Science
- Structure of the earth system
- Energy in the earth system
- Geochemical cycles

Science and Technology
- Abilities of technological design
- Understandings about science and technology

Science in Personal and Social Perspectives
- Populations, resources, and environments
- Natural hazards
- Risks and benefits
- Science and technology in society
- Natural resources
- Environmental quality

History and Nature of Science
- Science as a human endeavor
- Nature of science

Vocabulary

abyssal plain	Coriolis effect	neap tide	tidal range
adhesion	crest	old river	transpiration
algal bloom	dead zone	porosity	tributary
amplitude	desalination	rejuvenated river	trough
aquifer	divide	reservoir	valley glacier
cohesion	eutrophication	salinity	water budget
condensation	evaporation	seamount	watershed
continental glacier	geyser	sinkhole	water table
continental margin	groundwater	spring tide	wavelength
continental rise	hydroelectric power	sublimation	youthful river
continental shelf	levee	submersible	zone of aeration
continental slope	mature river	surface tension	zone of saturation

SCRIPTURE

Whoever drinks the water I give them will never thirst. Indeed, the water I give them will become in them a spring of water welling up to eternal life.

John 4:14

4.1.0 Water

LOOKING AHEAD

- For **Lesson 4.1.2**, obtain a potted plant for use in a transpiration demonstration. Arrange for students to visit a local water treatment plant. Contact a local water authority to determine the amount of water the local region uses compared with the amount of water it receives.
- For **Lesson 4.1.3**, arrange a field trip to a local glacier, if possible.
- For **Lesson 4.1.4**, obtain a groundwater model simulator. Arrange for a hydrogeology graduate student to speak to the class.
- For **Lesson 4.1.6**, invite a river-rafting guide to speak to the class.
- For **Lesson 4.1.7**, obtain materials needed for *Lab 4.1.7A Pond Exploration* and arrange a field trip to a local pond. Gather lake and pond water samples to be used later in the lesson and in **Lesson 4.1.9**. Arrange a guided tour of a lake that was formed by a receding glacier, if possible.
- For **Lesson 4.1.8**, arrange a field trip to a local lock and dam or hydroelectric dam, if possible. Invite an engineer to speak to the class about modern trends in dam construction. Obtain materials for *Try This: Building a Dam*.
- For **Lesson 4.1.9**, obtain a variety of unopened fertilizer bags. Invite a fish and wildlife expert to speak to the class.

SUPPLEMENTAL MATERIALS

- BLM 4.1.5A Global Water Challenges
- TM 4.1.1A Physical States of Water
- TM 4.1.1B Cohesion and Adhesion in Plant Roots
- TM 4.1.2A The Water Cycle
- TM 4.1.3A Muir Glacier 1915 and 2010
- TM 4.1.7A Riparian Food Web
- TM 4.1.7B Light and Dark Phases of Photosynthesis
- TM 4.1.8A Hydroelectric Dam
- TM 4.1.8B Hoover Dam
- TM 4.1.8C Types of Dams

(continued)

Chapter 4.1 Summary

Water is a remarkable molecule. It is one of the only substances on Earth to expand as it cools from a liquid state; virtually every other substance becomes denser as it cools, but water's unique quality produces floating ice. Seventy-one percent of Earth's surface is covered by water. God created water to be the primary ingredient of life. It provides stability against extreme temperature changes and dissolves nearly all vital nutrients necessary for the survival of organisms. Water's molecular structure is the reason it exhibits many unique physical and chemical properties.

Background

Lesson 4.1.1 – Water Properties

Water is a liquid substance composed of molecules that contain one atom of oxygen and two atoms of hydrogen. Pure water lacks color, taste, and smell. It becomes a solid at 0°C and a vapor at 100°C. The density of water is 1 g/cm^3 at 4°C.

The special physical and chemical properties of water are determined by water's molecular structure. Each water molecule has one oxygen atom that is bound to two hydrogen atoms that are separated by an angle of approximately 105°. The shape of a water molecule is bent, producing a slightly positive charge near the hydrogen end and a slightly negative charge near the oxygen end. This separation of charges makes water a polar molecule. Water molecules are attracted to one another, which results in hydrogen bonds. Typically, molecules with such low molecular weights would exist in a gaseous state, but hydrogen bonding is strong enough for water to maintain a liquid state. Hydrogen bonding also accounts for water's high specific heat. The specific heat is the amount of heat required to raise one gram of a substance one degree Celsius. Given water's high specific heat of 4.186 joule/gram °C, water can absorb a tremendous amount of heat. This ability is important for temperature regulation in water systems and during seasonal changes. For this reason, water is widely used for cooling and for transferring heat in thermal and chemical processes.

Water molecules bind not only to each other but also to many other substances in a process called *adhesion*. For example, the molecules in a thin glass tube adhere to the molecules of glass just above them, which draws other water molecules upward with them. The water surface then pulls the entire body of water to a new level until gravity is too great to overcome. This process is called *capillary action*. Without capillary action, the nutrients that plants need would remain in the soil.

Lesson 4.1.2 – The Water Cycle

It is important to recognize that the same water molecules that existed from the beginning of Creation have been transferred repeatedly from the oceans to the atmosphere by evaporation, dropped on the land as precipitation, and transferred back to the oceans by rivers and groundwater. This circulation is known as *the water cycle* or *hydrologic cycle*. At any given time, approximately 5 L out of every 100,000 L of water on Earth are in motion.

Lesson 4.1.3 – Glaciers

Glaciers are in constant motion. The speed of a glacier depends on the glacier's volume, the angle of the ground, the slope of the upper surface of the ice, the amount of water and debris the glacier holds, the air temperature, and the amount of friction present. Friction reduces the speed of the sides and bottom of a glacier. Friction enables the glacier's center structure to move faster than the sides. The surface of a glacier also moves faster than the bottom portion. If the ice at a glacier's edge melts faster than the glacier is moving, the edge of the glacier retreats. If the glacier moves faster than it melts, the edge advances. The glacier remains stationary only if the rate of movement and the rate of melting are equal.

Glaciers change topography through erosion, transportation, and deposition. Valley glaciers carve out cirques at their sources and grind away the bases of slopes and cliffs to transform V-shaped valleys into U-shaped valleys. This process often leaves the outlets of tributary valleys hanging above the new valley floor. Any streams that are present then fall in the form of waterfalls and cascades.

Lesson 4.1.4 – Groundwater

Each year 40,000 km^3 of water flows over the land into streams and rivers as runoff or soaks deep into soil and rock to become groundwater. People may think of groundwater as water that flows through underground rivers or collects in underground lakes. However, groundwater is not confined to a few channels or depressions. It exists almost everywhere underground between particles of rock and soil or in crevices and cracks in rock. At greater depths, these openings are much smaller because of the overlying weight. Therefore, less water is found at greater depths. Most groundwater is within 100 m of the earth's surface. Groundwater flows vertically and horizontally at a rate that depends on the glacial and bedrock geology. Water flows relatively quickly where there is dissolved limestone to form caverns and large openings.

The presence of springs depends on a region's geology. If an impermeable layer of rock underlies a layer of saturated soil or rock, a line of springs may appear on a slope where the clay layer forms outcrops. Springs also tend to appear along the fractures of igneous rocks, where the dissolving action of groundwater enlarges fractures in limestone to form small underground channels and caves, and along major faults where groundwater reaches the surface along the fault plane, which can help pinpoint a fault's position. Springs are a valuable source of water. Sometimes spring water is channeled into a reservoir. The geographic location of springs determined the sites of many ancient city-states, such as Troy, and dictated where North American pioneers settled.

Some groundwater flows to the surface to feed into lakes and streams. So, the quality of groundwater and surface waters is linked. Groundwater contamination can occur from natural sources, such as excessive salts or minerals, or human causes such as pesticides or gasoline spills. Groundwater contaminants include solvents, petroleum products, pesticides, nitrates, salts, sulfates, chromium, lead, viruses and bacteria, and radioactive compounds, such as uranium and tritium. Contaminated groundwater is dangerous because it can be drawn into wells. Increased groundwater pumping rates can sometimes magnify existing groundwater contamination problems or introduce groundwater contaminants into previously unpolluted wells.

Lesson 4.1.5 – Surface Water

The area of land that drains into a particular river system is known as *a watershed*. A watershed is limited by the elevation of the divide that separates it from any neighboring drainage systems. The size of the area, precipitation levels, evaporation, and the rate of water absorption by soil, rock, or vegetation all help determine how much water from the watershed reaches the river system.

Rainwater's natural runoff percentage depends on an area's soil characteristics, such as depth, permeability, and saturation capability. Urbanization, however, greatly affects runoff. As cities and suburban neighborhoods grow, they add acres of hard, impermeable surfaces to the landscape. Rainwater that lands on fields and forests soaks into the soil, but roofs, parking lots, and streets collect rainwater and force it through a drainage system. Unless the drainage system connects to a wastewater treatment plant, the rainwater and everything it picks up is funneled into streams and rivers.

Urbanization plays a major role in groundwater and stream pollution. Some of the pollutants include lead, zinc, copper, chromium, and arsenic. Urbanization also affects water temperature, pH, dissolved oxygen, alkalinity, hardness, and conductivity. Methods of controlling urban runoff include storage tanks, settling tanks, and retention ponds. When a problem is caused by runoff, defining who is liable for the public health and the environmental damages is challenging.

SUPPLEMENTAL MATERIALS

(continued from previous page)
- Lab 4.1.1A The Dissolving Power of Water
- Lab 4.1.2A Cloud Capacity
- Lab 4.1.4A Groundwater Contamination
- Lab 4.1.5A Hard Water
- Lab 4.1.7A Pond Exploration

- WS 4.1.1A Physical Properties of Water
- WS 4.1.2A Amount of Water
- WS 4.1.2B Comparing Water Budgets
- WS 4.1.3A Global Glaciers
- WS 4.1.3B Glaciers in the Water Cycle
- WS 4.1.6A River Creation
- WS 4.1.8A Dams

- Chapter 4.1 Test

BLMs, TMs, and tests are available to download. See Understanding Purposeful Design Earth and Space Science at the front of this book for the web address.

WORLDVIEW

- Water is the key resource for many animals and plants. Without water, most life-forms could not function properly and would probably cease to exist. Another kind of water is referenced in the Bible many times, which is the living water that only God can provide. David yearns for the living water in Psalm 63. In verse 1, he cries out, "I thirst for You, my whole being longs for You, in a dry and parched land where there is no water." Without God, life is empty, and the end result is death. Those individuals who are filled with the living water of Christ see purpose in their lives and are assured of eternal life in God's loving presence. It is the duty of God's children to share this water with others in order that they may no longer thirst for the life-sustaining power of God's living water.

Lesson 4.1.6 – Rivers

Rivers can be categorized into three basic stages: youthful, mature, and old. The channel of a youthful river is V-shaped, has a steep gradient of around 2–3 m of drop per horizontal kilometer, has rapids (generally just downstream of a tributary junction), and has waterfalls because the water has not developed a smooth path. A mature river meanders and has a floodplain not much wider than the meander belt. The river deposits sediment on the floodplain, creating fertile soil. An old river has a gradient of only about 20–40 cm of drop per horizontal kilometer, a wide floodplain, tight meanders, oxbow lakes, natural levees, and possibly swamps. Water flows slowly through old rivers.

A rejuvenated river is one that has experienced an increase in gradient and erosion power. Rivers are rejuvenated when tectonic uplift raises topographic relief. This change in topography results in steeper gradients and in greater potential energy in the flowing water. River volume is affected by precipitation in the drainage basin of the river. A river system may be enlarged when one river cuts through the divide that separates its drainage basin from that of another river and diverts the other river's water into its own channel.

Lesson 4.1.7 – Ponds and Lakes

The density of water and the concentration of dissolved oxygen in water changes with temperature fluctuations. Higher concentrations of dissolved oxygen exist in cooler waters. Water holds its maximum oxygen concentration at 5°C. Temperature fluctuations are most noticeable in shallow waters.

During cold weather months, the surface of a pond cools off rapidly and can freeze. A temperature inversion takes place: warmer water transfers to the lower portions of the pond and cooler water stays at the surface of the pond. Minerals that were originally stored at the bottom of the pond cycle to the top of the pond where they can benefit plants and animals.

Permanent ponds exist year-round. Vernal ponds exist only during a wet season and dry up during the hot summer months. When vernal ponds are filled with water, animal and plant life abounds. As soon as the water evaporates, plants and animals become dormant. Some even form a protective covering or burrow into the mud to hibernate until water again fills the pond. Larger plants will drop seeds that can germinate during the next wet season. Other plant species die back to their roots.

Glacial erosion often carves basins into bedrock to form lakes. Glacial moraine deposits can block preexisting stream valleys to form lake basins. Lakes can also form in calderas after volcanic craters collapse. Groundwater can dissolve limestone deposits to form caves that often contain underground lakes. Sometimes the cave roofs collapse to reveal deep lake basins. Tectonic activity can cause water to fill fault-generating rift valleys to form lakes.

Lesson 4.1.8 – Dams and Reservoirs

Some of the earliest known engineering works were levees built to prevent flooding. For example, ancient Egyptians built a series of levees along the Nile's banks that stretched for about 975 km from Aswan to the Mediterranean Sea. Levees were also built in ancient China and Mesopotamia. One of the largest levee systems today is located along the Mississippi River and its tributaries. This system stretches about 1,600 km from Missouri to the Mississippi delta. Levees have an average height of 7 m and can reach as high as 15 m.

Dams have been constructed for many years to provide a ready supply of water for irrigation and for other purposes. One of the earliest large dams was a marble structure built around the year 1660 in

Rajputana, India. Most modern dams are constructed for multiple purposes such as irrigation, flood control, navigation improvement, and power for hydroelectric plants.

Dams can be made of timber, rocks, earth, masonry, or concrete. Rock-fill dams consist of an embankment of loose rock with either an impermeable core, such as clay, or a watertight face on the upstream side. Earthen dams may be reinforced with a core of concrete or with a watertight upstream surface. Masonry and concrete dams can be arch dams or gravity dams, which depend on their own weight to resist the pressure of the water.

Before a reservoir can be built, the amount and distribution of rainfall, evaporation, runoff, soil or rock conditions, and elevation must be considered. If the ground of the reservoir is not sufficiently impervious to prevent excessive seepage, a lining of clay or other material can be added. The reservoir's retaining walls may be of loose rock, masonry, or earth. These materials form a good embankment but must be sealed with clay and covered with masonry to prevent erosion. Reservoirs can also be built as flood control along rivers, to maintain the water level on canals, to ensure water supply for hydroelectric plants, and to act as catch basins for silt on the tributaries of large rivers. Reservoir water can be lost to seepage or evaporation. Certain chemicals that form a film on the water surface and covered tanks made of prestressed concrete help prevent evaporation losses.

Lesson 4.1.9 – Eutrophication

In the early developing stages of a young lake, very little life can be supported. Over time, streams carry nutrients, such as nitrogen and phosphorus, into the lake. The nutrients encourage the growth of aquatic species. As plant and animal life begin to thrive, organic matter is deposited on the lake bottom. The lake becomes shallower and warmer as silt and organic debris pile up. Warm-water organisms begin to replace cold-water organisms. The lake's shallow areas become overgrown with marsh plants, which gradually fill in the original lake basin until the lake succumbs to bog and finally to land. This process is a natural part of God's design.

The natural life cycle of a lake can span thousands of years, depending on climate and the size of the lake. Today, pollutants from human activities are unnaturally accelerating this aging process. Over the past few decades, nitrates and phosphates from agricultural and industrial wastes and sewage have caused increased problems with eutrophication in lakes worldwide. Nitrates and phosphates overstimulate algae growth, robbing the water of dissolved oxygen. Other pollutants may poison fish populations, and their decomposition further depletes the water's dissolved oxygen content. The amount of oxygen in water is far more crucial than the amount of oxygen on land. Atmospheric air contains about 20% oxygen with about 200,000 of every 1 million air molecules being oxygen. Each cubic meter of air contains 270 g of oxygen, which is more than sufficient for land creatures. Each cubic meter of water holds only 5–10 g of dissolved oxygen. Water does not circulate as easily as air to replenish the oxygen.

Long-term solutions to eutrophication are being developed. Many wastewater plants are adding the process of denitrification to their treatment of sewage effluent. During denitrification, bacteria are used to remove the nitrates. The bacteria release harmless atmospheric nitrogen. Many farmers are reducing the amount of nitrogen fertilizers they use, which prevents excess fertilizer from being carried away by runoff. Forests and wetlands can serve as buffers between farmland and bodies of water, and water runoff can be minimized by properly grading and tilling farmland. In addition, soil tests and computerized tractors using satellite navigation equipment can be used to prevent the overfertilization of crops.

4.1.1 Water Properties

Water

OBJECTIVES

Students will be able to
- describe various life-sustaining properties of water.
- explain how polarity affects water's physical properties.

VOCABULARY

- **adhesion** the force of attraction between different molecules
- **cohesion** the molecular attraction between particles of the same kind
- **surface tension** the force that pulls molecules on the surface of a liquid together to form a layer

MATERIALS

- Wire screen, 2 canning jars, 5 squares of card stock, fine gauze, 2 bowls, 5 paper clips, 5 small squares of paper, liquid detergent, 2 eyedroppers, waxed paper, isopropyl alcohol, 5 clear drinking straws, 2 bowls, 2 hand lenses, 2 capillary tubes, two 100 mL beakers, 2 hot plates, 2 thermometers (*B*)
- Beaker, small coins, vinegar, rubbing alcohol (*C*)
- Clear container, powdered food coloring, oil (*D*)
- Shallow pan, fine powder, eyedropper, dish soap (*E*)
- TM 4.1.1A Physical States of Water
- TM 4.1.1B Cohesion and Adhesion in Plant Roots
- WS 4.1.1A Physical Properties of Water

PREPARATION

- Obtain materials for *Lab 4.1.1A The Dissolving Power of Water*.
- Obtain materials for *Try This: Which Property*.
- Set up the physical property of water stations. (*A*)
  Surface Tension: Cut 2 squares of wire screen and 5 squares of card stock.
  Float It: Cut 5 small squares of paper.

Introduction

Challenge students to list as many uses of water as they can think of. Invite a few student volunteers to read their lists to the class. Use the lists to guide students to identify the properties of water.

Display **TM 4.1.1A Physical States of Water**. Ask students what the chemical formula is for ice, liquid, and water vapor. (H_2O) Point out that all three states of water have the same chemical formula even though they are physically different.

Discussion

- Share that the percentage of water in people's bodies is dependent on their gender, age, and the geographical regions of their homes. Generally, women have a lower percentage of water than men, and babies have the most—about 78%. The following body parts contain varying percentages of water: brain and heart—73%, lungs—83%, skin—64%, muscles and kidneys—79%, and bones—31%. Lead students to infer the reasons behind these statistics.

- Draw the bent structure of a water molecule on the board. Explain that the two pairs of unbonded electrons determine the shape of the water molecule as well as the polarity of the molecule. Discuss polarity. Elaborate on how the polarity of the water molecule contributes to the cohesive and adhesive forces using **TM 4.1.1B Cohesion and Adhesion in Plant Roots**. Discuss the terms *cohesion* and *adhesion*. How are they similar? (**Answers will vary but should include they are both forces of attraction.**) How are they different? (**Answers will vary but should include that cohesion is the attraction between particles of the same kind and adhesion is the attraction between different molecules.**)

- Have each student explain to a classmate how the polarity of water molecules affects one of the physical properties of water.

- Ask the following questions:
  1. How is water different from other naturally occurring substances? (**Possible answers: Most other substances do not retain as much heat as water does; most have a more dense structure in their solid state; and many other liquid substances do not require as much energy to break molecular bonds.**)
  2. How do cohesive and adhesive forces apply to the movement of blood in capillaries? (**Water molecules are attracted to the capillary walls, an example of adhesion. As the blood travels through the capillary, the water molecules in the blood are attracted to each other and follow along through the capillaries as well, an example of cohesion.**)

Activities

Lab 4.1.1A The Dissolving Power of Water

- 100 mL graduated cylinders, 1 per group
- 100 mL beakers, 2 per group
- wax marking pencils, 1 per group
- spoons, 1 per group
- sugar, 50 g per group
- electronic balances, 1 per group
- metric rulers, 1 per group

Hot water works best in this experiment. Emphasize water's universal dissolving capability.

A. Complete *Try This: Which Property* with students. Discuss student results.

B. Divide the class into five groups, one for each of the physical property of water stations. Have groups rotate from station to station and record their observations on **WS 4.1.1A Physical**

172

Properties of Water. (Note: Sufficient materials are listed so two groups can simultaneously be at each station in the event that some groups move at a faster pace than others.)

Surface Tension: If time permits, have students complete the surface tension experiments they designed. Provide different sized jars and covers, such as wire screens with larger holes or fine gauze.

Float It: No additional instruction is needed.

Drop It: No additional instruction is needed.

Climbing Water: No additional instruction is needed.

Heat Retention: Remind students that the end of the thermometer should not touch the bottom of the beaker as this will give a false temperature reading. Explain how specific heat is related to the time it takes for water to come to a boil. Remind students that water can absorb a lot of energy because it has a high specific heat. Therefore, once the temperature in the beaker reaches the boiling point, the water molecules are able to absorb extra heat for several minutes before boiling occurs.

C. Fill a beaker with enough water to form a convex surface that is not overflowing. Have students predict how many small coins can be added to the beaker before the water overflows. Carefully drop coins individually into the beaker with the edge of the coin facing the water surface. Ask students why the water does not overflow with each coin addition. (**Answers will vary.**) Explain that the strong forces of surface tension prevent the water from overflowing. Repeat this demonstration using vinegar or rubbing alcohol. Have students predict the results when experimenting with the other liquids.

> **TRY THIS**
>
> **Which Property**
> - *balloon*
> - *wool cloth*
> **(polarity)**

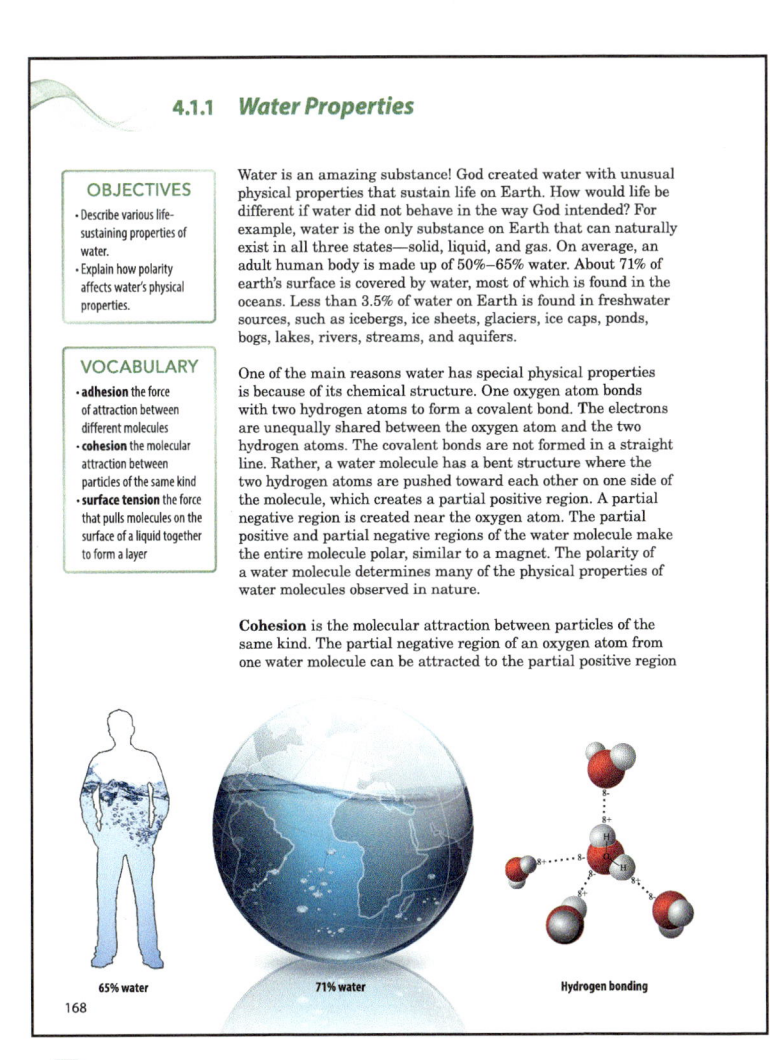

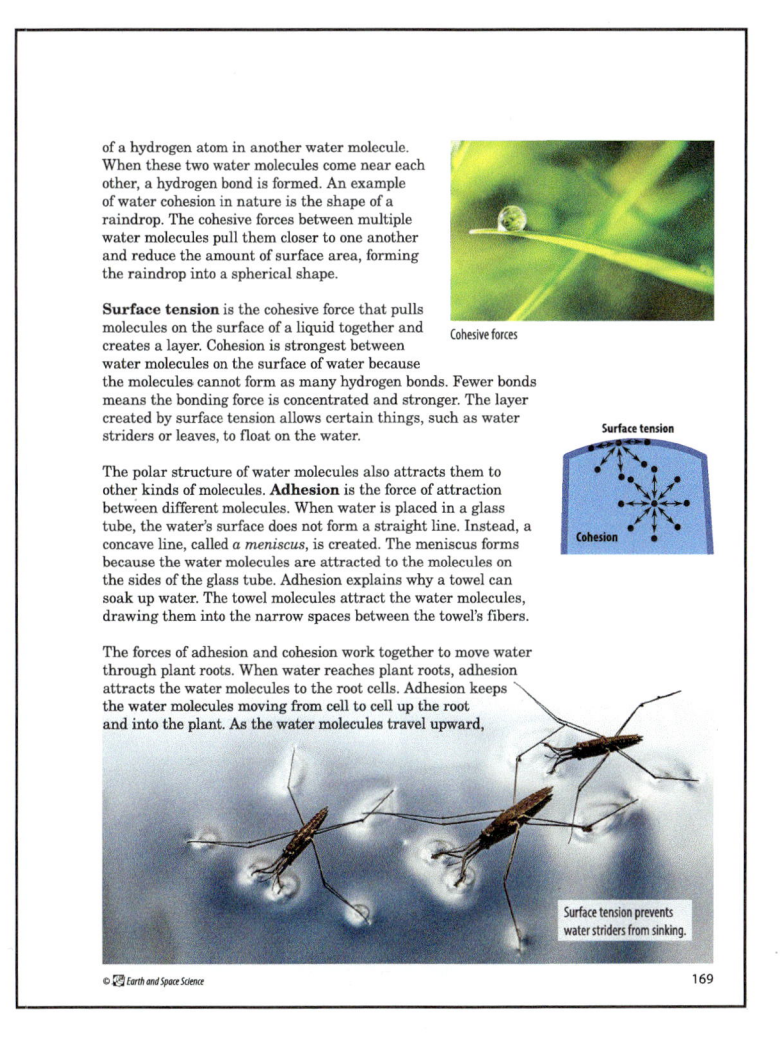

4.1.1 Water Properties

OBJECTIVES
- Describe various life-sustaining properties of water.
- Explain how polarity affects water's physical properties.

VOCABULARY
- **adhesion** the force of attraction between different molecules
- **cohesion** the molecular attraction between particles of the same kind
- **surface tension** the force that pulls molecules on the surface of a liquid together to form a layer

Water is an amazing substance! God created water with unusual physical properties that sustain life on Earth. How would life be different if water did not behave in the way God intended? For example, water is the only substance on Earth that can naturally exist in all three states—solid, liquid, and gas. On average, an adult human body is made up of 50%–65% water. About 71% of earth's surface is covered by water, most of which is found in the oceans. Less than 3.5% of water on Earth is found in freshwater sources, such as icebergs, ice sheets, glaciers, ice caps, ponds, bogs, lakes, rivers, streams, and aquifers.

One of the main reasons water has special physical properties is because of its chemical structure. One oxygen atom bonds with two hydrogen atoms to form a covalent bond. The electrons are unequally shared between the oxygen atom and the two hydrogen atoms. The covalent bonds are not formed in a straight line. Rather, a water molecule has a bent structure where the two hydrogen atoms are pushed toward each other on one side of the molecule, which creates a partial positive region. A partial negative region is created near the oxygen atom. The partial positive and partial negative regions of the water molecule make the entire molecule polar, similar to a magnet. The polarity of a water molecule determines many of the physical properties of water molecules observed in nature.

Cohesion is the molecular attraction between particles of the same kind. The partial negative region of an oxygen atom from one water molecule can be attracted to the partial positive region of a hydrogen atom in another water molecule. When these two water molecules come near each other, a hydrogen bond is formed. An example of water cohesion in nature is the shape of a raindrop. The cohesive forces between multiple water molecules pull them closer to one another and reduce the amount of surface area, forming the raindrop into a spherical shape.

Surface tension is the cohesive force that pulls molecules on the surface of a liquid together and creates a layer. Cohesion is strongest between water molecules on the surface of water because the molecules cannot form as many hydrogen bonds. Fewer bonds means the bonding force is concentrated and stronger. The layer created by surface tension allows certain things, such as water striders or leaves, to float on the water.

The polar structure of water molecules also attracts them to other kinds of molecules. **Adhesion** is the force of attraction between different molecules. When water is placed in a glass tube, the water's surface does not form a straight line. Instead, a concave line, called *a meniscus*, is created. The meniscus forms because the water molecules are attracted to the molecules on the sides of the glass tube. Adhesion explains why a towel can soak up water. The towel molecules attract the water molecules, drawing them into the narrow spaces between the towel's fibers.

The forces of adhesion and cohesion work together to move water through plant roots. When water reaches plant roots, adhesion attracts the water molecules to the root cells. Adhesion keeps the water molecules moving from cell to cell up the root and into the plant. As the water molecules travel upward,

65% water | 71% water | Hydrogen bonding

Cohesive forces

Surface tension

Cohesion

Surface tension prevents water striders from sinking.

NOTES

D. Demonstrate why water is called *the universal solvent*. Fill a clear container halfway with water. Review the concept of polarity. Add powdered food coloring. Explain that the powder has a polar molecular structure similar to water that allows the powder to dissolve. After the powder has dissolved, fill the remainder of the container with oil. Ask students why the oil does not dissipate or dissolve in the water like the powder did. (**Answers will vary but should include that oil has a nonpolar structure.**)

E. Fill a shallow pan with water. Sprinkle fine powder on the surface of the water. Use an eyedropper to add one drop of dish soap to the center of the pan and have students observe how quickly the powder retracts. Explain that this demonstrates the repellent action of a nonpolar substance in polar water as well as surface tension. Have students perform the same demonstration as time allows.

F. Direct student groups of two or three to pick one of the physical properties of water and to create a demonstration to present to the class.

G. Encourage students to research one literary reference to water that is found in poetry, prose, or the visual arts. Have them write a paragraph about the reference, including how the reference relates to one of the physical properties of water.

Lesson Review

1. If the human body is made up of 65% water, how many kilograms of water are in a 41.52 kg person? (**41.52 kg × 0.65 = 26.99 kg**)

2. How does the polarity of water affect the cohesive physical property of water? Give an example of water cohesion. (**Because water molecules are polar, they have both positively and negatively charged portions, so they attract other water molecules. The cohesive forces between water molecules pull them closer to one another and reduce the amount of surface area. Possible examples of cohesion: shape of a raindrop, water molecules traveling up a plant root**)
3. Why is it important that God created ice to be less dense than liquid water? (**If ice were more dense than liquid water, aquatic organisms would be crushed by sinking ice layers.**)
4. Select three properties of water and explain how they help sustain life. (**Possible answers: The properties of adhesion and cohesion help plant roots absorb water; the relative temperature stability of water regulates or stabilizes the temperatures of bodies of water and various organisms; the property of solubility helps sustain life because the water in blood dissolves oxygen, vitamins, minerals, and other nutrients, enabling the bloodstream to carry them throughout the body; and the structure of ice, which is less dense than liquid water, allows it to form at the surface of natural water sources rather than sinking to the bottom and destroying organisms below.**)

NOTES

Name: _____ Date: _____

Lab 4.1.1A The Dissolving Power of Water

QUESTION: Does the addition of sugar (solute) affect the volume of water (solvent)?

HYPOTHESIS: Answers will vary.

EXPERIMENT:

You will need:		
• 100 mL graduated cylinder	• wax marking pencil	• electronic balance
• two 100 mL beakers	• spoon	• metric ruler
	• 50 g sugar	

Steps:
1. Measure 50 mL of hot water and place it in a beaker. Measure the height of the water in centimeters.
   - Exact volume of water (mL): 50 mL
   - Height of water (cm): ~ 3.8 cm
2. Add 16 g of sugar (about 4 level teaspoons) to the empty beaker. Measure the height of the sugar in centimeters.
   - Height of sugar (cm): ~ 1.1 cm
3. Add 16 g of sugar (about 4 level teaspoons) to the beaker containing water, stirring vigorously.
4. Measure the height of the water and sugar mixture in centimeters.
   - Exact volume of mixture (mL): 52 mL
   - Height of mixture (cm): ~ 4.4 cm

ANALYZE AND CONCLUDE:
1. Calculate the volume difference between the mixture and the water.
   Volume of Mixture – Volume of Water = ~ 2 mL
2. Calculate the difference in height between the mixture and the water.
   Height of Mixture – Height of Water = ~ 0.6 cm
3. Compare the height of the sugar and the water height difference. Did the water rise as high as expected after the sugar was added? Explain the height variations.
   Students may expect the difference in water height to equal the height of the sugar. This is not the case. There is a lot of empty space between water molecules that solutes fill after they dissolve. Given that there are pre-existing empty spaces, the water height does not drastically increase after the sugar is added.

Lab 4.1.1A The Dissolving Power of Water

4. What does this experiment tell you about the molecular structure of sugar?
   The molecular structure of sugar is polar.
   Explain. Water can only act as a solvent on other substances with molecules that are polar. Since the water dissolved the sugar, sugar must be polar.
5. Design and test an experiment using other liquids or other solids. Receive approval for the proposed experiment from the teacher before proceeding. Explain the test results for the other liquids or solids tested with regard to polarity.
   Answers will vary but should include that substances that dissolved are polar and substances that did not dissolve are not polar.

4.1.2 The Water Cycle

Water

OBJECTIVES

Students will be able to
- explain the major components of the water cycle.
- determine the factors that affect a region's water budget.

VOCABULARY

- **condensation** the change of a substance from a gas to a liquid
- **evaporation** the change of a substance from a liquid to a gas
- **sublimation** the change of a substance from a solid to a gas without passing through the liquid state
- **transpiration** the loss of water by plants
- **water budget** the relationship between the input and the output of all the water on Earth

MATERIALS

- 2 potted plants, plastic bag, six 1,000 mL beakers, eyedropper (*Introduction*)
- Map of the major area watersheds (*C*)
- TM 4.1.2A The Water Cycle
- WS 4.1.2A Amount of Water
- WS 4.1.2B Comparing Water Budgets

PREPARATION

- Place a plastic bag over one portion of a potted plant's foliage at least one hour prior to presenting the lesson. (*Introduction*)
- Label six 1,000 mL beakers as follows: *Oceans, Icecaps, Groundwater, Lakes, Atmosphere,* and *Rivers*. (*Introduction*)
- Obtain materials for Lab 4.1.2A Cloud Capacity.
- Obtain a map of the major area watersheds from a government agency. (*C*)
- Arrange a field trip to the local water treatment facility. (*D*)

Introduction

Invite students to share what they know about the water cycle. Fill in any gaps and clarify any misconceptions using **TM 4.1.2A The Water Cycle**.

Display a potted plant in front of the class. Ask students whether they can see a plant breathe. (**Answers will vary.**) Now display the prepared potted plant. Point out the condensation that developed under the plastic bag and convey that it is evidence of transpiration.

Display the labeled beakers. Fill the Oceans beaker with 1,000 mL of water. Encourage students to predict how much water from the Oceans beaker will be transferred to the other five beakers. From the Oceans beaker, pour or use an eyedropper to transfer water to the other beakers as follows: Icecaps—24.0 mL, Groundwater—10.8 mL, Lakes—0.79 mL, Atmosphere—0.01 mL, and Rivers—0.01 mL. The Atmosphere and Rivers water quantities are so minute that not placing any water in these two beakers would be sufficient. Explain that Earth's total water supply can be found in the ratios displayed by the beakers.

Discussion

- Have students discuss real-world examples of water condensation, evaporation, and sublimation.

- Ask the following questions:
  1. What does a high relative humidity level indicate with regard to the condensation process? (**High relative humidity levels indicate there is more moisture in the air than is able to condense.**)
  2. How do humans and animals affect the water cycle? (**Humans affect the water cycle through storage of water in reservoirs, groundwater mining, irrigation, urbanization, combustion, and deforestation. Humans and animals consume water that is eventually eliminated as urine or evaporated via perspiration.**)

Activities

Lab 4.1.2A Cloud Capacity

- cotton balls, 1 per group
- scales, 2 per class
- eyedroppers, 1 per group
- 100 mL beakers, 1 per group

Students should complete this lab in groups of two. It is helpful if one student holds the cotton ball and one student operates the eyedropper.

A. Have students list the ways they used water in the past 24 hours. Explain that billions of liters (or gallons) of water are used for daily activities, such as taking showers, cleaning clothes, and flushing toilets. Challenge students to predict how much water is used for each of these tasks. Have students complete **WS 4.1.2A Amount of Water** and then present their daily use data.

B. Assign **WS 4.1.2B Comparing Water Budgets** for students to complete.

C. Display the map of major local watersheds. Have students use the map and other information the agency provides to piece together the local water budget. As a class, determine whether the local region has a positive or negative local water budget.

D. Take students on a field trip to a local water treatment plant. Instruct students to summarize their findings regarding the following treatment steps: intake, pretreatment, mixing, coagulation and flocculation, filtration, chlorination, and distribution. Have students write a comparison of the natural water cycle and the process at the treatment plant.

E. Encourage students to contact the local water authority to find the amount of water the local area uses compared with the amount of water it receives. Have students discover where the water supply is stored. Direct students to research the processes used to clean the water supply and the costs involved.

F. Challenge student pairs or groups to select one of the processes of the water cycle to teach for the class. Encourage them to be creative in their presentations.

G. Direct students to write an interesting story about a day in the life of a water molecule as it travels through the water cycle.

Lesson Review
1. What is the main contributor to the water cycle? (**the oceans**)
2. Name and explain the major components of the water cycle. (***evaporation***: **liquid water changes to water vapor;** ***transpiration***: **plants take in liquid water and release water vapor;** ***condensation***: **water vapor changes to liquid water;** ***precipitation***: **liquid water falls from the atmosphere;** ***sublimation***: **frozen water changes to water vapor without becoming a liquid**)
3. What factors affect a region's water budget? (**the amounts of precipitation, heat, wind, and vegetation**)
4. Why is Earth's water budget considered balanced? (**Earth's water budget is balanced because the water the earth uses equals the amount it receives.**)

NOTES

4.1.2 The Water Cycle

OBJECTIVES
- Explain the major components of the water cycle.
- Determine the factors that affect a region's water budget.

VOCABULARY
- **condensation** the change of a substance from a gas to a liquid
- **evaporation** the change of a substance from a liquid to a gas
- **sublimation** the change of a substance from a solid to a gas without passing through the liquid state
- **transpiration** the loss of water by plants
- **water budget** the relationship between the input and the output of all the water on Earth

The marvelous physical properties of water would not be very useful if all water stayed in one place. God created a water transportation system that stretches to every ecosystem on Earth. The continuous cycle that stores water and moves it from the atmosphere to the earth's surface and back to the atmosphere is called *the water cycle* or *hydrologic cycle*. The water cycle consists of five main processes: evaporation, transpiration, condensation, precipitation, and sublimation.

The main way that water enters the atmosphere is through **evaporation**, when liquid water is converted into water vapor. A large amount of energy, often in the form of heat, is required for evaporation to occur. During this process, heat is removed from the environment, creating a cooling effect. Scientists attest that about 90% of the water in the atmosphere evaporates from oceans, lakes, rivers, and streams. Most of the water in the atmosphere evaporates from oceans because of their large surface area. However, most of this water does not travel across land but instead falls back into the ocean water. A very small fraction of evaporated water is actually transported over land to fall in the form of precipitation.

Transpiration is the loss of water by plants. It represents the way that about 10% of the water in the atmosphere enters the water cycle. Transpiration rates are greatly affected by weather

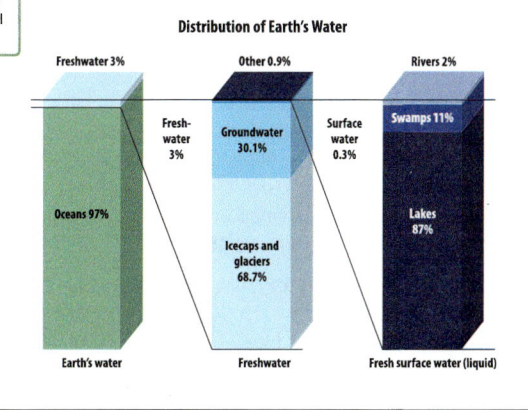

Distribution of Earth's Water

Freshwater 3% | Other 0.9% | Rivers 2%
Freshwater 3% | Groundwater 30.1% | Swamps 11%
Oceans 97% | Surface water 0.3% | Lakes 87%
 | Icecaps and glaciers 68.7% |

Earth's water | Freshwater | Fresh surface water (liquid)

conditions, such as temperature, humidity, wind, precipitation, soil composition, land slope, and plant type. As temperature increases, plant transpiration also increases, especially during the growing season. The warmer temperatures cause plant stomata to open and release water vapor. In contrast, when the temperature is cooler, plant stomata tend to remain closed, keeping the water contained in the leaves. When it is humid, the atmosphere contains excess water vapor, so transpiration rates are low. Transpiration rates tend to increase if wind replaces more saturated air with dryer air. Water that is transpired during dry periods can contribute to soil dryness. When there is not a enough water in the soil, plants transpire less water in an attempt to stay hydrated. Precipitation, the composition of the soil, and the land's slope are other factors that can impact the amount of water in the soil. Different plant types also affect transpiration rates. For example, in more arid regions, plants such as cacti and succulents tend to conserve water by transpiring less.

The next stage in the water cycle is **condensation**, which is the change of a substance from a gas to a liquid. Water molecules are grouped randomly when water is in a vapor form. As condensation occurs, the water molecules group in a more uniform pattern. Condensation is the opposite of evaporation. Evaporation requires a lot of heat. In contrast, condensation releases a lot of heat, so it occurs more frequently at higher elevations where the air is cooler. Depending on its temperature, air can hold a limited amount of water vapor. Warm air can hold more water vapor than cooler air. When air contains all the water vapor it can possibly hold, it is saturated. Scientists use the saturation point of air to determine what is called its *relative humidity*. Saturated air has a relative humidity of 100%, so air that is holding only half the water it can has a relative humidity of 50%.

Condensation is a crucial process of the water cycle because it creates clouds. When the invisible water vapor in the air cools to the point that visible drops of water form, it has reached the temperature called *the dew point*. At the dew point, clouds form because the water-saturated air condenses. Condensation can occur high in the sky to form clouds and at ground level to form fog and dew.

Clouds may eventually release the condensed water in the form of precipitation. Precipitation is water that falls from the atmosphere to the earth such as rain, snow, sleet, or hail. Most

BIBLE CONNECTION

Who Sends the Rain?
Evaporation. Transpiration. Condensation. Precipitation. Do these processes all just happen by chance? In Psalm 147, the psalmist makes plain that the water cycle is under God's dominion. Verse 8 says, "He covers the sky with clouds; He supplies the earth with rain and makes grass grow on the hills." These words were written long before the complex balance that forms Earth's water budget was understood by scientists, and yet the cycle is clearly depicted. Water vapor condenses in the clouds, falls as rain on the grass, and is transpired by the grass to return to the clouds once again. Jeremiah echoes this certainty when he says, "Do any of the worthless idols of the nations bring rain? Do the skies themselves send down showers? No, it is You, Lord our God. Therefore our hope is in You, for You are the One who does all this" (Jeremiah 14:22).

of the condensed water in clouds will not precipitate, or condense and fall, because updrafts of air keep the droplets suspended in the atmosphere. You may have seen wispy trails called *virga* extending from clouds. Such trails of moisture do not reach the ground. Most precipitation falls in the form of rain.

Sublimation is the change of a substance from a solid to a gas without passing through the liquid state. For example, when warm, dry winds flow over the mountains, snow is vaporized before it can melt. This process releases a very small amount of water vapor into the atmosphere. Without energy to fuel the process, sublimation could not occur. Sunlight often provides the energy needed for sublimation, which requires more energy than evaporation or transpiration because the step of changing to a liquid is skipped. It takes about five times more energy for water to move from ice to water vapor than it does for liquid water to become ice. Sublimation occurs more readily in areas with a low relative humidity, plenty of sunlight, low air pressure, and abundant dry winds. Many high altitude regions provide these requirements, including unique regional dry winds. For example, sublimation regions include the North American Rocky Mountains that experience Chinook winds; the Alps, where foehn winds frequent; the mountains in Libya, where ghibli winds occur; and the Andes Mountains, which experience zonda winds.

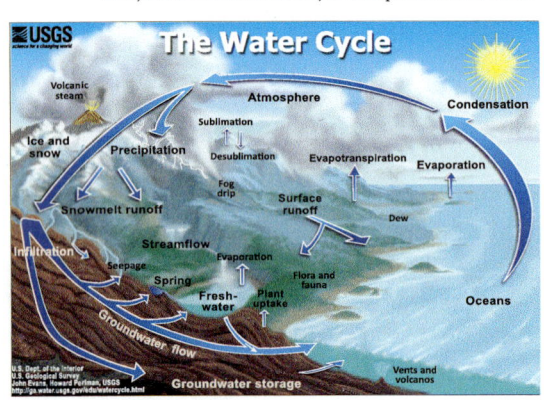

Wisps of precipitation that evaporate before they reach the ground are called *virga*.

The **water budget** is the relationship between the input and the output of all the water on Earth. Earth's water budget can be compared to a financial budget. For example, each month, you may earn a certain amount of money from a job or be given an allowance; money entering your budget is called *income*. The ways you spend your money are called *expenditures*. Expenditures can be divided into categories, such as tithing, saving, housing, food, clothing, and recreation. When your income equals your expenditures, then your budget is balanced. In Earth's water budget, precipitation and condensation represent income, and evaporation, transpiration, sublimation, and water runoff represent expenditures. Earth's water budget is balanced because the amount of water it uses equals the amount it receives.

Sublimation on Lhotse in the Himalayas

The water budget for Earth as a whole is balanced, but local water budgets are usually not balanced. For example, when more water falls on an area than evaporates or runs off, flooding occurs. Sometimes the amount of water that evaporates and transpires is greater than the precipitation that falls, which results in drought conditions. Regions and communities analyze their water budgets to manage water resources by comparing how much water is coming into an area with how much is being used. Local water budgets can change. Heavy rainfalls can cause flooding, and infrequent rain can cause drought. Long periods of heat can increase evaporation. Wind also increases evaporation and carries away the water in humid air. Vegetation also affects the local water budget. Plants keep water from running across the ground and plant roots hold water in the soil. As these factors change, so does the water budget.

Researchers study local and regional water budgets to gain a better understanding of how water is stored and how water flows in and out of a particular ecosystem. They can calculate the water availability and analyze and predict water shortages. A study of the water budget aids the decision-making processes and helps researchers address natural resource management issues.

LESSON REVIEW
1. What is the main contributor to the water cycle?
2. Name and explain the major components of the water cycle.
3. What factors affect a region's water budget?
4. Why is Earth's water budget considered balanced?

CHALLENGE

Design an experiment that tests an aspect of transpiration. Some questions to consider include the following: How much water does one plant transpire? Do different types of plants transpire different amounts of water? Do plants transpire different amounts of water under different conditions?

Name: _____ Date: _____

Lab 4.1.2A Cloud Capacity

QUESTION: How much water will a standard cotton ball hold?

HYPOTHESIS: Answers will vary.

EXPERIMENT:

You will need:	• scale	• 100 mL beaker
• cotton ball	• eyedropper	

Steps:
1. Fill the beaker with 50 mL of water.
2. Separately measure and record the mass of the cotton ball and one drop of water.
   - Mass of cotton ball: Answers will vary. grams
   - Mass of one drop of water: Answers will vary. grams
3. Predict how many drops of water can be added to the cotton ball before the water begins to drip.
   - Estimated number of water drops: Answers will vary. drops
4. Hold a small section of a cotton ball by pinching it with your index finger and thumb. The cotton ball should be held above the beaker.
5. Apply drops of water, one drop at a time, to the cotton ball.
6. Count how many drops of water are added to the cotton ball. Stop counting the number of drops added as soon as the cotton ball begins to drip.
   - Total number of water drops added: approximately 200 drops
7. Record the number of drops that other groups measured.
   Answers will vary.

ANALYZE AND CONCLUDE:
1. Was the estimated number of water drops equal to the actual number of water drops added to the cotton ball? Answers will vary.
2. Did each group record the same amount of water drops added to the cotton ball? Why did the results vary? Answers will vary. Groups will have different results because the water drops were not all the same size, the cotton balls were not all the same size, or students added water drops in one location rather than in multiple locations. If the drops were only added in one location, the cotton ball will not hold as much water.

Lab 4.1.2A Cloud Capacity

3. Graph the results on the chart below. Add a title for the graph.

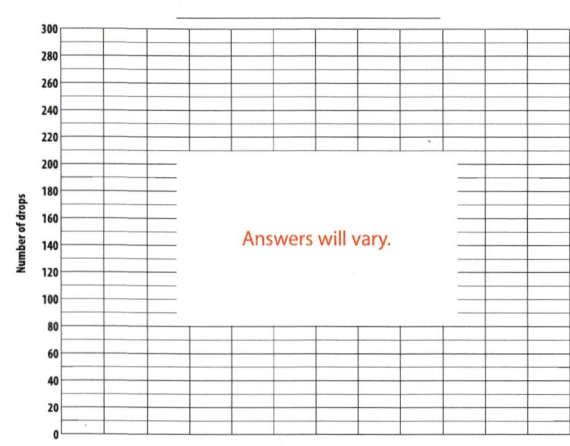

Answers will vary.

(Number of drops vs. Group number)

4. What forces allow the cotton ball to hold such a tremendous amount of water?
   adhesion and cohesion
5. Relate the saturation of the cotton ball to the saturation of a cloud. Answers will vary but should include that cotton balls can hold only a certain amount of water depending on how much space is available inside the cotton ball. Clouds can hold only the amount of water that can be supported by updrafts. Both cotton balls and clouds can hold a large amount of water.

Water

4.1.3 Glaciers

Introduction
Create a glacier for students to observe. In a shallow baking pan, crumble graham crackers to represent rocks and soil. Place two long, narrow pieces of cake on either side of the baking pan to represent sides of mountains. Leave enough room in the center of the pan where the graham crackers are still visible and the "glacier" can flow. Elevate one end of the baking pan about 45°. Add about 1 cup of shaved chocolate or pudding to the elevated end of the baking pan. Repeat this step every 5–10 minutes. The shaved chocolate or pudding represents the snow and ice that form a glacier. Students will observe the gradual movement of the "glacier" and the crevasses that develop. Point out to students that the bottom pieces of chocolate are more solid than the top pieces because they compacted as they moved down the pan, similar to snow transforming into ice as a glacier is formed. Also, direct students to observe how the graham crackers became adhered to the underside of the chocolate, similar to rocks that adhere to the underside of glaciers. Inform students that this model illustrates one of the two types of glaciers they will learn about in this lesson.

Discussion
- Display **TM 4.1.3A Muir Glacier 1915 and 2010**. Have students point out the differences in the two pictures. Explain that Muir Glacier is a receding glacier. Convey that other examples of receding glaciers include Pedersen Glacier in Alaska and Bossons Glacier in France. Encourage students to find images of these online. Ask students why scientists believe glacial receding has occurred. (**Possible answers: Earth's temperatures have risen; there is natural expansion and shrinking of glaciers.**) Is glacier recession something new or have glaciers receded in the past? (**Glacier recession is not new; many glaciers have receded in the past.**) Guide students to consider the effect glacial recession has on local and global environments. Glacial recession indicates the ice is melting or subliming. If enough ice melts, ocean water levels rise. The increased water level could cover inhabited land. Also, local fauna could lose their natural habitats. Ask students to think of ways that glacial receding is beneficial. (**Possible answers: provides drinking water, irrigates crops, generates hydroelectric power**)

- Begin a discussion about the difference between a snow patch and a glacier by asking students if all snow on a mountain is a glacier. (**No.**) Explain that a snow patch is an accumulation of snow and firn that does not melt seasonally. Ask students if they have seen snow on a mountain during the summer months. (**Answers will vary.**) Both snow patches and glaciers are formed on land and both are made of ice. Guide students to understand that the main difference between a snow patch and a glacier is that a glacier moves.

- Ask the following questions:
  1. What type of glacier did the model in the *Introduction* represent? (**valley glaciers**)
  2. What is firn? (**Grainy ice that forms when snow melts, compacts, and recrystalizes.**)
  3. Why is glacial ice often blue? (**Compacted glacial ice appears blue because when it becomes dense, it absorbs all colors in the light spectrum and reflects only blue.**)
  4. Explain the two glacier zones. (**The zone of accumulation is where snow converts to firn near the head of the glacier. The zone of ablation, near the foot of the glacier, is where glacial ice reduces by melting, evaporating, or calving.**)
  5. What is calving? (**the process by which portions of glaciers break off to create more glaciers**)

Activities
A. Complete *Try This: Ice Melt* with students. Discuss student predictions and the outcome.

B. Direct students to use internet sources to complete **WS 4.1.3A Global Glaciers**.

C. Have students use their Student Editions to complete **WS 4.1.3B Glaciers in the Water Cycle**.

OBJECTIVES
Students will be able to
- explain how glaciers form.
- analyze how glaciers contribute to the water cycle.
- contrast the two main types of glaciers.

VOCABULARY
- **continental glacier** a glacier that covers a large area of land in a continuous sheet
- **valley glacier** a long, narrow, U-shaped mass of ice that takes shape as ice moves down a mountain and through a valley area

MATERIALS
- Shallow baking pan; graham crackers; two long, narrow strips of cake; shaved chocolate or pudding (*Introduction*)
- TM 4.1.3A Muir Glacier 1915 and 2010
- WS 4.1.3A Global Glaciers
- WS 4.1.3B Glaciers in the Water Cycle

PREPARATION
- Obtain materials for *Try This: Ice Melt*. Pour 30 mL of water into one chamber of an ice cube tray. Repeat this step three more times. Next, place 2 drops of red food coloring in the first chamber, 2 drops of yellow food coloring in the second chamber, 2 drops of green food coloring in the third chamber, and 2 drops of blue food coloring in the fourth chamber. Freeze the water for two hours.
- If possible, arrange a field trip to a local glacier. (C)

TRY THIS
Ice Melt
- red, yellow, green, and blue food coloring
- ice cube trays, 1 per group
- graduated cylinders, 1 per group
- petri dishes, 4 per group

Assemble and freeze colored ice cubes. Have students predict which colored ice cube will melt the fastest in sunlight.

NOTES

D. If possible, take a field trip to a local glacier. Direct students to illustrate what they observe and to especially note any changes in the landscape.

E. Assign student groups to research a technological tool scientists use to observe and measure glacial movement. Have groups report on the tool they researched.

F. Direct students to create a travel brochure for a specific glacier. Guide students to include information about the type of glacier represented, the size and thickness of the glacier, the estimated age of the glacier, a map of the location, whether the glacier is expanding or receding, and tourist activities that are currently in place or created by students.

G. Divide the class into four groups. Each group will become experts on one of four topics: ice sheets, ice shelves, sea ice, or ice caps. After they thoroughly research the assigned topic, have students create new groups that include one individual from each of the four original groups. Each individual will then teach the new group of individuals about their original topic. Have new groups create a chart illustrating the similarities and differences among the four topics.

H. Encourage students to create a Venn diagram comparing continental and valley glaciers.

I. Have students create glacial analogies.

4.1.3 Glaciers

OBJECTIVES
- Explain how glaciers form.
- Analyze how glaciers contribute to the water cycle.
- Contrast the two main types of glaciers.

VOCABULARY
- **continental glacier** a glacier that covers a large area of land in a continuous sheet
- **valley glacier** a long, narrow, U-shaped mass of ice that takes shape as ice moves down a mountain and through a valley area

TRY THIS
Ice Melt
Predict which colored ice cube will melt the fastest. Pop out the colored ice cubes and place them on individual petri dishes. Set the petri dishes next to each other in direct sunlight and record how long it takes for each cube to melt. Was your prediction correct?

Many scientists believe the earth was cooler in the past and had much more ice than it does today. During an ice age, it is estimated that about one-third of the earth's land mass was covered with glaciers. Today, glaciers exist on every continent except Australia. A large mass of moving ice that forms on land and remains from year to year is called *a glacier*. Glaciers form at high altitudes or in polar regions where snow remains throughout the year. As the seasons change, snow partially melts, compacts, and recrystallizes. This process forms a grainy ice called *firn*. The pressure from the deep layers of snow squeezes the air from between the ice grains, reducing the pore space. The increased density changes the color and structure of firn from white snow to steel-blue, solid ice. Compacted glacial ice appears blue because when it becomes dense, it absorbs all colors in the light spectrum and reflects only blue. White glacier ice is less dense and contains many tiny air bubbles.

Snow converts to firn in a region near the head of a glacier called *the zone of accumulation*. Near the foot of a glacier in a region called *the zone of ablation*, glacial ice reduces by melting, evaporating, or calving. The process of calving involves portions of a glacier breaking off to create more glaciers. The accumulation zone is separated from the ablation zone by the equilibrium line.

Glaciers are constantly shrinking and growing. Glaciers increase in size when the rate of precipitation is faster than the rate of evaporation. When the layered snow evaporates at a faster rate than the precipitating snow, the size of a glacier decreases. The development of a glacier is affected by many factors, such as the amount of snowfall, the amount of sunlight, and the slope of the land. For example, if a mountain has a steep incline, snow has difficulty sticking to the surface and accumulating.

When a mass of ice reaches a thickness of about 40 m, it becomes so heavy that it begins to deform and move. Glacial movement is very slow. On average, glaciers move a few millimeters to a few meters each day. However, there are times when glaciers move rather quickly. During these surges, glaciers can move up to 6 km in a year. The stress of such movement causes glacial ice to form fractures called *crevasses* near the top of the glacier. Many crevasses form during glacial surges. In contrast, the lower portions of a glacier flow more like a thick fluid. The underside of a glacier tends to slip over the surface. Friction between the glacier base and the land reduces the speed at which the bottom of the glacier can move. This friction is why the underside of a glacier moves more slowly than its upper portion. As glaciers move, some glacial ice reenters the water cycle as glacial ice breaks off, melts, and sublimes.

Glaciers are an important component in Earth's water cycle because much more water is in storage on the planet than is moving through the water cycle at any one time. Freshwater on Earth is stored for short periods of time in lakes, longer stretches of time in groundwater, and even longer periods in glaciers. Much of the water that is now a liquid was once frozen in continental glaciers. Scientists estimate that in the past these larger glaciers lowered the sea level more than 100 m.

A **continental glacier** is a glacier that covers a large area of land in a continuous sheet. Today, continental glaciers cover about 11% of the earth's land mass. Continental glaciers are largely unaffected by underlying topography. They move out from a central region in all directions. Examples of continental glaciers are the Greenland and Antarctic ice sheets. Continental glaciers hold amazing amounts of

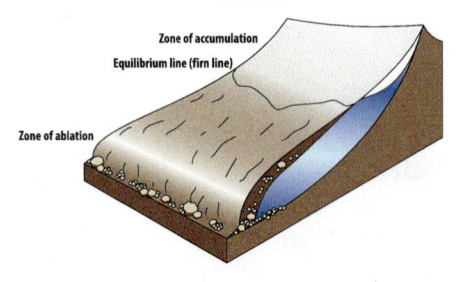

Glacier Zones
- Zone of accumulation
- Equilibrium line (firn line)
- Zone of ablation

Antarctica is a continental glacier.

Lesson Review

1. How do glaciers form? (**Glaciers form at high altitudes or in polar regions where snow remains throughout the year. As the seasons change, the snow partially melts, compacts, and recrystallizes. This process forms a grainy ice called *firn*. The pressure from the deep layers of snow squeezes the air from between the ice grains, reducing the pore space. Increased density changes the color and structure of firn from white snow to steel-blue, solid ice.**)

2. Explain how glaciers move. (**When a mass of ice reaches a certain thickness, it becomes so heavy that it begins to deform and move. Glacial movement is very slow. Lower portions of the glacier flow similarly to the way a fluid does. The underside of a glacier tends to slip over the surface. However, friction between the glacier base and the land reduces the speed at which the bottom of the glacier can move. This is why the underside of the glacier moves more slowly than the upper portion.**)

3. What part do glaciers play in the water cycle? (**More water is in storage on Earth than is moving through the water cycle at one time. Glaciers provide long-term storage of freshwater on Earth.**)

4. What is the difference between a continental glacier and a valley glacier? (**A continental glacier is a glacier that covers a large area of land in a continuous sheet, it moves away from a central region in all directions, and it is largely unaffected by underlying topography. A valley glacier is a long, narrow, U-shaped mass of ice that takes shape as ice moves down a mountain and through a valley area; it is much smaller than a continental glacier; and it originates in the snow or ice fields of a mountainous area.**)

NOTES

4.1.4 Groundwater

Water

OBJECTIVES

Students will be able to
- describe the different zones and features of groundwater.
- identify geological features formed by groundwater.
- illustrate how groundwater storage and movement are related to the water cycle.

VOCABULARY

- **aquifer** a permeable underground layer of rock
- **geyser** a hot spring that periodically erupts
- **groundwater** all the water found underground
- **porosity** a measure of the open space in rocks
- **sinkhole** a hole in the ground that forms when an underground cave collapses
- **water table** the boundary between unsaturated and saturated ground
- **zone of aeration** the underground region where pore spaces contain air and water
- **zone of saturation** the underground region where pore spaces are saturated with groundwater

MATERIALS

- Groundwater model simulator, pipettes, red and green dye, drinking glass, food coloring, straw (*Introduction*)
- Large, clear plastic containers, plastic or rubber tubing, water pumps, synthetic sponges, sand, gravel, brick fragments, topsoil, modeling clay, 5 gal buckets (*A*)
- TM 4.1.2A The Water Cycle

Introduction

Ask students where water can be found on Earth. (**Possible answers: oceans, lakes, rivers, glaciers, atmosphere, underground**) Display **TM 4.1.2A The Water Cycle**. Have students share why they think water stored underground is important. Convey that the movement and storage of groundwater is an integral part of the water cycle. Remind students that groundwater contains more than 30% of the accessible freshwater on Earth, and it is much more abundant than freshwater found in lakes and streams. Relate the amount of groundwater to a bank account. An unlimited amount of groundwater cannot be withdrawn but only the amount that was deposited.

Present a groundwater model simulator to the class. Draw attention to the following components: bedrock layer, zone of aeration, zone of saturation, aquifers, wells, and ponds. Have a student volunteer fill the reservoir with tap water. Ask students to observe as the water saturates the sand and gravel, travels to the wells, and enters the pond. Point out that the water cannot pass through the impermeable bedrock layer. Direct another student volunteer to pump one of the wells. Have students describe what they see. (**As water comes up through the well, the water level in the adjacent well drops.**) Have a third student volunteer fill the underground storage tank with red dye. Convey that the red dye represents pollution, such as petroleum. Guide the student volunteer to also pour green dye into the landfill. Both the red and green dyes will slowly leak into the aquifers and pond. Direct the second student volunteer to pump the well again and have students observe that the pollutants travel toward the pumped well and eventually toward the other wells. Guide students to document the order in which the pollutants traveled and to hypothesize how groundwater contamination would be detrimental to the human, animal, and plant populations in the area. (**Answers will vary.**)

To demonstrate the action of pumps, fill a drinking glass with colored water, insert a straw into the water, place your thumb over the end of the straw, and remove the straw from the glass. Share that the water column inside a pump gradually rises. Illustrate this idea by barely submerging the open end of the straw in the water, uncovering the top end, and quickly plunging the straw farther into the water. Cover the top of the straw again and quickly pull it up. This process may be performed multiple times to build up a water column.

Discussion

- Discuss how shared groundwater resources could possibly cause conflicts between neighboring cities, states, or countries. Ask students what qualities are desired in a public leader that would help rectify any potential disputes. (**Possible answers: patience, fairness, intelligence, honor**)

- Ask the following questions:
  1. How does water move underground? (**Groundwater flows through the open pore spaces between rock and soil particles.**)
  2. What is the difference between porosity and permeability? (**Porosity is the percentage of open spaces in rocks or sediments that hold groundwater. Permeability is the ability of water to flow through the open spaces in rocks or sediments.**)
  3. How does water enter an aquifer? (**Rainwater seeps through the permeable layers of rock.**)
  4. How does water exit an aquifer? (**through well pumps or springs**)
  5. How do you think the emergence of groundwater on Earth's surface could alter the rocks that surround it? (**The surrounding rocks may experience mechanical or chemical weathering from groundwater that emerges from the earth's surface.**)
  6. Why would someone draw water from the ground rather than from a nearby river? (**Possible answers: Groundwater may be less polluted and does not need to be filtered and purified as extensively as river water; the river may have dried up from overuse or during a drought.**)

Activities

Lab 4.1.4A Groundwater Contamination

- 25 cm × 16 cm sections of screen, 1 per group
- 500 mL beakers, 1 per group
- 250 mL beakers, 1 per group
- aquarium pebbles
- sand
- topsoil
- food coloring
- metric rulers, 1 per group
- timers, 1 per group

A. Assign students to research different types of aquifers one day before completing this activity. Divide the class into groups of three. Have groups design and create an aquifer model using the materials provided. Each model should clearly display the water table, zone of aeration, zone of saturation, bedrock, and a lake. Direct students to consider the type of material that will be used in each zone, the thickness of the zones, the thickness of the topsoil layer if there is one, and if there will be a confining layer. If the aquifer models do not initially pump water, instruct students to analyze what parts did not operate properly and reconstruct the model. Have groups explain to the class how each part works. (Note: A large hand soap pump works well as a water pump.)

B. Present the hydrogeology graduate student to share with the class why she is interested in the field of geology and to describe current fieldwork she is involved in.

C. Have students become groundwater advocates. Direct students to write letters to a local groundwater monitoring and protection agency detailing what they have learned about overuse;

PREPARATION

- Obtain a groundwater model simulator. Simulators may be purchased from a local or online distributor. (*Introduction*)
- Obtain materials for *Lab 4.1.4A Groundwater Contamination*.
- Assign students to research different types of aquifers one day before completing activity. (*A*)
- Obtain materials for the activity. (*A*)
- Arrange for a hydrogeology graduate student to speak to the class. (*B*)

Groundwater 4.1.4

Lakes, rivers, and glaciers hold an enormous amount of water. However, 30% of the earth's freshwater supply is locked up in the ground. **Groundwater** is water that seeps down into soil and rock crevasses to be stored underground.

When water falls on the land, it travels through the earth as far as possible until it reaches clay deposits. Water first seeps through the **zone of aeration**, which is the upper permeable layers of rock and soil. The pore spaces of this underground region contain both air and water, so the ground is not saturated. Beneath this area is the **zone of saturation**, which is an area of rock and soil with pore spaces completely filled with groundwater. The **water table** is the region that separates the zone of aeration from the zone of saturation. Water located in the zone of saturation does not remain stagnant because it moves and travels to new locations. However, water cannot move down through the saturated soil and rock located just above the clay layer. Clay particles have very little pore space, making clay impermeable.

The water in rivers can flow several meters per second, but groundwater may move only a few meters per day or a few centimeters in a decade. The movement of groundwater is a slow process because it must seep through varying pore spaces. Groundwater movement is measured according to flow velocity and permeability. Groundwater flows downhill along the water table. When the water table is steep, the gravitational force that pulls groundwater down is greater and the flow velocity, or speed, increases. Rock layers containing similarly sized sediment grains are porous and allow water to flow freely. In contrast, sediment grains of different sizes often block the flow of water. Loosely packed grains have high permeability, but tightly packed grains have fewer spaces and slow or even block the flow of water.

An **aquifer** is a permeable underground layer of rock that can hold a large amount of groundwater. The amount of water that an aquifer can hold depends on its **porosity**, which is the percentage of open spaces in rocks or sediments that will hold groundwater. The greater the porosity of the rocks and sediments, the more water an aquifer can hold. Aquifers are made up primarily of fractured sandstone and limestone rock. Groundwater travels through aquifers and eventually reaches land, where the water table cuts across Earth's surface. As water emerges from underground, it produces a small stream known as *a spring*. People can also tap into underground aquifers by digging or drilling wells. Water is not accessible unless the wells reach beneath the level of the water table. The level of the water table fluctuates seasonally because of different amounts of precipitation. For example, an extremely rainy season may cause the water table to rise, but during a dry season, springs or wells might dry up. If an aquifer dries up, water from precipitation will eventually refill the aquifer, which is a very slow process. In

OBJECTIVES
- Describe the different zones and features of groundwater.
- Identify geological features formed by groundwater.
- Illustrate how groundwater storage and movement are related to the water cycle.

VOCABULARY
- **aquifer** a permeable underground layer of rock
- **geyser** a hot spring that periodically erupts
- **groundwater** all the water found underground
- **porosity** a measure of the open space in rocks
- **sinkhole** a hole in the ground that forms when an underground cave collapses
- **water table** the boundary between unsaturated and saturated ground
- **zone of aeration** the underground region where pore spaces contain air and water
- **zone of saturation** the underground region where pore spaces are saturated with groundwater

CAREER
Hydrogeologist
Hydrogeologists study the distribution, movement, and quality of water found underground. They interpret data and information from maps and historical documents to develop a conceptual model of groundwater flow. The models are created to make predictions about future trends and impacts on groundwater movement and quality. Most of the work is completed in an office setting. However, designing and completing investigations in the field is also critical in order to test and develop models. Investigations in the field may incorporate various measurement and sampling techniques that can allow the hydrogeologist to collect data over a short or an extensive period of time. The work of a hydrogeologist leads to better management of natural resources and better protection of groundwater.

FYI
Yellowstone Volcanoes
The majority of the world's known geysers are found in Wyoming, most notably in Yellowstone National Park. There are about 500 geysers and 10,000 thermal features located in this area. What do the geysers indicate? Geysers indicate the presence of magma under the surface of the earth. The Yellowstone area experiences 1,000 to 3,000 earthquakes each year, which indicates activity belowground. Scientists believe there were three major volcanic eruptions in Yellowstone's past. One of the major eruptions is believed to have created the West Thumb of Yellowstone Lake. Some scientists consider Yellowstone to be a supervolcano capable of an eruption that would cover more than 386 cubic miles. An eruption is theoretically possible, but it is very unlikely to occur within the next thousand years. Geologic activity at Yellowstone has remained relatively constant for the past 30 years. Scientists continue to monitor the region closely for any sudden or strong movements or shifts in heat that would indicate increasing volcanic activity.

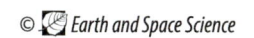

NOTES

subsidence (sinking of land); and chemical, salt, radon, or other pollution that pose possible threats to the underground water supply. Also, direct students to include in their letter a request for ways they can specifically protect the local and regional water supplies.

D. Encourage students to create presentations or posters to illustrate how groundwater is involved in the water cycle.

Lesson Review

1. Describe the different groundwater zones. How do rock and soil pore spaces affect the path of groundwater? (**zone of aeration: the underground region where pore spaces contain air and water; zone of saturation: the underground region where pore spaces are saturated with groundwater. Layers with similarly sized sediment grains are porous and allow water to flow freely. Sediment grains of different sizes can block the flow of water. Loosely packed grains are very permeable. Tightly packed grains have fewer spaces and slow or block the flow of water.**)
2. Explain two geologic features that are formed by groundwater. (**hot spring: hot groundwater that rises to the surface; geyser: hot spring that erupts; karst topography: chemical weathering by groundwater; sinkhole: limestone dissolved by groundwater under soil**)
3. Why do geysers have such intense heat? (**Geysers are located near magma pockets that transfer heat to the rocks surrounding the geyser chambers.**)
4. Describe why groundwater is an important component of the water cycle. (**Groundwater is an important component of the water cycle because water delivered in the form of precipitation is propelled through the water cycle as groundwater delivers the water back to the ocean or to land's surface where evaporation can occur.**)

---

fact, deep aquifers may take hundreds of years to regain their original water quantity. Aquifers may also be refilled with the use of rapid-infiltration pits or groundwater injection. To create a rapid-infiltration pit, water is spread over the land in furrows, pits, or ditches and allowed to seep into the aquifer. Recharge wells may also be constructed for direct water injection into the aquifer.

Spring water is usually the same temperature as the surrounding land area. Spring water found in cold climates may average 10°C, but warmer climates may have spring water that averages 24°C. Interestingly, spring water is usually warmer in the winter months and cooler in the summer months. Aquifers that are located near volcanic activity absorb a large amount of heat. Springs produced from extremely warm aquifers can form some interesting features on the earth's surface. When hot groundwater rises to the surface, it is called *a hot spring*. A **geyser** is a hot spring that periodically erupts. A narrow tube under a geyser connects the earth's surface to underground chambers. Heat from magma warms the rocks that surround the narrow tube and the water in the chambers. Over time, the heat causes the water to boil. When the water boils, tremendous pressure begins to build at the base of the chambers. The pressure forces water upward until it erupts through an opening in the earth's surface. During an eruption, liquid water is converted into very hot water vapor. The eruption continues until most of the tube and storage chambers are empty. This process can be repeated multiple times throughout a day. For example, Old Faithful in Yellowstone National Park in the United States erupts every hour or so throughout the day.

Another interesting feature forms when groundwater dissolves and carries away underground minerals such as salt, gypsum, or limestone. Land areas that have experienced large amounts of chemical weathering by groundwater are said to have a *karst topography*. Karst topography can be found around the world in such places as China, France, Mexico, Slovenia, and in the United States in parts of Florida, Indiana, and Kentucky. This type of topography is usually characterized by caverns, sinkholes, and gaps in rock that swallow streams. Occasionally, these streams will run underground for several kilometers until they come out through another gap in the rock.

Old Faithful

Limestone mountain karsts

If a water table drops too low, hollows in the limestone can collapse to form sinkholes. A **sinkhole** is a circular depression in the ground that forms when the roof of an underground cave collapses. Sinkholes also form when limestone under the soil dissolves. In September 2016, a massive sinkhole in Florida opened up underneath a storage pond near a fertilizer plant. More than 750 million liters of contaminated wastewater leaked into the Floridan aquifer, one of the state's main underground resources of drinking water. The Floridan aquifer is one of the highest water-producing aquifers in the world. It lies beneath the entire state of Florida and extends into southern Alabama, Georgia, and South Carolina. The sinkhole was about 14 m in diameter. Representatives from the fertilizer plant diverted the storage pond water to an alternate holding area and recovered the water by pumping the contaminated water through on-site production wells. To date, the largest natural sinkhole in the world is the Qattara Depression in Cairo, Egypt. It measures 80 km long by 120 km wide and is 133 m deep. Other naturally occurring sinkholes have become ponds or lakes.

Sinkhole in Didyma, Turkey

Water is essential for the survival of humans, animals, and plants. The more than 7.4 billion people who live on Earth use more water than can be supplied by lakes and rivers. As a result, people tap into the groundwater supply to meet their water needs. Areas with large populations or large areas devoted to agriculture can use up the water in aquifers faster than the water is replaced. When aquifers near the coasts are drained of water, saltwater seeps into them. Another issue affecting the groundwater supply occurs in regions where vegetation is lacking. Plant foliage that covers a large amount of the ground slows the movement of runoff and allows it to seep into aquifers. Plant roots prevent the collapse of pore spaces in soil and absorb harmful chemicals. Without vegetation, contaminated water can enter aquifers where salt and other pollutants can remain for years.

LESSON REVIEW
1. Describe the different groundwater zones. How do rock and soil pore spaces affect the path of groundwater?
2. Explain two geologic features that are formed by groundwater.
3. Why do geysers have such intense heat?
4. Describe why groundwater is an important component of the water cycle.

Name: _____ Date: _____

Lab 4.1.4A Groundwater Contamination

QUESTION: How do different sediments affect water permeability?

HYPOTHESIS: Answers will vary.

EXPERIMENT:

You will need:		
• 25 cm × 16 cm section of screen	• aquarium pebbles	• metric ruler
• 500 mL beaker	• sand	• timer
• 250 mL beaker	• topsoil	
	• food coloring	

Steps:
1. Roll the screen to form a long cylinder with a 5 cm diameter. The cylinder represents a well. Place and hold the cylinder in the center of a 500 mL beaker.
2. Add aquarium pebbles to the bottom of the beaker to form a 4 cm layer.
3. Pour water over the pebbles until the water level measures about 2 cm. This water represents groundwater.
4. Add a 2 cm layer of sand above the pebbles.
5. Add a 2 cm layer of topsoil above the sand. Carefully release the cylinder.
6. Add a few drops of food coloring to 150 mL of water in a separate beaker. The colored water represents contaminated water.
7. Pour 100 mL of the contaminated water over the topsoil around the well.
8. Record the amount of time it takes for the contaminated water to enter the groundwater. Answers will vary. minutes:seconds.
9. Record the amount of time it takes for the contaminated water to enter the well. Answers will vary. minutes:seconds.

ANALYZE AND CONCLUDE:

1. Did the contaminated water travel through each layer at the same rate of speed? Why? In which sediment layer did the water travel the fastest? the slowest?
   No. Each layer has a different amount of pore space that either inhibits or promotes the permeability of water molecules. Answers will vary.

2. Was the color of the well water the same color as the original contaminated water? Explain the differences. The color of the well water was less vibrant because the sediment layers acted as natural filters.

3. What type of improvements could be made to prevent contamination of the well water? Answers will vary but should include forming a barrier between the sand and pebble layers, inserting an alternate drainage site, or pumping out the contaminated water.

4.1.5 Surface Water

Water

OBJECTIVES

Students will be able to
- summarize how a stream forms.
- assess potential problems associated with contaminated runoff.

VOCABULARY

- **divide** a ridge or other elevated region that separates watersheds
- **tributary** a stream or river that flows into a larger stream or river
- **watershed** an area of land that drains into a particular river system

MATERIALS

- Local topographic map (*Introduction*)
- BLM 4.1.5A Global Water Challenges

PREPARATION

- Obtain materials for *Lab 4.1.5A Hard Water*.

TRY THIS

Water Softeners
- *No additional materials are needed.*

Introduction

Have students name the major rivers in the local area. Display a topographic map of the local region illustrating the streams that feed into the major rivers and the major watersheds that drain into the rivers. Remind students that contour lines that are drawn close together show a steep incline; lines drawn farther apart show a gentler slope. Ask students what might determine the size, shape, and location of the watersheds. (**Possible answer: Topography is often the major contributor to the size, shape, and location of watersheds.**)

Discussion

- Have students consider what pollutants may be commonly found in runoff. Possible pollutants include oil, grease, metals, and coolants from vehicles; fertilizers, pesticides, and other chemicals from farms, gardens, and homes; bacteria from pet waste and failing septic systems; soil from construction sites; soaps from car or equipment washing; and accidental spills from leaky storage containers.

- Discuss ways that individuals can reduce contaminated runoff.
  1. Maintain cars. Avoid dumping materials down storm drains. Recycle used oil, antifreeze, and other fluids when possible or take them to a facility that can safely dispose of them if they cannot be recycled. Fix oil leaks.
  2. When possible, wash cars at a commercial car wash rather than in the street or driveway. If washing at home, wash it on the lawn if allowed.
  3. Reduce the use of fertilizers and pesticides. Do not fertilize before a hard rainstorm. Consider using organic fertilizers. Water lawns in the morning or evening when temperatures are cool and the wind is low to maximize water absorption into the soil. Compost or mulch lawn clippings. Preserve existing trees or plant new ones because they absorb rainfall and help manage storm water.
  4. Replace part of the lawn with native, drought-resistant plants. Add compost to planting soil and cover with mulch to improve plant growth and reduce storm-water runoff.
  5. Maintain septic systems. If not maintained properly, these systems could pollute nearby lakes and streams.
  6. Reduce impervious surfaces at home and increase the vegetative ground cover. Reduce rooftop runoff by directing downspouts to vegetated areas, not to the storm drain on the street.
  7. Support local storm- or surface-water programs. These programs maintain a community's storm-water system, prevent flooding, and protect natural resources. Consider volunteering for stream restoration or other local volunteer projects focused on reducing runoff pollution.

- Consider why properly managing storm water is important. Poorly managed storm water contaminates groundwater systems that may be the source of drinking water, affects some coastal shellfish businesses to the point of possible closure, and harms or kills fish and other wildlife. Flooding can harm the stream and wetland habitats of fish and other wildlife. Flooding can also damage homes, businesses, and septic system drainage fields. Water shortages are possible in growing urban communities that cover the ground with impervious surfaces that prevent water from soaking into the ground and replenishing groundwater used for drinking water.

- Ask the following questions:
  1. What are watersheds used for? (**Possible answers: drinking water, water for agriculture and manufacturing, habitats for plants and animals**)
  2. If an organic garden is located downhill from a garden or farm that is treated with pesticides, is the garden located downhill really organic? Explain your answer using the concept of runoff. (**Possible answer: The downhill, organic garden will not be 100% organic because water that is not absorbed in the soil from the garden located up the hill will travel down the hill and may be absorbed in the soil of the downhill garden.**)

Activities

Lab 4.1.5A Hard Water

- plastic bottles with lids, 2 per group
- 100 mL graduated cylinders, 1 per group
- distilled water, 100 mL per group
- Epsom salts, 20 g per group
- dishwashing liquid
- metric rulers, 1 per group

Liquid detergents created for dishwashers do not work well in this lab because they are low-sudsing detergents.

A. Complete *Try This: Water Softeners* with students. Discuss their results.

B. Discuss current water challenges across the world using **BLM 4.1.5A Global Water Challenges**. Have student groups brainstorm ways in which these challenges could be addressed. Challenge student groups to create campaigns that encourage others to adopt better water usage practices. For example, in support of World Water Day, the class can choose one campaign to promote locally and globally, using social media outlets or by conducting fund-raising events.

C. Assign students to research the monthly rainfall amounts in the local region, the amount of water discharge produced, and the water level of a nearby lake. After gathering this information, direct students to report their findings in an essay that includes detailed annual graphs. Annual

NOTES

Surface Water 4.1.5

Have you ever watched water rushing down a street or pouring from your roof during a heavy rain or as snow melts? Precipitation that falls to the earth usually soaks into the ground or evaporates. But when the ground is fully saturated and can no longer absorb any more water, the water begins to flow over the land in the form of runoff.

Streams begin as runoff that travels down a hill or a mountain. The runoff erodes the soil to a depth of at least 30 cm to form a narrow ditch cut in the ground called *a gully*. Gullies often become empty and dry shortly after precipitation stops. Runoff from nearby slopes flows into the gully, producing even more erosion. Eventually, all of the flowing water widens and deepens the gully into a valley with a permanent stream.

The path that a stream follows is called *a channel*. A channel is generally long and narrow and is confined by banks and a streambed. Channels branch out and widen upstream as water carries sediment downstream. As streams gradually grow wider and deeper, the banks above the water level that do not have vegetation and the streambed below the water level erode.

The flowing water in streams and rivers constantly cuts into the stream's banks, washing away more sediment. Gravity always pulls running water to the lowest elevation. The surface features of a place or region, known as *the topography*, determine the way water drains from an area. Topography includes the shapes and patterns formed by height differences in the landscape. In some locations, rivers follow lines of rock, turning at sharp angles. Stream channels are somewhat straighter when they cut into bedrock because the channels tend to follow joints, faults, or other weak structural elements. Mountain streams form a deep V-shape with steep sides, and they flow quickly in relatively straight lines. Flatland streams have rounded banks. They spread out into broad floodplains and gradually meander to the river's mouth. A bluff may develop from this type of water passage. Bluffs consist of high banks with broad, steep, sometimes rounded cliff faces overlooking plains or bodies of water.

These water systems are examples of watersheds, which are also called *drainage basins*. A **watershed**

OBJECTIVES
- Summarize how a stream forms.
- Assess potential problems associated with contaminated runoff.

VOCABULARY
- **divide** a ridge or other elevated region that separates watersheds
- **tributary** a stream or river that flows into a larger stream or river
- **watershed** an area of land that drains into a particular river system

Guadalupe River watershed

Big Horn River and its tributaries in Wyoming and Montana as seen from space

is an area of land that drains into a particular river system. It includes the main river and all of the streams that flow into it. One of the largest watersheds in the world is the Amazon Basin, which covers over 7 million square kilometers. The largest watershed in the United States is the Mississippi River Basin, which covers more than one-third of the United States. A **tributary** is a stream or river that flows into a larger stream or river. In the United States, hundreds of tributaries stretch from the Allegheny Mountains in the east and the Rocky Mountains in the west as part of the Mississippi River watershed.

A **divide** is a ridge or other elevated region that separates watersheds. Every continent has a continental divide. Some divides span multiple continents. The continental divide in North America, also known as *the Great Divide*, runs through the Rocky Mountains and extends into portions of South America along the Andes Mountains. It separates the watersheds that flow into the Pacific Ocean from those that flow into the Atlantic Ocean and the Gulf of Mexico. Some divides form country borders, such as the Congo-Nile Divide in Africa.

Runoff dissolves minerals and soil as it flows across the land toward a stream. Runoff or stream water that flows over soft rocks like

Geographic divide in Africa

NOTES

amounts for each of the three data sets may be displayed on separate graphs or may overlap on one graph. An overlapping graph may best represent the relationship among the three data sets.

D. Encourage students to draw or model streams in the area. Have students show the formation of the stream and its contribution to the water flow in its watershed. Students should include labels with the water depth, water hardness, and the stream's impact on the watershed.

E. Direct students to create a presentation assessing the harmful effects of contaminated runoff and ways people can repair or prevent the problems associated with runoff.

F. Assign students to research a major flood or flood-prevention project from the following list. Have students explain why the flood occurred or why the project was needed; the effect floodwaters had on human, animal, and plant life; and the actions taken to prevent future flooding or how well the project worked to prevent flooding.
- **1902:** The Aswan Low Dam was built to maintain annual flood water in Egypt.
- **1927:** The Mississippi River flooded from Illinois to Louisiana, leaving more than 600,000 people homeless.
- **1938:** The central Chinese Nationalist Government caused the Yellow River to flood during the Second Sino-Japanese War to prevent Japanese forces from rapidly advancing.
- **1958:** A flood in Holland claimed 2,000 lives.
- **1974:** The United Kingdom built the Thames Barrier to protect London from flooding.
- **1988:** Monsoon rains in Bangladesh caused major flooding.
- **1996:** Volcanic eruptions in Iceland released meltwater from under the Vatnajokull glacier.

---

limestone may become what is called *hard water*, which is water that contains a high concentration of dissolved minerals such as calcium and magnesium. The hardness of water is measured by determining the amount of calcium carbonate found in a water sample. Water that flows over rocks that are not easily weathered does not pick up many, if any, dissolved minerals. This water is called *soft water*—water that does not contain dissolved minerals. Water is considered soft if the concentration of calcium carbonate is between zero and 60 mg/L.

Much of the rainfall in watersheds with forests and fields is absorbed into the soil. This water becomes groundwater and slowly makes its way into streams and rivers. In most rural areas, the majority of rainfall does not enter the streams and rivers all at once because of vegetation. The gradual infiltration helps prevent flooding, even during heavy rains.

In contrast, in urban areas, much of the absorbent topsoil and vegetation has been replaced by surfaces that water cannot penetrate, such as roads, parking lots, buildings, and other hard surfaces. Instead of being absorbed into the ground, rainfall flows over the hard surfaces to storm sewers, which redirect the runoff into local streams. These streams cannot handle large amounts of runoff all at once, so they often flood during storms. This type of runoff can harm streams. The water that flows across a parking lot or down a street during a heavy rain picks up pollution, such as oil, garbage, and possibly salt and sand used to melt snow and ice. The runoff carries these pollutants into streams. In summer, water that flows over hot pavement can raise stream temperatures, killing fish and other aquatic life.

Towns and cities can reduce the harmful effects of runoff in various ways through the elimination of pollution. For example, industrial waste and untreated sewage should not be dumped into water systems. To melt ice on roads during winter months, alternatives to salt can be used. In addition, reducing regional deforestation ensures that water can be absorbed by the soil and not create excess runoff. These are just a few ways towns and cities can reduce the harmful effects of runoff to benefit the entire ecosystem.

Effects of hard water

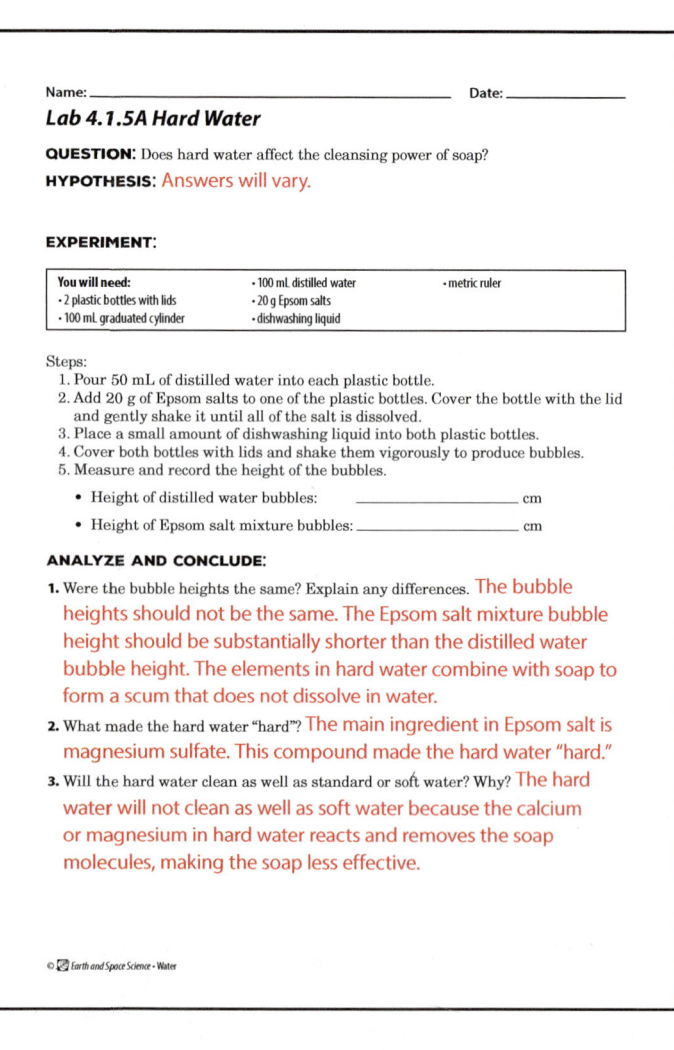

TRY THIS
Water Softeners
Design an experiment to test the effectiveness of water softeners.

LESSON REVIEW
1. How does a stream form?
2. What is the difference between hard and soft water?
3. How can runoff harm a stream?
4. How can urban areas reduce harm to streams from runoff?

Name: _____ Date: _____

Lab 4.1.5A Hard Water

QUESTION: Does hard water affect the cleansing power of soap?
HYPOTHESIS: Answers will vary.

EXPERIMENT:

You will need:	· 100 mL distilled water	· metric ruler
· 2 plastic bottles with lids	· 20 g Epsom salts	
· 100 mL graduated cylinder	· dishwashing liquid	

Steps:
1. Pour 50 mL of distilled water into each plastic bottle.
2. Add 20 g of Epsom salts to one of the plastic bottles. Cover the bottle with the lid and gently shake it until all of the salt is dissolved.
3. Place a small amount of dishwashing liquid into both plastic bottles.
4. Cover both bottles with lids and shake them vigorously to produce bubbles.
5. Measure and record the height of the bubbles.
   - Height of distilled water bubbles: _____ cm
   - Height of Epsom salt mixture bubbles: _____ cm

ANALYZE AND CONCLUDE:
1. Were the bubble heights the same? Explain any differences. The bubble heights should not be the same. The Epsom salt mixture bubble height should be substantially shorter than the distilled water bubble height. The elements in hard water combine with soap to form a scum that does not dissolve in water.
2. What made the hard water "hard"? The main ingredient in Epsom salt is magnesium sulfate. This compound made the hard water "hard."
3. Will the hard water clean as well as standard or soft water? Why? The hard water will not clean as well as soft water because the calcium or magnesium in hard water reacts and removes the soap molecules, making the soap less effective.

- **2005:** Hurricane Katrina caused devastating flooding in Louisiana, Mississippi, and Alabama.
- **2011:** Rio de Janeiro experienced major flooding and mudslides.
- **2016:** Hurricane Matthew decimated homes and forests in Haiti before it swept across the Bahamas, flooding coastal and inland regions from Florida to North Carolina.

G. Encourage students to create art pieces that represent how water positively affects their lives. Art pieces can include paintings, songs, photographs, collages, sculptures, or contemporary dances.

Lesson Review

1. How does a stream form? (**Streams begin as runoff that travels down a hill or a mountain. The water erodes the soil to a depth of at least 30 cm to form a gully. Runoff from nearby slopes flows into the gully, producing even more erosion. Eventually, all of the flowing water widens and deepens the gully into a valley with a permanent stream. Topography determines the size and shape of the stream as well as how much water discharge will be produced.**)
2. What is the difference between hard and soft water? (**Hard water contains a high concentration of dissolved minerals, such as calcium and magnesium. Soft water contains a low concentration of dissolved minerals.**)
3. How can runoff harm a stream? (**Runoff can cause flooding. It picks up oil, garbage, and other pollutants and carries them into a stream. Runoff that flows over hot pavement can increase stream temperatures, which can kill fish and other aquatic life.**)
4. How can urban areas reduce harm to streams from runoff? (**Possible answers: by not dumping industrial waste or untreated sewage into water systems, by using an alternative to salt on roads to melt ice, by reducing deforestation**)

NOTES

4.1.6 Rivers

Water

OBJECTIVES

Students will be able to
- compare the four main stages of rivers.
- relate the structure and function of rivers.

VOCABULARY

- **mature river** a meandering river located at a low elevation
- **old river** a slow-moving, flat river
- **rejuvenated river** a river with an increased stream gradient and power to erode
- **youthful river** a fast-flowing, irregular river with a steep, V-shaped channel

MATERIALS

- *Wind in the Willows* by Kenneth Grahame. Wordsworth Editions, 1998; globe (*Introduction*)
- Long plastic tray, sand, bricks, paper towels, flexible hose (*A*)
- Lab 2.1.5A Water Erosion and Time
- Lab 2.1.5B Water Erosion and Force
- WS 4.1.6A River Creation

PREPARATION

- Invite a river-rafting guide to speak to the class. (*B*)
- Have students bring in pictures of rivers. Set up three stations around the room, labeled *Youthful River*, *Mature River*, and *Old River*. (*C*)

Introduction

Read aloud the first chapter of *Wind in the Willows*. Reflect on Grahame's description of river water and how features of the watershed are described.

Have students try to name the 10 longest rivers in the world. Display a globe and identify the Nile River in northeastern Africa, the Amazon River in South America, the Yangtze River in China, the Mississippi-Missouri River in North America, the Yenisei River in Russia, the Yellow River in China, the Ob-Irtysh River in Russia and China, the Congo River in Africa, the Amur River in Russia, and the Lena River in Russia. These are generally considered the longest rivers in descending order. Explain that measuring the length of a river is complicated because rivers are not constant and it is sometimes difficult to determine a river's source. Challenge students to estimate the length of the Nile River, which is 6,650 km long.

Discussion

- Discuss the ways that rivers have influenced communities. Ask students why so many communities initially develop near rivers. (**Possible answers: Rivers aid in the transport of manufactured products; they connect communities; they draw animals to sources of water, providing a food source for individuals; energy and power supplies can be harnessed from the force of rushing river water; rivers help supply irrigation needs; and rivers supply recreation.**)

- Encourage students to discuss the role God has given humans regarding river sustainment. Have students describe gifts and abilities humans have that lend themselves to this role.

- Ask the following questions:
  1. How are rivers classified? (**Rivers are classified by their physical characteristics. For example, youthful rivers are fast-moving with steep, V-shaped channels and old rivers are slow-moving and flat.**)
  2. Do all rivers flow north to south? (**No. Rivers can flow in any direction to reach the lowest elevation.**)
  3. What does the health of a river indicate about the health of the surrounding ecosystem? (**If a river is unhealthy, it is likely that the surrounding ecosystem is unhealthy. An unhealthy river may indicate that pollution is in the area that could affect the plants and animals of the region. Living creatures require healthy water to survive.**)

Activities

A. Demonstrate how the types of rivers are created. Do not tell students what types of rivers are being created. Have students determine the type of river that is being demonstrated. Or have student groups complete **WS 4.1.6A River Creation**.
  Youthful River: Fill a long, plastic tray with sand several centimeters deep. Place four or five bricks under one end of the tray to create a steep incline. The other end of the tray should be placed on the edge of a sink to ensure the water will flow into the sink. Line the sink with paper towels to catch any sand that washes out of the tray. Connect one end of the hose to a faucet, place the other end of the hose at the elevated end of the tray, and increase the water flow to represent fast-moving water. Ask students to describe the type of channel that is produced. (**A deep, relatively straight channel is produced.**)
  Mature River: Smooth the sand and decrease the incline by removing one or two bricks. Lower the water flow to produce a meandering erosion pattern. Instruct students to observe and record details regarding the river bank erosion, occasional flooding of flat areas, and the development of oxbow lakes. (Note: It may be necessary to carve a meander pattern in the sand to establish the river flow for the demonstration.)
  Old River: Smooth the sand and set the tray to a low incline by using only one brick. Reduce the water flow considerably to produce large meanders. Students should observe the buildup

of sediment in the channel. Ask students what factors have changed in the course of the demonstration. (**Answers will vary but should include that the force of the water and the incline of the tray changed.**)

Rejuvenated River: Challenge students to propose a method to create a rejuvenated river using the current model. One method is to increase the incline of the old river setup. Have students observe and record any changes to the valley. Ask students to explain how a rejuvenated river forms. (**Possible answers: The river's base level falls because the amount of seawater decreases or the land rises, such as when a tectonic plate uplifts.**)

Refer students to **Lab 2.1.5A Water Erosion and Time** and **Lab 2.1.5B Water Erosion and Force**. Have students compare the activities there with the demonstrations. Ask students how erosion is connected to river formation. (**Erosion is the process that shapes rivers.**)

B. Present the river-rafting guide to share with the class his or her knowledge about different river patterns and explain why certain river types are better to raft than others.

C. Instruct student pairs to arrange pictures into the three categories. After the pictures have all been placed in a category, direct students to evaluate the results and to explain why they think a river should or should not be included in a particular category. Ask students why a fourth category, *Rejuvenated River*, was not included in the activity. (**They would be difficult to categorize based only on a picture.**)

D. Provide students with the following quote from architect Louis Sullivan from his essay "The Tall Office Building Artistically Considered": *Whether it be the sweeping eagle in his flight, or the*

NOTES

4.1.6 Rivers

OBJECTIVES
- Compare the four main stages of rivers.
- Relate the structure and function of rivers.

VOCABULARY
- **mature river** a meandering river located at a low elevation
- **old river** a slow-moving, flat river
- **rejuvenated river** a river with an increased stream gradient and power to erode
- **youthful river** a fast-flowing, irregular river with a steep, V-shaped channel

Rivers play an important role in society. Explorers charted routes along the paths of rivers. Rivers provided the energy for the first mills, and great port cities have been built on rivers. Transportation needs are met along river waterways. Drinking water is often supplied by rivers. Rivers keep underground aquifers full. Many recreational activities are facilitated by the flow of river water. All of these benefits are possible because the different types of rivers function in various ways. Think about the rivers in your area. Do they flow from the mountains or across flat land? Is the river path straight, or does it snake around in a winding path? Are rivers broken with rapids and waterfalls, or are they calm? Why are rivers so different?

In the late 19th century, a model was developed to describe stages in the development of rivers. At that time, rivers were thought to develop from a youthful stage to a mature stage to an old stage. Today, this model is no longer used. However, scientists still use the terms *youthful, mature,* and *old* to describe rivers. These terms do not identify the literal age of a river. Rather, the names describe river characteristics that depend on factors such as climate and the shape and erosion patterns of the land.

A **youthful river** is a fast-flowing, irregular river with a steep, V-shaped channel. Youthful rivers flow through hard rock, such as granite, that does not easily erode. The channels that develop from youthful rivers are deeper than they are wide. Given that the sides of these rivers are steep, water flows quickly and with force. The water in this type of river often tumbles over rocks in rapids and waterfalls. Youthful rivers carry relatively small volumes of water because they do not have very many tributaries. An example of a very important youthful river in Spain is the Ebro River. This river has the greatest discharge, or volume, and the largest drainage basin in Spain. Tributaries that flow into the Ebro River provide hydroelectric and thermoelectric power and are used for irrigation purposes.

A **mature river** is a meandering river located at a low elevation. Eroded channels that develop from mature rivers are wider than they are deep. Mature rivers are fed by many tributaries. Rain from watersheds flows from the tributaries into mature rivers. This enables mature rivers to carry more water than youthful rivers. Mature rivers are not as forceful as youthful rivers. They eventually lose their waterfalls and rapids, and they sometimes break their banks and flood the surrounding land. Instead of flowing in straight lines, mature rivers bend to form wide curves. Some of these curves may become cut off from the main river to form lakes or ponds, called *oxbow lakes*. The River Thames in southern England and the Mississippi River in the United States are examples of mature rivers.

An **old river** is a slow-moving, flat river. The slow pace of water prevents further land erosion. When water slows down, old rivers build up sediments in their channels and along their banks to form a broad, shallow plain. They are not as efficient as mature rivers at draining water from a watershed. Old rivers do not have as many tributaries as mature rivers because numerous smaller tributaries combine. The Tigris and the Euphrates rivers in the Middle East and the Indus River in South Asia are old rivers.

A **rejuvenated river** is a river in which the stream gradient, or slope, and power to erode have increased. Rejuvenated rivers develop when the river's base level falls because the amount of seawater decreases or the land rises. The land may rise when a tectonic plate uplifts, elevating a section of Earth's crust. When the land rises, the river channels become steeper. The river then cuts more deeply into the floor of the valley. For example, a rejuvenated river may cut a new V-shaped valley into an old river. The remains of the old floodplains may be present in steplike structures called *terraces*.

Rivers are a gift from God, but people have not always treated these gifts as they should. For a long time, many people thought that mighty rivers, such as the Mississippi, were self-cleaning machines that rushed pollution out to sea. In the 1960s, every major waterway in the United States was polluted with industrial waste and sewage. Laws were passed to reduce the pollution of waterways. These government actions have helped, but pollution is still an issue in many rivers.

HISTORY

The Role of Rivers
Rivers have played an important role in the history of many countries. They have defined both geographic and political boundaries. During the Civil War in the United States, many of the necessary resources needed by the Confederate Army were transported along the Mississippi River. In 1862, the Union Army decided to cut off all Confederate access to the Mississippi River. This action would eliminate a key supply route for the Confederate Army and affect its ability to defend certain positions. Ulysses S. Grant directed a fleet of gunboats to follow the Mississippi south while David G. Farragut's squadron in the Gulf of Mexico traveled north along the river. The two Union forces met at Vicksburg, Mississippi, and after a 40-day siege, successfully gained complete control of the Mississippi River. The Union's tactical decision to control the river was a major contributing factor to its victory over the Confederate Army.

NOTES

open apple-blossom, the toiling workhorse, the blithe swan, the branching oak, the winding stream at its base, the drifting clouds, over all the coursing sun, form ever follows function, and this is the law. Where function does not change form does not change. Assign students to write a paragraph explaining how the form of a river follows its function.

E. Encourage students to create presentations describing a local river. Students should include the following information in their presentation:
- Name of river's origin
- Latitude and longitude of river's origin
- Elevation at river's origin
- Direction of river's flow
- Name of outlet body of water
- Number and names of cities, states, or countries the river crosses

F. Direct students to research laws and initiatives that have been developed to protect rivers. For example, the United States has passed the National Environmental Policy Act, the Clean Water Act, the Safe Drinking Water Act, and the Water Framework Directive. Challenge students to find similar legislation in other areas of the world or initiatives passed by the United Nations.

G. Challenge student groups to create a list of questions about river protection in the local area. Direct groups to contact local environmental or governmental agencies that are involved in efforts to protect river environments. Have student groups ask the contacted agencies their questions. Then have groups share with the class what they learned from the experience.

An oxbow lake at Horseshoe Bend in Arizona

LESSON REVIEW
1. Compare the characteristics of the four stages of rivers.
2. How do people benefit from the fast-moving water found in youthful rivers?
3. Mature rivers bend to form wide curves. What is the name of the structure that is formed when these curves are cut off from the rest of the main river to form lakes or ponds?
4. Explain how the structure of a river and its functions are related.

Terrace development at the San Juan River in Goosenecks State Park, Utah

188

Lesson Review

1. Compare the characteristics of the four stages of rivers. (***youthful river***: **fast-flowing, irregular river; steep, V-shaped channel that is deeper than it is wide; not many tributaries.** ***mature river***: **meandering river located at a low elevation; channel wider than it is deep and less steep than those of youthful rivers; fed by many tributaries.** ***old river***: **slow-moving, flat river; erodes little land; less efficient than other rivers at draining water from a watershed; not many tributaries because numerous smaller tributaries combine.** ***rejuvenated river***: **river with increased stream gradient and power to erode; developed when river's base level falls or the land rises; river channel becomes steeper and cuts more deeply into the valley; can form terraces**)
2. How do people benefit from the fast-moving water found in youthful rivers? (**People have been able to use fast-moving youthful rivers to generate hydroelectric and thermoelectric power, to use them for irrigation, and to engage in recreational activities.**)
3. Mature rivers bend to form wide curves. What is the name of the structure that is formed when the curves are cut off from the rest of the main river to form lakes or ponds? (**oxbow lake**)
4. Explain how the structure of a river and its functions are related. (**Answers will vary but should include that a river's structure affects how it functions. For example, steep-sided youthful rivers are narrow, which forces the water in them to move downhill quickly. The force of such a river can be used to generate power. The number of tributaries a river has affects how much water it can carry. Youthful rivers don't have as many tributaries, so they can't carry as much water. Mature and old rivers are wider than young rivers, and they don't have waterfalls or rapids, which makes them more useful for transportation.**)

NOTES

4.1.7 Ponds and Lakes

Water

OBJECTIVES

Students will be able to
- compare features of ponds and lakes.
- determine how sunlight infiltration affects pond and lake ecosystems.
- explain how ponds and lakes form.

MATERIALS

- TM 4.1.7A Riparian Food Web
- TM 4.1.7B Light and Dark Phases of Photosynthesis

PREPARATION

- Obtain materials for *Lab 4.1.7A Pond Exploration* and arrange a field trip to a local pond. Prior to students arriving at the pond site, flag transects that are 5 m apart.
- Gather a variety of water samples from local ponds or lakes. (*A*)
- Obtain a local topographic map. (*B*)
- If possible, arrange a field trip to a lake that was formed from a receding glacier and request a guided tour to learn about the geologic formation of the lake. (*C*)

Introduction

Direct students to write a definition for the term *symbiotic*. Ask student volunteers to suggest words that are associated with this term and to write the words on the board. (**Answers will vary.**) Explain that the organisms that inhabit a pond or a lake exhibit symbiotic relationships. Display **TM 4.1.7A Riparian Food Web**. Invite three student volunteers to write a *P* next to the creatures that are producers, a *C* next to the creatures that are consumers, and a *D* next to creatures that are decomposers. Have students explain how organisms in a pond or lake environment are symbiotic.

Display **TM 4.1.7B Light and Dark Phases of Photosynthesis**. Relate the illustration to the concentration of dissolved oxygen and carbon dioxide in shallow and deep water during day and evening hours and during warm and cool months when sunlight may not always be prevalent.

Discussion

- Discuss the similarities and differences between a lake and a pond. Draw a Venn diagram on the board and have student volunteers fill in the information that is suggested by fellow classmates.

- Compare the shorelines and volumes of ponds or lakes of one geographic region to those of another. Have students graph the data. Discuss how the terminology regarding the two bodies of water may vary because one area may have less water than the other. Shoreline length and water volume are not universal indicators of what name the body of water has.

- Encourage conversation about students' experiences of swimming in lakes or ponds and the different temperatures felt the lower they swam or dove in the water. Have students illustrate their findings.

- Ask the following questions:
  1. Why is the temperature fluctuation in shallow water significant to native flora and fauna? (**Some plants and animals require a narrow range of temperatures to thrive.**)
  2. Do plants grow on the bottom of deep bodies of water? Why? (**No. Sunlight cannot infiltrate water at great depths, and without sunlight, plants cannot grow.**)
  3. How does the dumping of food scraps into a pond affect the ecosystem? (**Excess organic matter may cause an increase in the pond's bacteria as they feed on it and multiply, and more bacteria can reduce oxygen levels, which can harm pond plants and animals.**)

Activities

Lab 4.1.7A Pond Exploration

- tape measures, 1 per group
- jars, 1 per group
- thermometers, 1 per group
- vials, 1 per group
- dissolved oxygen test kits, 1 per group
- water hardness test strips, 2 per group
- litmus paper strips, 2 per group
- ammonia test strips, 2 per group
- nitrate test strips, 2 per group
- sieves, 1 per group
- buckets, 1 per group
- tweezers, 1 per group
- ice cube trays, 1 per group
- hand lenses, 1 per student
- pipettes, 1 per group
- D-frame aquatic dip nets, 1 per group

Place a dark sticker on the bottoms of the jars for students to observe while measuring turbidity, or clarity. Direct student groups to marked transects. Guide students to test and repeat for all portions involving test strips and to reuse their vials as needed.

A. Provide students with water samples that have been collected from a variety of ponds or lakes. Have students describe what they observe. Direct student pairs to place one drop of pond water on a microscope slide, to cover the slide with a coverslip, and to observe the water under a microscope. Have students draw any plant or microscopic creatures they observe.

B. Display a local topographic map. Direct students to select a nearby pond or lake and research how that pond or lake was formed. If no information is available, challenge students to form a hypothesis using clues from the surrounding geography. Have students describe features of the pond or lake as well as the nearby ecosystem it is part of.

C. If possible, take students on a field trip to a lake that was formed by a receding glacier (or present one that is available online). Introduce the ranger, who can explain the geologic formation of the lake. Guide students to study the plant growth patterns of the lake and compare them with the plant growth observed during the local pond survey.

Lesson Review

1. How are ponds and lakes alike? (**The water in ponds and lakes is either standing or slow moving. Both ponds and lakes are surrounded by land. Ponds and most lakes are freshwater ecosystems. Both are biodiverse habitats.**) How are they different? (**Ponds are generally smaller than lakes. Lakes are usually deeper than ponds, so light doesn't reach the bottom of a lake. The water temperature in ponds is more uniform.**)
2. Why is water depth an important physical characteristic of most ponds? (**Most ponds are shallow, which allows for sufficient sunlight to penetrate to the bottom. The sunlight increases the water temperature and enables plant growth and animal expansion throughout the pond.**)

NOTES

Ponds and Lakes 4.1.7

Ponds and lakes are places of wonder. Perhaps you have searched for frogs and turtles in the mud or looked for salamanders under rocks. The murky bank of a neighborhood pond is a great place to begin a lifelong discovery of God's creation.

Ponds and lakes contain standing or slowly moving water and are surrounded by land. Generally, ponds are smaller and shallower than lakes. However, this is not always true. Echo Lake in New Hampshire has a water surface area of about 5.6 hectares and is 3.35 meters deep. In contrast, the water in Island Pond in Vermont covers an area of 202.3 hectares and is 18.29 meters deep. Lakes and ponds can be found in a variety of sizes and depths.

Ponds and lakes are unique habitats, separate from the surrounding meadows or forests. Their ecosystems consist of freshwater organisms that depend on each other and the environment to survive. Algae, flagellates, plants, invertebrates, fungi, and fish are examples of the producers, consumers, and decomposers in these ecosystems. These organisms reside in one of the following four habitats: the shore, surface film, open water, or bottom water. The most plants and animals are found in the shallow depths of ponds and lakes where sunlight can penetrate.

OBJECTIVES
- Compare features of ponds and lakes.
- Determine how sunlight infiltration affects pond and lake ecosystems.
- Explain how ponds and lakes form.

Rooted plant life diminishes with lack of sunlight.

The temperatures of lakes and ponds are affected by the depth of sunlight infiltration. If a pond is very shallow, the sun's rays can reach to the bottom, enabling rooted plant growth throughout the pond. Without sunlight, rooted plant growth does not occur on the floors of ponds or lakes. Plant growth is further inhibited on lake shorelines that experience excessive wind and wave erosion.

Pond temperatures are relatively uniform throughout, but deep ponds or lakes can contain multiple thermal layers during the summer months in temperate regions. This layering, called *thermal stratification*, happens because the water's density changes with the temperature. Warm water is less dense than cold water. Spring winds help circulate lake water. As summer's heat warms a lake from the surface down, three layers of water form. The temperature of each water layer decreases moving downward. Wind circulates the less dense surface water, but the deeper water is relatively unmixed. Since water circulation moves oxygen through the water, the lake bottom has less oxygen in summer. Algae grows, preventing sunlight from reaching the lower layers in deep bodies of water. As temperatures cool in fall, the pond or lake water begins circulating again.

Thermal Stratification

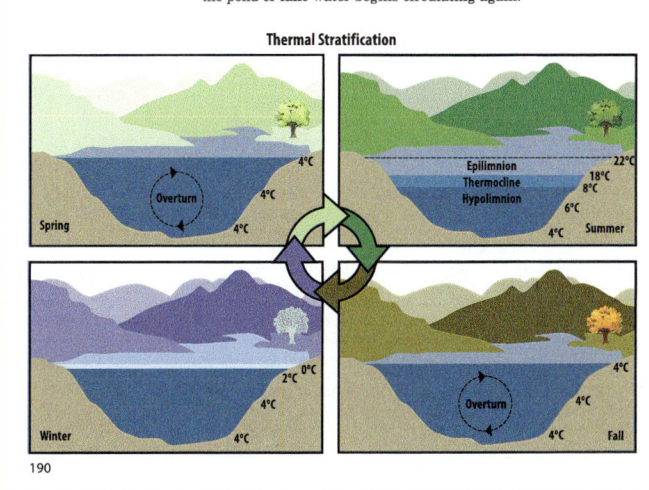

NOTES

3. Explain why there is a small percentage of plant and animal life at deeper water depths in ponds and lakes. (**Sunlight cannot penetrate to the lower depths, and without sunlight, plants and animals are unable to thrive.**)
4. Why is there less dissolved oxygen in pond or lake water at night than during the day? (**Plants do not produce oxygen when the sun is not visible. The oxygen that is produced by plants during daylight hours is consumed by animals and converted into carbon dioxide at night.**)
5. Describe two ways that ponds and lakes form. (**Possible answers: Depressions left by receding glaciers become filled with water; tectonic plates rise or rift and the resulting depression fills with water.**)

Dissolved oxygen and carbon dioxide quantities in shallow bodies of water also fluctuate with temperature variation. During the day, water plants and algae capture sunlight for photosynthesis. As they photosynthesize, plants give off a lot of oxygen, which means the oxygen level in the water peaks in the late afternoon. Plants stop producing oxygen when the sun sets, but animals continue to take in oxygen and expel carbon dioxide throughout the night. As a result, overnight the oxygen level in the water declines and the carbon dioxide level increases. Bacteria also give off carbon dioxide when they decompose organic substances in the water. Unlike shallow lakes and ponds, a deep lake's temperature, dissolved oxygen level, and carbon dioxide level remain about the same throughout the day.

Beaver dams can create ponds.

Not all ponds form in the same way. Many ponds formed thousands of years ago when glaciers receded, leaving holes in the ground. Some ponds form as rivers erode and change course. Some ponds form when landslides block the streamflow in steep valleys. Beavers can make ponds by damming up sections of a stream. Alligators sometimes hollow out small ponds as they search for water. Strong winds and meteorites can even carve out deep impressions in the soil, which later fill with water and form small ponds. Some ponds form in depressions in a forest during the wet season but evaporate during the dry season. People also dig new ponds that supply water for farm animals or for irrigation. Ponds generally have a short lifespan. Over time, they develop into marshes and then meadows.

Lake formation is very similar to pond formation, but requires a greater amount of force. Rivers can form oxbow lakes, and delta lakes can form at the mouth of a river when sediments dam up the current. In North America, thousands of both small lakes and deep lakes, such as New York's Finger Lakes, formed in deep depressions left by glaciers. The movement of the earth's plates has also formed enormous lakes. Scientists theorize that a crustal plate rose from the sea and isolated Florida's Lake Okeechobee and that tectonic rifting formed Lake Tanganyika in Central Africa. Deep, clear lakes often form in the craters of volcanoes, such as Quilotoa Crater Lake in Ecuador. Some lakes are formed when rivers are dammed. For example, dams along the Columbia and Colorado rivers resulted in a series of large lakes. Lake Nasser is a reservoir, or artificial lake, that lies behind the Aswan Dam in Egypt. Lakes have a longer lifespan than ponds. They can flourish for thousands of years.

The Great Lakes, which scientists believe were formed by glaciers, make up the world's largest surface area of freshwater. The United States and Canada share all of these lakes except for Lake Michigan, which lies entirely within the United States. The Great Lakes are unique among the world's freshwater lakes. All five lakes form a single watershed with one common outlet to the sea, which is the Saint Lawrence Seaway.

Although ponds and lakes only make up a small percentage of water on Earth, they are important sources of freshwater. In fact, many scientists consider lakes to be the best available freshwater source on Earth's surface. Ponds and lakes not only create habitats for plants and animals, but are also used by people for recreation, industry, and agriculture. These bodies of water and their contributions can be enjoyed for many years if the environment is purposefully protected.

LESSON REVIEW
1. How are ponds and lakes alike? How are they different?
2. Why is water depth an important physical characteristic of most ponds?
3. Explain why there is a small percentage of plant and animal life at deeper water depths in ponds and lakes.
4. Why is there less dissolved oxygen in pond or lake water at night than during the day?
5. Describe two ways that ponds and lakes form.

Kelimutu Crater Lake in Indonesia

Name: _____ Date: _____

Lab 4.1.7A Pond Exploration

QUESTION: How does the surrounding land contribute to the health of a pond-water system?

HYPOTHESIS: Answers will vary.

EXPERIMENT:

You will need:	• 2 water hardness test strips	• tweezers
• tape measure	• 2 litmus paper strips	• ice cube tray
• jar	• 2 ammonia test strips	• hand lenses
• thermometer	• 2 nitrate test strips	• pipette
• vial	• sieve	• D-frame aquatic dip net
• dissolved oxygen test kit	• bucket	

Steps:
Site Description

1. Topography surrounding the pond: Answers will vary.

2. Current latitude/longitude: Answers will vary.
3. Site length (meters): Answers will vary.
4. Minimum pond width (meters): Answers will vary.
5. Maximum pond width (meters), if measurable: Answers will vary.
6. Number of transects: Answers will vary.
7. What is the dominant vegetation type in the area (none, cultivated, meadow, scrub, or forest)? Is the vegetation located on both sides of the pond?
   Answers will vary.
8. Is the dominant substrate silt/sand (<2 mm), pebbles (2–8 mm), gravel (8–64 mm), cobblestone (64–256 mm), or boulders (>256 mm)?
   Answers will vary.
9. What is the estimated overhead forest cover (none, 1%–25%, 26%–50%, 51%–75%, or 76%–100%)? Answers will vary.

Lab 4.1.7A Pond Exploration

10. Describe the weather conditions, any notable or unusual site conditions, sampling problems, and sightings or observations—including plants, animals, invasive species, and human activities.
    Answers will vary.

11. Identify and describe any potential sources of pollution seen on or near the site.
    Answers will vary.

12. Sketch all pond features, transects, vegetation, and nearby permanent features, including roads, buildings, paths, and bridges. Label features according to a map legend and indicate directions.
    Drawings will vary.

Name: _____ Date: _____

Lab 4.1.7A Pond Exploration continued

Turbidity
Turbidity measures the cloudiness of water. The greater the turbidity of the water, the less life it can support. Water becomes turbid, or cloudy, when the suspended solids in the water increase. Soil erosion, urban runoff, waste discharge, algal blooms, and pond substrate disturbance all contribute to increased turbidity.
1. Place the jar with a sticker horizontally in the water and gently sweep the jar from side to side for 15 seconds.
2. Once the jar is full of water, remove it from the pond, and look through the top to observe the sticker on the bottom of the jar. Is the sticker 100% visible, 50% visible, or 0% visible? Answers will vary.

Temperature
Different species thrive at different water temperatures. If the temperature changes in a short amount of time, animals can experience stress, and decreased oxygen levels may be present. A temperature of 13°C is ideal during the fall and spring months.
1. Place a jar in the pond water at least 10 cm below the surface.
2. After 1 minute has elapsed, remove the jar and use a thermometer to record the temperature of the water in the jar. Answers will vary. °C

Dissolved Oxygen
Ponds with higher amounts of dissolved oxygen support more life. Oxygen levels are reduced when water becomes polluted with fertilizers, sewage, animal feces, and garden waste, which use oxygen to decompose.
1. Submerge a jar in the pond water for 20 seconds.
2. Withdraw the jar from the pond and use it to completely fill a small vial.
3. Take 2 oxygen tablets from the dissolved oxygen test kit, drop them into the vial, and place the lid on the vial. Water should overflow out of the vial.
4. Invert the vial for about 5 minutes until the oxygen tablets are fully dissolved.
5. The water color should change 5 minutes after the tablets have fully dissolved.
6. Follow the directions in the oxygen test kit to record the amount of oxygen dissolved in the water sample. Answers will vary. ppm

Water Hardness
A water hardness test measures the amount of calcium and magnesium present. If the level is high, this indicates there is too much algae. If the level is low, very little algae is present. To raise the calcium and magnesium levels, different salt solutions (not table salt) can be added to the water. A normal water hardness level ranges between 150 and 330 ppm.
1. Fill a vial with pond water.
2. Dip a water hardness test strip in the water and observe the color change.
3. Compare the strip's color to the key on the water hardness test strip container and record the value. Answers will vary. ppm

Lab 4.1.7A Pond Exploration

pH
Most aquatic animals thrive in a water pH range of 6.5–8.0. Acid rain and waste water discharge can affect the natural pH of a water source.
1. Fill a vial with pond water.
2. Dip a litmus test strip in the water and observe the color change.
3. Compare the strip's color to the pH key on the litmus test strip container and record the pH level. Answers will vary.

Ammonia
Decaying organic material and excess amounts of fish waste can contribute to a high ammonia level in pond water. If the ammonia level is too high, different types of bacteria can be added to the water. An ideal ammonia reading is <0.1 ppm.
1. Fill a vial with pond water.
2. Dip an ammonia test strip in the water and observe the color change.
3. Compare the strip's color to the ammonia key on the ammonia test strip container and record the value. Answers will vary. ppm

Nitrates
High levels of nitrates can be observed in water that contains a large quantity of plant life or algae. If the nitrate level is too high, buffers can be created between the water source and the pollution source, along with limiting the use of fertilizers. An ideal nitrate level is <20 ppm.
1. Fill a vial with pond water.
2. Dip a nitrate test strip in the water and observe the color change.
3. Compare the strip's color to the nitrate key on the nitrate test strip container and record the value. Answers will vary. ppm

Organisms
1. Use the jar to collect a water sample from just below the water surface.
2. Pour the water through a sieve and into a bucket.
3. Use tweezers to place any organisms into compartments in an ice cube tray. Use a pipette to add a little water to each compartment to keep the organisms moist.
4. Use a pipette to gather small organisms from the bucket that passed through the sieve and add them to the ice cube tray.
5. Examine the organisms with a hand lens and draw them in the space provided.
6. Return the organisms to the pond.
7. Repeat Steps 1–6 with water drawn from halfway between the surface of the pond and the bottom of the pond.
8. Repeat Steps 1–6 with water drawn from near the bottom of the pond.
9. Using a D-frame aquatic dip net, make several slow back-and-forth sweeps through the pond.
10. Use the space provided to draw and label the organisms in the net.

Name: _____ Date: _____

Lab 4.1.7A Pond Exploration continued

Surface Organisms Drawings

Drawings will vary.

Middle Organisms Drawings

Drawings will vary.

Bottom Organisms Drawings

Drawings will vary.

Lab 4.1.7A Pond Exploration

ANALYZE AND CONCLUDE:

1. Using the data, is the overall health of the pond good, fair, or poor? Explain. **Answers will vary.**

2. What factors contributed to the health of the pond? **sunlight exposure and pollution from urban runoff and waste discharge**

3. How could the health of the pond be improved? **Possible answers: by reducing soil erosion, fertilizer, and waste runoff**

4. What kind of impact would high turbidity have on the animals and plants in the pond habitat? **Water temperature would increase, the amount of dissolved oxygen would be reduced, and animals and plants would suffer with the decreased amount of dissolved oxygen.**

5. What kind of impact would the removal of vegetation around the water have on the pond habitat? **Without vegetation around the water, the water is subject to more intense sunlight that will warm the water, which will decrease the amount of dissolved oxygen and in turn stress the animal and plant life.**

Water

4.1.8 Dams and Reservoirs

Introduction
Ask students why flooding occurs. (**Answers will vary but should include excessive rain, supersaturated soil, rivers or streams that overflow their banks during a heavy rain, rapid ice or snow melting in the mountains, and coastal storms or tsunamis.**) Share with students that the Organization for Economic Cooperation and Development found that coastal flooding alone does $3 trillion in damage worldwide.

Review the definition of the term *floodplain* with students: a flat area along a river formed by sediments deposited when a river overflows. Expound if needed and clarify any misconceptions.

Show students a local topographic map. Guide them in locating the dams on the map and hypothesizing the purpose of each of the dams. Have students determine if any of the dams supply electricity to the school or nearby homes. If the local area does not have dams, substitute a map for an area that does.

Display **TM 4.1.8A Hydroelectric Dam** and explain to students the process of electric energy production. Share the following information:
- Water pressure increases as it flows down the penstock into the turbine that is connected to a generator.
- Large electromagnets attached to the rotor are located within coils of copper wire.
- The generator rotor spins the magnets, and a flow of electrons is created in the coils. This produces electricity, which can be stepped up in voltage through the station transformers and sent across transmission lines.
- The water that flows through the dam serves its purpose and joins the rest of the river.

Discussion
- Display **TM 4.1.8B Hoover Dam**. Have students imagine Hoover Dam collapsing. Present a topographic map of the region where Hoover Dam is located. (Note: Some Internet sites have interactive maps of this region.) Explain that Lake Mead, located behind the dam, contains about 10 trillion gallons of water. Guide students to identify which cities or towns would be initially affected by this event. Circle the cities or towns named. Relate that the population in these areas totals more than 100,000 people. Ask students how the communities might be affected. (**Possible answers: Water from Lake Mead would flood the cities beyond the dam. The cities that would initially be affected are Laughlin, Nevada; Bullhead City, Arizona; Needles, California; Lake Havasu City, Arizona; and Blythe, California.**) What damage would be done to lakes and dams below Hoover Dam by such an event? (**Terrific damage would be done to the lakes located downstream, including Lake Mohave, which is held in place by Davis Dam, and Lake Havasu, which is held in place by Parker Dam. The lakes and dams would likely be destroyed.**) Point out that these dams are used to produce hydroelectric power, irrigate farmlands, and supply drinking water to Los Angeles, California; Las Vegas, Nevada; Phoenix, Arizona; and San Diego, California. What impact would damage to these lakes and dams have on the area? (**Many people would be without electricity. Irrigation systems could collapse, which would affect farming in the area. Cities would be without drinking water.**)

- Challenge students to consider the ethical implications an engineering team must consider when constructing a dam or any other structure that has an impact on the environment.

- Ask the following questions:
  1. Why do levees only offer temporary flood protection? (**Rivers continue to build up sediments and increase in height, so levees have to be built up over time.**)
  2. Why are dams built? (**Dams are built to control flooding, to allow navigation of a river, or to create electricity.**)

OBJECTIVES
Students will be able to
- evaluate ways that flooding can be prevented.
- indicate advantages of using dams and hydroelectric power.
- explain disadvantages of using dams and hydroelectric power.

VOCABULARY
- **hydroelectric power** the electricity produced from the power of moving water
- **levee** a structure built to prevent a river from overflowing
- **reservoir** a natural or artificial lake used to store and regulate water

MATERIALS
- Topographic map, preferably local that includes dams (*Introduction*)
- Topographical map of Hoover Dam region (*Discussion*)
- TM 4.1.8A Hydroelectric Dam
- TM 4.1.8B Hoover Dam
- TM 4.1.8C Types of Dams
- WS 4.1.8A Dams

PREPARATION
- Obtain materials for *Try This: Build a Dam*.
- Arrange a field trip to a local lock and dam or a hydroelectric dam. (*B*)
- Invite an engineer to speak to the class about modern trends in dam construction. Have students prepare questions for the speaker. (*D*)

TRY THIS
Build a Dam
- *rectangular plastic containers, 1 per group*
- *gravel*
- *sand*
- *cardboard*
- *PVC pipe*
(*continued*)

TRY THIS

(continued from previous page)
- tape
- plastic wrap
- straws
- wooden dowels
- clothes pins
- wire or string
- screen

Dams must be able to hold back 5 L of water. Students should create a dam that can allow water to flow, stop, and flow again.

---

3. How does a hydroelectric dam produce electricity? (**Hydroelectric power is produced when a dam diverts the flow of a river to spin turbines, which generate electricity.**)

Activities

A. Complete *Try This: Build a Dam* with groups of students. Following the building process, ask students the following questions:
- Was the original design similar to the actual dam built? (**Answers will vary.**)
- If your team made changes during the construction phase, what caused you to make the changes? (**Answers will vary.**)
- If your team was given the opportunity to create a new dam, what would you do differently? (**Answers will vary.**)
- Could your team have built an effective dam using fewer materials? (**Answers will vary.**)
- What one additional material would your team have wanted to use to create an effective dam? (**Answers will vary.**)

B. Take students on a field trip to a lock and dam or to a hydroelectric dam. Assist students in preparing questions to ask the guide and have students complete **WS 4.1.8A Dams** at the conclusion of the tour. If such a trip is not possible, show an online virtual tour of such a facility.

C. Display **TM 4.1.8C Types of Dams**. Guide students to consider possible reasons for the design of each dam. Assign students to investigate how dam design and construction has changed in the last century. What modifications (if any) affect the environmental impact of a dam?

Dams and Reservoirs 4.1.8

In order to take advantage of rivers for transportation needs, energy sources, recreational activities, and drinking water resources, many cities and towns have been built on floodplains. With an increasing number of individuals building houses near rivers, flood prevention has become an important topic in many areas. However, it is important to realize that not all flooding is bad. In fact, flooding is a natural and even necessary occurrence in the life of many rivers. Flooding can even be beneficial because floodwaters leave behind sediments that produce flat, fertile plains for farming. To limit flood damage, scientists now believe it is best to leave floodplains undeveloped and in their natural state. If that is not possible, limiting or restricting how development occurs in floodplains helps lessen the devastation from floods. In regions where extensive development has already occurred on floodplains, scientists and engineers continue to research and develop plans that will prevent devastating floods.

God provided natural flood control systems in the environment. Natural flood control involves using forests, wetlands, and soil conservation methods to prevent excess runoff during heavy rains. Some human methods for controlling floods include levees and dams. A **levee**, also called *a dike*, is a structure built to prevent a river from overflowing. Most levees are built alongside a river. Levees offer temporary protection from flooding. As a river deposits sediment, the height of a levee must be raised. In contrast, a dam is a permanent structure built across a river

OBJECTIVES
- Evaluate ways that flooding can be prevented.
- Indicate advantages of using dams and hydroelectric power.
- Explain disadvantages of using dams and hydroelectric power.

VOCABULARY
- **hydroelectric power** the electricity produced from the power of moving water
- **levee** a structure built to prevent a river from overflowing
- **reservoir** a natural or artificial lake used to store and regulate water

TRY THIS
Build a Dam
Work with a team and fill a rectangular plastic container with gravel or small rocks that must not be removed. Using the provided materials, construct a dam that can hold back five liters of water and allow for a controlled release of some of the water. Test the dam for its ability to allow water to flow, stop, and flow again. Make adjustments as needed.

Levee breaks in New Orleans during Hurricane Katrina

---

to regulate water flow. By regulating the water flow, engineers can control flooding. Some of the largest structures on Earth are dams. Dams can be found on almost every continent.

Most dams are constructed for flood control or to allow navigation on a river. However, many large-scale dams were built in the 20th century to provide **hydroelectric power**, which is electricity produced from the power of moving water. To produce energy, the water level behind a dam is raised and then diverted over the turbines of a generator to create electricity. Hydroelectric power supplies nearly 16% of the world's electric power. This type of power is the most widely used renewable resource in the world.

Hetch Hetchy Reservoir formed behind the O'Shaughnessy Dam in Yosemite National Park.

Hydroelectric dams have many advantages compared to other energy sources. Unlike burning fossil fuels, this power source does not produce air pollution. Hydroelectric power is also one of the most economical sources of energy. Tens of thousands of dams around the world are used to generate hydroelectric power. Dams and hydroelectric power plants use technology that is well

BIBLE CONNECTION
God-Given Power
The first hydroelectric power plant was designed and built by Nikola Tesla and George Westinghouse in 1895. In Tesla's speech at the Niagara Falls power plant opening ceremony, he stated that such monuments "exemplify the power of men and the greatness of nations." Tesla placed his focus on the power and capabilities that people possess without giving proper attention to the one—God the creator—who provides people with the power to build such a marvel. More important, God supplied people with His power because individuals are created in His image. In Exodus 4:21, God tells Moses to "perform before Pharaoh all the wonders I have given you the power to do." Everything people have is a gift from God. With these gifts come great responsibility. Tesla shared later in his speech that the hydroelectric power station "signifies the subjugation of natural forces to the service of man." As good stewards of God's kingdom, people are directed to protect and utilize the resources wisely. These resources are not just for human gain.

D. Present the invited engineer to speak to the class about modern trends in dam construction.

E. Assign students to write an essay about levee failures. Direct students to include options to improve community protection in the essays.

F. Divide the class in half. Have one group of students represent citizens who are in favor of building dams. Have the other group of students represent those citizens who oppose building dams. Assign each group to write a statement defending its position. After reviewing students' statements, provide both groups with the opposing viewpoints and have each group write a response. Alternatively, have students participate in a formal debate.

G. Encourage students to investigate how lock and dam systems are being changed in many areas to benefit the environment and the barge industry. If possible, have students contact barge companies in the local area to find out how older short locks affect their industry. Assign students to contact local environmental agencies to find out about the environmental impact of systems with many short locks in contrast to systems with fewer longer locks. Challenge students to contact state and federal agencies about any plans to update local locks and dams. Have students research the cost of any planned projects and how such projects are funded.

H. Have students research how dam design has changed to be more protective of wildlife or to have a less harmful impact on the environment. Encourage students to share with classmates the changes they find. Changes may include the addition of fish ladders, construction of longer locks, or use of less intrusive construction methods.

NOTES

---

understood, and they cost less to maintain than nuclear or coal-burning power plants.

Frequently, a reservoir is created behind a dam. A **reservoir** is a natural or artificial lake used to store and regulate water. People use reservoirs for drinking water and for recreation.

However, dams change the natural environment, and sometimes these changes are not beneficial. God's creation is complex, and people must respect how the different facets work together. Constructing a dam across a river can disrupt the ecosystem, putting the native species at risk. Dams can harm fish or prevent them from reaching their spawning habitats. Scientists have determined that dams can lower the temperature of warm rivers by drawing in cooler water from deep reservoirs. Dams can also raise the temperatures of cold rivers by trapping the water in shallow reservoirs where energy from the sun is absorbed. Fish and other creatures cannot always adapt to such temperature fluctuations, so these changes may lead to lower population numbers.

Hydroelectric power station in Naberezhnye Chelny, Russia

Downstream from dams, the beds and banks of rivers often erode because the sediment needed to rebuild what erosion removes is stopped by the dam. Dams trap debris, which can clog reservoirs. Water pollution and heavy metals, such as boron, arsenic, and uranium, can accumulate in sediments and reservoir water. Polluted drinking water can create serious health hazards.

Caring for God's world is an intricate task. There are many factors that must be considered when people attempt to alter the flow of a waterway by building a dam, including the environmental impact, the economic impact, and the overall community impact. As with many human endeavors, the needs of the community should be weighed against the effects on God's creation.

LESSON REVIEW
1. Is flooding always harmful? Why?
2. Describe one method of flood prevention and explain why this method is the best.
3. What are the benefits of building dams for hydroelectric power?
4. How might dams have a negative effect on the environment?

NOTES

Lesson Review

1. Is flooding always harmful? Why? (**No. Flooding is a natural occurrence in the life of many rivers. It deposits sediments that can create flat, fertile plains that are ideal for farming.**)
2. Describe one method of flood prevention and explain why this method is the best. (**Answers will vary but should include limiting or restricting development in floodplain areas. If there are no communities on or near a floodplain, the possibility of a devastating flood would no longer be an issue.**)
3. What are the benefits of building dams for hydroelectric power? (**Possible answers: Hydroelectric power does not produce air pollution; it is one the least expensive sources of energy; hydroelectric power is a well-understood technology; and dams cost less to maintain compared to other types of power plants.**)
4. How might dams have a negative effect on the environment? (**Possible answers: Dams change the natural environment; dams trap debris that can clog reservoirs; health can be affected by drinking water polluted by heavy metals that accumulate in sediments instead of being diluted by flowing water; downstream riverbeds and banks erode over time; fish can be prevented from reaching spawning habitats; and river temperatures can be affected, which can affect plants and animals.**)

Water

4.1.9 Eutrophication

Introduction ✋

Write the following question on the board: *Can there ever be too much of a good thing?* Ask students what things might be good in moderation but can cause problems in great quantities. (**Possible answers: eating, exercising, hobbies, viewing digital screens, entertainment**)

Explain that ecosystems, especially those along rivers and lakes (riparian), function by balancing the types of organisms present, the quantity and flow of vital nutrients, and the physical environment. Too much of a good thing, such as nitrogen and phosphorus nutrients, can cause the balanced system to become unbalanced and adversely affect the entire system. This process is called *eutrophication*.

Distribute the ingredient lists from the fertilizers to student groups. Direct students to locate the three-digit code that indicates the ratio of nitrogen to phosphorus to potassium. Using the total mass of the fertilizer and the percentages of ingredients, have students calculate the mass of nitrogen and phosphorus in a bag of fertilizer. (**Answers will vary.**) Ask students what comprises the remainder of the fertilizer. (**Answers will vary.**) Explain that the remaining mass is fillers or inert ingredients that are included to help disperse the main ingredients. Challenge students to determine which type of fertilizer would be the least harmful to the environment if the fertilizer became runoff. (**Answers will vary.**)

Discussion

- Invite students who have observed the eutrophication of local lakes or ponds to describe what they saw that indicated this process was occurring. (**Possible answers: murky water covered with green, red, or brown algae; dead fish; strong, unpleasant odor**)

- Share with students that scientists have recognized the importance of the natural water purification systems God created. This recognition has led to the planting of forests and preservation of wetlands to act as buffer zones around farmland. Divide the class into four groups and assign each group to one of the categories below. Have groups quickly research their category to create a sentence describing how the assigned category is involved in the process of cleaning water naturally. Discuss group descriptions as a class and elaborate as needed. Point out that these natural methods were not designed for massive amounts of pollution.
  Evaporation. As water evaporates from the ground or from the surface of a body of water through the water cycle, debris is removed from the water. However, some tiny particles of dust or chemicals can remain to circulate through the water cycle.
  Bacteria. Organic waste products are converted by bacteria into simple compounds that do not harm the environment. Wetlands rich in decaying plant materials provide a habitat for huge numbers of decomposing bacteria. When farm runoff that contains excess nutrients is diverted into large wetlands, the bacteria can break down nitrates and phosphates.
  Wetlands. Wetlands act as natural filters for pollutants. They filter many harmful substances from the water that flows through them. Some communities use wetlands for natural wastewater treatment. Wetlands cannot handle large amounts of pollution and can only reduce the toxicity of some substances.
  Filtration. As water flows through the pore spaces of sand and gravel and through the upper layers of sediment in lakes and streams, many suspended particles are removed from the water. Some tiny pollutants or chemicals can remain in water that has been filtered in this way.

- Ask the following questions:
  1. What is eutrophication? (**the process by which nitrate or phosphate compounds overenrich a body of water and deplete it of oxygen**)
  2. How do algal blooms deplete the oxygen supply in water? (**When too much algae grows, it eventually covers the surface of the water, preventing light from reaching the water beneath**

OBJECTIVES

Students will be able to
- explain the process of eutrophication.
- analyze the effects of eutrophication.
- discover solutions related to eutrophication issues.

VOCABULARY

- **algal bloom** an explosive growth of algae caused by too many nutrients in the water
- **dead zone** an area that has been depleted of oxygen by eutrophication
- **eutrophication** the process by which nitrate or phosphate compounds overenrich a body of water and deplete it of oxygen

MATERIALS

- Ingredient lists from different fertilizers (*Introduction*)

✋ PREPARATION

- Gather the list of ingredients from several different types of fertilizers. (*Introduction*)
- Obtain materials for *Try This: Too Much of a Good Thing*.
- Invite a fish and wildlife expert to speak to the class. (*B*)

⚠ TRY THIS

Too Much of a Good Thing
- *pond water*
- *distilled water*
- *liquid fertilizer*
- *microscopes, 1 per group*
Note: Safety precautions must be taken when handling fertilizer.

Earth and Space Science

NOTES

the surface. As the algae die off, bacteria begin to multiply and use most or all of the available oxygen.)

Activities

A. Complete *Try This: Too Much of a Good Thing* with students. Experiments should include a control, samples with fertilizer, and samples with excess fertilizer. Procedures should include counting microorganisms at different intervals. This experiment is best completed over at least two weeks. Have students graph and defend their findings.

B. Present the fish and wildlife expert to speak to the class regarding local eutrophication issues including the immediate and long-term effects of eutrophication, how local land and water systems are being improved, and how students can contribute to improving the ecosystem.

C. Direct students to create a graphic organizer outlining the process of eutrophication.

D. Have students research the eutrophication problems experienced by major bodies of water around the world such as the Chesapeake Bay, the Baltic Sea, the Black Sea, the lagoon of Venice, the North Sea, and the Adriatic Sea. Guide students to determine how eutrophication has influenced the fishing, tourist, and recreational industries in these locations. Also, direct students to describe how local governments and citizens are addressing the problem.

Lesson Review

1. What two main chemical elements contribute to eutrophication? (**nitrogen and phosphorus**)

4.1.9 *Eutrophication*

OBJECTIVES
- Explain the process of eutrophication.
- Analyze the effects of eutrophication.
- Discover solutions related to eutrophication issues.

VOCABULARY
- **algal bloom** an explosive growth of algae caused by too many nutrients in the water
- **dead zone** an area that has been depleted of oxygen by eutrophication
- **eutrophication** the process by which nitrate or phosphate compounds overenrich a body of water and deplete it of oxygen

Aquatic plant and animal life requires the right amount of nutrients and minerals to thrive. During different periods in time, water sources can become saturated with an influx of nutrients and minerals. When this happens, plant and animal life is adversely affected. This overabundance of resources actually harms organisms rather than helps them. Their systems become overwhelmed and cannot function properly, and death is often a result. As the saying goes, "Everything in moderation."

The overenriching process begins not in the water but on land. Nitrogen and phosphorus are the two most common minerals required by plants, so most fertilizers contain these two minerals. Sewage also contains a significant amount of nitrogen and phosphorus. Rainfall can wash fertilizer and sewage into rivers, lakes, and other surface waters. Over time, a surplus of this fertilizer and sewage can build up in waterways and completely change the dynamics of the water system.

Eutrophication is the process where a body of water becomes overenriched with nitrate or phosphate compounds and is depleted of oxygen. The process begins with **algal bloom**, which is an explosive growth of algae caused by too many nutrients in the water. Algae feed on the extra phosphate and nitrate compounds and begin to reproduce at a dramatic rate. Large numbers of algae develop to form a layer that covers the water's surface. This layer blocks light from passing through the water. Without light to photosynthesize, aquatic plants beneath the water's surface quickly die. Bacteria and other microorganisms digest the dead plants and algae. With all this extra food, the bacteria reproduce quickly. The large bacteria population uses up much of the water's dissolved oxygen. Fish and other animal

TRY THIS
Too Much of a Good Thing
Using pond water, distilled water, and liquid fertilizer, design an experiment that tests the effects of phosphorus and nitrogen on water.

Bactrian camels relaxing in algal bloom

196

Phytoplankton swarming near the Swedish island of Gotland in the Baltic Sea

populations decrease in number because they do not receive sufficient amounts of oxygen to live.

Such an occurrence can happen in large bodies of water as well. A common example of algal bloom that occurs along the coast is called *red tide*. The seawater is often colored red because the dinoflagellate population increases. Dinoflagellates are aquatic, single-celled organisms that can release toxic substances that are harmful to marine life. Red tides kill many fish, so they are accompanied by the strong odor of dead fish. Brown tides may also develop with increases of *Aureococcus anophagefferens*, a particular type of algae that lives in areas where there is little sunlight and high concentrations of nitrogen and phosphorus.

Between the late 1850s and 1920, towns throughout North America and Europe designed sewer systems that pumped

2. Describe the roles algae and bacteria play in the eutrophication process. (**Algae feeds on the excess phosphate and nitrate compounds and begin to reproduce at a dramatic rate. A layer of thick algae forms a blanket that covers the surface of the water. This layer smothers plants and blocks light for photosynthesis from passing through the water, resulting in the death of plants. Bacteria digest the dead plants and algae and begin to reproduce quickly. Large bacteria populations use up the majority of the dissolved oxygen in the water, which deprives aquatic plants and animals.**)
3. How does eutrophication affect aquatic life? (**Aquatic plant and animal life is significantly reduced in areas where eutrophication has occurred primarily because of the lack of oxygen and sunlight.**)
4. What is an area that has been depleted of oxygen called? (**a dead or hypoxic zone**)
5. Name two methods that can reduce or prevent the effects of eutrophication. (**Possible answers: regulating fertilizer and waste runoff, creating more efficient flood control, restoring wetland and riparian systems that can capture excess nitrogen and phosphorus**)

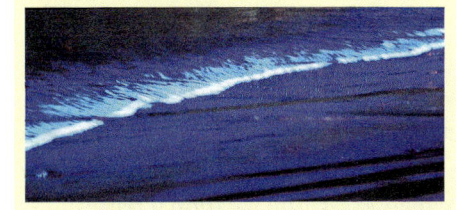

NOTES

4.2.0 Oceans

LOOKING AHEAD

- For **Lesson 4.2.3**, invite an oceanographer from a local college or university to explain the interaction between tides and currents and their significance on ocean travel.
- For **Lesson 4.2.4**, create mystery boxes for Lab 4.2.4A Predicting Ocean Floor Topography as described in the lesson.
- For **Lesson 4.2.5**, arrange a field trip to the USS *Nautilus* Memorial and Submarine Force Library and Museum or to a local maritime museum or to a submarine monument or visit a shipwright.
- For **Lesson 4.2.6**, obtain a 10-gallon aquarium and 20 pounds of sand. Invite an ocean conservationist to speak to the class.

SUPPLEMENTAL MATERIALS

- TM 4.2.1A Ocean Views
- TM 4.2.5A Ocean Floor Bathymetry
- TM 4.2.5B Deep-Sea Submersibles
- TM 4.2.5C Ocean Museum

- Lab 4.2.1A Ocean Water Density
- Lab 4.2.2A Wind and Waves
- Lab 4.2.3A Tides
- Lab 4.2.4A Predicting Ocean Floor Topography
- Lab 4.2.6A Distillation

- WS 4.2.2A Surface Currents
- WS 4.2.2B Currents Map
- WS 4.2.3A Tidal Range
- WS 4.2.4A Oceans vs. Land

- Chapter 4.2 Test
- Unit 4 Test

BLMs, TMs, and tests are available to download. See Understanding Purposeful Design Earth and Space Science at the front of this book for the web address.

Chapter 4.2 Summary

The world's oceans are an interconnected mass of saltwater that covers about 70% of Earth's surface. Continental masses separate the oceans into four major units: the Atlantic, Indian, and Pacific Oceans extend northward from Antarctica; and the Arctic Ocean, nearly landlocked by Eurasia and North America, caps the North Polar region. Some people consider the Antarctic Ocean, also known as *the Southern Ocean*, to be a fifth, separate ocean. The major oceans are subdivided into seas, gulfs, or bays. The Mediterranean Sea and the Black Sea, for example, are almost totally surrounded by land areas. Large saltwater bodies that are totally landlocked, such as the Caspian Sea, are actually salt lakes.

Despite technological advances, people know relatively little about the oceans. Public support for and interest in research is low. People do not fully understand or appreciate this large part of God's creation and the potential it holds.

Background

Lesson 4.2.1 – Ocean Water Properties

Ocean water accounts for 97% of Earth's water. Ocean water is a complex solution of pure water, solids, and gases, and it sustains a wide variety of life. A unique quality of ocean water is its salinity, which averages 34–37 parts per thousand. Because salts carry a charge, salinity is determined by measuring the water's conductivity. Salts are composed of ions. Chlorine is the most abundant ion in ocean water, followed by sodium, sulfate, magnesium, calcium, and potassium. (For a chlorine ion to become a chloride, it must gain an electron.) Together these ions constitute about 99% of ocean water's salinity. As salinity increases, osmotic pressure increases; this is significant because marine organisms depend on osmosis and specific degrees of salinity. An increase in salinity also causes freezing point depression.

The major water masses in the deep ocean are differentiated by their temperature and salinity. These properties determine relative densities, which in turn drive deep thermohaline circulation of the oceans. Thermohaline circulation refers to the deep-water circulation of the oceans and is primarily caused by differences in density between the waters of different regions. Most of the deep water acquires its characteristics in the Antarctic region and in the Norwegian Sea. Antarctic bottom water is the densest and coldest water in the ocean depths. It forms and sinks just off the continental slope of Antarctica and drifts slowly along the bottom as far as the middle of the northern Atlantic Ocean. The circulation of ocean waters is vitally important in dispersing heat energy around the globe. In general, heat flows toward the poles in the surface currents as the displaced cold water flows toward the equator in deeper ocean layers.

Lesson 4.2.2 – Ocean Motion

Ocean waves consist of a series of crests and troughs. The water particles follow vertical circular orbits—they briefly move forward as the wave crest passes and backward as the trough passes. After the wave has passed, the water particles return to their pre-wave position. The motion of the water particles decreases with depth; at a depth equal to about one-half of the wavelength, the water particles hardly move. Factors that affect the size and frequency of ocean waves include wind velocity and duration as well as the distance the wind has blown without changing direction, which is called *fetch*.

Ocean currents are part of a horizontal and vertical circulation system of ocean waters produced by gravity, wind friction, and water density variation in different areas of the ocean. The course of the currents is dictated by the Coriolis effect, density, and friction (including wind on the water's surface and friction between the layers of water). The Coriolis effect deflects currents about 45° from the wind direction and creates current cells called *gyres*. The rotational pattern causes the Northern Hemisphere gyres to displace their centers westward, forming strong western boundary currents against the eastern coasts of the continents, such as the Gulf Stream in the Atlantic Ocean. The reverse occurs in the Southern Hemisphere.

Vertical oceanic circulation brings up deep ocean water and moves surface water down. Where the wind causes surface currents to move away from shore or from each other, deeper water moves in to take the place of the diverging water in a process called *upwelling*. Upwelling generally brings extra nutrients near the surface, which in turn promote greater biomass in the ocean's upper regions. Where the wind causes ocean currents to move toward the shore or toward each other, the converging water is forced downward in a process called *downwelling*. Downwelling discourages biomass in the ocean's upper regions because nutrients are being moved into especially deep water. Variations in salinity and temperature also help account for vertical circulation; higher-saline water sinks, as does colder water. The smaller current systems found in enclosed areas are influenced more by the direction of water inflow than by than the Coriolis effect.

Lesson 4.2.3 – Tides

Tides are the alternate, regular rise and fall of sea level in the oceans caused by the moon and sun's gravitational force. The tide is sometimes considered the longest possible ocean wave, stretching all the way around the earth.

Currents redistribute an ocean's water as the tides change. As the tides rise and fall, a horizontal movement of water called *the tidal current* takes place. The incoming tide along the coast is called *a flood current*; the outgoing tide is called *an ebb current*. The strongest flood and ebb currents occur just before or near the times for a high or low tide. Near the coast, the direction of the current adjusts every 6 hours and 15 minutes, alternating between flowing toward the shore (flood current) and flowing away from the shore (ebb current). Tidal currents in the open ocean rotate to move in all directions over a period matching that of the local tide. When high tide is accompanied by high wind and low pressure, a tidal surge can occur.

The irregular shape of the ocean basins and coastlines complicates efforts at detailed predictions of ocean tides, although past tides at particular locations and circumstances can help predict future tides. People in some harbors use tide tables to determine the time and height of high tide and low tide using past observations. These tide tables are corrected for the varying positions of celestial bodies.

Earth's tides can be harnessed as a power source. To produce practical amounts of electricity, the difference between high and low tides must be around 5 m. This requirement is met by about 40 sites around the world. One of these sites is the Bay of Fundy between New Brunswick and Nova Scotia. The higher the tides, the more electricity can be generated from a given site, and the lower the cost of electricity produced. The technology that can convert tidal energy into electricity is similar to that used in traditional hydroelectric power plants. A dam is built across the tidal bay. Gates and turbines are installed along the dam. The gates are opened when a sufficient difference in water elevation is achieved on opposite sides of the gate. Water flows through the turbines, generating electricity.

Lesson 4.2.4 – Ocean Floors

The ocean covers about two-thirds of the earth's surface. Before the 19th century, the ocean floor was generally thought to be relatively flat and featureless. Oceanic exploration dramatically improved knowledge of the ocean floor. It has been determined that most of the geologic processes occurring on land are linked directly or indirectly to the dynamics of the ocean floor.

In the 19th century, deep-sea line soundings were routinely made in the Atlantic Ocean and Caribbean Sea. In 1855, a bathymetric chart was published that showed evidence of underwater

> **WORLDVIEW**
>
> - Psalm 93 uses the ocean as an indicator of God's power and permanence. In ancient times, the ocean represented power and chaos; it was the most powerful earthly force most people knew. No one could control the ocean, but it could do great damage. However, not even the ocean can undo what God has established. God is eternal, but the ocean began when God created it. The ocean is loud and strong, and people are rightly afraid of its power, yet God is stronger than the ocean. Since nothing is more powerful than God, people know that He must have all things under His control, even sad and difficult circumstances. Individuals also know that their salvation is forever secured by Jesus, the second person of the Trinity, who demonstrated His deity by controlling storms (Mark 4:39; 6:45–52).

mountains in the central Atlantic Ocean. This observation was later confirmed by survey ships laying the Trans-Atlantic telegraph line. Graphs mapping signals received from echo sounders showed a very rugged ocean floor and further confirmed the continuity and roughness of the mountain chain in the central Atlantic Ocean, which was later called *the Mid-Atlantic Ridge*.

Oceanic exploration grew in the 1950s. Oceanographic surveys revealed a great mountain range on the ocean floor that essentially encircled the earth. The global oceanic ridge covers an area that is more than 50,000 km long, 4 km high, and more than 1,500 km wide in places. Although the global oceanic ridge system is covered by water, it is the most prominent topographic feature on the earth's surface.

The ocean floor can be divided into various zones. The continental shelf extends from the coast to depths of about 100–200 m. On average, continental shelves are about 65 km wide. Continental shelves are usually covered with sand, silt, and mud, and they may feature small hills and ridges.

The continental slope has an average angle of 4° from the shelf break at depths of 100–3,200 m. The slope of the Pacific Ocean is steeper than that of the Atlantic Ocean. The Indian Ocean has the flattest continental slope. Continental slopes are marked by many canyons and mounds.

Abyssal plains have a depth of 3,000–6,000 m and are generally adjacent to a continent. They usually vary in depth from only 10–100 cm. The plains are the largest and most prevalent in the Atlantic Ocean, less prevalent in the Indian Ocean, and rare in the Pacific Ocean.

Lesson 4.2.5 – Ocean Exploration

Maps of the ocean help identify hazards on the seabed, determine routes for laying undersea cables, and assist in exploration for valuable minerals. Sonar technology grants scientists a picture of the ocean floor. Other mapping methods include radar and side scan sonar. Radar from satellites scans the ocean floor to detect large features. Side scan sonar uses a sled towed behind a research vessel. Side scan sonar can judge the ocean floor's texture but not an object's height.

Automated underwater vehicles (AUVs) function without continuous human control. They combine robotics, thrusters, sonar, and sensors with artificial intelligence. AUVs can remain submerged and operational for several months, which offers great promise for deep-sea exploration.

Remotely operated vehicles (ROVs) are often controlled by people on a ship. Robotic arms, cameras, lights, and sensors allow ROVs to "see" and "feel" for their shipboard operators. ROVs can remain submerged longer than vehicles carrying people, and they transmit data constantly.

Occupied submersibles carry people to the depths of the oceans. Their advantage over unoccupied submersibles is that the scientists can directly observe and react to new situations by changing experiments or devising new ones if necessary. Occupied submersibles can remain underwater for only short periods, however, and crew safety is always a concern.

Researchers are also developing an acoustic-daylight imaging system for tracking moving objects underwater. Background noise surrounds any submerged object forming a noise field, and the object's movement modifies this noise field. Researchers can track background noise as a source of illumination with an acoustics lens, which picks up the noise information and relays it to a computer. The computer uses the noise information to create a false-color image of the moving object.

Lesson 4.2.6 – Ocean Resources

A wide variety of resources is available in the ocean. Offshore coal mining began in the 1500s and is still being conducted through land-based mine shafts extended into the ocean or shafts lowered from artificial islands. Oil and natural gas are two other resources in the continental shelves. The number of offshore oil and natural gas rigs has increased as the demand for fossil fuels has risen. The largest offshore oil platform is the Berkut platform in Russia.

The offshore mineral mining industry is currently limited to shallower waters, and this industry is relatively small but growing. Sand and gravel are currently the most important offshore mineral resources. Many countries are dredging minerals mixed with sediments—tin in Indonesia and Thailand; gold in the Philippines and Alaska; ilmenite, rutile, zircon, and monazite in Australia, India, China, Sri Lanka, and the United States; and diamonds in southern Africa. Marine mineral resources in deeper waters, including some outside zones of national jurisdiction, offer further possibilities. Considerable research and development is necessary before mining can occur.

More than 8 million m^3 of freshwater are produced each day by the thousands of desalination plants in the world. About half of these plants use distillation, which accounts for about three-fourths of the water that is desalinated. Middle Eastern countries produce 75% of desalinated water, and the United States produces 10%. A majority of the remaining desalinated water is produced in Africa and Asia. This process is expensive, but it is necessary in regions that lack abundant freshwater supplies. Membrane processes are usually used for less salty water (brackish water). Another membrane process is electrodialysis, in which electricity drives the positive and negative ions of the dissolved salts through filters, leaving freshwater between filters.

NOTES

4.2.1 Ocean Water Properties

Oceans

OBJECTIVES

Students will be able to
- explain what affects ocean water salinity.
- examine what affects ocean water density and how density affects ocean water.
- illustrate how life-sustaining gases cycle through the ocean environment.
- model why heat and light from the sun extend only to certain depths.

VOCABULARY

- **salinity** the amount of dissolved salt in a given quantity of liquid

MATERIALS

- Pitcher, warm water, table salt, disposable cups, scale, stirring rod, wide-mouthed jar, fresh egg, small container (*Introduction*)
- 2 wide-mouthed jars, basin, blue and red food coloring, thin plastic sheet (*B*)
- TM 4.2.1A Ocean Views

PREPARATION

- Create a saltwater solution in a pitcher by mixing about 120 mL of warm water with 4 g of table salt. (*Introduction*)
- Obtain materials for *Lab 4.2.1A Ocean Water Density*.
- Obtain materials for *Try This: Bubbles Out, Bubbles In.*
- Immediately prior to the demonstration, place 2 wide-mouthed jars in a basin to catch spills. Fill one to the brim with hot water, and fill the other to the brim with cold water. Add several drops of blue food coloring to the cold water and allow it to disperse; add several drops of red food coloring to the hot water. (*B*)

Introduction

Pour tiny amounts of the prepared saltwater solution into disposable cups for students to taste. Explain that this is approximately what ocean water tastes like. Demonstrate that taste is not the only difference between saltwater and freshwater. Display a wide-mouthed jar of freshwater. Place an egg in the jar. The egg should sink because it is more dense than the freshwater. Challenge students to guess how many grams of salt will be needed to make the egg float. Select a student to tally the salt added. Then have students measure 1 g portions of table salt into a small container and add them one at a time to the jar until the egg floats. Carefully stir the solution after each addition of salt until the salt is dissolved. (Note: Test this ahead of time to ensure the egg will eventually float and that a supersaturated solution will not be necessary.)

Discussion

- Display **TM 4.2.1A Ocean Views**. Direct students to take 2–5 minutes to write down their thoughts and impressions related to each image. Emphasize that students do not need to write about every picture, nor do they need to write about the pictures in any order. Students should label each response with the letter that corresponds with the picture they are writing about. Invite students to share their thoughts about the pictures and their impressions about the ocean in general.

- Have students use an online Bible concordance or search engine to find verses that mention the ocean or the sea. Divide the class into small groups and have each group categorize at least 20 uses of the words *ocean* or *sea* in the Bible; groups should create their own categories. Guide a class discussion regarding how the Bible talks about the ocean. Possible topics include the ocean as God's creation, a symbol of judgment, a symbol of plenty or provision, a symbol of God's power, an indication of distance or expanse, or an indication of God's authority.

Activities

Lab 4.2.1A Ocean Water Density

- fine-point permanent markers, 1 per student
- metric rulers, 1 per student
- drinking straws, 3 per group
- modeling clay
- distilled water, 400 mL per group
- 250 mL beakers, 1 per group
- thermometers, 1 per group
- clocks
- metal beads, 12 per group
- buckets, 1 per group
- ice
- salt, 40 g per group
- colored pencils, 2 per student

Explain to students that the curved upper surface of a column of liquid, like the water in the beaker, is called *the meniscus*. When measuring the water in the beaker, students should measure from the bottom of the meniscus.

Option: Have students mix solutions with varying salinity and document each solution's density at room temperature and at 4°C–7°C. Then have students plot the data on a line graph.

A. Complete *Try This* activities with students.

B. Demonstrate how temperature affects density, which in turn affects water movement. Direct students' attention to the prepared jars. Place the thin plastic sheet over the mouth of the jar containing the cold water. Carefully place the two jars mouth to mouth with the cold water on top. Align the jars' mouths and have a student volunteer remove the plastic sheet. Ask students what they observed. (**The water in the two jars mixed.**) When the water stops moving, reinsert the film between the jars and separate them. Invite students to touch each jar. Ask students what they observe about the two jars. (**They are the same temperature.**) Repeat the experiment but put the hot water on top of the cold water. Ask students what they observed. (**The water in the two jars did not mix.**) After about 1 minute, replace the film between the jars and separate them. Invite students to touch each jar. Ask students what they observed about the two jars. (**One jar is still cold, and the other jar is still warm.**) Ask students why the water behaved as it did in each part of the experiment. (**Answers will vary but should include that in the first experiment, the cold water sank to the bottom of the jar of hot water and pushed the hot water up because cold water is more dense than hot water. The water continued to mix in the jar until is was the same temperature and density. In the second experiment, the less dense hot water remained on top of the more dense cold water.**) Challenge students to explain how temperature affects ocean currents.

C. Have students draw illustrations of how gases cycle through the ocean environment.

D. Assign students to create a model or diorama of the ocean's layers as determined by light penetration. Students should demonstrate the light intensity at each layer and identify several of the organisms that live in each layer. Challenge students to identify unique characteristics that enable organisms to live in each layer, to research any organisms that travel between light layers, and to explain how these organisms cope with the changes in pressure, temperature, and water density.

E. Assign students to research the life cycle of a sea mammal, such as the gray whale, that spends part of the year in the cold waters of the Arctic Ocean. Have students theorize why the salinity of ocean water is important for the mammal's habitat.

> **TRY THIS**
>
> **Bubbles Out, Bubbles In**
> • drinking glasses, 1 per group
> • cold tap water
> • access to a refrigerator
> (**The bubbles disappear because colder water can hold more dissolved gas. Warm ocean water would contain less oxygen than cold ocean water.**)
>
> **Freezing Point**
> *No additional materials are needed.*

4.2.1 Ocean Water Properties

OBJECTIVES
- Explain what affects ocean water salinity.
- Examine what affects ocean water density and how density affects ocean water.
- Illustrate how life-sustaining gases cycle through the ocean environment.
- Model why heat and light from the sun extend only to certain depths.

VOCABULARY
- **salinity** the amount of dissolved salt in a given quantity of liquid

If you were to observe Earth from space, you would notice how vast the oceans are compared to the continents. In fact, ocean water covers over 70% of the Earth; this water accounts for 97% of Earth's water. Ocean water is a careful recipe created to support a wide variety of life.

Ocean water is a complex solution of pure water, solids, and gases. It is 96.5% pure water. The other 3.5% is dissolved salts from at least 72 different elements. However, only six of these elements make up approximately 99% of the ocean's salt content.

Rivers are one source of the elements in oceans. The place where freshwater from a river meets salty ocean water is called *an estuary*, which often takes the shape of a wide, shallow bay. As river water flows over rocks, it dissolves some of the minerals in the rocks and carries them along. Each year, rivers carry 400 billion kg of these minerals to the ocean. Once in the ocean, elements may combine to create various salts. Sodium and chlorine, in their ionic forms, make up more than 90% of the elements that enter the ocean. These ions combine to create table salt, which explains why ocean water tastes salty. Minerals in ocean water also come from the rocks on the ocean floor or from openings in the sea floor called *hydrothermal vents*. These vents emit heated water from inside the crust that carries dissolved minerals.

Salinity is the measure of dissolved salt in a given amount of liquid. It is usually expressed in parts per thousand. The amount of salt in ocean water varies. Normal ocean water ranges from 34–37 parts per thousand. Ocean water in drier, hotter climates usually has more salt compared to ocean water in more humid, cooler climates. Less freshwater runs into the ocean and more water evaporates in drier climates. In contrast, places where large rivers drain into the ocean usually have low salinity because such rivers add large amounts of freshwater to the ocean.

People easily float on the Dead Sea because of its high salinity.

"The sea is His, for He made it, and His hands formed the dry land." Psalm 95:5

Hydrothermal vents

TRY THIS

Bubbles Out, Bubbles In
To observe that hot and cold water hold different amounts of dissolved oxygen, let a glass of cold water stand at room temperature for several hours. Observe the bubbles that form on the inside of the glass. Then carefully (without disturbing the bubbles) put the glass in a refrigerator and leave it there for several hours. What happens to the bubbles? Why? How might temperature changes in ocean water affect the creatures that live there?

TRY THIS

Freezing Point
Design an experiment to find out how the salt in ocean water affects the temperature at which it freezes.

FYI

Pile it On
If the salt in the oceans could be removed and spread evenly over Earth's land surface, it would form a layer more than 150 m thick, about the height of a 40-story building!

NOTES

Lesson Review

1. Where does the ocean's salt come from? (**from minerals dissolved in river water, eroded from rocks on the ocean floor, and carried up from the crust through hydrothermal vents**)
2. Explain why the salinity of ocean water is likely to be greater in a hot, dry climate than in a cool, wet climate. (**Heat causes evaporation, which takes away water and leaves salt; and dry climates do not have as many rivers contributing freshwater to the ocean, which decreases salinity. Cool, wet climates do not experience as much evaporation and add more freshwater through precipitation, both of which lower salinity.**)
3. What two factors especially affect ocean water's density? (**temperature and salinity**)
4. How does density affect ocean water? (**More dense water sinks and forces less dense water upward; this motion creates currents.**)
5. How do carbon dioxide and nitrogen cycle through the ocean environment? (**Carbon dioxide and nitrogen are used by phytoplankton to create food, and oxygen is a waste product. Other marine organisms use oxygen for respiration, and they expel carbon dioxide. Some of these same organisms eat the phytoplankton, which gives them needed nitrogen. The nitrogen moves up through the food chain as predators consume prey. When phytoplankton and other organisms die, they sink to the bottom of the ocean where bacteria decompose them and release the nitrogen back into the water. Currents bring the nitrogen back to the surface for the cycle to begin again.**)
6. Why do heat and light from the sun not travel to the lowest depths of the ocean? (**Both heat energy and light waves lose intensity as they pass through water, which prevents them from reaching the ocean's greatest depths.**)

Name: _____ Date: _____

Lab 4.2.1A Ocean Water Density

QUESTION: What things affect the density of ocean water?

HYPOTHESIS: Answers will vary.

EXPERIMENT:

You will need:		
• fine-point permanent marker	• distilled water	• bucket
• metric ruler	• 250 mL beaker	• ice
• 3 drinking straws	• thermometer	• 40 g of salt
• modeling clay	• clock	• colored pencils
	• metal beads	

Steps:
1. Use a permanent marker to mark 5 mm increments along the drinking straws.
2. Label the first straw *1*, the second straw *2*, and the last straw *3*.
3. Roll three small pieces of clay into balls, each about 1 cm in diameter. Plug one end of each straw with a clay ball. Use the smallest amount of clay possible that will completely seal the end of the straw. If the straws contain too much clay, they will not float.
4. Pour 200 mL of distilled water at room temperature into a 250 mL beaker. All measurements taken with this water will be recorded next to Warm Freshwater on the data table.
5. Place the thermometer in the beaker, allow it to sit for 2 minutes, and record the temperature on the data table. Remove the thermometer.
6. Place Straw 1 in the water with the plugged end down. If it does not stay upright in the water, drop some metal beads into it, one at a time, until the straw stays upright. The straw must float in the beaker without touching its bottom or sides.
7. Calculate the height of the straw's submerged portion by counting the number of underwater marks. If the water level falls between two lines, estimate the measurement. Remember to take the measurement from the bottom of the meniscus. Multiply the number of lines by 5 to calculate the total number of millimeters and round to the nearest whole number. Record the measurement on the data table. For example, if 3.25 lines are underwater, multiply by 5 mm to obtain 16.25 mm, and round to 16 mm.
8. Repeat Steps 6–7 with Straw 2 and Straw 3.
9. Average the measurements of all three straws and record it on the data table.
10. Place a thermometer into the beaker of water and place the beaker in a bucket of ice until the water temperature is 4°C–7°C. Remove the beaker from the ice and repeat Steps 6–9, recording the data next to Cold Freshwater on the data table. Do not change the number of metal beads in the straws.
11. Prepare a solution of 40 g of salt in 200 mL distilled water at room temperature.
12. Repeat the above procedures, recording the data under Warm Saltwater and Cold Saltwater.

Lab 4.2.1A Ocean Water Density

	Temperature	Straw 1 Height	Straw 2 Height	Straw 3 Height	Average Height
Warm Freshwater (WFW)					
Cold Freshwater (CFW)		Answers will vary.			
Warm Saltwater (WSW)					
Cold Saltwater (CSW)					

13. Create a line graph in the space provided to plot the table's information. Plot the temperature along the *x*-axis, and plot the average height along the *y*-axis. Use one color for the freshwater and a different color for the saltwater. Use the acronyms to label the points.

Possible Answer: [graph showing CSW, CFW, WSW, WFW points with Average Height on y-axis and Temperature on x-axis]

ANALYZE AND CONCLUDE:

1. How can the submerged straw indicate the density of the water sample? Objects float higher in water that is more dense. More straw above the water indicates greater water density.

2. Which sample of water had the greatest density? Which had the least? The cold saltwater had the greatest density. The warm freshwater had the least density.

3. How did temperature affect the density of each water sample? Colder temperatures made the water more dense.

4. Where in the ocean would you expect to find water that is least dense? The least dense water in the ocean would likely be found near the mouth of a stream near the equator where it is very warm.

5. How could you create water that is more dense? Water density can be increased by lowering the temperature to just above freezing, increasing the salinity, or both.

6. Why were three different straws used for each density measurement? Possible answer: The straws gave slightly different measurements because the amount of clay on each straw was slightly different, so averaging the measurements gives a more accurate result.

4.2.2 Ocean Motion

Oceans

OBJECTIVES

Students will be able to
- illustrate the parts of a wave.
- describe how waves are formed and what affects their growth in the deep ocean and close to shore.
- compare types and behaviors of currents near the shore.
- demonstrate the flow of major ocean currents.

VOCABULARY

- **amplitude** a wave's height or depth measured from the surrounding water level
- **Coriolis effect** the curving of moving objects from a straight path because of Earth's rotation
- **crest** the highest point of a wave
- **trough** the lowest point of a wave
- **wavelength** the distance between identical points on two back-to-back waves

MATERIALS

- Long plastic tub, prepared blue ice cubes, red food coloring (*Introduction*)
- Masking tape; drinking straws, 100 per group; metric rulers (*C*)
- WS 4.2.2A Surface Currents
- WS 4.2.2B Currents Map

PREPARATION

- Fill a long plastic tub about half full with tap water. Prior to freezing water in an ice cube tray, place a drop of blue food coloring in each compartment. (*Introduction*)
- Obtain materials for *Lab 4.2.2A Wind and Waves*.
- Obtain materials for *Try This: Saline Density*.

Introduction

Display the long plastic tub of water and have students gather around it. Ensure that it is in a stable location so it does not shake as students walk near it. Add several drops of red food coloring to the water at one end of the container. Select student volunteers to gently place blue ice cubes in the opposite end of the tub. Stress that the ice cubes should be added without creating waves. Direct students to record their observations.

Discussion

- Ask students the following questions related to the activity in *Introduction*:
  1. Why did the red coloring rise? (**Answers will vary but should include that the colder, more dense water moved beneath the warmer, less dense water and pushed it upward.**)
  2. Why did the red coloring eventually sink and mix with the blue? (**Answers will vary but should include the red coloring sank when its water reached the cooler end of the tub and became cool too, sinking beneath warmer water.**)
  3. How long do you think the current will continue circulating? (**until the temperature is the same throughout the tub**)

- Lead a discussion on how ocean life might be affected if all ocean water were of the same salinity and temperature. Guide students to understand that there would be less biodiversity because there would be no deep current to bring nutrients up from the bottom of the ocean or take oxygen down to the depths.

- Discuss how understanding currents could help search-and-rescue teams locate the site of a plane crash in the ocean or a ship sinking by using floating debris. Explain that the debris could be traced back to a likely point of origin by examining ocean currents, which would give rescue teams a location to begin searching for survivors.

Activities

Lab 4.2.2A Wind and Waves

- clear, deep plastic containers, 1 per group
- bricks, 3 per group
- sand
- bendable straws, 1 per group
- tape
- string
- metric rulers, 1 per group
- dry-erase markers, 1 per group
- food coloring, 4 colors per group

Option: Have students experiment by blowing softly for 2 minutes, blowing hard for 2 minutes, and blowing at varying intensities for 2 minutes. Direct students to create a data table similar to the one in the lab.

A. Direct students to complete *Try This: Saline Density*. Discuss the results of the activity.

B. Assign students to complete **WS 4.2.2A Surface Currents** and **WS 4.2.2B Currents Map**.

C. Direct groups of five students to construct simple wave machines using the following instructions: roll out 3.1 m of masking tape sticky side up and secure each end to a hard surface; attach the centers of 100 drinking straws across the tape with about 3 cm between each straw but leave about 5 cm free at each end of the tape. Place another strip of tape over the first, sticky side down, to sandwich the straws between the tape strips. If the two ends are held and the "machine" is tilted so all the straws have one end touching the ground, then the machine will look like a long fence with all the slats connected through their middles. To create waves, have two students each hold one end of the machine as another student plucks the end of one straw. Have students experiment by holding the machine more or less taut between them, plucking more than one straw at a time, and changing the force with which the straws are plucked. Ask students what happens to the straws as the wave passes. (**The straws move up and down, but they do not travel toward either end of the machine.**) What happens when more force is applied to the straws? (**The waves become bigger.**) Explain how this relates to ocean waves. Water molecules move up and down, but they do not move forward or backward. Greater force, such as wind or underwater vibrations, will create larger waves.

D. Assign students to select any major surface current and to create a presentation demonstrating how the current affects weather and travel over the ocean.

Lesson Review

1. Draw a wave and label its five parts. Then, define the different parts of a wave. (*crest*: highest point of a wave; *trough*: lowest point of a wave; *wavelength*: distance between identical points on two back-to-back waves; *wave height*: vertical distance between the wave crest and wave trough; *amplitude*: wave's height or depth measured from the surrounding water level)

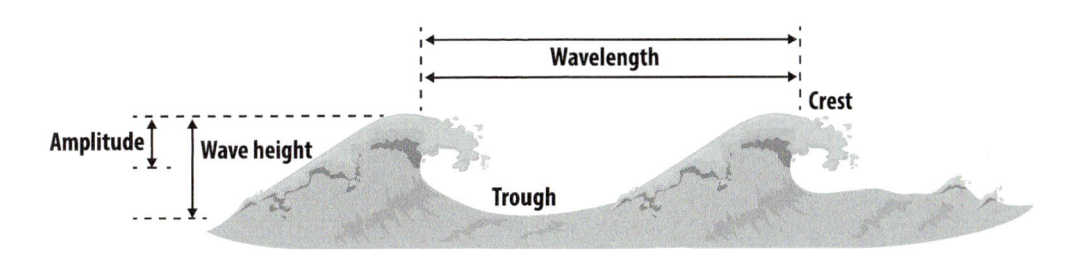

> ### 🅐 TRY THIS
> **Saline Density**
> • *four 500 mL beakers*
> • *yellow and blue food coloring*
> • *table salt*
>
> Emphasize that students should pour the water in very slowly.
>
> **(The water should turn green because the dense salty yellow water is being poured on top of less dense freshwater. The water should be layered with the blue water on top of the yellow water because the less dense water is on the bottom of the container.)**

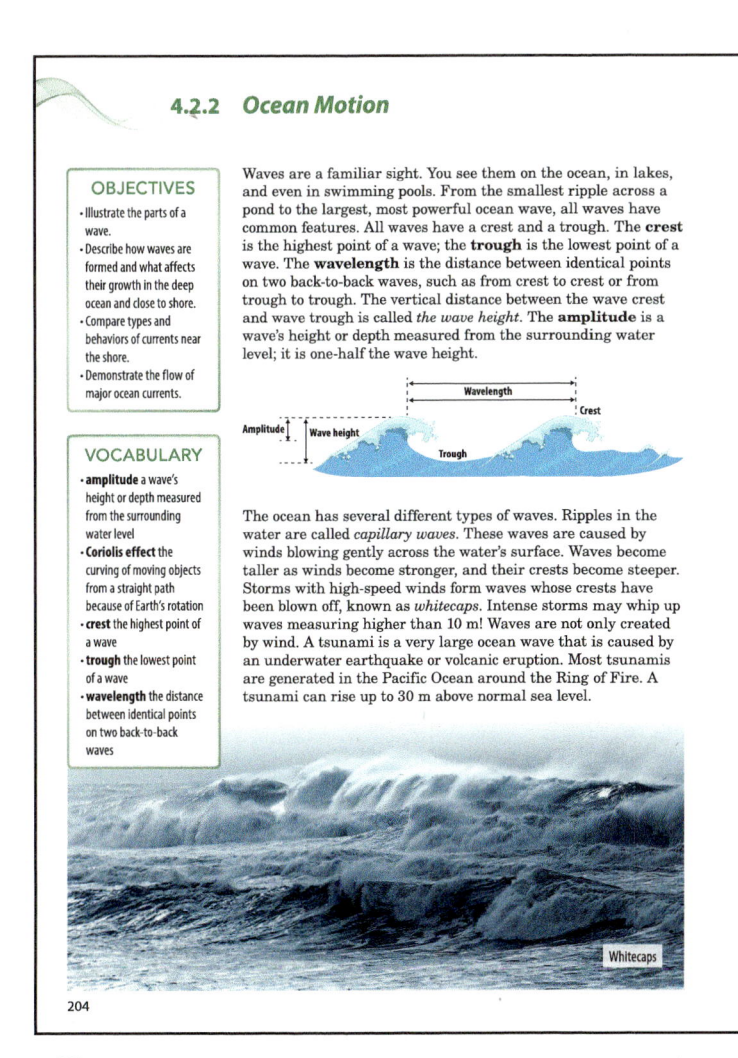

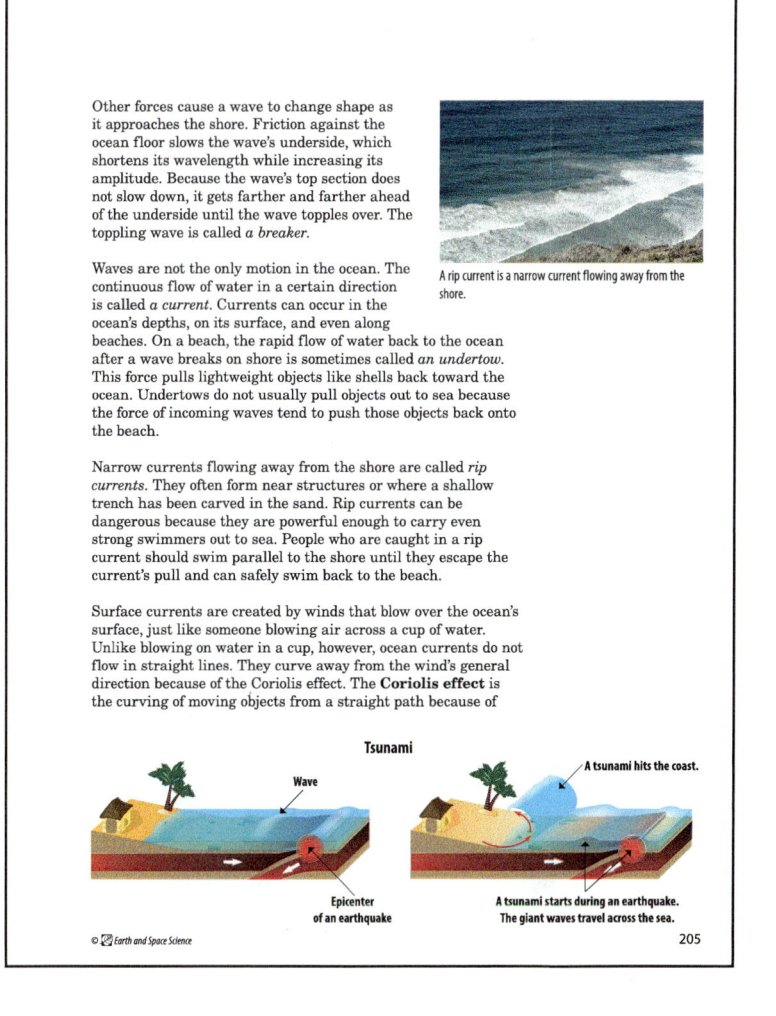

4.2.2 Ocean Motion

OBJECTIVES
- Illustrate the parts of a wave.
- Describe how waves are formed and what affects their growth in the deep ocean and close to shore.
- Compare types and behaviors of currents near the shore.
- Demonstrate the flow of major ocean currents.

VOCABULARY
- **amplitude** a wave's height or depth measured from the surrounding water level
- **Coriolis effect** the curving of moving objects from a straight path because of Earth's rotation
- **crest** the highest point of a wave
- **trough** the lowest point of a wave
- **wavelength** the distance between identical points on two back-to-back waves

Waves are a familiar sight. You see them on the ocean, in lakes, and even in swimming pools. From the smallest ripple across a pond to the largest, most powerful ocean wave, all waves have common features. All waves have a crest and a trough. The **crest** is the highest point of a wave; the **trough** is the lowest point of a wave. The **wavelength** is the distance between identical points on two back-to-back waves, such as from crest to crest or from trough to trough. The vertical distance between the wave crest and wave trough is called *the wave height*. The **amplitude** is a wave's height or depth measured from the surrounding water level; it is one-half the wave height.

The ocean has several different types of waves. Ripples in the water are called *capillary waves*. These waves are caused by winds blowing gently across the water's surface. Waves become taller as winds become stronger, and their crests become steeper. Storms with high-speed winds form waves whose crests have been blown off, known as *whitecaps*. Intense storms may whip up waves measuring higher than 10 m! Waves are not only created by wind. A tsunami is a very large ocean wave that is caused by an underwater earthquake or volcanic eruption. Most tsunamis are generated in the Pacific Ocean around the Ring of Fire. A tsunami can rise up to 30 m above normal sea level.

Other forces cause a wave to change shape as it approaches the shore. Friction against the ocean floor slows the wave's underside, which shortens its wavelength while increasing its amplitude. Because the wave's top section does not slow down, it gets farther and farther ahead of the underside until the wave topples over. The toppling wave is called *a breaker*.

Waves are not the only motion in the ocean. The continuous flow of water in a certain direction is called *a current*. Currents can occur in the ocean's depths, on its surface, and even along beaches. On a beach, the rapid flow of water back to the ocean after a wave breaks on shore is sometimes called *an undertow*. This force pulls lightweight objects like shells back toward the ocean. Undertows do not usually pull objects out to sea because the force of incoming waves tend to push those objects back onto the beach.

Narrow currents flowing away from the shore are called *rip currents*. They often form near structures or where a shallow trench has been carved in the sand. Rip currents can be dangerous because they are powerful enough to carry even strong swimmers out to sea. People who are caught in a rip current should swim parallel to the shore until they escape the current's pull and can safely swim back to the beach.

Surface currents are created by winds that blow over the ocean's surface, just like someone blowing air across a cup of water. Unlike blowing on water in a cup, however, ocean currents do not flow in straight lines. They curve away from the wind's general direction because of the Coriolis effect. The **Coriolis effect** is the curving of moving objects from a straight path because of

A rip current is a narrow current flowing away from the shore.

Tsunami

A tsunami hits the coast.

Epicenter of an earthquake

A tsunami starts during an earthquake. The giant waves travel across the sea.

NOTES

2. What forces create waves? (**wind and violent movements in the water, such as earthquakes**)
3. What happens to a wave as it nears shore? (**Friction against the ocean floor slows the wave's underside, which shortens its wavelength but increases its amplitude. Because the wave's top does not slow down, it gets farther ahead of the underside until the wave topples over.**)
4. How are rip currents different from undertows? (**Rip currents often form near structures or where a shallow trench has been carved in the sand. Rip currents can be dangerous because they are powerful enough to carry even strong swimmers out to sea. An undertow is the rapid flow of water back to the ocean after a wave breaks on shore. This force pulls lightweight objects like shells back toward the ocean. Undertows do not usually pull objects out to sea because the force of incoming waves tend to push those objects back up the beach.**)
5. What two forces produce ocean currents? (**wind and changes in water density**)
6. Why do ocean currents not flow in the exact same direction as the wind that creates them? (**Earth's rotation causes ocean currents to curve. This action is called *the Coriolis effect*.**)

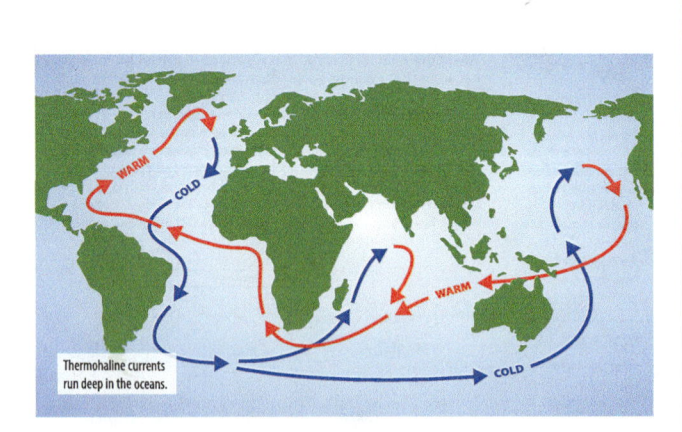

Name: _____ Date: _____

Lab 4.2.2A Wind and Waves

QUESTION: How are waves and surface currents created and sustained?

HYPOTHESIS: Answers will vary.

EXPERIMENT:

You will need:	• bendable straw	• dry-erase marker
• clear, deep plastic container	• tape	• 4 different colors of food coloring
• 3 bricks	• string	
• sand	• metric ruler	

Steps:
1. Label the two ends of the container A and B.
2. Lay a brick against one corner of end A so the brick's long end is parallel to the container's short end. Lay the second brick so one end is on the first brick and the other end is on the bottom of the container. The two bricks should look like a T with a big bump in the middle.
3. Add sand to the container until the level of sand is 3 cm from the top. Keeping the bricks in place, scrape the sand so most of it is mounded over the bricks at end A.
4. Bend the straw into an L shape.
5. Tape the straw to the inside of end B so it is centered between the corners and 1 cm down from the top edge. The short leg of the L should point at end A, and the straw's long leg should point straight up.
6. Tie the string around the third brick and lay the brick on its long edge parallel to end B. Drape the string over the side of the container and add water to the container until it is just below the straw.
7. Wait until there are no ripples in the water.
8. Measure the water level to the nearest millimeter from the tabletop.
9. One partner should blow very gently through the straw for 10 seconds to simulate wind as the other partner looks into the container at end A and uses a dry-erase marker to mark the maximum wave height on the nonsandy side.
10. Measure the wave height from the tabletop to the nearest millimeter. Find the amplitude by subtracting the calm water level from the wave height.
11. Wait until there are no ripples in the water.
12. Place 1 drop of food coloring in the water at each of the container's corners so each corner has a different color.
13. Repeat Steps 9–10, but create wind for 2 minutes without changing the force of blowing. Mark and measure the wave height every 30 seconds, but do not stop blowing. Record the amplitude for each wave measurement.
14. Wait until there are no ripples in the water.
15. Remove the straw, then quickly pull the string to flip the third brick over to simulate a tsunami.

Lab 4.2.2A Wind and Waves

ANALYZE AND CONCLUDE:

1. Look at the sand. How might waves affect a beach over many years? Over many years, a beach will eventually erode back toward the ocean unless new material is added.

2. Examine the Wave Data Table and explain the results. Answers will vary, but should include that the waves grew larger as the wind blew longer.

3. Draw the currents that formed during the 2 minutes of blowing. Show where the colors traveled.

Drawings will vary.

4. Where did the tallest waves form? along the beach

5. Where did the shortest waves form? in open water

6. How did the tsunami wave compare with the waves created by wind? The tsunami wave was larger than the waves created by the wind.

4.2.3 Tides

Oceans

OBJECTIVES

Students will be able to
- explain how the gravitational forces of the moon and sun produce ocean tides.
- contrast the two basic types of tides.
- relate tides to the position of the moon and sun.

VOCABULARY

- **neap tide** a tide that occurs when the sun and moon are at right angles to each other
- **spring tide** a tide that occurs when the sun, moon, and Earth are aligned
- **tidal range** the difference in water height between high and low tide

MATERIALS

- Football, ping-pong ball, basketball (*Introduction*)
- WS 4.2.3A Tidal Range

PREPARATION

- Locate interactive illustrations of inertia on the Internet. (*Discussion*)
- Obtain materials for *Lab 4.2.3A Tides*.
- Gather materials for *Try This: Moon Waves*.
- Invite an oceanographer from a local college or university to explain the interaction between tides and currents, and their significance to ocean travel. (C)

TRY THIS

Moon Waves
- tide charts, 1 per group
- lunar almanac charts, 1 per group
Charts may be found online.

Introduction

Relate the position of Earth, the moon, and the sun to the production of tides. Share that the earth will be represented by a football, the moon a ping-pong ball, and the sun a basketball. Explain to students that the points of the football are similar to the ocean bulges that are created because of the sun and moon's gravitational force or pull. As the moon's gravitational force pulls on the ocean water, an ocean bulge is created in the direction of the moon. The oceans bulge outward on the opposite side of the earth because the gravitational attraction of the moon is less than the inertial force on the far side of the earth. Assign three students to hold the balls. Direct students to stand in a straight line with the person holding the football standing in the middle. The individual in the middle needs to hold the football with the ends pointing toward the basketball and the ping-pong ball. Explain that a full moon or a new moon is being represented, which produces a spring tide. Direct students to create a formation that produces neap tides. The individual with the ping-pong ball should move to form a 90° angle with the football and the basketball, and the individual with the football needs to point one end of the football toward the ping-pong ball. Explain that a first- or third-quarter moon phase is being represented. Ask students what formation produces the largest tides? (**Answers will vary.**) Explain that the spring tide formation creates the largest tides because the gravitational force from the moon adds to the gravitational force exerted by the sun. The combination of forces creates a very high tide.

Discussion

- Discuss how a solar day and a lunar day differ. Explain that solar days account for the 24-hour period it takes for the earth to fully rotate on its axis. A lunar day is 24 hours and 50 minutes. The additional 50 minutes accounts for the time it takes the earth to catch up to the moon that is orbiting the earth. Relate the length of the lunar day to the fact that two high and two low tides occur every 24 hours and 50 minutes in some locations.

- Inform students that inertia is the resistance to a change in motion. Display an interactive illustration of inertia from the Internet. Discuss how the interplay of inertia and gravity works to form the tidal bulges on Earth. Explain how gravity and inertia counterbalance each other.

- Discuss the predictable nature of ocean tides. Lead the class in a discussion of what such predictable natural events tell people about the nature of God. Some people recognize God as Creator of the earth, but may not see Him as Sustainer. Ask students the following questions to encourage discussion: Does the predictability of the tides indicate whether God is present or absent? (**Answers will vary.**) What would happen if nature was not so predictable? For example, how would the world be different if the tides changed every day? (**Answers will vary.**) How would life on Earth change if forces such as gravity or inertia changed from day to day or from moment to moment? (**Answers will vary.**) What if such forces changed from place to place? (**Answers will vary.**) Does the regularity of the tides indicate anything about God's faithfulness? (**Answers will vary.**) Encourage students to find and share Scripture passages about God's faithfulness, such as Psalm 100, 111, or 136.

- Ask the following questions:
  1. Does the moon or sun create a larger tidal effect on Earth? (**the moon**)
  2. What would happen to Earth's tides if the moon did not exist? (**There would still be tides because of the sun's gravitational force, but the tides would be much smaller.**)
  3. How many spring tides and neap tides occur each month? (**Two spring tides and two neap tides typically occur each month.**)

Activities

Lab 4.2.3A Tides

- tape
- 11" × 17" pieces of paper, 1 per group
- compasses or round objects, 1 per group
- objects with equal masses, 3 per group
- springs, 2 per group
- metric rulers, 1 per group

Suggestions for objects include beanbags, wooden blocks, or cans.

A. Direct students to complete *Try This: Moon Waves* using tide and lunar almanac charts on the Internet. If possible, have students use a nearby location.

B. Assign students to complete **WS 4.2.3A Tidal Range**.

C. Introduce the oceanographer, who will explain the interaction between tides and currents and their significance to ocean travel.

D. Have students draw a diagram that shows how the moon influences tides. Diagrams should illustrate spring tides and neap tides.

NOTES

4.2.3 Tides

OBJECTIVES
- Explain how the gravitational forces of the moon and sun produce ocean tides.
- Contrast the two basic types of tides.
- Relate tides to the position of the moon and sun.

VOCABULARY
- **neap tide** a tide that occurs when the sun and moon are at right angles to each other
- **spring tide** a tide that occurs when the sun, moon, and Earth are aligned
- **tidal range** the difference in water height between high and low tide

Have you ever left your beach towel on the shore only to return later to find it soaked or lost to the tide? Tides touch coastlines all over the world. For years, philosophers and scientists attempted to explain the movement of tides. Pytheas of Massilia, an ancient Greek philosopher, first recognized that tides were related to the moon, but this relationship was not well understood until Sir Isaac Newton provided an explanation involving the gravitational attraction of the sun and moon on the oceans. Newton's laws indicate that the force between two objects depends on the mass of the two objects and the distance between them. For example, the center of the earth exerts a gravitational force on your body. At the same time, the mass of your body exerts a force on the earth. However, the mass of the earth is so much greater than the mass of your body that the earth is barely affected by your gravitational force. People feel the gravitational force from the earth every day. This force allows your feet to stay firmly planted on the ground.

In the same way, planetary masses exert forces on each other. Einstein further developed this idea in his general theory of relativity. Massive objects impact the geometry of space-time. Imagine a bowling ball placed in the middle of a trampoline. The bowling ball would cause the center of the trampoline to dip and bulge. If a smaller object was placed near the curvature of the trampoline, it would move in toward the bowling ball. The earth, moon, and sun behave in much the same way as the bowling ball. Smaller masses that are near the space-time curvatures created by a larger mass would be drawn toward the larger mass.

Low tide

The gravitational forces acting along the curves in space-time allow the moon to pull on every molecule on Earth. This force is most noticeable with matter in a liquid state. The moon exerts a force on all bodies of water, but it is only visible in very large bodies of water. You do not see tides appear in a glass of water because the sample is too small. Gravitational forces are most evident on large lakes and oceans.

The moon's gravitational force is a major contributor to the formation of tides. Inertia acts to counterbalance the pull of this force. Inertia is a property of matter that keeps matter at rest or moving in a straight line. On the half of the earth closer to the moon, the moon's gravitational force pulls strongly on the ocean water. As the gravitational force draws the water closer to the moon, inertia pulls to keep the water in place. The gravitational force is stronger than the inertia on this side of the earth, so the water ends up bulging on the side near the moon. On the opposite side of the earth, inertia is more powerful than the moon's gravitational pull because the moon is farther away. Inertia is what causes the movement of the water to create a bulge on this half of the earth. These two bulges form the earth's high and low tides. Because water is fluid, the two bulges stay aligned with the moon even as Earth rotates.

Space-time curvature

High tide

TRY THIS

Moon Waves
Compare a tide chart to an almanac chart of the positions of the moon. Plot the position of the moon relative to high and low tides for a lunar cycle.

NOTES

E. If the school is located near an ocean, encourage students to research the times of the next high and low tides. Ask students who might benefit from this information. (**Possible answers: beach visitors, mariners, fisherman, seaside industrial workers**)

Lesson Review

1. How do the gravitational forces of the moon and the sun produce tides? (**The gravitational forces acting along the curves in space-time allow the moon to pull on every molecule on Earth. This force is most noticeable with liquids like water. The moon exerts a force on all bodies of water, but it is only visible in very large bodies of water. Tides are produced as the earth spins and different regions of the world are facing the moon at various times throughout the day. The gravitational attraction is stronger on whatever area of the earth is facing the moon. The sun's gravitational pull also affects tides. It is not as strong as the moon's pull because the sun is farther away from Earth. Sometimes the gravitational forces of the moon and the sun combine to produce higher tides called *spring tides*.**)

2. Why do ocean bulges occur on Earth? (**On the half of the earth closer to the moon, the moon's gravitational force pulls strongly on the ocean water. As the gravitational force draws the water closer to the moon, inertia pulls to keep the water in place. The gravitational force is stronger than the inertia on this side of the earth, so the water ends up bulging on the side near the moon. On the opposite half of the earth, inertia is more powerful than the moon's gravitational pull because the moon is farther away. Inertia is what causes the movement of the water to create a bulge on this half of the earth.**)

3. Compare spring and neap tides. (**Spring tides occur when the sun, moon, and Earth are aligned. Spring tides have a wide daily tidal range, so the ocean rises and falls more than usual, and**

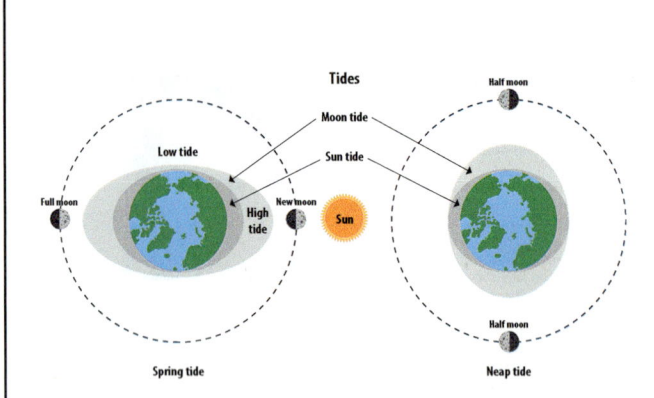

Page 210

Tides

The bulge of the tides appears slightly ahead of the moon's actual position. As water tides move across the bottom of the sea, friction is produced. Tidal friction, caused by the earth's rotation, prevents the ocean bulge from being located directly in line with the moon.

Tides are produced as the earth spins and different regions of the world are facing the moon at various times throughout the day. Two high tides and two low tides occur every 24 hours and 50 minutes. A lunar day lasts 50 minutes longer than a standard 24-hour day because the moon orbits the earth in the same direction that the earth is spinning on its axis. It takes an additional 50 minutes for the earth to catch up to the moon.

The sun also exerts a gravitational pull on Earth's oceans. You might think that the sun's pull would have an even greater influence on the earth, but it does not. Even though the sun is much more massive than the moon, the sun's gravitational force does not have as great of an impact because sun is so much farther away from Earth. In fact, the sun's tide-generating force is about half of the moon's force.

The gravitational forces of the sun and moon often act in conjunction with one another to produce higher tides. The **tidal range** is the difference in water height between high and low tide. When the sun, moon, and Earth are aligned, the sum of the gravitational forces produces a **spring tide**. Spring tides

Page 211

have a wide daily tidal range, so the ocean rises and falls more than usual. Spring tides occur twice a month, during the new moon and full moon. The term used to describe this type of tide is not related to the season. It was first used because tides were said to be *springing forth*.

During the moon's first and third quarters, the sun and the moon are at right angles to each other. At this time, the gravitational pull of the sun partially cancels the gravitational pull of the moon. Consequently, moderate tides are produced called **neap tides**. During neap tides, high tides are lower and low tides are higher, resulting in the lowest daily tidal range.

Tides vary from place to place. For example, the tides of the Atlantic Ocean are different from those of the Pacific Ocean. This difference occurs because the earth is not smooth. Ocean waters are not free to flow over the earth's entire surface in a continuous pattern. Large continents create natural barriers that ocean tides must travel around. Ocean water is also trapped by ocean basins and blocked by mid-ocean ridges and other underwater landscape features. Earth's rotation and the resulting Coriolis effect also influence the oceans' movement. These variations combine to produce different patterns of tides in different regions. Some ocean basins have complex waves that result in no tides. In other places tidal ranges may vary 1–3 m. In bays that are partly enclosed, such as Nova Scotia's V-shaped Bay of Fundy, the tidal range can reach 16 m.

Each day, two high tides and two low tides occur along most of the Atlantic Coast. Other regions, such as the Gulf of Mexico, experience one high tide and one low tide a day. Tides along the Pacific Coast have a mixed pattern, so the high and low tides differ in height along the coast.

LESSON REVIEW
1. How do the gravitational forces of the moon and the sun produce tides?
2. Why do ocean bulges occur on Earth?
3. Compare spring and neap tides.
4. When the sun and moon are at right angles to each other, how are tides affected?
5. Why do tides not appear the same in different regions of the world?

FYI

Tidal Bore
When ocean tides are exceptionally high, some of the water can rush into the narrow channels of rivers and estuaries along the coast. This wall of water is called *a tidal bore*. It is a true tidal wave. Tidal bores only occur at spring tides, never at neap tides. As the ocean tide flows in, it meets the resistance of the river water that is stagnant or flowing in the opposite direction. Significant waves can be created that travel up the river at speeds of up to 24 kph. The largest tidal bores on Earth can be found on the Qiantang River in China. These bores can reach heights of nearly 9 m.

they occur twice a month, during the new moon and full moon. Neap tides happen during the moon's first and third quarters, when the sun and the moon are at right angles to each other. Neap tides are moderate. During neap tides, high tides are lower and low tides are higher, resulting in the lowest daily tidal range.)

4. When the sun and moon are at right angles to each other, how are tides affected? (**Moderate tides are produced. The high tides are lower and the low tides are higher.**)
5. Why do tides not appear the same in different regions of the world? (**The earth is not smooth. Ocean waters are not free to flow over the earth's entire surface in a continuous pattern. Large continents create natural barriers that ocean tides must travel around. Ocean water is also trapped by ocean basins and blocked by mid-ocean ridges and other underwater landscape features. The earth's rotation and the resulting Coriolis effect also influence the oceans' movement. These variations combine to produce different patterns of tides in different regions.**)

NOTES

Name: _____ Date: _____

Lab 4.2.3A Tides

QUESTION: How are the tides on Earth affected by the position of the moon?

HYPOTHESIS: Answers will vary.

EXPERIMENT:

You will need:	• compass or round object	• metric ruler
• tape	• 3 objects with equal masses	
• 11" × 17" piece of paper	• 2 springs	

Steps:
1. Tape the paper down on a flat surface.
2. Draw a circle on the piece of paper using the compass or round object. The circle should have at least a 25 cm diameter. This image represents the earth.
3. Label the three objects A, B, and C.
4. Attach A to B with one spring and attach B to C with the other spring.
5. Arrange the objects on the piece of paper with B in the middle of the circle, A touching the line just outside of the circle to the left of B, and C touching the line just outside of the circle to the right of B. Draw a small dot on one side of each of the objects and label the dot in relation to the objects.
6. Slide A to the left and observe the movement of B and C.
7. Draw a small dot on one side of each of the objects in the new position and label the dot in relation to the objects.
8. Measure the distance each object moved.
   - Distance B moved: Answers will vary. cm
   - Distance C moved: Answers will vary. cm

ANALYZE AND CONCLUDE:
1. Describe the movement of B and C when A is pulled. B moves toward A. If pulled hard enough, C will also move, but it tends to remain stationary as the spring connecting B and C stretches.
2. Do B and C move equal distances? Why or why not? B and C do not move equal distances because the force exerted by the pull of A is stronger on B, given their closer proximity.
3. If A represents the moon's gravitational force and the circle represents Earth, what do B and C represent? Earth's tides

---

Lab 4.2.3A Tides

4. If another object (D) with a weak gravitational force is placed in this configuration directly above B, how will the movement of B and C be affected? B will not move as far toward A, and it will also move slightly toward D. C will behave similarly to the first configuration.
5. What type of tides would be produced in the configuration described in question 4? neap tides

4.2.4 Ocean Floors

Oceans

OBJECTIVES

Students will be able to
- model ocean floor features.
- relate floor features in the Atlantic Ocean and the Pacific Ocean.

VOCABULARY

- **abyssal plain** a large, nearly flat region beyond the continental margin
- **continental margin** the part of the earth's surface beneath the ocean that is made of continental crust
- **continental rise** the base of the continental slope
- **continental shelf** a broad, relatively shallow underwater terrace that slopes outward from the shoreline
- **continental slope** the steepest part of the continental incline located at the edge of the continental shelf
- **seamount** an underwater volcanic mountain that rises at least 1,000 m above the abyssal plain

MATERIALS

- Model-building materials (*B*)
- WS 4.2.4A Oceans vs. Land

PREPARATION

- Obtain materials for *Lab 4.2.4A Predicting Ocean Floor Topography*. Prepare the mystery boxes before class. Shoe boxes with lids work well, and ocean floor models can be made from clay, building blocks, or plaster. Poke several small holes large enough for a probe to be inserted in a straight line along the length of the lid about 2 cm apart. Create a different ocean floor model in each box.

TRY THIS

Ocean Floor Features
No additional materials are needed.

Ocean Vacation
No additional materials are needed.

Introduction

Have students complete **WS 4.2.4A Oceans vs. Land**. Guide students to answer as many questions as possible without receiving assistance from classmates. Have students place a star next to answers that reference oceans. Ask students if they were able to answer more land questions or more ocean questions. Discuss why there may be such a lack of knowledge regarding oceans. (**Possible answers: It is easier to observe landforms; many people do not spend as much time in the water compared to the amount of time spent on land; and it can sometimes be more dangerous to explore in the ocean than on land.**)

Discussion

- Discuss the importance of studying the ocean floor. Lead students to discuss how studying the ocean can give people a better appreciation of God's creation. Point out that studying organisms that can survive without sunlight provides scientists with an understanding of the different processes some organisms utilize to thrive. Ask how a study of the ocean floor can provide insight into how people can better care for the planet. (**Answers will vary.**)

- Ask the following questions:
  1. Considering what you know about the ocean, what regions of the ocean floor do you think have been explored the most? (**the continental margin that contains the shelf, slope, and rise**)
  2. What regions of the ocean floor have been the least explored? (**abyssal plains and trenches**)
  3. Where do you think Earth's tallest mountains are located? (**on the ocean floor**)
  4. Using what you know about living things, why do you think there are fewer organisms living at the bottom of the ocean compared to the upper regions of the ocean? (**The lack of sunlight in that region means plants cannot use photosynthesis. Plants provide food for many living things in the ocean, so there is a lack of nutritional sources at the bottom of the ocean.**)

Activities

Lab 4.2.4A Predicting Ocean Floor Topography
- mystery boxes, 1 per group
- metric rulers, 1 per group
- probes, 1 per group

A. Assign *Try This* activities for students to complete. Discuss their findings and drawings.

B. Divide the class into groups of three to create large models of individual ocean floors. Models may be built with modeling clay, papier-mâché, or any suitable material. Features should include continental margins, continental shelves, oceanic ridges, continental slopes, continental rises, trenches, abyssal plains, and seamounts. Remind students that the heights and depths of some of the features should vary significantly. Have students label the features and compare characteristics among the other ocean models.

C. Assign student pairs to compose 10 questions about the basic features of the ocean. Instruct students to conduct a survey to assess general knowledge of ocean features by posing their questions to five people of different ages outside of the classroom. As a group, discuss the results with facts.

D. Direct student groups to use their survey results to make an ocean education booklet, poster, or computer presentation to educate the public about basic ocean features.

E. Challenge students to research how many scientists believe plate tectonics formed the distinctive features of the ocean floor. Have students present their findings and discuss any differing opinions or theories.

Lesson Review

1. What three regions comprise the continental margin? (**continental shelf, continental slope, and continental rise**)
2. Where does the transition from continental crust to oceanic crust occur? (**the continental rise, which is at the base of the continental slope**)
3. What are oceanic ridges? (**underwater mountain chains that form where tectonic plates pull apart**)
4. How do many scientists believe oceanic ridges are formed? (**Many scientists believe that as tectonic plates under the ocean pull apart, magma from Earth's mantle moves up through the crust. Over time, this hardened magma forms the ridges.**)
5. Describe the continental margin in the Atlantic Ocean. (**The continental shelves are wide, and the continental slope drops gradually to a vast underwater plain.**)
6. Where are a majority of Earth's major ocean trenches located? (**the Pacific Ocean**) Why? (**The Pacific Ocean has areas undergoing subduction, which is the process of an oceanic plate being pushed under a continental plate. Many ocean trenches were formed by subduction**).

NOTES

4.2.4 Ocean Floors

OBJECTIVES
- Model ocean floor features.
- Relate floor features in the Atlantic Ocean and the Pacific Ocean.

VOCABULARY
- **abyssal plain** a large, nearly flat region beyond the continental margin
- **continental margin** the part of the earth's surface beneath the ocean that is made of continental crust
- **continental rise** the base of the continental slope
- **continental shelf** a broad, relatively shallow underwater terrace that slopes outward from the shoreline
- **continental slope** the steepest part of the continental incline located at the edge of the continental shelf
- **seamount** an underwater volcanic mountain that rises at least 1,000 m above the abyssal plain

How much of the earth's surface do you think you could visit if you could only travel by car? Less than one-third of Earth's surface can be explored this way because about 70% of Earth's landscape is under ocean water. Oceanographers have mapped about 5%–15% of the ocean floor, but in this small fraction of the space, a variety of landscapes has been discovered.

You may think continental land stops abruptly where it meets the ocean. However, a gradual incline of sediment stretches to the bottom of the ocean floor. The flattest part of this incline is called the **continental shelf**. A broad, relatively shallow underwater terrace, the continental shelf forms on the edge of a continental landmass as the land gradually slopes away from the shoreline. The waters on the outer edge of the continental shelf are rarely greater than 150 m deep.

Unlike the ocean floor's deeper regions, the continental shelf is filled with life. God uses living things to transform some ocean landscapes on the continental shelf. In tropical regions of the world, tiny coral animals build small shelters out of calcium carbonate, a hard substance similar to limestone. Layers of coral homes form massive underwater rock structures called *coral reefs*. For example, the Great Barrier Reef in Australia stretches for 2,000 km along Australia's northeastern coast. A barrier reef is a wall of coral that lies several kilometers off the coast. Eventually, some of these towers build up hundreds of meters high to form islands called *cays*.

At the edge of the continental shelf is the **continental slope**. The continental slope is the steepest part of the continental incline. The slope represents the transitional area between the continental shelf and the continental rise. It lies at depths between 100 m and 2,500 m. At a depth of about 130 m, the underwater landscape drops steeply down to the ocean floor several kilometers deep.

Great Barrier Reef

At the base of the continental slope is the **continental rise**. This region marks the transition from continental crust to oceanic crust. As sediment erodes from the continental crust, it travels down the shelf and slope and forms a thick deposit layer. Essentially, mass wasting occurs below the water's surface in this region. The rate of the erosion process is accelerated during storms as the waves and currents stir up layers of sediment.

The continental shelf, continental slope, and continental rise are known collectively as the **continental margin**, which is the most studied part of the earth's continental crust beneath the ocean. Because it is made of continental crust, the continental margin is considered to be part of the continent, even though it is underwater. Continental margins on the leading edges of tectonic plates are usually narrow and have steep continental slopes and poorly developed continental rises. Margins on the trailing side of tectonic plates are broad with gentle continental slopes and well-developed continental rises.

In the Atlantic Ocean, where earthquakes and volcanos are relatively rare, the continental shelves are wide. In the Pacific Ocean, where earthquakes and volcanoes are more common, the continental shelves are narrow and are separated from the ocean floor by deep trenches. Long, narrow underwater valleys are called *trenches*. Most of these trenches are 8,000–10,000 km deep. The deepest ocean trench is the Mariana Trench in the Pacific Ocean, which reaches a depth of 11,034 m. The majority of Earth's ocean trenches are found in the Pacific Ocean because the

TRY THIS

Ocean Floor Features
Choose the Pacific, Atlantic, Indian, or Antarctic Ocean to study. Tally the number of ridges and trenches in the ocean. Note the different depths of the ocean. Compare your results with results collected by other students who studied another ocean. Create a chart comparing the features of the four main oceans.

TRY THIS

Ocean Vacation
Design a postcard featuring a section of one of the ocean floors. Draw a scenic picture on the front of the postcard. On the back of the postcard, write a letter to a friend describing your underwater vacation to that section of the ocean. Include details about scenery, temperature, light availability, ocean life, and other interesting facts.

Pacific has areas that are undergoing subduction. Subduction is the process of an oceanic plate being pushed under a continental plate. Many ocean trenches were formed by subduction.

In the Atlantic Ocean, the continental slope moves gradually down to form a fast underwater plain. The irregular landscape found in the Pacific Ocean often forms mountain ranges instead of plains. Underwater mountain chains that form where tectonic plates pull apart are called *oceanic ridges*. Many scientists believe that magma from Earth's mantle oozes through the crust to further develop the mountain ranges. Over time, as more magma is deposited, the ridges form mountain ranges. A rift, or small indentation, can be found on both sides of oceanic ridges. The most prominent oceanic ridge can be found in the Atlantic Ocean rather than the Pacific Ocean.

Beyond the continental margin lies a massive expanse of nearly flat land called the **abyssal plains**. Some of the larger plains can be thousands of kilometers long and hundreds of kilometers wide. They are normally located next to a continent. The abyssal plains—the flattest places on Earth—cover 54% of the earth's surface and about 75% of the deep ocean. If it was possible to roam across the abyssal plains, you could potentially travel 1,300 km without climbing or dropping more than 1 m during the entire trip.

Abyssal plains are formed from thick sediment layers that originate from the continental margins. In the Atlantic Ocean basin, the plains are uninterrupted. The Pacific Ocean is laden with more trenches, making the presence of vast abyssal plains much less common. Occasionally, ocean crust irregularities called **seamounts** protrude out of the plain surface. Seamounts are underwater volcanic mountains that rise at least 1,000 m above the abyssal plain. These features dot all major ocean basins but do not appear above the water's surface. An estimated 10,000 seamounts are located in the Pacific Ocean alone.

LESSON REVIEW
1. What three regions comprise the continental margin?
2. Where does the transition from continental crust to oceanic crust occur?
3. What are oceanic ridges?
4. How do many scientists believe oceanic ridges are formed?
5. Describe the continental margin in the Atlantic Ocean.
6. Where are a majority of Earth's major ocean trenches located? Why?

Name: _____ Date: _____

Lab 4.2.4A Predicting Ocean Floor Topography

QUESTION: Is the line and sinker method an accurate way to determine the appearance of the ocean floor?

HYPOTHESIS: Answers will vary.

EXPERIMENT:

| You will need: | • mystery box | • metric ruler | • probe |

Steps:
1. Do not remove the lid from the mystery box.
2. Starting at one end of the box, measure the distance in centimeters between the edge of the box and each hole. Record measurements in the table below.

Hole Identification	Distance (cm)	Hole Identification	Distance (cm)

Answers will vary.

3. Insert the probe into each hole. When the probe hits a surface, place a finger where the probe meets the box, remove the probe, and measure the length of the probe from the tip to your finger. Record measurements in the table below.

Hole Identification	Distance (cm)	Hole Identification	Distance (cm)

Answers will vary.

Lab 4.2.4A Predicting Ocean Floor Topography

4. Plot the distance (*x*-axis) versus height (*y*-axis) to create a line graph below and connect the points. Label and title the graph.

Answers will vary.

ANALYZE AND CONCLUDE:

1. Relate the image of the line to the actual box floor features. Answers will vary but should include that the line gives a basic outline of the box floor.

2. In the space below, draw a hypothesized view of the box floor. Imagine that the box is the ocean floor. Label the regions that represent the continental shelf, continental slope, and oceanic ridge. Drawings will vary but should include the ocean floor features included on the box floor.

3. Remove the box lid. Does the box floor look identical to the line graph? Why? Answers will vary but most line graphs will not look identical to the box floor because only a few slits were measured rather than a full scan of the box.

4. Using the lab results, is the line and sinker method the best way to map the floor of the box? What other methods might an oceanographer use when mapping the ocean floor? No. A better method to map the ocean floor is to use echo sounders or multibeam sonar.

Oceans

4.2.5 Ocean Exploration

Introduction
Display **TM 4.2.5A Ocean Floor Bathymetry**. Guide students to a region where mountains are visible and a region where mountains are not visible on the ocean floor. Ask students why the second region does not appear to contain mountains (**Answers will vary.**) Is this because there are no mountains in that particular region? (**Answers will vary.**) Explain that only a fraction of the ocean floor has been mapped, so a particular region may not show mountains because it has not been mapped yet. Direct students' attention to the Mariana Trench. Challenge students to predict how many species have been discovered in regions where no light can penetrate. (**Answers will vary.**) Student answers will probably be low. There are more than 17,000 species that thrive in the deep sea where sunlight does not penetrate.

Present **TM 4.2.5B Deep-Sea Submersibles**. Display a photo of *Deepsea Challenger*. Discuss how technologies have changed and advanced the field of ocean exploration.

Discussion
- Inquire why students think more time and money have been contributed to space exploration than to ocean exploration. (**Answers will vary but may include the lack of visibility in the ocean, the visibility of stars and planets, and the popularity of science fiction books and movies.**) Discuss the difficulties of both space and ocean exploration. Which region is most easily accessible? (**Answers will vary.**) What can be done to encourage more ocean exploration? (**Answers will vary.**) Why are some individuals more interested in the moon that is 384,472 km away than in the oceans on Earth? (**Possible answers: The moon is visible and government support exists for space travel.**)

- Poll students to determine how many have snorkeled or gone scuba diving. Invite students to share their experiences with the class. If available, present scuba gear to the class. Explain the importance of each piece of equipment.

- Display **TM 4.2.5C Ocean Museum**. Discuss why the ocean could be considered the largest museum on Earth.

- Discuss inventions that were originally intended for another purpose. For example, submersibles were initially created to act as military crafts, but later became important vessels for ocean exploration.

- Ask the following questions:
  1. What assumptions are made about the ocean? (**Answers will vary but may include that the ocean is a gloomy place with nothing to offer and there is not much life at the very bottom of the ocean.**)
  2. What would you want to learn about an ocean floor site being explored for the first time? How would you collect data? (**Answers will vary.**)

Activities
A. Conduct a field trip to a local maritime museum, submarine monument, or visit a shipwright. If possible, visit the USS *Nautilus* Memorial and the Submarine Force Library and Museum.

B. Have students create a time line that includes the significant contributions to ocean exploration. Some significant contributions include the following:
  - **1818:** Sir John Ross utilizes the line and sinker method to map the Northwest Passage.
  - **1872–1876:** The HMS *Challenger* voyage launches the science of oceanography.
  - **1920:** Alexander Behm introduces echo sounding technology.
  - **1934:** William Beebe and Otis Barton obtain a glimpse of previously unseen ocean regions

---

OBJECTIVES

Students will be able to
- discuss the history of ocean exploration.
- identify technologies that advanced the study of the oceans.

VOCABULARY

- **submersible** a small underwater vessel

MATERIALS

- TM 4.2.5A Ocean Floor Bathymetry
- TM 4.2.5B Deep-Sea Submersibles
- TM 4.2.5C Ocean Museum

PREPARATION

- Obtain a photo of *Deepsea Challenger*. (*Introduction*)
- Obtain scuba gear. (*Discussion*)
- Arrange a field trip to a local maritime museum, submarine monument, or shipwright. (*A*)

NOTES

- **1948:** Auguste Piccard dives in a bathyscaphe.
- **1951:** Challenger Deep is discovered in the Mariana Trench.
- **1960:** Jacques Piccard and Don Walsh dive to the bottom of Challenger Deep.
- **1961:** The theory of seafloor spreading is proposed.
- **1964:** The *Alvin*, a manned deep-sea submersible, is launched.
- **2012:** James Cameron pilots the submersible *Deepsea Challenger* to a world-record depth for a solo descent in Challenger Deep.

C. Assign students one of the following deep-sea submersibles to research: *Bathysphere, Bathyscaphe, Alvin, Johnson Sea-Link, Deep Discoverer, Deepworker, Nautile, Hercules ROV, Jason ROV, MIR I and II, Pisces IV and V, ROPOS,* and *Deep Flight*. Using the information collected, instruct students to create a promotional brochure inviting others to travel in the submersible.

D. Direct students to identify three to five things that ocean exploration has revealed in the past five years. Topics may include medical advancements, energy resources, or historical revelations.

E. Have students explore the history of undersea habitats like *Sealab* and *Aquarius*. Guide students to determine why these habitats have been discontinued.

F. Encourage students to compile a list of detailed reasons why ocean exploration is important. Assign student pairs to formulate a letter to the president, secretary of the treasury, or country leader explaining why funds for ocean exploration and research should be included in the budget.

Ocean Exploration 4.2.5

Around 100 BC, Posidonius, a Greek philosopher and geographer, set sail to the middle of the Mediterranean Sea. He was not a merchant carrying precious cargo, a fisherman, or a warrior sailing to meet an enemy. Posidonius was seeking to answer the age-old question of how deep the ocean is. In the middle of the Mediterranean Sea, Posidonius's crew fastened a stone to 2 km of rope and threw it overboard. The men were overjoyed when the rope stopped unwinding, indicating that the stone had reached the bottom of the sea. By subtracting the remaining length of the rope from the total rope length, the explorers were able to determine how deep the Mediterranean Sea was in that particular location. For 2,000 years, the line and sinker method was the only way to plumb the depths of the oceans.

In 1872, the modern science of oceanography emerged when the HMS *Challenger*, a British naval ship, set out on a 1,000-day oceanography journey to cross the Atlantic, Antarctic, and South Pacific Oceans. The journey covered more than 68,000 nautical miles. Researchers were able to identify many new marine organisms and gather temperature, ocean current, water chemistry, and ocean floor deposit data at 362 oceanographic stations.

In the 1800s, small underwater vessels known as **submersibles** were created to act as military crafts. The first submersible was a one-person wooden US submarine named the *Turtle*,

OBJECTIVES
- Discuss the history of ocean exploration.
- Identify technologies that advanced the study of the oceans.

VOCABULARY
- **submersible** a small underwater vessel

CAREER

Oceanography
Oceanography is the study of all aspects of the world's oceans and seas. This field can be divided into four major branches. Physical oceanography focuses on properties of ocean water (density, pressure, and temperature), the movement of ocean water (tides, waves, and currents), and the water-atmosphere interaction. Chemical oceanography studies the chemical composition of ocean water and biogeochemical cycles. Marine geology delves into the topography of the ocean floor. Marine ecology focuses on the plants and animals in the ocean. The study of oceanography aids scientists in making more accurate predictions regarding long-term weather changes. Scientists also gain a better understanding of the effects of ocean pollutants and discover more efficient ways of preserving natural resources.

---

which was used in the American Revolutionary War against the British. Submersibles were also used during the American Civil War. Many of these early vessels were not very effective. They often were destroyed as they attempted to demolish enemy craft. Andrew Campbell and James Ash built a submarine, the *Nautilus*, in 1886. The *Nautilus* was driven by electric motors that were powered by a storage battery. On January 21, 1954, another vessel named *Nautilus* was deployed by the US Navy. This submarine was the first to be nuclear powered, and it was the first ocean craft capable of prolonged submersion. In addition, it could reach speeds of more than 20 knots (37 kph) and could maintain the speed for an extended amount of time.

Another technological advancement was the invention of echo sounders, which were instruments used by ships to determine the depth of water. The first echo sounders were developed by oceanographers in the 1920s. Echo sounders measured the time it took for a produced sound to return, or echo, from the ocean floor. Once again, military needs encouraged advancements in this technology as echo sounders were designed to detect submarines during World War II. An added benefit of the advancements was the mapping of undersea areas that had been hidden. This technology gave scientists a picture of the ocean floor. On the civilian front, echo sounders have been used

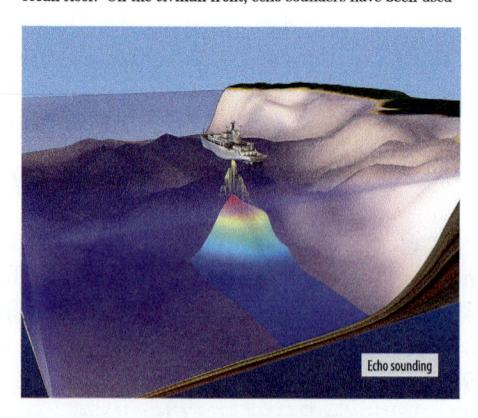

Echo sounding

G. Challenge students to design ways to research the ocean.

H. Assign student pairs or groups to plan an ocean civilization. Have students create models of their planned spaces. Remind students to plan for all aspects of life in such an environment, including food, shelter, energy, travel, and waste removal. Encourage students to check plans for future settlements in space as possible sources of information.

Lesson Review

1. Discuss the importance of three events in the history of ocean exploration. (**Posidonius's use of line and sinker to determine ocean depth; voyage of HMS *Challenger* begins science of oceanography; building the *Turtle* for use in war leads to development of submersibles for exploration; creation of echo sounders for use in war leads to them being used to map the ocean floor; use of multibeam sonar to map ocean floor**)
2. Explain how a line and sinker were used to determine the depth of the ocean. (**sailors attached a rock to a measured line of rope; rock was thrown overboard; when line stopped moving downward, it had reached the bottom of the ocean; sailors measured how much rope was left and subtracted that amount from the full measurement of the rope to determine how deep the ocean was in that spot**)
3. What two inventions significantly advanced the study of oceans? (**Possible answers: submersibles, echo sounders, multibeam sonar, satellites, remotely operated vehicles**)
4. Describe how echo sounders work. (**A sound is produced just below the surface of the water, and the time it takes for the echo to return from the bottom of the ocean is measured.**)
5. What technological advancement improved the process of echo sounding? (**multibeam sonar**)

NOTES

FYI

ROV *Hercules*
Very few remotely operated vehicles (ROVs) in the world are specifically designed for scientific purposes. One such vehicle is *Hercules*, first launched in 2003. This machine can descend to depths of 4,000 m where it recovers shipwreck artifacts and gathers biological and geological information in the deep ocean. *Hercules* was designed with two manipulator arms, a high-definition camera, and acoustic sensors. One important feature of this ROV is the ability to measure depth, which can be used when mapping the ocean floor. *Hercules* can also measure water temperature, salinity, pressure, and oxygen concentration. The machine's cylindrical titanium pressure housings can undergo extreme water pressure. The electrical components that are not in pressure housings are immersed in mineral oil because mineral oil does not significantly compress under high pressure, does not cause corrosion, and does not conduct electricity. *Hercules* is operated by external pilots, who send signals through a long fiber-optic cable.

to locate fish, measure the thickness of Arctic ice, and record information for oceanographic charting.

Multibeam sonar, developed in the 1960s and 1970s, was used to accomplish the mapping of the ocean floor. This technology uses sound waves, which reflect off the ocean floor. A computer then creates images of the ocean landscape. This method of mapping is precise but slow. Scientists estimate that it will take an additional 125 years to map the entire ocean floor using this method. In addition to multibeam sonar, radar from satellites has also been used to detect large features on the ocean floor, including mountains. Currently, the most accurate, detailed method of exploring the ocean floor is the use of underwater cameras attached to a remotely operated vehicle (ROV). This method is useful in ocean depths up to 4,000 m where artificial light can penetrate.

Recently, the National Oceanic and Atmospheric Administration (NOAA) mapped more than 1 million km² of ocean floor in the Atlantic Ocean, Pacific Ocean, and Gulf of Mexico. Although this is a significant number, the ocean covers 335,258,000 km², which means the study of the ocean floor has only begun. Obviously there is still much to learn about this part of God's creation.

Ocean exploration has greatly advanced over the past century. Still, most of the ocean remains a mystery. For comparison, so far 12 people have been to the moon, but only three people, one of whom was James Cameron, have visited the Mariana Trench. In fact, Cameron broke a record in 2012 when he reached a depth of 10,898 m at the Mariana Trench's Challenger Deep area. While the ocean is a very important part of God's creation, support for ocean research has been lacking. Perhaps as more of this hidden world is mapped, support for ocean research and exploration will be expanded.

LESSON REVIEW
1. Discuss the importance of three events in the history of ocean exploration.
2. Explain how a line and sinker were used to determine the depth of the ocean.
3. What two inventions significantly advanced the study of oceans?
4. Describe how echo sounders work.
5. What technological advancement improved the process of echo sounding?

BIOGRAPHY

Jacques-Yves Cousteau
One of the major contributors to the study of the ocean was Jacques Cousteau. His love of the ocean and diving sparked an interest in expanding ocean exploration. However, prior to 1943, divers did not have any way to remain underwater for an extended period of time. Cousteau and Émile Gagnan developed the first fully automatic compressed-air Aqua-Lung to solve this problem. With this invention, divers could remain underwater longer. Cousteau also invented a small, easy-to-operate submarine and underwater cameras. Cousteau and his colleagues filmed their explorations so people could see what was beneath the ocean's surface. These films allowed people to see a part of God's creation that had been hidden. Later in his career, Cousteau founded a number of research and conservation organizations to help preserve the ocean world. During his career, which lasted to the early 1990s, he changed the way people viewed and understood the ocean.

4.2.6 Ocean Resources

Oceans

OBJECTIVES

Students will be able to
- summarize different types of ocean resources.
- explain the process of desalination and other distillation methods.
- analyze the risks and benefits related to harvesting ocean resources.
- investigate alternative ocean energy sources.

VOCABULARY

- **desalination** the process of removing salt from ocean water to obtain freshwater for drinking, irrigation, or industrial uses

MATERIALS

- 10-gallon aquarium, sand, tomato, paper towels, salt (*Introduction*)
- Large world map (*A*)

PREPARATION

- Cover the bottom of the 10-gallon aquarium with a thick layer of sand. (*Introduction*)
- Gather materials for *Lab 4.2.6A Distillation*.
- Invite an ocean conservationist to speak to the class. Assign each student to write three questions for the speaker prior to the presentation. (*B*)

Introduction

Display a 10-gallon aquarium that contains a thick layer of sand covered by water. Inform students that the aquarium represents a portion of the ocean. Ask students what process forms much of the sand. (**erosion**) Explain that sand collects on the bottom of the ocean after land erosion has occurred and the currents have transported the eroded materials to different ocean regions. Scoop up and remove a substantial portion of sand from the middle portion of the aquarium. Have students observe how the ocean floor is disrupted. Ask students what effect they think this disruption would have on surrounding environments. (**Answers will vary but should include that the floating sand blocks sunlight, which could reduce plant life.**) Point out that such a disruption could cause organisms that rely on a specific habitat to thrive to die out, unwanted organisms to be transferred to other regions of the ocean, and pollution to spread to other parts of the ocean.

Share with students that the earth has a natural desalination process that is driven by the water cycle. Ask student volunteers to define the words *evaporation*, *condensation*, and *precipitation*. Have students determine the order these processes should follow for clean water to be collected.

Remind students that even though there is a natural desalination process, only 1% of the water on Earth is available for drinking. Ask students where they might find a large volume of water that could be turned into drinking water. (**the ocean**) Is it safe to drink large amounts of seawater? (**No.**) Explain that drinking large amounts of seawater can lead to severe dehydration, fatal seizures, heart arrhythmias, and kidney failure. To demonstrate this idea, place a tomato on a paper towel and spoon a large amount of salt on top of the tomato. After about 10 minutes, have students describe their observations. (The tomato will be completely dehydrated, similar to human cells that are exposed to saltwater, and the paper towel will be saturated with tomato juice.) Discuss the different methods of mechanical desalination.

Discussion

- Discuss the ethical responsibility of scientists and manufacturers who seek to use ocean resources.

- Ask the following questions:
  1. Who would be interested in desalination processes? (**Possible answers: farmers who need desalinated water to nourish their crops, ranchers who need desalinated water for their livestock, individuals who live in areas where freshwater is not available**)
  2. How is tidal power similar to hydroelectric power? (**Both systems use the movement of water to spin a turbine to produce electricity.**)
  3. What energy production system utilizes the natural temperature differences between deep, cold water and warm surface water? (**ocean thermal energy conversion**) What two processes drive turbines in this system? (**condensation and evaporation**)
  4. What are the most valuable resources extracted from the oceans? (**Answers will vary but should include sand, gravel, salt, water, oil, and natural gas.**)

Activities

Lab 4.2.6A Distillation

- boiling chips
- saltwater, 100 mL per group
- 250 mL Erlenmeyer flasks, 1 per group
- conductivity probes, 1 per group
- distilled water
- rubber tubing, 1 piece per group
- rubber stoppers with hole, 1 per group
- clamps, 1 per group
- ring stands, 1 per group
- metric rulers, 1 per group
- Bunsen burners or hot plates, 1 per group
- test tubes, 1 per group
- 600 mL beakers, 2 per group
- ice

A. Display a large map of the world's oceans. Using different symbols to represent metals, diamonds, sulfur, sand, gravel, and fossil fuel reserves, instruct students to find out and then to mark where these natural resources can be found in the oceans.

B. Introduce the ocean conservationist to the class. Allow time for student questions.

C. Have students research and write a brief report about an organism in the ocean that scientists believe may potentially be able to fight certain diseases. Students should include who discovered the medical benefits of the particular organism, in what part of the ocean the organism can be found, and how the treatment can be derived from the organism.

D. Challenge students to imagine they are stranded on an island with no obvious sources of freshwater. Direct students to design a distillation method to separate salt from ocean water.

E. As a class, research what countries use desalination processes to obtain pure drinking water. Have students determine the reasons why these countries use this method to obtain freshwater.

F. Assign students to create a map of the desalination process. Maps should include an ocean water pumping station; pretreatment station; evaporator, condenser, and membrane station; wastewater removal pump; posttreatment station; and storage container.

G. Instruct students to create a computer presentation or poster explaining alternatives to the use of oil and natural gas.

NOTES

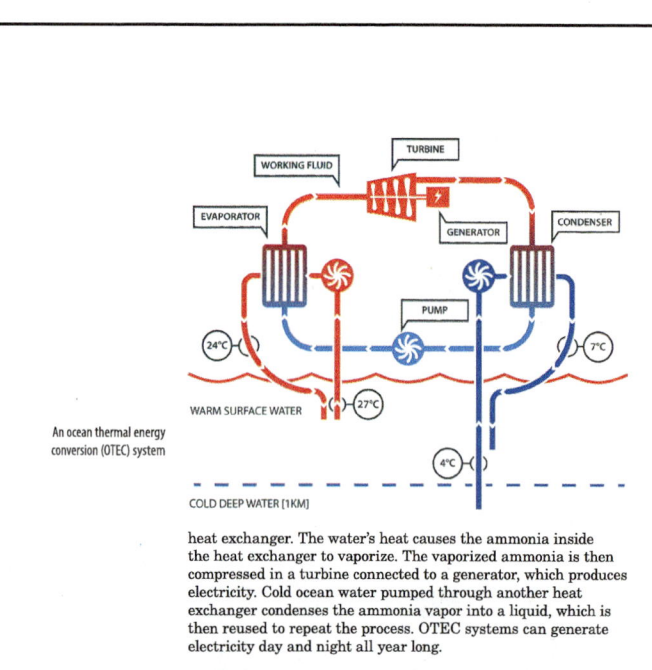

NOTES

H. Divide the class into two groups. Assign one group to represent people in favor of offshore drilling, and the second group to represent those opposed to offshore drilling. Have students research both sides and write a statement advocating their position. Before discussing both statements as a class, ask students whether those in favor and those opposed to the process have a bias in their arguments. (**Yes.**) Discuss the short-term and long-term effects presented in the statements. Ask if Christians have a responsibility to join this conversation. (**Answers will vary.**) How should Christians approach individuals from both sides of the discussion? (**It is important that such discussions begin with grace and humility.**)

I. Encourage students to compare and contrast the uses of ocean resources in two different regions of the world. Have students present their findings to the class.

Lesson Review

1. Summarize different ocean resources. (*minerals*: **gold, diamonds, zinc, copper;** *sand; gravel; salt; water; fossil fuels*: **oil, natural gas;** *alternative energy*: **wave power, tidal power, OTEC**)
2. Why are people not able to extract more ocean minerals? (**Current technology is expensive, and it is often difficult to extract minerals in very deep water.**)
3. What kind of climate promotes desalination processes? (**hot, dry climates**)
4. How is the direct contact membrane distillation process different from the standard distillation process? (**The direct contact membrane distillation process uses a hydrophobic membrane. Less heat is required for the direct contact membrane process than for the standard distillation process.**)

5. Why might some people oppose offshore drilling? (**Drilling through ocean rocks can cause pollution or major oil spills, which can harm or kill living things.**)
6. Name two renewable resources that are alternatives to energy derived from oil or natural gas. (**Possible answers: wind power, tidal power, OTEC**)

NOTES

Name: _____ Date: _____

Lab 4.2.6A Distillation

QUESTION: What method of distillation most efficiently separates salt from water?

HYPOTHESIS: Answers will vary.

EXPERIMENT:

You will need:		
• boiling chips	• distilled water	• metric ruler
• 100 mL saltwater	• rubber tubing	• Bunsen burner or hot plate
• 250 mL Erlenmeyer flask	• rubber stopper with hole	• test tube
• conductivity probe	• clamp	• two 600 mL beakers
	• ring stand	• ice

Steps:
1. Place a few boiling chips and 100 mL of saltwater in the Erlenmeyer flask.
2. Set the conductivity probe to the highest setting using mg/L units.
3. Rinse the probe with distilled water.
4. Place the probe in the saltwater solution and record the conductivity reading. _____ mg/L
5. Rinse the probe again with distilled water.
6. Place the rubber tubing in the hole of the rubber stopper and insert the stopper into the mouth of the Erlenmeyer flask.
7. Position the Erlenmeyer flask in the clamp on the ring stand at least 2.5 cm above the Bunsen burner or place the flask directly on the hot plate. See illustration for proper set up.
8. Extend the rubber tubing to fit inside the test tube. Do not allow the rubber tubing to form any kinks; this could cause a potentially dangerous pressure problem.
9. Fill one 600 mL beaker with ice and water to create an ice-water bath.
10. Place the test tube in the ice-water bath.
11. Heat the saltwater inside the Erlenmeyer flask just to boiling. Do not allow the boiling water to reach the rubber stopper. If it looks like the bubbles may reach the rubber stopper, remove the heat source and adjust the burner settings.
12. Observe the desalinated water collecting in the test tube.
13. When most of the saltwater has boiled off, turn off the heat source and allow the apparatus to cool for 10 minutes. Do not allow all of the saltwater to evaporate.
14. Fill the second 600 mL beaker with room-temperature water and carefully transfer the test tube from the ice water to the room-temperature water.
15. Adjust the conductivity probe to the lowest setting using mg/L units.
16. Repeat Steps 3–5. Record the conductivity reading of the desalinated water inside the test tube. _____ mg/L

Lab 4.2.6A Distillation

ANALYZE AND CONCLUDE:

1. Which sample had a higher conductivity reading? Why? The saltwater sample has a higher conductivity reading because when salt dissolves in water, the sodium and chlorine elements separate into their ionic form. This presence of negative ions and positive ions produces a charge. Pure water does not contain charged particles, so it produces a zero conductivity reading.

2. What is the most important variable that contributes to an effective separation in a distillation? the amount of heat

3. What factors may have contributed to a positive conductivity reading for the desalinated water sample? Possible answers: Some of the saltwater may have contaminated the rubber tubing and leaked into the test tube; the test tube may not have been completely void of contaminants.

4. What other methods would be more efficient to produce a desalinated water sample? direct contact membrane distillation or reverse osmosis

UNIT 5

Meteorology

Chapter 1: *The Atmosphere*
Chapter 2: *Weather*
Chapter 3: *Climate*

Key Ideas

Unifying Concepts and Processes
- Systems, order, and organization
- Evidence, models, and explanation
- Change, constancy, and measurement
- Evolution and equilibrium
- Form and function

Science as Inquiry
- Abilities necessary to do scientific inquiry
- Understandings about scientific inquiry

Earth and Space Science
- Structure of the earth system
- Earth's history
- Earth in the solar system
- Energy in the earth system
- Origin and evolution of the earth system

Science and Technology
- Abilities of technological design
- Understandings about science and technology

Science in Personal and Social Perspectives
- Personal health
- Natural hazards
- Risks and benefits
- Science and technology in society
- Personal and community health
- Natural and human induced hazards
- Science and technology in local, national, and global challenges

History and Nature of Science
- Science as a human endeavor
- History of science
- Nature of scientific knowledge
- Historical perspectives

Vocabulary

aeroplankton	cloud	heat	plasma
air mass	coalescence	humidity	radiation
anemometer	conduction	ice age	shelterbelt
atmosphere	convection	interglacial period	solar wind
atmospheric pressure	deposition	ion	temperature
barometer	El Niño	isobar	thunder
CFCs	fog	lightning	UV
climate	front	microclimate	virga
climate change	greenhouse gas	ozone	

SCRIPTURE

He draws up the drops of water, which distill as rain to the streams; the clouds pour down their moisture and abundant showers fall on mankind.

Job 36:27–28

5.1.0 The Atmosphere

> **LOOKING AHEAD**
>
> - Obtain magnetite mineral for **Lesson 5.1.3**.
> - For **Lesson 5.1.4**, obtain materials for *Lab 5.1.4A Sunscreen*.
>
> **SUPPLEMENTAL MATERIALS**
>
> - Lab 5.1.1A Percentage of Oxygen
> - Lab 5.1.1B My Daily Breath
> - Lab 5.1.4A Sunscreen
>
> - WS 5.1.1A Air Pollutants
> - WS 5.1.2A Layers Chart
> - WS 5.1.2B Layers
> - WS 5.1.3A Comparing Theories
>
> - Chapter 5.1 Test
>
> *BLMs, TMs, and tests are available to download. See Understanding Purposeful Design Earth and Space Science at the front of this book for the web address.*

Chapter 5.1 Summary

The atmosphere is arguably the earth's most precious resource. It blocks out harmful rays from the sun and deflects meteorites. It contains oxygen and other essential gases that all living things need for survival. The water cycle and nitrogen cycle take place in the atmosphere. The atmosphere keeps the earth warm and protects it from the sun's extreme heat. God created an atmosphere that works in harmony with the water, land, and creatures of the biosphere in order to provide all living things with a healthy environment.

Background

Lesson 5.1.1 – Composition of the Atmosphere

The atmosphere is made of several important gases. Nitrogen, which is 78% of the atmosphere by volume, dilutes oxygen and prevents rapid burning at the earth's surface. Oxygen composes 21% of the atmosphere by volume, and argon accounts for 0.9% of the atmosphere by volume. Minerals that contain potassium produce terrestrial argon when the radioactive isotope potassium-40 decays. The argon leaks from the rocks and into the atmosphere. Trace gases such as neon, helium, methane, nitrous oxide, ozone, and carbon dioxide make up the final 0.1%. Carbon dioxide makes up 0.036% of the atmosphere by volume, but the burning of fossil fuels adds more carbon dioxide to the atmosphere. Many scientists attribute global climate change to the carbon dioxide that is added to the atmosphere by these fuels. Water vapor composes from 0%–4% of the atmosphere depending on the location and the time of day. Water vapor is essential for life processes and helps prevent some heat loss from the earth. The amount of water vapor depends on the rates of condensation and evaporation. Warm air holds more water vapor than cool air.

Trace gases, which make up very small amounts of the atmosphere, include neon, krypton, hydrogen, and xenon. Ozone is a product of unaltered oxygen molecules and oxygen molecules that have split into single atoms by the sun's radiation. Although the ozone layer in the stratosphere absorbs the sun's harmful rays, too much ozone in the troposphere can damage lung tissue and plants.

Lesson 5.1.2 – Layers of the Atmosphere

The lower part of the atmosphere is called *the homosphere* because it retains a consistent composition. The homosphere reaches an approximate height of 80 km, and although the composition of this layer is consistent, its temperature varies widely.

Above 80 km, air currents no longer mix the atmosphere's gases. This level is called *the heterosphere*, and gases of the heterosphere are separated into layers according to their density. The lowest layer of the heterosphere is made of a layer of diatomic oxygen (O_2). Above the oxygen layer is a layer of helium, and hydrogen atoms drift into the upper layer of the heterosphere. The heterosphere has no definite boundary, but at a height approaching 10,000 km, the density of the hydrogen atoms is approximately the same as it is in space.

Lesson 5.1.3 – Magnetosphere

The earth's magnetic field permeates not only the planet but a vast volume of surrounding space. It reacts to changes in the solar wind. The interaction between the solar wind and the magnetosphere's plasma acts like an electric generator, creating electric fields within the magnetosphere. These fields circulate the plasma within the magnetosphere and accelerate some electrons and ions to higher energies.

The sun's ultraviolet rays ionize the upper atmosphere, creating the electrically conducting ionosphere and a source of plasma for the magnetosphere. The magnetosphere stores the solar wind's energy and releases it in sudden surges. Mars, which has little or no magnetic field, is believed to have lost much of its former atmosphere to space. Scientists believe this loss was caused, at least in part, by the direct impact of the solar wind on Mars' upper atmosphere. Venus is also thought to have lost nearly all of its water to space, largely because of solar wind-powered ablation.

Interactions between the sun and the magnetosphere can change the electrical and chemical properties of the atmosphere and ozone layer and alter high-altitude temperatures and wind patterns. Strong solar wind can produce powerful magnetic storms in space near the earth. In addition, the interaction of ions from solar wind with the magnetosphere causes auroras around Earth's magnetic poles. The magnetic field captures these ions and conducts them downward to the poles. The ions collide with oxygen and nitrogen atoms; electrons are knocked away, leaving the ions in an excited state. The ions then emit various wavelengths of radiation, creating reds and greenish blues. When solar activity is particularly intense, auroras can extend to the middle latitudes; they have been sighted as far south as 40° north latitude.

Lesson 5.1.4 – Ozone Layer

Ozone is oxygen that has been rearranged from normal diatomic oxygen (O_2) to triatomic oxygen (O_3). The stratosphere holds approximately 90% of all ozone in the atmosphere.

The ozone layer thins over the Antarctic every year from August–November. The lower stratosphere over the South Pole is the coldest spot on Earth, and the cold winds that blow around the region block out warmer air. The small amount of water vapor in the stratosphere freezes, forming thin clouds of ice crystals. The ice crystals convert safe molecules such as $ClONO_2$ and HCl to reactive molecules such as $HOCl$ and Cl_2, both of which release Cl atoms. Warm air eventually melts the ice crystals, and by November the hole is nearly refilled.

Since evidence of seasonal ozone decline was discovered over Antarctica in 1985, the depletion of the ozone layer and the role of human activity in this decline has become a worldwide concern. The hole allows some UV radiation to reach Earth. The increased UV radiation could harm plants and marine organisms and reduce crop yields and forest productivity. Because tiny plant and marine organisms are essential to the marine food web, small increases in ultraviolet exposure could significantly change ecosystems. The ozone hole could also potentially harm humans. UV radiation depresses the immune system, activates the herpes virus (which can lead to cold sores), and increases the risk of skin cancer, skin diseases, eye diseases, and other health risks. A depletion in atmospheric ozone by 1% is estimated to lead to a 3% increase in skin cancer.

It has been theorized that the ozone thinned because of chlorofluorocarbons (CFCs), which are used as refrigerants. When UV light waves strike CFC molecules, a carbon-chlorine bond breaks and produces a chlorine atom. The chlorine atom then reacts with ozone to yield oxygen and chlorine monoxide. CFCs, which were invented in 1928, were considered a great improvement over other refrigerants. The Dupont Company was the largest producer of CFCs in the United States, producing 50% of America's CFCs and 25% of the world's.

In 1987, 46 countries signed the Montreal Protocol, which imposed immediate reductions in CFC production and a complete phaseout of CFCs by 1995 in developing countries. By that time, more than 1 billion metric tons of CFCs had been produced, and more than 18 million tons had been released into the atmosphere. In 1992, NASA was releasing 61.6 metric tons of chlorine into the stratosphere from space shuttle launches alone. As late as 1995, Congress was still holding hearings on whether the ozone hole was real and if CFCs were to blame. The state of Arizona declared the Montreal Protocol invalid within its boundaries. The treaty has reduced CFC production by 90% since its inception. With the precautions that are in place, the ozone layer is expected to recover in 50–100 years.

WORLDVIEW

- After spending time researching data and evidence on a particular topic, people usually reach a conclusion, perhaps a conclusion that is held very tightly. Whether it is about the origins of the magnetosphere or the relationship between CFCs and the ozone, such a conclusion can arouse strong passions. Research, along with personal passion and pride, can create zealots. Zeal and knowledge are good, but often how one's passion is expressed is more important than the actual scientific conclusions a person is passionate about. And knowing the Who of Jesus Christ is more important than all other knowledge. Proverbs 17:27 says, "The one who has knowledge uses words with restraint, and whoever has understanding is even-tempered." People can easily become heated in a discussion, especially if they are passionate and convinced they are right. Once a debate begins, it is easy to lose sight of what is most important. It can become a contest to prove, "I am right and you are wrong!" However, this attitude brings division to the Body of Christ and does not give glory to God. As advised in Proverbs 17:14, "Starting a quarrel is like breaching a dam; so drop the matter before a dispute breaks out." Instead of holding tightly to human scientific conclusions, hold tightly to the Savior of the universe. In that way, believers let their light shine before men, that others may see their good deeds and praise the Father in heaven (Matthew 5:16). Do not let quarrels and dissensions divide the house of God.

5.1.1 Composition of the Atmosphere

The Atmosphere

OBJECTIVES

Students will be able to
- list the components of the atmosphere and their concentrations.
- explain the purpose of oxygen, nitrogen, and carbon dioxide.
- illustrate how the levels of oxygen and nitrogen are maintained.

VOCABULARY

- **aeroplankton** microscopic organisms that float in the atmosphere
- **atmosphere** a mixture of gases that surrounds the earth
- **atmospheric pressure** the pressure exerted by Earth's atmosphere at any given point

MATERIALS

- Hydrogen peroxide, 2 L plastic bottle, baker's yeast, balloon, candle, matches glass jar (*Introduction*)
- WS 5.1.1A Air Pollutants

PREPARATION

- Obtain materials for *Lab 5.1.1A Percentage of Oxygen* and *Lab 5.1.1B My Daily Breath*.
- Gather materials for *Try This: See the Atmosphere*.

TRY THIS

See the Atmosphere
- beakers, 1 per group
- paper towels, 1 per group
- large pans, 1 per group

(Only the outside of the beaker should get wet; the paper towel should remain dry. This activity demonstrates that air takes up space and maintains its space as the beaker is submerged. The air prevents the water from coming into the beaker.)

Introduction

Begin oxygen-creating reaction prior to the lesson. Pour 470 mL of hydrogen peroxide into the plastic 2 L bottle. Quickly add about 6 mg of baker's yeast, then stretch the mouth of the balloon over the opening of the bottle. Gently shake the bottle back and forth to increase the reaction rate. The balloon will catch the oxygen produced from the reaction. Leave the balloon on the bottle until ready to use. Write *Oxygen* and *Atmosphere* on the board. Ask students what they know about oxygen and atmosphere and how they might be related. (**Answers will vary.**) Light a candle and ask students what the candle needs in order to burn. (**Answers will vary but should include oxygen.**) Ask students what will happen if a jar is placed over the candle. (**The candle will go out.**) Why? (**Because the candle used up all the oxygen that is left in the jar.**) Demonstrate by placing the jar over the candle and wait for it to go out. Finally, demonstrate the importance of the proper concentration of oxygen in the atmosphere. Ask for a student volunteer. Relight the candle. Have the student hold the jar over the candle with one edge of the rim on the table and the other edge lifted up. Remove the balloon from the bottle, holding the end closed. Slowly release the oxygen under the lifted edge of the jar. The flame should glow brighter with the increased concentration of oxygen. Oxygen is one of many gases in the atmosphere and is always maintained at a specific level. How is that possible? (**Answers will vary.**) Why is it necessary? (**Life on Earth depends on oxygen.**)

Discussion

- Discuss the purposeful design of the atmosphere. All living things use resources from the atmosphere, but the resources are never used up. Why is oxygen important to living things? (**Animals, bacteria, microorganisms, and plants require oxygen for respiration to produce energy in their bodies.**) Why is nitrogen important to living things? (**Living things need nitrogen for their amino acids, which build the proteins that all animals need.**) Nitrogen is a vital ingredient needed by plants to grow, and plants use this nitrogen to make proteins needed by all animals. Explain that nitrogen also prevents too much oxygen from accumulating, which could spark worldwide fires. Why is carbon dioxide important to living things? (**It is needed by green plants for photosynthesis, and it traps heat, warming the earth.**) Which gas in the atmosphere do you think is most important? (**Answers will vary but should include that all of the gases are important and serve a purpose. We need all of them to live.**) How do you think the existence of cycles gives evidence of a Creator? (**Answers will vary.**) What do you think would happen to life on Earth if the cycles did not exist? (**Answers will vary but should include that life on Earth would end.**)

- Ask the following questions:
  1. What is the atmosphere? (**A mixture of gases that surrounds the earth.**)
  2. When do people come in contact with the atmosphere? (**Constantly. The atmosphere reaches down to the ground.**)
  3. Explain how the atmosphere is replenished with nitrogen. (**The atmosphere is replenished through the nitrogen cycle—nitrogen-fixing bacteria convert atmospheric nitrogen into nitrogen compounds needed by plants. When dead animals decay and living animals excrete waste products, nitrogen is released back into the atmosphere.**)

Activities

Lab 5.1.1A Percentage of Oxygen

- clear, shallow bowls, 1 per group
- wide-mouthed glass jars, 1 per group
- small candles with stands, 1 per group
- matches
- food coloring
- marking pens, 1 per group

Lab 5.1.1B My Daily Breath
- 2 L bottles, 1 per group
- plastic tubing
- disinfectant wipes
- large basins, 1 per group
- graduated cylinders, 1 per group

Use the disinfectant wipes to sanitize the plastic tubing before students blow into it.

A. Complete *Try This: See the Atmosphere* with students. Discuss the questions posed in the Student Edition.

B. Have students complete **WS 5.1.1A Air Pollutants**. Discuss students' findings on pollutants and suggested solutions.

C. Divide the class into groups. Have each group prepare a chart or computer presentation of one of the cycles: oxygen or nitrogen. Assign groups to present their work to the class.

D. Encourage students to research how atmospheric pressure affects such things as the boiling point of water, baking and cooking, and health conditions such as congestive heart failure or asthma. Have students choose a topic. Direct students to create an informational handout or public service announcement to educate the public about how atmospheric pressure causes the effect and what can be done to compensate for the effect of atmospheric pressure. Assign students to present their handouts or public service announcements to the class.

NOTES

5.1.1 Composition of the Atmosphere

OBJECTIVES
- List the components of the atmosphere and their concentrations.
- Explain the purpose of oxygen, nitrogen, and carbon dioxide.
- Illustrate how the levels of oxygen and nitrogen are maintained.

VOCABULARY
- **aeroplankton** microscopic organisms that float in the atmosphere
- **atmosphere** a mixture of gases that surrounds the earth
- **atmospheric pressure** the pressure exerted by Earth's atmosphere at any given point

Imagine flying thousands of kilometers above the earth. What would you see? The oceans would look dark blue, the continents would look brown and green, and clouds would stretch across the surface of the globe in swirling bands and clusters. The **atmosphere** is a mixture of gases that surrounds the earth and is held in place by gravity.

God wrapped this thin layer of gases around the earth like a protective cocoon to help maintain the living things on Earth. Although the atmosphere reaches hundreds of kilometers above the earth, a large percentage of it is squeezed into the first 40 km. In fact, you travel through the earth's atmosphere whenever you take a step.

If you were to compare the atmosphere near the ground with the atmosphere near its upper limit, you would notice differences in pressure. **Atmospheric pressure** is the pressure exerted by the weight of Earth's atmosphere at any given point on Earth. The atmospheric pressure is greater at lower altitudes than at higher altitudes because the gases in the atmosphere are most highly concentrated at sea level, where they are "packed down" from the weight of the gases above them. The gases thin out with an increase in altitude. People may have a hard time breathing in the mountains because there are fewer air molecules. Almost 90% of the atmosphere's mass lies within 40 km of the earth's surface. The constant motion of the air keeps the atmospheric gases near the earth's surface well mixed. If this were not the case, you would run the risk of walking into a patch of air with no oxygen.

The lower portion of the atmosphere contains about 78% nitrogen, 21% oxygen, and 1% of other trace gases such as carbon dioxide. The lower atmosphere also contains water vapor. Humid air holds up to 4% water vapor; dry air holds less than 1%. Many different kinds of particles and pollution also float around in the atmosphere—dust, mineral particles, soil, ash, smoke, salt crystals from sea spray, carbon monoxide, smog, pollen, fungal spores, small organisms, and insect eggs. Some of the solid particles can float in the atmosphere for months or even years. The small organisms that float in the atmosphere are called **aeroplankton**. The solid particles in the atmosphere are called *atmospheric dust*.

Most atmospheric gas is nitrogen. Living things need nitrogen. Animals need nitrogen to form amino acids to build proteins. Nitrogen is a key ingredient in chlorophyll, which plants use in photosynthesis. Nitrogen is recycled from the atmosphere to the soil and back into the atmosphere through the nitrogen cycle. Through this cycle, nitrogen-fixing bacteria convert nitrogen in the atmosphere into nitrogen compounds that plants can use. Without these nitrogen-fixing bacteria, life on Earth could not exist. The cycle is completed when nitrogen is

High concentrations of atmospheric dust can be a problem in large cities like Beijing, China.

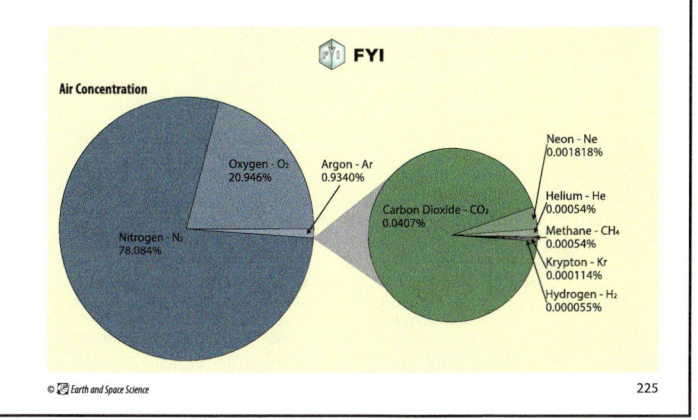

TRY THIS
See the Atmosphere
The atmosphere has mass and takes up space, similar to liquids. This concept can be difficult to believe because the atmosphere cannot be seen and people can walk right through it without any effort. To help you "see" the atmosphere, crumple a paper towel and stuff it into the bottom of a beaker so it cannot fall out. Fill a large pan or tub with water. The sides of the pan need to be taller than the beaker. Invert the beaker and carefully push it straight down to the bottom of the pan. Now, remove the beaker without tilting it. Was any part of the beaker wet? If so, which parts? Why? What happened to the paper towel? Why?

FYI
Air Concentration
- Nitrogen - N₂ 78.084%
- Oxygen - O₂ 20.946%
- Argon - Ar 0.9340%
- Carbon Dioxide - CO₂ 0.0407%
- Neon - Ne 0.001818%
- Helium - He 0.00054%
- Methane - CH₄ 0.00054%
- Krypton - Kr 0.000114%
- Hydrogen - H₂ 0.000055%

NOTES

Lesson Review

1. Why is there more atmospheric pressure at lower altitudes than at higher altitudes? (**The lower atmosphere has greater density because the gases are more highly concentrated at sea level. Also, the upper atmosphere is pressing down on the lower atmosphere.**)
2. What are the main components of the atmosphere? What is the usual concentration of each component? (***nitrogen*: 78%, *oxygen*: 21%, *argon*: 0.934%**)
3. What else is in the atmosphere besides gases and water vapor? (**dust, mineral particles, soil, ash, smoke, salt crystals, carbon monoxide, smog, pollen, fungal spores, insect eggs, aeroplankton. The collective term for the solid particles is atmospheric dust.**)
4. Explain the purpose of oxygen, nitrogen, and carbon dioxide. (***oxygen*: needed for respiration and combustion; *nitrogen*: needed as part of the amino acids to build proteins; *carbon dioxide*: needed for plants to photosynthesize and produce oxygen**)
5. If living organisms use oxygen and nitrogen, how is it possible that the concentration of each of these gases is maintained at a constant level? (**Both elements are constantly recycled. Oxygen is used up through respiration and combustion but plants continue to produce it. Special bacteria use the nitrogen from the atmosphere and change it into a form that plants can use. Nitrogen is released back into the atmosphere when dead animals decay and animals excrete waste. God created the cycles to maintain balance.**)

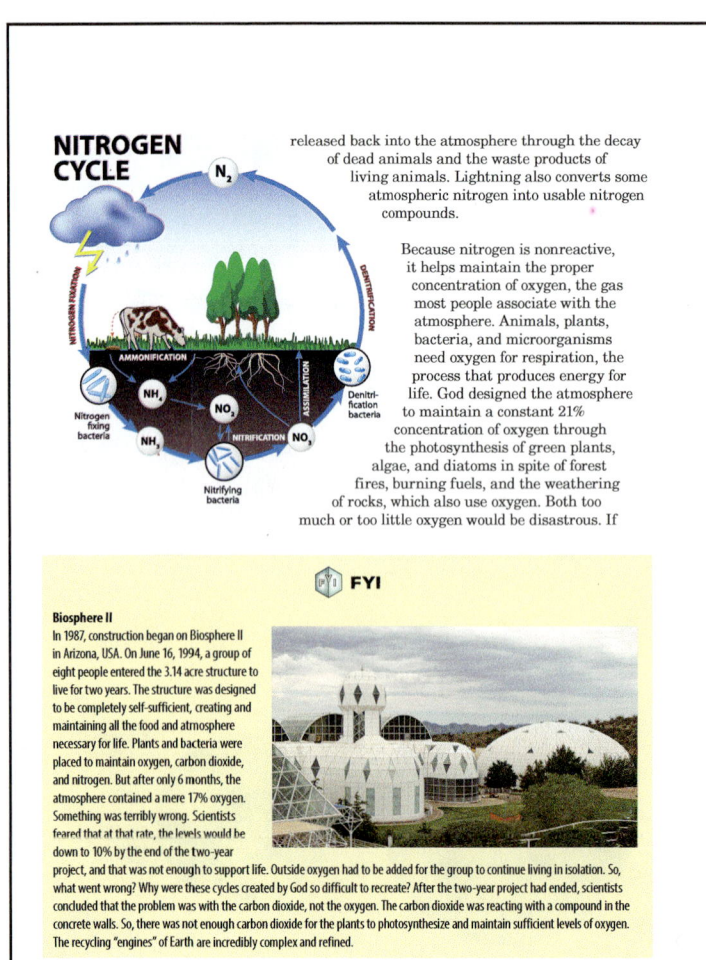

Name: _____ Date: _____

Lab 5.1.1A Percentage of Oxygen

QUESTION: What is the percentage of oxygen in the atmosphere?
HYPOTHESIS: Answers will vary.

EXPERIMENT:

You will need:	• small candle with stand	• marking pen
• clear, shallow bowl	• matches	
• wide-mouthed glass jar	• food coloring	

Steps:
1. Put about 2 cm of water in the bowl. Invert the jar and place it into the bowl to ensure the water level just reaches over the rim of the jar. Add or remove water as needed. Remove jar and make a small mark on the jar to show the water level.
2. Using centimeters, measure the height of the jar to that mark. Answers will vary.
3. Add a couple drops of food coloring to the water. Gently stir.
4. Set stand and candle in the center of the bowl. Light the candle.
5. Invert the jar and place it over the candle; leave it until the flame goes out.
6. Mark the side of the jar to show where the new water level is.
7. Measure the distance between the first and second mark. Answers will vary.
8. Divide the answer in Step 7 by the answer in Step 2. This will give you the percentage of oxygen. Answers will vary but should be close to 21%.

ANALYZE AND CONCLUDE:
1. Why did the flame go out? The flame used up all the oxygen in the jar.
2. What happened to the water level? Why? The water level rose because the amount of air inside the jar was now less than the amount of air outside the jar; there was less air pressure to keep the water level down.
3. The atmosphere is 21% oxygen. Do your results align with this statement? Why? Answers will vary. If student answers have a deviation >5, possible reasons for the error should be given.
4. How do your results compare with other lab groups? Why? Answers will vary. Unwanted variables are always a possibility in the lab. Students should evaluate possible reasons for the different results.

Lab 5.1.1A Percentage of Oxygen

5. If you were to do the lab correctly several more times, would you always get the same results? Why? Answers will vary but may include yes, because the concentration of oxygen is always the same. Oxygen is constantly replenished through photosynthesis and constant motion of the air keeps the atmospheric gases well mixed.

Name: _____ Date: _____

Lab 5.1.1B My Daily Breath

QUESTION: How much of the atmosphere do I breathe each day?
HYPOTHESIS: Answers will vary.

EXPERIMENT:

You will need:	• plastic tubing	• large basin
• 2 L bottle	• disinfectant wipes	• graduated cylinder

Steps:
1. Breathe normally for 30 seconds, counting your breaths. Multiply by 2 and record your answer. Answers will vary.
2. Calculate how many breaths you take each day (Answer from Step 1 × 60 × 24). Record your answer. Answers will vary.
3. Fill the 2 L bottle with water and screw on the cap.
4. Fill the basin half full with water.
5. Invert the bottle and place it in the basin. Take the cap off underwater.
6. Keeping the open end of the bottle underwater, insert one end of the plastic tubing about 3 cm into the bottle.
7. Breathing normally, exhale once through the other end of the tube so the bottle collects the air.
8. Still keeping the opening of the bottle underwater, remove the tubing and replace the cap.
9. Empty the bottle into the graduated cylinder.
   How many mL of water are left? _____
10. Subtract your answer in Step 9 from 2000 mL to determine the volume of air in one breath. _____
11. Calculate the volume of air you breathe in a day. (Step 2 × Step 10) Answers will vary.

ANALYZE AND CONCLUDE:
1. How much of that air is oxygen? (Step 11 × 21%) Answers will vary.
2. How much of that air is nitrogen? (Step 11 × 78%) Answers will vary.
3. Compare your results to other students' results. Why do you think the results are different? Answers will vary.

Lab 5.1.1B My Daily Breath

4. What errors could have been made during the lab to give incorrect results? Answers will vary.

5. How much oxygen do you breathe in a week? Answers will vary.
6. How much nitrogen do you breathe in a week? Answers will vary.

239

5.1.2 Layers of the Atmosphere

The Atmosphere

OBJECTIVES

Students will be able to
- compare the layers of the atmosphere.
- identify the benefits of the atmosphere.
- relate the benefits of the atmosphere to how God sustains life on Earth.

VOCABULARY

- **ion** an atom with an electrical charge that has gained or lost one or more electrons

MATERIALS

- Five-layer cake (*Introduction*)
- WS 5.1.2A Layers Chart
- WS 5.1.2B Layers

PREPARATION

- Bake a five layer cake to represent the atmosphere. For example, use food coloring to tint cake batter for layers from a dark blue to a very light blue or cut the layers to make them different thicknesses. Frost the cake one uniform color. (*Introduction*)

Introduction

Place the cake on a table. Ask students what they see of the atmosphere. Is the atmosphere the same from the bottom to the top? (**Answers will vary.**) Slice open the cake to reveal the layers. Discuss the differences in the layers. Point out that the atmosphere has layers like the cake and that they are not the same. Each layer has its own distinct characteristics. (Note: As an alternative to a layer cake, use a clear container with layers formed by different grains, colored sand, or other materials.)

Discussion

- Discuss the unique features of each layer of the atmosphere. Point out how God's design can be seen in the benefits found in each layer.

- Read **Romans 1:18–20**. Discuss how creation gives evidence of a Creator. Even the ancient Greek philosophers Plato and Aristotle saw evidence of atmospheric layers without knowing God. Plato gave credit to a deity for creating the cosmos. Aristotle credited natural processes for the "design." What other evidence exists that Earth and its inhabitants were created? (**Answers will vary.**) How can believers discuss the evidence in the natural world for God as the Creator with nonbelievers? (**Answers will vary but should include acting in a godly manner and avoiding arguments.**)

Activities

A. Have students complete **WS 5.1.2A Layers Chart** and **WS 5.1.2B Layers**. Extend the activity by having students use the information to make an illustration or a model of the layers. Encourage students to research the layers for additional information for their illustrations or models.

B. Challenge students to identify the atmospheric layer using the following descriptions. Create a large sign for each layer of the atmosphere. Place the signs around the room where students can see them easily. Randomly read aloud one characteristic from the following list. Direct students to stand near the layer sign the characteristic describes. Repeat for each description.

Troposphere
- The layer of air is the closest to the earth.
- The layer has an average height of about 12 km.
- Its name comes from the Greek root meaning "change."
- All of Earth's weather happens in this layer.
- Nearly all the clouds occur in this region.
- This layer acts like an air conditioner for the earth.

Stratosphere
- This region is the second layer of the atmosphere.
- The layer is 12–50 km from the earth's surface.
- Almost the entire ozone layer is located here.
- Commercial jets fly here to take advantage of the calm, thin air.

Mesosphere
- The third layer, which lies 50–80 km above the earth.
- Temperatures decrease in this layer, reaching the lowest temperature in the homosphere.
- Meteors burn up in this layer.

Thermosphere
- This region is the first layer of the heterosphere.
- In this layer, ultraviolet radiation turns molecules into ions.
- Radio waves bounce off layers of air here and return to Earth.
- The lower region contains the ionosphere.

Exosphere
- This region is the outermost layer of the atmosphere.
- In this layer, air molecules thin out and become outer space.
- The end of this layer cannot be measured.

C. Divide class into groups. Assign each group a layer of the atmosphere. Have the groups pretend to be travel agents and design a poster or an advertisement selling their layer to potential travelers.

D. Challenge students to compare and contrast flying a commercial jet and a fighter jet. Have students answer the following questions: In what layers of the atmosphere do the planes fly? What accommodations need to be made to fly in different layers of the atmosphere? What challenges do pilots face in each case?

NOTES

Lesson Review

1. What are the differences between the homosphere and the heterosphere? (*homosphere*: **lower part of the atmosphere where the air is mixed evenly throughout;** *heterosphere*: **atmosphere above 80 km; air molecules are separated; fewer air molecules; much colder than the homosphere**)
2. Describe the layers of the heterosphere. (*troposphere*: **from the earth to 12 km, temperature decreases as the altitude increases;** *stratosphere*: **12–50 km above the earth, contains jet streams and the ozone, the temperature increases as the altitude increases;** *mesosphere*: **50–80 km above the earth, lowest temperature in heterosphere, meteors burn up in this layer**)
3. Why does the temperature decrease with increasing altitude in the homosphere? (**As the warm air rises, it expands and cools.**)
4. Where is the ionosphere located? List two benefits of the ionosphere. (**The ionosphere starts in the mesopause and extends up into the thermosphere. It provides a beautiful light show with the auroras, and it helps transmit short wave radio signals.**)
5. Why does the temperature increase in the mesosphere? (**The ozone layer traps much of the sun's heat which warms up the mesosphere.**)

5.1.2 Layers of the Atmosphere

OBJECTIVES
- Compare the layers of the atmosphere.
- Identify the benefits of the atmosphere.
- Relate the benefits of the atmosphere to how God sustains life on Earth

VOCABULARY
- **ion** an atom with an electrical charge that has gained or lost one or more electrons

HISTORY
Shell Theory
Greek philosopher Plato believed that the earth was round and that the stars had circular orbits. Contrary to the culture of his time, he believed a divine being created the earth for the good of mankind. Plato's student, Aristotle, rejected the idea of a divine creator. He believed the earth could neither be created nor destroyed; it would last forever. Aristotle hypothesized the spherical earth was surrounded by three shells made of water, air, and fiery matter. Even before the help of modern technology, the evidence of God's design was apparent: the atmosphere had layers.

The atmosphere is divided into two main parts. The lower part of the atmosphere is called *the homosphere* because it is the same throughout (in Greek, *homo* means "same"). The homosphere extends from the surface of the earth upward about 80 km. Even though the components of the homosphere are the same throughout, its temperature varies widely. Above 80 km, air currents no longer mix the atmospheric gases. The upper level of the atmosphere is called *the heterosphere* because the gases there are not mixed, but are separated (in Greek, *hetero* means "different"). Earth's gravity pulls heavier gases, such as oxygen and carbon dioxide, closer to its surface, but in the heterosphere, lighter gases move farther away from the earth and can eventually escape into space.

The homosphere and heterosphere can be further divided according to temperature, composition, and density. The three layers within the homosphere are the troposphere, stratosphere, and mesosphere.

The troposphere is the layer of the atmosphere closest to the earth. All life exists in the troposphere. Its average height is about 12 km. The word *troposphere* comes from the Greek root meaning "change." This name is appropriate because almost all weather happens in the troposphere. Hurricanes, tornadoes, rainstorms, blizzards, hailstorms, monsoons, wind, lightning, thunderstorms, and all other weather patterns occur in the troposphere. Most clouds also form in the troposphere. The changing weather mixes up the gases into a homogeneous, or uniform, solution. The troposphere acts like Earth's air conditioner. The sun heats up the earth's surface, which then heats the air it touches. The warmed air expands and rises, but as it moves higher, the air expands further and cools. Consequently, the temperature within the troposphere drops as the altitude increases.

The boundary between the turbulent troposphere and the next layer is called *the tropopause*. The air temperature remains fairly constant in the tropopause at about −60°C. Gaps in the tropopause form a layered structure. Jet streams, which are ribbons of high-speed winds that occur at altitudes of 6–14 km, form in these gaps. The airline industry benefits from these high-speed winds because flying planes along the jet streams saves fuel and speeds up flight times, bringing travelers to their destinations faster.

The stratosphere is the layer of the atmosphere above the troposphere. It is 12–50 km above the earth. The stratosphere has very little vertical air movement. Jets that fly in the stratosphere take advantage of the calm, thin air and leave white trails of fine, white crystals across the sky called *contrails*. The top of the stratosphere is called *the stratopause*, where the temperature reaches about −15°C. The stratosphere contains the very important ozone layer. The temperature of this layer increases with altitude because ozone molecules absorb incoming solar energy. This energy absorption maintains earth's temperatures and prevents much of the sun's harmful ultraviolet radiation from reaching the earth's surface.

Jet flying through the stratosphere

The layer above the stratosphere is the mesosphere, which is 50–80 km above the earth. With temperatures that can drop to −100°C, the mesosphere is the homosphere's coldest layer. The mesosphere has very few oxygen molecules to absorb heat, which is why it is so cold. The mesosphere is the layer where meteors, or shooting stars, burn up and leave beautiful streaks across the night sky. The upper part of the mesosphere is called *the mesopause*, where the temperature begins to increase again.

The outer two layers of the atmosphere are part of the heterosphere. Beyond the mesopause is the thermosphere. It reaches more than 227°C at night and steadily rises to 1,700°C during the day, probably because free atoms of oxygen there absorb heat from the sun. However, because there are so few oxygen molecules in the thermosphere, there are not enough of them to provide warmth. If you were in the thermosphere, you would be very cold.

The lower region of the thermosphere, at 80–300 km above the earth, is called *the ionosphere*. Here the air is extremely thin. Ultraviolet radiation from the sun bombards the molecules here, smashing them apart into **ions**, which are electrically charged atoms that have gained or lost one or more electrons. Ions create an electrical layer that

228

229

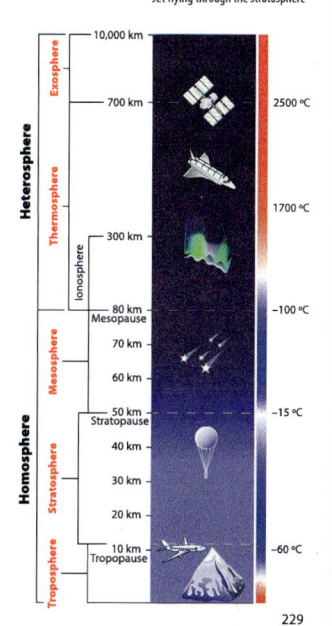

Meteor

reflects radio waves. Short wave radio signals bounce off the ionized air and return to Earth, making it possible to broadcast radio signals beyond the horizon. Also, when the ionized particles collide, beautiful lights are created. In the Northern Hemisphere, they are called *the Aurora Borealis* and in the Southern Hemisphere they are called *the Aurora Australis*.

Finally, the outermost region of the heterosphere is called *the exosphere*. The exosphere is the atmosphere's hottest layer. The few molecules of air that exist in the exosphere have a huge temperature range. During the daytime, temperatures can reach upwards of 2,500°C. However, the nighttime temperature drops to –270°C, which is almost absolute zero, the coldest possible temperature. This final region of the atmosphere continues to thin out and ultimately becomes outer space. Satellites that allow communication through television and other means are located in the exosphere.

LESSON REVIEW
1. What are the differences between the homosphere and the heterosphere?
2. Describe the layers of the heterosphere.
3. Why does the temperature decrease with increasing altitude in the homosphere?
4. Where is the ionosphere located? List two benefits of the ionosphere.
5. Why does the temperature increase in the mesosphere?

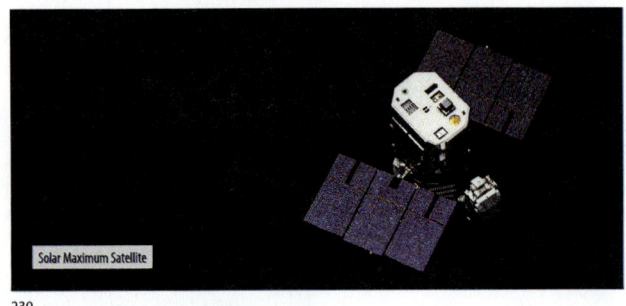

Solar Maximum Satellite

The Atmosphere

5.1.3 Magnetosphere

Introduction
Show the class a magnetite mineral. Discuss how lodestones were once used as navigational devices. A lodestone is a form of magnetite, which is a mineral containing a magnetic oxide of rock. When a lodestone is cut into an elongated shape, it is polarized with north and south ends that act like a magnet. Lodestones were first used by ships sailing out of sight of land. By hanging the stone on a string, the lodestone would align itself with true north, which allowed sailors to keep their bearings on the open sea. Later, lodestones were made into needles and used in the first magnetic compasses.

The magnetic force found in lodestones extends to the outer atmosphere. Use two bar magnets and the repulsion between their ends to introduce the concept of the magnetosphere. Emphasize that the earth acts like a gigantic magnet with two magnetic poles. Lines of force extend between the north magnetic pole and the south magnetic pole. These two poles produce a magnetic field around the earth, affecting the atmosphere and protecting life on Earth.

Discussion
- Discuss the two theories about the electric current that creates the magnetosphere. The dynamo theory supports an old earth; the rapid-decay theory supports a young earth. Explain that there is scientific evidence to support both theories, but neither can be proven. Remind students that both theories are supported by scientists who are Christians. Read the following Scriptures: **Ecclesiastes 10:12; Philippians 2:1–5; Colossians 3:12–16; Hebrews 12:14–15; and 2 Peter 1:5–8**. How should a Christian approach a debate on this topic? (**Answers will vary but should include with humility.**) Encourage students that such discussions require Christians to have the courage to confront their personal assumptions and the wisdom to know which questions matter.)

- Ask the following questions:
  1. Why is the earth a gigantic magnet? (**The motion in the earth's core creates an electric current, which generates a magnetic field around the earth.**)
  2. What would the magnetosphere look like without solar wind? (**It would be symmetrical, with similar field lines circling the earth.**)
  3. What causes solar wind? (**the continuous flow of plasma from the sun**)
  4. How do geomagnetic storms affect human life? (**Geomagnetic storms can disrupt radio signals, create errors in navigation systems, and cause power surges and blackouts.**)
  5. What technology did Van Allen use to discover the radiation in the magnetosphere? (**Balloons were sent up into the atmosphere to measure radiation. Later, a satellite was equipped with a Geiger counter that could measure radiation levels and an altimeter to measure altitude.**)

Activities
A. Complete *Try This: Deflection Protection* with students. Discuss students' answers to the questions.

B. Have students complete **WS 5.1.3A Comparing Theories**.

C. Challenge students to create a computer representation, poster, or model of the magnetosphere.

D. Encourage students to read informational articles about solar wind and the magnetosphere and then make a scrapbook depicting what they learned.

E. Direct students to research a myth about the auroras and present findings to the class. Students may choose one of the following options or find one of their own:
  - The Inuit tribes around the Hudson Bay and the spirits of the dead
  - Medieval Europeans and the reflections of heavenly warriors
  - People in northern Norway avoided "intimidating" the northern lights
  - People groups who thought the northern lights were an omen

---

OBJECTIVES
Students will be able to
- illustrate the structure of the magnetosphere.
- summarize how solar wind affects the magnetosphere.
- compare the two theories about the electric current in the core.
- infer the benefits of the magnetosphere.

VOCABULARY
- **plasma** a superheated gas composed of electrically charged particles
- **solar wind** the continuous flow of plasma from the sun

MATERIALS
- Magnetite mineral, bar magnets, (*Introduction*)
- WS 5.1.3A Comparing Theories

PREPARATION
- Gather materials for *Try This: Deflection Protection*.

TRY THIS
Deflection Protection
- foam balls, 7.5 cm, 1 per group
- fine insulated wire, 3 m per group
- metric rulers, 1 per group
- electrical tape
- 1.5 volt batteries, 1 per group
- iron filings

(The filings will line up with the loops and follow the pattern of the magnetic field. Solar wind particles also move around the magnetic field. The compass point will change direction because the magnetic field will affect it.)

NOTES

Lesson Review

1. How is the magnetic field generated around the earth? (**It is generated by the motion of the earth and the convection currents in the core.**)
2. Explain the structure of the magnetosphere. (**Earth has two magnetic poles. Lines of force extend between the poles and produce the magnetosphere. The magnetosphere is shaped like a comet because the solar wind presses against the magnetosphere to compress it. It spreads out on the other side of the earth.**)
3. Describe the two theories about the electric current in the earth's core. (**The dynamo theory states that the magnetic field is constant but will fluctuate. The magnetic reversals in the rocks take thousands of years to create. The dynamo theory fits the old earth model. The rapid-decay theory states that the magnetic field is steadily decaying or growing weaker. The changes in the magnetic poles are the result of catastrophic events. The rapid-decay theory fits the young earth model.**)
4. How does solar wind affect the magnetosphere? (**Solar wind pushes against the magnetosphere on the sun side of the earth and causes it to spread out on the other side.**)
5. Explain how technology was used to discover radiation in the magnetosphere. (**James Van Allen sent satellites up into the magnetosphere on a small rocket. The satellites were equipped with Geiger counters and altimeters. These measured the radiation levels and the altitudes, or locations of the radiation.**)
6. How does the magnetosphere benefit the earth? (**The magnetosphere protects life from damaging solar wind, and it helps with navigation by keeping a compass pointing true north.**)

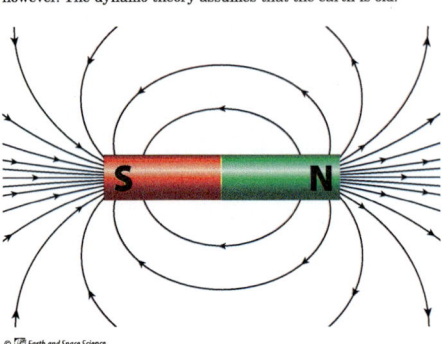

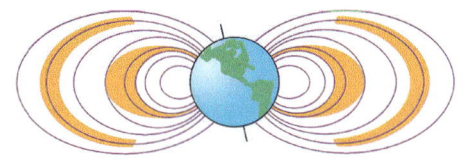

Model of Van Allen radiation belts

The rapid-decay theory assumes the magnetic field is the result of how the earth was created. According to this theory, the process of creation generated the motion necessary to create the electric current that results in the magnetosphere. As the motion of the earth and its core continues, the flow of electricity is resisted and the magnetosphere decays, or becomes weaker. Studies conducted from 1971–2000 indicate that the magnetic field weakened by 1.4%. Evidence of magnetic field reversal exists on every continent, which the rapid-decay theory suggests has happened as the result of catastrophic events such as the Flood. Using the mathematical equations in the rapid-decay theory and projecting backward in time, the earth would be no more than 10,000 years old, which supports young-earth theory.

In science, when making and testing theories with indirect observations and data, scientists take the theory and make

TRY THIS

Deflection Protection

The earth is surrounded by a magnetic field that is invisible but powerful. The solar wind produced by the sun contains particles that are dangerous to human life. As the solar wind approaches the earth, the magnetosphere stops much of it from reaching the surface. Punch a small hole, about 1 cm in diameter, through the center of a 7.5 cm foam ball. Feed 3 m of fine insulated wire through the hole, and form continuous loops of varying sizes around the ball. Make the loops look similar to the magnetic field lines. Allow the ends of the wire to extend about 15 cm from each pole. Remove about 1 cm of insulation from each wire end and use electrical tape to attach the exposed wire ends to a 1.5 volt battery. Place this setup on a large piece of paper and sprinkle iron filings through the loops. What happens to the iron filings? How is this similar to what might happen to solar wind particles as they approach the earth? Then, hold a compass away from the model and note where north is. A compass always points north because it is affected by the magnetosphere. Now place the compass near the model and move it from pole to pole. What happens to the compass? Why?

HISTORY

Van Allen Radiation Belts

In 1958, James Van Allen detected belts of trapped radiation within the magnetosphere. These belts are now known as *the Van Allen radiation belts*. Van Allen had explored the earth's upper atmosphere with balloons that could measure atmospheric radiation levels. Van Allen and his team placed a Geiger counter and an altimeter on Explorer I, America's first spacecraft, to take radiation readings at different heights. During the flight, the Geiger counter began to register thousands of particles passing through the satellite every minute. Scientists learned that space was radioactive. When spacecraft are exposed to fast photons that can penetrate several millimeters of metal for long periods, the electronic circuits and solar cells of spacecraft can deteriorate. Space missions are generally planned to minimize exposure to the most intense regions of the radiation belts.

predictions. The dynamo theory suggests the magnetic field will fluctuate over time, but overall the strength will remain the same. The rapid-decay theory predicts the earth's magnetosphere will decay, or continue to get weaker and weaker, over time. Both theories have been used to predict the magnetic fields for other planets. The dynamo theory predicted Mercury should not have a magnetic field, but it does. It predicted that Mars should have one similar to the earth's, but the magnetic field of Mars is very weak in comparison. The rapid-decay theory predicted the magnetic fields for Neptune and Uranus years before they were actually measured. In those cases, the measurements matched the predictions.

As the earth soars through space, it is bombarded by **solar wind**—the continuous flow of **plasma**, or superheated gases composed of electrically charged particles, from the sun. The earth's magnetic field repels the solar wind, protecting the earth from the wind's harmful effects. The magnetic field of a dipolar magnet is very symmetrical, so it would be reasonable to assume the field around the earth would be symmetrical. But the effect of solar wind creates an area surrounding the earth that is shaped like a comet. The winds on the sun side press against the magnetosphere to compress it. Then it spreads out on the other side of the earth.

Scientists have used the sensitive instruments on satellites to determine that the magnetosphere is filled with various plasmas that distort the magnetic field. Bands of high energy radiation

called *Van Allen radiation belts* stretch away from the earth for several thousand kilometers to tens of thousands of kilometers. James Van Allen discovered these belts of radiation in 1958 through the use of radiation detectors called *Geiger counters* on the satellites *Explorer 1* and *Explorer 3*.

Sometimes huge explosions occur on the sun, causing solar flares and plasma to be hurled in all directions. When this happens, solar wind becomes stronger. Gusts of solar particles enter Earth's atmosphere and disturb the earth's magnetic field. This extra disturbance from solar wind is called *a geomagnetic storm*. Geomagnetic storms can disrupt radio signals, create errors in navigation systems, and even cause power surges and blackouts on Earth.

LESSON REVIEW
1. How is the magnetic field generated around the earth?
2. Explain the structure of the magnetosphere.
3. Describe the two theories about the electric current in the earth's core.
4. How does solar wind affect the magnetosphere?
5. Explain how technology was used to discover radiation in the magnetosphere.
6. How does the magnetosphere benefit the earth?

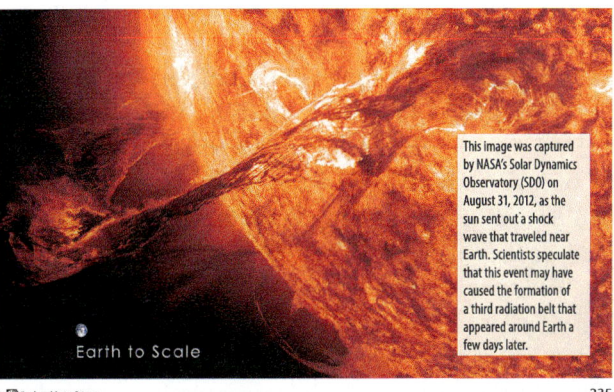

This image was captured by NASA's Solar Dynamics Observatory (SDO) on August 31, 2012, as the sun sent out a shock wave that traveled near Earth. Scientists speculate that this event may have caused the formation of a third radiation belt that appeared around Earth a few days later.

Earth to Scale

5.1.4 Ozone Layer

The Atmosphere

OBJECTIVES

Students will be able to
- explain the purpose of the ozone layer.
- compare the oxygen in the ozone to the oxygen used for respiration.
- theorize the relationship between CFCs and the hole in the ozone.

VOCABULARY

- **CFCs** synthetic compounds consisting of carbon, fluorine, and chlorine
- **ozone** a three-atom form of oxygen gas (O_3) that protects Earth from UV radiation
- **UV** ultraviolet radiation from the sun

MATERIALS

- Globe, waxed paper, flashlight (*Introduction*)

PREPARATION

- Obtain materials for *Lab 5.1.4A Sunscreen*.

Introduction

Direct students' attention to the globe. Have a student hold the piece of waxed paper a few inches above the globe. Shine the flashlight on the waxed paper and ask students to observe how the light shines on the globe. (**The waxed paper diffuses the light.**) Then, tear a hole in the waxed paper and shine the light on the globe as before. Does the hole make any difference in how the light shines on the globe? (**Yes, the light coming through the hole is brighter than the light coming through the waxed paper.**) Assuming the Earth's ozone layer is similar to the waxed paper, how does the ozone layer protect the earth? (**The ozone layer prevents much of the sun's rays from reaching Earth.**)

Discussion

- Discuss the different theories about the hole in the ozone layer: it was caused by human activity, such as CFCs and other emissions from industrialization; it is an annual seasonal thinness of the layer caused by the polar vortex; and it is a naturally occurring thinness caused by the polar vortex, but since the area is thinner than the rest of the layer, CFCs are more damaging there. Ask students how personal bias might affect scientific research. (**Answers will vary but should include that scientific research can be affected by the bias of scientists, organizations, and other entities, which may affect interpretations of data or findings.**)

- Discuss the UV index and why it is important to protect skin and eyes from harmful UV radiation. Explain the "shadow rule": when a person's shadow is shorter than the person is tall, a person is more likely to get sunburned because the intensity of UV radiation is directly related to the angle of the sun.

- Ask the following questions:
  1. How was the ozone layer discovered and studied? (**Gordon Dobson discovered temperature differences above the troposphere as he was studying meteor trails. He designed a spectrometer to study the ozone and later placed more spectrometers around the world.**)
  2. What are CFCs? (**man-made chemicals containing carbon, fluorine, and chlorine**) What prompted the invention of CFCs? (**They were created to replace toxic gases that were being used as refrigerants. Several fatal accidents occurred when the toxic gases leaked from early refrigerators.**)
  3. What did scientists discover about CFCs and the ozone? (**Scientists discovered that CFCs can destroy ozone.**)

Activities

Lab 5.1.4A Sunscreen

- transparency sheets, 5 per group
- marking pencils
- sunscreen, SPF 4, 15, 30, 45
- craft sticks, 4 per group
- black paper or plastic to cover windows
- UV meter cards, 5 per group
- photo-sensitive paper, 5 per group
- tape
- sunlamps, 1 per group
- stopwatches, 1 per group

Transparency sheets and photo-sensitive paper should be the same size. Keep photo-sensitive paper in protective package until room is darkened. Note: SPF stands for *Sun Protection Factor*.

A. Assist students in creating a time line of major scientific findings on the ozone layer hole from 1985 to the present using the Internet. Assign students to use what they learn from the time line to write a persuasive essay on their view of the relationship between the ozone hole, the polar vortex, and human activity.

B. Challenge students to research the risks of sunscreen use. Then, have students evaluate whether sunscreen or wearing protective clothing such as hats and long sleeves are better for skin.

C. Have students research the benefits of vitamin D and answer the following questions: *How much vitamin D does a person need? What does vitamin D do in the body? What is the healthiest way to get vitamin D—sunlight, foods, supplements?* Challenge students to evaluate the benefits and risks of UV exposure with regard to vitamin D needs.

D. Assign student groups to research a type of skin cancer. Direct groups to find information on the following: causes, prevention, treatment, geographic areas with higher frequencies, and any current research or studies. Have groups share their findings in class.

Lesson Review

1. Where is the ozone layer located in the atmosphere? (**It is located in the stratosphere.**)
2. What is the purpose of the ozone layer? (**The purpose of the ozone layer is to absorb UV radiation and protect life from high amounts of radiation that can cause death.**)
3. What is UV radiation? (**It is high frequency wave lengths of light energy from the sun. These are invisible but very dangerous.**)
4. How is the oxygen used for respiration different from ozone? (**The oxygen we breathe is made of two oxygen atoms and the oxygen in the ozone is made of three oxygen atoms. The ozone molecule can absorb UV radiation. The diatomic oxygen molecule cannot.**)
5. Explain the two theories on the hole in the ozone layer. (**The hole is a seasonal thinning of the ozone layer over Antarctica that corresponds with the polar vortex every year. The hole in the ozone was caused by the use of CFCs in refrigeration.**)
6. Why were CFCs banned? (**CFCs were banned because scientists discovered they can destroy ozone molecules. Some scientists believe CFCs caused or increased the hole in the ozone.**)

NOTES

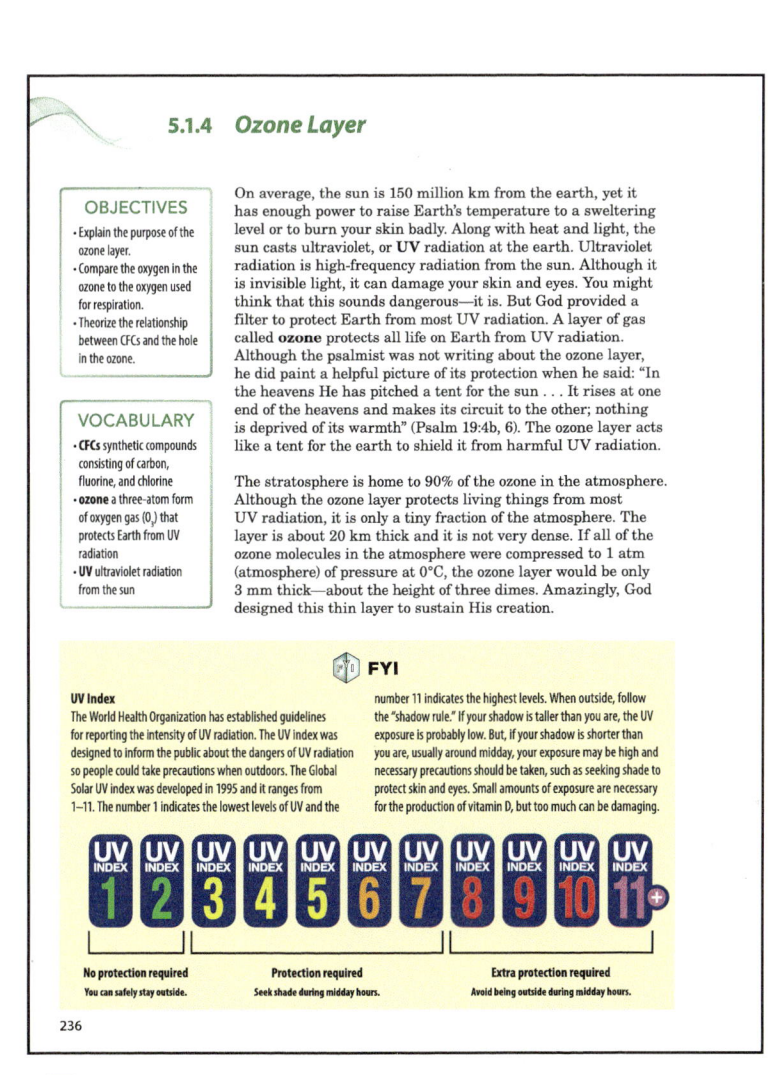

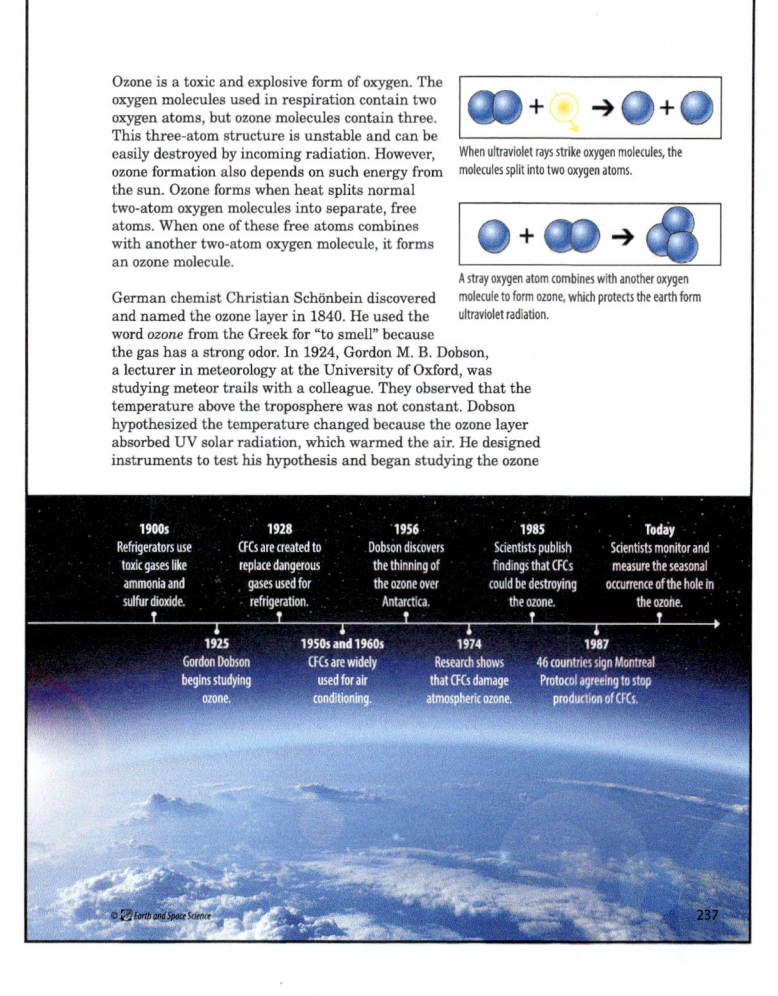

layer in 1925. By 1956, Dobson had refined his instruments and 44 Dobson spectrometers were distributed throughout the world to measure the level of ozone in the atmosphere at different locations. Soon after, he discovered the yearly variation of ozone at Halley Bay in Antarctica. He found that the ozone layer would thin out annually between the months of August and November, the time of a natural phenomenon called *the polar vortex*.

Meanwhile, after several gas leaks in early refrigerators resulted in fatalities, a search for a less toxic refrigerant began. Chlorofluorocarbons, or **CFCs**, were first synthesized in 1928 as a safe alternative to toxic gases like ammonia and sulfur dioxide that were used as refrigerants. Less than 10 years later, millions of refrigerators were sold with the new, safer refrigerant. CFCs were also widely used for air conditioning in homes, businesses, and cars.

During the 1970s, scientists began questioning whether emissions from high-altitude jets or spacecraft would damage the ozone layer. Two chemists, Professor F. Sherwood Rowland and Dr. Mario Molina, showed that CFCs could damage atmospheric ozone. This information, along with decades of data collected over Antarctica, convinced many people that such advances were destroying the ozone layer. Concern about the use of CFCs increased in the 1980s when scientists in the British Antarctic Survey published a paper claiming to have discovered a hole in the ozone. The seasonal thinning of the ozone over Antarctica, first discovered by Gordon Dobson, then became known as *the hole in the ozone*. Although this was not a new discovery, it did appear that the area was thinner than it had been in the past. When the study was published, many people thought that the industrialized world was going to destroy the ozone layer completely. In 1987, as a result of the research, 46 countries signed the Montreal Protocol, an agreement to stop using ozone-damaging chemicals by 1996. The agreement has now been signed by nearly 200 countries.

Has the ban on CFCs decreased the size of the ozone hole? Is it too soon to tell? There are different opinions on whether the ban has affected the size of the seasonal ozone hole. Some scientists believe the ban has helped the ozone hole shrink. Others believe the area's size fluctuations are strictly seasonal and have little to do with CFCs. As in all areas of science, personal bias affects how events are interpreted.

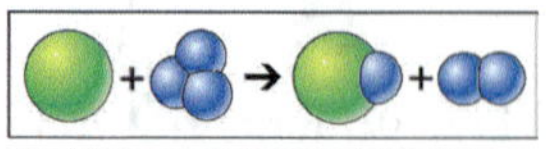

Scientists continue to monitor the ozone layer. In 2016, the thin area over Antarctica reached a record size in spring, but shrank back to a below-average measurement by fall. As scientists continue to study the ozone layer and the seasonal phenomenon of its thinning, more details of God's creation will be discovered and better understood.

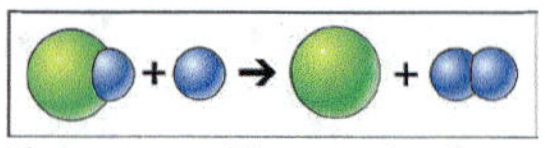

When CFCs are exposed to heat and radiation, chlorine atoms are freed. These atoms then attach to an oxygen atom in ozone, destroying the ozone's filtering ability.

When an oxygen atom of chlorine monoxide combines with a free oxygen atom to form an oxygen molecule, the chlorine atom is released, which destroys another ozone molecule. In this way, the chlorine atoms in CFCs can destroy many ozone molecules.

LESSON REVIEW
1. Where is the ozone layer located in the atmosphere?
2. What is the purpose of the ozone layer?
3. What is UV radiation?
4. How is the oxygen used for respiration different from ozone?
5. Explain the two theories on the hole in the ozone layer.
6. Why were CFCs banned?

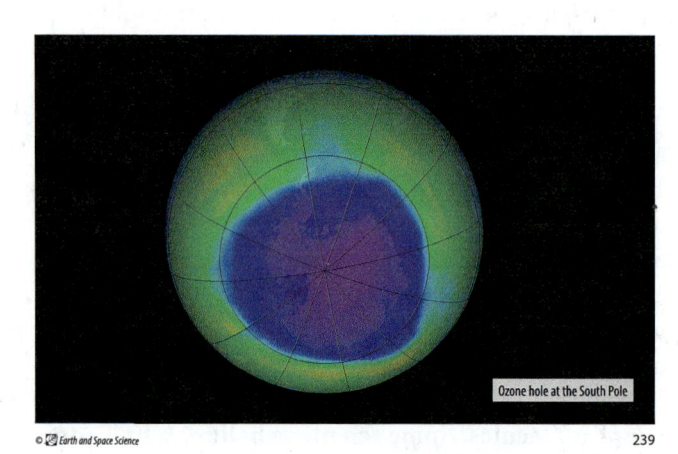

Ozone hole at the South Pole

Lab 5.1.4A Sunscreen

QUESTION: Do higher SPF sunscreens filter out more UV radiation?

HYPOTHESIS: Answers will vary.

EXPERIMENT:

You will need:		
• 5 clear transparency sheets	• 4 craft sticks	• 5 sheets of photo-sensitive paper
• marking pencil	• black paper or plastic to cover windows	• tape
• sunscreens, SPF 4, 15, 30, 45	• 5 UV meter cards	• sunlamp
		• stopwatch

Steps:
1. Label one transparency sheet *Control*. Label each of the remaining sheets with a different SPF rating.
2. Using a separate craft stick for each sunscreen, apply thin layer of sunscreen to the appropriate transparency sheet. Spread the sunscreen evenly over the transparency; it does not need to reach the edge of the sheet. 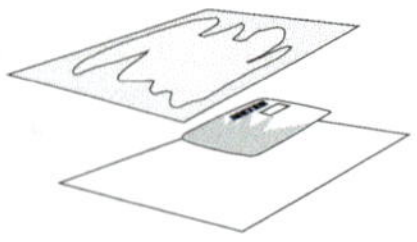
3. Cover the windows with black paper or black plastic and turn out the lights so the room is as dark as possible.
4. Place a UV meter card on a piece of photo paper and cover these with a transparency. If the sunlamp is large enough to expose all 5 sheets at once, set up all 5 sheets. If not, expose 1 sheet at a time and keep the unused photo paper inside the protective package until ready for use.
5. Expose the paper and UV cards to the sunlamp for 5 minutes.
6. Turn off the sunlamp, and using as little light as possible, quickly observe each segment of paper and UV meter card.
7. Record the color of the paper and the UV meter reading in the data table.

SPF	Color of Paper	UV Meter Reading	Rank
Control			
4		Answers will vary.	
15			
30			
45			

Lab 5.1.4A Sunscreen

ANALYZE AND CONCLUDE:
1. How does the SPF rating influence the UV protection? The higher the SPF rating, the greater the protection from the UV radiation.

2. How might the thinning of the ozone layer influence the sun's effect on human life? A thinning ozone layer would increase the harmful effects of the UV radiation.

3. How do your results compare to the SPF ratings? Answers will vary.

4. If your results are not what you expected, what might have gone wrong in the experiment? Answers will vary.

5. Other than using sunscreen, what could you do to protect yourself from UV radiation? Answers will vary but may include use the shadow rule, wear sunglasses, wear a hat with a wide brim, and stay in the shade.

6. Do you think the brand of sunscreen or the thickness with which it is applied can affect how well the sunscreen protects from UV radiation? Why? Answers will vary.

7. Design an experiment that tests different brands of sunscreen or an experiment that tests different thicknesses of the same sunscreen. Answers will vary.

5.2.0 Weather

Chapter 5.2 Summary
Weather may be the most discussed topic in the world. In many places, the weather changes constantly, so consideration of the weather pervades people's lives. Understanding the weather is directly applicable to students' daily lives. In this chapter, students will study atmospheric conditions that cause weather and different weather patterns and phenomena. They will observe various cloud formations and learn about different types of precipitation and storms. Students will also learn how humans try to forecast and control the weather, and they will observe weather changes and use these observations to make their own weather predictions.

Background
Lesson 5.2.1 – Heat Transfer
Heat is transferred throughout the biosphere in essentially three ways: radiation, conduction, and convection. Radiation is the transfer of heat through space or matter by electromagnetic waves. All radiation travels through space at the speed of light—300,000 km/sec. Conduction is the transfer of heat from one substance to another by direct contact between the two substances. For example, as the air above the earth's surface comes into contact with the warm ground, the air is warmed. Temperatures close to the ground are usually higher than temperatures a few meters above the ground. Conduction only plays a minor role in heating the land, ocean, and atmosphere because soil, water, and air are poor conductors of heat. During convection, currents form in a fluid when the molecules in the heated portion speed up and spread out. The warmer part of the liquid becomes less dense and rises, carrying heat with it. When a warmer fluid moves to a cooler region, it cools, becomes more dense, and sinks again. Convection currents are largely responsible for the weather.

The sun's angle above the horizon determines how quickly a section of the earth's surface is heated. The higher the sun, the less slanted the rays that intercept each square meter. The efficiency of these slanted rays to deliver energy to the surface increases with the height of the sun. Earth receives the most heat when the sun is the highest above the horizon. Since the earth is tilted, the heating rate depends on latitude. This heating rate is the highest during the summer. During the winter, the sun never rises very high above the horizon as it moves throughout the day, so its heating ability is lower.

Lesson 5.2.2 – Air Pressure
The weight of the atmosphere exerts pressure on everything on the earth's surface. The ratio of the weight of air to the area of a surface is called *atmospheric pressure*. At an altitude of 5,000 m, the atmospheric pressure is half the value of the atmospheric pressure at sea level and there is only half the available oxygen. Differences in air pressure become apparent for people who have a cold and travel by plane. Changes in air pressure cause pain in the ears and sinuses, a condition called *barotrauma*. Usually the human body can equalize differences in air pressure between the middle ear and the environment, but when a person has a cold, the air pressure on either side of the eardrum becomes uneven, causing pain, temporary hearing loss, and possibly a ruptured eardrum.

Atmospheric pressure is also called *barometric pressure*, and it is measured with a barometer. Mercury is the most common liquid used in a barometer because its density allows the vertical column to be a practical size. If water were used instead of mercury, the column would have to be 10.3 m high. Mercury barometers are often corrected for ambient temperature and the local value of gravity in order to increase their accuracy in a certain location. They are used to calibrate aneroid barometers. Aneroid barometers can be made smaller than mercury barometers, so aneroid barometers are commonly used in aircraft altimeters. Aneroid barometers are also commonly used in barographs, the instruments that mechanically record changes in barometric pressure over time.

LOOKING AHEAD
- Obtain houseplants for **Lesson 5.2.1** to use in *Lab 5.2.1A Heat*.
- For **Lesson 5.2.2**, obtain a 50–90 cm glass tube that is closed at one end for *Lab 5.2.2A Barometric Pressure*.
- For **Lesson 5.2.3**, invite a meteorologist to speak to the class about why meteorologists measure wind speed. If applicable, have the meteorologist explain any distinctive local winds and their effect on the area.
- For **Lesson 5.2.5**, obtain bicycle pumps for *Lab 5.2.5A Cloud Formation*.

SUPPLEMENTAL MATERIALS
- BLM 5.2.2A Barometric Pressure
- BLM 5.2.9A Weather Report

- TM 5.2.5A Cloud Types
- TM 5.2.8A Hurricanes
- TM 5.2.9A Weather Map
- TM 5.2.10A Severe Weather

- Lab 5.2.1A Heat
- Lab 5.2.2A Barometric Pressure
- Lab 5.2.3A Anemometer
- Lab 5.2.5A Cloud Formation
- Lab 5.2.6A Relative Humidity
- Lab 5.2.7A Air Mass Interaction

- WS 5.2.1A Daily Temperature
- WS 5.2.4A Global Winds
- WS 5.2.5A Cloud Observations
- WS 5.2.6A Average Precipitation
- WS 5.2.7A Air Masses
- WS 5.2.8A Stormy Weather
- WS 5.2.8B On-the-Scene Report
- WS 5.2.9A Weather Predictions
- WS 5.2.9B Weather Map
- WS 5.2.9C Weather Report Evaluations

- Chapter 5.2 Test

BLMs, TMs, and tests are available to download. See Understanding Purposeful Design Earth and Space Science at the front of this book for the web address.

WORLDVIEW

- The sound of ocean waves can sooth and relax. Wind can bring relief on a warm summer day. But when the waves are high enough to flood city streets and the force of the wind is strong enough to blow over buildings, the enjoyment turns into disaster. In a similar way, the foundation of life can be shaken to the core. The apostle Paul used the analogy of waves and wind to describe the power of false doctrine. He exhorted the church at Ephesus to "no longer be infants, tossed back and forth by the waves, and blown here and there by every wind of teaching and by the cunning and craftiness of people in their deceitful scheming" (Ephesians 4:14). It is important that believers are prepared to stand against the waves of deception, so fortifying the mind is imperative. Paul encouraged the Roman Christians to resist conforming to the pattern of the world, but to be transformed by the renewing of their minds so they would be able to test and approve what God's will is (Romans 12:2). When the wind of life's challenges blows, and it is unclear whether it is a summer breeze or a tropical cyclone, follow James's advice: "If any of you lacks wisdom, you should ask God, who gives generously to all without finding fault, and it will be given to you. But when you ask, you must believe and not doubt, because the one who doubts is like a wave of the sea, blown and tossed by the wind" (James 1:5–6).

Lesson 5.2.3 – Local Winds

Wind is the flow of air relative to the earth's surface. A wind is named according to the direction from which it blows, so a "west wind" blows from the west. Meteorologists measure wind speed with an anemometer. *Anemos* is Greek for "wind" and *metron* is Greek for "measure." The most common kind of anemometer consists of three or four cups attached to short rods that are secured at right angles to a vertical shaft. Wind velocity can also be measured by the pressure of the air blowing into a pitot tube, which is an L-shaped tube with an end open toward the wind and the other end connected to a pressure-measuring device. Wind can be measured electrically by its cooling effect on a heated wire, which causes the electric resistance of the wire to change.

Near large bodies of water, the land and water display a difference in temperature during the day because of their specific heats. Rising air can cause clouds and precipitation; descending air provides fair weather. The peninsula of Florida in the United States illustrates this effect. Florida receives a large amount of precipitation every year. A sea breeze develops on both sides of Florida, eventually reaches the center of the peninsula, and produces a convergence line down the middle of the peninsula. This line of convergence pushes the air upward and produces clouds, precipitation, and thunderstorms.

In contrast, the mountain-valley breeze is on a diurnal cycle. These winds are thermally induced, although they do not form by the classic thermal mechanism. The cold mountain night breeze forms because dense cold air forms at the ground through radiative cooling and flows down toward the valley. The wind in a valley breeze flows from a cooler region to a warmer region at the surface. A mountain slope facing the sun receives sunlight much more directly than the flatlands in the valley, so the slopes heat up more.

Lesson 5.2.4 – Global Winds

The Coriolis effect occurs because different parts of the earth rotate at different speeds. The equator rotates the fastest, at about 1,675 kph, and the poles experience no rotation. The Coriolis effect deflects wind direction according to the rotation of the earth. Wind traveling long distances deflects to the right in the Northern Hemisphere and to the left in the Southern Hemisphere. The Coriolis effect is always perpendicular to the wind direction.

Lesson 5.2.5 – Clouds

Clouds and fog are visible masses of water or ice particles in the atmosphere. Countless particles of dust, ice, salt, and other solid particulates provide surfaces for water vapor to condense into clouds. On average, these tiny particles are only 0.001 mm in diameter, and their tiny size enables them to remain suspended in the air for long periods of time. Water molecules attach to the particles forming droplets, which combine to make clouds.

Cloudiness is an important element of climate. Cloud cover should be taken into consideration when studying global temperatures because it can affect how much radiation reflects from Earth back into space. Fog is actually a cloud of small water droplets near ground level. To be considered fog, the cloud must generally reduce horizontal visibility to less than 1,000 m. Fog forms when water vapor condenses on particles in the air, which occurs when the relative humidity of the air exceeds saturation by 1%. In highly polluted air, the particles may be large enough to cause fog at humidities of 95% or less.

Lesson 5.2.6 – Precipitation

Humidity is a primary element of weather. Absolute humidity is the mass of water vapor per unit volume of natural air; relative humidity is the ratio of the actual water vapor content of the air to its total capacity at the given temperature. The rate of evaporation decreases as the ratio of water to air increases, the air approaches saturation, and the saturation point increases with temperature.

Even when the relative humidity of cold air is very high, it cannot hold as much water as warm air because it has a low absolute humidity. Cold air with a high relative humidity feels colder than dry air of the same temperature because the water vapor in the air increases the conduction of heat away from the body. Warm air with a high relative humidity feels warmer because of the reduced cooling effect offered by evaporation. A low relative humidity diminishes the effects of extreme temperatures.

Several factors control rainfall distribution: temperature, winds, ocean currents, proximity to water, and mountain ranges. In convection currents, the ascending air is cooled by expansion, resulting in the formation of clouds and rain. In the broad belts of descending air, great deserts have formed because descending air is warmed by compression and absorbs instead of releases moisture.

Sleet consists of small, partially melted bits of ice. Raindrops become sleet when they pass through a layer of the atmosphere with a temperature below the freezing point. Snowflakes that melt as they fall through a warm layer turn into sleet when they pass through a freezing layer. Sleet is often mixed with snow and rain. Sleet falls only at temperatures at or below freezing, usually in the winter. Unlike sleet, hail may fall during any season. Hail pellets, composed of ice or of ice and snow, usually occur as a cold front or thunderstorm passes. Small hailstones have a soft center and a single outer coat of ice; large hailstones usually have alternate hard and soft layers.

Lesson 5.2.7 – Air Masses and Fronts

An air mass is a large body of air within Earth's atmosphere with a similar temperature and humidity throughout the body at any one height. Air masses, which vary in height, form over what are called *source regions*, which are large bodies of water or landmasses with relatively uniform topography. A body of air that remains over a source region for days or weeks reaches an equilibrium with the earth's surface. Air masses generally do not remain stationary for long. Though they never move in exactly the same way, they follow patterns that meteorologists use to predict the weather. The boundary between two air masses is called *a front*. Moving fronts indicate that the weather is about to change. On a weather map, fronts show a large change in temperature and a shift in wind direction. A front can also be described as the boundary between adjacent systems with different conditions. As an air mass moves away from its source region, it brings its particular weather conditions to new areas, where exposure to new elements slowly modifies its characteristics. Fronts are zones of rapid transition from cold to warm or dry to moist air. Turbulence at boundaries often breeds low-pressure storms.

Lesson 5.2.8 – Storms

The term *storm* describes a wide variety of atmospheric disturbances, characterized by low barometric pressure, cloud cover, precipitation, and strong winds. Some storms can be frightening and unpredictable. Even with today's technology, the weather cannot be predicted with total accuracy. Thunderstorms are local in nature and accompanied by brief, heavy rain showers and sometimes hail. Thunderstorms often appear in the late afternoon or early evening because they are caused by the convection currents that rise following the sun's warming of a large body of moist air near the ground. As this air rises, it cools by expansion, and the water vapor condenses to form a cumulus cloud. A cumulus cloud can reach a height of 6.5 km and spread out in the shape of an anvil.

Lightning, one of the most powerful energy sources on Earth, discharges up to 30 million volts, but its short duration prevents lightning from being harnessed as an energy source. Scientists are continually researching methods to protect people and property from lightning strikes. For example, by using laser beams as lightning rods, scientists are researching ways to divert lightning away from crucial sites.

Hurricanes form in conditions of low central pressure relative to the surrounding pressure. The resulting pressure gradient combines with the Coriolis effect, which causes air to circulate around the core of lowest pressure. Hurricanes rotate in a counterclockwise direction in the Northern Hemisphere and in a clockwise direction in the Southern Hemisphere. The friction of the earth's surface slows the air and causes it to spiral gradually inward toward lower pressures. The rising currents near the center cool through expansion when they reach the lower pressures of higher altitudes. This cooling raises the relative humidity. Tropical cyclones form over warm tropical oceans and are not associated with fronts. A tropical cyclone of severe intensity is called *a hurricane* in the Atlantic Ocean, *a typhoon* in the Pacific Ocean, or *a tropical cyclone* when it occurs in the Indian Ocean region.

Lesson 5.2.9 – Forecasting Weather
Since the beginning of time, people have attempted to predict the weather. The ancient Egyptians recorded weather signs and used them to predict coming weather events. Aristotle wrote the treatise *Meteorologica*, detailing his view of many aspects of weather and climate. In the 4th century BC, the Chinese divided their calendar into 24 parts according to observed weather changes. Once the thermometer, barometer, hygrometer, and anemometer were invented, modern meteorology was born. With the invention of the telegraph in 1848, meteorologists were able to rapidly share weather information, which led to longer-term forecasting.

Doppler radar systems, which can probe deep into thunderstorms to determine when tornadoes form, have been installed in states where tornadoes are common. So far these systems have greatly reduced the number of false alarms given to residents in tornado-prone areas. Meteorologists also hope to take advantage of the ever-increasing power of computers to run ensemble forecasting programs. Ensemble forecasting helps determine whether a weather system is in a predictable pattern by running several computer simulations with slightly different data.

Lesson 5.2.10 – Controlling Weather
Weather modification, such as cloud seeding, originally involved cumulus clouds, but the short life span and instability of such clouds complicated seeding operations. Orographic clouds, which form over mountainous areas, are the preferred cloud for seeding experiments because they last longer than cumulus clouds.

Weather modification and precipitation enhancement are controversial topics. Some people believe that weather modification defies ecological ethics because it interferes with a natural process and produces unpredictable and uncontrollable results. Others believe that weather modification holds great potential to enhance water resources. The issue of human modification of weather becomes more complex when intentionality is considered. Planned weather modification, such as cloud seeding, would intentionally influence atmospheric forces for beneficial purposes. Inadvertent weather modification is not intentional; for example, climate change may be brought on by industrialization, urbanization, irrigation, and changes in land-use patterns. Much of the evidence in support of weather modification relies on statistical indicators instead of physical support.

The science of weather modification is still mostly in a research and experimentation stage. More work is needed to establish a sound scientific basis. In 1985, the American Meteorological Society adopted a policy statement to address planned and inadvertent weather modification. The policy acknowledges that under favorable conditions and with existing weather modification technology, the precipitation yield of cold orographic cloud systems could be increased and noted that under certain conditions, decreases in orographic precipitation is also possible.

Weather

5.2.1 Heat Transfer

Introduction
Place the prepared refrigerated glass of water and the ice cubes on a table. Mention that the water is cold. Ask students if there is any heat in the system on the table. (**Answers will vary.**) Place the ice cubes in the water. Explain that now there is heat. Convey that heat is the transfer of energy from one substance to another. The molecules in a substance have energy. When two substances with different temperatures or different amounts of energy come into contact with each other, energy is transferred. Ask whether the energy transferred from the water to the ice cubes or the ice cubes to the water. (**The energy transferred from the water to the ice cubes.**) Reiterate that when energy is transferred, there is heat.

Discussion
- Demonstrate and discuss the three types of heat transfer.
  Radiation: Place a thermometer on the table. State the temperature to students. Place a heat lamp above the thermometer and have students observe the rising temperature.
  Conduction: Heat a pan of water over a hot plate.
  Convection: Fill a beaker with water. Add purple potassium permanganate crystals to the water and place the beaker on a tripod over a Bunsen burner. Gently warm the water. Have students observe the motion of the dissolving purple crystals in the water. Purple streaks should begin moving in a circular motion as the water warms. Explain that as the water in the bottom of the beaker warms, it becomes less dense and rises. When it reaches the top of the beaker, the water cools, becomes denser, and starts to sink. Ask students how long the water will continue to circulate. (**until there is no longer a temperature difference in the beaker's water**) Remind students that warm and cold air react in the same way.

- Ask the following questions:
  1. Can two cold objects have heat? Why? (**Yes, because heat is the transfer of energy. If one cold object is warmer than the other cold object, energy can be transferred from one object to the other.**)
  2. What is radiation? Give an example. (**Radiation is the transfer of energy through space by electromagnetic waves. The sun radiates the earth.**)
  3. What is conduction? Give an example. (**Conduction is the transfer of heat by direct contact. One example is a pan on a hot burner. The hot pan conducts heat to the food inside the pan.**)
  4. How is convection related to weather? (**The gases in the atmosphere are constantly moving. As they are heated up by radiation and conduction, they expand and rise. As they rise, they start to cool and then condense and sink. Convection currents cause different weather patterns.**)
  5. What causes the liquid to rise or fall in a thermometer? (**The liquid in the thermometer rises when the substance it is measuring transfers energy to it. When the substance being measured has less energy, the thermometer transfers energy to the substance, so the liquid falls.**)

Activities

Lab 5.2.1A Heat
- wide-mouthed jars, 1 per group
- thermometers, 4 per group
- pie pans, 1 per group
- sand
- tape
- small houseplants, 1 per group
- black construction paper
- stopwatches, 1 per group
- graph paper
- colored pencils

A. Complete *Try This* activities with students.

OBJECTIVES
Students will be able to
- compare the methods of heat transfer.
- correlate the earth's seasons to its orbit.
- infer how a thermometer measures temperature.

VOCABULARY
- **conduction** the transfer of heat from one substance to another substance through direct contact
- **convection** the transfer of heat that occurs in moving fluids, liquids, or gases that is caused by the circulation of currents from one region to another
- **heat** the transfer of energy from one substance to another
- **radiation** the transfer of energy through space by electromagnetic waves
- **temperature** the measure of energy in the molecules of a substance

MATERIALS
- Refrigerated glass of water, ice cubes (*Introduction*)
- Thermometer, heat lamp, pan, hot plate, beaker, purple potassium permanganate crystals, tripod, Bunsen burner (*Discussion*)
- WS 5.2.1A Daily Temperature

PREPARATION
- Refrigerate glass of water; make ice cubes. (*Introduction*)
- Obtain materials for *Lab 5.2.1A Heat*.
- Obtain materials for *Try This* activities.

TRY THIS
Convection
- *beakers, 2 per group*
- *triple beam balances, 1 per group*
(**No. Convection happens because warm water is less dense than cold water. The warm water rises and the colder water sinks. The warm water has less mass than the cooler water.**)

🛈 TRY THIS

Make a Thermometer
- *eyedroppers, 1 per group*
- *food coloring*
- *clear, narrow-necked plastic bottles, 1 per group*
- *isopropyl alcohol*
- *clear straws, 1 per group*
- *clay*

(**The water level rises when it is in the sun because the heat adds energy to the water molecules and makes them less dense. The water level goes down when the water cools off because the cooler liquid has less energy and the water is more dense.**)

B. Have students complete **WS 5.2.1A Daily Temperature**. Direct students to record data at the same time of day for five days.

C. Direct students to research temperature facts: highest temperature ever recorded, lowest temperature ever recorded, the greatest temperature change in one day, the most rapid temperature change, and any other facts students may find interesting. Assign them to write a brief article about the facts they researched for a science magazine or a science blog. In their articles, have them theorize why such a weather event occurred.

D. Challenge students to research the history of thermometers and the different types of thermometers. Discuss why is it important to know the temperature of substances.

E. Assign student pairs to research cities at different latitudes. Have pairs find information about their cities' seasonal conditions at eight set points in Earth's orbit (divide the orbit into eight equal segments). Challenge students to create illustrations reflecting where their cities are located at these eight points in Earth's orbit. Allow time for pairs to present their information and illustration to the class. Discuss how the sun's path and the angle of sunlight hitting Earth's surface vary depending on the latitude of each city and Earth's position in its orbit.

Lesson Review

1. What is heat? (**Heat is the transfer of energy from one substance to another.**)
2. Compare the three methods by which heat can be transferred. (***radiation***: **energy is transferred through space by electromagnetic waves;** ***conduction***: **energy is transferred through direct**

5.2.1 *Heat Transfer*

OBJECTIVES
- Compare the methods of heat transfer.
- Correlate the earth's seasons to its orbit.
- Infer how a thermometer measures temperature.

VOCABULARY
- **conduction** the transfer of heat from one substance to another substance through direct contact
- **convection** the transfer of heat that occurs in moving fluids, liquids, or gases that is caused by the circulation of currents from one region to another
- **heat** the transfer of energy from one substance to another
- **radiation** the transfer of energy through space by electromagnetic waves
- **temperature** the measure of energy in the molecules of a substance

Weather—the daily variations of temperature, humidity, cloud cover, wind, and precipitation in the atmosphere—influences your life constantly. Your ball game is rained out. A blizzard strikes, so school is cancelled. Your hat gets blown away at the bus stop. The weather helps determine whether you will wear shorts or jeans, whether you will go camping in April or July, or maybe even whether you feel cheerful or grouchy. Some hobbies or interests even depend on the weather. For example, whether you can snow ski or water ski depends on the weather. In some parts of the world, the weather is somewhat consistent. In other areas, the weather seems to change daily.

Although weather might seem unpredictable, the processes that create it are orderly. One piece of the weather puzzle involves heat. The heat on Earth comes from the sun, of course. But have you ever thought about how the sun actually heats the earth? **Heat** is the transfer of energy from one substance to another. God created heat to travel from place to place in an orderly fashion, so it always moves from warmer objects to cooler objects. Heat can travel from one object or location to another in three different ways: radiation, conduction, and convection. Heat from the sun is transferred through space to the earth by **radiation**, which is the transfer of heat through space by electromagnetic waves. When heat radiates from the sun through space, it reaches the air molecules in the earth's atmosphere. This radiated heat then warms the air molecules, which then warm the molecules of other substances they come in contact with.

Soccer players

The sun provides radiant heat.

They conduct the heat to such substances as the ground. In this way, the ground is warmed by conduction.

Conduction is the transfer of heat from one substance to another substance through direct contact. When you put an ice pack on your forehead, the heat from your forehead is conducted to the ice pack. In reality, the ice pack cools the forehead because the energy is transferred from your forehead to the ice pack. This heat transfer continues until the ice pack and your forehead are the same temperature. Conduction is why the ice melts and your head feels cooler. A pan on a hot burner is another example of conduction. The pan is very hot because metal conducts heat well. If you placed your hand near the pan, you would feel the radiated warmth as the heat energy from the pan was conducted to your fingers. The heat of the stove burner moves through conduction. When heat radiates from the sun through the atmosphere to the earth's surface, the warmed air comes in contact with the warm ground, and the temperature of the air rises even more. In this way, the earth is heated by conduction.

Have you ever heard the expression "warm air rises"? This phrase describes **convection**, the

Heat will be conducted if the girl touches the pan.

FYI

Gliding On the Currents
Hawks and other birds of prey often sail through the sky without flapping their wings. They glide on the warm convection currents that are rising from the earth. By gliding on these currents as they search for prey, these birds can conserve the energy they would otherwise have to use for flying.

contact; *convection*: energy is transferred through moving fluids and caused by the circulation of currents. As warm air rises, it expands and cools. Then the cool air sinks and condenses.)

3. What creates the seasons? Explain. (**The seasons are the result of the earth's tilt away from or toward the sun as it rotates around the sun. The tilt gives an uneven distribution of radiation on the earth.**)

4. If the Northern Hemisphere is experiencing summer, what season is it in the Southern Hemisphere? Why? (**The Northern Hemisphere experiences summer when it is tilted toward the sun. It is receiving more radiation at this point in orbit. At the same time, the Southern Hemisphere is tilted away from the sun and is receiving less radiation, so it is winter.**)

5. How does a liquid thermometer measure temperature? (**A thermometer measures the change in energy. It has its own energy when placed into a substance. If the thermometer is warmer than the substance, it will lose heat to the substance and the liquid in the thermometer will go down. If the thermometer is cooler than the substance, energy from the substance will be transferred to the thermometer and the liquid in the thermometer will rise.**)

NOTES

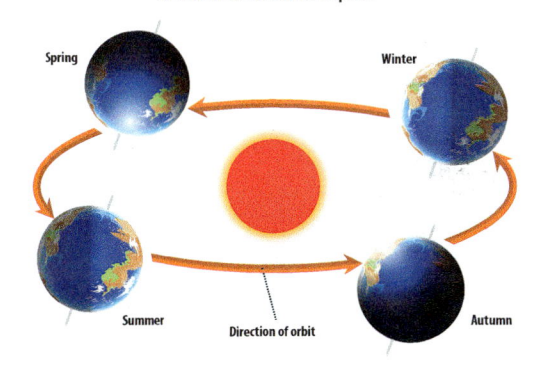

Name: _____ Date: _____

Lab 5.2.1A Heat

QUESTION: Which object will absorb the most heat?

HYPOTHESIS: Answers will vary.

EXPERIMENT:

You will need:	• sand	• stopwatch
• wide-mouthed jar	• tape	• graph paper
• 4 small thermometers	• small houseplant	• colored pencils
• pie pan	• black construction paper	

Steps:
1. Fill the jar with water and put a thermometer in it.
2. Fill the pie pan with sand and put a thermometer in the middle of the pan. Do not cover the thermometer with sand.
3. Tape a thermometer to the underside of the houseplant's leaves or its stem.
4. Place the black construction paper on the windowsill in direct sunlight and place a thermometer in the center of the paper.
5. Place the jar of water, pan of sand, and houseplant on the windowsill. Be certain that all four objects receive equal amounts of sunlight.
6. Record the temperature on each thermometer every 2 minutes for 10 minutes.

Temperature

Item	2 Minutes	4 Minutes	6 Minutes	8 Minutes	10 Minutes
Water					
Sand		Answers will vary.			
Houseplant					
Paper					

Lab 5.2.1A Heat

7. Make a line graph showing the temperature increase of the four objects. Plot temperature on the *y*-axis and time on the *x*-axis. Include a title for your graph. Use different colored pencils to represent the temperature change of each substance.

Answers will vary.

ANALYZE AND CONCLUDE:

1. Which substance absorbed the most heat? Which substance absorbed the least? **Answers will vary but should include that the sand or the black paper absorbed the most heat and the houseplant or the water absorbed the least heat.**

2. Which substances heated up the slowest? What areas of the earth might these represent? **The water and the plant heated up the slowest. The water can represent the oceans or other large bodies of water. The plant can represent forests or other areas with vegetation.**

3. Which substances heated up the fastest? What areas of the earth might these represent? **The sand and the black paper heated up the fastest. These would represent desert areas or areas with little vegetation.**

4. Which type of heat transfer caused the items in the lab to heat up? Explain. **Both radiation and convection were involved. The sun radiated the air. The warm air then convected heat to the objects.**

Weather

5.2.2 Air Pressure

Introduction
Fill two aluminium soda cans about one-eighth full of water. Place both cans in a pan and heat the pan until the water is boiling. Using tongs, take one can out of the pan and place it right side up in an ice water bath. Ask what happened to the can. (**Nothing.**) Remove the second can from the pan and quickly place it upside down in an ice water bath. Ask what happened to the can. (**The can imploded.**) Why? (**Answers will vary.**) Explain that the change in air pressure caused the can to implode. When the can was in the boiling water, the steam pushed the air out of the can. As soon as the can was placed in the ice bath upside down, the steam condensed back to liquid. However, no air could get in to the can to replace the space occupied by the steam. When the steam condensed, the air pressure outside the can was greater than the air pressure inside the can because there was not any air left in the can. Convey that air pressure is all around, but people do not feel it because the pressure inside their bodies that is pushing out is equal to the air pressure around them that is pushing in.

Discussion
- Discuss how air pressure affects baking and cooking. Remind students that the boiling point of water goes down at high altitudes because there is less air pressure pushing down on the top of the water, so the steam escapes sooner. Ask students if this causes food to cook slower or faster. (**Answers will vary but should include that food cooks slower.**) Explain that it can take longer to cook foods like pasta because water boils at a lower temperature.

- Ask students the following question: *If air is dense at sea level, why is it warmer there than on a mountain?* (**Answers will vary.**) On an interactive whiteboard, trace a 2" and a 4" circle side by side. Then, use a marker to draw 15 thick horizontal lines through both circles to represent heat. The smaller circle should be almost filled, but the larger one will not be. Explain that the small circle represents the air at sea level; the larger circle is the air on a mountain. The air pressure at sea level is greater, so heated air is forced into a smaller space. On a mountain, the air pressure is lower, so heated air can spread out. Both circles have the same amount of heat, but the mountain's lower air pressure lets the heat spread out. The higher air pressure keeps the heat in place at sea level.

- Ask the following questions:
  1. What factors create change in atmospheric pressure? (**rotation of the earth, solar energy, gravity, and altitude**)
  2. Why do your ears pop when you are going up a mountain? (**The air pressure on the outside of my body is now less than the pressure on the inside of my body. The popping is caused by the pressure difference.**)
  3. What type of weather is related to the two types of pressure systems? (**low-pressure: cloudy skies, storms, rain showers, and blizzards; high-pressure: clear skies and sunny conditions**)
  4. Explain how changing atmospheric pressure affects the height of the mercury in a mercury barometer. (**When the air pressure outside the tube increases, the mercury is forced up into the tube. When the air pressure outside the tube decreases, the mercury in the tube moves down.**)

Activities

Lab 5.2.2A Barometric Pressure
- glass tubes, 50–90 cm in length, closed at one end, 1 per group
- colored water
- beakers, 1 per group
- ring stands with clamp, 1 per group
- cardboard strips, 4 cm × 10 cm, 1 per group
- marking pencils, 1 per group
- tape
- metric rulers, 1 per group

A eudiometer tube can be substituted for the glass tube. **BLM 5.2.2A Barometric Pressure** can be used in place of students making a scale. Check an online source to find the weather history.

OBJECTIVES
Students will be able to
- describe the factors that affect air pressure.
- compare low-pressure and high-pressure areas.
- demonstrate how atmospheric pressure relates to weather.
- explain how a barometer measures atmospheric pressure.

VOCABULARY
- **barometer** an instrument used to measure atmospheric pressure

MATERIALS
- 2 aluminum cans, pan, hot plate, tongs, 2 bowls of ice water (*Introduction*)
- 2 wooden paint stirrers, heavy book, newspaper (B, Option 1)
- Ice pick, aluminum can, bowl (B, Option 2)
- BLM 5.2.2A Barometric Pressure

PREPARATION
- Obtain glass tubes for *Lab 5.2.2A Barometric Pressure*.
- Gather materials for *Try This: Caving In*.

TRY THIS

Caving In
- boiling water
- funnels, 1 per group
- 2 L bottles with caps, 1 per group

NOTES

A. Complete *Try This: Caving In* with students. Discuss the answers.

B. Choose one of the following options as an activity for student groups:
   Option 1: Heavy Paper
   - 2 wooden paint stirrers
   - heavy book
   - newspaper

   Direct students to lay a paint stirrer over the edge of the table so that half of it is on the table and half of it is over the edge, then have students set a heavy book on the paint stirrer. Ask students what will happen if a group member quickly strikes a blow to the half of the paint stirrer that is protruding over the table. (**The stick will break because the book is too heavy to allow the stick to move.**) Have groups demonstrate this action. Then, have groups repeat the activity, replacing the heavy book with a full sheet of newspaper. Ask students what will happen if a group member quickly strikes a blow to the half of the paint stirrer that is protruding over the table. (**Answers will vary but should include that the paper will fly off the table and the stick will fall on the floor.**) Direct groups to demonstrate this action. (Note: If students strike hard enough and quickly enough, the stick will break.) Explain that the newspaper distributes the air pressure so that the paper seems heavy like the book.

   Option 2: Water Flow
   - ice pick or other sharp object
   - aluminum can
   - bowl

   Direct student groups to punch a hole near the bottom of an aluminum can filled with water. Have groups allow the water to flow for a moment into the bowl. Then, direct a group member

---

5.2.2 Air Pressure

OBJECTIVES
- Describe the factors that affect air pressure.
- Compare low-pressure and high-pressure areas.
- Demonstrate how atmospheric pressure relates to weather.
- Explain how a barometer measures atmospheric pressure.

VOCABULARY
- **barometer** an instrument used to measure atmospheric pressure

The weight of the atmosphere presses down on everything on Earth's surface. The pressure exerted by Earth's atmosphere at any given point is called *atmospheric pressure*. Atmospheric pressure varies from place to place on Earth. These differences are caused by Earth's rotation on its axis, solar energy, gravity, and altitude.

The earth's rotation and convection currents move air masses of varying pressures into different regions, which changes the air pressure in other areas. Air masses with different densities create low- and high-pressure areas, which cause Earth's weather patterns. An area where warm air is rising is referred to as *a low-pressure area*. As air rises in low-pressure areas, strong winds develop, and water vapor from warm air condenses to form clouds. Low-pressure systems bring unsettled weather such as cloudy skies, storms, rain showers, or blizzards. An area where cold air is sinking is called *a high-pressure area*. In high-pressure areas, the sinking air prevents the formation of many clouds. Usually, high-pressure systems have clear, sunny skies.

Solar energy heats the atmosphere near the equator more than it heats the air near the poles. This heat energy causes air molecules to move around and spread out. The volume of the air expands, which makes the air less dense. The resulting warm air has a lower atmospheric pressure. Because it is less dense, warm air rises. Cold, dense air has a higher atmospheric pressure than

TRY THIS
Caving In
Hot air is less dense than cold air and therefore has lower air pressure. Boil two liters of water. Using a funnel, carefully fill a 2 L bottle with the hot water. To avoid the risk of getting burned, do not hold the bottle when pouring the water. Leave the water in the bottle for two minutes. Pour it out and screw the lid on the bottle immediately. What happened to the bottle? Why?

Air travels as wind from areas of higher pressure moves to areas of lower pressure.

Cooler air (higher pressure) — Warmer air (lower pressure)

Mercury barometer

warm air. Because cold air is more dense than warm air, cold air sinks beneath warm air. This movement generates convection currents of air.

Gravity is another factor in atmospheric pressure change. Gravity pulls gas molecules to the earth's surface, which compresses 99% of the atmosphere's mass into a height of only 40 km above the earth's surface. The air at sea level is more dense than the air at high altitudes.

Since higher altitudes have less dense air, there is less weight pressing down on objects. This produces a lower atmospheric pressure. Your body is naturally acclimated to the air pressure of its usual surroundings. When you experience a change in altitude, the air pressure around you changes as well, but the pressure inside your body takes time to adjust. For example, your ears may pop when you drive into the mountains or ride a skyscraper's elevator. This sensation happens because the cavities in your body, such as those in your lungs, ears, and sinuses, contain air. Since these areas do not adjust automatically to pressure changes, you may experience ear pain or a headache at times with a change in altitude.

Atmospheric pressure is measured using an instrument called a **barometer**. There are two basic types of barometers. A mercury barometer is a tube filled with liquid mercury that opens into to place a hand over the opening on the top of the can to stop the water flow. Ask students why the water stopped flowing. (**Answers will vary.**) Explain that air pressure holds the water in place at the can's bottom and prevents it from flowing out when there is no air pressure from above the water. Have the students remove their hands from the top of the can to allow the water to flow again. Ask why it is flowing now. (**because air can now flow in the opening and push the water through the hole**)

C. Have students research what preparations are needed to make a trip to high altitude. For example, soldiers are frequently deployed to high altitude areas. What preparations do they make to get ready for such a deployment? What measures do soldiers use to counteract the effects of being at high altitude? Why is it important for military personnel to be able to counteract the effects of high altitude very quickly? Discuss student findings.

D. Divide the class into groups to research the history of a weather instrument used on naval vessels. Direct students to include information on how the instrument worked and what it was used for. Encourage groups to present their findings to the class.

Lesson Review

1. Explain how the rotation of the earth, solar energy, gravity, and altitude affect air pressure.
(**Earth's rotation moves air masses of varying pressures into different regions, which changes the air pressure in other areas; solar energy warms air molecules, which causes them to move and spread out; gravity pulls gas molecules toward the earth's surface, which causes the air at sea level to be dense; and high altitudes have less dense air, which results in lower air pressure.**)

NOTES

Aneroid barometer

a container of mercury. Rising atmospheric pressure forces the mercury higher in the column; falling pressure allows the mercury to sink lower. A falling barometer indicates that the atmospheric pressure is falling and that a low-pressure system is forming, strengthening, or approaching. One way mercury barometers indicate atmospheric pressure is by the measurement of the height of the mercury in millimeters. For example, a height of 760 mm is considered standard atmospheric pressure. A second type of barometer is the aneroid barometer. The word *aneroid* means "not liquid." An aneroid barometer consists of a metal container from which most of the air has been removed. When the atmospheric pressure rises, the sides of the container bend inward. The sides bulge outward when atmospheric pressure drops. These changes cause a pointer to move along a dial, indicating the change in air pressure.

LESSON REVIEW
1. Explain how the rotation of the earth, solar energy, gravity, and altitude affect air pressure.
2. What effect does the change in altitude have on your body? Why?
3. Explain the difference between a low-pressure area and a high-pressure area.
4. How is atmospheric pressure related to weather?
5. What is a barometer?
6. How does an aneroid barometer work?

FYI

Cave Breathing
Temperature differences create wind as warm air rises and cooler air sinks. A change in temperature also creates a change in atmospheric pressure. Wind Cave in South Dakota is so large that it has its own atmospheric pressure. When the earth's surface pressure increases, air is forced into the cave and its pressure increases as well. When the surface pressure lowers, or the temperature rises, air rushes out of the cave and the atmospheric pressure lowers. This movement of air in and out of the cave makes it appear the cave is breathing, but the air movement is the result of changes in atmospheric pressure.

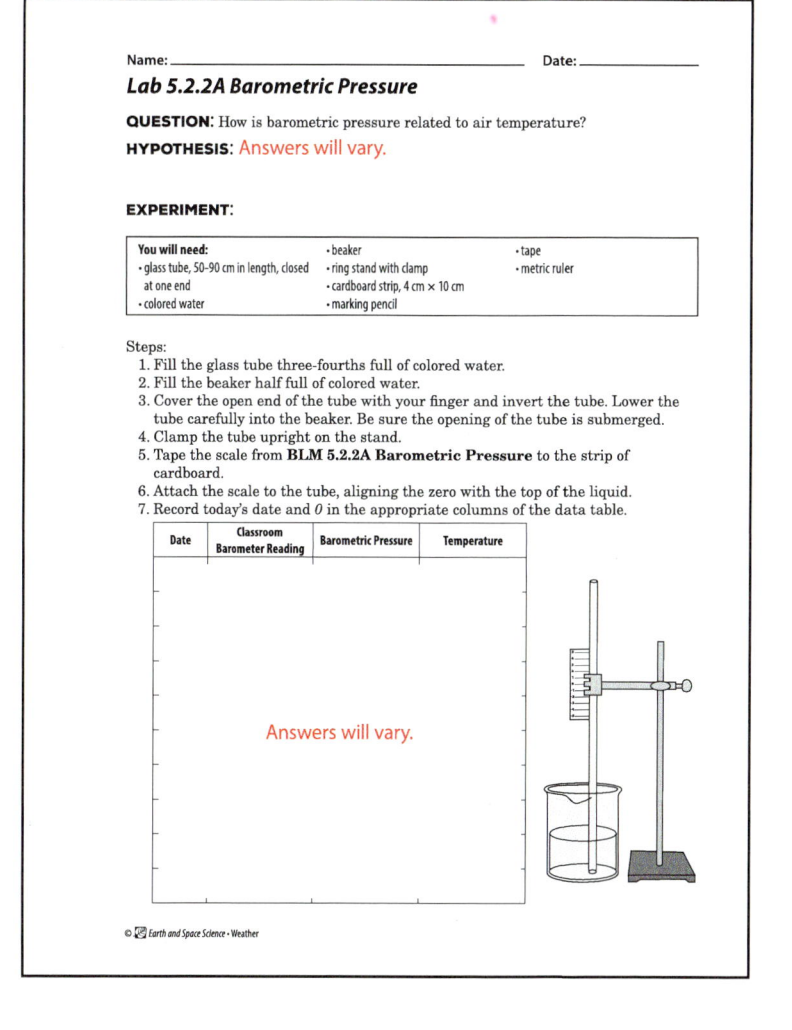

246

Name: _____ Date: _____

Lab 5.2.2A Barometric Pressure

QUESTION: How is barometric pressure related to air temperature?
HYPOTHESIS: Answers will vary.

EXPERIMENT:

You will need:		
• glass tube, 50-90 cm in length, closed at one end	• beaker	• tape
• colored water	• ring stand with clamp	• metric ruler
	• cardboard strip, 4 cm × 10 cm	
	• marking pencil	

Steps:
1. Fill the glass tube three-fourths full of colored water.
2. Fill the beaker half full of colored water.
3. Cover the open end of the tube with your finger and invert the tube. Lower the tube carefully into the beaker. Be sure the opening of the tube is submerged.
4. Clamp the tube upright on the stand.
5. Tape the scale from **BLM 5.2.2A Barometric Pressure** to the strip of cardboard.
6. Attach the scale to the tube, aligning the zero with the top of the liquid.
7. Record today's date and *0* in the appropriate columns of the data table.

Date	Classroom Barometer Reading	Barometric Pressure	Temperature

Answers will vary.

NOTES

2. What effect does a change in altitude have on your body? Why? (**When the altitude changes, the air pressure around me changes as well, but the pressure inside my body takes time to adjust. It can cause ear pain and headaches.**)
3. Explain the difference between a low-pressure area and a high-pressure area. (**A low-pressure area is characterized by warm air expanding, rising, and decreasing in density. A high-pressure area is characterized by cold air condensing, sinking toward the surface of the earth, and becoming more dense.**)
4. How is atmospheric pressure related to weather? (**As air rises in low-pressure areas, strong winds form, and water vapor from warm air condenses to form clouds. Low-pressure systems bring unsettled weather such as cloudy skies, storms, rain showers, or blizzards. In high-pressure areas, the sinking air prevents the formation of many clouds and usually there are clear skies and sunny conditions.**)
5. What is a barometer? (**A barometer is an instrument used to measure atmospheric pressure.**)
6. How does an aneroid barometer work? (**It is a metal container that has no air in it. When the atmospheric pressure increases, the sides of the container bend inward. When the atmospheric pressure decreases, the sides bulge outward. A pointer is attached and moves along a dial to indicate the changes.**)

Lab 5.2.2A Barometric Pressure

8. Research today's temperature and barometric pressure. Record the data in the appropriate location. (Note: If the barometric pressure is given in inches of Hg instead of millimeters, use the following conversion factor: 25.4 × inches of Hg = mm of Hg.)
9. Continue recording data for several days.
10. Make two separate graphs of the data you collected. Use one graph for the barometric pressure and the second graph for the daily temperature. For each graph, place the days on the x-axis because this is the independent variable. The barometer and temperature readings should be placed on the y-axis because these are dependent variables.

ANALYZE AND CONCLUDE:

1. What day had the highest classroom reading? What day had the highest barometric pressure? Are these on the same day? Why? **Answers will vary.**

2. What day had the highest temperature reading? What was it? **Answers will vary.**

3. What day had the lowest class reading? What day had the lowest barometric pressure? Are these on the same day? Why? **Answers will vary.**

4. Which day had the lowest temperature? What was it? **Answers will vary.**

5. Is there a direct relationship or an inverse relationship between barometric pressure and temperature? Why? **There is an inverse relationship between barometric pressure and temperature. Low pressure systems are the result of warmer air. High pressure systems are the result of cooler air.**

6. Challenge: Graph the classroom readings. Place the days of the week along the x-axis. The y-axis should be numbered from –10 mm through 10 mm. Compare the change in pressure in the classroom to the actual barometer changes. Are the lines on the classroom graph and the barometer graph the same? Why? **Answers will vary.**

Weather

5.2.3 Local Winds

Introduction
To demonstrate wind formation, set the block of ice inside the aquarium at one end and the pan of sand at the other. Cover the aquarium with the sheet of plexiglass and suspend the heat lamp over the end of the aquarium containing the pan of sand. Then, light the candles. Allow a few minutes for the heat circulation to begin inside the aquarium and for the candlewicks to burn a bit. Blow out the candles and place them inside the aquarium on top of the ice block. Quickly replace the plexiglass sheet. Have students observe what happens to the smoke, which should move from the cold end to the warm end.

Discussion
- Discuss severe weather involving strong winds. Encourage students who have experienced severe weather with high wind speeds to describe what they witnessed. Ask students what might cause the strong winds of such storms. (**Answers will vary.**) Discuss what can be done to prevent wind damage.

- Ask the following questions:
  1. Why does the wind change direction along a coastline? (**The land and the water heat up and cool off at different rates. The varying rates create two different pressure systems, and the high-pressure system moves into the low-pressure area.**)
  2. What causes the wind to change direction in a valley? (**When mountains lack thick vegetation to hold in heat, they lose heat at night more quickly than nearby valleys. This heat loss causes the local winds to change direction.**
  3. Which landscape heats up faster, a bare mountain slope or a slope covered with plants? (**a bare mountain slope**) Which one loses heat faster? (**a bare mountain slope**)
  4. What is an anemometer? (**An anemometer is an instrument that measures wind speed.**) How does it work? (**An anemometer has several arms with a small cup attached to each arm. The arms are attached to a shaft. When the wind blows, the cups catch the wind and spin the shaft. The rotations are measured to calculate the wind speed.**)
  5. What makes an anemometer an important tool for meteorologists? (**Meteorologists graph wind speeds and use this information to predict changing weather patterns.**)

Activities

Lab 5.2.3A Anemometer
- small paper cups, 4 per group
- scissors, 1 pair per group
- markers, 1 per group
- strips of stiff cardboard, equal in length, 2 per group
- staplers, 1 per group
- pushpins, 1 per group
- pencils with eraser, 1 per group
- metric rulers, 1 per group
- stopwatches, 1 per group

A. Complete *Try This: Wind Speed* with students.

B. Introduce the invited meteorologist. Have students ask their prepared questions about why meteorologists measure wind speed. If the area experiences distinctive local winds, allow time for the meteorologist to explain their effect on the area to the class.

C. Divide the class into groups. Give each group a compass and finely ground grass or very fine silt. Take students outside to a paved area to measure the direction of the local winds.
  - Direct a student from each group to drop a large pinch of grass clippings, and have the other students in the group observe the path of the clippings.
  - Challenge each group to determine the wind's direction with the compass, using the fallen clippings as a reference point.
  - Discuss the factors that determine the direction from which the wind is blowing.

OBJECTIVES
Students will be able to
- summarize the nature of wind.
- compare the types of local winds in different geographical areas.
- explain why meteorologists measure wind speed.

VOCABULARY
- **anemometer** an instrument used to measure wind speed

MATERIALS
- Aquarium, block of ice, foil pan, sand, plexiglass sheet, heat lamp, 2 pillar candles, matches (*Introduction*)
- Compasses, finely ground grass clippings or very fine silt (*C*)
- Wind speed indicators (*D*)

PREPARATION
- Obtain materials for *Lab 5.2.3A Anemometer*.
- Gather materials for *Try This: Wind Speed*.
- Freeze water in a small dishpan to form a block of ice. Pour several cups of sand in a foil pan. (*Introduction*)
- Invite a meteorologist to speak to the class. Have students prepare questions about why meteorologists measure wind speed. (*B*)

TRY THIS

Wind Speed
- string, 30 cm per group
- glue
- table tennis balls, 1 per group
- protractors, 1 per group

NOTES

D. Assign pairs of students to choose a location to measure the wind speed at the same time of day for a week. Have students use wind speed indicators and record the wind speed at the their locations. Direct students to graph their results. Discuss the differences in wind speed from location to location that are noted by the pairs of students.

E. Encourage students to research wind shear. Assign them to write a news story on wind shear that answers the following questions: *What is wind shear? Why are some areas more prone to wind shear? What types of accidents have occurred because of wind shear? What safety precautions are in place to prevent accidents? How is technology being used to detect wind shear?*

Lesson Review

1. Explain what wind is and how it is created. (**Wind is the movement of air from a high-pressure area to a low-pressure area. It is created by unequal heating of the atmosphere.**)
2. What factors affect local winds? (**different geographical surfaces, like land and water; different landscapes in close proximity, like mountains and valleys; and different amounts of vegetation**)
3. Describe sea breezes and land breezes. (**A sea breeze occurs in the daytime when the land near the water heats up faster than the water. The air over the land heats and rises, creating a low-pressure system. The cooler air over the water rushes in to take its place. A land breeze occurs in the evening as the land cools faster than the water. The warm air over the water rises and the cool air over the land rushes in underneath it.**)
4. What is the difference between mountain and valley breezes? (**A valley breeze occurs in the daytime as the warm air in the valley rushes up the side of the mountain. A mountain breeze occurs in the evening when the cool air above the mountain sinks down into the valley.**)

Local Winds 5.2.3

If you have ever flown a kite at the beach, you know that lake or ocean shores are usually windy, but do you know why? When air is heated, its density decreases. As warm air rises, its air pressure lowers, creating a low-pressure area. Cooler, denser air from a high-pressure area then moves underneath the warm air. Earth's surface absorbs heat from the sun more quickly than a body of water can, so at the beach when the warm air over the land meets the cooler air over the water, it results in wind. Wind is the movement of air under high pressure toward an area of low pressure. There are two types of wind: local winds and global winds. Both types of wind result from atmospheric differences caused by the unequal heating of the atmosphere. Global winds will be discussed in the next lesson.

Sea breezes are local winds that form during the day when cooler air over the water moves in to replace the warmer air that rises up from the land. At night, land temperatures drop more quickly than water temperatures. Air from the cooler land is then drawn toward the warmer air over the water, which forms local winds called *land breezes*.

Local winds are also influenced by the landscape. During the day, mountains heat up more quickly than valleys because the sun's radiation reaches the slopes before it moves down into the valleys. In addition, bare mountain slopes can absorb solar energy better than areas that are covered with plants like in most valleys. The surface of the mountain heats the air around it. The warmer air expands, creating a low-pressure area near the top of the mountain. The low-pressure pulls the cooler air up from the valley, creating a local wind known as *a valley breeze*.

OBJECTIVES
- Summarize the nature of wind.
- Compare the types of local winds in different geographical areas.
- Explain why meteorologists measure wind speed.

VOCABULARY
- **anemometer** an instrument used to measure wind speed

TRY THIS

Wind Speed
Cut a piece of strong string 30 cm long. Glue one end of the string to a table tennis ball. Glue the other end of the string to the center point on the base of a protractor. Hold the protractor upside down and away from your body so the ball can swing freely. The ball should hang straight down to 90°. If it does not, make whatever adjustments are needed. To measure the wind speed, hold the instrument so it points into the wind. Mark the angle where the ball swings. Use the following chart to find out how strong the wind is. Compare your results with those of other students.

Angle	90°	85°	80°	75°	70°	65°	60°	55°	50°	45°	40°	35°
Kph	0	9.3	13.2	16.3	19.0	21.6	24.0	26.6	29.0	31.5	34.4	37.6

Sea and Land Breezes

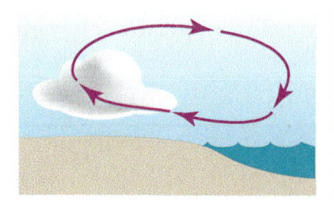

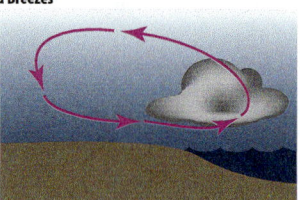

Without thick vegetation to hold in heat, mountains lose heat faster than valleys at night. This heat loss causes the local winds to change direction. *Mountain breezes* are local winds that form at night when cooler, denser mountain air moves down into the valleys. If you have ever camped on a mountain, you know that the temperature drops quickly after the sun sets.

As local winds blow over mountain peaks and sink over the other side, the winds form high-pressure systems. These systems create warm temperatures that can produce intense warm winds. Some examples of local winds that sweep down mountain slopes are the chinook on the east side of the Rocky Mountains, the foehn in Switzerland's alpine valleys, and the zonda in the Argentine Andes. Local winds can also be formed by low-pressure areas that result from the intense heating of inland areas. The sirocco, a hot wind formed in this manner, blows to the Mediterranean Sea from the Sahara Desert, and the khamsin brings hot winds to southern Egypt from the Sahara.

Meteorologists measure wind speed using an instrument called an **anemometer**. An anemometer looks like a weather vane. Small cups, all pointing in the same direction, are attached to spokes, which are connected to a freely rotating shaft. As the wind hits the cups, it turns the shaft. Wind speed is determined by counting the number of rotations that occur in a set time. Meteorologists measure the wind's speed and direction to help graph the movement of high- and low-pressure systems, which helps them predict changing weather patterns.

Anemometer

5. Explain how a chinook wind is formed. (**A chinook is formed when local winds blow over mountains and sink over the other side. The winds form high-pressure systems with clear skies. This creates warm temperatures, producing warm winds.**)
6. Why do meteorologists measure wind speed? (**Meteorologists measure the wind's speed and direction to help them graph the movement of high- and low-pressure systems. This information helps them predict changing weather patterns.**)

NOTES

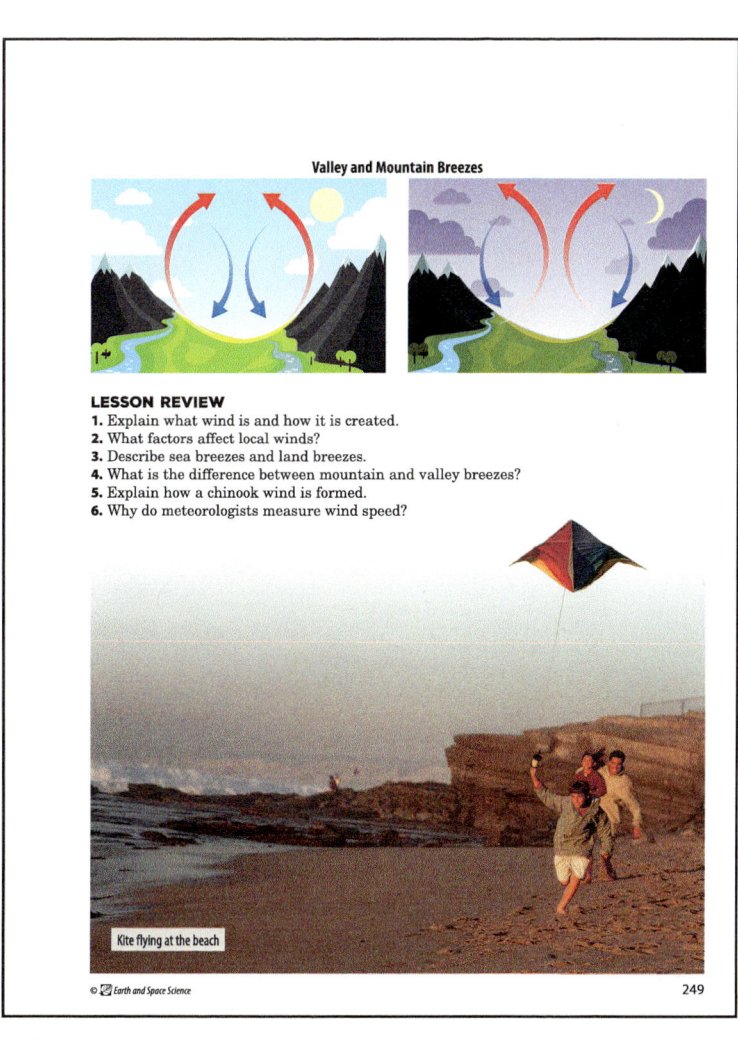

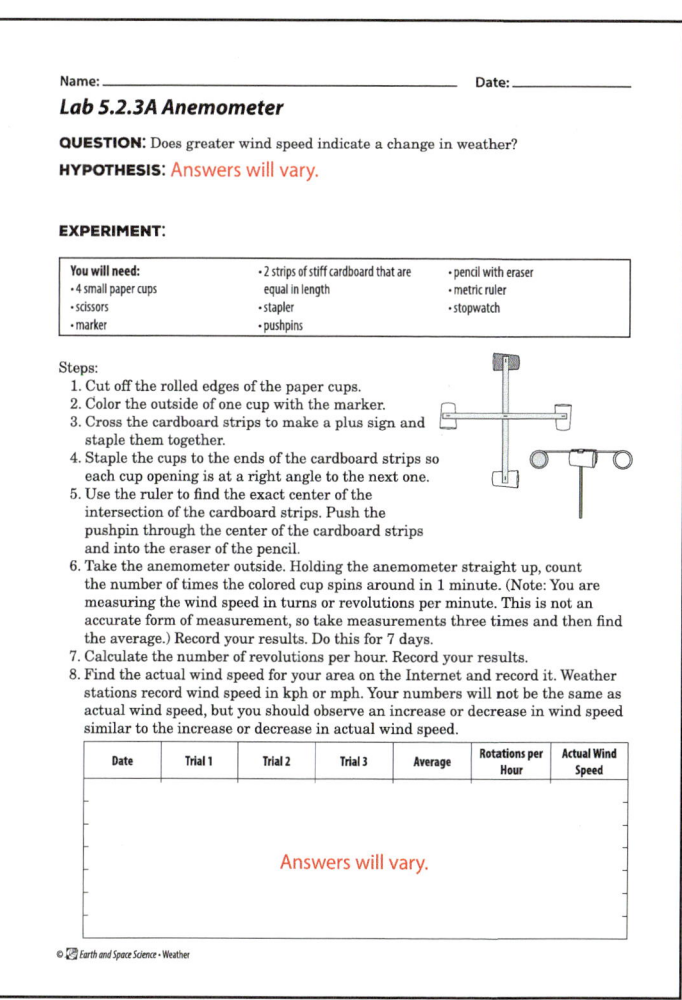

Lab 5.2.3A Anemometer

ANALYZE AND CONCLUDE:

1. On which day did you have the highest number of revolutions per hour? Was it on the same day as the highest reading for the actual wind speed?
   Answers will vary.

2. On which day did you have the lowest number of revolutions per hour? Was it on the same day as the lowest reading for the actual wind speed?
   Answers will vary.

3. What problems may have affected the accuracy of your anemometer?
   Answers will vary.

4. What improvements could you make to get more accurate results?
   Answers will vary.

5. On the days with higher wind speed, did you see a change in weather? If yes, what was the change? Answers will vary.

6. What causes wind and why do meteorologists measure it? Wind is the result of unequal heating of the atmosphere. Changes in high pressure or low pressure areas cause wind and can be used to predict changing weather patterns.

Weather

5.2.4 Global Winds

Introduction
Explain the Coriolis effect using a globe. Ask students how Earth moves. (**It spins.**) Do people spin with it? (**Yes.**) What would happen if people could jump and stay suspended in the air for several hours? (**Answers will vary.**) Where would a person land? (**Answers will vary.**) Demonstrate the Coriolis effect by touching your location on the globe, having your finger jump off, and spinning the globe half a turn.

Discussion
- Pin a cardboard disk to a bulletin board and smoothly spin it. As the disk is spinning, draw a straight line from its center to its outer edge with a marker. Explain to the class that the spinning disk deflects the path of the marker just like the spinning of the earth deflects the motions of air currents and other objects. Using a different color marker, make one dot near the center of the circle to represent a point near the North Pole. Make a second dot near the edge of the circle to represent a point near the equator. Ask students to predict where the line will end up if you start at the North Pole and aim for the equator point as the cardboard disk is spinning. (**Answer will vary but should include it will end up to the west of the point.**) Demonstrate what happens. Now start from the point on the equator and aim for the point near the pole. Ask students to predict where the point will end up. (**Answers will vary but should include it will end up east of the point.**) Demonstrate to check students' predictions.

- Divide the class into three groups and have each group explain to the class the characteristics and actions of one of the global wind loops: polar easterlies, westerlies, or trade winds. Discuss the information that the groups present on the three global wind loops.

- Ask the following questions:
  1. If you jumped up and stayed suspended in the air for 12 hours, where would you land when you came back down? (**Answers will vary but should include west of where I started.**)
  2. There are two places on the earth that you could jump up, stay suspended for 12 hours, and land in the same spot when you came back down. Where are these two places? (**North Pole and South Pole**)

Activities
A. Complete *Try This: The Speed of the Earth* with students.

B. Have students complete **WS 5.2.4A Global Winds**. Point out that the globe on the worksheet is drawn from a different perspective than the globe in the Student Edition.

C. Assign students to trace the routes used by early explorers and to evaluate whether global winds influenced the use of these routes. Provide time for students to present to the class the information they find.

D. Encourage students to investigate how global winds influence the movement of heat, precipitation, or the path of hurricanes and typhoons around the world.

E. Challenge students to explain how air pollution could reach the Arctic Circle. The North Pole remains relatively untouched by human activity, but its air is polluted with the same kind of pollution as in certain North American and European cities. Have students draw a map to illustrate their explanations for Arctic air pollution.

OBJECTIVES
Students will be able to
- identify and explain three global wind patterns.
- apply the Coriolis effect to global travel.

MATERIALS
- Globe (*Introduction*)
- Pushpin, cardboard, bulletin board, markers (*Discussion*)
- WS 5.2.4A Global Winds

PREPARATION
- Cut a disk from cardboard. (*Discussion*)
- Obtain materials for *Try This: The Speed of the Earth*.

TRY THIS

The Speed of the Earth
- chalk
- long pole

Use the length of the pole to guide the size of the circles.

(**The person at the equator moved the fastest. The person at the North Pole moved the slowest. The person at the equator had a lot farther to walk, so he or she had to move faster. Different parts of the earth travel at different speeds. The latitude lines near the poles have the smallest circumference and the equator has the largest circumference. The equator area has to move faster to rotate with the poles.**)

NOTES

Lesson Review

1. Why is the air warmer at the equator than at the poles? (**The sun is almost directly overhead for most of the year. The direct rays heat the earth's surface rapidly. Regions the sun's rays strike with a greater angle do not heat up as quickly.**)
2. Describe the trade winds. (**At the equator, warm air rises and moves toward both the North and South Poles. As it rises and travels away from the equator, the air begins to cool. When these winds reach about 30° latitude, they start sinking, change direction, and then return to the equator to start the process all over again.**)
3. Describe the pressure and wind pattern in the polar latitudes. (**The high pressure at each pole moves cold air toward the equator. As the air sinks, it travels near Earth's surface and begins to warm up. As it warms, winds called *polar easterlies* expand and rise. Eventually, at about 60° latitude, they rise to the upper troposphere where they cause a high-pressure area. The air then changes direction and heads back toward the poles.**)
4. Explain how and where the westerlies form. (**The westerlies are created by the actions of the trade winds and the polar easterlies. They form between 30° and 60° latitude and blow toward the poles at the surface and toward the equator in the upper troposphere.**)
5. Other than different air temperatures and pressure areas, what else affects the movement of global winds? (**Possible answers: the rotation of the earth, or the Coriolis effect**)
6. An airplane is traveling straight south from the equator to Antarctica. What direction would its path seem to bend? Why? (**The path would bend to the east. Because of the Coriolis effect, the point on the equator is moving much faster than the destination in Antarctica.**)

5.2.4 Global Winds

OBJECTIVES
- Identify and explain three global wind patterns.
- Apply the Coriolis effect to global travel.

The unequal heating of the earth's surface forms not only local winds but also large global wind systems. Near the equator, the sun is almost directly overhead most of the year. The direct rays of the sun heat the earth's surface there rapidly and create a low-pressure area. In contrast, regions where the sun's rays strike at a greater angle do not heat up as quickly, which is why temperatures near the equator are warmer than those at the poles.

At the equator, warm air rises and moves toward both the North and South Poles. As it rises and travels away from the equator, the air begins to cool. When these winds, which are called *the trade winds*, reach about 30° latitude, they start sinking. The trade winds then change direction rapidly and return to the equator to start the process all over again. The opposite happens at the poles. The high pressure at each pole moves the cold air toward the equator. As the air sinks, it travels near Earth's surface and begins to warm up. As it warms, winds called *polar easterlies* expand and rise. Eventually, at about 60° latitude, they rise to the upper troposphere where they cause a high-pressure area. The air then changes direction and heads back toward the poles.

A third loop of wind is created by the actions of the trade winds and the polar easterlies. Between 30° and 60° latitude, the winds blow toward the poles at the surface and toward the equator in the upper troposphere. These winds, called *westerlies*, can be very strong and can cause unstable weather patterns in the middle latitudes.

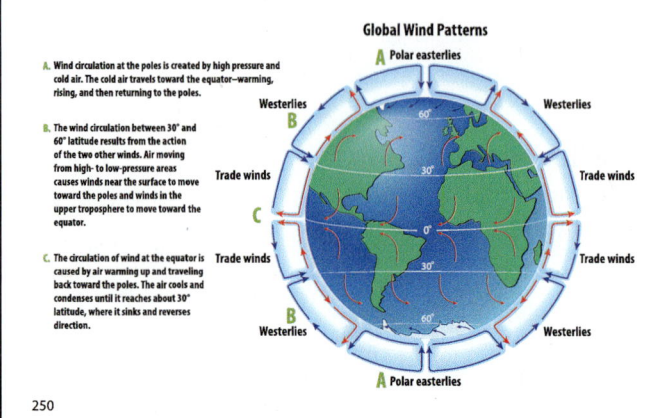

Weather patterns would not be difficult to predict if you only had to consider global winds. However, the rotation of the earth adds another dimension to the movement of the winds. The latitudes of the earth move at different speeds because Earth is a sphere. The equator is the widest part of the planet, so Earth's circumference is greatest at 0° latitude. The latitudes near the poles, on the other hand, have the smallest circumference. If you plotted points along a line of longitude from the North Pole to the South Pole, the points near the equator would have to travel much faster to make a full circle in 24 hours than the points near the poles.

Because the latitudes move at different speeds, winds bend. For example, the polar easterlies bend in the opposite direction of the earth's rotation. Air over the poles moves slower than air at lower latitudes. As the wind leaves the poles and flows toward the equator, the earth moves ahead of the wind, which makes it seem like the air from the poles is moving backward. The westerlies move close to the earth between 30° and 60° latitude. These winds begin at a fast-moving latitude and move toward a part of the earth that is moving more slowly. The winds get ahead of the earth's rotation and curve from west to east. In contrast, the trade winds move toward the fastest moving part of the earth so they bend back to the west as they travel.

The rotation of the earth curves the paths of winds, flying objects, and sea currents that travel through different latitudes because of the Coriolis effect. The Coriolis effect was first described by the French mathematician Gustave-Gaspard Coriolis in the early 19th century. Because of the Coriolis effect, an object that moves in a straight line above the earth's surface and not parallel to the equator appears to curve because the earth is spinning beneath it. Similarly, something that is relatively stationary near the equator will tend to turn because the earth is moving faster than places to the north or south of it. Tropical storms rotate counterclockwise in the Northern Hemisphere and clockwise in the Southern Hemisphere. A common myth is that water in toilet bowls, bathtubs, and sinks rotates counterclockwise in the Northern Hemisphere and

Hurricanes are steered by global winds.

Global Wind Patterns

A. Wind circulation at the poles is created by high pressure and cold air. The cold air travels toward the equator—warming, rising, and then returning to the poles.

B. The wind circulation between 30° and 60° latitude results from the action of the two other winds. Air moving from high- to low-pressure areas causes winds near the surface to move toward the poles and winds in the upper troposphere to move toward the equator.

C. The circulation of wind at the equator is caused by air warming up and traveling back toward the poles. The air cools and condenses until it reaches about 30° latitude, where it sinks and reverses direction.

FYI

Windy Jargon

Trade winds: These winds occur between the equator and 30° latitude. The trade winds helped ships sail on the trade routes west from Europe to the Americas.

Polar easterlies: These winds originate at the poles and bend from east to west.

Westerlies: These winds originate between 30° and 60° latitude, blow from the west, and bend to the east.

Horse latitudes: At around 30° latitude, the winds become very weak. In the 18th and 19th centuries, sailors on ships that got stranded in this latitude might throw any horses they were carrying overboard to save the drinking water.

clockwise in the Southern Hemisphere as it drains. However, the Coriolis effect is very slight on something as small as water going down a drain. Other rotational forces easily overcome the Coriolis effect in such instances.

Scientists calculating the trajectory of a vehicle launched into space must take the Coriolis effect into account. The Coriolis effect is also considered when airlines determine flight paths. Suppose a commercial jet is scheduled to fly from Norway to Nigeria. If the plane were to fly directly south, it would completely miss Nigeria. Nigeria, which is close to the equator, is on a latitude that is moving much faster than one near Norway. When the plane reached the equator, the earth would have moved to the east and because of the Coriolis effect, the plane's path would bend. Consequently, the plane would end up over the Atlantic Ocean rather than in Nigeria.

Airline flight paths

LESSON REVIEW
1. Why is the air warmer at the equator than at the poles?
2. Describe the trade winds.
3. Describe the pressure and wind pattern in the polar latitudes.
4. Explain how and where the westerlies form.
5. Other than different air temperatures and pressure areas, what else affects the movement of global winds?
6. An airplane is traveling straight south from the equator to Antarctica. What direction would its path seem to bend? Why?

TRY THIS

The Speed of the Earth

Outside on a paved surface, use chalk to draw four large concentric circles. Consider the center circle to be a line of latitude near the North Pole. The largest circle is the equator, and the other circles are lines of latitude in between. Have one person stand on each line of latitude, lining up as though creating a line of longitude. Have each person grasp a long pole to maintain the straight line. Mark the starting point for the line of people. Now, direct the whole line to move once around their circles. The person at the North Pole will dictate the speed. The line of people should stay straight as it moves around its circles and should arrive back at the starting point at the same time. Which person moved the fastest? Which person moved the slowest? Why was there a difference in the speeds? How does this activity relate to the movement of the earth?

5.2.5 Clouds

Weather

OBJECTIVES

Students will be able to
- describe how a cloud is formed.
- categorize different cloud types.
- summarize how clouds affect weather.

VOCABULARY

- **cloud** a visible collection of tiny water droplets or ice crystals in the atmosphere
- **fog** a low-level cloud caused by condensation of warm water vapor as it passes over a cold area
- **virga** a streak of precipitation that evaporates before reaching the ground

MATERIALS

- Index cards (B)
- TM 5.2.5A Cloud Types
- WS 5.2.5A Cloud Observations

PREPARATION

- Obtain materials for *Lab 5.2.5A Cloud Formation*.

Introduction

Invite several students to draw clouds on the board. After they are finished, ask them what cloud shapes are missing. Encourage a couple more students to draw any missing shapes. Have students describe the different shapes that were drawn. Explain that in 1803, Luke Howard gave clouds Latin names according to the clouds' shapes and that these names are still used today.

Discussion

- Using **TM 5.2.5A Cloud Types**, discuss the different types of clouds. Have students identify each type. The altitude of the cloud is not measurable, therefore it may not be possible to identify whether or not the clouds are low-, mid-, or high-level clouds. Explain that some cloud types can occur at multiple levels, such as nimbostratus and stratocumulus, which can form at both the mid-level and low-level of the atmosphere.

- Discuss how clouds form and what makes fog different from clouds. Explain that clouds form when water vapor condenses on small particles of dust, ice, salt, and other solids and that the water droplets combine to form clouds. Point out that fog forms in a similar way near the ground and is common in areas near bodies of water and in valleys. Ask students why fog is common in these areas. (**The air near a body of water or in a valley tends to be cool, so when warm water vapor passes over it, fog forms.**)

- Ask the following questions:
  1. Originally, cirrus clouds were named for their shape. What else does *cirrus* indicate today? (**that they are high-level clouds**)
  2. What types of clouds are found at low and mid levels? (**stratus and cumulus**)
  3. Explain the difference between a nimbostratus and a stratocumulus cloud. (**A nimbostratus cloud is a low-level or mid-level cloud that brings rain. A stratocumulus cloud is a low-level or mid-level cloud, but it is more puffy or lumpy than a nimbostratus cloud and it isn't a rain cloud.**)
  4. Which type of cloud can bring violent storms? (**cumulonimbus**)
  5. In what ways can clouds affect temperature? (**Clouds carry water from the ocean, which can fall as precipitation and lower the temperature. They can cool the earth because they reflect sunlight, and they can warm the earth because they can trap heat.**)

Activities

Lab 5.2.5A Cloud Formation
- safety goggles, 1 per student
- isopropyl alcohol
- 2 L bottles, 1 per group
- duct tape
- bicycle pumps, 1 per group

A. Using **WS 5.2.5A Cloud Observations**, instruct students to record cloud observations daily. Encourage students to compare their observations. Ask students why there are differences. (**Clouds can change during the day, and observations may have been made at different times. Clouds can be difficult to classify because each type can come in a variety of shapes. Without using an altimeter, the altitude cannot be determined, so students may assign them to different levels.**) Direct students to save the completed worksheet for use in Lesson 5.2.9.

B. Have students use index cards to make flash cards for the different cloud types. Challenge students to work in pairs and to quiz each other using the flash cards.

C. Divide the class into groups. Assign each group to make a presentation on the different cloud types and their altitudes. Have the groups present their information to other science classes.

Lesson Review

1. Explain how water becomes a cloud. (**Clouds are visible collections of tiny water droplets or ice crystals in the atmosphere. Water vapor condenses on small particles of dust, ice, salt, and other solids. These particles can remain suspended in the air for a long time because they are so small. The water droplets combine to form clouds.**)
2. What do the three Latin root words for types of clouds mean? (***Cumulus*** **means "lumpy,"** ***stratus*** **means "layered," and** ***cirrus*** **means "lock of hair."**)
3. Give the name and meaning of the prefixes and suffixes that can be added to cloud root names. (***Nimbus*** **means "rain,"** ***fracto*** **means "fracture," and** ***alto*** **means "high."**)
4. Describe two types each of high-level, mid-level, and low-level clouds. (**Possible answers:** ***high-level clouds***: **cirrus—are thin, wispy, or feathery and contain ice crystals; cirrocumulous—are thin, patchy, and form wavelike patterns; cirrostratus—are widespread, are like thin veils or strands of cotton, and create halos around the sun and moon.** ***mid-level clouds***: **altostratus—form gray or blue sheets; altocumulus—are white or gray patches of puffy clouds, form in layers, and create coronas.** ***low-level clouds***: **cumulus—are white and puffy, form in daytime, disappear at night, and signal fair weather; stratus—are gray, form layers, and produce fine drizzle; nimbostratus—are dark gray and produce light continuous rain; stratocumulus—are gray, form irregular masses, and spread into rolling layers or puffy waves; cumulonimbus—are associated with thunderstorms, can have flattened tops because of wind, can rise to great heights, have low bases, produce virga and sometimes tornadoes.**)
5. How do clouds affect weather? (**Clouds carry water inland from the oceans, cool the earth by reflecting sunlight, and keep areas warm by absorbing or trapping heat to warm the earth.**)

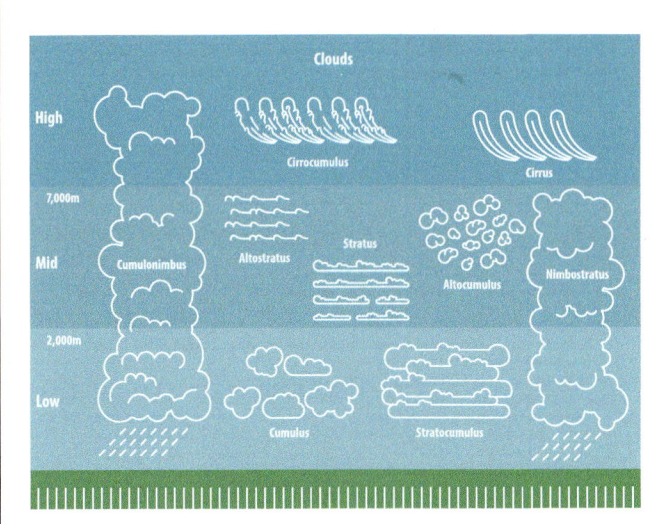

NOTES

BIBLE CONNECTION

Rain in the Bible
Clouds can bring rain and refreshment or storms and destruction. Rain appears often in the Bible. In Noah's day, God used the clouds to bring forth great rain and through the rain, judgment: "The floodgates of the heavens were opened. And rain fell on the earth forty days and forty nights" (Genesis 7: 11b–12). During Elijah's time, Israel experienced a severe drought and famine because God withheld the rain until the unfaithful people of Israel acknowledged Him as Lord. Elijah sent his servant to go out to the sea to look for a cloud. He did this seven times until finally, the servant reported, "A cloud as small as a man's hand is rising from the sea." Elijah told King Ahab to hitch up his chariot and get off the mountain before the rains stopped him. "The sky grew black with clouds, the wind rose, a heavy rain started falling and Ahab rode off to Jezreel" (1 Kings 18:44–45).

sometimes form around the sun or the moon when the ice crystals of cirrostratus clouds refract light at certain angles.

The middle layers of clouds are generally stratus and cumulus clouds. These clouds form between 2,000 m and 8,000 m. Altostratus clouds are sheets of gray or blue across the sky. When sunlight pierces through these clouds, it looks like a light shining through frosted glass. In contrast, altocumulus clouds are white or gray patches of puffy clouds that form layers across the sky. When seen through this type of cloud, the sun can form a corona, or a disk, that is pale yellow or blue on the inside and a reddish hue on the outside.

Low-level clouds form below 2,000 m. Cumulus clouds are the familiar white, puffy clouds people sometimes see animal shapes in. They form during the day but disappear at night and are generally signs of fair weather. Gray stratus cloud layers make the sky look heavy and ominous, but they produce only a fine drizzle of rain. Dark gray nimbostratus clouds, however, are true rain clouds that produce light but continuous rain. Stratocumulus clouds are not rain clouds; they are gray, irregular masses that spread out into rolling layers or puffy waves. Cumulonimbus clouds are sometimes called *thunderheads* because these clouds are frequently associated with thunderstorms. High winds often flatten their tops, which can rise to great heights. Cumulonimbus clouds have low bases under which ragged clouds can form. Sometimes cumulonimbus clouds produce **virga**, which is precipitation that evaporates before it reaches the ground. Some cumulonimbus clouds produce violent tornados.

Relative altitude of cloud formations

Clouds affect both weather and climate. They carry water inland from the oceans and thereby bring rain and snow to regions far from large bodies of water. Clouds have a cooling effect on the earth because they reflect sunlight. However, they also have a warming effect because they absorb or trap heat, which keeps the earth warm. Both of these features affect regional weather patterns and global temperatures.

LESSON REVIEW
1. Explain how water becomes a cloud.
2. What do the three Latin root words for types of clouds mean?
3. Give the name and meaning of the prefixes and suffixes that can be added to cloud root names.
4. Describe two types each of high-level, mid-level, and low-level clouds.
5. How do clouds affect weather?

Storm supercell

Name: _____   Date: _____

Lab 5.2.5A Cloud Formation

QUESTION: Will a cloud form?

HYPOTHESIS: Answers will vary.

EXPERIMENT:

You will need:	• isopropyl alcohol	• duct tape
• safety goggles	• 2 L bottle	• bicycle pump

Steps:
1. Put on safety goggles.
2. Pour 2 tbsp of alcohol into the bottle.
3. Screw the lid on the bottle, hold it horizontally, and then slowly rotate it. Some of the alcohol will evaporate. At this point, the bottle contains liquid alcohol, free alcohol molecules, and water vapor.
4. Remove the lid and place a piece of duct tape over the opening of the bottle, creating a tight seal.
5. Poke a small hole in the tape over the opening.
6. Pressurize the bottle by pumping air through the hole in the duct tape. Hold the air pump tightly against the duct tape, sealing the opening. Continue pumping until you cannot pump any more air in the bottle. This increases air pressure and temperature.
7. Keeping your face away from the opening of the bottle, pull the pump away from the bottle. What happened? A cloud formed in the bottle.

ANALYZE AND CONCLUDE:

1. Did a cloud form? Yes.
2. How does a cloud form in this system? A cloud is formed when the pressure is released or lowered and the water vapor cools and condenses on the alcohol molecules.
3. If a cloud did not form, what could be the problem with the system? Possible answers: There are not enough particles in the bottle; the bottle was not sealed well so the pressure didn't change when the pump was removed; the water vapor did not cool enough to condense.
4. How is the lab activity similar to real clouds forming? In both cases, water vapor cools and condenses on microscopic particles to form clouds.

Weather

5.2.6 Precipitation

Introduction
Show the umbrella to class. Explain that the word *umbrella* comes from the Latin word *umbra*, meaning "shade" or "shadow." Umbrellas were first used by the Egyptians and Assyrians as shades from the sun. In ancient Greece and Rome, only women used umbrellas. This custom was true in England until the 1750s, when Jonas Hanway boldly carried an umbrella in public in London. Although he was publicly criticized for using an umbrella, which was regarded as being taboo because it was too French, the use of umbrellas eventually caught on throughout England.

Discussion
- Discuss relative humidity and how it is measured. Ask students to define humidity. (**the amount of water vapor in the air**) What is the difference between humidity and relative humidity? (**Answers will vary.**) Explain that relative humidity is the amount of water vapor in the air compared with the amount that the air can hold at that particular temperature. Present the following question to students and discuss the answer: *What is the relative humidity if the air has 12 grams of water in every cubic meter but can hold 18 grams?* (**[12 ÷ 18] × 100 = 66%**) Why if two locations have the same temperature, but different relative humidities, does one feels hotter than the other? (**Answers will vary but should include that the location that feels hotter has a higher relative humidity. When people sweat at that location, they are not cooled as much. The air cannot absorb as much of the sweat.**)

- Discuss the differences between sleet, freezing rain, hail, and snow. Sleet starts as liquid rain, but passes through freezing air and freezes before it hits the ground. Freezing rain also starts as a liquid but passes through air that is almost freezing. It does not freeze until it touches a surface and loses just enough energy to freeze on contact. Hail forms when convection currents carry rain droplets up to altitudes where they freeze. Hail may be carried back up through freezing air several times, adding layers of ice. Snow starts as ice particles, not liquid precipitation.

- Ask the following questions:
  1. How is frost formed? (**If a dew point below 0°C is reached, frost forms because the water vapor changes directly into a solid.**)
  2. Why do cloud droplets stay in the clouds and not fall like raindrops? (**Cloud droplets are too small to be affected by gravity. The air currents keep them aloft.**)

Activities

Lab 5.2.6A Relative Humidity

- thermometers, 1 per group
- hot plates, 1 per group
- beakers, 3 per group
- ice water
- room-temperature water
- 7 cm × 7 cm squares of cotton fabric, 3 per group
- rubber bands, 1 per group
- heat-resistant gloves, 1 pair per group
- small pieces of cardboard, 1 per group

A. Complete *Try This: Rain Gauge* with students. Have students place their rain gauges outside and record the precipitation amounts for a few weeks. Allow time for groups to present their totals.

B. Assign students to complete **WS 5.2.6A Average Precipitation**.

C. Have students test for acid rain. Guide students to collect samples of precipitation and to use pH indicator papers to compare the pH of rainwater or snow to tap water. Discuss the test results. If tests indicate the rain is acidic, encourage students to suggest possible causes for the results.

D. Direct students to make recipe cards including all the "ingredients" necessary for each type of precipitation.

OBJECTIVES
Students will be able to
- compare relative humidity in warm and cool temperatures.
- describe the process of cloud droplets becoming precipitation.
- correlate precipitation types to atmospheric conditions.

VOCABULARY
- **coalescence** the process of coming together
- **deposition** the changing of a gas directly into a solid
- **humidity** the amount of water vapor in the air

MATERIALS
- Umbrella (*Introduction*)
- Graph paper, colored pencils (*B*)
- pH indicator paper (*C*)
- WS 5.2.6A Average Precipitation

PREPARATION
- Obtain materials for *Lab 5.2.6A Relative Humidity*. Cut cotton fabric into 7 cm squares.
- Obtain materials for *Try This: Rain Gauge*.

TRY THIS
Rain Gauge
- tall narrow jars, 1 per group
- large wide-mouthed jars, 1 per group
- metric rulers, 1 per group
- permanent markers, 1 per group

NOTES

E. Challenge students to research average rainfall in three climates: desert, tropical, and temperate. Have students record the latitude and longitude, average relative humidity, average temperatures, and rainfall amounts, including which climates have the greatest and least average rainfall.

Lesson Review

1. Explain the difference between humidity and relative humidity. (**Humidity is the amount of water vapor in the air. Relative humidity is the amount of water in the air compared with the amount that the air can hold at that particular temperature.**)
2. What is the relative humidity if the air has 16 grams of water in every cubic meter, but it can hold 17 grams in every cubic meter? (**[16/17] × 100 = 94%**)
3. Two different locations have a temperature of 36°C. One has a relative humidity of 25% and the other has a relative humidity of 85%. Which location will feel hotter? Why? (**The location with 85% humidity will feel hotter because it has so much moisture in the air that sweat will not evaporate very much to cool a person down.**)
4. What has to happen for a cloud droplet to precipitate? (**Air currents keep tiny droplets aloft, so cloud droplets must come together to form larger drops through the process of coalescence.**)
5. Compare the atmospheric conditions that result in rain, sleet, freezing rain, and snow. Describe these types of precipitation. (**Rain may begin as frozen crystals in high cold clouds or as water vapor in lower warm clouds. Sleet may start as rain that freezes as it passes through cold clouds or as ice or snow that melts and refreezes before hitting the ground as ice. Freezing rain starts as rain, falls through air that is almost freezing, then loses energy when it hits the ground and freezes on contact. Snow starts as ice particles in cold clouds and stays frozen all the way to the ground.**)

Precipitation 5.2.6

On a humid day, the air feels sticky, and you feel sticky too. When it is very humid, it can feel harder to breathe and the heavy, damp air makes you feel hot. The air feels uncomfortably sticky because it has more water vapor than drier air. Humidity makes you feel hot because the extra water vapor prevents your perspiration from evaporating. The evaporation of sweat is what cools the body because the process removes heat energy.

The amount of water vapor in the air is called **humidity**. Humidity increases as more water molecules evaporate into the air. Relative humidity is the amount of water vapor in the air compared with the amount that the air can hold at that particular temperature. For example, if a cubic meter of air is holding 16 grams of water vapor at a temperature with the potential to hold 20 grams of water vapor, then the relative humidity is 80% ([16 ÷ 20] × 100 = 80%). The air is holding 80% of what it can hold. Warm air can hold more water vapor than cool air, so humid days tend to be hot days. The higher the relative humidity, the warmer a person feels. All precipitation can be traced back to this water in the air.

The most common forms of precipitation are rain, snow, sleet, and hail. Precipitation falls to the ground for the same reason that any other object falls to the ground—gravity. Rain does not constantly fall from clouds because the size of a cloud droplet is only 0.0001–0.005 cm in diameter; this is so tiny that air currents hold the droplets aloft. In a process known as **coalescence**, cloud droplets join into drops large enough to fall. Once the water is at least 0.5 cm in diameter, it is considered a raindrop. The average raindrop contains a million times more water than the average cloud droplet. Rain may begin as frozen crystals in high, cold clouds or as water vapor in lower, warm clouds.

Sleet forms when precipitation falls through a layer of freezing air. Scientists have differing views on how sleet forms. According to cold cloud theory, it can start as snow, then melt and refreeze. Warm cloud theory suggests it starts as rain, passes through a layer of freezing air, and freezes before it hits the ground. If the rain does not freeze until it lands, it is called *freezing rain*. These raindrops fall through a layer of air that is close to freezing, which lowers the temperature

OBJECTIVES
- Compare relative humidity in warm and cool temperatures.
- Describe the process of cloud droplets becoming precipitation.
- Correlate precipitation types to atmospheric conditions.

VOCABULARY
- **coalescence** the process of coming together
- **deposition** the changing of a gas directly into a solid
- **humidity** the amount of water vapor in the air

257

Cold Cloud Precipitation

WARM AIR →

Frozen precipitation melts and reaches the ground as rain.

Frozen precipitation melts in warm air. Rain falls and freezes on cold surfaces as freezing rain.

Frozen precipitation melts in shallow warm air, then refreezes into sleet before reaching the surface.

Snow falls through cold air and reaches the surface.

of the raindrops to within a degree or two of 0°C. When the drops reach the surface, they lose enough energy to freeze on contact.

In contrast, hail is precipitation in the form of ice lumps. Hail forms when convection currents carry rain droplets up to altitudes where they freeze. Hailstones may be carried back up through freezing air several times, which adds layers of ice. Large hailstones can be very destructive, even damaging crops and property. In April 1986, the heaviest hailstone ever recorded, which weighed 1 kg, fell on Bangladesh.

Snow is made of ice particles that fall as small pellets, individual crystals, or a combination of crystals called *snowflakes*. In order for snow to form, the air must be supersaturated with water vapor at a temperature below 0°C. The lower the temperature, the smaller the snowflakes produced. Low temperatures result in small snowflakes because cold air holds less water vapor; there is not as much moisture to form big, fluffy snowflakes. For this reason, snow that falls in very cold temperatures is hard to pack into snowballs. The snowflakes do not have enough moisture to hold together well. Snowflakes become larger as the air becomes warmer.

Dew forms when the water vapor above the ground cools in the night air and condenses on

Freezing rain on branches

258

TRY THIS

Rain Gauge
Obtain a tall, narrow jar and a large, wide-mouthed jar such as a peanut butter jar. The opening of the large jar should be the same diameter as the jar itself. With a metric ruler and black marker, draw centimeter markings on the large jar. Add water to the large jar to a height of 2.5 cm and pour this water into the tall, narrow jar. Mark this level as *2.5 cm*. Divide the space below this mark into ten 0.25 cm units. Place the jars outside where water will not splash into them or run into them from a roof. Use the large jar to measure heavy rainfall and the tall jar to measure light rainfall.

the ground. The air temperature at which water vapor begins to condense is called *the dew point*. If a dew point below 0°C is reached, frost forms. Frost is caused by **deposition**, which is the changing of a gas directly into a solid. In the case of frost, water vapor becomes ice without first becoming a liquid.

LESSON REVIEW
1. Explain the difference between humidity and relative humidity.
2. What is the relative humidity if the air has 16 grams of water in every cubic meter, but it can hold 17 grams in every cubic meter?
3. Two different locations have a temperature of 36°C. One has a relative humidity of 25% and the other has a relative humidity of 85%. Which location will feel hotter? Why?
4. What has to happen for a cloud droplet to precipitate?
5. Compare the atmospheric conditions that result in rain, sleet, freezing rain, and snow. Describe these types of precipitation.

Close-up of a snowflake

FYI
Rain Records
The greatest recorded annual rainfall is 2,300 cm from August 1860 to July 1861 in Cherrapunji, Meghalaya, India.

The greatest rainfall in 24 hours is 182.5 cm in 1952 in Cilaos, Reunion, in the Indian Ocean.

The greatest average annual rainfall is in Mt. Waialeale, Hawaii, with 1,140 cm each year.

One of the driest places on Earth is the Atacama Desert in Chile, where less than 5 mm of rain falls each year.

Name: _____ Date: _____

Lab 5.2.6A Relative Humidity

QUESTION: How does temperature affect relative humidity?
HYPOTHESIS: Answers will vary.

EXPERIMENT:

You will need:		
• thermometer	• ice water	• rubber band
• hot plate	• room-temperature water	• heat-resistant gloves
• 3 beakers	• three 7 cm × 7 cm squares of cotton fabric	• small piece of cardboard

Steps:
1. Place the thermometer on a flat surface for 5 minutes. Record the temperature. This temperature reading is the dry bulb temperature for all three trials.
2. Pour ice water into one beaker and room-temperature water into another beaker.
3. Heat some water in the third beaker over the hot plate; do not exceed 34°C.
4. Secure a square of cloth around the bulb of the thermometer with the rubber band, and moisten the cloth in the hot water.
5. Wearing heat-resistant gloves, remove the thermometer from water. Hold the top of the thermometer and rapidly fan the thermometer bulb with the cardboard for 1 minute.
6. Quickly unwrap the cloth and record the wet bulb temperature.
7. Subtract the temperature of the wet bulb from the temperature of the dry bulb and record the difference. _____
8. Use the data table to determine the relative humidity of the cloth.
9. Repeat Steps 4–8 with the beakers of ice water and room-temperature water.

Water	Water Temperature	Dry Bulb Temperature	Wet Bulb Temperature	Temperature Difference	Relative Humidity
Hot					
Ice water		Answers will vary.			
Room Temperature					

ANALYZE AND CONCLUDE:
1. Which water temperature had the lowest relative humidity and which had the highest? Ice water had the lowest relative humidity and hot water had the highest.

Lab 5.2.6A Relative Humidity

2. Why did the temperatures produce different relative humidities? The temperatures for the wet bulbs were lower than the dry bulb temperature because the water in the cloth evaporated and carried away some of the heat. The cold water lowered the temperature more than the room temperature water or hot water, and as a result, it had a lower relative humidity. Cold air does not hold as much moisture as warm air.

3. Why might you feel uncomfortable on a hot, humid day? Humid air slows the evaporation rate of sweat on the skin. The evaporation of the sweat is how the body cools.

4. If your results are not what you expected, explain why you think that is. Answers will vary.

Relative Humidity (%)

Dry Bulb Temp °C	\multicolumn{10}{c}{Difference Between Dry Bulb and Wet Bulb Temperature °C}
	1	2	3	4	5	6	7	8	9	10
0	81	64	46	29	13					
2	84	68	52	37	22	7				
4	85	71	57	43	29	16				
6	86	73	60	48	35	24	11			
8	87	75	63	51	40	29	19	8		
10	88	77	66	55	44	34	24	15	6	
12	89	78	68	58	48	39	29	21	12	
14	90	79	70	60	51	42	34	26	18	10
16	90	81	71	63	54	46	38	30	23	15
18	91	82	73	65	57	49	41	34	27	20
20	91	83	74	66	59	51	44	38	31	24
22	92	83	76	68	61	54	47	41	34	28
24	92	84	77	69	62	56	49	44	37	31
26	92	85	78	71	64	58	51	47	40	34
28	93	85	78	72	65	59	53	48	42	37
30	93	86	79	73	67	61	55	50	44	39
32	93	87	80	74	68	63	57	52	47	42
34	93	87	81	75	69	64	59	54	49	44

5.2.7 Air Masses and Fronts

Weather

OBJECTIVES

Students will be able to
- classify air masses.
- explain how fronts are formed.
- evaluate a front to determine what weather will result.

VOCABULARY

- **air mass** a large body of air with consistent temperature and humidity
- **front** the boundary between two air masses

MATERIALS

- Ring stand, 2 small paper bags, thin dowel rod—50 cm long, tape, candle, matches (*Introduction*)
- TM 5.2.5A Cloud Types
- WS 5.2.7A Air Masses

PREPARATION

- Obtain materials for *Lab 5.2.7A Air Mass Interaction*.

Introduction

Demonstrate warm air rising and cooler air sinking by setting up a ring stand with the ring about 30 cm above the table. Punch a hole 2 cm from the bottom on both sides of two small paper bags. Cut the bags to measure 10 cm tall. Slide an inverted, open bag onto either end of a very thin dowel rod. Balance the dowel rod on the ring stand. If necessary, use tape to lightly secure the rod to the ring stand. Light a candle and place it under one of the bags. Be sure the bag is about 15 cm from the flame. As the air in the bag over the candle warms, the bag will rise. Discuss the results with students.

Discussion

- Discuss the density of a warm air mass and a cold air mass. Remind students that warm air rises. Ask students how density affects the motion of a warm air mass. A cold air mass? (**Less dense, warm air will rise. More dense, cold air will sink.**) Point out that other natural environments experience this process, such as local and global winds or ocean currents. Ask what this process is called. (**convection**) Discuss the similarities and differences of the process in these natural environments. In all these cases, warmer substances are less dense and rise, and cooler substances are more dense and sink. However, unlike the other environments that experience convection, air masses do not follow a circular pattern. They travel in a more linear pattern across the earth and change altitude depending on temperature changes.

- If possible, display an Internet weather map for the local area that shows a front. Discuss how fronts form and why they are responsible for changes in the weather. Have students suggest how the weather will change if an incoming front is cold. Point out that cold fronts often produce strong or violent storms, such as thunderstorms or tornadoes. Ask students to describe how an occluded front forms. (**A fast-moving cold air mass overtakes a warm air mass that is going the same direction and lifts it off the ground, cutting it off from the earth's surface.**) What type of weather does an occluded front produce? (**low temperatures, many clouds, and lots of precipitation**) How does a stationary front form and what weather can it produce? (**A stationary front forms between two air masses that meet and stall, with neither replacing the other. This type of front produces weather similar to the weather formed by a warm front.**)

- Ask the following questions:
  1. Why does an air mass stay together as it moves? (**because the air mass has a consistent temperature and moisture level**)
  2. What type of air mass is cold and dry? (**continental polar**) Warm and humid? (**maritime tropical**) Cold and humid? (**maritime polar**)
  3. What is a squall line? (**a line of heavy thunderstorms that develops in front of a high-speed cold front**)

Activities

Lab 5.2.7A Air Mass Interaction

- identical, wide-mouthed jars, 2 per group
- hot and cold water
- red and blue food coloring
- spoons, 1 per group
- large pans, 1 per group
- plastic cards, slightly larger than mouth of the jars, 1 per group

Remind students about the convection currents in wind patterns and ocean currents. A similar motion is found with the interaction of air masses.

A. Using **TM 5.2.5A Cloud Types**, review cloud types and discuss which cloud types are associated with each type of front.

B. Assign **WS 5.2.7A Air Masses** for students to complete. Discuss student answers.

C. Have students sketch a cold front, a warm front, a stationary front, and an occluded front. Direct them to use red to indicate warm air and blue to signify cold air. Drawings should include the type of weather and clouds produced by each front and arrows to show the direction of the wind. For example, a squall line of thunderstorms can be drawn in front of the cold front, and two parallel arrows can be used to show the stasis of stationary fronts.

D. Divide the class into groups. Have each group teach one of the following topics to the class: types of air masses, cold fronts, warm fronts, occluded fronts, or stationary fronts.

E. Encourage students to research what technologies are being used by meteorologists locally, nationally, and internationally to track air masses and fronts as they develop.

Lesson Review

1. What is an air mass? (**a large body of air with a consistent temperature and moisture level**)
2. Describe each type of air mass. (***maritime polar***: **forms over polar oceans, cold and humid;** ***maritime tropical***: **forms over tropical oceans, warm and humid;** ***continental polar***: **forms over polar land, cold and dry;** ***continental tropical***: **forms over tropical land, warm and dry;** ***arctic air mass***: **forms over Arctic, very cold**)
3. How is a front formed? (**A front is formed when two air masses of different densities and temperatures encounter each other. The boundary between the two air masses is a front.**)

NOTES

5.2.7 Air Masses and Fronts

OBJECTIVES
- Classify air masses.
- Explain how fronts are formed.
- Evaluate a front to determine what weather will result.

VOCABULARY
- **air mass** a large body of air with consistent temperature and humidity
- **front** the boundary between two air masses

When you listen to the weather report online or watch it on TV, the meteorologist might say that a cold air mass is moving down from the Arctic or that a warm air mass will be moving in, bringing a change in the weather. An **air mass** is a large body of air with a consistent temperature and moisture level. Air masses usually cover hundreds or thousands of square kilometers. Because the air mass has a consistent temperature and moisture level, it will tend to stay together as it moves.

Air mass names refer to the regions where they formed. A region's topography influences the moisture level of an air mass. Continental air masses form over land and are characterized by dry air; maritime air masses form over water and are characterized by moist air. The regions' temperature affects the temperature of the air mass. Polar air masses form in the polar regions and tropical air masses form over warm tropical areas. These designations are combined to form four different classifications for air masses that more accurately describe their characteristics. A cold air mass that forms over water is called *a maritime polar* (mP), and a warm air mass that forms over water is called *a maritime tropical* (mT). A cold air mass that forms over land is called *a continental polar* (cP), and a warm air mass that forms over land is called *a continental tropical* (cT). Another type of air mass, called *an arctic air mass* (A), can form over the Arctic in the winter. These air masses are very cold and very dry.

Air masses can change as they travel and encounter different conditions. If they stay over an area for a length of time, the mass can take on the characteristics of that area. For example, a maritime polar can become a continental polar if it stays over land for a length of time. A continental tropical can become a

FYI

Air Masses

Air Mass Type	Weather Map Symbol	Source	Characteristics
Arctic	A	Arctic basin	Extremely cold
Maritime Polar	mP	Polar ocean regions	Cold and humid
Maritime Tropical	mT	Tropical ocean regions	Warm and humid
Continental Polar	cP	Polar land regions	Cold and dry
Continental Tropical	cT	Tropical land regions	Warm and dry

Air Masses

maritime tropical if it stays over water for a length of time. How much a mass changes depends on how fast it is moving.

At some point, a moving air mass will encounter another air mass. The boundary between two air masses of different densities and temperatures is called a **front**. Fronts are usually responsible for weather changes. For a front to form, one air mass must overtake another air mass. The type of front that forms is determined by the way the colder air mass is moving.

Cold fronts form when a cold air mass overtakes a warm air mass. The dense, cold air lifts the warm air. If the warm air is moist, cumulus and cumulonimbus clouds form. Cold fronts produce strong but short-lived storms. Sometimes a line of heavy thunderstorms called *a squall line* develops ahead of a high-speed cold front. Slower moving cold fronts produce less cloud cover and precipitation because the cold air lifts the warm air very slowly.

Sometimes fast-moving cold air overtakes a warm air mass that is moving in the same direction and lifts it entirely off the ground. This occurrence is called *an occluded front*. The warm air is completely cut off, or occluded, from the earth's surface and held up in the atmosphere's upper levels. The advancing cold front joins with the cool air mass already beneath the warm air mass. Occluded fronts produce low temperatures, many clouds, and lots of precipitation.

Warm fronts form when a warm air mass overtakes a cold air mass. Being less dense, the warm air rises above the cold air

NOTES

4. Explain the difference between a cold front and a warm front. (**Cold fronts form when a cold air mass overtakes a warm air mass. The dense, cold air lifts the warm air. Cumulus and cumulonimbus clouds are formed and sometimes cause strong, but short-lived storms. Warm fronts form when a warm air mass overtakes a cold air mass. Being less dense, the warm air rises above the cold air, and forms a gradual slope that is often marked by clouds. Cirrus clouds condense at the height of the slope, and cirrostratus clouds form in the middle. If the warm air mass is very humid, nimbostratus clouds accumulate at the leading edge of the front, producing heavy rains or snowfall over large areas.**)

5. What type of weather results from an occluded front? (**Occluded fronts produce low temperatures, many clouds, and a lot of precipitation.**)

---

and forms a gradual slope that is often marked by clouds. Cirrus clouds condense at the height of the slope, and cirrostratus clouds form in the middle. If the warm air mass is very humid, nimbostratus clouds accumulate at the leading edge of the front, producing heavy rains or snowfall over large areas.

A stationary front forms between two air masses that meet and stall with neither replacing the other. Stationary fronts produce weather that is similar to the weather formed by warm fronts.

LESSON REVIEW
1. What is an air mass?
2. Describe each type of air mass.
3. How is a front formed?
4. Explain the difference between a cold front and a warm front.
5. What type of weather results from an occluded front?

Cold front
Occluded front
Stationary front
Warm front

262

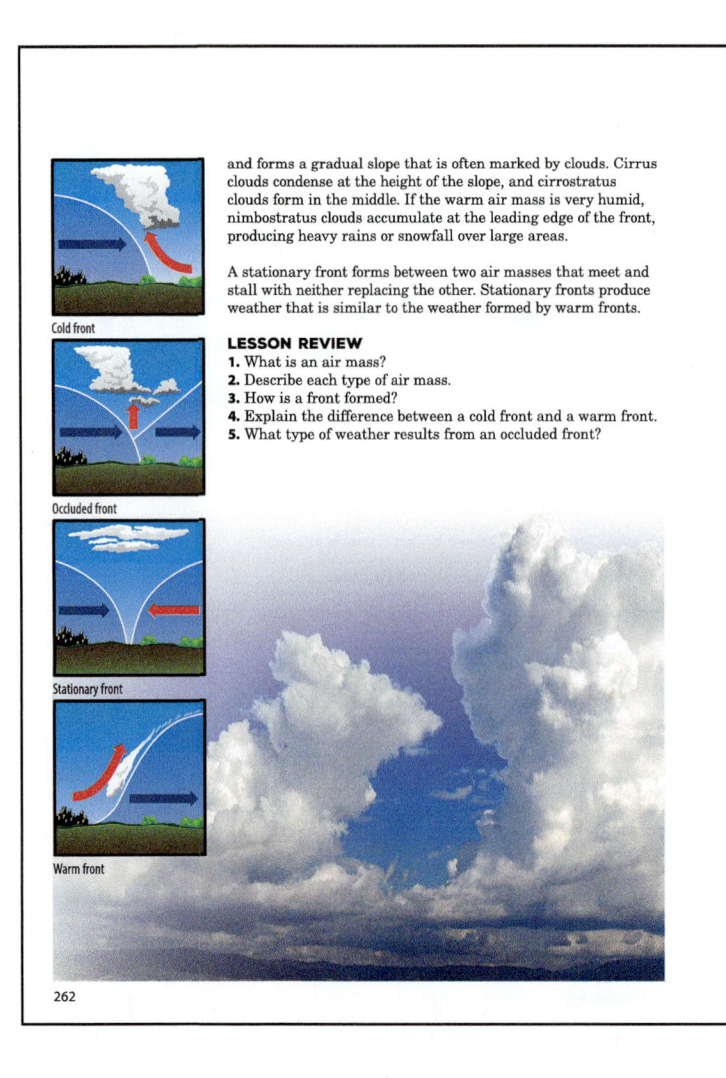

Name: _____ Date: _____

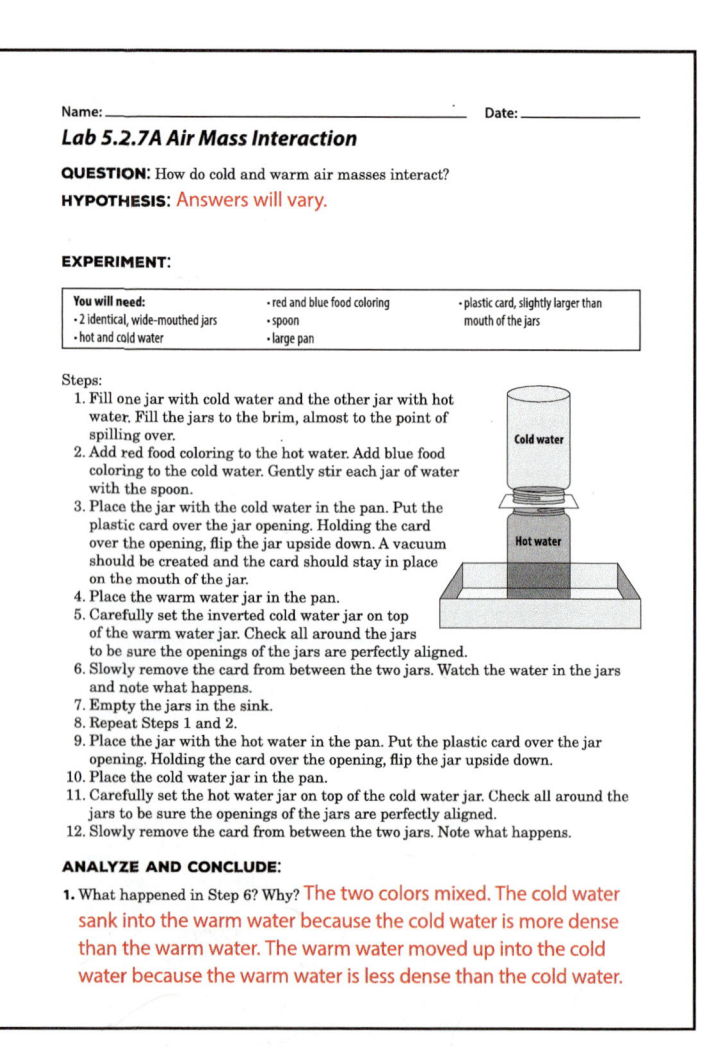

Lab 5.2.7A Air Mass Interaction

QUESTION: How do cold and warm air masses interact?
HYPOTHESIS: Answers will vary.

EXPERIMENT:

You will need:	• red and blue food coloring	• plastic card, slightly larger than
• 2 identical, wide-mouthed jars	• spoon	mouth of the jars
• hot and cold water	• large pan	

Steps:
1. Fill one jar with cold water and the other jar with hot water. Fill the jars to the brim, almost to the point of spilling over.
2. Add red food coloring to the hot water. Add blue food coloring to the cold water. Gently stir each jar of water with the spoon.
3. Place the jar with the cold water in the pan. Put the plastic card over the jar opening. Holding the card over the opening, flip the jar upside down. A vacuum should be created and the card should stay in place on the mouth of the jar.
4. Place the warm water jar in the pan.
5. Carefully set the inverted cold water jar on top of the warm water jar. Check all around the jars to be sure the openings of the jars are perfectly aligned.
6. Slowly remove the card from between the two jars. Watch the water in the jars and note what happens.
7. Empty the jars in the sink.
8. Repeat Steps 1 and 2.
9. Place the jar with the hot water in the pan. Put the plastic card over the jar opening. Holding the card over the opening, flip the jar upside down.
10. Place the cold water jar in the pan.
11. Carefully set the hot water jar on top of the cold water jar. Check all around the jars to be sure the openings of the jars are perfectly aligned.
12. Slowly remove the card from between the two jars. Note what happens.

ANALYZE AND CONCLUDE:

1. What happened in Step 6? Why? The two colors mixed. The cold water sank into the warm water because the cold water is more dense than the warm water. The warm water moved up into the cold water because the warm water is less dense than the cold water.

Lab 5.2.7A Air Mass Interaction

2. What happened in Step 12? Why? The two colors did not mix. The less dense, warm water stayed in the top jar and the more dense, cold water stayed in the bottom jar.

3. How is this experiment like a warm air mass and a cold air mass coming in contact with each other? Warm air and cold air masses behave the same way as warm water and cold water. Warm air rises just like the warm water because they are both less dense than the cold air or cold water. The cold water sank just like the cold air sinks. It is more dense than the warm air.

4. What problems did you encounter during the lab? Answers will vary.

5. If your results were not what you expected, what might have gone wrong? Possible answers: The jars did not have enough water, the temperature difference in the jars was not great enough, or the jars were not aligned correctly.

5.2.8 Storms

Weather

OBJECTIVES

Students will be able to
- identify conditions necessary for a blizzard.
- model the stages of a thunderstorm and the formation of lightning.
- compare the nature of a tornado to the nature of a hurricane.

VOCABULARY

- **lightning** the electrical discharge of energy from storm clouds
- **thunder** the sound that results from the rapid heating and expansion of air that accompanies lightning

MATERIALS

- Balloons, freezer, cooler, ice, heat lamp (*Introduction, Option 1*)
- Glass rod, animal fur friction pad, plastic comb (*Introduction, Option 2*)
- TM 5.2.8A Hurricanes
- WS 5.2.8A Stormy Weather
- WS 5.2.8B On-the-Scene Report

PREPARATION

- Inflate a balloon and store it in a freezer before class. Bring it to class in a cooler with ice. (*Introduction*)
- Obtain materials for *Try This: Tornado in a Bottle.*

TRY THIS

Tornado in a Bottle
- 2 L bottles, 2 per group
- food coloring
- tornado tubes or electrical tape
- 9.5 mm washers, 1 per group

(**The water is swirling in a vortex and has energy just like a tornado. It is different in that it is water swirling in a contained space, not wind swirling and moving in unpredictable directions.**)

Introduction

Choose from the following options to introduce students to the physical forces found in thunderstorms:

Option 1: Thunder
- balloons
- freezer
- cooler
- ice
- heat lamp

Demonstrate the expansion of air that leads to thunder by first asking students what happens to air when it is heated. (**It expands.**) Take the prepared cold, inflated balloon out of the cooler and have students observe its expansion in the warmer air. You may speed up this process by holding the balloon under a heat lamp. Ask what would happen if the balloon continued to expand. (**Answers will vary.**) Have a student inflate another balloon until it pops. Explain that lightning heats up the air so much and so fast that the expansion causes sound waves, which are heard as thunder.

Option 2: Lightning
- glass rod
- animal fur friction pad
- plastic comb

Explain that air moving inside a storm cloud rubs tiny water droplets and ice together, and they become charged with static electricity. The positive electrical charges collect near the top of the cloud. The negative charges collect near the bottom. This separation of electrical charges is very unstable. The charges equalize through lightning. Inform students that you will demonstrate the formation of lightning. Darken the room. Rub the glass rod vigorously back and forth with the animal fur. Hold the comb firmly in the other hand and bring it slowly near the rod. When the gap between the rod and the comb is small enough, a tiny spark should jump across.

Discussion

- Discuss the formation of thunderstorms. Ask students where thunderstorms usually occur. (**They usually occur in the tropical and midlatitude zones.**) Why? (**Answers will vary.**) Remind students that thunderstorms often occur in these regions in spring and in the late afternoon during summer because of the warmth of the ground and the movement of fronts carrying precipitation. Highlight the three stages of thunderstorms.

- Using **TM 5.2.8A Hurricanes**, discuss how a hurricane forms and the damage a hurricane can cause when it reaches land. If the school is in a hurricane-prone area, have students share their experiences with these storms. Then, discuss the formation of a tornado and the damage it can cause. If the school is located in an area that experiences tornadoes, encourage students to share any personal experiences. Use a Venn diagram to compare the nature of both types of storms. Ask students which lasts longer, a tornado or a hurricane. (**hurricane**)

- Discuss Matthew 7:24–27, "Therefore everyone who hears these words of Mine and puts them into practice is like a wise man who built his house on the rock. The rain came down, the streams rose, and the winds blew and beat against that house; yet it did not fall, because it had its foundation on the rock. But everyone who hears these words of Mine and does not put them into practice is like a foolish man who built his house on sand. The rain came down, the streams rose, and the winds blew and beat against that house, and it fell with a great crash." In this analogy, Jesus uses storms to represent the difficulties in life. Believers are not promised to have sunshine and blue skies all the time. Psalm 34:19 declares, "The righteous person may have many troubles, but the Lord delivers him from them all." Harsh storms and brutal winds are a reality for any person, the wise and the foolish. If a person's spiritual foundation has been set by using biblical principles (the rock), then that foundation is reliable and the person's continuing walk and reliance on Jesus is what will get them through the storms. Ask students what storms God has carried them through. (**Answers will vary.**) How should life's storms be handled? (**Answers will vary.**)

- Ask the following questions:
  1. What types of clouds are associated with thunderstorms? (**cumulus and cumulonimbus**)
  2. Can nimbostratus clouds produce rainstorms? (**Yes.**)
  3. What conditions have to be present for a snowstorm to be a blizzard? (**The winds have to be above 56 kph for at least 3 hours and visibility must be reduced to 0.4 km or less.**)
  4. What is thunder? (**It is the sound that results from the rapid heating and expansion of air that accompanies lightning.**)
  5. What is a stepped leader? (**It is the beginning of a lightning bolt when the negative charges in the cloud stretch toward the ground.**)
  6. Why are tornadoes more dangerous than thunderstorms? (**Tornadoes have very high-speed rotating winds. These powerful winds can touch down on the ground and destroy buildings and trees. The path tornadoes take are very unpredictable. Thunderstorms do not travel along the ground. They do not spin and their winds do not have the high wind speeds of tornadoes.**)

Activities

A. Complete *Try This: Tornado in a Bottle* with students.

B. Have students complete **WS 5.2.8A Stormy Weather**.

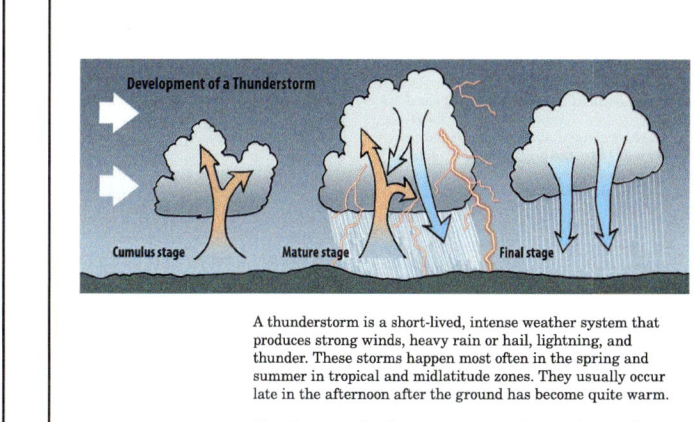

NOTES

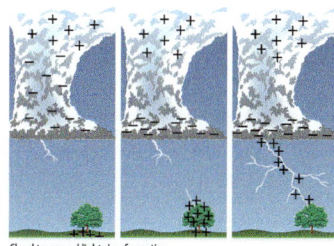

NOTES

C. Divide the class into groups. Assign groups to use **WS 5.2.8B On-the-Scene Report** to research either a tornado or hurricane that has severely impacted a community. Have groups create an on-the-scene report about the storm and present it to the class. Direct students to use the information used for WS 5.2.8B to write a script. Have students assign themselves the following parts: reporter at the scene, reporter in the news station, witness 1 and witness 2. A props person and a camera person can also be assigned if needed.

D. Encourage students to choose a recent tornado, hurricane, or blizzard and research the costs to the government and the affected area for emergency help, cleanup, and possible rebuilding. As a class, use the information gathered to compare the costs of such storms. Have students consider what measures might be taken to reduce damage from these storms.

E. Challenge students to create a service project to help storm victims of an area recently devastated by a tornado or a hurricane.

Lesson Review

1. Where do rainstorms usually form? (**along cold fronts**) What type of cloud is involved in a thunderstorm? (**cumulonimbus**)
2. If it has been snowing for five hours and the wind is blowing at 45 kph and visibility is 2 km, is this a blizzard? Why? (**No. Even though it has been snowing for five hours, it is not a blizzard because the winds must be at least 56 kph and the visibility has to be reduced to 0.4 km.**)
3. Describe the three stages of a thunderstorm. (***cumulus stage***: **warm air that is quite a bit warmer than its surroundings rises from ground, a cumulus cloud forms as humid rising air**

---

charges. As they hit each other, electrons are knocked off and charges begin to accumulate. The positive charges collect near the top of the cloud and the negative charges collect near the bottom. The negative charges attract positive charges on the ground. Then, the beginning of a lightning bolt, which is called *a stepped leader*, forms when the negative charges stretch toward the ground. A stepped leader reaches down from the cloud and a positively charged return stroke reaches up from the ground. Finally, when the pressure between the positive and negative charges becomes too high, the charges are released as lightning. As the air heats up from the lightning, the rapid increase in temperature causes the air to expand. The sound that results from the rapid heating and expansion of air is called **thunder**.

Thunderstorms can produce violent tornadoes. A tornado is a rotating funnel of air that has a high wind speed and a low central air pressure. Tornadoes are usually attached to the base of a thundercloud. High winds may cause rising warm air to rotate. When this happens, the cloud begins to spin at incredible speeds, forming a tornado. Many funnel clouds gyrate through the troposphere without touching down, but when the spinning section of the cloud reaches down to the earth's surface, the funnel cloud becomes a tornado and it can cause severe damage. Tornadoes can rise and touch down in a random path across the

Tornado

FYI
Enhanced Fujita Scale
In 1971, Dr. Tetsuya Theodore Fujita created a scale to estimate tornado wind speeds according to damage done by a tornado. The Enhanced Fujita Scale (EF) has replaced the original scale, and it takes more variables into account. The EF scale considers 28 different damage indicators, including building types and trees.

EF Scale	Class	Description	Wind Speed
EF0	weak	Gale	105–137 kph
EF1	weak	Moderate	138–177 kph
EF2	strong	Significant	178–217 kph
EF3	strong	Severe	218–266 kph
EF4	violent	Devastating	267–322 kph
EF5	violent	Incredible	>322 kph

TRY THIS
Tornado in a Bottle
Fill a 2 L bottle about two-thirds full of water. Add a little food coloring. Use a tornado tube connector or electrical tape with a 9.5 mm washer to connect this bottle to another 2 L bottle. Place the two bottles on a table so the filled bottle is on top. Rapidly rotate the bottles in a circle a few times, and then place the bottles on the table. Observe the formation of a funnel-shaped vortex. Notice the shape of the vortex and the flow of the water as it empties. How is this like a real tornado? How is it different from a real tornado?

The hurricane's eye wall pulls up warm air from the ocean's surface. The rising warm air draws in more air from the surface to take its place. The warm air is less dense, so the air pressure drops at the surface, creating an area of low pressure, or the eye. The cold, dense air from above sinks down and warms. As more air rises, more air rushes toward the center, creating the spiraling winds.

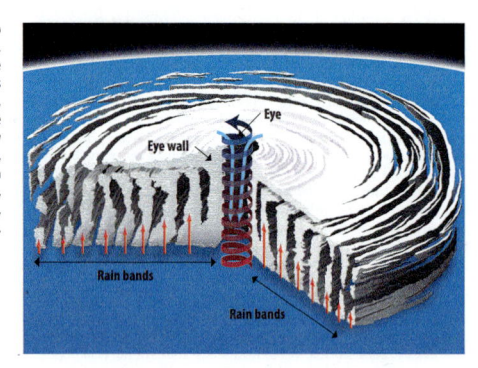

earth. For example, tornadoes can destroy houses on one side of a street without even touching the houses on the other side.

Several warning signs help meteorologists predict that a thunderstorm is spawning a tornado. They watch for strong and persistent rotation in the base of the cloud, or hail and heavy rain followed by dead calm or an intense wind shift. Loud, continuous rumbling and white flashes at ground level can also be signs of a tornado. Another sign that a tornado may form is a low-hanging appendage that looks like a crashing ocean wave reaching toward the ground and rotating.

Another very destructive storm is a hurricane. Hurricanes are large, rotating tropical weather systems with sustained wind speeds from 119 kph to 240 kph. These storms pack more power than hundreds of thunderstorms. A large hurricane holds enough energy to supply the United States with electricity for six months. Unlike tornadoes and short-lived thunderstorms, hurricanes can last for several days.

Hurricanes form over warm ocean waters near the equator. When the moisture in warm, rising air condenses to produce a large amount of energy, a low-pressure area is left above the surface of the water. The high pressure surrounding this low-pressure area moves in to take its place; it will warm up and also rise. As warm air keeps rising, cooler, denser air continues condenses into cloud droplets, there is very little rain, but there may be some lightning; *mature stage*: begins with fully formed cumulonimbus cloud, has heavy rain, updrafts and downdrafts, strong winds, thunder and lightning, and possibly hail; *final stage*: the storm dissipates, wind and hail die down, and the rain is much lighter.)

4. Give a detailed description of how lightning forms. (**As water droplets or ice crystals move around in a cloud, they hit each other and electrons are knocked off. Charges begin to accumulate. The positive charges collect near the top of the cloud and the negative charges collect near the bottom. The negative charges attract positive charges on the ground. A stepped leader reaches down from the cloud, and a return stroke reaches up from the ground. When the pressure between the positive and negative charges becomes too high, the charges are released as lightning.**)

5. What weather conditions can create a tornado? (**Tornadoes come from thunderstorms and are usually attached to the base of a thundercloud. High wind speeds can cause rising warm air to rotate.**)

6. How is a hurricane formed? (**Hurricanes form near the equator over warm ocean water. When the moisture in warm, rising air condenses to produce a large amount of energy and heat, a low-pressure area is left over the surface of the water. The high pressure surrounding this low pressure moves in to take its place, then warms and rises. As warm air keeps rising, cooler, denser air continues to move in to take its place. The air is drawn into a narrow column in the center of the storm and releases more heat to sustain the storm's power. The eye of the hurricane remains calm as the winds swirling around it gain more energy.**)

NOTES

5.2.9 Forecasting Weather

Weather

OBJECTIVES

Students will be able to
- identify various weather instruments and the data they collect.
- interpret weather map symbols.

VOCABULARY

- **isobar** a line that connects points of equal atmospheric pressure

MATERIALS

- Recording of five days of local weather forecast (*Introduction*)
- BLM 5.2.9A Weather Report
- TM 5.2.9A Weather Map
- WS 5.2.5A Cloud Observations
- WS 5.2.9A Weather Predictions
- WS 5.2.9B Weather Map
- WS 5.2.9C Weather Report Evaluations

PREPARATION

- Record local weather forecasts for five days prior to the introduction of the lesson. (*Introduction*)
- Verify that students have completed and have available **WS 5.2.5A Cloud Observations** to be used with **WS 5.2.9A Weather Predictions**. (*A*)
- Print samples of weather maps from the Internet or other sources. (*B*)

Introduction

Play the recorded forecasts for the class as a way of introducing this lesson. Note the improving accuracy of the weather forecasts for a particular day as that day approaches. Have students observe the weather conditions outside to determine the accuracy of the different predictions.

Explain the different types of forecasts. Discuss the accuracy of the different types of weather forecasts. Inform students of the current short-range, medium-range, and extended forecasts and have students chart their accuracy in the coming days. Short-range forecasts predict weather conditions over the next 12–48 hours, medium-range forecasts predict the weather for a period of 2–7 days, and extended forecasts predict weather conditions beyond 7 days.

Discussion

- Display **TM 5.2.9A Weather Map**. Point out the different parts of the map. Have students identify cold fronts, warm fronts, isobars, and high- and low-pressure areas. Discuss the direction the fronts are moving and how the pressure changes as it moves away from the center of the high- and low-pressure areas. Ask students what information can be found on a weather map. (**Answers will vary but should include cloud cover, wind speed, wind direction, air masses, weather fronts, precipitation, and storms.**) If a map has an L surrounded by isobars, how does the pressure change as the isobars get farther away from the L? (**The pressure will get higher, or less "low," as distance increases from the center of the low-pressure system.**) If a line with triangles on one side and semicircles on the other side is on a weather map, what does it mean? (**It means there is a stationary front. Neither air mass is really moving.**)

- Discuss the various tools used in forecasting the weather.
  Weather balloons: used to make wind maps of the upper level of the atmosphere for airplane flights and to calculate wind speeds and direction at different atmospheric levels
  Radiosondes: instruments sent into the atmosphere on special balloons to measure the temperature, pressure, and humidity at various heights in the atmosphere
  Radar equipment: tracks storms and hurricanes; reveals the location, shape, and intensity of cloud cover; specially designed radar in planes indicates storms or turbulent areas
  Weather satellites: carry high resolution radiometers that map clouds, measure atmospheric moisture, determine wind speeds, and calculate the sea surface temperatures for the entire Earth both day and night; provide satellite images used to determine weather patterns across the globe
  Computers: process thousands of observations from weather stations, aircraft, ships, and satellite signals from around the world every few hours; developing technology will be able to process quadrillions of calculations every second; create accurate images of current weather conditions around the world
  Weather maps: observations from local weather stations and data collected from aircraft, satellites, radar, and radiosondes plotted on weather maps using standard symbols decided by the international meteorological community

- Ask the following questions:
  1. If the meteorologist said a severe thunderstorm was heading to the area, what type of front would viewers expect to see on the weather map? (**a cold front**) What would it look like? (**It would be a line with triangles on it pointing toward this location.**)
  2. Which type of forecast is the most accurate: short-range, medium-range, or extended? (**short-range**) Why? (**because there is less time for small changes to create wide variations**)
  3. Why is weather forecasting important? (**Answers will vary but should include that forecasting warns people of dangerous weather, which gives them a chance to take cover or move to safety. It also helps pilots avoid flying into storms.**)

Activities

A. Assign **WS 5.2.9A Weather Predictions** for students to complete using data from **WS 5.2.5A Cloud Observations** and historical data from an online source.

B. Have students complete **WS 5.2.9B Weather Map** using printed sample weather maps.

C. Divide the class into groups. Direct each group to create a fictitious weather report using **BLM 5.2.9A Weather Report**. Extend the project with **WS 5.2.9C Weather Report Evaluations**. Have groups present their weather reports and discuss their probable accuracy.

D. Challenge students to investigate how modern weather satellites are used.

E. Assign students to construct a time line of the history of weather forecasting.

F. As a class, make a list of the different technologies that have affected weather forecasting. Some examples include radar, computers, satellites, and even smart phones, which can alert people of incoming dangerous weather. Ask students to consider whether more technological advancement is needed in weather prediction or in effectively informing the public.

Lesson Review

1. What instruments were used at the onset of modern weather forecasting? (**barometer, thermometer, hygrometer, and anemometer**) What information do they collect? (**atmospheric pressure, temperature, humidity, and wind speed**)

5.2.9 Forecasting Weather

OBJECTIVES
- Identify various weather instruments and the data they collect.
- Interpret weather map symbols.

VOCABULARY
- **isobar** a line that connects points of equal atmospheric pressure

If you want to know what kind of weather to expect later in the day, you probably turn on the TV to catch the weather forecast, listen to a weather report on the radio, or check the weather map on the Internet. Before these conveniences were available, people relied on their own observations to predict the coming weather.

Even in Bible times people tried to predict the weather using physical signs without knowing the science behind it. Once when Jesus was speaking to some Pharisees, He said, "When evening comes, you say, 'It will be fair weather, for the sky is red,' and in the morning, 'Today it will be stormy, for the sky is red and overcast.' You know how to interpret the appearance of the sky" (Matthew 16:2–3a). A reddish evening sky is caused by light interacting with dry dust particles, so a red evening sky indicates that dry weather is coming. A gray evening sky means that the atmosphere is heavy with water droplets that will probably fall the next day. If clouds look red in the morning, the sun is rising in clear skies to the east with clouds approaching from the west. This indicates that a storm system to the west is moving your way. Perhaps you have heard the saying "Red sky at night, sailors delight; red sky in morning, sailors take warning." This adage relies on the same principle. People have used observation to predict the weather for thousands of years.

Weather prediction becomes more efficient and accurate with each passing decade. Modern weather forecasting began after the barometer, which measures atmospheric pressure, and the thermometer, which measures temperature, were invented in the 17th and 18th centuries. The hygrometer, which measures humidity, and the anemometer, which measures wind speed, were also used in early forecasting. Today meteorologists use these and other instruments to help predict the weather. For example, instruments called *radiosondes* measure the temperature, pressure, and humidity at different heights of the atmosphere. Weather balloons calculate wind speed and direction by tracking the position of the radiosondes they carry. Radio transmitters then send the information to a receiver on the ground.

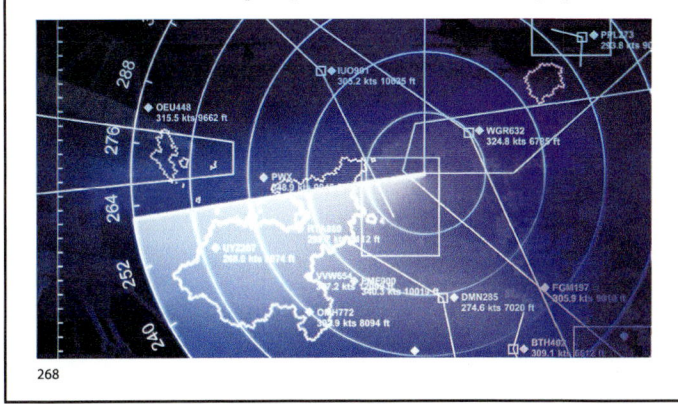

Doppler radar

Radar helps meteorologists track storms. Airplanes have specially designed radar to help pilots avoid storms or turbulent areas. Radar beams are concentrated electric pulses that shoot outward. These pulses are returned when they hit objects such as rain, hail, or even bugs. Radar screens show the location, velocity, shape, and sometimes size of the objects that the radar beams strike. The newest type of radar, Doppler radar, allows meteorologists to track precipitation within a storm and identify areas of rotation, which indicate tornadoes.

Weather satellites are also important weather forecasting tools. The first weather satellite was sent into orbit in 1960. Newer satellites carry instruments that map clouds, measure atmospheric moisture, determine wind speeds, and calculate ocean surface temperatures for the whole world, day and night. Because clouds outline storms, satellite images can show weather patterns across the world.

Computers have also helped more reliably forecast the weather. Accurate weather forecasts require observations from many satellites and other sources. Supercomputers can process thousands of observations from weather stations, aircraft, ships, and satellites from around the world every few hours. Developing technology will have the ability to process quadrillions of calculations every second. This wealth of data will enable computers to create accurate images of current weather conditions around the world.

TV meteorologist

NOTES

2. What information do weather balloons and radiosondes collect? (**Weather balloons calculate wind speeds and directions by tracking the position of the radiosondes they carry. Radiosondes measure the temperature, pressure, and humidity at different heights—into the atmosphere.**)
3. How is radar helpful to pilots? (**It helps pilots avoid storms or turbulent areas.**)
4. Which type of front has triangles and semicircles on the same side of the line? (**occluded front**)
5. What are isobars? (**Isobars are lines that connect points of equal atmospheric pressure.**)
6. If an *H* is in the middle of several isobars, what does it mean? (**The H marks the center of a high-pressure area.**)

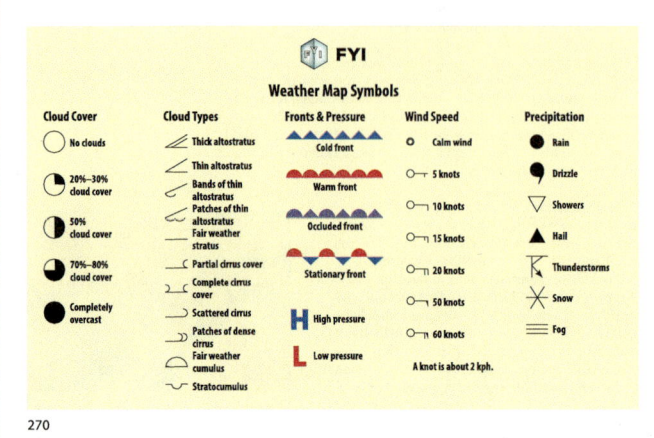

Weather

5.2.10 Controlling Weather

Introduction
Display **TM 5.2.10A Severe Weather**. Ask students what destruction they see. (**flooding, destroyed homes, crop damage, wrecked cars, snow-covered vehicles, downed power lines**) Discuss what caused the damage for each image. Have students imagine being able to control certain types of weather. Explain that scientists not only study the weather to be able to predict what type of weather is coming, but also to find ways to control the weather. Ask students why scientists want to control weather. (**Possible answers: to prevent severe weather, flooding, droughts, hurricanes, tornadoes**) What type of weather would you like to control or change? (**Answers will vary.**)

Discussion
- Read **1 Kings 18:16–46**. Discuss the struggle between Elijah and the prophets of Baal. This passage is the earliest account of humans taking action to change the weather. Discuss this passage in light of James 5:16b–18.

- Discuss the ethical nature of controlling the weather. Weather control can be used or misused. Point out that God gave people the ability to discover and understand the physical nature of weather. Ask what it would be like if scientists could completely control weather. (**Devastating storms could be avoided, but opinions about what weather is best would vary greatly. For example, nobody would agree on when it should rain, what the temperature should be, or if it should snow.**) Point out that setting up good conditions for one region might require drought in another area because scientists could only redirect the moisture that already exists, not create it. Weather control could even be misused as a weapon of war. Ask if weather is something that should be in God's hands only. (**Answers will vary.**)

- Compare cloud seeding methods. Explain how cloud seeding produces rain. Clouds are seeded with crystals of dry ice, or frozen carbon dioxide, to allow rain droplets to form. The tiny crystals attract water molecules, which eventually grow large enough to form rain or snow. Point out that other current methods of cloud seeding use chemicals. Ask students what problems exist with current cloud seeding methods. (**Possible answers: They are expensive; chemical cloud seeding may harm the environment.**) What other method of cloud seeding is being tried? (**Researchers are using laser beams.**) Explain that laser beams ionize air molecules. Polar water molecules are attracted to the charged particles and can condense on these particles. Ask what benefits there would be to using laser beams for cloud seeding. (**Possible answers: Laser beams do not add harmful chemicals to the environment; laser beams are easier to control than chemicals; laser beams are less expensive than chemical seeding.**)

- Ask the following questions:
  1. Does a shelterbelt control weather? (**No. It only prevents the existing weather from damaging crops. It does not actually prevent the wind from blowing.**)
  2. How did FIDO work? (**The heat from oil burners lowered the relative humidity and caused water droplets to evaporate. This cleared the fog so the pilots could see the runway to land.**)
  3. Aside from cloud seeding, what other research is being done with laser beams? (**Scientists are also studying the possibility of using laser beams to control the path of lightning strikes.**)
  4. Can scientists stop lightning? (**No.**) Do you think they should? Why? (**Answers will vary.**)

Activities
A. Complete *Try This: Making Snow* with students.

B. Divide the class into groups. Assign each group to design a device to remove fog. The device can be attached to a car or be designed for personal use. Guide them to use the science from FIDO for this invention. Have students draw or create a mock-up of the design and write a scientific explanation of how it works. Provide time for students to present their inventions to the class.

OBJECTIVES
Students will be able to
- explain how cloud seeding works.
- summarize the pros and cons of controlling weather.

VOCABULARY
- **shelterbelt** a barrier of trees or shrubs designed to protect crops from wind damage

MATERIALS
- TM 5.2.10A Severe Weather

PREPARATION
- Obtain materials for *Try This: Making Snow*. Sodium polyacrylate is the absorbent substance in diapers. Tear open diapers and remove this material.

TRY THIS
Making Snow
- sodium polyacrylate, 5 g per group
- clear plastic cups, 1 per group
(**No. Answers will vary.**)

© Earth and Space Science

NOTES

C. Assign students to respond to the following prompt by writing an essay or a skit: *What do you think would happen if weather could be controlled completely?*

D. Encourage students to research Seasonal Affected Disorder, a seasonal depression brought on by lack of sunlight. Have students explain how controlling weather conditions could help people with SAD.

Lesson Review

1. How do shelterbelts control the effects of weather? (**They divert the wind, which can erode soil and destroy crops.**)
2. What weather element was a problem for WW II aircraft in Britain? (**fog**) How was it controlled? (**Oil burners were used to clear fog from English airfields. The heat from the burners lowered the relative humidity and caused the water droplets to evaporate.**)
3. Explain the process of cloud seeding. (**Clouds are seeded with crystals of dry ice, or frozen carbon dioxide, to provide particles for rain droplet formation. The tiny crystals attract water molecules, which eventually grow large enough to form rain or snow.**)
4. Explain how laser beams may be used to control the weather. (**Someday, laser beams may be used to seed clouds and direct the path of lightning.**)
5. What are the benefits and dangers of controlling the weather? (**Some of the benefits of controlling the weather would be controlling the path of lightning, reducing the amount and size of hailstones, and bringing rain to areas experiencing drought. The dangers of controlling the weather are that it could damage the environment and the balance of Earth's systems, it could cause flooding or droughts, and it could cause political problems between nations.**)

5.2.10 Controlling Weather

OBJECTIVES
- Explain how cloud seeding works.
- Summarize the pros and cons of controlling weather.

VOCABULARY
- **shelterbelt** a barrier of trees or shrubs designed to protect crops from wind damage

The struggle between Elijah and the prophets of Baal is one Biblical account of people taking action to change the weather. Elijah asked God to bring fire from heaven and a rainstorm to prove His power over the lifeless god Baal.

Attempts to control the weather can be made because weather behaves in an orderly way. When looking at the methods that God uses to form weather, it is sometimes possible to control its effects. For example, for centuries farmers have grown **shelterbelts** to divert the wind, which can erode soil and destroy crops. But serious efforts to control the weather were not made until recently. During World War II, the thick British fog made safely landing planes very difficult. Winston Churchill initiated the development of a system to clear fog from runways so Allied aircraft could land. The Fog Investigations and Dispersal Operations (FIDO) was formed. FIDO used oil burners near the runways to clear fog from English airfields. The heat from the burners lowered the relative humidity and caused the water droplets to evaporate. This effort saved the lives of many Allied military personnel.

A breakthrough in weather control was made in 1946 when scientists discovered cloud seeding. Seeding clouds can reduce the amount or size of damaging hail, reduce the density of fog, or cause clouds to produce precipitation. In order to form rain droplets, water vapor needs particles, such as dust or salt. Clouds are seeded with crystals of dry ice, or frozen carbon dioxide, to provide particles for rain droplet formation. The tiny crystals attract water molecules, which eventually grow large enough to form rain or snow. Today, silver iodide crystals and calcium

Shelterbelt of trees

WWII aircraft

chloride are also used, depending on the type of clouds. However, scientists warn that chemical cloud seeding can harm the environment.

Cloud seeding has not always turned out well. In 1947, scientists associated with Project Cirrus tried to weaken hurricane winds by seeding the storm clouds. After the first hurricane seeding, the storm changed course, hitting Savannah, Georgia, and causing widespread damage. Weather modification experiments are still conducted today by countries around the world.

Scientists continue to conduct research on other methods of cloud seeding that are less expensive and less harmful to the environment. Jérôme Kasparian, physicist at the University of Geneva, Switzerland, is experimenting with the use of laser beams to create cloud condensation nuclei to encourage the formation of precipitation without adding harmful chemicals into the environment. Laser beams are also easier to control and less costly than chemical seeding procedures. The powerful laser beam bursts remove electrons from molecules in the atmosphere.

FYI
Suitable for Seeding
What makes a cloud suitable for seeding? There are three tests a cloud has to pass in order to qualify. The cloud must be able to sustain a large updraft of moist air. It must be free of ice because the presence of ice means the cloud does not need to be seeded. And the temperature within the cloud needs to be less than 0°C.

Lightning over city

Negev Desert, Israel

The newly charged particles then attract polar water molecules, which condense and form water droplets. Laser beams can also be used to direct electric discharges. Researchers have been able to control short distance electrical discharges and even redirect them around objects. This technology eventually may be used to redirect long distance discharges, like cloud to ground lightning.

Many scientists consider the benefits of controlling the path of lightning bolts, reducing the amount and size of hailstones, or bringing rain to areas experiencing drought worth pursuing. Lives could be saved, crops could be protected or watered, water supplies could be increased, and property damage could be reduced. Other scientists argue that controlling the weather is a dangerous attempt to play God that may damage the environment because Earth's systems are delicately balanced. Manipulating weather conditions could change the course of storms or cause flooding in some areas and droughts in others. Controlling the weather could also become political as nations work to provide their citizens with the weather they need at the expense of other countries. Though the outcome is uncertain, both the research and the debate will continue.

LESSON REVIEW
1. How do shelterbelts control the effects of the weather?
2. What weather element was a problem for WWII aircraft in Britain? How was it controlled?
3. Explain the process of cloud seeding.
4. Explain how laser beams may be used to control the weather.
5. What are the benefits and dangers of controlling the weather?

TRY THIS

Making Snow
It is your turn to control the weather. Put 5 g of sodium polyacrylate in a cup. Add 55 mL of water to the cup. Watch what happens. The powder has such a high concentration of sodium that it draws the water into the powder. This process is called *osmosis*. Did you really make snow? Could this process be used to make snow? Why?

5.3.0 Climate

LOOKING AHEAD

- For **Lesson 5.3.1**, invite a climatologist to speak to the class.

SUPPLEMENTAL MATERIALS

- TM 5.3.1A Cape Town Climograph
- TM 5.3.2A Köppen Classification System
- TM 5.3.3A Greenhouse Gas Emissions
- TM 5.3.3B Measuring Climate Change

- Lab 5.3.1A Earth's Axial Tilt
- Lab 5.3.2A Microclimates
- Lab 5.3.3A Greenhouse Gases

- WS 5.3.1A Rain Shadow Effect
- WS 5.3.1B La Paz Climate Analysis
- WS 5.3.2A Local Climate
- WS 5.3.3A My Contribution

- Chapter 5.3 Test
- Unit 5 Test

BLMs, TMs, and tests are available to download. See Understanding Purposeful Design Earth and Space Science at the front of this book for the web address.

Chapter 5.3 Summary

The earth's climate affects all people in many ways. Favorable climates allow humans to grow the food they need to survive, and these good climates sustain diverse ecosystems from which medicine, foods, and other products are harvested. Climates influence cultures and determine livelihoods. Because of this, climatologists are searching ice cores at the poles and analyzing fossils in order to understand how Earth's climate has changed in the past and to predict future changes. Even small variations in climate are meaningful events in God's delicately-balanced world. Some people consider this time in history to be uncertain with respect to climate. However, more study is required. Students need a solid understanding of how climates are created, how humans are affecting the global climate, and steps that can be taken to maintain the earth's climatic balance.

Background

Lesson 5.3.1 – Climate Factors

A region's climate depends primarily on latitude because the angle of the sun's rays and the global winds are easy to predict along various latitudes. Plotting average daily temperatures on a world map reveals zones that roughly follow the parallels of latitude. Near the equator, the average daily temperature is highest, and it is lowest at the poles. However, these zones do not occur in straight parallel lines. Other factors are also relevant in determining climate.

Proximity to land and water masses, altitude, topography, winds, ocean currents, and prevalence of cyclonic storms are important factors that contribute to the climate of a particular region. With the exception of regions near the equator, continental climates are characterized by dry, sunny weather, low humidity, and seasonal extremes in temperature. Death Valley in California, for example, boasts the highest temperature on record (56.7°C) and Antarctica the lowest (−89.2°C).

Marine climates have small annual and diurnal temperature variation but receive heavy rainfall on the windward side of coastal highlands and mountainous islands. Prevailing winds play a dominant role in coastal climates. Eastern coasts generally receive the heavier rainfall in the trade-wind belts, and the western coasts in westerly belts. The climate of both resembles continental climates when the wind is blowing from the interior of the continent.

Lesson 5.3.2 – Climate Regions

Climate regions are characterized largely by latitude or temperature differences. The tropical region can be found roughly between 0° latitude and 15° latitude in both the Northern and Southern Hemispheres. It spans the equator and is characterized by high temperatures and small seasonal and diurnal changes. The doldrums—a relatively calm area near the equator where warm air rises—are responsible for much of the climate characteristics found in tropical regions. This region receives the most direct sunlight and is hot year-round. Some areas are hot and wet; others are hot and dry; and some experience wet and dry seasons. Moist tropical regions often experience afternoon thunderstorms—the clouds build up in the presence of hot temperatures in the morning and release rain in the afternoon.

The moderate and continental regions have warm or hot summers and cool or cold winters. Here, there are trade winds—the winds that blow easterly from both 30° latitudes toward the equator, and horse latitudes—a relatively calm area located near both 30° latitudes where cool air sinks. Winds from the west bring precipitation year-round (especially to west coasts). Most of the precipitation falls in the winter. The higher latitude areas of this region, especially those far inland, experience winter snow. Regions closer to the tropics have hot, dry summers and wet winters.

Polar regions are located at latitudes of 60° to 90° in both the Northern and Southern Hemispheres where the sun is always low in the sky. These regions receive almost no sunlight in the winter. Temperatures are cold year-round. Winters are severe in the Arctic Circle. Summers are short

5.3.0 Climate

and mild with daily maximum temperatures of 15°C to 18°C. The lowest extreme temperature of −89.2°C was recorded in this part of the world. Antarctica stays below −50°C for more than half the year.

Lesson 5.3.3 – Climate Changes

Climates have changed throughout history. Long-term climate changes can be attributed to ice ages. Ice age cycles alter the amount and distribution of sunlight on the planet. The effect of the cycles can be measured by analyzing deep-sea core samples.

Many scientists have been interested in discovering what causes ice ages. One such scientist was Milutin Milankovitch, a Serbian astronomer, who attributed the cause of Ice Ages to Earth's tilt with regard to the sun and Earth's elliptical orbit. Summers are cooler when the earth is farthest from the sun and when its axis is less steeply tilted toward the sun. This will cause unmelted winter snow and ice to accumulate yearly that may trigger an ice age. Whether human-induced climate change will alter the naturally occurring climatic cycles is under scientific debate.

Some scientists believe that the presence of coal beds in North America and Europe along with evidence of glaciation in these same areas indicates that the continents experienced alternating warm and cold climates. Global climate is a complex system influenced mainly by natural forces such as the sun, the oceans, the atmosphere, topography, and plant and animal life. The complex interactions among these various spheres are difficult to predict.

Given that climate change occurs on long time scales, people assume the weather to which they are accustomed is normal. Despite yearly fluctuations of climatic elements, climate has changed very little during the period of recorded history.

Climate change is a natural event. However, since the Industrial Revolution, humans have been contributing to excess carbon dioxide and other greenhouse gases that are believed to have a significant impact on the climate. The effects of greenhouse gases are difficult to predict, even on a short-term level. Scientists agree that the ocean has been rising 2 mm a year for the past several decades, but most are unwilling to predict a sudden acceleration in the sea level. It is predicted that thermal expansion, the rising of seawater with increased temperatures, will raise the sea level 30 cm in the next century.

WORLDVIEW

- Climates may change, but God remains the same. We are reminded of this promise in Hebrews 13:8 and Lamentations 5:19. The Lord will reign forever. His throne will endure from generation to generation. God knows when the seasons will come and go, how global temperatures and precipitation will fluctuate, and how long His people are to remain on this Earth. One day, God will create a new heaven and a new Earth. We do not need to worry about anything. God has everything under control. But, as long as the earth endures, it should be treated as a precious gift (Genesis 8:22). This means that we must be good stewards of the world that has been placed in our hands to cherish and protect.

5.3.1 Climate Factors

Climate

OBJECTIVES

Students will be able to
- differentiate between weather and climate.
- relate factors that influence climate.

VOCABULARY

- **climate** the weather pattern an area has over a long period of time

MATERIALS

- Globe, flashlight (*Introduction*)
- TM 5.3.1A Cape Town Climograph
- WS 5.3.1A Rain Shadow Effect
- WS 5.3.1B La Paz Climate Analysis

PREPARATION

- Obtain a globe and a flashlight for the latitude activity. (*Introduction*)
- Gather materials for *Lab 5.3.1A Earth's Axial Tilt*.
- Invite a climatologist to speak to the class. Have students prepare questions to ask. (*F*)

TRY THIS

Changing Times
No additional materials are needed.

Introduction

Provide students with the following hypothetical situation: On March 15, Brussels' temperature was 11°C and the amount of rainfall that fell that day was 4 mm. Ask students if this scenario describes the weather of Brussels or the climate? (**the weather**) What is the reasoning behind your choice? (**The scenario only describes one day. To determine the climate of Brussels, the average temperatures and precipitation levels would need to be collected over a long period of time.**) Why do some people confuse weather and climate? (**Answers will vary but should include that both address solar radiation, temperature, humidity, wind speed, and precipitation.**)

Display a globe with a tilted axis and have a student shine a flashlight on the globe in line with the equator. Demonstrate how an acute angle of sunlight is created near the poles. Ask students what kind of an affect indirect sunlight that is spread over a larger area would have in latitudes far north or south. (**Indirect sunlight that is spread out would make latitudes near the poles feel colder.**) Discuss two regions that have the same latitude. For example, regions of Quebec, Canada, and London, England, both have a latitude of 51.51°N. However, their climates are very different. Have students speculate why two locations with the same latitude would have different climates. (**Possible answers: One region may have a higher elevation; it may be located near a mountain or a large body of water.**)

Discussion

- Draw a chart with two columns. Write *Geographic Features* above the first column and *Effect on Climate* above the second column. Have students list different geographic features under the first column. Some students may think geographic features include manufactured structures such as roads or buildings. Clarify any misconceptions. As a class, discuss the effects on climate and fill in the second column. Ask how elevation and the proximity of large bodies of water and mountains affect temperature and precipitation. (**Regions near large bodies of water tend to have more moderate temperatures and more precipitation. The windward side of mountains will have higher levels of precipitation, and the leeward side will have decreased precipitation. As the elevation increases, the average temperature decreases.**)

- Discuss how climate affects culture. Ask how people's lives are determined by the local climate. (**Answers will vary but should include climate determines types of housing accommodations, food choices, types of occupations, fashion choices, and recreational activities.**) Direct students to share how the local climate has affected their lives. (**Answers will vary.**)

- Ask the following questions:
  1. What is the main difference between weather and climate? (**time**)
  2. Can a region experience an unusual change in weather but retain the same climate? (**Yes.**) Ask students to provide examples of this idea. (**Answers will vary but should include the Sahara experiencing a temporary snowfall but retaining a desert status that has a hot, dry climate.**)
  3. Does wind produce a dry climate or a wet climate? (**Wind type determines if an area will have a dry or wet climate. Calm doldrums produce warm, low-pressure air that constantly rises and loses moisture near the equator. These areas have a wet climate. The wind moving into subtropical areas is usually dry and sinking. These areas have a dry climate. Warm air masses collide with cold air masses in mid-latitudes producing average amounts of precipitation.**)

Activities

Lab 5.3.1A Earth's Axial Tilt

- large sheets of paper, 4 per group
- flashlights, 1 per group
- metric rulers, 1 per group
- protractors, 1 per group

A. Direct students to perform *Try This: Changing Times*.

B. Complete **WS 5.3.1A Rain Shadow Effect** with students. Discuss their answers.

C. Display **TM 5.3.1A Cape Town Climograph**. Explain to students that a climograph is a double graph showing temperature and precipitation averages for a location. The four main factors that affect climates are latitude, elevation, proximity to large bodies of water or large mountain ranges, and whether the location is along the coast or inland. Ask students the following questions:
- What is the average precipitation each month? (**48 mm**)
- Are there any patterns observed? (**Winter months of March through October have increased precipitation and decreased temperatures.**)
- What is the range of precipitation? (**80 mm**)
- What is the average temperature? (**17°C**)
- Calculate the temperature range. (**10°C**)
- What are the hottest months? (**January and February**) What are the coldest months? (**June, July, and August**)
- Determine possible factors that create observed climate conditions. (**Cape Town, South Africa, is located along the coast of Africa on the westward side of a mountain range.**)

After students have answered the above questions, assign **WS 5.3.1B La Paz Climate Analysis**.

D. Have students create a Venn diagram to compare weather and climate by filling in the areas that are unique to climate and weather and how climate and weather are similar.

NOTES

5.3.1 Climate Factors

OBJECTIVES
- Differentiate between weather and climate.
- Relate factors that influence climate.

VOCABULARY
- **climate** the weather pattern an area has over a long period of time

What are the long-term weather conditions where you live? Whether you live in Norway or Nepal, Albania or Argentina, you have some idea of what kind of weather to expect in any given month. The weather tends to follow the same general pattern year after year. **Climate** is the weather pattern an area has over a long period of time.

Climate describes many factors, including average monthly precipitation, average high and low daily temperatures, average humidity, cloud cover during different seasons, wind speed, maximum wind gusts, and levels of solar radiation. In general, climate describes the average temperature range and average annual precipitation of a region. However, a study of climate also analyzes extreme fluctuations of these factors over long periods of time.

The climate in a specific area is determined by a combination of many factors. One of the most important factors is latitude, which is the area's distance from the equator. Latitude is measured in degrees. Areas closest to the equator, at a latitude of 0°, receive the most direct sunlight. Annual temperatures do not vary much from place to place in regions near the equator, but there is a wide variation of annual rainfall amounts. As a result, this region of the world is not described as having four seasons. A better way to describe the climate near the equator is by designating alternating wet and dry seasons.

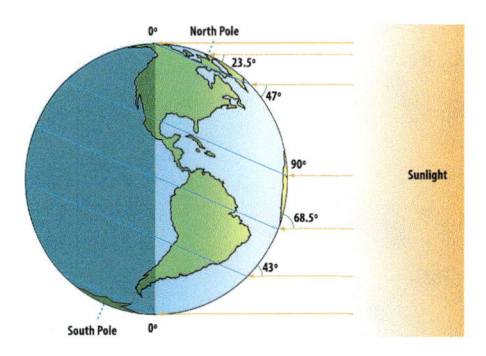

The sun strikes different latitudes at different angles. This illustration shows the sun's angle at the summer solstice.

Given that the earth is tilted on its axis, the sun's rays strike latitudes located closer to the poles at indirect angles. Also, the curvature of the earth forces the sun's rays to spread out over a larger distance. Much less sunlight reaches these latitudes, and the sunlight that does come in contact with the land is often reflected by ice. Indirect sunlight dispersion and energy reflection make these regions of the world relatively cold. The average daily temperatures near the poles are lower than the average daily temperatures near the equator. In these regions, the sun appears for only a few hours in the winter and never sets in the summer. The lowest average temperatures usually occur near the poles at a latitude of 90°, where the angle of the sun is the least direct.

Wind is another factor that affects climate. For example, the doldrums produce warm, low-pressure air that constantly rises and loses moisture. Areas in this latitude subsequently have heavy rains. The wind moving into subtropical areas is usually dry and sinking, so these areas have little precipitation. In middle latitudes, warm air masses collide with cold air masses. These areas have average amounts of precipitation.

The direction of the wind also affects an area's climate. For example, the Sahara—the world's largest desert—is one of the

Sahara Desert

NOTES

E. Direct students to contact the National Oceanic and Atmospheric Administration (NOAA) to inquire about the climate computer models they use to forecast future climate patterns. Have students select one model and create a report or computer presentation that illustrates the climate patterns that are predicted for the future.

F. Introduce a climatologist to speak to the class. Provide time for students to ask their questions.

Lesson Review

1. What is the difference between weather and climate? (**Weather is the daily variations of temperature, humidity, cloud cover, wind, and precipitation in the atmosphere. Climate is the pattern of weather an area has over a long period of time.**)
2. Why do regions at varying latitudes have different climates? (**The position of the sun relative to the earth, the earth's tilt, and the curvature of the earth are responsible for different climates at varying latitudes. Areas closest to the equator receive the most direct sunlight. Sun's rays strike latitudes located closer to the poles at indirect angles. The curvature of Earth forces the sun's rays to spread out over a large distance.**)
3. Describe a rain shadow effect. (**As air travels up the windward side of a mountain, the air mass becomes colder and condenses, releasing moisture in the form of precipitation. The dry, cold air passes over the mountain and travels down the leeward side of the mountain where it increases in temperature and absorbs ground moisture. The result is a rain shadow effect.**)
4. Explain how ocean currents affect the climate of coastal areas. (**Warm ocean currents transfer heat to the air above. Cold ocean currents absorb heat from the air above. The air is then transported across land that will change the temperature of coastal areas and precipitation.**)

Page 278

driest places on Earth, even though it is bordered on the west by the Atlantic Ocean. The trade winds that blow across the Sahara dissipate cloud cover. This allows more sunlight to penetrate and heat the surface of the earth. Even with such a large body of water nearby, the winds keep the Sahara very dry.

Topography also influences climate. Elevation and the presence of mountains or large bodies of water have a great impact on climate. Altitude—the height above sea level—is one important factor. When altitude increases, the average temperature decreases. As a general rule, for every 100-meter increase in altitude, the temperature drops 0.65°C. Even at the equator, low temperatures allow high mountain peaks to be covered with snow year-round.

Mountains also change the temperature and humidity levels of passing air masses. This creates different climates between the top of a mountain and nearby valleys. As air carried by wind travels up the windward side of the mountain, it cools with increasing elevation and much of the contained moisture condenses and falls in the form of precipitation. Increased precipitation occurs at the highest elevations. An increase in precipitation means that lush vegetation often grows on the windward side of a mountain. On the other side of the mountain, the leeward side, a rain shadow effect can be observed. The dry air that released most of the moisture content travels down the mountain. As it travels down the leeward side, the air temperature increases and any moisture on the ground is absorbed by the air mass. The result is an arid climate, often a desert. A great example of the rain shadow effect occurs over the Tibetan Plateau where moist air travels over the Himalayan Mountains, releasing all moisture content before reaching the dry plateau.

Ocean currents that can carry warm or cold water for thousands of kilometers also influence the climate of coastal regions. In general, currents carry warm water from the equator to the poles and bring cold water back toward the tropics. The air above the ocean water either releases or absorbs heat that is then transported over land. Without the circulation of ocean water, regions near the equator would be unbearably hot, and regions near the poles would be even colder. Currents regulate global climate and allow

TRY THIS

Changing Times
Design an experiment to show how the climate in an area may change. Include only one climate factor as your variable.

Rain shadow over the Andes Mountains

Page 279

for coastal regions to have minimal temperature fluctuations compared to inland areas. Typically, coastal regions are warmer in the winter and cooler in the summer compared to inland areas at similar latitudes. On the west coast of the United States, cold ocean currents produce cool air, which the wind blows to shore. As a result, California's coastline is cool, even though temperatures a few kilometers inland often reach temperatures above 38°C in the summer.

Ocean currents not only influence temperature, they also affect precipitation. Warm, moist air from warm ocean currents produces more precipitation. The ocean currents near Peru and the northern coast of South America affect rainfall in places as far away as Australia. Ocean currents also influence the development of storms, especially where cold polar air meets moist, warm air. Winter storms can quickly strengthen near the eastern coast of the United States and Canada when polar air masses collide with warmer air from the Gulf Stream.

LESSON REVIEW
1. What is the difference between weather and climate?
2. Why do regions at varying latitudes have different climates?
3. Describe a rain shadow effect.
4. Explain how ocean currents affect the climate of coastal areas.
5. What are five main factors that affect climate?

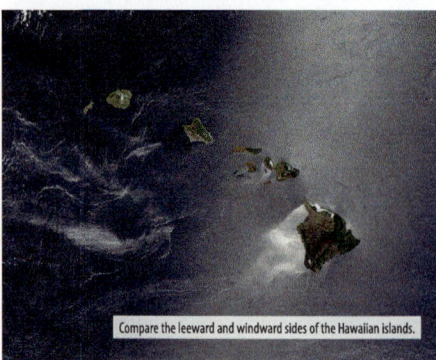

Compare the leeward and windward sides of the Hawaiian islands.

5. What are five main factors that affect climate? (**Earth's tilt, amount and direction of wind, topography, mountains, and ocean currents**)

NOTES

Name: _____ Date: _____

Lab 5.3.1A Earth's Axial Tilt

QUESTION: How does latitude determine climate?

HYPOTHESIS: Answers will vary.

EXPERIMENT:

You will need:	• flashlight	• protractor
• 4 large sheets of paper	• metric ruler	

Steps:
1. Tape one large sheet of paper to a flat surface.
2. Shine the flashlight in a vertical position, 20 cm above the piece of paper.
3. Trace the outline of the ellipse of light made by the flashlight. Label the sheet of paper *90 Degrees*.
4. Repeat Steps 1 through 3 with the flashlight positioned at a 70°, 45°, and 20° angle. Make sure that the flashlight remains 20 cm above the pieces of paper.

ANALYZE AND CONCLUDE:

1. What angle cast the smallest beam of light? 90°

2. What angle cast the largest area of light? 20°

3. Predict which angle would produce the most heat. Explain your prediction.
The 90° angle would produce the most heat. The same amount of heat is produced by the flashlight regardless of the angle. However, the impact is more intense if the heat is spread out over a smaller surface area.

4. How is this activity related to climate? Regions that are near the equator receive more direct sunlight in a localized area, similar to the flashlight positioned at a 90° angle. Portions of the earth that are located near the poles receive less intense sunlight because the axial tilt causes sunlight to be dispersed over a larger surface area. The result is a cooler climate because of the cooler temperature.

5. Why is the latitude of a region not a complete indicator of climate patterns?
Other factors including wind patterns and topography contribute to regional climates.

5.3.2 Climate Regions

Climate

OBJECTIVES

Students will be able to
- categorize climates according to the Köppen climate classification system.
- identify the main contributing factors for each climate region.
- distinguish between climate and microclimate.

VOCABULARY

- **microclimate** the unique climate conditions that exist over small areas of land within larger climate regions

MATERIALS

- TM 5.3.2A Köppen Classification System
- WS 5.3.2A Local Climate

PREPARATION

- Obtain materials for *Lab 5.3.2A Microclimates*.
- Contact another school in a different climate subtype or region to create open dialogue between students regarding climate conditions. (*B*)

Introduction

Review with students what climate is. Have students describe climate conditions that are common to the local area. Then have them brainstorm about the climate conditions in other larger regions that may be different from the local area. Ask students to describe any similarities and differences that they note. (**Answers will vary but should include temperature, precipitation, and vegetation patterns.**) List on the board the characteristics that students mention. Explain that some scientists divide the world into large regions that have similar average temperatures, average precipitation amounts, and natural vegetation patterns. These areas are called *climate regions*. Generally, there are five major climate regions: tropical, dry, moderate, continental, and polar.

Identify the climate regions characterized in the list students made. Fill in any climate regions that students have not mentioned. Ask students what makes Indonesia different from Egypt. (**Answers will vary.**) What makes Sweden different from Portugal? (**Answers will vary.**) Clarify that the boundaries of a country or state do not determine a climate region.

Discussion

- Display **TM 5.3.2A Köppen Classification System**. Ask students to determine the approximate percentage of land covered by dry climates. (**about 20%**) What is the most extensive climate region? (**tropical climate region**) What climate region is predominately in the Northern Hemisphere? (**continental climate region**)

- Discuss who may benefit from climate classification systems, other than scientists and students. (**Answers will vary but should include farmers, government agencies measuring population needs, and extreme winter sports enthusiasts.**)

- Ask the following questions:
  1. Why do scientists and other individuals classify climates? (**Similarities and differences between climates can be recognized using classification systems. This helps individuals study climates using a well-organized, systematic approach. Also, recognizing similarities among different climates gives individuals a better understanding of the subject matter.**)
  2. What type of climate is found 15° north and south of the equator? Describe the two main types of ecosystems that can be found in this climate. (**Tropical climates have rain forests and savannas.**)
  3. Explain why no two areas on Earth have exactly the same climate conditions. (**Climate conditions are determined by a variety of factors. Different areas have diverse combinations of these factors, which makes it impossible for two areas on Earth to have the exact same climate conditions.**)
  4. What are the defining characteristics of dry, moderate, and polar climates? (**Dry climates have high average temperatures because of a lack of air moisture. Moderate climates can have thunderstorms, tropical cyclones, frost, and tornadoes. Polar climates are determined by arctic air masses.**)
  5. Siberia has which type of climate—one that is characterized by annual temperatures that can increase or decrease by as much as 60°C? (**continental climate**)

Activities

Lab 5.3.2A Microclimates

- thermometers, 1 per group
- hand lenses, 1 per group
- metric rulers, 1 per group
- trowels, 1 per group
- paper towels
- soil color charts, 1 per group

A. Complete **WS 5.3.2A Local Climate** with students. Discuss student answers.

B. Contact another school in a different climate subtype or region. Have students communicate with students from the other school and share local weather conditions, interesting flora and fauna, and activities unique to the particular climates.

C. Have students identify possible drawbacks that exist in the Köppen classification system. (**The system does not include temperature extremes, precipitation intensity, or net radiation.**) Using what students know about climate factors, direct students to create a climate classification system. Instruct students to identify what is lacking in their proposed systems and specify how boundaries were determined. If the boundaries are not clear, distinct lines, students should suggest ways to remedy the issue. Students may want to consider using average climate conditions or include climate extremes.

D. Encourage students to design a graphic organizer showing all five climate regions. For each region, have students include the average amount of precipitation, the average temperature, and the predominant type of unaltered vegetation.

Lesson Review

1. What are the three factors that determine a climate region according to Köppen's climate classification system? (**average temperature, average precipitation, and natural vegetation patterns**)
2. What climate region has the largest annual temperature ranges? (**continental climates**)
3. Identify the three contributing factors that determine tropical climates. (**trade winds, the intertropical convergence zone, and monsoons**)

NOTES

5.3.2 Climate Regions

OBJECTIVES
- Categorize climates according to the Köppen climate classification system.
- Identify the main contributing factors for each climate region.
- Distinguish between climate and microclimate.

VOCABULARY
- **microclimate** the unique climate conditions that exist over small areas of land within larger climate regions

The study of different climates is important and fascinating. To help individuals examine climates more closely, scientists have devised various climate classification systems. It can be difficult to classify climate regions because no two locations have exactly the same climate conditions. Most of the classification systems attempt to group regions that have similar climate patterns and geographic features.

The most widely used classification system was derived by German climatologist Wladimir Köppen in 1900. Köppen grouped climates by measuring and observing average temperature, average precipitation, and natural vegetation patterns on each of the seven continents. His classification system is divided into five main climate regions—tropical, dry, moderate, continental, and polar. Each region has a letter designation of A, B, C, D, or E. The five main regions are further divided into a number of subtypes to include humid subtropical, mediterranean, and continental subarctic.

Tropical climates, region A, can be found at low latitudes on either side of the equator, roughly between 0° latitude and 15° latitude in both the Northern and Southern Hemispheres. Intense, constant sunlight is indicative of this region. As a result, high temperatures occur each month, usually above 18°C, and there are no distinct seasons. The warm air produces high humidity levels and heavy precipitation. True selva, canopy rain forest, and jungle-like vegetation flourishes in this climate region.

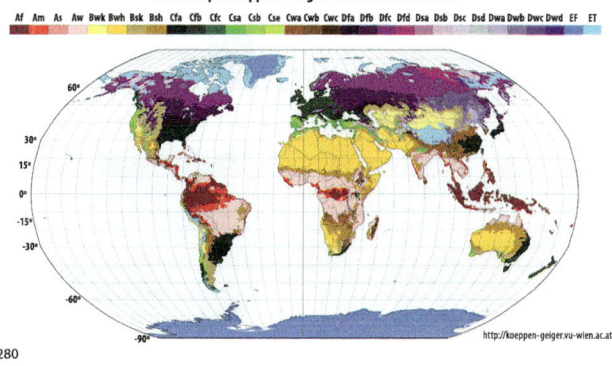

Tropical climate conditions are determined mainly by fluctuating trade winds during warm and cool months, the intertropical convergence zone, and the monsoon wind system in eastern Asia. Trade winds persistently blow from east to west toward the Equator. As the winds encounter the coastlines of tropical regions that are often lined with mountain ranges, precipitation occurs. Tropical disturbances, including tropical cyclones, can also be carried in the trade winds that result in excess moisture production. The intertropical convergence zone is a band that wraps around the earth close to the equator. In this zone, trade winds converge and air ascends. Widespread cloud cover, frequent thunderstorms, and heavy rainfall occurs in this zone. As the intertropical convergence zone moves away from a particular area, the active, warm air is replaced by stable, dry air. In some tropical climate subtypes, the effects of alternating monsoon wind systems are noticeable. During the warmer months of the year, warm, moist, maritime tropical air moves in one direction across the land to produce precipitation. The winds change direction during the cooler months of the year, bringing cool, dry air over the land.

Intertropical convergence zone

There are three tropical climate subtypes. These subtypes include wet equatorial climate, tropical monsoon and trade-wind littoral climate, and tropical wet-dry climate. Rain forests are common in the tropical regions. Most rain forest climates are found in Central Africa, the Amazon Basin, and Southeast Asia. Tropical climates also include savannas, which have very wet, warm months and very dry, cool months.

A – Tropical
Cayenne, French Guiana

Dry climates, region B, are classified as arid or semiarid. This region can be found mostly between 50°N and 50°S, mainly in the 15°–30° latitude band in both the Northern and Southern Hemispheres. Low precipitation, great annual precipitation variability, low relative humidity, high evaporation rates, little cloud cover, and intense sunlight are characteristics of dry climates. On average, dry climates receive

World Map of Köppen–Geiger Climate Classification

http://koeppen-geiger.vu-wien.ac.at

NOTES

4. How does the intertropical convergence zone affect land masses in tropical regions? (**In the intertropical convergence zone, trade winds converge and air ascends. This produces active, warm air with widespread cloud cover, frequent thunderstorms, and heavy rainfall. Stable, dry air moves in as the intertropical convergence zone moves away.**)
5. What is the difference between climate and microclimates? (**Climate is the pattern of weather over a large area for a long period of time. Microclimates are unique climate conditions that exist over small areas of land within larger climate regions.**)
6. Compare and contrast dry and polar climate regions. (**Dry and polar regions receive very little precipitation. Polar regions are permanently covered with snow and ice and temperatures are mostly below 10°C. Dry regions have low relative humidity, high evaporation rates, little cloud cover, and intense sunlight. Annual rainfall in dry regions ranges from 0 cm to 25 cm. Average temperatures in dry regions range from 21°C to 32°C.**)

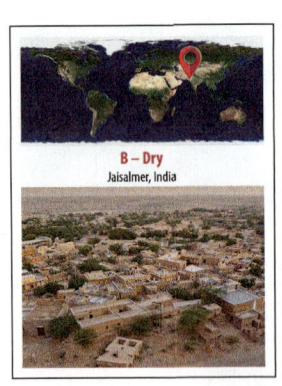

B – Dry
Jaisalmer, India

0–25 cm of rainfall each year. However, average values have little meaning because one year 25 cm of rain may fall and in another year there may be no rainfall. Some researchers claim that portions of the Atacama Desert in Chile have never received any rain.

Dry climates are surrounded by dry air that creates high average temperatures. Average monthly temperatures range between 21°C and 32°C. Temperatures vary greatly each day with an increase or decrease of up to 35°C. The highest air temperatures on Earth have been recorded in dry climates. As of 2016, Death Valley, California held the world record for the highest air temperature of 56.7°C. The actual surface temperatures in this area can reach 94°C. There is very little vegetation in dry climates. Vegetation must be hardy and drought-tolerant with thick bark, no leaves, and large water storage capabilities to be able to survive in such harsh conditions. Examples of dry climate vegetation include cacti, Joshua trees, and creosote bushes.

The three dry climate subtypes are tropical and subtropical desert climate, mid-latitude steppe and desert climate, and tropical and subtropical steppe climate. Tropical and subtropical desert climates are the driest places on Earth. Many of these regions are located far into the interior of continents where moisture-bearing winds are not present. Also, high mountains along the western coasts block the winds, so rain falls on the windward sides of the mountains, leaving the leeward sides dry.

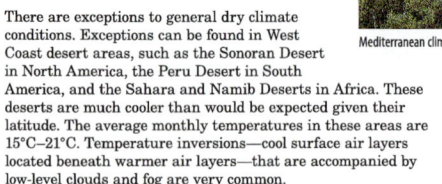

There are exceptions to general dry climate conditions. Exceptions can be found in West Coast desert areas, such as the Sonoran Desert in North America, the Peru Desert in South America, and the Sahara and Namib Deserts in Africa. These deserts are much cooler than would be expected given their latitude. The average monthly temperatures in these areas are 15°C–21°C. Temperature inversions—cool surface air layers located beneath warmer air layers—that are accompanied by low-level clouds and fog are very common.

Moderate and continental climates, regions C and D, are located between 25° and 70° in both the Northern and Southern Hemispheres. A majority of the climate conditions and variations in seasons observed in these two climate regions are determined by the location and intensity of the Westerlies. During the summer months, the polar front and jet stream move toward the poles, and warm, moist tropical air masses extend to higher latitudes. In winter months, the wind pattern circles toward the equator, the tropical air masses move back to lower latitudes, and cold polar outbreaks infiltrate these two climates. The majority of vegetation that grows in these regions includes evergreen and deciduous trees, bushes, shrubs, and grasses.

Moderate climate subtypes include humid subtropical climates, Mediterranean climates, and marine west coast climates. Climate conditions vary among subtypes. It is common for different moderate climate subtypes to have severe thunderstorms, tropical cyclones, frost, tornadoes, and strong monsoonal wind gusts. Summers can be moderate to hot. Winters are usually mild. The average temperatures vary by subtype. Some subtypes have an average

Sonoran Desert

Mediterranean climate

C – Moderate
Joinville, Brazil

282 283

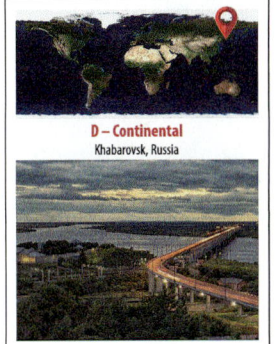

D – Continental
Khabarovsk, Russia

temperature of 27°C during the summer with a maximum value between 30°C and 38°C. In other subtypes, the temperature rarely exceeds 20°C during the summer. The average temperature during winter months is between 5°C and 13°C. Precipitation also varies from one subtype to another. Many portions of this climate region have evenly distributed precipitation throughout the year. More precipitation is typically produced in the summer. Annual precipitation totals vary from 50 cm to 250 cm of rain and can even reach totals of over 500 cm of rain. A smaller percentage of subtypes have a reduced annual precipitation total, usually 35–90 cm of rain. Areas located far into the interior of the land have even less precipitation and opposite seasonal patterns—dry summers and rainy winters.

The largest annual temperature ranges on Earth occur in continental climates. Inland temperatures can change by 30°C during different seasons. An extreme temperature range occurs in central Siberia where temperatures can increase or decrease by as much as 60°C during different times of the year. Coastal areas have more balanced temperature ranges. Continental climate summers are relatively hot. Some interior regions reach temperatures near 25°C. Winters are cold with average temperatures remaining below freezing for several months. Different subtypes have longer winters than others. It is common to have average temperatures of –40°C to –50°C at higher latitudes.

Annual precipitation totals vary across the humid continental climate subtype and the continental subarctic climate subtype. Precipitation totals for humid continental climates range from 50 cm to 125 cm. Most of the precipitation that falls on continental subarctic climate regions occurs during summer months, and totals are usually less than 50 cm. With extremely low temperatures during the winter months, snow tends to remain on the ground for the majority of the year. Summers are generally short and mild.

Weather is subject to change in lower Northern Hemisphere latitudes because the region is located between polar and tropical air masses. These land masses can experience severe thunderstorms, tornadoes, and high winds that can create blizzard conditions. The higher latitudes are dominated by more continental polar air that results in short, clear days with low humidity.

Polar climates, region E, are influenced by arctic and polar air masses at latitudes of 60°–90° in both the Northern and Southern Hemispheres. These regions are significantly cold with average temperatures generally remaining below 10°C. The highest latitudes never have an average monthly temperature

Siberia

FYI

Urban Climate

In metropolitan areas, climate conditions are significantly different from the surrounding rural communities. This is because of the installation of tall buildings and pavement that affect wind flow, precipitation runoff, and the overall energy balance. Asphalt, specifically, absorbs, stores, and redirects more solar energy than vegetation and soil in rural areas. Also, high pollutant concentrations have a considerable impact on temperature, visibility, and precipitation in urban areas. Sometimes, weather patterns can allow for the buildup of pollutants to remain stagnant over a city for days. This creates a temperature inversion where air temperature increases with increasing elevation. Excess pollutants can cause acute stress and even death. Londoners experienced the effects of too much pollution in December 1952 where 3,500 people died from respiratory issues.

The centers of cities are especially warm. They are known as *heat islands*. It is common for the daily minimum temperature in these heat islands to be 6°C to 11°C warmer than surrounding rural areas. At night, the city remains warm because the stored heat radiation is gradually released. The average relative humidity in these areas is usually several percentages lower than their rural counterparts because cities lack the moisture released by plants, and runoff increases without soil absorption. Average wind speeds are 20% to 30% lower than the country because of increased frictional drag between city buildings and low-level wind meetings. Atmosphere particles hovering over cities reduce solar radiation penetration and often cause water vapor to condense in the form of fog. Precipitation has also been estimated to be higher in urban environments.

Temperature inversion over Dubai

E – Polar
Kulusuk, Greenland

that exceeds 0°C. It is common for polar climates to remain below freezing for nine months of the year. These regions are permanently covered with snow and ice. The cold air does not contain very much moisture, so this region also receives very little precipitation. Interestingly, polar climates receive more precipitation than desert climates.

There are two polar subtypes—tundra climate and snow and ice climate. The Arctic Circle has a tundra climate. Winters are severe in this part of the world, and the short tundra summer temperatures are mild with a daily maximum of 15°C to 18°C. The lowest extreme temperature of –89.2°C was recorded at the Vostok II research station in Antarctica. Lower latitudes in this climate region have few trees and it is covered in lichen, moss, algae, and rocky areas without vegetation. Some grasses and low shrubs can also survive in portions of the polar climate.

There are small climate regions that were not included in the original Köppen classification system. These regions are not easily categorized and account for the highland areas that are scattered throughout the other climate regions. In general, higher elevations result in a decrease in pressure, temperature, and atmospheric humidity. Highland areas have similar annual temperature ranges and precipitation amounts compared to the surrounding lowlands. They are usually slightly cooler and receive somewhat more precipitation. Sequoias and bristle cone pines thrive in highland climates until they reach the tree line, where trees cease to exist.

Not all climate regions are uniform. Unique climate conditions that exist over small areas of land within larger climate regions are called **microclimates**. A few examples of microclimates include gardens, valleys, and parks. These regions are usually located within a few meters above Earth's surface and are surrounded by vegetation. Microclimate conditions are determined by temperature, wind, humidity, and solar radiation near the ground surface. Changes in topography can create microclimates. Razor-backed mountains on the Hawaiian Islands separate lush rainforests from dry savannas. Large bodies of water, such as oceans and the Great Lakes of North America, moderate the weather conditions on the shore. The temperature contrast between the cold air moving over the Great Lakes and the warmer water temperature in the Great Lakes often causes heavy snow squalls downwind of the Great Lakes. This snowfall is called *lake effect snow*.

Some plant and animal species may have specific living conditions that are well-suited to microclimates. For example, some species of ants rely on specific microclimate conditions to regulate their body temperature. Various microclimates located within close proximity to one another promotes beneficial biodiversity.

LESSON REVIEW
1. What are the three factors that determine a climate region according to Köppen's climate classification system?
2. What climate region has the largest annual temperature ranges?
3. Identify the three contributing factors that determine tropical climates.
4. How does the intertropical convergence zone affect land masses in tropical regions?
5. What is the difference between climate and microclimates?
6. Compare and contrast dry and polar climate regions.

Antarctica

BIBLE CONNECTION

The Negev

Moses described the popular seasonal migrations to the Negev in Genesis 20. For thousands of years, people have traveled with their flocks and herds to the southern region of Israel located between the hill country of Judah to the north and the deserts of Zin, Shur, and Paran to the south. Abraham, Isaac, and Jacob grazed their livestock in the semiarid Negev during the winter and migrated north to Judah, located near Bethel and Shechem, for the summer months. Open, rugged and sparsely populated, the region supports scrub brush but no forests. It has two seasons: a mild winter with periodic rains and a hot, dry summer. Given that less than 200 mm of rain falls annually in the Negev, the area is unsuitable for farming.

Name: _____ Date: _____

Lab 5.3.2A Microclimates

QUESTION: What determines a microclimate?
HYPOTHESIS: Answers will vary.

EXPERIMENT:

You will need:	• metric ruler	• soil color chart
• thermometer	• trowel	
• hand lens	• paper towels	

Steps:
1. Select a small location on the school grounds that has unique climate conditions compared to the surrounding area. Draw a map of the school grounds and mark the particular location you will be examining.

Drawings will vary.

2. Provide details regarding the weather conditions on the day of experimentation. Is there any cloud cover, strong winds, or high humidity levels?
   Answers will vary.

3. Using a thermometer, measure the air temperature in the small location. Make sure the thermometer is not in direct sunlight as this will give an inaccurate air temperature measurement. After five minutes has elapsed, record the thermometer reading.
   Air temperature: Answers will vary.

4. Apply a dry paper towel to the soil surface. Classify the soil surface as extremely moist, somewhat moist, dry, or extremely dry. Answers will vary.

Lab 5.3.2A Microclimates

5. Dig a small 15 cm hole and apply another dry paper towel to the soil below ground. Classify the soil found below ground as extremely moist, somewhat moist, dry, or extremely dry. Answers will vary.

6. Note the color of the soil using the soil color chart, if available.
   Answers will vary.

7. Describe the vegetation in the small location.
   Answers will vary.

8. Measure and record the height of the tallest, unaltered plant. If the tallest, unaltered plant is a tree, estimate the height of the tree in meters.
   Plant height: Answers will vary. cm

9. Repeat Steps 2 through 6 over a period of four days and record data in the table.

	Weather Conditions	Air Temperature (°C)	Soil Surface Moisture Content	Below Ground Soil Moisture Content	Soil Color
Day 2					
Day 3		Answers will vary.			
Day 4					
Day 5					

ANALYZE AND CONCLUDE:
1. Research the daily temperatures and precipitation values for the surrounding climate and record in the table.

	Air Temperature (°C)	Amount of Precipitation (cm)
Day 2		
Day 3	Answers will vary.	
Day 4		
Day 5		

Name: _____ Date: _____

Lab 5.3.2A Microclimates continued

2. How does the microclimate compare to the surrounding overall climate?
   Answers will vary.

3. Would the observed conditions be the same during a different season? Explain.
   Answers will vary but should include that conditions differ with varying seasons (such as the monsoon season), the position of the earth relative to the sun, and reversal of trade winds.

4. If anomalies are observed in the microclimate, what is responsible for the unusual conditions? Answers will vary.

5. Compare air temperature values with other groups and graph the results.

Answers will vary.

(Temperature vs. Location 1, Location 2, Location 3, Location 4)

Climate

5.3.3 Climate Changes

Introduction
Have students define weather and climate and explain how the two are different. Remind students why some people may confuse the two topics. Challenge students to think about why the idea of climate change may not seem important to individuals who use the terms *weather* and *climate* synonymously. (**Individuals who equate weather and climate may not think climate change is important because weather is constantly changing.**)

Discuss general short-term and long-term changes. Ask students to provide decisions that people make that have short-term effects. (**Possible answers: types of food to eat, where they want to park their car, what line to select at the supermarket**) What decisions can people make that have long-term effects? (**Possible answers: People may choose to attend a university or trade school that will affect their long-term future; they make choices regarding how they treat their body, such as ingesting excessive sugar or overusing drugs.**)

Ask students if they think most of people's decisions regarding climate will have short-term or long-term effects. (**Answers will vary but should include that most decisions have short-term effects.**) It is important to mention that some decisions that appear to have short-term effects may lead to long-term effects. One example is the emission of excess carbon dioxide and other greenhouse gases that may not produce a noticeable temperature change yet, but may lead to a significant global temperature change in the future that could affect crop production, the survival of vulnerable plant and animal species, and the global economy.

Discussion
- Have students share what they know about climate change. Ask students if their families have made any personal changes in response to what they have learned about climate change. (**Answers will vary.**) Do they think climate change is an important issue to address? (**Answers will vary.**) If an issue does not immediately affect someone, should it be ignored? (**Answers will vary.**)

- Have students name greenhouse gases. Ask students to predict which gases are emitted in the largest quantity as a result of people's activities. (**carbon dioxide, methane, nitrous oxide, and fluorinated gases**) Predict which economic sector contributes the most greenhouse gas emissions. (**electricity and heat production**) What countries do you think are responsible for a majority of the greenhouse gas emissions? (**China, United States, European Union, India, Russian Federation, and Japan**) Display **TM 5.3.3A Greenhouse Gas Emissions**.

- Display **TM 5.3.3B Measuring Climate Change** and discuss with students various ways that scientists attempt to measure climate change.

- Share with students that God provided the earth for humans to dwell in and enjoy. This provision came with a great responsibility to maintain the earth. Some individuals have placed a higher value on the environment than on the people who dwell in it. Ask students if this choice is in agreement with the Bible. (**No.**) Some environmental activists greatly criticize the Industrial Revolution for having a negative impact on global climates. Ask students how the Industrial Revolution has improved the quality of human lives. (**The Industrial Revolution has improved methods of transportation, communication, agriculture, and manufacturing.**) Some political activists are trying to regulate carbon dioxide production. Ask students if they think personal liberties are at risk with such regulations. (**Answers will vary.**) Who will be impacted the most? (**Answers will vary but should include the poor, very young, and elderly.**) In what ways would individuals be impacted the most? (**Possible answers: increased food production costs, and more expensive methods of powering vehicles and heating and cooling homes**) How would you solve this dilemma? (**Answers will vary but should include implementing more cost-effective choices such as reducing waste, recycling, and driving low-cost, fuel-efficient vehicles.**)

OBJECTIVES
Students will be able to
- summarize short-term and long-term climate change.
- examine theories that attempt to explain what causes ice ages.
- outline natural processes and human activities that contribute to climate change.

VOCABULARY
- **climate change** any long-term change in Earth's climate
- **El Niño** periodic changes in oceanic and atmospheric conditions in the Pacific Ocean that cause unusually warm surface water
- **greenhouse gases** a portion of atmospheric gas molecules that deflects infrared radiation back to Earth's surface that was initially on a path to escape into space
- **ice age** a period of time when ice collects in high latitudes and moves toward lower latitudes
- **interglacial period** a warm period that occurs between glacial periods when large ice sheets are absent

MATERIALS
- TM 5.3.3A Greenhouse Gas Emissions
- TM 5.3.3B Measuring Climate Change
- WS 5.3.3A My Contribution

PREPARATION
- Gather materials for *Lab 5.3.3A Greenhouse Gases*.
- Obtain materials for *Try This: Ocean CO₂*.

TRY THIS

Ocean CO₂
- jars, 1 per student
- cabbage water
- straws, 1 per student

Place 2–3 cups of chopped cabbage in a pot. Cover with water. Bring the water to
(continued)

ⓘ TRY THIS

(continued from previous page)
a boil. Allow water to cool for 30 minutes. Pour cool water through a strainer. **(Oceans become more acidic with the addition of CO_2.)**

- Ask the following questions:
  1. How do climatologists measure climate change? **(Climatologists make inferences from indirect data collected from the measurement of tree rings, ice cores, and geologic patterns.)**
  2. What is the most recent observation made by scientists regarding Earth's temperature? **(Earth's temperature appears to be increasing at a much faster rate than previously recorded.)**
  3. Would the inhabitants of Earth benefit from the elimination of all greenhouse gases? Why? **(A majority of the inhabitants of Earth would not benefit from the elimination of all greenhouse gases because without any greenhouse gases, the average temperature of Earth's surface would be about −18°C. The extremely low temperature would impact all areas of life.)**
  4. What human activities contribute to the production of greenhouse gases other than the operation of vehicles? **(Answers will vary but should include deforestation and the use of fossil fuels to provide power and energy to homes and businesses.)**

Activities

Lab 5.3.3A Greenhouse Gases

- thermometers, 3 per group
- glass jars, 2 per group
- paper towels
- sun lamps, 1 per group (optional)

If this laboratory experiment is performed on a cloudy day that lacks adequate sunshine, a sun lamp may be used.

5.3.3 Climate Changes

OBJECTIVES
- Summarize short-term and long-term climate change.
- Examine theories that attempt to explain what causes ice ages.
- Outline natural processes and human activities that contribute to climate change.

VOCABULARY
- **climate change** any long-term change in Earth's climate
- **El Niño** periodic changes in oceanic and atmospheric conditions in the Pacific Ocean that cause unusually warm surface water
- **greenhouse gases** a portion of atmospheric gas molecules that deflects infrared radiation back to Earth's surface that was initially on a path to escape into space
- **ice age** a period of time when ice collects in high latitudes and moves toward lower latitudes
- **interglacial period** a warm period that occurs between glacial periods when large ice sheets are absent

The weather changes all the time. It can be hot and dry one day and cool and rainy the next. Sometimes the weather may even change within a day. Changes in the weather are obvious. Climate changes are much less noticeable given that they occur over a longer period of time. Although climates change very slowly, they do change. It has been determined that for most of history climate change resulted from natural processes. Today, many scientists have come to the conclusion that human activity is also influencing climate change.

Climate change is the term currently used to describe any long-term change in Earth's climate. Climate change not only refers to the increase in Earth's temperature, but also changes in precipitation and wind patterns. Climatologists attempt to measure climate change using fragmented and indirect evidence such as tree rings, ice cores, and geologic patterns. Most observations and measurements have been collected beginning in the 19th century. Few geologic records exist prior to the late 18th century. Predictions for climate conditions during this period of time are largely hypothetical.

Changes in climate have both short-term and long-term effects. An example of a short-term change occurs as a result of pollutants, ash, or dust in the atmosphere. Air pollutants gather over industrial areas, block sunlight penetration, cool

FYI

Number of Ice Ages

Prior to the 1970s, many old-earth scientists originally thought there had been four ice ages in Earth's past. That belief was rejected and replaced with the idea that there have been thirty or more ice ages. In the past 800,000 years, old-earth scientists speculate that there have been eight ice ages that each lasted 100,000 years. However, these scientists have difficulty explaining any recent ice ages based on the temperature and precipitation rates observed today. Young-earth scientists attempt to reconcile this issue by proposing that only one ice age occurred and that it followed the biblical Flood. A worldwide Flood would have caused major changes in the earth's crust, earth movements, and tremendous volcano activity that would have greatly disturbed the climate. The shroud of volcanic dust and aerosols that would have been produced during this time would have been trapped in the stratosphere for several years following the Flood. That dust could have reflected some of the solar radiation back into space and caused cooler summers.

regions, and cause greater precipitation. Ash and dust from volcanic eruptions can have the same effect as pollutants. Major eruptions have been known to induce freezing climates because the dust and the gases block solar energy from reaching the earth's surface. Volcanic clouds cause temperatures to drop. The greatest volcanic eruption on record happened on the island of Sumbawa in 1815. The following year became known as *the year without a summer*. Parts of Europe and the New England area of the United States experienced heavy snow in June and frost in July and August. In 1991, Mount Pinatubo in the Philippines released a haze of sulfur dioxide gas that spread throughout much of the world. The two summers following the eruption were cooler than usual.

Short-term climate change induced by volcanic plumes of ash and dust

Another short-term climate change, known as **El Niño**, results from periodic changes in oceanic and atmospheric conditions in the Pacific Ocean that cause unusually warm surface water. During El Niño years, easterly trade winds in the Pacific Ocean decrease or change direction, allowing warm water to flow eastward. The shift in wind patterns brings drought to parts of Africa and Australia and extraordinary typhoons to Polynesia. California and the southern portion of the United States usually suffer severe winter storms during El Niño conditions. Canada and other parts of the United States usually experience warmer winters with less precipitation than usual.

Long-term climate change often has more dramatic results. Indirect geologic evidence indicates that during certain periods the earth's climate was much colder than it is now. At times, much of the earth was covered with sheets of ice. An **ice age** is a period of time when ice collects in high latitudes and moves toward lower latitudes. Ice ages have periods of cold and periods of warmth. During cool glacial periods, the sheets of ice advance and cover a larger area. Sea level drops because a large amount of ocean water freezes. **Interglacial periods** are warm periods that occur between glacial periods when large ice sheets are absent. During these times, some of the ice melts and the sea level rises.

Scientists have several theories on why ice ages occur, even though they may disagree on the number of ice ages that have occurred. In the 1920s, Serbian astronomer Milutin Milankovitch linked ice ages to changes in the earth's orbit and

A. Complete *Try This: Ocean CO₂* with students. Remind students that cabbage water is a pH indicator. Review with students that cabbage water is red in the presence of acids and blue-green in the presence of bases.

B. Direct students to complete **WS 5.3.3A My Contribution.** Discuss student answers.

C. Guide students to conduct a school survey regarding current views of climate change. Have students share their findings with the class. Discuss with students any ideas that appear to have been formulated with a lack of knowledge. Direct students to choose one idea and write an essay providing more detailed information. Have students share their essays with original survey participants and conduct the survey again. Encourage students to share if any of the original survey answers changed.

D. Instruct students to research the top five ranking local facilities for all greenhouse gases and the top-ranking local facilities for carbon dioxide. Have students report any efforts being made by the facilities to reduce greenhouse gas emissions.

E. Encourage students to become plant and animal advocates. Assign students one of the following plants or animals to research that are specially adapted to a particular climate: Quiver tree, Orange-spotted filefish, Polar bear, Adélie penguin, North Atlantic cod, Staghorn coral, and the Golden toad. Have students determine the major challenges and consequences that these plants and animals face when climates change even slightly. Some challenges may include reduced food sources, temperature sensitivities, and habitat loss.

NOTES

TRY THIS

Ocean CO₂
Fill a jar halfway with cabbage water. Observe the color. This water represents the ocean. Place a straw in the water and gently blow without causing any water to spill out of the jar. Note the color change. How is increased atmospheric carbon dioxide affecting oceans?

the tilt of the earth's axis. According to his theory, the amount of solar radiation that reaches the earth varies with changes in the earth's orbit. Over an estimated cycle of 100,000 years, Earth's orbit becomes less circular and more elliptical. When the orbit is more elliptical, seasons are more severe—summers are hotter and winters are colder. Seasonal extremes occur less during more circular orbits. In addition, over a 41,000-year period, scientists hypothesize that the earth's tilt on its axis varies between 22.1° and 24.5°. The variation in the earth's tilt could affect seasonal differences. Winters would have more precipitation and warmer temperatures. Summers would have less precipitation and cooler temperatures. Snow in high latitudes would not melt in the summer because temperatures would be cooler than average, which could result in increased glacial formation. According to the Milankovitch theory, scientists predict that the next ice age will occur in another 50,000 to 100,000 years.

Some theories attribute climate change to solar activity and the movement of plate tectonics. Observations and measurements have indicated that the sun has been increasing in brightness over time. This is important to understand because the sun contributes heat energy to the earth. An increase in heat energy would lead to an increase in the atmospheric temperature on Earth. Also, sunspots appear and disappear over time. Some scientists relate the reduced sunspot activity from 1645 to 1715 with the theorized Little Ice Age. During the Little Ice Age, Europe and the North Atlantic region felt the greatest

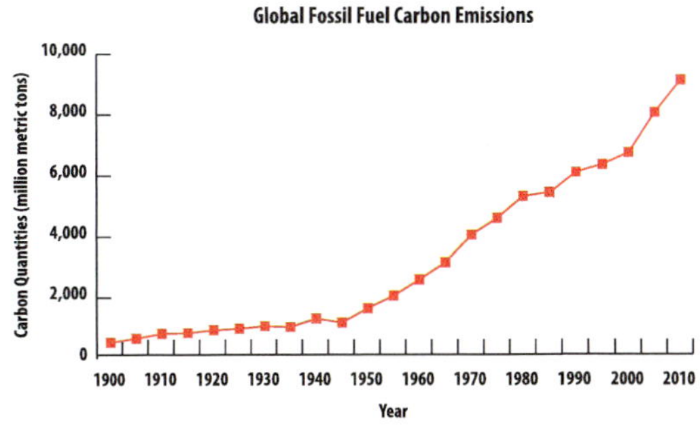

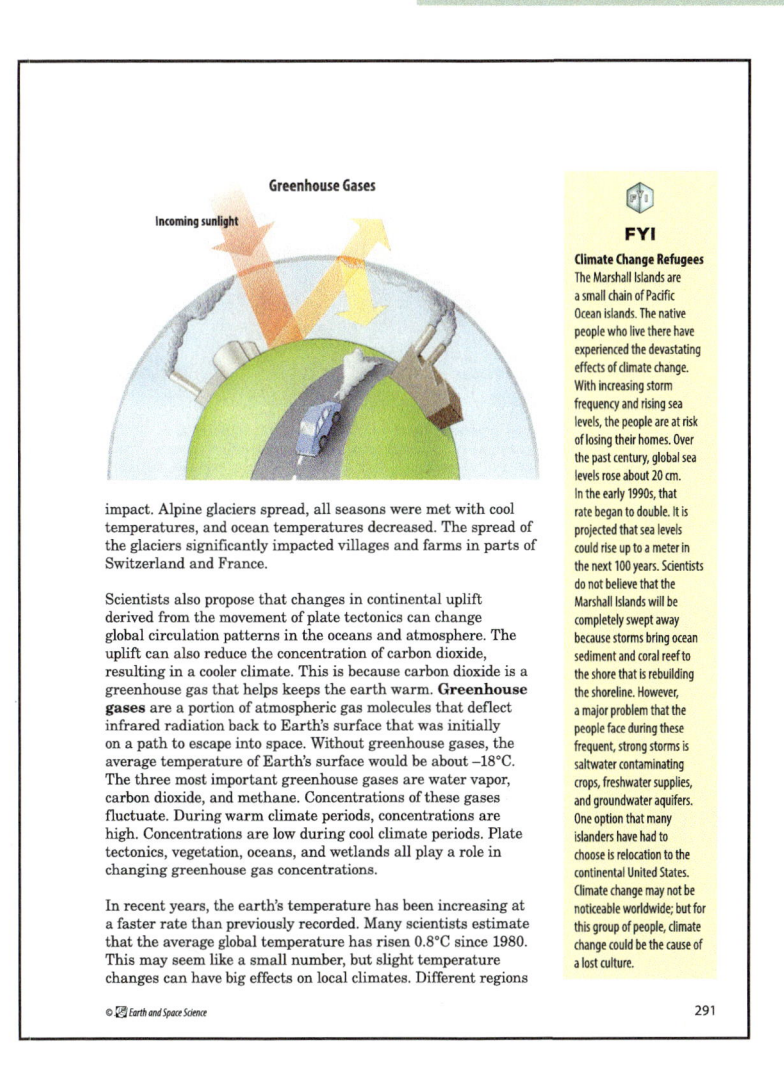

Global Fossil Fuel Carbon Emissions

impact. Alpine glaciers spread, all seasons were met with cool temperatures, and ocean temperatures decreased. The spread of the glaciers significantly impacted villages and farms in parts of Switzerland and France.

Scientists also propose that changes in continental uplift derived from the movement of plate tectonics can change global circulation patterns in the oceans and atmosphere. The uplift can also reduce the concentration of carbon dioxide, resulting in a cooler climate. This is because carbon dioxide is a greenhouse gas that helps keeps the earth warm. **Greenhouse gases** are a portion of atmospheric gas molecules that deflect infrared radiation back to Earth's surface that was initially on a path to escape into space. Without greenhouse gases, the average temperature of Earth's surface would be about –18°C. The three most important greenhouse gases are water vapor, carbon dioxide, and methane. Concentrations of these gases fluctuate. During warm climate periods, concentrations are high. Concentrations are low during cool climate periods. Plate tectonics, vegetation, oceans, and wetlands all play a role in changing greenhouse gas concentrations.

In recent years, the earth's temperature has been increasing at a faster rate than previously recorded. Many scientists estimate that the average global temperature has risen 0.8°C since 1980. This may seem like a small number, but slight temperature changes can have big effects on local climates. Different regions

Greenhouse Gases

FYI

Climate Change Refugees
The Marshall Islands are a small chain of Pacific Ocean islands. The native people who live there have experienced the devastating effects of climate change. With increasing storm frequency and rising sea levels, the people are at risk of losing their homes. Over the past century, global sea levels rose about 20 cm. In the early 1990s, that rate began to double. It is projected that sea levels could rise up to a meter in the next 100 years. Scientists do not believe that the Marshall Islands will be completely swept away because storms bring ocean sediment and coral reef to the shore that is rebuilding the shoreline. However, a major problem that the people face during these frequent, strong storms is saltwater contaminating crops, freshwater supplies, and groundwater aquifers. One option that many islanders have had to choose is relocation to the continental United States. Climate change may not be noticeable worldwide; but for this group of people, climate change could be the cause of a lost culture.

NOTES

F. Instruct students to collect local temperature data from the past 50 years. Assign students to graph the data and note any dramatic changes. Have students determine the causes of the dramatic changes and note any periods of extensive heat, drought, or flooding. Direct students to determine how the climate changes have affected the local habitat, water availability, and survival of vegetation. Challenge students to predict the temperature trend in the next 50 years.

Lesson Review

1. Name and describe one short-term climate change and one long-term climate change. (**Volcanic eruptions are one example of a short-term climate change. The dust and gases produced by the eruption block solar energy that can cause temperatures to drop and induce freezing climates. An ice age is an example of a long-term climate change. Ice ages are periods of time when ice collects in high latitudes and moves toward lower latitudes.**)
2. What two Earth attributes did the Milankovitch theory suggest are the causes of ice ages? (**changes in the earth's orbit and the tilt of the earth's axis**)
3. How do changes in solar activity contribute to climate change? (**The sun contributes heat energy to the earth. Scientists hypothesize that the sun has been increasing in brightness over time. With greater luminosity, the earth is gradually getting warmer. Also, the frequency of sunspots causes atmospheric temperatures to fluctuate.**)
4. Explain how greenhouse gases affect the earth's atmospheric temperature. (**Greenhouse gases absorb infrared radiation that is deflected by Earth's surface and redirect the radiation to the surface of the earth. This process makes the surface of the earth warmer.**)
5. What are the two primary sources of carbon dioxide released by humans? (**fossil fuel combustion and deforestation**)

Lab 5.3.3A Greenhouse Gases

ANALYZE AND CONCLUDE:

1. Why is it important that all three thermometers be the same temperature at the beginning of the laboratory experiment? Limiting thermometer variations reduces inaccuracy. Also, it is easier to compare temperature readings if they all begin at the same point.

2. What greenhouse gas was tested in this experiment? water vapor

3. Graph the three sets of temperature readings over time in the area below.

Dry, Covered Thermometer	Moist, Covered Thermometer	Uncovered Thermometer
Time (min)	Time (min)	Time (min)

Answers will vary but all three lines should have a positive slope.

(Y-axis: Temperature (°C))

4. How do the three sets of temperature readings compare? The temperature increases over time for all three thermometers with the greatest increase occurring under the glass with the moistened paper towel.

5. Explain why the thermometer under the jar with the moistened thermometer had the greatest temperature increase over time. The water vapor absorbed more solar radiation. Given that the heat could not escape the glass jar, it was redirected to the thermometer and caused the temperature to increase.

Earth and Space Science

UNIT 6

Chapter 1: *Natural Resources*
Chapter 2: *Pollution Solutions*

Key Ideas

Unifying Concepts and Processes
- Systems, order, and organization
- Evidence, models, and explanation
- Change, constancy, and measurement
- Evolution and equilibrium
- Form and function

Science as Inquiry
- Abilities necessary to do scientific inquiry
- Understandings about scientific inquiry

Earth and Space Science
- Structure of the earth system
- Geochemical cycles

Science and Technology
- Abilities of technological design
- Understandings about science and technology

Science in Personal and Social Perspectives
- Personal health
- Populations, resources, and environments
- Natural hazards
- Risks and benefits
- Science and technology in society
- Personal and community health
- Population growth
- Natural resources
- Environmental quality
- Natural and human induced hazards
- Science and technology in local, national, and global challenges

History and Nature of Science
- Science as a human endeavor
- Nature of science
- Nature of scientific knowledge
- Historical perspectives

The Environment

Vocabulary

biomass	furrow	nuclear energy	smog
coal	geothermal energy	ore	solar energy
contour farming	incineration	peat	stewardship
cover crop	leachate	petroleum	strip cropping
crop rotation	natural gas	quarry	terrace farming
desertification	natural resource	renewable resource	
fossil fuel	nonrenewable resource	sludge	

SCRIPTURE

Therefore I tell you, do not worry about your life, what you will eat or drink; or about your body, what you will wear.... But seek first His kingdom and His righteousness, and all these things will be given to you as well.

Matthew 6:25, 33

6.1.0 Natural Resources

LOOKING AHEAD

- For **Lesson 6.1.1**, fill a jar for each student pair with 90 beans of one variety and 10 beans of another variety. Gather 1 blindfold for each pair of students.
- For **Lesson 6.1.2**, obtain soil moisture sensors for *Lab 6.1.2A Plants and the Atmosphere*. Also, arrange a field trip to a local water treatment plant.
- For **Lesson 6.1.3**, obtain one medium red foam ball, four small blue foam balls, and four large white foam balls. Also, gather samples of peat, lignite coal, subbituminous coal, bituminous coal, and anthracite. Obtain a large global map.
- For **Lesson 6.1.4**, gather materials needed for the *Introduction* activity. Arrange a field trip to a rock quarry, mine, or cement plant.
- For **Lesson 6.1.5**, obtain a large piece of poster board. Also, if possible, arrange a field trip to a wetland. Or, contact a classroom located near a wetland.
- For **Lesson 6.1.6**, invite a local farmer to speak to the class. If possible, arrange a field trip to a no-till farm.
- For **Lesson 6.1.7**, obtain materials for a Telkes solar oven.
- Begin growing bean plants for *Lab 6.2.3A Acid Rain*.

SUPPLEMENTAL MATERIALS

- BLM 6.1.4A Minerals Uses

- TM 6.1.1A Resource Replacement Time Scale
- TM 6.1.2A The Carbon Cycle
- TM 6.1.2B Global Water Shortage
- TM 6.1.3A The Formation of Coal
- TM 6.1.4A Mineral Uses
- TM 6.1.4B Metallic Properties

- Lab 6.1.2A Plants and the Atmosphere
- Lab 6.1.3A Oil Reserve Model
- Lab 6.1.4A Mining Minerals
- Lab 6.1.4B Mining Conditions

(continued)

Chapter 6.1 Summary

People often take for granted the amazing natural resources God has provided to make life possible. Homes are framed with wood and constructed with cement, wallboard, and other materials from the earth or sea. Without fertile soil and the exact mix of certain gases in the atmosphere, people would not have food to enjoy, and many of the products that are made of plastic, metal, and manufactured materials that require resources such as oil, minerals, ores, and water would not exist.

A person today could spend months, perhaps years, touching only pavement, tile, and other human-made products. Nevertheless, these manufactured items would depend entirely on the abundance and renewal of resources with which God has blessed the earth. Foods originate with fertile soil. Once ingested, oxygen is required to convert those foods into usable energy. Carbon dioxide in the atmosphere keeps the earth warm, and atmospheric nitrogen provides the only raw materials for the protein required in the diets of living organisms.

Background

Lesson 6.1.1 – Renewable and Nonrenewable Resources

Natural resources form a continuum from the most renewable to the least renewable. Various types of natural resources have different cycling times. Renewable resources generally have short cycling times. Solar energy and wind energy are the most renewable because they are continually produced. Living plants and animals have relatively short reproduction cycles and are therefore classified as renewable resources. Nonrenewable resources have very long cycles. For example, minerals and fossil fuels take thousands of years to renew. Resources are considered nonrenewable when the rate of use exceeds the cycling capacity.

Lesson 6.1.2 – Air, Water, and Trees

Cyanobacteria and photosynthesizing protists, including some types of algae, provide between 70% and 80% of the world's oxygen supply. Trees and other plants supply oxygen as well. Forest fires, burning fuels, and the weathering of rocks use up oxygen. However, the percentage of oxygen in the atmosphere has remained constant for hundreds and perhaps thousands of years.

Animals release carbon dioxide through respiration. Although only a small percentage of the atmosphere is made of this gas, carbon dioxide acts as a protective blanket around the earth. Without any carbon dioxide in the atmosphere, the earth would be too cold to support life.

Water is another renewable resource provided by God. This resource is becoming even more precious as the human population grows and as more water is used for food production. In light of freshwater shortages, many regions are turning to desalination for drinking water.

Trees are a crucial resource. They play an important role in the maintenance of the environment and in manufacturing. Trees efficiently improve air quality and reduce pollution. A mature tree can annually absorb 22 kg of small particulates and gases such as carbon dioxide emitted from automobiles or factories. It is estimated that one mature tree can produce roughly 118 kg of oxygen each year—enough oxygen for two people.

Many manufactured products include material from trees. Cellulose products include paper and refined products such as rayon, cellophane, solid rocket fuel, and industrial explosives. Chemical and resin products include soap, turpentine, chewing gum, maple syrup, rubber, and charcoal. Fruits, nuts, chocolate, and many spices come from trees. Lumber, plywood, and particle board are important for building. However, the primary use of wood throughout the world is fuel.

Lesson 6.1.3 – Fossil Fuels

Fossil fuels enable the use of many technologies including space exploration, gas grills, automobiles, plastics, and many more. They have made people's lives much more comfortable. However, many individuals do not realize the amount of influence fossil fuels have on their lives. For example, during the Arab oil embargo in the early 1970s, the United States' gross national product declined and unemployment doubled. The Gulf War was fought 20 years later partly because of disputes about oil supplies in the Middle East. Concerns over fossil fuel consumption may continue given that the hydrocarbon components of petroleum, natural gas, and coal make useful energy sources.

Petroleum is found in large quantities below the surface of the earth. Once the petroleum forms, it flows upward in the earth's crust because it has a lower density than the shales, sands, and carbonate rocks that constitute Earth's crust. Much of this oil flows out at the earth's surface or onto the ocean floor. Surface deposits also include bituminous lakes and escaping natural gas. Workers extract petroleum for fuel and for raw material in the chemical industry. To refine petroleum, chemists use fractional distillation to heat petroleum, which causes the vapors of the different components to condense on collectors at different heights. These various components are then further processed into various petroleum products.

In recent years, the world's availability of petroleum has steadily declined, and its relative cost has increased. Some estimate that petroleum may become an uncommon commercial material by the middle of the 21st century. Additional oil reserve discoveries could still be made, and new efficient technologies may be developed to use the oil. Knowing that the world's oil supply is reaching its limit, many scientists have been researching ways to extract oil from the ground more efficiently.

Natural gas is often found in association with petroleum. Its chief component is methane. Ethane, propane, butane, and other hydrocarbon compounds make up the remainder. Because it is highly flammable, natural gas is mainly used as a fuel.

Coal is burned to create electrical energy and heat. Ideally, coal should be high in heat content but low in impurities. If coal contains high sulfur contents, it can be corrosive, degrade product quality, and increase air and water pollution.

Two-thirds of the world's coal supply is found in the United States, Russia, and China. Although coal is a nonrenewable resource, most experts believe that the world's coal supply could last several hundred years.

Lesson 6.1.4 – Rocks and Minerals

Each year, the rivers of the world sweep 24 billion tons of weathered rock and sediments into the sea. Humans remove about 3 billion tons of rock and sediment each year. A primary purpose for quarrying gravel, sand, and limestone is to use it as building materials for roads and buildings.

In some Central and South American countries, gemstone and precious metal mines are created by deforesting rain forests, and the chemicals used in the extraction and processing of the precious stones pollute the soil and streams surrounding the mines. When metal concentrations in ores are high, little waste is left at the mining site. Many ores, such as iron ore, require at least 25% of the metal present in the ore for successful economic retrieval. However, some metals, such as copper, can be extracted with as little as 1% of the metal in the ore. Unfortunately, this type of mining operation leaves enormous amounts of unwanted rubble.

Every stage of processing a mineral involves a high risk of pollution. Liquid effluents must be treated properly. Tailings must be sealed to contain their dust. Tailings must also be sealed to prevent poisonous liquids from leeching into the ground. Toxic gases and dust must be removed

---

SUPPLEMENTAL MATERIALS

(continued from previous page)
- WS 6.1.1A Renewable and Nonrenewable Resources
- WS 6.1.1B Computer Energy Audit
- WS 6.1.2A Paper, Plastic, or Cloth

- Chapter 6.1 Test

BLMs, TMs, and tests are available to download. See Understanding Purposeful Design Earth and Space Science at the front of this book for the web address.

WORLDVIEW

- God has bestowed His people with immeasurable gifts. These gifts include natural resources. The air that people breathe, the sunlight that warms their skin, and the water that nourishes their bodies are all renewable resources that were provided by God. He not only provided these resources, but also ensures that they will be replenished through cycles. So, why are some people concerned about Earth's resilience? In Matthew 6:25, we are reminded that we do not need to worry about what to eat, drink, or wear. God already knows our needs. This does not mean that people should not take care of the earth and misuse resources. We have been blessed, and should desire to be a blessing to others.

from smelters before they reach the outside air. When environmental controls for one or more stages of a mining process are lax or nonexistent, serious pollution results. This danger is most prominent in developing countries where financial resources and a lack of proper technology can contribute to the pollution of the local environment surrounding a mining operation.

Modern society uses enormous amounts of aluminum, copper, chromite, aluminum, nickel, and other metals. The consumption of minerals rises much faster than population growth. However, it is believed that mineral reserves are adequate for the next century. With proper care, the pollution caused by mineral extraction and processing can be contained. The minerals in many manufactured products can be recycled, further preserving Earth's future.

Lesson 6.1.5 – Land

Widespread agriculture promotes human population growth. This population growth dictates the use of land, a valuable resource. The issue is not whether there is enough space for people to live, but whether there is enough farmland to grow sufficient food. Little new farmland is available to be cultivated. Most new agricultural land is of poor quality and is quickly degraded by erosion.

The rate of deforestation is high. Since 1900, approximately 90% of tropical forests in coastal West Africa have disappeared, and about 17% of the Amazon rain forest has been lost in the past 50 years. Industrial logging, plantation agriculture, cattle ranching, mining, and small scale agriculture and fuel collection are major factors.

Desertification is another issue associated with land use. Unfortunately, humans are expanding the deserts and arid regions of the world yearly by overgrazing domesticated animals and using improper agricultural methods. About 14% of the world's population lives in dry areas. These communities survive primarily by raising livestock, which in turn overgraze these regions and turn them into deserts.

Surface soil salinization is a problem in irrigated areas. Usually this is due to inadequate drainage of the irrigated land. When water cannot flow freely, it evaporates, leaving dissolved salts. Over many years, the accumulation may be significant enough to render the soil unsuitable for crops. Heavy soils with poor drainage are at risk for salinization. For example, salinization has damaged land that was irrigated in the western United States, the Nile Basin, and the Tigris-Euphrates region.

Another type of land degradation is the loss of wetlands. Wetlands serve as filters for pollution, especially eutrophication-causing phosphates and nitrates. They also act as buffer zones against flooding. Wetlands are generally lost to agriculture, tourist development, and hydroelectric dams. As the need for cheap energy increases, fertile valleys and wetlands are flooded as reservoirs when dams are built. The Aswan Dam in Egypt causes the annual loss of 40 million tons of soil, which has since been replaced by chemical deposits.

Solutions for conserving all the diverse land resources are difficult and complex. Radical economic, social, and political reforms may be necessary to preserve land in developing countries, while industrialized countries may use wealth and technology to preserve their land. Sound forest management combined with the use of recycled products and consumer conservation can provide wood products on a sustainable basis. Draining canals and drainage tiles may reduce salinization. Educating people on the economic and environmental value of wetlands may help curb urban sprawl and tourist development in these areas.

Lesson 6.1.6 – Soil

Soil is necessary for most plants, animals, and people to live. Certain agricultural methods may lower soil fertility because crops can deplete the soil of nutrients, cultivation may degrade soil structure, and exposure to wind and rain can encourage erosion. Proper soil management can maintain and even improve soil fertility.

Several methods have been implemented to combat soil erosion. Cover crops and crop rotation conserve soil fertility by stabilizing the soil and replenishing nutrients. Minimum-tillage systems help maintain soil structure. Fertilizers can help restore soil fertility by adding nutrients over and above those available in the soil. Amending soil with organic matter helps improve the soil's structure. Soil acidity can be decreased by adding calcium carbonate or increased by adding sulfuric acid.

Contour farming involves plowing rows that follow the contours of the land. These furrows are perpendicular to the slope of the land. If furrows are plowed parallel to the slope, water runs rapidly through them and carries away precious topsoil. Contour farming is often used to farm the slopes of volcanic mountains, which have nutrient-rich soil. Contour plowing prevents runoff and erosion from steep volcanoes. Row crops are planted across the slope as much as possible. This reduces erosion, controls water flow, and increases infiltration.

Terrace farming not only conserves soil but also utilizes land that would otherwise not be useful for farming. Farmers construct steplike ridges that follow the slope of a hill. The runoff from the hill improves soil retentiveness and increases arability. Often the runoff is diverted into a larger irrigation system.

Lesson 6.1.7 – Alternative Energy Sources
One alternative energy source is solar energy. The first solar-powered motor was patented by a French mathematics teacher named *Augustin-Bernard Mouchot*. However, the convenience of using coal and oil kept solar energy from becoming a major energy source until the energy crisis of the 1970s. Solar energy is now one of the most promising energy sources for the future. Most spacecraft use solar power for their instrumentation and communication. Future solar energy technology could allow for solar-run automobiles and power plants.

The amount of solar energy varies. This presents a challenge because most applications require readily available, constant energy. The solution to this dilemma lies in forecasting the amount of available solar energy in the future, capturing that energy, and storing it for later use. Another challenge is to convert the sun's radiation into useful energy. Although sunlight falls on Earth in great amounts, it is not concentrated. Many solar energy systems require large, space-consuming collectors.

Another energy source is biomass, which uses the process of photosynthesis as an energy source. Biomass involves the conversion of crops or agriculture waste products into liquid or gaseous fuels. A United Nations study suggests biomass could supply 55% of the world's energy needs by 2050.

Geothermal energy, although sufficient to produce a fraction of the energy needs, is a useful form of energy. Geothermal energy is obtained from the internal heat of the planet and can be used to generate steam for a steam turbine that generates electricity. One type of geothermal system pumps water into hotspots and uses the steam to generate electricity. Another method utilizes volcanic magma to boil water. A third system recovers heat from dry rock by circulating water around it and transferring the heat to a generator.

Nuclear energy is the energy stored in the nucleus of an atom. The energy can be released from the atom through fission or fusion. These processes involve a small amount of mass being converted to energy. Fission involves the splitting of atoms. Fusion involves the joining of nuclei. Uranium, the fuel used in fission reactions, is scattered throughout the earth's surface.

NOTES

6.1.1 Renewable and Nonrenewable Resources *The Environment*

OBJECTIVES

Students will be able to
- distinguish between renewable and nonrenewable resources.
- recognize the importance of energy and resource conservation.

VOCABULARY

- **natural resource** any substance, organism, or energy form found in nature that can be used by living things
- **nonrenewable resource** a resource that cannot be replaced once it is used or can only be replaced over an extremely long period of time
- **renewable resource** a resource that is constantly available or that can be replaced in a relatively short period of time through natural processes

MATERIALS

- Two varieties of dried beans, jars, blindfolds (*Introduction*)
- Matches (*B*)
- TM 6.1.1A Resource Replacement Time Scale
- WS 6.1.1A Renewable and Nonrenewable Resources
- WS 6.1.1B Computer Energy Audit

PREPARATION

- Fill each jar with 90 beans of one variety and 10 beans of another variety. (*Introduction*)

Introduction

Divide the class into pairs. Provide each pair with a prepared jar of 100 dried beans, 90 beans that represent nonrenewable resources and 10 beans that represent renewable resources. Explain to students that more demand is placed on resources as human populations increase. Ask students to speculate what challenge this situation may pose. (**If demand becomes too great, there is the potential that nonrenewable resources would be depleted.**) Have one student in each pair wear a blindfold to represent an impartial consumer. This individual will draw the beans from the jar. Complete Scenario 1 on **WS 6.1.1A Renewable and Nonrenewable Resources** as a class. This scenario represents a human population with no growth and a constant yearly energy consumption rate. Share with students that Scenario 2 represents a developed nation with population growth and increased energy consumption. Ask students if industrialized or developing nations consume more energy. (**industrialized nations**) Discuss ideas for extending energy resources in a growing society.

Discussion

- Display **TM 6.1.1A Resource Replacement Time Scale**. Remind students that renewable and nonrenewable resources are largely defined by the amount of time it takes to replace the resource. It may be interesting to have students speculate the amounts of time before revealing the answers.

- Discuss how fossil fuels are formed beneath the earth's surface.

- Ask the following questions:
  1. Why do people use fewer renewable resources than nonrenewable resources? (**They are less expensive, they provide more energy.**)
  2. Should more renewable resources be utilized now or in the future when the nonrenewable resources are depleted? (**It is better to be proactive and utilize renewable resources before nonrenewable resources are depleted.**)
  3. Are renewable resources unlimited? (**Theoretically. However, in the case of trees, if all of the mature species are depleted, the resource will no longer be available.**)

Activities

A. Challenge students to compute the energy used by computers in the school by conducting an energy audit. Have students complete **WS 6.1.1B Computer Energy Audit**.

B. Direct a student volunteer to light a match. Ask the class the following questions:
   - What does the match produce? (**light, heat, and smoke**)
   - How could the match be useful? (**The match could start a fire or ignite a gas burner that could then be used to cook food or keep people warm.**)
   - How is the match similar to resources such as coal and petroleum? (**Coal and petroleum can be used to generate heat and light. They also produce pollution and are quickly used up.**)

   Instruct the student to try to light the same match. Ask why the match does not light for a second time. (**All of the material on the end of the match head reacted during the initial lighting.**) What type of resource would a match be? (**nonrenewable resource**)

C. Have students identify items in the classroom whose original material is naturally replaceable or not replaceable. Lead a discussion about what different materials could be exchanged for the items that are not naturally replaceable. Have students write a summary of the discussion.

D. Instruct students to develop a plan for conserving water and energy at school or at home. Direct them to illustrate their plan in the format of a brochure.

E. Assign students to research a particular form of energy. Some choices include fossil fuels, biofuels, geothermal energy, solar power, wind power, biomass, and hydroelectric power.

F. Have students develop a computer presentation or video to encourage use of renewable resources.

G. Have students investigate unlimited resources such as solar, wind, or ocean-wave energy. Ask students to summarize the latest research. Direct students to answer the following: *What can these resources be used for?* (**Solar energy can be converted into electricity or used to heat air, water, or other fluids. Wind energy can be converted into kinetic power to grind grain and pump water. It can also be converted into electricity to power homes and businesses. Ocean-wave energy can generate electricity in desalination plants, power plants, and water pumps.**) *Describe possible challenges associated with using these type of resources.* (**Solar energy can be expensive to convert and sunlight is not always available. Wind energy may not be cost effective, and turbines pose a threat to birds and bats. Ocean-wave energy equipment can be costly to install and is dangerous in harsh working environments.**) *What countries are working on such energy sources?* (**Possible answers: China, United States, Japan, United Kingdom, India, Germany**)

Lesson Review
1. What type of resource cannot be replaced once it has been used up? (**nonrenewable resources**)
2. What are the main sources of renewable energy? (**the sun, air, water, and living things**)
3. Fossil fuels are an example of what type of resource? (**nonrenewable resource**)
4. How can individuals help manage renewable resources wisely? (**Individuals can manage renewable resources by ensuring there are enough resources that can reproduce or be naturally replaced. People need to make careful choices regarding the use of renewable resources.**)

NOTES

6.1.1 Renewable and Nonrenewable Resources

OBJECTIVES
- Distinguish between renewable and nonrenewable resources.
- Recognize the importance of energy and resource conservation.

VOCABULARY
- **natural resource** any substance, organism, or energy form found in nature that can be used by living things
- **nonrenewable resource** a resource that cannot be replaced once it is used or can only be replaced over an extremely long period of time
- **renewable resource** a resource that is constantly available or that can be replaced in a relatively short period of time through natural processes

God created the earth with everything that is needed for His people to live and thrive. These sustaining materials are called **natural resources**. A natural resource is any substance, organism, or energy form found in nature that can be used by living things. Certain natural resources, such as air and water, are necessary for life. Some natural resources, such as coal and natural gas, make people's lives more convenient and comfortable. Many natural resources provide energy. Natural resources may be found in the atmosphere, oceans, or deep within the earth's crust.

Natural resources are often classified as renewable or nonrenewable. **Renewable resources** are resources that are constantly available or that can be replaced in a relatively short period of time through natural processes. For example, when energy from the sun powers a solar-powered calculator, the energy used is immediately replaced by more energy from the sun. Air, water, and living things are considered *renewable resources*. Renewable resources include energy resources such as solar energy, wind energy, and energy gleaned from the heat inside of the earth. A resource can maintain a renewable status as long as the resource is not completely depleted before it can reproduce or be naturally replaced. For example, trees are a renewable resource. However, if too many trees are cut down before more trees can be produced or replaced, the renewable resource no longer exists.

Nonrenewable resources are resources that cannot be replaced once they are used or can only be replaced over an extremely long period of time. Once these resources have been completely depleted, there will be no more available for people and other organisms. Examples of nonrenewable resources are fossil fuels (coal, oil, and natural gas) and minerals. These substances exist in limited quantities. Fuel alternatives will need to be in place when the fossil fuel stores have been depleted.

Many people do not know what it is like to live without readily accessible energy sources. When cars run low on gas, more gasoline is usually available. When it is dark, people can quickly turn on the lights. If the outside temperatures are cold, people turn up the heat. Because nonrenewable resources have been available for a long time, most people do not think about how many resources and how much energy are being

Natural, renewable resource

CAREER

Conservationist
A conservationist works primarily with landowners and government agencies to develop ways to protect natural resources. These resources include soil and water. Conservationists help develop ways to efficiently utilize land without harming the environment.

There are two primary branches of conservation that individuals may opt to pursue. They may choose to work directly in the field as a conservation worker. These individuals serve in interventionist roles. For example, some scientists clear the land of any invasive plant species that may damage the surrounding ecosystem. Another example of a conservation worker is an archaeological conservationist. These individuals work in the laboratory where they carefully clean site finds in an attempt to preserve and store them for future study.

The second primary branch of conservation is conservation science. Individuals who pursue this avenue do not get to participate in many hands-on activities. Rather, they focus on theory, conduct research, and develop conservation materials and methods to be used in the field.

consumed. Although energy can never be created or destroyed, nonrenewable energy resources can be completely expended. Likewise, energy can be used wisely or unwisely.

It is important to think carefully about how choices regarding natural resources relate to the well-being of God's creation. Even small choices matter. For example, those who can afford to pay high electric bills should be careful not to waste electricity. Although more gasoline is usually available at the next gas station, making good energy choices, such as walking, biking, or skating whenever possible instead of riding in a car, helps to maintain the delicate resource balance in nature.

The issues related to renewable and nonrenewable resources are not always obvious. Making good energy choices involves some careful thought and a lot of personal responsibility.

LESSON REVIEW
1. What type of resource cannot be replaced once it has been used up?
2. What are the main sources of renewable energy?
3. Fossil fuels are an example of what type of resource?
4. How can individuals help manage renewable resources wisely?

6.1.2 Air, Water, and Trees

The Environment

OBJECTIVES

Students will be able to
- examine the natural processes that renew oxygen, nitrogen, and carbon.
- discover the implications of an uneven global distribution of water.
- correlate the value of tree contributions with the need to protect this resource.

MATERIALS

- Molecular model kit (*Introduction*)
- TM 6.1.2A The Carbon Cycle
- TM 6.1.2B Global Water Shortage
- WS 6.1.2A Paper, Plastic, or Cloth

PREPARATION

- Obtain materials for *Lab 6.1.2A Plants and the Atmosphere*.
- Gather materials for *Try This: Transpiration*.
- Arrange a field trip to a local water treatment plant. (C)

TRY THIS

Transpiration
- sealable plastic bags, 1 per group
- rubber bands or tape
- gardening shears
- graduated cylinders, 1 per group

(The calculation should only include daylight hours.)

Introduction

Review the water and nitrogen cycles with students. Ask students to describe carbon. (**Answers will vary but should include that carbon is an element that is the building block of most living organisms.**) Explain to students that carbon also cycles through nature. Have students name places where they think carbon can be found. (**Possible answers: in the atmosphere, rocks, hydrocarbons found in fossil fuels, and the sun**) Use a molecular model kit to create various carbon compounds found in nature such as carbon dioxide, methane (CH_4), and propane (C_3H_8). Have students create a carbon cycle map. Display **TM 6.1.2A The Carbon Cycle** and encourage students to add any steps to their map that they may not have included. Draw special attention to carbon that circulates through water. Challenge students to list countries that experience regular water shortages. Display **TM 6.1.2B Global Water Shortage**. Ask students to explain why they think the water needs in the highlighted countries will increase or decrease in the future. Read **Matthew 25:35, 40**. Ask students if they think Christians have an obligation to help individuals in these countries regarding diminishing water supplies and to state their reasons why. (**Answers will vary.**)

Discussion

- Ask students how long they could survive without renewable resources. (**Answers will vary.**) Have students hold their breath for as long as they can without losing consciousness. Share with students that the amount of time they could hold their breath was the length of time they could survive without a renewable resource. This resource is air. Review the process of photosynthesis with students and how oxygen is generated.

- Remind students that air, water, and trees are considered *renewable resources*. Have students define renewable resources. (**Renewable resources are resources that are constantly available or that can be replaced in a relatively short period of time through natural processes.**) Ask students if renewable resources can be consumed without limit. Why? (**Renewable resources can be replaced through natural processes. However, if the resources are constantly consumed in large amounts without allowing for the appropriate amount of time to be replaced, the resource may be depleted.**) What is an example of a renewable resource becoming used up? (**Possible answer: cutting down an entire forest before mature trees can have the opportunity to reproduce**)

- Ask the following questions:
  1. Will the amount of air on Earth change over time? (**No. However, air quality can change.**)
  2. Why is there a greater demand on renewable resources? (**Increasing populations are creating a greater demand on renewable resources.**)
  3. Why are trees an important renewable resource? (**Trees replenish the atmosphere's oxygen and contribute to the water cycle through transpiration. They also shelter a majority of Earth's land species.**)
  4. What products from trees do humans consume or use? (**Possible answers: paper, wood, furniture, chewing gum, hair dye**)

Activities

Lab 6.1.2A Plants and the Atmosphere

- paintbrushes, 2 per group
- identical small plants, 2 per group
- petroleum jelly
- plastic wrap
- clear, plastic bags, 2 per group
- thermometers, 1 per group
- soil moisture sensors, 1 per group

A. Complete *Try This: Transpiration* with students. Ask students if they think other trees or plants would transpire the same amount of water. How could this exercise benefit people in countries with water shortages? (**Possible answer: Individuals could create a system where a portion of** a tree transpires water into a collection tank that is connected to a crop watering system. The water would be released directly to the roots, and the smaller percentage of water produced could help prevent groundwater contamination.)

B. Assign **WS 6.1.2A Paper, Plastic, or Cloth** for students to complete. Encourage students to discuss their findings with their family and make an informed decision about what kind of grocery bags to use.

C. Conduct a field trip to a local water treatment plant. Direct students to select two or three processing agents to research and to determine the potential impact these agents or their wastes may have on the environment. Have them discuss whether the result is worth the risk. (**Answers will vary.**)

D. Divide the class into four or five groups. Assign each group one of the following roles: farmer, landscaper, restaurant owner, nurse, and firefighter. Share with students that a severe drought has swept through the nation. Have students research the effects of drought on their assigned role. Direct students to answer the following: *What potential problems would be faced with the water reduction? How would you handle these problems created by the drought?* Have students present their findings to the class in the form of a drama, newspaper article, or computer presentation. Discuss whether certain measures should be limited to extreme circumstances, such as flooding or droughts, or if the measures would be good stewardship practices that should be conducted daily.

E. Guide students to create a story about a world without air, water, or trees.

NOTES

6.1.2 Air, Water, and Trees

OBJECTIVES
- Examine the natural processes that renew oxygen, nitrogen, and carbon.
- Discover the implications of an uneven global distribution of water.
- Correlate the value of tree contributions with the need to protect this resource.

Imagine hearing the following scenario on the evening news: *Scientists have calculated that the earth has enough oxygen left to sustain the population for only 29 more years. Please try not to breathe too deeply. And hold your breath a few times a day. If everybody does their part, the air might last for 32 years.* Does this sound like a ridiculous statement? Yes, because air is a renewable resource.

The atmosphere is a mixture of gases that surrounds the earth. These gases comprise air, which is one of Earth's most precious renewable resources. Although the amount of air available on Earth will never change, humans can influence the quality of air that many living things require. For example, preventing the deterioration of ozone gas layers in the upper atmosphere helps shield the earth from harmful ultraviolet radiation. Also, solid particles in the atmosphere, such as those from smoke, must be kept at low levels in order to avoid interfering with sunlight penetration. Toxic materials in the air must also be kept at low levels so that living things do not inhale or absorb them. It is important that the atmosphere maintain the right balance of gases.

Two primary atmospheric gases—oxygen and nitrogen—are important ingredients for the majority of life on Earth. Humans, animals, plants, aerobic bacteria, and most other microorganisms require oxygen for respiration to produce the energy they need to thrive. Oxygen is renewed by green plants, bacteria, algae, and plankton that produce oxygen as a by-product of photosynthesis. As long as green plants and other oxygen-producing organisms are alive, the atmosphere will have a supply of oxygen.

The atmosphere is a renewable resource.

Water is a renewable resource.

A large percentage of the atmosphere is made of nitrogen—a noncombustible gas. Plant and animal cells process nitrogen to make proteins that stabilize cell boundaries. Nitrogen-fixing bacteria pull nitrogen out of the air to make nitrogen compounds. Plants absorb the nitrogen compounds to make amino acids—the building blocks of proteins. Animals ingest the amino acids in plants to manufacture proteins. Lightning also converts atmospheric nitrogen into useful nitrogen compounds. When animals decay, bacteria recycle nitrogen gas back into the atmosphere.

Even though only a small percentage of the atmosphere is made of carbon dioxide, this gas works in conjunction with other greenhouse gases to act like a blanket for the earth. It traps heat and keeps the biosphere warm. Without carbon dioxide and the other greenhouse gases, the earth would be freezing cold.

Similar to nitrogen and water, carbon circulates in various forms through nature. Plants need carbon dioxide in order to carry out photosynthesis. Carbon dioxide is renewed by humans and animals who produce this gas as a by-product of respiration. The burning of fossil fuels adds more carbon dioxide to the air. This extra carbon dioxide traps additional heat and is maybe contributing to an increase in global temperature. God created

TRY THIS

Transpiration
Place a sealable bag over a small, leafy tree branch early in the morning. The branch needs to be in direct sunlight for at least two hours. Secure the opening of the bag around the tree branch with a rubber band or tape to prevent any air from entering the bag. After two hours have elapsed, clip off the branch and gently tap the branch to collect the water in one corner of the bag. Carefully pour the water into a graduated cylinder and record the volume of water that was transpired. Estimate the percentage of tree leaves that are contained in the bag. Calculate how much water the entire tree will transpire in one day. (Should the calculation cover a 24-hour period or only daylight hours?)

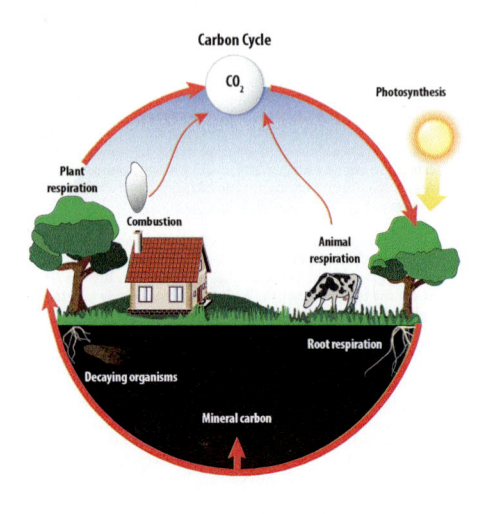

NOTES

Lesson Review

1. What two primary atmospheric gases are important for most of life on Earth? (**oxygen and nitrogen**) Why? (**Most organisms require oxygen for respiration to produce the energy that they need to thrive. Plant and animal cells process nitrogen to make proteins that stabilize cell boundaries.**)
2. Why is the carbon cycle important? (**Plants need carbon dioxide in order to carry out photosynthesis, which provides oxygen for other organisms. Carbon dioxide is also an important greenhouse gas used to maintain warm temperatures on the earth.**)
3. How are the oxygen, nitrogen, and carbon cycles renewed? (***oxygen***: **by green plants, bacteria, algae, and plankton;** ***nitrogen***: **by nitrogen-fixing, which creates nitrogen compounds that are then absorbed by plants that are then eaten by animals, which later die and release the nitrogen in the decaying process;** ***carbon***: **by humans and animals, which produce this gas as a by-product of respiration**)
4. What are implications of the uneven global water distribution? (**If a region has a chronic water shortage and a significant population growth, more freshwater will be needed or the available freshwater could be depleted at a faster rate.**)
5. What value do trees offer as renewable resources and what threatens the supply of them? (**Trees replenish the atmosphere's oxygen and contribute to the water cycle through transpiration. They supply the materials for things such as paper, building supplies, and furniture. Growing populations demand land use, which has been provided by the clearing of forests.**)
6. How can people protect atmospheric resources? (**People can reduce greenhouse gas emissions that could potentially cause the ozone layer to deteriorate. Also, industries can find ways to reduce pollutants in the atmosphere.**)

Name: _____ Date: _____

Lab 6.1.2A Plants and the Atmosphere

QUESTION: How do plants help renew the atmosphere?

HYPOTHESIS: Answers will vary.

EXPERIMENT:

You will need:		
• 2 paintbrushes	• petroleum jelly	• thermometer
• 2 identical small plants	• plastic wrap	• soil moisture sensor
	• 2 clear, plastic bags	

Steps:
1. Have two group members use the paintbrushes to coat both sides of all the leaves on one of the plants with petroleum jelly.
2. Saturate the soil of both plants with equal amounts of water.
3. Cover the soil of both plants with plastic wrap and place a plastic bag over each plant.
4. Place both plants on a windowsill in direct sunlight and observe for several days.
5. After several days have elapsed, measure the temperature of both soils.
   - Non-petroleum plant soil: Answers will vary. °C
   - Petroleum plant soil: Answers will vary. °C
6. Predict which plant will have the greater soil moisture content.
   Answers will vary.
7. Use the soil moisture sensor to measure the water content percentage in both soils.
   - Non-petroleum plant soil: Answers will vary. %
   - Petroleum plant soil: Answers will vary. %

ANALYZE AND CONCLUDE:

1. Describe the appearance of the petroleum-covered plant after several days have passed. Explain why the plant appears this way. The leaves of the petroleum covered plant turned yellow and fell off. Green plants receive carbon dioxide through tiny pores called *stomata* on the underside of leaves. They convert the carbon dioxide into glucose and oxygen through a process called *photosynthesis*. The petroleum prevented carbon dioxide from entering the stomata.

Lab 6.1.2A Plants and the Atmosphere

2. Which plastic bag held the most water? Where did the water come from? The plastic bag surrounding the plant that was not covered with petroleum jelly held the most water. The water was produced by the plant via transpiration.

3. Correlate the soil water contents to the amount of water produced on the inside of the plastic bags. The petroleum-covered plant's soil contained more water compared to the amount of water in the nonpetroleum plant's soil because it could not be transpired by the plant to be released on the inside of the plastic bag.

4. Explain why the plastic bag around the plant that was not covered with petroleum appears inflated. The plant that was not covered with petroleum produced water and oxygen gas through a process known as photosynthesis. Extra oxygen caused the plastic bag to inflate.

5. How do green plants help renew the atmosphere? Green plants and other photosynthesizing organisms regulate the atmosphere's gases in order to maintain a balance between oxygen and carbon dioxide. Without green plants, the biosphere would not have enough oxygen to sustain most living things. Also, too much carbon dioxide could build up in the atmosphere, which could lead to increased global temperatures. Green plants also contribute to the water cycle.

6.1.3 Fossil Fuels

The Environment

OBJECTIVES

Students will be able to
- summarize the formation of fossil fuels.
- highlight the properties and uses of fossil fuels.
- examine the advantages and disadvantages of processing fossil fuels.
- describe four types of coal.

VOCABULARY

- **coal** a solid fossil fuel formed from decomposed plant remains
- **fossil fuel** a source of energy formed from the buried remains of dead plants and animals
- **natural gas** a mixture of methane and other gases formed from decomposed marine organisms
- **peat** a substance made of partially decayed plant matter
- **petroleum** a liquid fossil fuel formed from microscopic plants, animals, and marine organisms

MATERIALS

- Red, blue, and white foam balls (*Introduction*)
- Samples of peat, lignite coal, subbituminous coal, bituminous coal, and anthracite (*Discussion*)
- Large global map (*A*)
- TM 6.1.3A The Formation of Coal

PREPARATION

- Obtain samples for the *Introduction* activity.
- Gather materials for *Lab 6.1.3A Oil Reserve Model*.

Introduction

Ask students why they think petroleum, natural gas, and coal are given the name fossil fuels. (**They all are derived from dead plant or animal material that received energy ultimately from the sun.**) Have students name the last time they used fossil fuels. (**Answers will vary but should include when they put on a particular type of clothing or rode in a car.**)

Remind students that all fossil fuels are made up of hydrocarbons. Make a model of a hydrocarbon by connecting a medium-size red foam ball that represents carbon with four small blue foam balls that represent hydrogen. Ask students how people can get energy from a hydrocarbon. (**Through the process of combustion in the presence of oxygen, hydrocarbons release energy.**) Display two pairs of large, connected white foam balls that represent oxygen. Remove the hydrogen foam balls and attach two hydrogen foam balls to each oxygen foam ball. Attach two oxygen foam balls to the carbon foam ball. Clap your hands or say "pop" each time a hydrogen atom is removed to indicate energy is released. Ask students what the products are from the combustion reaction. (**water and carbon dioxide**) Which product is observed escaping out of car tailpipes? (**Water vapor. Carbon dioxide is invisible.**) Have students review the process.

Discussion

- Display samples of peat, lignite coal, subbituminous coal, bituminous coal, and anthracite or **TM 6.1.3A The Formation of Coal**. Ask students how each type of coal formed. (**Lignite was created when immense pressure was applied to peat. Additional pressure was applied to lignite to create subbituminous coal that eventually converted into bituminous coal and finally anthracite.**)

- Have students trace the reverse energy path of a steam locomotive. (**Mechanical energy converted from thermal energy from chemical energy produced by combustion of coal, which is a fossil fuel that is derived from dead plants or animals that received energy from the sun.**)

- Discuss the use of fossil fuels. Ask students how people use energy in their daily lives. (**Possible answers: People use energy when they watch television, wash clothes, heat their homes, operate a computer, drive to school, and turn on a lamp.**) How many of these operations use fossil fuels? (**A large percentage of these operations require the use of fossil fuels.**) What would happen if one fossil fuel was completely consumed? (**Answers will vary but should include people could use alternative energy sources, or the prices of other fossil fuels may increase dramatically with supply and demand.**)

- Ask the following questions:
  1. Are fossil fuels renewable or nonrenewable? Why? (**Fossil fuels are nonrenewable because the conversion of decayed material into fossil fuels is a very slow process.**)
  2. What are the three major types of fossil fuels? (**petroleum, natural gas, and coal**)
  3. Where can fossil fuels be found? (**Petroleum is found in underground pools or reservoirs. Some petroleum is located near the surface of the earth in the form of oil shale and tar sands. Natural gas is found under land and ocean floors, usually above petroleum layers. Coal is found deep within the earth's crust.**)

Activities

Lab 6.1.3A Oil Reserve Model

- sand
- small rocks
- cardboard boxes, 1 per group
- balloons, 1 per group
- syringes, 1 per group
- viscous liquid
- sheets of graph paper, 1 per group
- metric rulers, 1 per group
- probes, 1 per group

Shoeboxes work well for this lab exercise. Chocolate pudding and molasses work well as viscous liquids. Any colored liquid will suffice. It may be a good idea to fill the balloons prior to students conducting the lab. After students finish the lab, discuss the high cost of obtaining oil and other costs associated with its production.

A. Assign students one of the three major fossil fuels (petroleum, natural gas, or coal). Have them research global regions that have the highest densities of the assigned fossil fuel. Display a large global map. Guide students to work together to create a fossil fuel key listing the fuels' properties, and have students place color-coded stickers over the regions.

B. Assign students to illustrate the formation of a fossil fuel.

C. Divide the class into three groups. Assign each group one of the fossil fuels. Have each group create a chart listing the advantages and disadvantage of the assigned fossil fuel. Ask students to share results with the class.

NOTES

6.1.3 Fossil Fuels

OBJECTIVES
- Summarize the formation of fossil fuels.
- Highlight the properties and uses of fossil fuels.
- Examine the advantages and disadvantages of processing fossil fuels.
- Describe four types of coal.

VOCABULARY
- **coal** a solid fossil fuel formed from decomposed plant remains
- **fossil fuel** a source of energy formed from the buried remains of dead plants and animals
- **natural gas** a mixture of methane and other gases formed from decomposed marine organisms
- **peat** a substance made of partially decayed plant matter
- **petroleum** a liquid fossil fuel formed from microscopic plants, animals, and marine organisms

Did you know that the energy that powers the car or bus that takes you to school originally came from the sun? Gasoline is derived from oil, which is a fossil fuel. **Fossil fuels** are sources of energy formed from the buried remains of dead plants and animals. The process for the formation of fossil fuels is very slow. This is why fossil fuels are nonrenewable resources.

Like the energy that people obtain from food, fossil fuels can be traced back to producers that harness energy from the sun in a process known as *photosynthesis*. This process models the first law of thermodynamics that states that energy cannot be created or destroyed. Plants convert energy from the sun into stored energy. As the plants decay, pressure and heat cause the matter to form layers that are eventually converted into fossil fuels. When the fossil fuels are burned, the thermal energy is released in the form of heat. Therefore, the energy in the natural gas that your furnace burns and the energy in the gasoline that fuels your car ultimately came from the sun!

The three major types of fossil fuels are petroleum, natural gas, and coal. **Petroleum**, a crude oil, is a liquid fossil fuel formed from microscopic plants, animals, and marine organisms. It can be found in underground pools or reservoirs. This oily, flammable mixture is separated into liquid fuels and other products in refineries. The liquid fuels include gasoline, jet fuel, diesel fuel, kerosene, and fuel oil. Other products separated from petroleum include lubricants, waxes, and tars such as asphalt.

Petroleum is a nonrenewable fossil fuel.

Petroleum meets a large portion of the world's energy needs. Most petroleum is drawn from oil wells, but some petroleum is located near the surface of the earth in the form of oil shale and tar sands. Oil shale is a fine-grained rock containing a solid, waxy mixture of hydrocarbon compounds. Tar sands are layers of sand soaked with thick petroleum.

Some of the world's largest mining projects mine oil from the tar sands in northeastern Alberta, Canada. Clay, sand, and water along with bitumen, a thick black oil, are the components of tar sand. Alberta's tar sands are the world's largest. This form of oil can be mined using two methods. Open-pit mining involves digging huge holes in the earth's surface. Oil miners use heat and diluents (thinning agents) to remove oil from the tar sand. The remaining sand is put back into the open-pit mine, which is reclaimed—brought back to a condition where it can be used for other purposes. The second method involves injecting steam into the tar sand deposit. Oil miners pump out and process the heated oil. This method is useful for deep deposits.

Petroleum is not only an energy resource, it is also a valuable ingredient in many manufactured products. Petroleum is used to

NOTES

D. Have students illustrate and describe the different kinds of coal in a graphic organizer.

E. Direct students to write a brief description about how their lives would be affected without electricity.

Lesson Review

1. Describe how petroleum and natural gas are formed. (**Petroleum is formed from microscopic plants, animals, and marine organisms. Natural gas is a mixture of methane and other gases formed from decomposed marine organisms. When these organisms died and sank to the ocean floor, they were covered by layers of sediments and compressed into sedimentary rock. Pressure, heat, and bacterial activity transformed these organic substances into oil and natural gas.**)

2. Contrast the properties and uses of the three major fossil fuels. (**Petroleum is an oily, flammable mixture separated into liquid fuels and other products in refineries. It is used for energy and manufacturing of plastics, nylon and rayon fabrics, and vitamin capsules. Natural gas is a group of gases—methane, butane and propane. These are used for heating, generating electricity, filling lighters, cooking, and fuel. Coal is solid and used to generate electricity and provide heat.**)

3. Examine the advantages and disadvantages of processing fossil fuels. (**Petroleum is inexpensive, reliable, efficient, and convenient to transport. Disadvantages include the production of atmospheric pollution and the potential for oil spills. Natural gas burns cleaner than other fossil fuels and creates less pollution. Natural gas is also plentiful, and it is convenient to transport and use. Burning natural gas does have a drawback—it produces carbon dioxide,**

---

make products such as plastics, nylon and rayon fabrics, vitamin capsules, and other products.

The reliance on petroleum has advantages and disadvantages. It is an inexpensive, trusted source of energy. High-grade oil and gasoline burn relatively efficiently. Petroleum is convenient to transport and new drilling technologies continue to improve oil production. One disadvantage of burning petroleum is the production of atmospheric pollution, such as smog and carbon dioxide, that contributes to climate change. Another issue is the potential for oil spills that harm the environment.

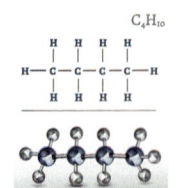

Natural gas is a mixture of methane and other gases formed from decomposed marine organisms. When these organisms died and sank to the ocean floor, they were covered by layers of sediments and compressed into sedimentary rock. Pressure, heat, and bacterial activity transformed these organic substances into oil and natural gas. Geologists typically find oil and natural gas together. The gas layers usually rise above the oil layers because gas is less dense than oil.

Almost all natural gas provides fuel for heating and for generating electricity. People commonly burn methane to warm houses and cook on gas stoves. Butane and propane are also natural gases. Butane is used in lighters and camp stoves. People burn propane as heating fuel and as cooking fuel for outdoor grills. Some motor vehicles are powered by natural gas. These vehicles produce less pollution than gasoline and diesel vehicles. Natural gas burns cleaner than other fossil fuels and creates less pollution. Natural gas is also plentiful, and it is convenient to transport and use. Burning natural gas does have a drawback—it produces carbon dioxide, a major greenhouse gas. However, it produces less carbon dioxide than petroleum or coal for the same amount of heat energy produced.

Coal, a solid fossil fuel formed from decomposed plant remains, was the first fossil fuel to be used by humans. Prior to this time, wood was the primary source of fuel throughout the world. Coal mining in Europe dates back to at least the 13th century. This fossil fuel was once the leading energy source for home heating in the United States and Canada. Many power plants still use coal to generate steam needed to produce electrical energy in power plants.

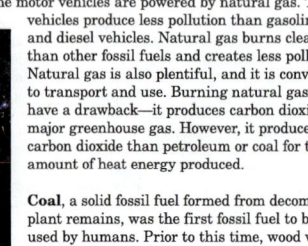

The formation of coal can be traced to wetlands such as swamps, bogs, and moors. **Peat** is a substance made of partially decayed plant matter. Over time, heat and pressure turn the peat into coal. There are four types of coal that contain various types and amounts of carbon and other impurities. Each type produces a different amount of heat energy. When placed under tremendous pressure from overhead rock layers, peat is transformed into lignite. Lignite is a soft coal with a woody texture. It is very moist and crumbles when exposed to air. Subbituminous coal forms when lignite is placed under pressure. This type of coal is less moist and has more carbon than lignite. The application of both pressure and heat to lignite produces bituminous coal. The majority of coal is bituminous coal, which has less moisture and more carbon than subbituminous coal. It also has a high sulfur content. When bituminous coal is burned, sulfur dioxide is released into the atmosphere. Sulfur dioxide is a serious form of air pollution that causes a variety of lung diseases. When sulfur dioxide mixes with the water in the atmosphere, acid rain precipitates. Bituminous coal under extreme pressure forms anthracite coal, the highest ranked coal and hardest form of coal. Anthracite coal contains the highest energy content.

Burning coal has both advantages and disadvantages. Coal is the most abundant fossil fuel, is relatively inexpensive to produce, and is easy to transport. Devices called *scrubber systems* work like a shower to dissolve sulfur oxides before they are released into the atmosphere. This process helps to reduce pollution. However, these systems do not reduce all pollution. Burning coal produces more carbon dioxide per unit of heat energy than burning either petroleum or natural gas. Coal mines can also pollute the land, and strip-mining coal out of the ground can harm a region's environment.

LESSON REVIEW
1. Describe how petroleum and natural gas are formed.
2. Contrast the properties and uses of the three major fossil fuels.
3. Examine the advantages and disadvantages of processing fossil fuels.
4. Name four types of coal and explain how they are formed.

FYI

Rapid Fossil Fuel Formation
Young-earth scientists do not agree with old-earth scientific evidence that supports the idea that fossil fuels require millions of years to form. Rather, their geologic data suggests fossil fuels can form rapidly. If given the proper anaerobic conditions, fossil fuels can be created within a few hours. So, where did today's fossil fuels come from? There are two explanations. One states that God created fossil fuels during the week of creation. Those fuel reserves were located deep in the earth. When the Flood occurred, the shifting of tectonic plates forced the reservoirs to crack open, and fuel permeated the upper layers of the earth's crust. The other young-earth explanation addresses additional fossil fuel quantities. An enormous amount of vegetation was buried very quickly during the Flood. The vegetation became the coal that can be observed today. Each of these schools of thought more closely adheres to the idea that the earth is less than 10,000 years old.

a major greenhouse gas. Coal is the most abundant fossil fuel, is relatively inexpensive to produce, and is easy to transport. The disadvantages of burning coal include more carbon dioxide produced per unit of heat energy than burning either petroleum or natural gas, coal mines that can pollute the land, and strip-mining that can harm a region's environment.)

4. Name four types of coal and explain how they are formed. (**Peat is transformed into lignite when placed under extreme pressure from overhead rock layers. It is soft, has a woody texture, is very moist, and crumbles when exposed to air. Subbituminous coal forms when lignite is placed under pressure. It is less moist and has more carbon than lignite. Applying pressure and heat to lignite produces bituminous coal. It has less moisture and more carbon than subbituminous coal, and it has a high sulfur content. Bituminous coal under extreme pressure forms anthracite coal, the highest ranked coal and hardest form of coal. Anthracite coal has the highest energy content.**)

NOTES

Name: _____ Date: _____

Lab 6.1.3A Oil Reserve Model

QUESTION: How do individuals locate oil reserves?

HYPOTHESIS: Answers will vary.

EXPERIMENT:

You will need:		
• sand	• balloon	• graph paper
• small rocks	• syringe	• metric ruler
• cardboard box	• viscous liquid	• probe

Steps:
1. Place sand and small rocks in a cardboard box such as a shoebox.
2. Fill a balloon with the viscous liquid using a syringe, and place the balloon in a layer of sand and rock.
3. Attach graph paper to the lid of the box, and poke holes large enough to insert a probe at each gridline cross section. Mark an X on one side of the lid, and secure the lid on the box.
4. Map the location of the "oil reserve" and exchange boxes with another group of students.
5. Do not adjust or open the lid of the other students' box.
6. Tap the lid of the box to estimate by sound where the oil reserve is located. Mark the areas on the graph paper that you think should be probed.
7. Use a ruler and mark 1 cm increments, beginning at the bottom, on the probe.
8. Probe gently into the areas you designated to be probed until the oil reserve has been tapped.
9. Monitor how many times the probe was inserted into the box and to what depth. Each centimeter costs $200,000.00, and each time the probe is removed and reinserted, a cost of $100,000.00 is applied.

ANALYZE AND CONCLUDE:

1. Calculate how much your oil tapping adventure cost. Answers will vary.

2. What changes would you make to the procedure? Possible answers: I would spend more time tapping and listening to the change in sound; I would probe in a general area rather than all spread out.

3. How did the oil reserve's estimated location compare to the actual location determined by the group that created the box? Answers will vary.

6.1.4 Rocks and Minerals

The Environment

OBJECTIVES

Students will be able to
- relate the properties of minerals with their uses.
- compare mining and quarrying.
- indicate issues associated with mining.

VOCABULARY

- **ore** a naturally occurring mineral from which a useful metal or mineral is recovered
- **quarry** a location where rocks are removed from the ground

MATERIALS

- Coins, gold and copper jewelry, graphite, enamel and porcelain, insecticides and matches, sandpaper, glass (*Introduction*)
- Copper sheets, aluminum wires, hammers (*Discussion*)
- Colored juice, beakers, molds, freezer (*A*)
- BLM 6.1.4A Mineral Uses
- TM 6.1.4A Mineral Uses
- TM 6.1.4B Metallic Properties

PREPARATION

- Prepare mineral samples for the *Introduction* display.
- Obtain materials for *Lab 6.1.4A Mining Minerals*. Purchase a birdseed mix containing sunflower seeds and millet.
- Gather materials for *Lab 6.1.4B Mining Conditions*.
- Arrange a field trip to a rock quarry, mine, or cement plant. (*C*)

Introduction

Have student groups create a list of minerals and brainstorm some different uses of the listed minerals. Present students with a variety of minerals that they may or may not have included in the original list. Some possible examples of minerals include the following or can be shown by displaying **TM 6.1.4A Mineral Uses**:
- Coins made of copper, nickel, or gold
- Gold and copper jewelry
- Graphite in pencils
- Enamel and porcelain made from feldspar
- Insecticides and matches made of sulfur
- Sandpaper derived from emery
- Glass made from quartz

Share more examples of minerals and their uses from **BLM 6.1.4A Mineral Uses**.

Discussion

- Provide sheets of copper and aluminum wires for students to examine. Explain that copper is an extremely malleable substance and the element aluminum is very ductile. Have students fold the copper sheets into multiple layers and pound them flat with a hammer. Display **TM 6.1.4B Metallic Properties**. Ask students to name the property being illustrated in each image. (*french horn*: **malleability**; *copper wires*: **ductility**; *horseshoe*: **good heat conductor**; *telephone line*: **good electricity conductor**; *molten aluminum*: **high melting point**; *metal anvil*: **hardness**)

- Discuss how human reliance on minerals affects the earth. (**The more humans rely on minerals, the more mining activities will occur that disrupt ecologic balance. Some mining efforts can promote the contamination of surrounding landscapes with asbestos, mercury, and other toxins.**) Encourage students to discuss how people can be good stewards of minerals. (**Answers will vary but should include how to extract and use minerals wisely.**)

- Lead a discussion about a different type of mining technology that incorporates the application of bacteria. Share with students that some bacteria are capable of isolating metals from their ores. When the bacteria are genetically modified, they can process metals with less environmental consequences. For example, to extract copper, the bacteria *Tiobacillus ferroxidans* and sulfuric acid are sprayed onto the rock. The solution reacts with the rock, excess liquid is removed, and the copper is extracted at a cost of about one third of the cost of using conventional processing methods. As an added benefit, the process produces nickel as a by-product.

- Ask the following questions:
  1. Name two malleable minerals and two ductile minerals. (**Possible answers: *malleable minerals*: aluminum, gold, silver, copper, steel, iron; *ductile minerals*: gold, copper**)
  2. Why are minerals harder to extract than rocks? (**Minerals often have to be separated from their ores by means of magnets, electricity, or by crushing and separating.**)
  3. What are some issues associated with mining? (**Returning the mining site to its original state is not always feasible. Wet residues or wastes are stored in ponds that may break, resulting in the release of toxic chemicals into rivers, lakes, or oceans. Dry wastes are stored in piles that could create dust, landslides, or, if exposed to water, could release acids and toxic metal compounds into the ground.**)

Activities

Lab 6.1.4A Mining Minerals

- 450 g of birdseed mix
- shallow pans, 1 per group
- gold beads or sequins, 2 per group
- silver beads or sequins, 4 per group
- red beads or sequins, 8 per group
- green beads or sequins, 16 per group

Share with students that the sunflower seeds represent iron, the other seeds represent waste, the gold beads represent gold, the silver beads represent silver, the red beads represent copper, and the green beads represent zinc.

Lab 6.1.4B Mining Conditions

- blueberry muffins, 1 per group
- paper towels
- paper clips
- toothpicks
- pins

Nut or raisin muffins or chocolate chip cookies may be used in place of blueberry muffins.

A. Have students create an analogy that describes the formation of an alloy. Direct groups to pour a small amount of colored juice into a beaker filled with water. Explain to students that the mixing of juice with water represents copper being mixed with zinc to make brass. Next, have students pour the mixture into a mold and place it in the freezer. Explain to students that molten

NOTES

6.1.4 Rocks and Minerals

OBJECTIVES
- Relate the properties of minerals with their uses.
- Compare mining and quarrying.
- Indicate issues associated with mining.

VOCABULARY
- **ore** a naturally occurring mineral from which a useful metal or mineral is recovered
- **quarry** a location where rocks are removed from the ground

Historically, people have utilized minerals in various aspects of daily living. Ancient people valued salt for its taste and as a food preservative. They used metals in tools and vessels. Gold, silver, and other minerals represented currency. Eventually, coins were stamped from these valuable metals. Rare gemstones were prized for their durability and beauty.

A mineral is a natural, inorganic solid found in the earth's crust. Many modern conveniences and necessities would not be possible without minerals. Water pipes, electric wires, toothpaste, soap, refrigerators, bowls, glasses, spoons, vitamins, clock radios, and doorknobs are manufactured with minerals. Almost everything that cannot be grown is a mineral or comes from a mineral.

Metals are extremely useful minerals. Metals are good conductors of heat and electricity. This means that they allow heat and electricity to easily pass through them. Because they are able to conduct electricity, copper and other metals are fashioned into electric wires. A variety of metals is used to make pots and pans, which efficiently conduct heat to cook food. Metals also have high melting points that prevent them from melting under ordinary temperatures.

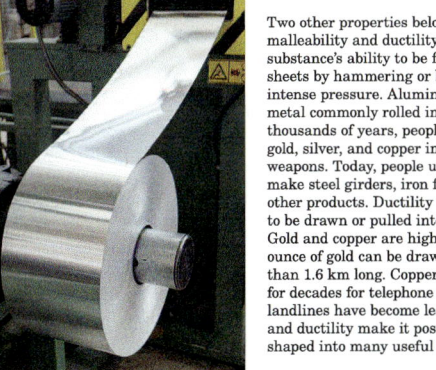

Two other properties belonging to metals are malleability and ductility. Malleability is a substance's ability to be flattened into thin sheets by hammering or by the application of intense pressure. Aluminum is a malleable metal commonly rolled into kitchen foil. For thousands of years, people have hammered gold, silver, and copper into jewelry, tools, and weapons. Today, people use malleable metals to make steel girders, iron fences, and countless other products. Ductility is a substance's ability to be drawn or pulled into the form of a wire. Gold and copper are highly ductile metals. An ounce of gold can be drawn into a wire more than 1.6 km long. Copper wires have been used for decades for telephone lines although now landlines have become less popular. Malleability and ductility make it possible for metals to be shaped into many useful forms.

Some metals are found in nature not bound to other elements. However, many metallic and some nonmetallic minerals form ores. An **ore** is a naturally occurring mineral from which a useful metal or mineral is recovered. Ores are often found in veins—ore-filled fissures. The veins vary in thickness from 1 cm to over 100 m. Some veins are deep within the ground. Others lie close to the earth's surface. Veins of ore at Earth's surface are exposed to weathering. They are often redeposited in gravel or sand. To be useful, the metals must be extracted from their ores. This is done with various methods that depend on the type of ore. Possible methods involve separating metal from ore by use of magnets or electricity, or by crushing and separating.

Press molding machines demonstrate the malleability of aluminum.

Ductile metal fences keep livestock secure.

The process of removing useful minerals from the ground is called *mining*. Mining minerals from the ground has a greater impact on the land than quarrying because valuable minerals must be separated from unusable ones. Metals must be separated from their ores. This type of mining operation leaves large amounts of waste. Dumping the rubble back into the hole is not an easy solution. Before they are mined, rocks and minerals are tightly compressed. However, once they are broken and processed, spaces are created between fragments, increasing the volume. Wet residues are kept in ponds. The ponds where the wet wastes are kept may break, releasing toxic chemicals into rivers, lakes, or oceans. Dry waste rock is stored in piles. If the discarded mineral fragments are exposed to water, they release acids and toxic metal compounds into the ground. Dry rubble can blow around as dust. Large, unstable piles of rubble may also create landslides.

Some minerals, such as gold and silver, must be processed with toxic chemicals after they are removed from the ground. The chemicals used for processing, such as potassium cyanide, can pollute nearby soil and streams. Today, the disposal of wastes is carefully regulated to ensure that damage is reduced.

Alloys, which are made of two or more metals or a metal and a nonmetal, are some of today's most important industrial materials. Alloys are often preferred over pure metals given that alloys are usually much stronger and harder than the individual metal components. Around 3,000 BC, the first alloy—bronze—was discovered. Bronze, an alloy of copper and tin, was so much

Iron ore

NOTES

brass is poured into molds to make musical instruments, decorations, doorknobs, and plumbing materials.

B. Direct students to choose one item from home that can be traced back to a mineral. Guide students to research how the mineral was mined, what issues there are in mining this mineral, the processes required to extract the mineral from an ore, any other processes necessary to make the mineral suitable for use, and how the mining for the mineral may have changed in the past 50 years due to technological advances. Have students present the mineral item and their findings to the class.

C. Visit a nearby rock quarry, mine, or cement plant to witness the nature of mineral extraction and the processing industry. Encourage students to ask on-site staff what steps they are taking in an effort to be environmentally responsible.

D. Assign students to research one to three minerals that are necessary for good health. Have students create a poster or computer presentation explaining the quantity or percentage of the mineral that the body requires, where the body obtains the mineral, how the mineral is used in the body, and the effect of too little or too much mineral in the body.

E. Direct students to write a recipe for pewter or bronze. Students should include the amount of each ingredient and thorough preparation instructions.

---

harder than copper that it transformed civilization, ushering in the Bronze Age. About 3,000 years ago, iron metallurgy was discovered. This began the Iron Age.

Today, people benefit from hundreds of alloys made from numerous metallic elements. Many alloys are formulated to resist corrosion and have low melting points. Iron-aluminum alloys are magnetic. Copper alloys have desirable thermal and electrical properties. Some alloys are very strong even at high temperatures, and others are unusually resistant to wear.

Metals are not the only useful minerals. Many nonmetallic minerals are also useful. The "lead" in your pencils is not lead but a mixture of clay and the mineral graphite. Feldspar is used to make glass, porcelain, and tableware. Garnet and emery are adhered to paper to make sandpaper. Barite is a mineral rich in barium, which supplies the bright green color in flares and fireworks. The mineral talc is used as a filler in cosmetics, paint, and plastics. Calcite is the primary mineral in limestone and marble. Muscovite has many uses in the electrical industry. Quartz is used to make ceramics, industrial cleaners, and laboratory equipment. Sulfur is used to make paper, rubber, and fertilizers. It is also used to preserve dried fruits. Fluorite is used in the process of smelting iron.

Minerals are not only practical, many are beautiful as well. Throughout history, people have fashioned precious gemstones and metals into jewelry and other decorations. Talc is used to make porcelain paper. Native Americans have used turquoise to make jewelry for centuries. Precious gemstones such as diamond, sapphire, and emerald, and semiprecious stones such as peridot, aquamarine, amethyst, citrine, and onyx are cut and polished as gemstones.

Rocks, which are composed of one or more minerals, are also valuable resources. Limestone, an important building stone, is the main component in cement, glass, and soil conditioners. Sand and gravel are ingredients of concrete and asphalt. Sandstone is an important building stone and it can be made into glass and other products. Clay

The Golden Gate Bridge is made of steel, an iron-carbon alloy.

and shale are materials mixed in cement, bricks, ceramics, and tiles. Gypsum provides strength to wallboard and is a stiffening agent in bakery products.

Quarries are locations where rocks are removed from the ground. Many old, abandoned quarries are still visible as holes, but many countries are now requiring mining companies to restore the sites to a usable status. Although quarries can leave holes and strip vegetation, they do not poison the land. Plants eventually grow again on the site. Abandoned quarry sites typically cannot be reworked into farming land, but they can eventually support interesting ecosystems.

LESSON REVIEW
1. The minerals used in pots and pans exhibit what physical properties?
2. Why do malleable and ductile physical properties help make minerals useful?
3. Where are rocks removed from the ground?
4. How does quarrying differ from mining?
5. How can individuals reduce the environmental impact associated with mining?

F. Have students research the history of mining in the local area. Ask students if there are any new innovations in mining.

Lesson Review

1. The minerals used in pots and pans exhibit what physical properties? (**good heat conduction and malleability**)
2. Why do malleable and ductile physical properties help make minerals useful? (**These two physical properties help make metals moldable into useful forms.**)
3. Where are rocks removed from the ground? (**quarries**)
4. How does quarrying differ from mining? (**Quarrying is removing rocks from the ground. The process leaves holes and strips vegetation but does not poison the ground. Mining is the process of removing useful minerals from the ground. Mining minerals from the ground has a greater impact on the land than quarrying because valuable minerals must be separated from unusable ones.**)
5. How can individuals reduce the environmental impact associated with mining? (**Possible answers: by mining only ores found close to the earth's surface; utilizing new methods of mining that do not have as great an environmental impact; preventing residue and waste leakage into rivers, lakes, or oceans; storing dry waste in more secure areas; utilizing less toxic chemicals when processing minerals after they have been separated from their ores**)

NOTES

Name: _____ Date: _____

Lab 6.1.4A Mining Minerals

QUESTION: How are minerals extracted from the ground?
HYPOTHESIS: Answers will vary.

EXPERIMENT:

You will need:	• 2 gold beads or sequins	• 16 green beads or sequins
• birdseed mix	• 4 silver beads or sequins	
• shallow pan	• 8 red beads or sequins	

Steps:
1. Pour the birdseed in the shallow pan.
2. Add the beads to the pan, and stir the mixture well.
3. Spend five minutes searching through the mixture and "mining" the various seeds and beads. The sunflower seeds represent iron, the other seeds represent waste, the gold beads represent gold, the silver beads represent silver, the red beads represent copper, and the green beads represent zinc. Place the mined products on a piece of paper in separate piles.
4. At the end of five minutes, count each kind of product from the piles made, and record the quantity in the table below.

	Gold	Silver	Copper	Zinc	Iron	Waste
Quantity						

5. Count and record the remaining millet grains scattered in the pan. _____

ANALYZE AND CONCLUDE:
1. Multiply the quantity of each mineral by the designated value of each mineral in the table below. Answers will vary.

Mineral	Quantity	Value	Total for each
Gold		$8.00	
Silver		$6.00	
Zinc		$3.00	
Copper		$2.00	
Iron		$1.00	

2. Calculate the total for all minerals. Answers will vary.
3. Multiply the quantity of millet grains by $5.00. This represents the cleanup cost. Subtract this amount from the total for all minerals. Answers will vary.

Lab 6.1.4A Mining Minerals

4. If any additional millet grains remained in the pan, multiply this quantity by $5.00. This represents the environmental damage penalty. Answers will vary.
5. What is your total profit? (Total for all minerals – cleanup cost – environmental damage penalty) Answers will vary.
6. Was mining for gold or iron more profitable? Why? Iron because there is a larger quantity even though the cost is less.
7. What is the most profitable, responsible way for miners to mine? Why? Mining iron was more profitable because there is a larger quantity even though the cost is less.

Name: _____ Date: _____

Lab 6.1.4B Mining Conditions

QUESTION: How does mining affect the land?
HYPOTHESIS: Answers will vary.

EXPERIMENT:

You will need:	• paper towels	• toothpicks
• blueberry muffin	• paper clips	• pins

Steps:
1. Place the muffin on a paper towel.
2. Describe the surface of the muffin. Answers will vary.
3. Draw a diagram of the muffin and speculate where the "deposits" (blueberries) are located, both on the surface and inside the muffin.

Answers will vary.

4. Using paper clips, toothpicks, and pins, remove the "deposits" from the muffin. Remember that you will have to reclaim your "land." This means the "land" will need to be returned to its original condition after mining. You may have to decide whether to mine some deposits.
5. When you have mined as much as possible, lay aside the deposits. Try to return the "land" to its original state.
6. Describe the surface of the mined muffin. Answers will vary.

ANALYZE AND CONCLUDE:
1. Compare the condition of the muffin before and after mining. How did the surface areas change as a result of mining? Answers will vary.

Lab 6.1.4B Mining Conditions

2. What problems were encountered when trying to reclaim the land? Not all areas could be patched, and not all crumbs fit neatly back in place.
3. Relate these problems to the problems that mining companies face. It is often difficult to return the unused portion of the ore to the mined area.
4. What factors should determine if mining should be allowed in areas where the land cannot be reclaimed? Answers will vary.
5. Were portions of the land mined only to reveal that no deposit lay underneath? Probably.
6. Were some deposits too difficult to reach without destroying the land? Answers will vary.
7. Is it practical to refrain from mining anything from the earth? No.
8. What solutions can you suggest to lessen the land destruction caused by mining? Answers will vary.

The Environment

6.1.5 Land

Introduction
Distribute the prepared poster board pieces to each student. Have students imagine that they have inherited a piece of land and enough money to do whatever they want with the land. Share with students that they may only designate one purpose for their land. Students may build a resort, ranch, farm, amusement park, mine, residential neighborhood, nature preserve, and so on. Allow students time to sketch and label their poster board. Collect the poster board pieces and arrange them in order according to the numbers on the reverse side of the poster board. Ask students the following questions: (**Answers will vary.**)
- How many preserved their land? What prompted them to do this?
- How many planned agriculture? If none of the students planned agriculture, ask them where they expect to get their food. Did they expect others to dedicate their land to agriculture?
- How many cleared wetlands and forests?
- How will their development contribute to pollution? Remind students that land adjacent to bodies of water may produce pollution that will contaminate the water sources and affect other land. Trace the pollution down the river and through the lakes. Mark any affected land.

Discussion
- Ask the following questions:
  1. What are some of the biggest threats to farmland? (**Answers will vary but should include expanding deserts, erosion, and increasing populations and cities.**)
  2. Why is deforestation not the best choice when creating new farmland? (**Rain forest soil has few nutrients and erodes easily. All forests play an important role in the environment, they are habitats for many organisms, they produce oxygen, and trees are storehouses of carbon.**)
  3. Why are wetlands valuable? (**Wetlands are important habitats and nurseries for waterfowl, fish, and invertebrates. They filter out pollution, and they help prevent flooding by soaking up extra water.**)

Activities
A. Display a current, detailed map of the local region. Have students develop a color code to identify the following areas on the map: residential, industrial, agricultural, recreational, and undeveloped land. Direct student volunteers to outline and label these areas with the appropriate color. Have students determine what area is predominant in the local region. Ask students what are some pollution sources. (**Answers will vary.**) How do they affect other regions? (**Answers will vary but students should identify pollution sources such as fertilizers and pesticides from farmlands or residential lawns, air pollution from heavy traffic, or chlorine from waste treatment plants.**) Have students determine some changes that could help balance the health of natural ecosystems with economic growth. (**Possible answer: more low-impact industry growth to reduce development and pollution in natural locations**) Direct student groups to develop an idea that preserves natural ecosystems and wildlife and also takes into account people's economic needs. (**Possible answers: Restrict the use of fertilizers and pesticides for homes near streams and rivers; those people building new residential homes in ecologically sensitive areas must maintain 50% of their property as a natural wildlife habitat.**)

B. Have students answer the following question in a few paragraphs: *What should be considered before developing new land?* (**Answers will vary but should include the idea that the development of new land depends on what the land will be used for, what need the new development will fulfill, and how the new land will affect surrounding areas.**)

C. Direct students to create a time line, outlining the progression and advancements in agriculture.

OBJECTIVES
Students will be able to
- explain three processes that cause land deterioration.
- identify ways that people can prevent land deterioration.

VOCABULARY
- **desertification** the making of new deserts by degrading land that used to be healthy and productive

MATERIALS
- Poster board (*Introduction*)
- Local map (*A*)

PREPARATION
- Prior to class, lay out pieces of poster board to represent a land mass. There may be one large land mass or two smaller, similar land masses. Sketch a few lakes, wetlands, forests, and streams on the land mass, and identify areas where natural resources might lie deep in the ground. Cut the land mass into equal size pieces—one for each student, and place numbers on the back of the pieces so the pieces may be put back together in order. (*Introduction*)
- If possible, arrange a field trip to a wetland. (*E*)

Earth and Space Science

NOTES

D. Have the class create a newscast describing the problem of desertification and water scarcity in Africa, Australia, or the United States. Assign student groups one of the following newscast roles:
- *Science Correspondent*: This group should focus on geographic, meteorological, or human factors contributing to desertification. Students should include graphics to illustrate the issues.
- *Community Correspondent*: This group should focus on the personal or human costs of desertification and water scarcity and provide ways to prevent desertification.
- *Economic Correspondent*: This group should focus on the impacts of desertification on the local and national economies as well as the global economy.
- *Political Correspondent*: This group should focus on the various political tensions created and made worse by these problems.
- *Production Group*: This group is responsible for integrating all the parts of the script and directing and recording the final production.
- *Set Designers*: This group is responsible for designing the set and obtaining appropriate stage props.
- *Expert Panel*: This group should get together at the end of the newscast. Their task is to suggest possible solutions to these issues.

E. If possible, take students on a field trip to visit a local wetland. Have students observe the plants and animals that are unique to this environment. Encourage students to hypothesize what might happen if the wetland became devoid of water. **(Answers will vary but should include that flooding may increase in surrounding areas, many of the species may become endangered, and there would be more pollution.)** Another option would be to communicate with a classroom

---

6.1.5 Land

OBJECTIVES
- Explain three processes that cause land deterioration.
- Identify ways that people can prevent land deterioration.

VOCABULARY
- **desertification** the making of new deserts by degrading land that used to be healthy and productive

There have been numerous inventions over the past few centuries that have changed people's lives. For example, it may be hard to imagine what life would be like without indoor plumbing, cars, computers, plastic, television, refrigerators, glass windows, sewing machines, aluminum cans, or cell phones. About 11,000 years ago, an invention changed the course of history. That invention was farming.

Farming and agriculture originated in the Middle East. Individuals who lived in the area known as *the Fertile Crescent* began clearing native vegetation in order to cultivate selective crops. Agriculture also began around the same time in East Asia with the domestication of millet and rice. Farming spread throughout the region of the Tigris and Euphrates Rivers more than 6,000 years ago, and soon people could raise enough food to support cities. Beginning 2,000 years later, agriculture spread throughout Europe. People in Central America began to grow grain more than 5,000 years ago. The developments in food production transformed the land and changed the world.

Land is a renewable resource. Although people cannot make more land, if properly managed, land can be renewed for further use. This renewal requires care, because improper treatment can cause land to become useless. Today, it is more important than ever not to let this happen. The world's population is growing quickly, in part, because many people have been blessed with better health care and more efficient food production. At the same time, however, the amount of farmland per person is shrinking. Expanding deserts and exponential population growth

Wheat field farming

308

HISTORY

Presidential Protection

"We have become great because of the lavish use of our resources. But the time has come to inquire seriously what will happen when our forests are gone, when the coal, the iron, the oil, and the gas are exhausted, when the soils have still further impoverished and washed into the streams, polluting the rivers, denuding the fields and obstructing navigation."
– Theodore Roosevelt

Theodore Roosevelt, 26th president of the United States, was a great proponent of conservation. When he became president in 1901, Roosevelt created the United States Forest Service to protect wildlife and preserve public lands. The intent was to conserve forests for continued use and utilize natural resources to ensure their sustainability. He established 150 national forests, 51 federal bird reserves, 4 national game preserves, 5 national parks, and 18 national monuments by issuing the American Antiquities Act. During his presidency, Theodore Roosevelt protected approximately 230 million acres of public land. Today, there is a national wildlife refuge in every state.

are destroying farmland. Because less land is now available to raise food, people should make good use of the land they have to ensure that everyone has enough to eat.

People in the United States and Europe do not feel the pinch of decreasing farmland as much as people who live in developing countries. About 80% of the world's people live in developing countries. Many of these countries are located in tropical regions where poverty, unequal land distribution, and growing populations are forcing the people to destroy rain forests and convert the land into cropland and pasture. Tropical forests, which make up half of the earth's forested land, have more plants and animals than any other ecosystem. Most of the nutrients are in the upper canopy of the trees, making the soil thin and poor. When rain forests are cut down for logging and agricultural purposes, the species-rich forest disappears and heavy rains wash away the topsoil. This makes the land fertile for only two or three growing seasons. Poor farmers must then clear more forests. The deforestation also puts the land at risk for floods and landslides. For example, in recent years, forests above the city of Ormoc in the Philippines were logged. The rains from a

Desertification by overgrazing

© *Earth and Space Science*   309

that is located near a wetland. Have students interact with one another by sharing photographs of local plants and animals. Encourage students to ask questions regarding how the wetland has changed over time.

Lesson Review

1. What factors contribute to the destruction and deterioration of farmland? (**expanding deserts and growing populations**)
2. Why is land considered a renewable resource? (**Land is the medium where individuals can cultivate food.**)
3. Explain why rain forest deforestation is a growing problem. (**Rain forest soil is not very fertile. When trees are cleared in these regions, the topsoil is easily washed away. The land is fertile for only two or three seasons. This forces individuals to cut down even more trees.**)
4. Why do desertification and salinization occur? (**Desertification occurs when land is degraded and forms new deserts. This may occur if livestock is allowed to overgraze or if too many forests are cut down to collect fuel. Salinization occurs when too many water-soluble salts accumulate in the soil because not enough water replenishes the soil after it evaporates.**)
5. What can individuals do to prevent land deterioration? (**Answers will vary but should include people can use recycled items, support land conservation efforts, reduce overgrazing, implement improved irrigation methods, rotate crops, restore wetlands.**)

NOTES

1991 typhoon caused a mudslide in the unforested area that killed 7,000 people.

One solution to the tropical deforestation problem is to encourage people who live in and near these forests to find ways to utilize rain forest resources instead of chopping down vegetation. Tropical plants offer countless foods and medicines. People who do not live in these forests can help prevent deforestation by using recycled items, reducing use of palm oil, and supporting conservation organizations.

Salinization

Another land use issue is **desertification**, the making of new deserts by degrading land that used to be healthy and productive. The degradation of land may be influenced by human activities such as letting livestock overgraze the land and cutting down forests to collect fuel. About 25% of the world's people live in dry areas. Most of the people in these areas make their living by raising livestock. If the livestock overgraze the dry land, the land turns into a desert. This activity along with other degradation processes converts 12 million hectares of land into deserts every year. Planting native vegetation in the affected areas can prevent desertification. The vegetation anchors soil and retains water. Other ways of preventing desertification include improving irrigation methods, rotating crops, rotating grazing patterns, and terracing (creating multiple steplike levels of flat ground on hillsides).

Salinization, the process in which water-soluble salts accumulate in soil, is another process that can ruin land. This process can happen in two ways. Salts occur naturally in soil. If the water that evaporates from the soil is not replenished, the concentration of salt increases. Too much salt is toxic to plants. Such land may have to be abandoned. This type of salinization occurs in arid and semiarid regions such as the plains and prairies. Salinization may also be caused by irrigation. If the water drawn from rivers or underground reservoirs has a high salt content, it will increase the salinity of that irrigated land over time.

Wetlands are another important land resource. A **wetland** is an area of land where the water level is near or above the soil surface for all or most of the year. Wetlands are important habitats and nurseries for waterfowl, fish, and invertebrates. They filter pollution and help prevent flooding by soaking up

310

excess water. Until recently, few people understood the value of wetlands. As a result, many wetlands were drained for farmland or converted into development areas.

Today, fewer wetlands are being destroyed. Some are being restored, and new wetlands are being created. A few groups of people worry that restoring wetlands creates a breeding ground for mosquitoes. However, mosquitoes are rarely a problem in properly restored wetlands. This is because the birds that wetlands attract eat the mosquitoes.

You can see that land resources are valuable. As people learn more about the design of the earth and how it works together as a whole, they can make better decisions about land use. It is important that everyone has enough room to live and enough food to eat. If cared for properly, the beauty of God's creation can be enjoyed by many future generations.

LESSON REVIEW
1. What factors contribute to the destruction and deterioration of farmland?
2. Why is land considered *a renewable resource*?
3. Explain why rain forest deforestation is a growing problem.
4. Why do desertification and salinization occur?
5. What can individuals do to prevent land deterioration?

Wetlands are an important natural resource.

6.1.6 Soil

The Environment

OBJECTIVES

Students will be able to
- analyze and explain why soil erosion occurs.
- investigate methods that help prevent soil erosion.

VOCABULARY

- **contour farming** the plowing of furrows around a hill perpendicular to its slope to reduce erosion
- **cover crop** the fast-growing vegetation planted on bare farmland to prevent erosion
- **crop rotation** the successive planting of different crops to prevent erosion and to improve fertility
- **furrow** a ditch in farmland
- **strip cropping** the planting of alternating bands of crops and cover vegetation in a planned rotation of equal widths
- **terrace farming** the construction of steplike ridges built into the slope of the land

MATERIALS

- Cardboard boxes, soil, pans, watering cans, metric rulers, grass and bean seeds, rocks, tile fragments (*Introduction*)

PREPARATION

- Pack soil into a cardboard box to form a slope. Remove a short side of the box, leaving the end open. (*Introduction*)
- Invite a local farmer, possibly one from an organic farm, to speak to the class. Encourage students to develop questions to ask the farmer prior to the visit. (*A*)
- If possible, arrange a field trip to a no-till farm. (*B*)

Introduction

Present students with a three-sided cardboard box containing soil that is packed into a slope to represent a field. Ask students what would happen if the soil was watered. (**The topsoil would run off.**) How could this be prevented? (**Answers will vary but should include by planting vegetation.**) Place the pan below the box and pour water over the soil from several angles. Have students observe the soil erosion. Provide student groups with a cardboard box. Direct them to pack the box with soil to form a slope. Set soil height requirements for each side of the slope. For example, the height limit for the top of the slope should be 15 cm and the height limit for the bottom of the slope should be 8 cm. Make adjustments for smaller boxes. Provide students with a variety of grass and bean seeds, rocks, and tile fragments. Have students create an agricultural blueprint for their soil slope to prevent as much soil erosion as possible. Guide students to place rocks and plant seeds after you approve their blueprints. Water the bare soil in your box as frequently as students water their seeds. When the seeds produce the first visible plants, direct students to measure the depth of the soil to determine which plan produced the least amount of erosion. Have students analyze each design and discuss why some designs had less erosion than others.

Discussion

- Ask the following questions:
  1. What are some things that cause erosion? (**deforestation, leaving farmland unplanted, and plowing furrows into steep hills**)
  2. What are some ways to reduce erosion? (**maintaining forests, planting native vegetation, planting cover crops, strip cropping, contour farming, terrace farming, crop rotation, and no-till farming**)
  3. How can trails cause rapid soil erosion? (**Paths frequented by vehicles or hikers can turn into streams during wet weather. These streams form gullies and cause serious erosion.**)
  4. What are cover crops, and how do they slow soil erosion? (**Bare land is exposed to wind and water, which allows for rapid erosion. Cover crops are plants used to cover bare land. Their roots secure the soil.**)

Activities

A. Introduce the local farmer whom you have invited to share the type of farming techniques employed, the type of crops produced, and the type of industry served by his or her farm.

B. If possible, take students to a no-till farm. Request that a farm staff member demonstrate the type of equipment that is used on the farm. Ask the staff member to explain why this type of farming was chosen and how the no-till farming technique has improved the farm's crop yield and has reduced soil destruction.

C. Assign one of the following farming techniques for students to research: strip cropping, contour farming, terrace farming, crop rotation, or no-till farming. Have students discover the advantages and disadvantages of the assigned technique, what group of people benefits most from the particular type of farming, and different technologies and farming methods that may have developed from a greater understanding of the particular farming technique.

D. Challenge students to illustrate erosion causes and solutions with labels added for descriptions.

Lesson Review

1. Why does soil erosion occur? (**Deforestation, exposed soil, and the plowing of furrows can lead to soil erosion.**)
2. Describe the two steps of soil erosion. (**Soil particles become detached from topsoil layers. These particles are then carried away from the land by wind or water.**)

3. Why would a farmer choose to plant cover crops in strips between rows of crops? (**The cover crop reduces the amount of bare land and greatly reduces soil erosion.**)
4. What type of land benefits most from terrace farming? (**land with steep slopes**)
5. What method involves the successive planting of different crops to prevent erosion and improve soil fertility? (**crop rotation**)
6. Name the two techniques farmers combine to reduce erosion by 75%. (**strip cropping and contour plowing**) How does the combination reduce erosion? (**Strip cropping is the planting of alternating bands of crops and cover vegetation. The narrow bands are plowed at right angles to the slope of a mountain, and the lower furrows trap the soil eroded from higher furrows.**)

NOTES

6.1.6 Soil

OBJECTIVES
- Analyze and explain why soil erosion occurs.
- Investigate methods that help prevent soil erosion.

VOCABULARY
- **contour farming** the plowing of furrows around a hill perpendicular to its slope to reduce erosion
- **cover crop** the fast-growing vegetation planted on bare farmland to prevent erosion
- **crop rotation** the successive planting of different crops to prevent erosion and to improve fertility
- **furrow** a ditch in farmland
- **strip cropping** the planting of alternating bands of crops and cover vegetation in a planned rotation of equal widths
- **terrace farming** the construction of steplike ridges built into the slope of the land

Have you ever closely examined soil? Without soil, most creatures living on land would not last very long. Soil is a very precious gift. God designed soil to be a source of nutrients for living things—even those that live in the water.

Soil is formed over a long period of time by the interaction of living and nonliving things. Rain, wind, sleet, and ice break rocks apart through weathering. The process of erosion transports these rock particles to other places. The resulting rock particles are an important part of soil. Nutrients are added to soil by bacteria, decomposing plants, and animals. The formation of soil is an extremely slow process. Soil is formed at a rate of about 10 mm every 100 to 400 years.

Although the process of erosion naturally forms soil, problems arise when erosion speeds up so much that the topsoil washes or blows away faster than it can be replaced. Erosion may speed up as a result of human activity. Soil's ability to be renewed depends on human use.

For example, trees and other plants prevent erosion because their roots hold soil in place. When hills and other land surfaces are deforested, the topsoil is exposed to water and wind. The topsoil can then be washed or blown away. Leaving farmland bare allows wind and running water to speed up erosion. Plowing **furrows**—ditches in farmland—down sloped land allows water to move quickly over soil, forming gullies. A gully is a narrow ditch cut in the earth by runoff. This process, known as *gully erosion*, wears away land quickly and can make the area useless for farming.

Soil erosion happens in two steps. First, soil particles become detached from topsoil layers. Then, these particles are carried away from the land. Reducing soil erosion must address both of these steps. When land is not being used to grow crops, cover vegetation should be planted. A **cover crop** is fast-growing vegetation planted on bare farmland to prevent erosion. These crops also help to increase nutrients in the soil and provide organic matter. When rain falls on vegetation or wind blows across plant-covered ground, soil particles are less likely to detach from topsoil.

Another method farmers often employ is the planting of grains after the fall harvest to prevent erosion during the winter. Alternating strips of crops and grass can also help by reducing the amount of bare land exposed to erosion. Corn may be planted in bands next to alfalfa, which is an excellent cover crop. The cover crop greatly reduces soil erosion. The planting of alternating bands of crops and cover vegetation in a planned rotation that are of equal widths is called **strip cropping**.

Combining strip cropping with contour plowing can reduce erosion by 75%. **Contour farming** is the plowing of furrows around a hill perpendicular to its slope to reduce erosion. Because the furrows are at right angles to the slope, lower furrows trap the soil eroded from higher furrows. If furrows

Cover crop

Contour farming

BIBLE CONNECTION

The Sabbath Year 
In Leviticus 25, God commanded Moses to provide instructions to the Israelites regarding the use of their land. The Israelites may not have rotated their crops or introduced terrace farming, but they did give their fields a year of rest. God said, "For six years sow your fields, and for six years prune your vineyards and gather their crops. But in the seventh year the land is to have a year of Sabbath rest, a Sabbath to the Lord" (Leviticus 25:3–4a). The purpose of resting the land was to invigorate the land for future production. The Israelites learned that they did not need to worry during this year, because the Lord always provided.

FYI

Farmers Using Global Positioning Systems
Global positioning systems are no longer just for providing travel directions. Farmers around the world are participating in "precision farming" or "site-specific farming" with the aid of GPS and geographic information systems (GIS). They use these systems to map fields, plan farms, sample soils, guide tractors, measure variable rate applications, scout crops, and map yields. GPS is especially useful when farmers attempt to work in low visibility conditions such as dust, rain, fog, or darkness. Specifically, farmers employ GPS to help them more precisely apply pesticides, fertilizers, and herbicides. This practice ensures the farmer has better control over the dispersion of specific chemicals, and they no longer have to use human "flaggers" to guide aircraft sprayers. As the crop dusters fly over the field, chemicals are applied only to certain areas. This reduces the amount of chemicals needed and minimizes chemical drift. Overall, GPS has helped farmers reduce expenses, increase their production yields, and become more environmentally friendly.

were plowed parallel to the slope, water would run down through them, carrying away precious topsoil.

Terrace farming is often used to prevent erosion on steeper slopes. **Terrace farming** is the construction of steplike ridges that are built into the slope of the land. The terraces help keep water from flowing downhill. Short, level steps allow heavy rains to soak into the soil rather than run off and cause erosion. Bench terracing converts the land into a series of level strips like a staircase. This method requires extensive planning, but it is very effective. Countries that experience severe erosion hazards and require flood irrigation methods benefit from bench terracing.

Crop rotation is another soil conservation method. **Crop rotation** is the successive planting of different crops to prevent erosion and to improve fertility. Some crops, such as corn, expose the soil to erosion. Cover crops, such as alfalfa, do not. Farmers may choose to plant corn one year and alfalfa the next. The small gullies formed during the corn years are filled in with soil during the alfalfa years. Some crops, such as soybeans, replenish important nutrients like nitrogen to the soil that were depleted by previous crops such as corn.

A recent method of farming is called *no-till farming*. Rather than plowing the land, the farmer leaves the stubble from the previous year's crop on the surface and plants the new crop using special planters that do not greatly disturb the soil. The stubble prevents wind and water erosion and helps trap snow for additional moisture.

Soil erosion is an ever present issue. Scientists continue to investigate methods to prevent it. Can you think of other ways to prevent soil erosion?

LESSON REVIEW
1. Why does soil erosion occur?
2. Describe the two steps of soil erosion.
3. Why would a farmer choose to plant cover crops in strips between rows of crops?
4. What type of land benefits most from terrace farming?
5. What method involves the successive planting of different crops to prevent erosion and improve soil fertility?
6. Name the two techniques farmers combine to reduce erosion by 75%. How does the combination reduce erosion?

Terrace farming

The Environment

6.1.7 Alternative Energy Sources

Introduction
Share with students that many people are accustomed to easy access to energy such as turning on a light or driving a car. Have students imagine how their lives would change if all of the fossil fuels were depleted. Ask them how they would produce electricity. (**Answers will vary.**)

Have students suggest other processes that God has woven into the design of creation that people might be able to harness and use for energy. (**Possible answers: wind, geothermal energy, hydroelectric energy, and biomass**) Ask students which of these energy sources they have observed or currently use in their homes. (**Answers will vary.**)

Guide students to list ways that people use nuclear energy. (**Possible answers: generating power, radiation treatments of cancer, weapons**) Why is the use of nuclear energy controversial? (**Possible answers: There are safety hazards associated with harnessing and maintaining nuclear energy. Also, contamination due to improper disposal of nuclear waste is of concern.**)

Discussion
- Have student groups determine ways that they could accomplish daily tasks using alternative energy sources. Discuss ideas as a class.

- Lead a discussion to review different alternative energy sources and to have students briefly explain how they work. As a class, decide which alternative energy source would be the best choice to harness energy in the local region. Ask whether this energy source is already being used, and why. (**Answers will vary.**)

- Ask the following questions:
  1. Why are alternative energy sources preferred over fossil fuels? (**The production of alternative energy does not produce greenhouse gases. These sources are also renewable.**)
  2. Why are more fossil fuels still being used instead of alternative energy sources? (**Fossil fuels are relatively inexpensive and more convenient to harness and maintain. Systems, technology, and processes are already in place to retrieve and use them. It can be difficult to change a system that has been operating for many years.**)
  3. Why are solar energy and other alternative forms of energy not more widely used? (**Answers will vary but should include that other alternative forms of energy may be expensive to harness and maintain, some forms of energy produce harmful by-products, and some energy sources may not always be available.**)

Activities
A. Complete *Try This: Solar Energy Conversion* with students. Have students relate this activity to solar photovoltaic cells.

B. Complete *Try This: Turbines* with students. Discuss industries that utilize turbines.

C. As a class, research how to build a Telkes solar oven, and test designs by trying to cook a grilled cheese sandwich or pizza.

D. Direct student groups to design an experiment to test the energy supply of different types of biomass, to include various types of nuts, grass clippings, and other organic material. If time permits, encourage student groups to conduct their experiment.

E. Inform students they will be participating in a simulation of nuclear fission that creates a chain reaction. Give one foam ball to each of two students. Instruct all students to gather together to create a uranium nucleus. Toss a foam ball, representing a neutron, at the group of students. Have

OBJECTIVES
Students will be able to
- examine ways people harness alternative energy sources.
- summarize uses of alternative energy sources.
- compare the benefits and drawbacks of using alternative energy sources.

VOCABULARY
- **biomass** organic matter that contains stored energy and is used to produce fuel
- **geothermal energy** the energy collected from heat trapped in the earth's crust
- **nuclear energy** the energy that comes from changes in the nuclei of atoms of radioactive elements
- **solar energy** the radiation from the sun that causes chemical reactions, generates electricity, and produces heat

MATERIALS
- Nuts, grass clippings, other organic material (D)
- Foam balls (E)

PREPARATION
- Obtain materials for *Try This* activities.
- Gather materials to build a Telkes solar oven. (C)

TRY THIS
Solar Energy Conversion
- *tape*
- *black construction paper*
- *150 mL beakers, 2 per group*
- *plastic wrap*
- *thermometers, 2 per group*

(**The temperature change in the wrapped beaker was greater because the paper absorbed more solar energy that was then transferred to the water.**)

TRY THIS

Turbines
- tape
- cardboard strips or metal or plastic fins, 4 per student
- corks, 1 per student
- pushpins, 2 per student
- craft sticks

(wind or geothermal resources)

the group then split into two fission products. Direct the two students holding foam balls to separate from the group. Guide these two students to toss their foam "neutron" balls at the two fission product groups and continue the process. A total of four fission product groups will result along with six neutrons. Discuss the power of nuclear fission. Have students write a summary of the activity and discussion.

F. Assign students to locate the closest nuclear power plant. Direct them to provide details regarding the plant's name, the plant's location, when the plant began operation, how much electricity the plant produces annually, and if there have been any issues associated with the plant's operation.

G. Assign student groups to determine possible methods to contain nuclear waste. Students should consider both short-term and long-term solutions. Discuss ideas as a class.

Lesson Review

1. Describe how solar cells work. Why do consumers value solar cells? (**Sunlight enters the cell and electrons flow across the layers of silicon and metal, solar inks, solar dyes, or conductive plastics to create an electric current. The electric current is harnessed to be a source of energy and electricity. Solar cells do not have any moving parts, do not require fuel to produce electric power, and they can last for many years.**)

2. How can wind generate power? (**Wind spins the blades of a wind turbine that is connected to a hub that is mounted on a turning shaft. The shaft travels through a gear transmission box. A generator converts the mechanical energy into electrical energy.**)

6.1.7 Alternative Energy Sources

OBJECTIVES
- Examine ways people harness alternative energy sources.
- Summarize uses of alternative energy sources.
- Compare the benefits and drawbacks of using alternative energy sources.

VOCABULARY
- **biomass** organic matter that contains stored energy and is used to produce fuel
- **geothermal energy** the energy collected from heat trapped in the earth's crust
- **nuclear energy** the energy that comes from changes in the nuclei of atoms of radioactive elements
- **solar energy** the radiation from the sun that causes chemical reactions, generates electricity, and produces heat

God wove enough energy resources into the design of the solar system to provide all of the energy people would ever need. Communities have already begun to tap into some of this energy. Fossil fuels have been a tremendous blessing over the past years. However, they are nonrenewable resources, so people need to find other ways to heat homes, power vehicles, and run factories.

Consider the sun's energy. Think about how hot the sun feels when you are sitting on a beach in the middle of the summer, or the scorching temperatures the sun produces in deserts. **Solar energy** is radiation from the sun that causes chemical reactions, generates electricity, and produces heat. The earth receives more energy from the sun each year than the combined energy total contained in all of the known fossil fuel reserves. In fact, every year the sun emits enough radiation to supply the earth with 15,000 times more energy than the world consumes. This makes solar energy a good energy choice.

Solar energy can be captured and used in many ways. Simple forms of direct solar energy collection include the positioning of windows in a house to absorb the maximum amount of sunlight and hanging clothes outside to dry. Technology has helped people capture solar energy more efficiently. Sunlight can be changed directly into electrical energy by using solar cells. Solar cells have been made out of layers of silicon and metal, solar inks, solar dyes, and conductive plastics. When sunlight enters the cells, electrons flow across the layers, creating an electric current. One solar cell produces a very small amount of electrical energy. For example, it takes a few solar cells to power a calculator or a watch. Thousands of solar cells are needed to provide enough electrical energy for a house or a business. A building that uses solar energy typically has large panels mounted on the roof to provide the electrical energy. Solar panels consist of many solar cells wired together.

Solar cells are quiet and reliable. They do not have any moving parts and do not require fuel to produce electric power, and they can last for many years without maintenance. The use of solar cells does not create any pollution, and solar energy is free. Why, then, do so few people harness solar energy this way? One reason is cost. Although solar energy is free, solar cells are expensive to make. Fortunately, the cost of solar power has been decreasing, and many believe that solar power will become more cost effective compared to other sources of energy.

Solar energy can also be captured by using unglazed solar collectors. Unglazed solar collectors consist of boxes covered with a dark metal or plastic that absorbs solar energy. The energy is transferred to a fluid in the collector. The fluid is then pumped through tubes that run through pipes. This action raises the temperature of the fluid. In warm, sunny regions, many people have solar water heaters. Even if an individual does not live in a warm and sunny climate, solar energy can be harnessed in any climate. Solar collectors can also be used to generate electricity.

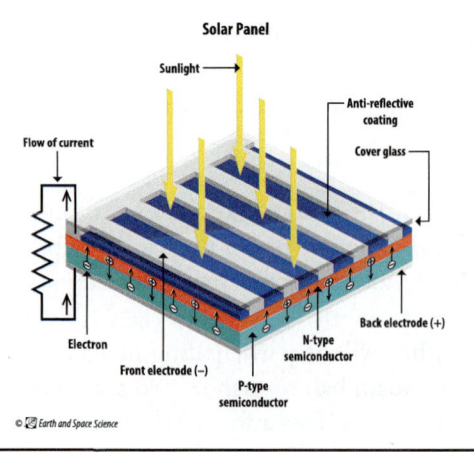

TRY THIS

Solar Energy Conversion
Tape a band of black construction paper around one beaker. Fill it and one other beaker with 100 mL of room temperature water. Cover both beakers with plastic wrap. Poke a small hole in the plastic wrap of both beakers and insert a thermometer into each beaker. Record the initial temperatures. Place both beakers on a windowsill in direct sunlight for 20 minutes. Record the final temperatures. Why was there a noticeable temperature difference between the water samples in the two beakers?

BIOGRAPHY

Mária Telkes
Mária Telkes, a Hungarian-born American physical chemist and biophysicist, designed the first solar-powered heating system to be used in the home. In 1948, the world's first home heating system was constructed in Dover, Massachusetts that used solar collectors and Glauber's salts to store heat. Telkes also created a solar distiller during World War II that separated salt from seawater, creating water that was safe to drink.

Solar Panel — Sunlight, Anti-reflective coating, Cover glass, Flow of current, Electron, Front electrode (−), P-type semiconductor, N-type semiconductor, Back electrode (+)

3. What concerns do some individuals have regarding wind turbines? (**They are noisy. Some believe the land where turbines have been built could be used for more productive things such as agriculture. Others are concerned about the threat turbines pose to birds and bats.**)
4. Explain how individuals harness geothermal energy. (**People have positioned turbines near steam that is produced by hot groundwater that is in close proximity to magma. The steam turns the turbines to generate electrical energy. Hot water can also be pumped through pipes that travel to buildings to produce heat.**)
5. Name some benefits associated with the use of biomass. (**Using biomass helps conserve fossil fuels, reduces greenhouse gas emissions and other air pollutant emissions, and saves space in sanitary landfills.**)
6. Describe the nuclear reaction that is most commonly used today to harness nuclear energy. (**Nuclear fission occurs when the nuclei of radioactive atoms split and a tremendous amount of energy is released.**)

NOTES

TRY THIS

Turbines
Tape cardboard strips to a cork or insert metal or plastic fins into slits on the cork that were cut using a single-edge blade. Insert pushpins into the two ends of the cork to act as axles. Fashion a U-shaped holder using craft sticks. Allow water to flow underneath the turbine. What other renewable resources could move the turbine?

Another alternative energy source is wind power. A substantial percentage of the solar energy reaching Earth is converted into wind energy through convection currents that develop as a result of temperature differences. Wind power can be used to generate electricity. The wind spins the blades on a wind turbine that is connected to a hub that is mounted on a turning shaft. The shaft travels through a gear transmission box, which increases the turning speed. A generator converts the mechanical energy into electrical energy. Wind power is very inexpensive in areas where strong winds blow. In Denmark, 42% of the population's energy consumption is generated by wind turbines. A drawback of wind power is that not all regions receive enough wind to make good use of wind power. Wind turbines are also noisy. Some people are concerned about the amount of space taken up by wind turbines and the threat they may pose to flying bats and birds.

Imagine obtaining energy to run your vehicle or heat your home from plants. **Biomass** is derived from organic matter, such as crops or waste products, that contains stored energy and is used to produce fuel. This energy ultimately comes from the sun. Plants use energy from the sun during photosynthesis to make food, and animals obtain this energy by eating plants. You have used biomass energy if you have ever burned wood or paper. Food

Wind power

is also biomass. About 90% of energy consumption in developing countries comes from burning wood or charcoal. Many people also burn dried animal dung. Other sources of biomass include yard clippings, crop waste, and sawdust or bark from lumber mills. Using biomass not only helps conserve fossil fuels, it also helps reduce greenhouse gas and other air pollutant emissions and saves space in sanitary landfills.

New ways of using biomass are now being discovered. Biogas—methane from manure and organic waste—is an energy source in some developing countries such as portions of India and China. Another way to use biomass is through a process called *fermentation*, which converts sugars and starch into alcohol. This alcohol can be separated and burned as a fuel or mixed with gasoline to produce gasohol. One bushel of processed corn can produce about 9.5 liters of ethanol. It would take more than 90% of the land in the United States to make fuel that its citizens use in their cars. Although biomass is renewable, it requires a tremendous amount of land that could be used to grow food for people.

Biomass power plant with wood chip fuel storage

Certain bacteria can act on biomass to make methane gas, which is then converted into methanol. Both ethanol and methanol are now being blended with gasoline to reduce pollution levels, and biomass has the potential to be converted into gases that are burned to produce electricity.

Another alternative energy source is known as **geothermal energy**—energy collected from heat trapped in the earth's crust. In some regions, the rainwater that penetrates the ground comes in close proximity to magma. The energy from the magma heats the groundwater and turns it into steam. The water can reach temperatures of more than 150°C—much hotter than boiling water. This steam and hot water escape through geysers or wells. Sometimes engineers use the steam to turn turbines to generate electrical energy. They also pump hot water through pipes that travel to buildings to produce heat. A place called *The Geysers* in California generates enough electricity to power a city the size of San Francisco.

If you have ever visited Niagara Falls or some other large waterfall, you have seen how much energy falling water holds. Hydroelectric power is electricity produced from the conversion of potential energy in falling or fast-flowing water to mechanical energy. A dam built across a river holds back the water, which is then directed past the turbines of a generator to produce electricity. Hydroelectricity is the world's most widely used renewable resource. It supplies nearly 16% of the world's electric power. This type of power is well suited for hilly countries that receive high amounts of rainfall. Dams, however, disrupt the migratory paths of salmon and steelhead. Dams also lower water quality and cause erosion problems.

Geothermal power plant in Kenya

Another alternative energy source is **nuclear energy**—energy that comes from the changes in the nuclei of atoms of radioactive elements such as uranium. Usually nuclear energy is produced through a process called *fission*, in which the nuclei of radioactive atoms are split and a tremendous amount of energy is released. Nuclear power plants use radioactive atoms as fuel. The energy released by fission produces steam to run electric generators. Nuclear reactors are also used to propel ships, submarines, and spacecraft.

Although the production of nuclear power does not produce the same kinds of pollution compared to the burning of fossil fuels, it does produce dangerous wastes. These radioactive wastes must be removed from the power plant and stored until they are no longer radioactive. The deactivation of radioactive waste can extend for thousands of years.

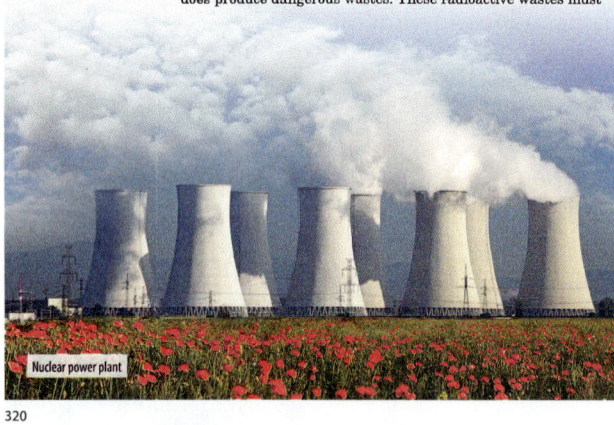

Nuclear power plants generate a lot of heat. In response, large amounts of water are required to cool power plants. If the cooling system fails, the plant can overheat and the reactor could melt. The meltdown would allow a large amount of radiation to escape into the environment. This happened in Chernobyl, Ukraine, in 1986 and in Fukushima, Japan, in 2011.

Nuclear power plants are more expensive to build than fossil fuel facilities. However, nuclear power is relatively inexpensive to maintain. As more fossil fuels are depleted, nuclear energy will probably become a common choice. Uranium, the fuel used in nuclear power plants, is also relatively inexpensive compared to coal and other fossil fuels. Scientists estimate that a pellet of uranium that has a mass of several grams can produce as much energy as 564 L of oil or 807 kg of coal. Uranium is a nonrenewable resource, but many speculate that uranium supplies will last longer than oil supplies. This projection is subject to change with new data observations.

Nuclear fuel rods

Nuclear energy can also be produced through a process known as *fusion*—the joining of two or more nuclei with small masses into a nucleus with a larger mass. Very high temperatures are required for this reaction to occur. As a result, fusion reactions are limited to laboratory experiments and are currently not useful as an energy source.

LESSON REVIEW
1. Describe how solar cells work. Why do consumers value solar cells?
2. How can wind generate power?
3. What concerns do some individuals have regarding wind turbines?
4. Explain how individuals harness geothermal energy.
5. Name some benefits associated with the use of biomass.
6. Describe the nuclear reaction that is most commonly used today to harness nuclear energy.

6.2.0 Pollution Solutions

Chapter 6.2 Summary
Despite people's everyday activities having negatively impacted the environment, there are many ways to recover from the damage and to reduce or prevent future damage. God created a biosphere that recycles and reuses nutrients, water, soil, and air; likewise, it is prudent for people to reuse and recycle the many products used in order to lessen the amount of pollution created.

Background
Lesson 6.2.1 – Land Pollution
Solutions to the solid waste problem revolve around changing attitudes and practices—people could choose to replace their throw-away mentality with one of conservation and repair; manufacturers could reduce unnecessary packaging of their products; and metal, plastics, paper, and glass could be recycled. Crushed glass could replace a percentage of the stone and sand used in asphalt for road construction; and new technology developed by the U.S. Bureau of Mines to convert organic waste into oil could become the norm.

Various disposal methods apply to solid waste; all have advantages and disadvantages. Sanitary landfills involve depositing refuse in shallow layers. Within 24 hours of the trash being dumped, the layers are compacted and sealed by soil or a chemically inert material. Liners made of clay, plastic, or composite are placed between each layer. However, clay is easily fractured, and certain organic chemicals can diffuse through clay. Some chemicals can also degrade clay. The best landfill liners are of tough plastic, but certain household chemicals will permeate even these liners. Vinegar, shoe polish, and margarine can cause stress cracks in plastic liners. Composite liners include a plastic liner and compacted soil. Under certain circumstances even these can crack and leak. The leachate from these landfills can pollute groundwater and harm living things; leachate from hazardous materials added to the landfill greatly multiplies this risk.

Incineration reduces the volume and weight of solid waste, but it does not destroy many hazardous chemicals. Also, burning may actually release chemicals that were safely embedded in trash. Carcinogens, such as dioxin, furan, and toxic metals found in both gas emissions and solid residues, are the most troubling concern of incinerating trash. Gas emissions from incineration must be properly controlled and solid residues buried properly. Although many incinerators convert garbage to energy, it has been impossible to destroy all carcinogenic chemicals before the exhaust is released.

Hazardous wastes are generated primarily by chemical production, manufacturing, and other industrial activities. Hazardous waste landfills are required to have double liners and collection systems for leached materials, but even landfills compliant with current regulations may still leak. Unfortunately, many home hazardous wastes such as oil-based paints, solvents, motor oil, and used car batteries can end up in local landfills not designed to accommodate hazardous waste products. Encouragement by community leaders to properly dispose of these materials and products could minimize this problem.

Radioactive wastes are another environmental threat. The running of commercial nuclear power plants and the manufacture of nuclear weapons generate radioactive waste. Nuclear power plants need to replace their fuel once a year. The used fuel in fuel rods contains uranium, plutonium, and other radioactive isotopes that remain dangerous for thousands of years. Many proposals for disposing of radioactive wastes such as rocketing it into space, dumping it in Antarctica, or burying it in the ocean have all been rejected for economic or environmental reasons. The current plan for radioactive waste disposal is burying it in bedrock or salt mines. Salt mines are considered safe storage places because the existence of undissolved salt indicates that running water does not pass through these places. Some salt mines once considered for storage sites, however, were found to contain water. Canada is considering the burial of radioactive wastes deep in the Canadian Shield, a large area of igneous and metamorphic rock that forms North America.

LOOKING AHEAD

- For **Lesson 6.2.1**, plan a field trip to a recycling plant or conservation agency.
- Invite a missionary or health-care worker that has worked in a country with poor water conditions to speak to the class. Begin growing bean plants and obtain feathers and water quality testing kits for **Lesson 6.2.3**.
- For **Lesson 6.2.4**, plan a field trip to the local department of conservation. Have students collect packaging waste from home for a week.

SUPPLEMENTAL MATERIALS

- BLM 6.2.2A The Air We Breathe

- Lab 6.2.1A Sanitary Landfill
- Lab 6.2.2A Air Pollution and Plants
- Lab 6.2.2.B The Air We Breathe
- Lab 6.2.3A Oil Spill
- Lab 6.2.3B Acid Rain

- WS 6.2.1A Recycling Journal
- WS 6.2.4A Do the Math

- Chapter 6.2 Test
- Unit 6 Test

BLMs, TMs, and tests are available to download. See Understanding Purposeful Design Earth and Space Science at the front of this book for the web address.

WORLDVIEW

- The apostle John wrote in Revelation 21:1–5a, "Then I saw 'a new heaven and a new earth,' for the first heaven and the first earth had passed away, and there was no longer any sea. I saw the Holy City, the new Jerusalem, coming down out of heaven from God, prepared as a bride beautifully dressed for her husband. And I heard a loud voice from the throne saying, 'Look! God's dwelling place is now among the people, and He will dwell with them. They will be His people, and God Himself will be with them and be their God….' He who was seated on the throne said, 'I am making everything new!'" It is true that this picture of a new heaven and a new earth is glorious, and it gives Christians great hope for a world free of death and destruction and full of beauty and perfection. It can be tempting to look at this future only and forego the responsibility to the earth here and now. It is easy to operate from the mind-set that this earth will all be destroyed some day and replaced, so why bother taking care of it? Even though God is going to replace this earth, Christians are not released from current responsibility. God has first given us this world to steward with all the resources and trials it avails. As His children, it is our responsibility to manage His resources to the best of our ability, to do all things as unto the Lord. This includes reducing waste, reusing products, and recycling all that can be recycled.

Lesson 6.2.2 – Air Pollution

There are several types of air pollution. Ozone, a summertime air pollution, is created when other pollutants in the air react with sunlight. Ozone is beneficial in the ozone layer, but at ground level it is a health hazard.

Particulates are solid air pollutants that are 0.025–0.1 mm in diameter. Particulates are produced by many sources, including the burning of diesel fuels; the burning of fossil fuels; the mixing and application of fertilizers and pesticides; road construction; industrial processes such as steel making, mining, and agricultural burning; and the use of fireplaces and wood stoves. Particulates can damage the lungs and are the main source of haze, which reduces visibility.

Visible air pollution is called *smog*. Chemists refer to smog as photochemical smog because of sunlight's role in creating it. Smog is a mixture of many different chemical compounds, the concentration of which varies throughout the day. Cars give off nitrogen oxide (NO), and in the presence of solar energy, nitrogen oxide reacts with oxygen in the atmosphere to form ozone (O_3). Nitrogen oxide also reacts with ozone to produce nitrogen dioxide (NO_2). Ozone reacts with atmospheric oxygen and hydroxide (OH) molecules to form highly reactive molecules called *free radicals*. Many different hydrogen carbons and free radicals then react with nitrogen dioxide to produce aldehydes and other smog components.

Smog can be related to temperature inversion, a condition in which the atmosphere's temperature increases with altitude instead of decreasing. This condition may arise when a cold front is passing, or from overnight radiative cooling of surface air. During a temperature inversion, air pollution that has been released into the atmosphere's lowest layer is trapped and can be removed only by strong horizontal winds. High-pressure systems can combine temperature inversions and low speed winds, resulting in severe smog that cannot disperse.

Outdoor air pollution has far-reaching effects. For example, acid rain often falls hundreds of kilometers away from where it formed. Acid rain created in the industrial regions of the Ohio Valley falls on the northeastern region of the United States and on eastern Canada. Various organisms suffer when acid accumulates in lakes; for example, many populations of trout and salamanders in North America and Scandinavia have been wiped out. Acid rain also wears away stone surfaces on buildings and monuments, and it can affect the visibility in a region. Biologists now suspect acid rain deteriorates forests.

Indoor air pollution is a hazard in both developed and less developed countries. In developed countries, insulation traps pollution, such as carbon monoxide from space heaters or gas stoves, inside buildings. Carbon monoxide drives oxygen from the bloodstream, causing apathy, fatigue, headache, disorientation, and decreased muscular coordination and visual acuity.

Various types of air pollution can cause asthmatic attacks, hyperreactive pathways, respiratory infections, and reversible changes in lung function. Nonrespiratory effects from air pollution can result when pollutants are inhaled in through the lungs and then absorbed into the bloodstream. For example, airborne lead can cause nervous disorders, hyperactivity, kidney damage, and poisoning, especially in children. In less developed countries, a lack of running water and indoor sanitation can abet respiratory infections.

Lesson 6.2.3 – Water Pollution

Without clean water, life on Earth could not exist. If the capacity of a body of water to dissolve, disperse, or recycle is exceeded, all additional substances or forms of energy become pollutants.

Oil spills first became a major problem in the 1960s with the intensified petroleum exploration on the continental shelf and the use of large tankers. Oil spills are difficult to clean because the slicks quickly disperse in many directions. Cleaning methods include skimming, which separates the oil from the water and deposits it in tanks, and sorbents, which absorb the oil. Skimming is effective only in calm waters, and sorbents present disposal problems. Bacteria that can digest petroleum and other hydrocarbons are being investigated as possibilities for oil spill cleanups.

Water also becomes polluted when silt, soil, and other suspended particles wash from plowed fields, construction and logging sites, urban areas, eroded river banks, and deforested slopes. By volume, sediment pollution is the greatest water pollutant. Sediments choke streams and fill other bodies of water.

Pollution also tarnishes groundwater. Sources of groundwater pollution include landfill sites, leaking septic tanks, underground gasoline storage tanks, industrial waste storage tanks, mining tails, oil fields, agricultural pesticides and fertilizers, and toxic chemicals stored outdoors. Radioactive waste has also contaminated groundwater supplies. Such pollutants can run through an aquifer spreading for hundreds of kilometers before being released into a stream or river. Unfortunately, it is nearly impossible to trace the path of pollutants in groundwater, and it is nearly impossible to clean. In Elmira, Ontario, pesticide manufacturing resulted in the contamination of the community's aquifer. Wells were shut down, and a pipeline was constructed to bring water from about 20 km away. Agent Orange, the defoliant used in the Vietnam War, was manufactured there.

Thermal pollution is usually caused by the discharge of water that has been used as a coolant in fossil-fueled or nuclear power plants. Although the warmer water can favor a diversity of life in water that otherwise would have been too cold, the addition of heat may make it less suited to desirable species. The added energy throws the ecosystem out of balance.

Sewage is comprised of water used for laundering and bathing as well as human waste. When sewage systems overflow, during heavy rain for example, untreated sewage can pollute bodies of water. Microorganism-laden water can present many health risks such as typhoid and dysentery, which are major problems in many parts of the world.

Lesson 6.2.4 – Stewardship

The question of what good stewardship looks like is answered in Genesis 1:28. God says that human-kind should fill the earth; subdue it, which in part means to cultivate the land; and take charge of, or rule, the creatures living on the earth. Some people believe that this means to use the earth and its resources in any way desired. They believe that humans can never destroy God's creation. Others believe that the requirements of good stewardship can be met by properly disposing of any waste that is generated. Still others are more proactive, trying to avoid waste in the first place. Within these three perspectives is a wide continuum considering the various degrees of consequences for humans, animals, plants, and water; economic ramifications; and the interpretation of biblical mandate.

Ultimately good stewardship is acting in such a way that best cares for God's creation. Christians have biblical and practical reasons for conserving nature and using it wisely. Not all pollution and waste is avoidable, but minimizing it can bring huge benefits.

Countless cooperative and individual efforts have contributed to improvements in the environment over the past decades. Environmental legislation can be an agent of good stewardship, not so much because of its restrictions and mandated compliance but because of the improvements to environmental performances that it encourages. For example, in 1969 Canada passed the Clean Air Act; in 1970 and 1977 the Clean Air Act and amendments to the act were passed by the U.S.

government to protect the air and improve its quality. This legislation stimulated research into new technologies, such as unleaded gasoline, and today lead particulates in the air have declined 95%; other major air pollutants have also decreased. In 1972, the Clean Water Act was enacted to maintain fishable and swimmable water quality, and in 1974 the Safe Drinking Water Act was passed to ensure a safe supply of public drinking water.

Technology is another agent of better stewardship. For example, the switch from fuels such as wood to gas and electricity reduced residential sulfur dioxide emissions by 80% between 1940 and 1970; alternative energy sources can reduce them even more. Soda cans used to contain more metal. A thousand cans once required 164 pounds of metal; now they require under 33 pounds, and recycling technologies can save even more metal. Scrubbers reduce air pollution, as do other newer devices on smokestacks and the exhaust systems of cars.

Individual efforts for stewardship include reducing, reusing, and recycling. Consumers can buy products with less packaging to reduce waste. At this point the use of fossil fuels cannot be completely eliminated because manufacturing depends on them for countless products, and they are used for transportation and generating electricity in houses and businesses. However, pollution levels can be reduced by conserving fossil fuels and recycling products that produce air or water pollution during their manufacturing. This not only cuts down on pollution but also conserves energy by reducing the demand for fossil fuels, which will not last forever. Part of reducing is refusing to buy or use certain products in the first place.

Most waste that is generated can be recycled. Composting yard waste and food scraps is an easy way to reuse and recycle products, which reduces the volume placed in landfills.

Students often feel powerless to solve environmental problems, but they can do many things to help care for creation. Although these solutions may seem small and ineffective, when many people do them they can have a big ecological impact. God often uses small acts to transform the world.

Pollution Solutions

6.2.1 Land Pollution

Introduction
Distribute identical pieces of candy or small trinkets to students. Have students vote on the product they would buy. Note which product received the most votes. Often students will vote for those with the most packaging. Discuss why this is so. Ask the following questions: *Why do you think better packaging means a better product?* (**Answers will vary.**) *What will ultimately happen to the packaging?* (**Answers will vary.**) *What do companies do to entice consumers to purchase their products?* (**Answers will vary.**) Show students several products that are overpackaged and comparable products that are not overpackaged. Ask whether overpackaging is a problem. Why? (**Answers will vary.**)

Discussion
- Lead a discussion about why manufacturers began increasing the packaging materials for their products. Convey that, in part, the results stemmed from public health concerns because of products being tampered with, theft of goods inside easier-to-open packaging, new approaches in marketing, and shelf longevity of a product. Generate conversation about how if more people lived in a godly manner, some of the extra packaging would not be needed.

- Make a chart on the board for students to compare the advantages and disadvantages of each type of waste disposal. Discuss which options are best and why.

- Present the topic of how companies and individuals could lessen the amount of toxic chemical pollution by finding ways to remove the toxicity altogether by using other chemicals or by abiding by the recommendations for recycling the chemicals or disposing of them properly.

- Ask the following questions:
  1. How is a sanitary landfill better than a dump? (**The sanitary landfill is designed to protect the environment. The dump is not.**)
  2. Why do items decay slower in a sanitary landfill than under natural conditions? (**They are not exposed to oxygen or moisture, which increases the rate of decay.**)
  3. Why do people not want to live near a dump or landfill? (**Possible answers: It smells bad; the garbage could contaminate the soil and water; garbage can catch fire if the methane gas builds up.**)
  4. Do you think incineration is a good solution to solid waste? (**Possible answers: Yes, it reduces the amount of space needed for trash; it reduces the possibility of land pollution; often the energy produced from the burning can be used to create steam to generate electricity or heat buildings.**) Does the solution create any new problems? (**Yes, there is potential for air pollution.**)
  5. How is radioactive waste different from chemical waste? (**Radioactive waste remains radioactive for thousands of years and there is no way to counteract its effects.**)

Activities

Lab 6.2.1A Sanitary Landfill
- 1 gal milk jugs or 2 L bottles with caps, 1 per group
- clay
- litmus paper
- soil
- graduated cylinders, 1 per group
- small pieces of garbage (food scraps, paper clips, aluminum foil, rubber bands, paper, pennies, newspaper, plastic scraps)
- filters, 1 per group
- cups or boxes large enough to support jugs, 1 per group
- beakers, 4 per group
- hot plates, 1 per group

A. Complete *Try This: Waste Not* with students. Discuss answers with students.

OBJECTIVES
Students will be able to
- compare the different types of land pollution.
- evaluate the different types of waste disposal.

VOCABULARY
- **incineration** the burning of solid waste materials
- **leachate** a solution formed when pollutants from sanitary landfills are dissolved in rainwater and seep into the groundwater
- **sludge** the solid waste leftovers from sewage treatment

MATERIALS
- Small wrapped candies or small trinkets (*Introduction*)
- WS 6.2.1A Recycling Journal

PREPARATION
- Obtain materials for *Lab 6.2.1A Sanitary Landfill*. Cut 8 cm off the bottom of the gallon milk jugs or 2 L bottles.
- Gather materials for *Try This: Waste Not*.
- Plan a field trip to a recycling plant or conservation agency. (C)

TRY THIS
Waste Not
- *mass scale*

NOTES

B. Have students use **WS 6.2.1A Recycling Journal** to record for two days the items they throw away or recycle. After the two days, discuss the results. Ask how they could recycle more and throw away less. (**Answers will vary.**)

C. Take students to a local recycling plant or conservation agency to find out what your community is doing to reduce pollution and recycle resources. Have students compare these efforts to the practices of godly stewardship.

D. Divide the class into groups. Have each group compete to create the best plan for the school to reduce waste. Direct each group to design a campaign poster that includes the following categories:
- **Refuse.** Do not buy items that use energy inefficiently, waste materials, or use toxic products.
- **Reduce.** Buy items with less packaging.
- **Reuse.** Use plastic bags, bottles, cardboard boxes, and plastic boxes multiple times instead of discarding them.
- **Recycle.** Collect recyclable materials such as glass, plastic materials, newspapers, and cans. Buy products made with recycled materials.

E. Challenge students to research how their community disposes of solid waste to determine what methods are being used to reduce land pollution and to assess which items can be recycled.

6.2.1 Land Pollution

OBJECTIVES
- Compare the different types of land pollution.
- Evaluate the different types of waste disposal.

VOCABULARY
- **incineration** the burning of solid waste materials
- **leachate** a solution formed when pollutants from sanitary landfills are dissolved in rainwater and seep into the groundwater
- **sludge** the solid waste leftovers from sewage treatment

Taking out the trash probably is not your favorite chore—moving heavy bags of smelly garbage outside is nobody's idea of fun. Sometimes the bag leaks, and your hands get grimy and slimy. Soon after you carry one bag out of the house, another one fills up. It is a dirty, continual job. Have you ever thought about where all of this garbage goes? When you throw things away, they do not disappear. The trash is just moving from one place to another. Trash pollutes the land.

A pollutant is a substance or condition in the air, soil, or water that harms living things. Pollutants can be solid wastes including those from industrial and agricultural sources; household wastes such as glass, paper, rubber, plastic, and textiles; scraps from butchered animals and food; and **sludge**, which is solid waste leftovers from sewage treatment.

Solid wastes often end up in sanitary landfills, large holes in the ground in which wastes are compacted and covered with soil. Sanitary landfills are designed to store wastes from living areas in a way that protects human health and the environment. The holes are lined with plastic, clay, or other materials, which are intended to keep leachate from leaking into the soil. **Leachate** is a solution formed when pollutants from sanitary landfills are dissolved in rainwater and seep into the groundwater. The wastes in sanitary landfills are compacted into the smallest space possible and covered with soil about once a day. It takes a long time for waste in sanitary landfills to decay. For example, paper can take 60 years to decay, and many plastics and synthetic materials take several hundred years. Because the materials used to line sanitary landfills seal out most oxygen and moisture, both of which speed decomposition, even substances that usually decay quickly do not do so in landfills.

Although sanitary landfills are a common method of waste disposal, they present problems. The linings often leak, allowing leachate into the soil before the wastes completely break down. When the wastes in a sanitary landfill do begin to decompose, methane gas is often produced. Methane gas burns easily, sometimes fueling underground fires in sanitary landfills. Some sanitary landfills have even exploded from this gas. Landfill developers sometimes place pipes in the landfills to safely remove methane gas, after which it is sold as a fuel.

Taking out the trash

The biggest problem with using sanitary landfills is finding the space for them; besides, nobody wants to live next to a landfill. Sanitary landfills, then, are not a perfect solution to the problem of solid wastes; they take up land space and sometimes leak leachate into the groundwater.

An alternative to sanitary landfills is **incineration**, the burning of solid waste materials. This reduces the volume and weight of trash, but it does not destroy many hazardous chemicals. Special incinerators can burn garbage more safely and heat water, turning it into steam that generates electricity or heats buildings.

Hazardous wastes are another type of pollutant. Factories often pollute the land and water with hazardous wastes, which are products that can cause serious health problems or even death. For example, factories that produce fuels generate hazardous wastes as by-products. If these wastes are not stored properly, they can seep into the soil and pollute the land. The best solution, of course, is for factories to reduce the amount of hazardous wastes that they produce. Another solution is treating these

Sanitary landfill

FYI

Garbage Slide
In July of 2000 in Payatas Estate in Quezon City, Philippines, at an open dumpsite, a landslide occurred that killed at least 140 people. The mountain of garbage was nearly 7 stories high before it failed. The landslide dropped 15 m of trash on nearby homes, burying 350 people in tons of garbage.

Lesson Review

1. What is a pollutant? (**a substance or condition in the air, soil, or water that harms living things**) What are the different types of land pollution? (**solid wastes from industry and agriculture, household wastes, scraps from butchered animals or food, and sludge**)
2. Describe how a sanitary landfill prevents land pollution. (**Sanitary landfills are designed to store wastes from living areas in a way that protects human health and the environment. The holes are lined with plastic, clay, or other materials, which are intended to keep leachate from leaking into the soil. Pipes have been built into some landfills to safely remove methane gas from the sanitary landfills; this gas can even be used as a fuel.**)
3. What problems are associated with sanitary landfills? (**Leachate can leak into the soil. It takes a long time for waste to decay. When decay happens, methane can be produced and can cause fires or explosions.**)
4. What are the benefits of incineration? (**It reduces the size of landfills and contamination of soil. The burning of trash can produce steam, which can be used to generate electricity and heat buildings.**)
5. Why are researchers exploring new methods of radioactive waste disposal? (**Radioactive waste takes thousands of years to decay. It can leak out of the disposal barrels and contaminate the surrounding land and water. Radiation can be deadly, so finding better means of disposal is very important.**)
6. What is the best way to manage waste? (**The best way to manage waste is not to create it in the first place.**)

NOTES

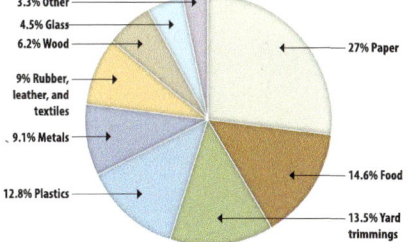

Name: _____ Date: _____

Lab 6.2.1A Sanitary Landfill

QUESTION: Will the garbage pollute the water?

HYPOTHESIS: Answers will vary.

EXPERIMENT:

You will need:		
• 1 gal milk jug or 2 L bottle with cap	• graduated cylinder	• filter
• clay	• small pieces of garbage (food scraps, paper clips, aluminum foil, rubber bands, paper, pennies, newspaper, plastic scraps)	• cup or box large enough to support jug
• litmus paper		• 4 beakers
• soil		• hot plate

Steps:
1. Turn the jug upside down.
2. Spread a thin layer of clay into the capped portion of the jug.
3. Mix food scraps, paper, and other garbage together. Place the mixture on top of the clay.
4. Sprinkle 2 cm of soil over the garbage mixture.
5. Prop the jug cap-side down into a cup or box.
6. Use the graduated cylinder to pour 25 mL of water over the mixture in the jug. Record your observations.
   Answers will vary.

7. Record your observations again after 24 hours. Answers will vary.

8. After another 24 hours, test the pH of a sample of clean water. This is the control. Record the pH. Answers will vary.
9. Pour another 25 mL of water over the mixture. Uncap the jug, and allow the water to drain through the clay into a beaker. Test the pH of the water. Record the pH level. Answers will vary.
10. Pour half of the water through a filter, and let it drain into a second beaker. Record your observations. Answers will vary.

11. Pour the remaining water into another beaker, and boil it until the water is gone. As a control, boil the same amount of clean water in another beaker until it is gone. Record your observations. Are there any dissolved solids? Answers will vary.

Lab 6.2.1A Sanitary Landfill

ANALYZE AND CONCLUDE:

1. Which garbage items have decomposed? Why? Answers will vary.

2. How does the garbage smell? What does this suggest? Awful. The garbage is decomposing.

3. What is the pH of the clean water? about 7
   What is the pH of the garbage water? Answers will vary.

4. How can a leaking sanitary landfill present a problem for the surrounding community? The garbage can change the pH of the soil and water, making them harmful to living organisms.

5. Why would putting hazardous materials such as chemicals and oil into a landfill cause a problem? These can leak into the soil and harm life.

6. What is the purpose of a landfill liner? Is it foolproof? A liner should prevent chemicals and other toxic substances from getting into the soil and water. No.

7. What are some alternatives to landfills? Answers will vary.

8. Consider the advantages and disadvantages of each alternative. Which is the best alternative? Why? Answers will vary.

Pollution Solutions

6.2.2 Air Pollution

Introduction
Have students speculate when air pollution began. Share that serious air pollution began with the industrial revolution starting as early as the 1700s, but polluted air was a problem for cities and towns in ancient times. Ancient people burned wood to cook food and keep warm, and the smoke from the wood hung over towns and cities. In 61 AD, the Roman philosopher Seneca described the city of Rome with "heavy air" and "the stink of the smoky chimneys thereof."

Ask students where most air pollution comes from. (**Answers will vary.**) Then, coat a microscope slide with petroleum jelly. Inform students how you ground the coal to produce coal dust, and put the dust in a beaker. Light it with a match. With forceps, hold the slide over the burning coal. Afterward, show students the slide. Have them identify how fossil fuels are commonly burned. (**Answers will vary but should include that gasoline and diesel fuel are burned in automobiles or many factories burn coal.**)

Guide students to consider whether air pollution is always visible. (**Answers will vary.**) To demonstrate that air pollution is not always visible and that it spreads, have students act as pollution detectors. Cut a large onion at one student's desk. Have students raise their hands as soon as they can smell the onion. Note which students' eyes begin to water. Set up a fan next to the onion and turn it on. Remark which students can then smell the onion "fumes" and which students' eyes water. Compare the fumes to pollution—they are not seen but they provoke a negative reaction.

Discussion
- Discuss the presence of air pollution inside buildings. Ask what types of pollutants are in the air at school and at home. (**Answers will vary.**) Shake a used furnace filter over a petroleum jelly-coated microscope slide. Have students examine the particulates with a hand lens or microscope. Ask students what they think made the debris. (**Answers will vary.**) How does it end up in the filter? (**Answers will vary.**)

- As a class, brainstorm ways industries and individuals can reduce air pollution. Consider the following types of pollution and ask students to come up with solutions:
  1. Toxic chemicals released in the air (**Possible answer: People should reduce the use of paints, solvents, and pesticides as much as possible.**)
  2. Burning of fossil fuels to purify water (**Possible answer: People could take shorter showers and turn off the water when they are not using it.**)
  3. Burning of fossil fuels for transportation (**Answers will vary but should include carpooling, riding a bike, walking, and taking public transportation.**)
  4. Burning of fossil fuels to generate electricity (**Answers will vary but should include always turning off lights and other electrical devices when they are not in use and using a microwave instead of a conventional oven.**)

- Ask the following questions:
  1. Why do researchers believe air pollution and weekend rain are related? (**Weekend storms are enhanced by the air pollution from vehicles driven by people on their way to and from work during the week. All raindrops form around particles, and the particles from air pollution can build up until they seed the clouds and cause rain.**)
  2. What types of substances make up particulate matter? (**certain kinds of smoke, such as diesel smoke and wood smoke; acids, such as nitrates and sulfates; allergens, like pollen and mold spores; dust; and fine ash**)
  3. What are some problems caused by particulate matter? (**Smaller particles can reach the lungs and irritate the respiratory system. In some people, this irritation triggers asthma attacks. Particulates also cover plants and trees, blocking out the sunlight they need for photosynthesis.**)

OBJECTIVES
Students will be able to
- identify causes of air pollution.
- describe the effects of air pollution.

VOCABULARY
- **smog** a dense, brownish haze formed when hydrocarbons and nitrogen oxides react in the presence of sunlight

MATERIALS
- Microscope slides, petroleum jelly, coal, beaker, matches, forceps (*Introduction, Discussion*)
- Onion, knife, cutting board, fan (*Introduction*)
- Used air filter, microscope slide, petroleum jelly, hand lens or microscope (*Discussion*)
- BLM 6.2.2A The Air We Breathe

PREPARATION
- Grind a piece of coal into dust. (*Introduction*)
- Obtain a used air filter or used furnace filter. (*Discussion*)
- Gather materials for *Lab 6.2.2A Air Pollution and Plants* and *Lab 6.2.2B The Air We Breathe*.

© *Earth and Space Science*

NOTES

4. Describe how smog is formed and the problems it can cause. (Gasoline contains hydrocarbons that are released into the air. Some nitrogen oxides are also produced from the combination of nitrogen and oxygen at high temperatures. The hydrocarbons and the nitrogen oxides react in the presence of sunlight to form a dense, brownish haze called *smog*. Smog can cause breathing problems, especially for people with asthma.)
5. What is the difference between the ozone layer in the atmosphere and ozone pollution? (The ozone layer is designed to protect life from harmful UV radiation. Ozone at ground level is dangerous. Ozone can aggravate asthma and other lung ailments. Ozone pollution is the result of oxides of nitrogen and hydrocarbons reacting together. Heat and sunlight speed ozone formation, so ozone pollution is more prominent in the summer.)
6. Why do you think smoking cigarettes is banned in many public places? (Answers will vary but should include that cigarettes have 4,000 different chemicals in them, and they cause 480,000 deaths in the U.S. every year.)

Activities

Lab 6.2.2A Air Pollution and Plants

- lima beans
- wide-mouthed jars with lids, 3 per group
- paper towels
- marking pencils
- automobile and adult
- stopwatches, 1 per group

6.2.2 Air Pollution

OBJECTIVES
- Identify causes of air pollution.
- Describe the effects of air pollution.

VOCABULARY
- **smog** a dense, brownish haze formed when hydrocarbons and nitrogen oxides react in the presence of sunlight

It has happened to everyone at one time or another. The sun shines Monday through Friday. You eagerly look forward to the weekend, planning a ball game or a hike or a picnic or a day at the beach. But when Saturday comes, rain drenches your plans. You grumble that it seems to rain more on weekends than during the week. And you know what? It does!

Researchers have found that in some areas, Saturdays have an average of 22% more precipitation than Mondays. This is not a coincidence. It happens because of air pollution. Weather data dating back to 1946 seems to indicate that the air pollution from vehicles driven by people on their way to and from work during the week enhances weekend storms. All raindrops form around particles, and the particles from air pollution can build up until they seed the clouds and cause rain. Weekly pollution levels often match the weekly precipitation levels, peaking on Saturday and dipping by Monday.

Harmful gases and tiny solid and liquid particles contaminate the air. Such particles, called *particulate matter*, are tiny bits of matter that float around in the atmosphere. They include certain kinds of smoke, such as diesel smoke and wood smoke; acids, such as nitrates and sulfates; allergens, like pollen and mold spores; dust; and fine ash. Your nose hairs, mucus, and breathing tubes catch larger particles, but smaller particles can reach your lungs and irritate your respiratory system. In some people, this irritation triggers asthma attacks. Particulates also cover plants and trees, blocking out the sunlight they need for photosynthesis.

Rained-out sporting event

HISTORY

Industrial Revolution
Do you think of air pollution as a recent problem? It is not. In ancient times, smoke from burning wood and trash caused air pollution in cities, but the problem of air pollution was small compared with what it is now. The Industrial Revolution changed England. In the late 18th century, people were leaving their farms to work in factories. The smoke and soot from these factories caused large-scale air pollution. At that point people did not realize that air pollution had serious consequences.

Air pollution worsened during the first half of the 20th century, but people did not pay much attention to the problem until 1952. In London that year, a temperature inversion contributed to smog that killed 4,000 people in four days and hospitalized thousands more. It was called *the Killer Fog*. The particles from this extreme smog entered people's lungs and caused dangerous complications.

Much air pollution comes from the industrial burning of fossil fuels and the burning of fossil fuels in motor vehicles. Here is how this happens. Gasoline contains carbon-based compounds called *hydrocarbons*. Although most gasoline is burned in the vehicle's engine, some is not; so hydrocarbons are released into the air. During the burning process, some nitrogen oxides are also produced from the combination of nitrogen and oxygen at high temperatures. The hydrocarbons, together with the nitrogen oxides, react in the presence of sunlight to form a dense, brownish haze called **smog**. Chemicals in smog irritate the eyes and cause breathing problems, especially for people with asthma or other respiratory ailments.

Smog can build up over cities because of a phenomenon called *temperature inversion*. A temperature inversion occurs when cool air near the earth's surface is trapped under a layer of warm air. Usually cool air is heated by the earth's surface and it rises, taking air pollutants with it; but during a temperature inversion, the warmer layer of air acts like a lid, trapping the pollutants and the cool air near the surface of the earth.

Air pollutants can also cause acid rain. Acid rain is the result of sulfur dioxides and nitrogen oxides reacting in the atmosphere with water and returning to Earth as precipitation. Acid rain contaminates forests and pollutes water and land.

Industrial smoke stacks polluting the air

Lab 6.2.2B The Air We Breathe

- scissors, 1 pair per group
- heavy paper
- glue
- clear packing tape
- hole punches, 1 per group
- string
- microscopes, 1 per group

Print **BLM 6.2.2A The Air We Breathe**. Hand out Collector Cards, 1 per group. Each group should hang the card in a different area of the school.

A. Divide the class into groups. Direct each group to design a product to detect, reduce, or filter air pollution and make a graphic or model of the prototype. Convey to groups that they need to write a script for an infomercial and present the product to the class.

B. Have students make a chart to compare the air quality index or smog index of five large cities from different countries. Direct students to chart air quality in each city for 10 days and evaluate which city has the best air quality and which city has the worst.

C. Guide students to research a specific type of air pollution and to creatively present the information to the class. Specific types of pollution may include carbon monoxide, sulfur dioxide, toxins, ozone, chlorofluorocarbons, radon, nitrogen oxides, and carbon dioxide.

D. Assign students to research black lung disease and to present their research to the class. Have students include causes, symptoms, and treatments.

NOTES

Because of weather patterns, acid rain often falls many kilometers away from the source of the air pollution.

Another form of air pollution is ozone pollution. Do not confuse ozone pollution with the ozone layer in the upper atmosphere, which shields the earth from harmful ultraviolet radiation. Ozone is a reactive gas made of three oxygen atoms (O_3). It forms when hydrocarbons react with a group of air pollutants called *oxides of nitrogen*. Hydrocarbons are emitted not only by motor vehicles but also by dry cleaning solutions and many other sources. Heat and sunlight speed ozone formation, so ozone pollution is heaviest in the summer. Perhaps you have heard ozone warnings on the news in the summer. People are encouraged not to drive unnecessarily, mow the lawn, or fill up the gas tank because these things all contribute to the formation of ozone. Breathing ozone can trigger breathing problems, especially for those with asthma or other lung ailments. Ozone adversely affects certain types of vegetation as well.

Sometimes the air indoors can be just as polluted as the air outdoors. Indoor air pollution is a problem inside houses and office buildings. One reason for this is that modern buildings are well insulated for energy efficiency. This means that there are not any small cracks or spaces through which the indoor air pollution can escape, so it stays trapped inside. Natural gas burned in cooking stoves or space heaters produces nitrogen dioxide and carbon monoxide, and wood burned in fireplaces adds particulates to these gases. Synthetic fibers from carpeting, foam insulation in furniture, particle board, and wall coverings are other sources of indoor air pollution. This is because some of these materials release gases such as formaldehyde. Office buildings with office equipment, carpeting, and poor ventilation usually have high indoor air pollution. Commercial products such as furniture polishes, cleaning agents, and glues also add to indoor air pollution. Smoking can be a major cause of indoor air pollution; that is why smoking is now banned in many public places.

Laws such as the United States' Clean Air Act of 1970 helped define acceptable air quality levels. These laws have encouraged industries and individuals to find new technologies and use energy sources that produce less pollution. New knowledge and a good use of God's gifts can help solve the problem. Hydrogen powered cars may offer cleaner alternatives for the future. Solar power, wind power, and fusion are other possible energy sources that could provide energy without causing pollution. More efficient industrial processes, smaller cars, smaller houses, turning off extra lights, and walking or riding public transportation could reduce the amount of fossil fuels burned. This decreases air pollution. Most cities now have air quality indexes that rate daily air pollution on a numerical scale and report it along with weather forecasts. This is a way of informing people, especially people with respiratory problems, about the risks of air pollution from day to day.

LESSON REVIEW
1. How does air pollution cause rain?
2. What problems result from air pollution?
3. What type of pollution is caused by burning fossil fuels? Why is it bad?
4. Why are temperature inversions a problem?
5. Why is ozone a problem mostly during the summer?
6. List three indoor air pollutants. What causes these types of pollution?

FYI

Smoking
Smoking is a major form of air pollution. Every year more than 480,000 people in the U.S. die from the effects of smoking. About one in every five deaths in the United States and Canada is the result of smoking. Tobacco accounts for 33% of all cancer deaths and 87% of lung cancer deaths.

Why are cigarettes so harmful? Cigarettes contain 4,000 different chemicals. They include acetone (nail polish remover), ammonia (toilet cleaner), arsenic (rat poison), formaldehyde (a fluid used to preserve bodies), hexamine (barbecue lighter), hydrogen cyanide (gas chamber poison), methane (swamp gas), methanol (rocket fuel), nicotine (insecticide), and toluene (industrial solvent).

NOTES

E. Challenge students to research legislation such as the Clean Air Act of the United States and how the air pollution problem has improved since then. Guide them to develop a campaign to improve the indoor air quality at home or at school.

Lesson Review

1. How does air pollution cause rain? (**The particulate matter can seed the clouds.**)
2. What problems result from air pollution? (**mostly lung problems and breathing problems; eye irritation; plants being covered with air pollution and therefore not able to photosynthesize**)
3. What type of pollution is caused by burning fossil fuels? Why is it bad? (**The burning of fossil fuels can release hydrocarbons and nitrogen oxides that react with sunlight to create smog. This can trigger asthma attacks and other lung ailments.**)
4. Why are temperature inversions a problem? (**Usually cool air is heated by the earth's surface and it rises, taking air pollutants with it; but during a temperature inversion, the warmer layer of air traps the pollutants and the cool air near the surface of the earth.**)
5. Why is ozone a problem mostly during the summer? (**The heat from the sun increases the production of ozone.**)
6. List three indoor air pollutants. What causes these types of pollution? (**Answers will vary but should include space heaters produce carbon monoxide; carpeting, foam insulation, particle board, and wall coverings can produce formaldehyde; commercial products emit toxic fumes; and smoking can produce many toxins.**)

Name: _____ Date: _____

Lab 6.2.2A Air Pollution and Plants

QUESTION: How will car exhaust affect bean growth?
HYPOTHESIS: Answers will vary.

EXPERIMENT:

You will need:
- lima beans
- 3 wide-mouthed jars with lids
- paper towels
- marking pencil
- automobile and adult
- stopwatch

Steps:
1. Fold each paper towel 3–4 times to fit in the bottom of a jar. Saturate each one with water and place one in the bottom of each jar.
2. Place about 5 beans in each jar.
3. Cap one of the jars. Label it *Control*.
4. Have an adult start a car and expose the second jar to the exhaust fumes for 30 seconds. (Be careful not to breathe in any of the fumes.) Cap the jar immediately to trap the gases. Label this jar *Air Pollution—30 seconds*.
5. Expose the third jar to the exhaust fumes for two minutes. Cap the jar immediately to trap the gases. Label this jar *Air Pollution—2 minutes*.
6. Illustrate and describe your observations of the beans after 2 days, 3 days, and 5 days. Estimate the height of the tallest sprout in each jar and record this data.

Time	Control	Air Pollution—30 seconds	Air Pollution—2 minutes
2 days			
Height			
3 days			
Height			
5 days			
Height			

Answers will vary.

Lab 6.2.2A Air Pollution and Plants

ANALYZE AND CONCLUDE:

1. Which set of beans grew the tallest? Answers will vary.
2. Which set of beans looked the healthiest? Were they also the tallest?
   Answers will vary.
3. To what type of pollution did you expose the second and third jars?
   Possible answers: air pollution, car exhaust, carbon monoxide, hydrocarbons
4. Does the length of time that the seeds were exposed to pollutants affect growth? Why? Yes, because the more air pollution, the less the beans were able to grow.
5. Do you think this lab accurately tests the harmful effects of air pollution? Why?
   Answers will vary.
6. Create a new experiment to test the harmful effects of air pollution on plants. Be sure to have a control for comparison and only one variable to test.
   Answers will vary.

Name: _____ Date: _____

Lab 6.2.2B The Air We Breathe

QUESTION: What types of airborne pollutants are in the school?

HYPOTHESIS: Answers will vary.

EXPERIMENT:

You will need:	• glue	• string
• scissors	• clear packing tape	• microscope
• heavy paper	• hole punch	

Steps:
1. Cut the three holes in the collector card as marked. Glue the card to a heavy piece of paper. Cut out the holes.
2. Tape a piece of packing tape across the back of the card so the adhesive shows through all the holes. Be careful not to touch the adhesive showing through the holes.
3. Place the card on the microscope to observe the adhesive. Illustrate what you see.

Drawings will vary.

4. Write your name, location where you will hang the card, the date, and the time.
5. Punch a hole in the top. Use the string to hang the card somewhere in the school.
6. After a week, retrieve the card and look at the adhesive under the microscope.

What do you see? Illustrate your results. _____

Drawings will vary.

Lab 6.2.2B The Air We Breathe

ANALYZE AND CONCLUDE:

1. Does the tape look different than when you started? Answers will vary.
2. What types of particulate matter do you think you see on the tape?
   Answers will vary.

3. Using the information in the textbook and other research, hypothesize what types of air pollution was collected. Explain. Answers will vary.

4. What could be done to reduce the air pollution in the room or area you tested?
   Answers will vary.

6.2.3 Water Pollution

Pollution Solutions

OBJECTIVES

Students will be able to
- compare types of water pollution.
- identify different types of natural water purification.

MATERIALS

- Tray, cardboard egg carton devoid of holes, paper towels, flax meal, food coloring (*Introduction*)
- Water quality testing kits (*B*)

PREPARATION

- Obtain feathers for *Lab 6.2.3A Oil Spill*.
- Gather the growing bean plants for *Lab 6.2.3B Acid Rain*.
- Obtain materials for *Try This: Dilution Solution*.
- Obtain water quality testing kits. (*B*)
- Invite a missionary or health-care worker to speak to the class about poor water conditions. (*C*)

TRY THIS

Dilution Solution
- beakers, 5 per group
- food coloring

Ask students if the water in the fourth beaker is clean. Why? **(No, the water in the fourth beaker is not clean because there is still some food coloring in it. It has been diluted, but the coloring has not been removed.)**

Introduction

Set a tray on the table. Remove the lid from an empty cardboard egg carton and place the carton on top of several paper towels inside the tray. Slightly tilt the carton and begin pouring water into one of the end cups. Draw students' attention to the water flowing from cup to cup. Share that this demonstrates how bodies of water are connected. Stop pouring once all the cups are full. Add 1 tsp of flax meal or other fine substance to one of the end cups. Have students watch the water carry the meal from cup to cup. At the opposite end of the carton, add about 10 drops of food coloring. Stir it slightly. Have students watch it slowly move from cup to cup. Ask what this demonstrates about water pollution. (**A little pollution can travel a long distance and contaminate lots of water.**) Leave the egg carton out for a couple of hours. The paper towels should become saturated with the colored water. Draw students' attention to the paper towels and convey that what happened demonstrates the contamination of groundwater.

Discussion

- Discuss the importance of a clean water supply. Ask how polluted water affects people, plants, and animals. (**It can make people and animals sick. Plants may not grow well. Crops may be unfit to eat.**) What are some ways water becomes polluted? (**Answers will vary.**) What can be done to reduce pollution in the local area? (**Answers will vary.**)

- Genesis 2:15 states that the Lord God took the man and put him in the Garden of Eden to work it and take care of it. God calls His people to steward the earth. Discuss God's plan for His people and read **Ephesians 2:10** and **1 Corinthians 10:31**. Ask students what environmental actions could exhibit good stewardship to honor God. (**Answers will vary but should include reducing usage of water and energy, using cleaner energy, and disposing of toxic chemicals properly.**)

- Ask the following questions:
  1. What is acid rain and how does it affect water? (**Acid rain is rain with sulfur dioxides and nitrogen oxides in it. Acid rain contaminates water supplies, making them too acidic.**)
  2. How is an oil spill cleaned up? (**The spill is contained by booms, and oil is skimmed from the surface or absorbed with certain materials. Chemicals can break down the oil. Scientists are exploring the possibility of using bacteria to clean up oil spills.**)
  3. Why is it bad to pour used oil down the drain or onto the ground? (**The oil can hurt the soil and the plants. It can also make its way into the groundwater supply.**)
  4. Is an increase in water temperature a type of pollution? (**Yes, it is called *thermal pollution*.**)
  5. When is soil considered a pollutant? (**when it interrupts the flow of water, clogs fish gills, and smothers coral reefs**)
  6. The water cycle is designed to purify water through evaporation. What other natural process is being used to help remedy water pollution? (**Scientists are using certain bacteria to clean polluted water and soil. It is called *bioremediation*.**)

Activities

Lab 6.2.3A Oil Spill

- shallow tubs, 1 per group
- feathers, 1 per group
- blue food coloring
- paper towels
- vegetable oil
- dish soap
- sponges, 1 per group
- bowls, 1 per group
- plastic straws, 2 per group
- cotton balls
- syringes or pipettes, 1 per group

Lab 6.2.3B Acid Rain

- young bean plants, 3 per group
- jars with lids, 3 per group
- masking tape
- marking pens, 1 per group
- white vinegar

A. Complete *Try This: Dilution Solution* with students.

B. Challenge students to test pH and chlorine levels of different water sources using a water quality testing kit. Have students compare the quality of the sources and hypothesize the causes of these differences.

C. Divide the class into groups. Have each group research a country that lacks proper water treatment methods to answer the following questions: *What waterborne diseases are commonplace? What symptoms do they cause? What are missionaries or Peace Corps workers doing to help people in these situations?* Extend the activity by presenting the invited missionary or health-care worker familiar with these conditions to speak to the class.

D. Have students research new technologies in air moisture recovery. Research is being done to collect water vapor and fog from the atmosphere. Direct students to answer the following: *Are any of these technologies in use? How effective and cost efficient are they? Where are they being used? Could this technology affect weather patterns? Why?*

NOTES

6.2.3 Water Pollution

OBJECTIVES
- Compare types of water pollution.
- Identify different types of natural water purification.

Have you ever been really, really thirsty? Maybe you were on a long bike ride and drained your water bottle halfway through the ride. Maybe you forgot to stop at the drinking fountain before you ran laps in physical education class. Or maybe you spent a summer day at the beach and remembered too late how quickly the hot sun dries you out. At times like these, nothing looks better than a glass of refreshing, clear water.

Water is a precious gift from God. As the population grows and agriculture and industry require more water, freshwater is becoming less available. Some regions have plenty of water; other regions are dry.

Water all over the world is being polluted. Acid rain pollutes groundwater and bodies of water such as streams, rivers, and lakes. When the acidity of such a body of water is increased, the organisms that live there can get sick or die. Oil can also pollute the water. Offshore oil wells that drill the oil in ocean beds can have accidents and spill oil into the water. When oil tankers are damaged, oil can leak into the water. For example, on March 24, 1989, the oil tanker *Exxon Valdez* struck Bligh Reef in Prince William Sound, Alaska, spilling more than 42 million liters of oil. The spill harmed or killed migratory shore birds and waterfowl; sea otters; and other species such as porpoises, sea lions, and whales. A larger oil spill occurred in April 2010 when an offshore oil rig exploded in the Gulf of Mexico, which spilled more than 635 million liters of oil. Oil spills are difficult to clean because they spread quickly. Oil companies contain the spill by floating

Clean drinking water

Controlled burn following the Deepwater Horizon Drilling platform oil spill

barriers, called *booms*, and skimming oil from the surface of the water or absorbing it with certain materials. Chemicals can also be used to break down the oil. And because some bacteria consume oil, scientists are exploring the possibility of using bacteria to clean up oil spills. As bad as oil spills are, daily operations of oil tankers leak more oil overall than oil spills do.

The United States and Canada depend heavily on groundwater for drinking water and crop irrigation. Groundwater is easily polluted. One liter of motor oil dumped into a drain can pollute almost a million liters of groundwater. Anything dumped on or into the ground including pesticides, fertilizers, factory wastes, chemicals, oil, and radioactive wastes will eventually find its way into groundwater and pollute it.

Water can also be "polluted" by temperature changes. Power plants use water to cool steam in the turbines. After the water is used, it is released into rivers, lakes, or bays. But it is much warmer than when it went into the plant. Many fish and animals are not created for warmer water, and the temperature increase can kill them. This type of pollution is called *thermal pollution*.

NOTES

Lesson Review

1. List ways that oil can get into water sources. (**Offshore oil wells that drill the oil in ocean beds can have accidents and spill oil into the water. When oil tankers are damaged, oil can leak into the water. Oil can be dumped down a drain or directly onto the soil.**)
2. Why is an oil spill a problem? (**It can kill migratory shore birds and waterfowl, sea otters, and other species. It can also catch fire and create air pollution.**)
3. What is thermal pollution? What causes it and how can it be remedied? (**Thermal pollution is an increase in water temperature. Power plants use water to cool steam in the turbines. After the water is used, it is released into rivers, lakes, or bays. Cooling ponds and cooling towers can be used to combat the high temperatures.**)
4. What causes an excess of algae? Why is it a problem? (**Extra nutrients in the water increase the growth of algae. When the algae die, the decomposition uses oxygen and chokes the rest of the ecosystem.**)
5. How are sea animals affected by solid waste? (**Soil can interrupt the flow of streams, clogging fish gills and even smothering coral reefs. Plastic wastes can choke or strangle fish, sea turtles, sea mammals, and sea birds.**)
6. Describe the natural water purification processes. (**When water evaporates, the impurities remain behind, and the purified water returns to liquid through condensation so living things can use it. Bacteria convert organic waste products into simple compounds that do not harm the environment. Decomposing bacteria live in wetlands rich in decaying plant materials. When farm runoff is diverted into large wetlands, these bacteria break down the excess nutrients so algae do not grow too quickly. When water seeps through gravel and sand, suspended particles are filtered out.**)

Name: _____ Date: _____

Lab 6.2.3A Oil Spill

QUESTION: How does an oil spill affect waterfowl?
HYPOTHESIS: Answers will vary.

EXPERIMENT:

You will need:	• paper towels	• bowl
• shallow tub	• vegetable oil	• 2 plastic straws
• feather	• dish soap	• cotton balls
• blue food coloring	• sponge	• syringe or pipette

Steps:
1. Fill the tub halfway with water. Add 3 drops of food coloring.
2. Set the feather in the water. What happens? It floats.
3. Remove the feather and set it on paper towels to dry.
4. Pour 40 mL of oil in the water. What does the oil do? The oil spreads out on top of the water.
5. Place the feather on top of the oil, holding one end. Gently move the feather around to coat it with oil. Does it float? Describe how the feather looks and feels. It still floats some. Part of the feather is staying in the oil. The feather is oily and clumpy.
6. Remove the feather and set it on the paper towels. Put water in the bowl and wet the sponge. Add a drop of dish soap to the sponge and clean the feather.
7. When finished cleaning, rinse the feather and let it dry.
8. Take the straw(s) and cut or connect them so their measurement equals the width of the tub.
9. Place the straw at one end of the pan and carefully push the oil toward the other end.
10. On the side of the straw that has only a little oil left, use a cotton ball to absorb the oil.
11. Use the syringe to remove the large patch of oil.

ANALYZE AND CONCLUDE:
1. Does the feather feel as light as it did at the beginning of the experiment? No.
2. How well do you think the feather would float after being soaked in oil and cleaned? Possible answer: not very well

---

Lab 6.2.3A Oil Spill

3. How do you think water fowl are affected by oil spills? Water fowl soaked in oil would probably die because they wouldn't be able to fly with oil on their wings.
4. Were you able to remove all the oil from the water? Why? No. Answers will vary.
5. What are some other ways you could remove the oil from the water? Answers will vary.
6. Do you think oil spills in the ocean are more difficult to clean than the oil spill in the experiment? Why? Yes, because the ocean is much bigger and the water is constantly moving. It would be very difficult to get all the oil. Many animals and plants would die before the cleanup was complete.
7. Oil burns and water does not. Would burning the oil be a good solution to get rid of it? Why? No, because the fumes would create air pollution and many plants and animals could die in the flames.

Name: _____ Date: _____

Lab 6.2.3B Acid Rain

QUESTION: How does acid rain affect plants?
HYPOTHESIS: Answers will vary.

EXPERIMENT:

You will need:	• masking tape	• white vinegar
• 3 young bean plants	• marking pen	• water
• 3 jars with lids		

Steps:
1. Label each plant: A, B, or C.
2. Label each jar: A, B, or C.
3. Add 120 mL of white vinegar to Jar A.
4. Add 60 mL of white vinegar to Jar B.
5. Do not add any vinegar to Jar C.
6. Add 120 mL of water to each jar and stir gently to mix the solution.
7. Set the bean plants in a sunny location.
8. Water each plant with 60 mL of solution from its corresponding jar.
9. Record observations daily for 5 days. If the soil gets dry, water again with 60 mL of solution. Record any additional watering.

Day	Plant A	Plant B	Plant C
Start, 1			
2			
3		Answers will vary.	
4			
5			

Lab 6.2.3B Acid Rain

ANALYZE AND CONCLUDE:
1. Which plant is the healthiest? Why? Plant C because it received only water, no acid
2. Which plant is the least healthy? Why? Plant A because it received the highest concentration of acid
3. Were your results what you expected? Why? Answers will vary.
4. How is the experiment similar to acid rain? The acid added to the water simulated industrial acids added to precipitation. Acid rain can contaminate water supplies.
5. What causes acid rain? Acid rain is the result of sulfur dioxides and nitrogen oxides reacting in the atmosphere with water and returning to Earth as precipitation.

© Earth and Space Science

6.2.4 Stewardship

Pollution Solutions

OBJECTIVES

Students will be able to
- explain why it is important to be a good steward.
- infer methods of good stewardship.

VOCABULARY

- **stewardship** the attentive management of something entrusted to one's care

MATERIALS

- Disposable gloves and trash bags (C)
- WS 6.2.4A Do the Math

PREPARATION

- Obtain materials for *Try This: Reduce*.
- Plan a field trip to the local department of conservation. Before going, have students prepare a list of questions to ask during the field trip. (D)

TRY THIS

Reduce
- large mass scale

Ask students how much waste they were able to reduce. How much waste could they reduce in a whole year? How much waste would be reduced if everyone in the class did this for a whole year? Is there value in reducing waste? Why? (**Answers will vary. Yes, there is value. It is good stewardship and an act of honoring God. More resources will be available to others.**)

Introduction

Read a psalm of praise and thanksgiving such as **Psalm 100** or **Psalm 111**. Have students create a list of blessings God has given, both tangible and intangible. Guide them to focus on the natural resources that are available.

Discussion

- Discuss ways to reduce, reuse, and recycle at school. Ask students how they could encourage others to do this too. Read **Matthew 7:3–5; 18:15** and **Philippians 2:1–8**. Use the principles from these verses to determine the correct way to encourage others to be good stewards of natural resources.

- Ask the following questions:
  1. Why should people follow God's example of recycling? (**We are made in His image. We should reflect Him in all that we do.**)
  2. How does turning off hot water reduce the burning of fossil fuels? (**Heating water often requires the burning of fossil fuels.**)
  3. Electricity is a clean energy source, yet reducing its use also reduces the burning of fossil fuels. Why? (**Electric power plants often use fossil fuels to generate electricity.**)
  4. How is good stewardship an act of honoring God? (**Answers will vary but should include that obedience to God shows love for Him. Caring for God's creation—what He entrusted us with—shows that we care about Him and what is His.**)
  5. In what ways do people benefit from good stewardship? (**Answers will vary but should include the air is cleaner, natural resources are not used up as quickly, the land looks better, and people are healthier.**)

Activities

A. Complete *Try This: Reduce* with students. Discuss students' results.

B. Have students complete **WS 6.2.4A Do the Math**. Extend the activity by increasing the number of lightbulbs in the first question. Increase the number of people in the second and third questions. Add the mass of junk mail collected in students' homes to the fourth question.

C. Take students outside on the school grounds to collect trash. Students should wear disposable gloves and take trash bags. Take precautions so students do not get cut on broken glass or sharp metal. If you are able to recycle, have students sort the trash accordingly. Discuss why this is an important activity.

D. Take the class on a field trip to a local department of conservation. Follow up the field trip by having students write about what they learned, how they can participate in conservation, and any ideas they have about improving conservation efforts.

E. Divide the class into groups to create a plan to compost and recycle waste from the lunchroom. Instruct students that they will be responsible for separating the trash at the end of the lunch period and to create a schedule for all students to participate.

Lesson Review

1. What is stewardship? (**Stewardship is attentively managing something entrusted to us.**)
2. How does God recycle? (**God recycles important gases, water, and nutrients through cycles like the water cycle, oxygen cycle, nitrogen cycle, and carbon cycle.**)
3. Why should you be a good steward? (**It is the job of God's people to manage and care for creation. People should sustain the earth because they were created to reflect God. Creation can continue to praise Him, and it is an act of worship. Good stewardship ensures that natural**

resources continue to be available and they are in good condition for people, plants, and animals.)

4. How can you be a good steward? (**Possible answers: by taking shorter showers; turning off the water while brushing teeth; reducing, reusing, recycling; sharing rides; using public transportation; combining errands; walking or biking instead of riding in a car**)
5. Who benefits from wise stewardship? (**everyone who is alive now and future generations**)

NOTES

6.2.4 Stewardship

OBJECTIVES
- Explain why it is important to be a good steward.
- Infer methods of good stewardship.

VOCABULARY
- **stewardship** the attentive management of something entrusted to one's care

Have you ever been disgusted by the sight of someone tossing trash from a car window; cigarette butts littering the sidewalk; or plastic containers floating down a river? People can recognize pollution when they see it, and most know enough not to litter or to dump paint thinner down the drain. It is easy to avoid doing these things, but there are many decisions to make about how to keep God's creation healthy.

Imagine the earth as God created it. Beautiful ecosystems, clean resources, wild places, and amazing creatures decorated the earth. People depend on a healthy, clean biosphere, so it is important to be good stewards of the earth. For the Christian, the importance of **stewardship** runs even deeper. God created a beautifully balanced atmosphere and biosphere and He sustains it by the natural processes He set in place. It is the job of His people to manage and care for that creation.

In the same way God provides for and sustains the world, people should also sustain the earth because they were created to reflect God. It is the privilege of all people to enjoy the fruitfulness of creation, and it is the responsibility of everyone to maintain it, to steward what has been entrusted to all. "The Lord God took the man and put him in the Garden of Eden to work it and take care of it" (Genesis 2:15). When pollution is reduced and resources are maintained, the healthy creation can continue to praise Him—the animals, plants, mountains, streams, stars, and, of course, people.

Littered waterway

The idea of stewardship is biblical, but the execution demands both a group effort and individual efforts. Society has done a better job with stewardship in recent decades. Believe it or not, littering used to be acceptable, but now society frowns on littering. Today people do not litter nearly as much, and many organized groups participate in cleanup projects. As a result, the land is much more litter-free than it used to be.

New technology has been created to improve recycling processes and use cleaner fuel. Solar energy, for example, does not produce as much pollution as burning fossil fuels. Scrubber systems in coal-burning plants work like showers to reduce pollution. Pollution devices in smokestacks and cars are also reducing pollution.

Picking up litter

Industries and individuals can also care for creation. One way to do this is to conserve water. Because heating water often requires the burning of fossil fuels, cutting back on hot water reduces air pollution. Industries can reduce water usage by replacing water-cooled equipment with air-cooled equipment; installing automatic shut-off nozzles; and using specialized water brooms instead of hoses to clean floors or other hard surfaces. Taking shorter showers, for example, will conserve water. Turning off the water while brushing teeth also helps.

Another way to be a good steward is to imitate how God sustains His creation. Ecosystems recycle important gases, water, and nutrients through the oxygen cycle, nitrogen cycle, carbon cycle, and water cycle so they can be used over again. Image bearers should follow the example of God in whose image they are made. Conserving and recycling finished products, conserving energy, reducing the demand for fossil fuels, and decreasing pollution levels are means of honoring God.

You can be a better steward by sharing rides, using public transportation, and combining errands. You can walk or bike whenever possible instead of taking the car.

Conserving the environment will help maintain a livable atmosphere for many generations to come. Christians should take the lead in preserving God's wondrous creation and the delicately natural systems He uses to sustain the world. He said, "For every animal of the forest is Mine, and the cattle on

TRY THIS

Reduce
Prepackaged foods generate waste in excess packaging. Collect all packaging waste from your household for a week. Weigh it. The following week, collect all packaging waste again, but use less prepackaged food and pack lunches in reusable containers. Weigh it and compare the weight to the first week. By how much were you able to reduce the waste? If you continued to do this for a whole year, how much waste could you reduce? How much waste would be reduced if everyone in the class did this for a whole year? Do you think there is value in reducing waste? Why?

FYI

Solar Bike Path

In November of 2014, the town of Krommenie, Netherlands, built a section of bike path using solar panels. The path collects solar energy to power the city and provides a safe bike path for commuters. The use of the solar road can reduce the use of fossil fuels and the formation of pollution. New technology often comes with a high price but is necessary to move forward. At a cost of $3.7 million, SolaRoad developed and installed a 70 meter stretch of solar panel pathway. After 6 months of use, it generated 3,000 kilowatt hours (kWh), enough energy to power one household. With energy costs at about $2/kWh, the money spent on the solar panel road could have powered 173 households. Although the cost of research and development seems high, pursuing alternate forms of renewable energy is good stewardship. This example demonstrates reduction in fossil fuel use as more commuters are riding bikes on a safe bike path and less coal is burned to generate electricity.

a thousand hills" (Psalm 50:10). Every natural resource belongs to God. Praise Him for His provision and honor Him with your stewardship.

LESSON REVIEW

1. What is stewardship?
2. How does God recycle?
3. Why should you be a good steward?
4. How can you be a good steward?
5. Who benefits from wise stewardship?

Solar energy reduces pollution.

UNIT 7

Chapter 1: *Solar System*
Chapter 2: *Planets*
Chapter 3: *Sun, Earth, and Moon*

Key Ideas

Unifying Concepts and Processes
- Systems, order, and organization
- Evidence, models, and explanation
- Change, constancy, and measurement
- Evolution and equilibrium
- Form and function

Science as Inquiry
- Abilities necessary to do scientific inquiry
- Understandings about scientific inquiry

Earth and Space Science
- Earth in the solar system
- Origin and evolution of the earth system

Science and Technology
- Abilities of technological design
- Understandings about science and technology

Science in Personal and Social Perspectives
- Natural hazards
- Science and technology in society

History and Nature of Science
- Science as a human endeavor
- Nature of science
- Historical perspectives

Astronomy

Vocabulary

annular eclipse
aperture
aphelion
apogee
asteroid
astronomical unit
aurora
chromosphere
comet
convective zone
corona
crater
dwarf planet
eclipse
electromagnetic radiation
electromagnetic spectrum
ellipse
equinox
friction
full moon
gas giant
geocentric
heliocentric
Kuiper belt
lunar calendar
lunar eclipse
lunar regolith
mare
meteor
meteorite
meteoroid
nebula
new moon
penumbra
perigee
perihelion
phase
photosphere
planet
prominence
radiative zone
reflecting telescope
refracting telescope
rille
solar calendar
solar eclipse
solstice
terrestrial planet
umbra
waning moon
waxing moon

SCRIPTURE

The heavens declare the glory of God; the skies proclaim the work of His hands.

Psalm 19:1

7.1.0 Solid System

LOOKING AHEAD

- For **Lesson 7.1.3**, invite a scientist from a local university or observatory to demonstrate and explain how refracting and reflecting telescopes work. Schedule a field trip to an observatory. Obtain a refracting telescope and a reflecting telescope. Schedule a nighttime sky viewing for students and parents.
- For **Lesson 7.1.4**, invite a professor from a local university to share what scientists can learn about the universe from various types of nonoptical telescopes.
- For **Lesson 7.3.2**, assign *Try This: Moonrise* about seven days prior to the lesson.

SUPPLEMENTAL MATERIALS

- BLM 7.1.1A Faith and Science
- BLM 7.1.5A Sun and Planets Measurements

- TM 7.1.2A Moon Phases
- TM 7.1.6A Auroras

- Lab 7.1.2A Sundial
- Lab 7.1.3A Atmospheric Gases
- Lab 7.1.3B Refracting Telescope
- Lab 7.1.5A Indirect Evidence
- Lab 7.1.6A Diameter of the Sun

- WS 7.1.1A Galileo and the Sun
- WS 7.1.1B Galileo Goes to Jail
- WS 7.1.2A Astronomical Observations
- WS 7.1.5A Solar System Diagram
- WS 7.1.6A Sun Diagram

- Chapter 7.1 Test

BLMs, TMs, and tests are available to download. See Understanding Purposeful Design Earth and Space Science at the front of this book for the web address.

Chapter 7.1 Summary

This chapter provides an overview of the solar system and the ways in which it has been understood and explored. The chapter begins with a brief history of astronomy that traces the Western understanding of the organization and controlling forces of the solar system. The lesson on calendars demonstrates ways in which various cultures have applied observations of the sun and moon to keeping track of time; it also traces the roots and development of the Gregorian calendar, which is widely used throughout the world today. From here, the chapter examines optical and nonoptical telescopes and their contributions to understanding the solar system and universe. The chapter concludes with respective lessons on the structure of the solar system and the sun, celestial bodies that people have observed in various ways for centuries.

Background

Lesson 7.1.1 – History of Astronomy

Although many ancient cultures have a strong history of astronomical studies, modern astronomy has its roots in ancient Babylon and Greece. The ancient Babylonians believed that the arrangement of the planets carried warnings of natural disasters; therefore, they studied the movements scientifically for superstitious reasons. They kept detailed astronomical records that enabled them to predict the movements of the visible planets. The ancient Greeks inherited some of the Babylonians' knowledge. The Greeks also kept records, developed mathematics that enabled Eratosthenes to calculate the circumference of Earth, and worked out a geocentric planetary model. Aristotle theorized that all things on Earth are imperfect and that the heavenly bodies are made of a more exalted and perfect substance than all earthly objects. His ideas were accepted from around 350 BC until the works of Copernicus in the early 1500s. During the Middle Ages, Arabs continued building on the Greeks' knowledge, and they improved several astronomical instruments.

In 1543, Nicholas Copernicus published his theory that planets rotated around the sun instead of Earth. Copernicus' theory, however, was perceived to contradict the philosophical and religious beliefs that man, being made by God in His own image, was superior to the rest of creation and would therefore be placed at the center of the universe. People thought that Copernicus's theories suggested that humans are simply part of nature, not superior to it, and they strongly objected to this perceived demotion. Johannes Kepler's observations and his study of Danish astronomer Tycho Brahe's earlier data led him to improve on Copernicus's theory by showing that each planet moved around the sun not in a circle but in an ellipse. Kepler developed three laws of planetary motion that have enabled people to chart the routes of celestial bodies and direct the flight of man-made satellites. Galileo formulated the law of falling bodies and discovered parabolic trajectories, but his discoveries with the telescope challenged the current theories of the universe. He proved that there was more than one center of motion in the universe and that the sun is the center of the solar system. Isaac Newton developed laws of motion and gravitation, which helped refine calculations made using Kepler's laws. William Herschel advanced a method of measuring time by the apparent movement of the stars around Earth, which improved the tracking of individual stars and the systematic observation of celestial bodies. He also discovered Uranus and hypothesized that nebulae are composed of stars.

Lesson 7.1.2 – Calendars

Throughout history, civilizations have observed celestial bodies for purposes ranging from farming to business to religious ceremony. Even before ancient people could explain such observations, they created calendars using the movements of these objects to help them mark time. The ancient Chinese used a star calendar and determined the seasons and the passage of time by observing the relationship between the positions of the moon and stars. The ancient Egyptians were the first to create a calendar that synchronized the lunar and solar years. The ancient Babylonians developed the concept of the Zodiac, which divided the year into 12 parts. The Maya predicted eclipses, and their calendar had two simultaneous cycles. One cycle had 20 named days, each numbered for a 260-day cycle. The second cycle had a 365-day cycle that used solar days. The Gregorian calendar is a solar calendar, and it is the calendar widely used today. However, certain orthodox churches continue to use the earlier Julian calendar, and other ethnic and religious groups use various other lunar calendars to mark their cultural or religious observances.

Lesson 7.1.3 – Optical Telescopes

Astronomers have made tremendous strides in observing stars and galaxies. The first step to broadening the view of the solar system was the invention of the telescope. Telescopes were invented in the early 1600s by Dutch spectacle makers. Hans Lippershey is often specifically credited with inventing the telescope. Galileo, however, was the first to use the telescope systematically to look beyond Earth's atmosphere. Optical telescopes include refracting and reflecting varieties. Refracting telescopes are the simplest, using lenses to gather light. Refracting telescopes have the disadvantage of chromatic aberration, meaning they do not focus all colors at the same point. Reflecting telescopes solve this problem by reflecting the light instead of bending it. In the 1800s, telescopes were combined with spectography. A spectrograph is an instrument that separates the light or radiation shed by a source into a spectrum of colors. Various lines in the spectrum indicate the presence of different gases in the object that is shedding the light. (Students will learn more about this process in the chapter about stars.) Optical telescopes can also be combined for greater resolution. Working together, the Keck Observatory telescopes in Hawaii have the resolving power of a single telescope with a mirror 85 m in diameter. Despite advances in telescopes that are launched into Earth's orbit, large optical telescopes on Earth are still being built, including the Giant Magellan Telescope and the European Extremely Large Telescope, both of which are being built in Chile. These two telescopes are scheduled to be completed in 2024.

Lesson 7.1.4 – Nonoptical Telescopes

A variety of nonoptical telescopes, designed to detect signals from discrete portions of the electromagnetic spectrum, are used to observe objects throughout the universe. These telescopes reveal information that optical telescopes cannot reveal. Radio telescopes pick up radio waves. The main dish of most radio telescopes is mounted on a support frame that allows it to point at any part of the sky. The main dish reflects radio waves to an antenna called *the feed*, which sends the waves to a receiver. To create more detailed images, radio telescopes are often built in arrays, but they can also be constructed as enormous single dishes set in the ground. A component of radio astronomy is the use of radar to produce images of celestial objects in the solar system that are difficult to see with optical telescopes. Such imaging is possible because radio waves undergo little to no distortion when they pass through gas and dust.

Infrared telescopes are helpful for viewing cool objects, which give off most of their radiation as infrared light, making them difficult to see with other telescopes. In order to prevent the infrared sensors from picking up heat from their own electrical components, these parts are cooled well below the freezing point of water. Ultraviolet telescopes are used to study gases throughout space. These gases provide information about planetary atmospheres, comets, star formation within galaxies, and supernovas.

Conventional X-ray telescope mirrors have shallow angles and are coated with heavy metals, such as gold, to prevent the X-rays from penetrating the mirror. These telescopes are used to explore dark energy and dark matter, which many scientists assume to be present because of how galaxies are held together internally and clustered with each other. No one knows much about dark energy or dark matter, but X-rays are produced either by them or near them, which enables scientists to map their locations and study them.

Gamma-ray telescopes pick up the short-lived bursts of cosmic gamma rays. These rays do not reach Earth's surface, so they can be detected only directly from space. Gamma rays expel electrons

WORLDVIEW

- Psalm 19 proclaims, "The heavens declare the glory of God; the skies proclaim the work of His hands," and "Their voice goes out into all the earth, their words to the ends of the world." Nearly everyone can look to the night sky and marvel at the complexity and immensity of what they see. Those who study astronomy and its related disciplines have the privilege of seeing these qualities on an immeasurably grander scale. Such knowledge should inspire awe in the powerful God who created such order and beauty. Even more amazing than the complexities of the solar system, however, is the attention that God pays to people on Earth. Compared with the solar system, not to mention the universe, we appear to be insignificant. Yet we are the only part of God's creation made in His image and with whom He interacts on a personal level. The second half of Psalm 19 speaks of the blessings of God's law. God's law is a far more personal revelation of Himself than the creation. The law shows us what God cares about, and it reveals how we can be in a relationship with Him. Together, God's creation and law reveal an almighty yet personally relational God.

as they move through detectors, and these electrons can be recorded as flashes of light. Scientists use gamma-ray emissions to better understand how a variety of celestial objects function, including supernovas, black holes, and pulsars.

Lesson 7.1.5 – Solar System

The solar system is comprised of the sun, the eight major planets, and many other smaller objects. Although there is a general consensus on how these objects are defined, some astronomers differ on exactly what a planet is, exactly what objects comprise the solar system, and how the solar system came into existence. Most astronomers use five broad groups to categorize the objects in the solar system. The first group of celestial objects includes the terrestrial planets, which differ in detail but are basically all Earth-like. These planets are Mercury, Venus, Earth, and Mars. These planets have silicate rock surfaces and iron cores; all of these planets have been significantly warmed by the sun.

The second group of celestial objects consists of a large army of asteroids largely located in the asteroid belt between the orbits of Mars and Jupiter. Many of the asteroids' orbits have been established, yet there are far more to investigate. Sometimes Jupiter or another large object will knock an astroid out of its orbit or change its orbit. Therefore, not all asteroids are found in the main asteroid belt; some have orbits that lie in the vicinity of the terrestrial planets and cross the orbits of those planets. Among these asteroids are comets and meteoroids. Comets differ from asteroids in that they are mostly ice and rock, as opposed to the largely rocky composition of asteroids. Meteoroids are thought to be chunks broken off from either asteroids or comets.

Four gaseous planets make up the third group of celestial objects. Jupiter, Saturn, Uranus, and Neptune all formed in the very cold regions of space where the sun supplied little heat. The abundant hydrogen and helium in this cold region moved very slowly, so the gases were captured by these masses' heavier-element cores to form giant planets.

The fourth group of celestial objects is the dwarf planets. Dwarf planets are composed mostly of rock and ice. A few, such as Ceres, are located in the asteroid belt, but most exist beyond Neptune. Pluto was once classified as a true planet. In 2006, it was reclassified into the newly created dwarf planet category because it did not appear to have cleared its orbit of other objects, as other true planets had. Other dwarf planets include Eris, Haumea, and Makemake.

The Oort cloud is an unconfirmed region of the solar system, thought to be up to 100,000 astronomical units from the sun. (One astronomical unit is the average distance between the sun and Earth.) The Oort cloud is considered by some astronomers to be the fifth group of objects in the solar system, and it is named after the astronomer who hypothesized its existence, Jan H. Oort. Scientists who accept Oort's theory believe that the Oort cloud contains about 1 trillion objects, which are thought to be the source of the long-period comets. Long-period comets are those that take more than 200 years to orbit the sun. Some long-period comets are thought to take up to 30 million years to complete a single orbit.

Over the years, several theories have been proposed to explain the scientific beginning of the solar system. Such theories attempt to explain not only the beginnings of the sun, planets, moons, asteroids, and comets, but also the chemical and physical differences of the planets and their orbital regularities. The theories also account for the relative angular momentum of the sun and planets that springs from their rotational and orbital motions. Two prominent theories on the beginning of the solar system are the young-earth and old-earth theories, and there are Christians who hold to each one. However, there are differences of opinion on how the solar system was formed even within each of these major camps.

As with all theories, those who develop them start from a certain set of presuppositions. Many proponents of the young-earth theory presuppose that the Genesis account of Creation is factual and that God created the universe to appear much like it does today. Many proponents of the old-earth theory presuppose that the Genesis account is not intended to factually teach how God made the universe and that God used natural processes over billions of years to accomplish His creation. The basic difference is that young-earth Christians look for ways science supports the Bible, and old-earth Christians look for ways the Bible can be explained in terms of science.

Lesson 7.1.6 – Sun

The sun dominates Earth's sky during the day because it is relatively close to Earth at an average distance of 149,597,871 km away. That seems far, but the nearest star is 250,000 times farther from Earth than the sun is. Like the earth, the sun spins on its axis, but its gases rotate at different rates depending on whether they are above or below the convective zone. The gases at the sun's equator rotate faster than those at its poles.

The sun is composed of several layers. The innermost layer is the core. The density of the core is great, but it is still a gas because of its high temperature (15 million °C). Nuclear fusion reactions take place in the core. The energy from this intensely heated core radiates up through the next level, the radiative zone. Streams of superheated gas transport energy up through the convection zone to the sun's surface, and convection currents take the gas back into the sun. The top layer of the sun, sometimes considered the first atmospheric layer, is the photosphere, a relatively thin layer of hot gas that radiates visible light to the entire solar system. The chromosphere, the next layer of the sun's atmosphere, extends above the photosphere in a layer of reddish-pink gas. The red color springs from the abundance of hydrogen. The chromosphere is visible on Earth only when the bright light of the photosphere is blocked from view by an eclipse or special instruments. The corona is the outermost layer of the sun's atmosphere. The corona is faint and not normally seen with the unaided eye except during solar eclipses. This layer of superheated gas extends millions of kilometers into space. The sun loses particles constantly; protons and electrons evaporate from the sun and reach Earth at velocities of 500 km/s. Magnetic fields in the corona hold in most of the sun's mass, but particles often slip through holes in the fields.

Sunspots are relatively dark and cool spots on the photosphere. Sunspots undergo 11-year cycles, increasing sharply in number and then decreasing until the sun is again nearly free of spots. Solar flares occur in the corona above sunspots. These eruptions vary in size and duration, but they all send out tremendous bursts of radiation and particles. In contrast, prominences are not as violent as solar flares. They occur at the convergence of magnetic field boundaries.

NOTES

7.1.1 History of Astronomy

Solar System

OBJECTIVES

Students will be able to
- chronologically order significant contributions to the history of astronomy.
- evaluate the significance of major contributions to the study of astronomy.
- determine the effect of technology on people's understanding of the universe.

VOCABULARY

- **geocentric** centered on or around Earth
- **heliocentric** centered on or around the sun

MATERIALS

- Orange, gray, yellow, blue, and red sheets of paper (*Introduction*)
- BLM 7.1.1A Faith and Science
- WS 7.1.1A Galileo and the Sun
- WS 7.1.1B Galileo Goes to Jail

PREPARATION

- Mark off a 4-meter diameter circle on the classroom floor and mark the center. In a large space, mark off 4 concentric circles with the following radii: 1.5 m, 3 m, 4 m, 6 m; the last circle may be omitted if space does not permit it. Divide the circles into quarters by marking lines through the shared midpoint. (*Introduction, B*)
- Obtain materials for *Try This: Astrolabe*.

TRY THIS

Astrolabe
- plastic protractors, 1 per student
- string
- scissors
- metal washers, 1 per student
- drinking straws, 1 per student
- tape
- card stock
- hole punch

Introduction

Lead students to the prepared large, concentric circles. Inform students that the circles represent the approximate orbits of the four inner planets around the sun. Ask students to name the four inner planets. (**Mercury, Venus, Earth, Mars**) Inform students that one way to remember the order of these planets is through a mnemonic device, such as "Monica Visits Every Monday." Select a volunteer to stand in the center and hold an orange sheet of paper to represent the sun. Select other volunteers to stand in a straight line on the circles radiating away from the sun in the following order: Mercury—gray paper, Venus—yellow paper, Earth—blue paper, Mars—red paper. Demonstrate the approximate motion of the planets by instructing volunteers to complete counterclockwise orbits according to the following directions:
- **Mercury**: Complete 2 orbits by the time Earth completes 0.5 orbit
- **Venus**: Complete 0.75 orbit by the time Earth completes 0.5 orbit
- **Mars**: Complete 0.25 orbit by the time Earth completes 0.5 orbit

Direct students to pay particular attention to the relationship between the orbits of Earth and Venus. (Note: The first quarter line is the first line counterclockwise from the line where volunteers lined up.) Have students note at which quarter Venus is located during four consecutive Earth orbits. (*Earth orbit 1*: **Venus on second quarter**; *Earth orbit 2*: **Venus on fourth quarter**; *Earth orbit 3*: **Venus on second quarter**; *Earth orbit 4*: **Venus on fourth quarter**) Ask how the position of Venus changes in relation to Earth as they orbit the sun. (**Sometimes Venus seems to be ahead of Earth, and sometimes Venus seems to be chasing Earth.**) How would the movement of Venus appear to change from the perspective of someone on Earth? (**Venus would occasionally seem to move backward.**)

Discussion

- Guide a discussion regarding why people believed in a geocentric model of the solar system. Ask why ancient civilizations might have assumed that Earth is the center of the solar system. (**Possible answers: The sun, moon, planets, and stars all appear to rotate round Earth; they had no way of knowing that Earth moved; they believed that people are the highest order of all things, and it made sense to them that the most important thing would be in the center of the universe.**)

- Ask the following questions:
  1. What problem with a geocentric view of the solar system lead Ptolemy to propose the epicycle? (**Possible answer: The planets sometimes appeared to move backward.**) What would that orbit look like? (**The planets would move in small circles as they orbit the earth in one large circle.**)
  2. What technology demonstrated the heliocentric view of the solar system? (**telescope**) How has that invention affected astronomy? (**Answers will vary but should include that it confirmed the solar system was heliocentric, it revealed details about the planets and stars, and it continues to allow scientists to learn more about the universe.**)

Activities

A. Complete *Try This: Astrolabe* with students. Display a globe or a map marked with lines of latitude. Ask students what they would need to know in order to determine their latitude using their astrolabe. (**how far up in the sky a star or the sun is supposed to be on any particular day**) Challenge students to infer what information could be determined from knowing how the stars change their positions throughout the seasons. (**Possible answers: Earth's tilt on its axis; Earth's movement through space**) Why might ancient astronomers have held to a geocentric view of the universe? (**Observations from Earth seem to indicate that the sun and stars rotate around Earth.**) Inform students that the astrolabe can be used to measure the height of an object such as a tree using the following steps:

- Move away from the object until a 45° angle can be obtained by sighting through the straw at the top of the object.
- Measure the distance from the ground to your eye. (For example, 1.5 m)
- Measure the distance from where you took the sighting to the base of the object. (For example, 7.5 m)
- Add the two distance measurements together to get the height of the object. (For example, 9 m)

B. Lead a discussion on the question of whether faith and science are in conflict using **BLM 7.1.1A Faith and Science** Have students complete **WS 7.1.1A Galileo and the Sun** and **WS 7.1.1B Galileo Goes to Jail** as part of the discussion activity.

C. Select a student volunteer to stand in the center of the prepared 4-meter diameter circle. Select a second volunteer to stand on the circle's perimeter, pointing an illuminated flashlight at the first volunteer's waist. Dim the lights. Direct the first volunteer to stand still as the second volunteer slowly moves around the circle's perimeter, keeping the flashlight focused on the first volunteer. Have the first volunteer describe his or her observations. Ask students which model of the solar system this represents. (**geocentric**) Next, direct the volunteers to trade places so the person with the flashlight is in the center. Have the volunteer in the center stand still. Direct the volunteer on the perimeter to face the center and to walk around the circle. Have the volunteer on the perimeter compare what he or she observed from both positions. Ask students what model of the solar system the second demonstration represents. (**heliocentric**)

NOTES

7.1.1 History of Astronomy

OBJECTIVES
- Chronologically order significant contributions to the history of astronomy.
- Evaluate the significance of major contributions to the study of astronomy.
- Determine the effect of technology on people's understanding of the universe.

VOCABULARY
- **geocentric** centered on or around Earth
- **heliocentric** centered on or around the sun

Even before modern science, people used observations as a means to understand the orderly nature of creation. For example, even in ancient times people knew about five of the planets: Mercury, Venus, Mars, Jupiter, and Saturn. How could the early astronomers distinguish planets from stars? Unlike stars, planets change position in the sky. (The Greek word for *planet* means "wanderer.") People observed planets temporarily joining star patterns before moving to other parts of the sky. These observations alerted them to the existence of these five faraway bodies.

The ancient Greeks were the first to try to explain why the celestial bodies move as they do. Eudoxus (c. 395–c. 337 BC) suggested that the stars and planets appear to move over time because they were attached to a transparent sphere rotating on a separate axis. More than a century later, the famous Greek philosopher and scientist Aristotle (384–322 BC) published his belief that eclipses showed that Earth is a sphere. He said that Earth does not move and that the other celestial bodies move in perfect circles around Earth. Approximately 100 years after Aristotle published his theories, Eratosthenes (276–195 BC) mathematically proved that Earth is a sphere by using geometry. Eratosthenes calculated Earth's circumference using the lengths of the midsummer shadows at Syrene (now Aswan) and Alexandria, along with the distance between the two cities.

Aristotle

TRY THIS

Astrolabe
Sailors navigated by using astrolabes to measure the angle between the horizon and stars. You can create your own astrolabe. Obtain a plastic protractor with a hole in the center of the straight part. Cut a 25 cm length of string and tie a large knot in one end. Thread the string through the protractor's hole. Tie a small weight, such as a washer, to the string's free end. Tape a drinking straw along the flat edge of the protractor; make sure the straw extends past the edges of the protractor by several centimeters. Cut a 10 cm × 10 cm square of card stock and punch a hole in it just big enough for the straw to pass through the hole. Slide the card stock over the straw until the card stock touches the protractor. Hold the protractor so the curved side is down and look through the straw at the top of a tall object, such as a tree. Pinch the string against the protractor and write down the number where the string crosses the protractor. Subtract this number from 90 to obtain the object's elevation in degrees.

In 140 AD, the Greek astronomer Ptolemy (100–170 AD) wrote the *Almagest*. Ptolemy's book was the first successful attempt to explain how all the observations of planetary movements fit into a single system. In this work, Ptolemy described a universe in which the sun, moon, and planets revolve around Earth. He even tried to explain why some planets seem to move backward at times. Ptolemy thought that each planet moved in a type of orbit called *an epicycle*, meaning each planet moved in small circles as they orbited Earth in one large circle. The word **geocentric** describes the view that Earth is the center of the universe. This belief influenced scientific thought until the 16th century.

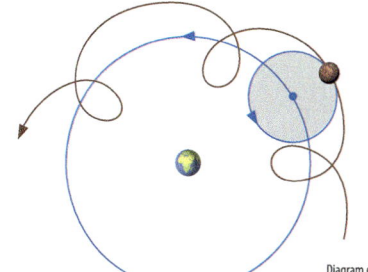

Diagram of Ptolemy's view of the planetary motion

NOTES

D. Assign students to research the history of astronomy in African, American, Asian, or Middle Eastern area cultures. Have students hypothesize why the history of international astronomy is primarily associated with European cultures.

E. Have students create a time line of significant discoveries, including the Antikythera mechanism, or theories in the history of astronomy from the earliest recorded astronomical observations to the present day. Guide students to incorporate discoveries or theories from non-Western cultures.

F. Direct students to research contributions to astronomy made since the year 2000.

Lesson Review

1. Starting with Copernicus and ending with Newton, list the scientists who made significant contributions to developing the heliocentric view of the universe and identify each person's contribution. (***Copernicus*: provided a believable explanation for the heliocentric view, *Kepler*: described how planets orbit the sun, *Galileo*: proved by observation with the telescope that planets orbit the sun, *Newton*: mathematically demonstrated that gravity is the force that moves the planets in their orbits**)
2. What was significant about Ptolemy's explanation of the universe? (**Ptolemy's book was the first successful attempt to explain how all the observations of planetary movements fit into a single system.**)
3. Explain the importance of the Arabian contributions to astronomy. (**The astrolabe enabled scientists to make more precise measurements, and algebra and the number system enabled scientists to make precise and complicated calculations.**)

BIBLE CONNECTION

Tell Me the Dream
"In the second year of his reign, Nebuchadnezzar had dreams; his mind was troubled and he could not sleep. So the king summoned the magicians, enchanters, sorcerers, and astrologers to tell him what he had dreamed. When they came in and stood before the king, he said to them, 'I have had a dream that troubles me and I want to know what it means…. Tell me the dream, and I will know that you can interpret it for me.' The astrologers answered the king, 'There is no one on earth who can do what the king asks!' " (Daniel 2:1–3, 9b–10a).

The ancient Babylonians did not distinguish between astrology and astronomy as people do now; the astrologers in Nebuchadnezzar's court were also astronomers. Astrology is the false belief that the movement of the various bodies in the solar system can influence human life. Astronomy is the scientific study of matter beyond Earth's atmosphere. Although these men predicted the movements of the planets, their interpretations of what their false gods meant through the movements was guesswork. Only God could reveal Nebuchadnezzar's dream.

When the Greek and Roman civilizations faded, the knowledge that they had accumulated was not lost. The Arabs built on the Greeks' knowledge of astronomy. Many bright stars have Arabic names. For example, the name *Betelgeuse* (the bright orange star in the constellation Orion) is Arabic in origin. The Arabs perfected an instrument called *an astrolabe* to determine the altitude of the sun and stars and to create accurate charts of their apparent movements. They also invented algebra, which used the works of the Greeks and Indians, and they introduced the number system that includes the number zero. These systems enabled the precise calculation of astronomical distances and movements.

Until the 1500s, Western civilizations were still using Ptolemy's system. But Polish astronomer Nicolaus Copernicus (1473–1543) decided that there must be a simpler system. In place of Ptolemy's geocentric view of complex epicycles, Copernicus proposed that the sun was the center of the solar system. The word **heliocentric** describes this view. The heliocentric view of the universe easily explains the movements of the planets and even explains why some planets appear to move backward at times. In his book

Arabic astrolabe

340

HISTORY

Music of the Spheres
You may know Pythagoras as the person who developed the Pythagorean Theorem, which is $a^2 + b^2 = c^2$. He also suggested the idea of the "Music of the Spheres." Pythagoras believed that the universe was made of 10 spheres nested inside each other. These spheres were named for objects they were thought to carry: Counter-Earth, Earth, Moon, Sun, Mercury, Venus, Mars, Jupiter, Saturn, and Fixed Stars. Pythagoras also believed that these vast objects produced a tone as they sailed along their courses, much like the whistle of a ball spun on a string. He thought that if a person could hear them, these spinning spheres would be in beautiful harmony, because he thought that the distances between the planets must correspond to the harmonic ratios between musical notes.

De Revolutionibus, Copernicus pointed out that if Earth and the planets orbited around the sun, the planets would appear to move backward in their orbits when Earth passed them. (Think of passing a truck on the highway. If you stare at the truck, it can seem as though you are sitting still and the truck is moving backward.)

Copernicus's work was expanded by German mathematician and astronomer Johannes Kepler (1571–1630). Kepler benefitted from the work of Danish nobleman Tycho Brahe (1546–1601), who created and updated precise tables of the planets' motion and improved astronomical instruments. As Kepler studied Brahe's data in light of the heliocentric view, he discovered certain laws of planetary motion. For example, planets travel in elliptical orbits with the sun as one of two focus points of those orbits. Kepler's laws have been an important foundation to the study of the solar system. They have helped chart the orbits of planets, the moon, and spacecraft. Although Kepler explained the laws that govern the motion of planets, he could not explain what caused the planets to move.

In the early 17th century, Galileo Galilei (1564–1642) improved the telescope and used it for astronomy. His observations supported the heliocentric view. For example, he observed that Venus does not revolve around Earth. Galileo also made new discoveries about the universe, including the discovery of four of Jupiter's moons. The existence of these moons disproved the idea that only Earth could attract celestial bodies into orbit.

Sir Isaac Newton (1642–1727) identified why the planets orbit the sun and why the moon orbits Earth. Newton suggested that

Nicolaus Copernicus

Johannes Kepler

Galileo Galilei

341

4. In what way did Eratosthenes's calculation of Earth's circumference affect people's understanding about Earth? (**Calculating Earth's circumference proved Aristotle's theory that Earth is a sphere.**)
5. Explain one way in which the technology of the telescope affected people's understanding of the universe. (**Possible answers: The telescope enabled people to see that the sun and not Earth is the center of the solar system; the telescope showed people that there are multiple galaxies in the universe; the telescope enabled people to see that the universe is expanding.**)

NOTES

---

celestial bodies keep their orbits through a force called *gravity*. He went on to mathematically describe the motions of the planets in terms of their mass, momentum, and force. Newton's laws are still applied today as scientists launch spacecraft throughout the solar system. Newton's work tied together the knowledge gathered by Copernicus, Brahe, Kepler, Galileo, and many others.

German astronomer Sir William Herschel (1738–1822), the astronomer who discovered Uranus, was famous for building large, powerful telescopes. Some of the objects that Herschel studied were fuzzy patches of light, which others thought were some sort of luminous fluid called *nebulae*. Herschel's telescopes enabled him to see that a few of these nebulae were actually clusters of stars. However, he was not able to build a telescope strong enough to verify that all the nebulae were star clusters. Herschel called these clusters *island universes* of stars.

As stronger telescopes were built, scientists learned more. In the 1920s, American astronomer Edwin Hubble (1889–1953), discovered that Herschel's island universes are galaxies outside the Milky Way galaxy. Until that time, the Milky Way was all anybody knew about the universe. Most people assumed that even Herschel's island universes were still part of the star system. Hubble not only discovered that the universe is

Sir Isaac Newton

Sir William Herschel's 12-meter reflecting telescope

Galaxy NGC 1309 is 100 million light-years from Earth.

HISTORY

A Changing Understanding  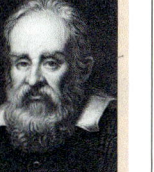
Galileo's use of the telescope gave him a unique view into the solar system and convinced him of the validity of Copernicus's theory that the Earth moved around the sun. Other scientists disputed his finding because they did not have access to this new technology and because the idea of a heliocentric solar system went against the accepted scientific views of the day. When church leaders disputed Galileo's findings, they echoed the opinions of leading scientists, who were reluctant to trust such revolutionary data from one source. Heliotropism was not heretical; it had been used by Pope Gregory XIII to revise the calendar although it was still considered unproven. Ultimately it was not Galileo's scientific findings that got him sentenced to house arrest by the Church; it was his attacks on the biblical arguments against Copernicus's theory and on the Pope. Politics rather than science got Galileo into trouble. In 1992, the Roman Catholic Church officially cleared Galileo's name.

Edwin Hubble at telescope

wider than the galaxy, but he discovered that the universe is expanding. This discovery has become an important part of discussions about the physical beginning of the universe.

Today, advanced telescopes, supercomputers, and spacecraft continue to expand people's knowledge of the universe. But even as new discoveries answer some questions, other discoveries lead to more questions. Many wonders of God's universe wait to be explained and explored.

LESSON REVIEW
1. Starting with Copernicus and ending with Newton, list the scientists who made significant contributions to developing the heliocentric view of the universe and identify each person's contribution.
2. What was significant about Ptolemy's explanation of the universe?
3. Explain the importance of the Arabian contributions to astronomy.
4. In what way did Eratosthenes's calculation of Earth's circumference affect people's understanding about Earth?
5. Explain one way in which the technology of the telescope affected people's understanding of the universe.

7.1.2 Calendars

Solar System

OBJECTIVES

Students will be able to
- determine the usefulness of the sun in measuring time during a day and during a year.
- summarize the development of the Gregorian calendar.
- evaluate the benefits of lunar and solar calendars.

VOCABULARY

- **lunar calendar** a calendar that uses the phases of the moon
- **solar calendar** a calendar that uses the amount of time Earth takes to orbit the sun

MATERIALS

- Meterstick, flashlight, sticky note (*Introduction*)
- TM 7.1.2A Moon Phases
- WS 7.1.2A Astronomical Observations

PREPARATION

- Obtain materials for *Lab 7.1.2A Sundial*.
- Obtain materials for *Try This: Cast a Shadow*.

TRY THIS

Cast a Shadow
- metersticks, 1 per group
(**Shadows move as the sun changes position in the sky, so by tracking their progress, approximate times can be set; except at the equator, shadows change position throughout the year for any given time of day, so by tracking the changes around the year, patterns can be noted; Earth's position in relation to the sun changes throughout the day and year.**)

Introduction

Have students complete **WS 7.1.2A Astronomical Observations**. Discuss student answers.

Display **TM 7.1.2A Moon Phases**. Ask students how many distinct phases they see. (**15**) Inform students that they will learn some information about the moon in this chapter and more detailed information about the moon in the next chapter.

Select a student volunteer to hold a meterstick vertically on a table in front of a cleared surface. Dim the lights in the room and shine a flashlight on the meterstick. Invite a second volunteer to place a sticky note along one edge of the shadow. Move the flashlight slowly from side to side and challenge students to tell you when to stop so the shadow lines up exactly with the sticky note. Point out that people used sundials to mark time in a similar way.

Discussion

- Lead a discussion regarding the ease of marking time using the moon and the sun. Encourage students to support their ideas by referring to the activities in *Introduction*.

- Guide a discussion regarding the possible strengths and weaknesses of lunar and solar calendars.

- Discuss the Gregorian calendar. Lead students to consider why it is widely used today.

- Ask the following questions:
  1. Why do you think many people today use BCE (Before the Common Era) instead of BC and use CE (Common Era) instead of AD? (**Possible answer: Many people do not want to acknowledge God's existence, but the current dating system is so widely used that it would be inconvenient to establish a new starting point for numbering years.**)
  2. What information was necessary before people could create an accurate solar calendar? (**the precise length of a solar year**)
  3. Why would it have been a problem to have a shifting solar calendar that told people to plant corn in April? (**April might be in the spring for a while, but eventually it would move to the summer, which would be too late for planting.**)

Activities

Lab 7.1.2A Sundial

- paper plates, 1 per group
- metric rulers, 1 per group
- protractors, 1 per group
- flashlights, 1 per group
- clay and toothpick, optional

Allow students sufficient time to create their sundials, then dim the lights to better enable them to see the sundials' shadows.

Some students may struggle to make the connection between the sundial and the position of the sun in relation to Earth. For these students, use a small ball of clay to fix a toothpick perpendicular to a point on a globe and shine a flashlight at the globe's center. This modification works best if the globe is mounted so its polar axis reflects the tilt of Earth's polar axis. Have a student slowly rotate the globe and observe the way the shadow moves. Move the flashlight around the globe and have students observe how the shadow changes; rotate the globe so the toothpick always faces the flashlight.

A. Complete *Try This: Cast a Shadow* with students.

B. Have students develop presentations showcasing an ancient structure or device used either for making astronomical observations or for marking time.

C. Assign students to research calendar systems developed by ancient cultures, such as those of the Egyptians, Maya, or Arabs. Challenge students to design a new calendar system for the modern world.

D. Have students research a culture, either ancient or contemporary, that did not use calendars or that used only very basic calendars. Encourage students to present their findings to the class.

E. Direct students to analyze recent suggestions for improving the Gregorian calendar. Have students describe the suggested changes and explain why the changes are thought to be necessary. Challenge students to consider the difficulties such changes may cause to governments and people. Ask students if they think such changes will be attempted in their lifetimes. Why? **(Answers will vary.)**

Lesson Review

1. Why do you think sundials were more common than solar calendars in ancient times? **(Possible answers: Shifting shadows are easily measured during a day, but tracking the changes in shadows from day to day are not as easy to recognize; in ancient times, most people did not realize that the earth orbits the sun.)**
2. Summarize the important people and events in the development of the Gregorian calendar. **(The calendar started as an ancient Roman lunar calendar. Julius Caesar and Sosigenes changed the lunar calendar to a solar calendar with 365 days and a leap year every fourth year to account for the extra 0.25 day in a solar year. Julius Caesar adjusted the year 46 BC to realign the calendar and seasons. Pope Gregory XIII asked astronomers to provide a more accurate length of a solar year, and they calculated 365.242 days. Pope Gregory adjusted the calendar so years divisible by 4 and centuries divisible only by 400 would be leap years. Later, astronomers declared that years divisible by 4,000 would not be leap years. Pope Gregory also adjusted the length of the year 1582 to realign the calendar and seasons.)**
3. What benefits are associated with a lunar calendar? **(Possible answers: Dates are easily determined according to phases of the moon; people can simply look at the moon to identify how far they have progressed through a month.)**
4. What benefits are associated with a solar calendar? **(Possible answers: Named months are always connected with specific seasons; it is possible to calculate specific dates from year to year.)**
5. Why might some cultures use both a solar and a lunar calendar? Use an example to support your answer. **(Answers will vary but should include that some cultures, such as those of the Hindus, Muslims, and Jews, preserve ancient traditions or religious observances that were originally scheduled using a lunar calendar. Because these dates are associated with phases of the moon, it would be difficult to accurately preserve their exact observances with a solar calendar. However, the Gregorian solar calendar has become nearly universally accepted, and its use makes it easier to do business internationally.)**

NOTES

7.1.2 Calendars

OBJECTIVES
- Determine the usefulness of the sun in measuring time during a day and during a year.
- Summarize the development of the Gregorian calendar.
- Evaluate the benefits of lunar and solar calendars.

VOCABULARY
- **lunar calendar** a calendar that uses the phases of the moon
- **solar calendar** a calendar that uses the amount of time Earth takes to orbit the sun

Ancient people did not fully understand the movements of celestial bodies in the solar system or why the patterns of the stars in the sky changed. However, they did notice the patterns and cycles, and they kept records of the orderly patterns of God's creation. By using their records, they tracked the passage of time and seasons.

Before the invention of mechanical clocks, a simple and common device was used to track time. This device, the sundial, uses a vertical piece to cast a shadow on a horizontal surface. As the sun moves across the sky above the sundial, the sundial's shadow moves. A sundial enabled people to divide daylight time into smaller units. In this way, people could organize their days by time. For example, in one of Jesus's parables in Matthew 20, laborers expected to be paid according to the amount of time they worked.

On a larger scale, keeping track of the sun's movement across the stars helped early farmers determine the best times of year to plant and harvest different crops. Such records led to the creation of calendars. A calendar is a system for organizing time. Different civilizations developed various calendars according to their observations of the sky. It takes Earth a year to revolve once around the sun. The moon takes about a month to revolve around Earth. And in one day, the Earth spins one time on its axis. Most nations now use a calendar in which one year has 365 or 366 days. Each of the 12 months has 28–31 days. However, early calendars were different.

A simple sundial

A complex sundial

Astronomers in ancient Babylon (what is now Iraq) tracked the positions of the planets and the moon as early as 700 BC. Their studies resulted in a calendar with 12 months; each month began with a new moon. A calendar that uses the phases of the moon is called a **lunar calendar**. The Babylonians also had a number system based on the number 60. The concept is similar to the metric system, which uses 10. The idea of a 60-second minute and a 60-minute hour came from the Babylonian number system.

The most widely used calendar today is called *the Gregorian calendar*. It is known as a **solar calendar** because it uses the amount of time Earth takes to orbit the sun. However, the Gregorian calendar had its roots in the lunar calendars of ancient Rome. In fact, most ancient calendars were lunar calendars. The Gregorian calendar calculated 355 days in a year. Since the calendar had about 11 fewer days than a true solar year, the calendar dates would shift through the seasons. For example, over time spring would shift from March to October. To keep the calendar in line with the seasons, extra months were added. The system eventually proved to be confusing and ineffective. Julius Caesar consulted astronomer Sosigenes of Alexandria to reform the calendar. They created the Julian calendar after calculating that each year has 365.25 days. Days were added to the year 46 BC to return the seasons to their original positions in the year. An extra day was added every four years to keep the seasons in place according to the calendar. A year in which an extra day, February 29th, is added to the calendar is called *a leap year*.

The Julian calendar was not quite accurate. In the mid-1500s, Pope Gregory XIII turned to astronomers for the solution. They determined that there are actually 365.242 days in a year. The Pope dropped 10 days from the year 1582 to bring the calendar back into line with the seasons. To keep the Gregorian calendar accurate, he restricted leap years to years that are divisible by 4, but century years are only leap years if they are evenly divisible by 400. For example, the year 2000 was a leap year, but the year 2100 will not be. This adjustment brought the average year length closer to 365.242 days. Later, years evenly divisible by 4,000 were declared not to be leap years. These adjustments will keep the Gregorian calendar accurate for another 20,000 years.

The Gregorian calendar is not the only calendar in use around the world. In some cultures, the Gregorian calendar is used for business purposes, but a lunar calendar is used to determine certain holidays and religious observances. For example, India

TRY THIS

Cast a Shadow
On a sunny day, fix an upright meterstick outdoors. Track the shadow at regular intervals throughout the day. Graph the data. How can shadows help people keep track of time? What can shadows demonstrate about Earth's relationship to the sun? Could shadows help people keep track of time throughout the year as well as time throughout the day?

Julius Caesar

Pope Gregory XIII

FYI

The Jewish Calendar
The Jewish calendar uses the cycles of the moon. The names of the months in the Jewish calendar are Babylonian; they were adopted during the Babylonian exile. The Jewish calendar starts with the creation of Adam.

On the Jewish calendar
- Abraham was born in 1948.
- The Israelites' exodus from Egypt was in 2448.
- Jesus was born about 3754–3756.
- Columbus came to America in 5252.

HISTORY

Accurate Clocks
For thousands of years, people have observed the sky and devised ways to mark the movements of the sun and the moon. For example, an arrangement of stones found at Nabta in southern Egypt are arranged to line up with the sun on the longest day of the year. The stone arrangement is estimated to be 6,000–7,000 years old. Stonehenge, built near Salisbury, England, around 2000 BC, lines up with the sun on the longest and shortest days of the year.

Stonehenge

The Maya, who lived in Mexico and Central America, developed advanced systems of mathematics and astronomy. They had a number system that included the number 0, and they designed buildings so that at set times the sun would shine through precise openings in walls or would cast shadows in predetermined places. The Maya also developed a complex calendar that included an accurate solar year.

Temple of Kukulcan, Chichen Itza, Mexico

BIBLE CONNECTION

The King's Birthday
In the 6th century, a scholar designed the current system of dating from the birth of Jesus. AD means "anno Domini," which is Latin for "in the year of our Lord." BC means "before Christ." There is no year 0 between BC and AD; the calendar transitions from 1 BC directly to 1 AD. Today many scholars believe that Christ was born several years before 1 AD.

and most Muslim nations use the Gregorian calendar for civil purposes. In India, religious observances are determined according to the ancient Hindu lunar calendar that includes adjustments to keep its months aligned with the seasons. Muslims follow a purely lunar calendar established by the Koran for religious observances. Because this calendar does not make any adjustments for the solar year, Muslim religious days do not always occur in the same season. Israel officially uses a lunar calendar that includes an extra month at precise intervals to keep it aligned with the seasons.

LESSON REVIEW
1. Why do you think sundials were more common than solar calendars in ancient times?
2. Summarize the important people and events in the development of the Gregorian calendar.
3. What benefits are associated with a lunar calendar?
4. What benefits are associated with a solar calendar?
5. Why might some cultures use both a solar and a lunar calendar? Use an example to support your answer.

Name: _____ Date: _____

Lab 7.1.2A Sundial

QUESTION: How can shadows be used to tell time?

HYPOTHESIS: Answers will vary.

EXPERIMENT:

You will need:		
• paper plate	• metric ruler	• flashlight
	• protractor	

Steps:
1. Draw a line through the approximate center of the paper plate using the ruler.
2. Use the protractor and ruler to mark and draw a line that runs through the center of the first line and is perpendicular to it. The lines should create a giant plus sign that divides the plate into four equal sections. Label each line *North*, *South*, *East*, or *West*, so the plate looks like a simple magnetic compass without a needle.
3. Mark 30° segments around the plate using the protractor. If you live in the Northern Hemisphere, label the *North* line 12. If you live in the Southern Hemisphere, label the *South* line 12. If you live near the equator, you may write 12 on either the *North* or the *South* line. Continue numbering each line so the plate looks like a clockface.
4. Carefully poke a pencil through the center of the plus sign and adjust it so it stands straight up.
5. In a darkened room, use a flashlight to make the pencil's shadow appear on the plate.
   - Make the pencil's shadow as long and as short as possible. Observe the flashlight's positions when the shadow is longest and shortest.
   - Make the pencil's shadow move clockwise around the marks on the plate. Observe the flashlight's position in relation to the shadow.
   - Shine the flashlight on the pencil from different angles. Observe how the shadow changes as the angle of the light changes.

ANALYZE AND CONCLUDE:
1. How was the flashlight positioned to create the shortest pencil shadow?
   The flashlight was shining straight down over the pencil's end.

2. How was the flashlight positioned to create the longest pencil shadow?
   The flashlight was shining perpendicular to the pencil's side.

Lab 7.1.2A Sundial

3. Why should a sundial be situated so the mark for noon points north in the Northern Hemisphere and points south in the Southern Hemisphere? Earth's curve causes objects standing perpendicular to Earth's plane to be pointed northward away from the sun in the northern hemisphere and southward away from the sun in the southern hemisphere, which causes their shadows to fall in the same directions.

4. At the equator, what would a sundial's shadow look like? A shadow may not be visible.

5. How might the length of shadows change at different times during the year? Shadows will be longer when the sun is lower in the sky during winter, and shadows will be longer when the sun is higher in the sky during summer.

6. How could shadows be used to help calculate the length of a solar year? Possible answer: A person could note where a shadow falls when the sun is highest in the sky on a particular day. When the shadow returns to that exact location, one solar year has passed.

7. What challenges exist when using shadows to help measure a solar year? Possible answers: overcast skies, creating a shadow with a crisp outline, accurately measuring small changes in a shadow's length or position

7.1.3 Optical Telescopes

Solar System

OBJECTIVES

Students will be able to
- evaluate the usefulness of refracting and reflecting telescopes.
- summarize the key problems and solutions for accurate Earth-based observations of celestial bodies.

VOCABULARY

- **aperture** an opening through which light passes
- **reflecting telescope** a telescope that uses a series of mirrors to magnify objects
- **refracting telescope** a telescope that uses a series of lenses to magnify objects

MATERIALS

- Refracting telescope, reflecting telescope (D)

PREPARATION

- Obtain materials for *Lab 7.1.3A Atmospheric Gases* and *Lab 7.1.3B Refracting Telescope*.
- Obtain materials for *Try This: Reflecting Telescope*.
- Invite a scientist from a local university or observatory to demonstrate and explain how refracting and reflecting telescopes work. (B)
- Plan a field trip to an observatory. (C)
- Obtain a refracting and a reflecting telescope and schedule a nighttime sky viewing for parents and students. (D)

TRY THIS

Reflecting Telescope
- concave mirrors, 1 per student
- table
- outside window
- flat mirror
- convex lenses, 1 per student

(**Possible answers:** *advantages*—wider field of view, more stable, *(continued)*

Introduction

Guide students to imagine that they can see an emergency vehicle parked about 1 km away, and the driver turns on the emergency lights and siren at the same time. Ask students whether they will see the lights or hear the siren first. (**see the lights**) Why? (**Light travels faster than sound.**) Inform students that light travels so quickly that it seems instantaneous in normal experiences. Challenge students to calculate the time in minutes that light takes to travel from the sun to Earth given that light travels 299,792,458 m/s in a vacuum and that the distance between the sun and Earth is 149,597,871 km. (**299,792,458 m/s × 1 km/1,000 m = 299,792.458 km/s; 149,597,871 km ÷ 299,792.458 km/s = 499.3 s; 499.3 s × 1 min/60 s = 8.32 min**)

Discussion

- Discuss the speed of light. Do people see light from the sun or stars in real time? Why? (**No. Light takes time to travel.**) How is people's knowledge of the universe affected by the fact that light takes time to travel? (**Possible answers: We cannot see the universe as it really is in real time; new stars may be born that we cannot see, and old stars may have died without our knowledge; celestial bodies that are new to us may be very old; we can know where celestial bodies have been, but we cannot know for certain where they are now.**) Do you think optical telescopes can currently see to the farthest reaches of the universe? Why? (**Possible answers: No, because more powerful telescopes continue to reveal new details of the universe, so it is probable there is even more to discover; No, because the light from the farthest stars and galaxies may be too dispersed to see or may be blocked by other objects.**)

Activities

Lab 7.1.3A Atmospheric Gases

- pins, 1 per group
- aluminum foil
- tape
- books, 2 per group
- flashlights, 1 per group
- hot plates, 1 per group

Check to ensure that the tape peels off books easily without causing damage.

Lab 7.1.3B Refracting Telescope

- concave lenses, 1 per group
- convex lenses, 1 per group
- scissors, 1 pair per group
- aluminum foil
- straight pins, 1 per group

Be sure that each group has a convex lens that is larger and less powerful than its concave lens. The thinner the middle of the concave lens, the more powerful the students' telescopes will likely be. The concave lenses should be as thick in the middle as possible to obtain a clearer image. Pins should create holes no larger than 1 mm in diameter.

Option: Obtain lenses in a variety of thicknesses and diameters and have students compare the images produced using different combinations of lenses.

A. Complete *Try This: Reflecting Telescope* with students.

B. Introduce the scientist to demonstrate and explain how refracting and reflecting telescopes work.

C. Take students on a field trip to visit an observatory.

D. Take students and parents on a sky-viewing event to view celestial bodies using the refracting and reflecting telescopes. Have students compare the telescopes' apertures, greatest effective power of magnification, fields of view, clarity, and ease of use. The same activity can be done during a class period by observing distant objects or the moon if it is visible.

E. Direct students to research methods used in large optical telescopes to reduce the twinkling appearance of stars.

F. Have students research a person who influenced the development or use of optical telescopes, such as Laurent Cassegrain, James Gregory, Bernhard Schmidt, Bernard Lyot, William Lassell, Alvan Clark, Edwin Hubble, Jerry Nelson, Adam Riess, or David Charbonneau.

> **TRY THIS**
>
> *(continued from previous page)*
> **collects more light so objects appear brighter;** *disadvantages*—**not as easy to transport or to aim at different parts of the sky, blurry objects)**

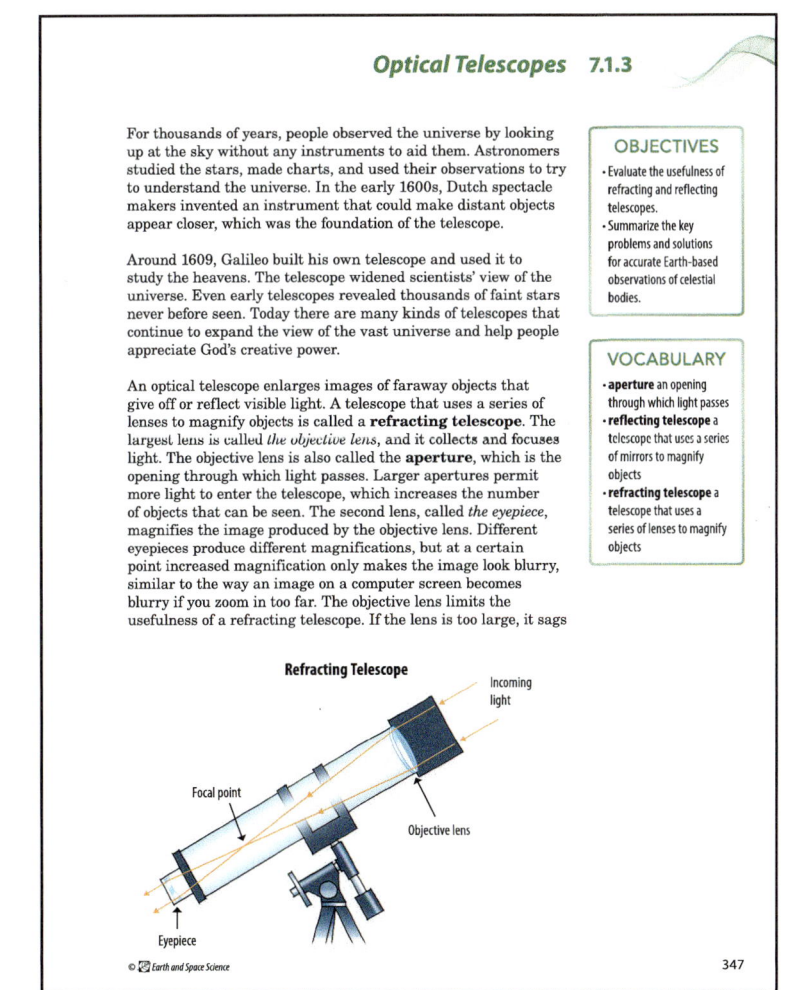

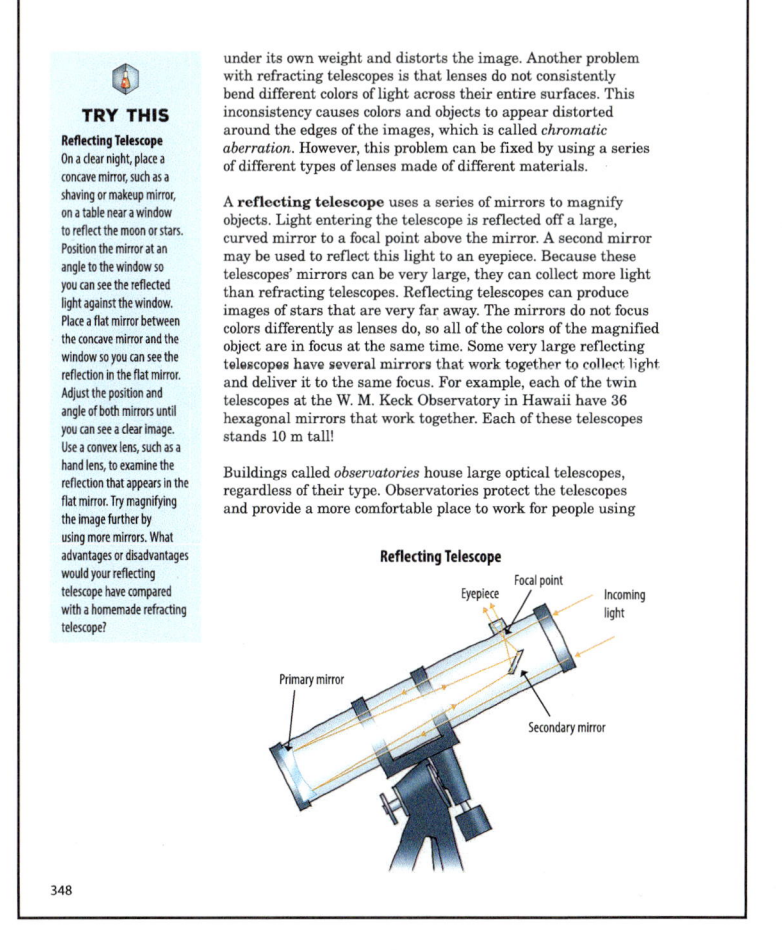

Lesson Review

1. What is the basic difference between refracting and reflecting telescopes? (**Refracting telescopes use a series of lenses to magnify objects; reflecting telescopes use a series of mirrors to magnify objects.**)
2. If you were building an optical telescope today, which type of telescope would you build? Why? (**Answers will vary but should include that refracting telescopes are limited by the size of the objective lens and experience chromatic aberrations, but reflecting telescopes don't have these problems.**)
3. Why are mountaintops popular locations for observatories? (**Building observatories on mountaintops helps limit artificial light pollution and it elevates them above some of Earth's atmosphere, which can distort images.**)
4. What advantage does the Hubble Space Telescope have over telescopes located on Earth? (**The Hubble Space Telescope's view is not obstructed by Earth's atmosphere.**)

Optical Telescopes 7.1.3

For thousands of years, people observed the universe by looking up at the sky without any instruments to aid them. Astronomers studied the stars, made charts, and used their observations to try to understand the universe. In the early 1600s, Dutch spectacle makers invented an instrument that could make distant objects appear closer, which was the foundation of the telescope.

Around 1609, Galileo built his own telescope and used it to study the heavens. The telescope widened scientists' view of the universe. Even early telescopes revealed thousands of faint stars never before seen. Today there are many kinds of telescopes that continue to expand the view of the vast universe and help people appreciate God's creative power.

An optical telescope enlarges images of faraway objects that give off or reflect visible light. A telescope that uses a series of lenses to magnify objects is called a **refracting telescope**. The largest lens is called *the objective lens*, and it collects and focuses light. The objective lens is also called the **aperture**, which is the opening through which light passes. Larger apertures permit more light to enter the telescope, which increases the number of objects that can be seen. The second lens, called *the eyepiece*, magnifies the image produced by the objective lens. Different eyepieces produce different magnifications, but at a certain point increased magnification only makes the image look blurry, similar to the way an image on a computer screen becomes blurry if you zoom in too far. The objective lens limits the usefulness of a refracting telescope. If the lens is too large, it sags

OBJECTIVES
- Evaluate the usefulness of refracting and reflecting telescopes.
- Summarize the key problems and solutions for accurate Earth-based observations of celestial bodies.

VOCABULARY
- **aperture** an opening through which light passes
- **reflecting telescope** a telescope that uses a series of mirrors to magnify objects
- **refracting telescope** a telescope that uses a series of lenses to magnify objects

Refracting Telescope
Focal point — Objective lens — Incoming light — Eyepiece

under its own weight and distorts the image. Another problem with refracting telescopes is that lenses do not consistently bend different colors of light across their entire surfaces. This inconsistency causes colors and objects to appear distorted around the edges of the images, which is called *chromatic aberration*. However, this problem can be fixed by using a series of different types of lenses made of different materials.

A **reflecting telescope** uses a series of mirrors to magnify objects. Light entering the telescope is reflected off a large, curved mirror to a focal point above the mirror. A second mirror may be used to reflect this light to an eyepiece. Because these telescopes' mirrors can be very large, they can collect more light than refracting telescopes. Reflecting telescopes can produce images of stars that are very far away. The mirrors do not focus colors differently as lenses do, so all of the colors of the magnified object are in focus at the same time. Some very large reflecting telescopes have several mirrors that work together to collect light and deliver it to the same focus. For example, each of the twin telescopes at the W. M. Keck Observatory in Hawaii have 36 hexagonal mirrors that work together. Each of these telescopes stands 10 m tall!

Buildings called *observatories* house large optical telescopes, regardless of their type. Observatories protect the telescopes and provide a more comfortable place to work for people using

TRY THIS
Reflecting Telescope
On a clear night, place a concave mirror, such as a shaving or makeup mirror, on a table near a window to reflect the moon or stars. Position the mirror at an angle to the window so you can see the reflected light against the window. Place a flat mirror between the concave mirror and the window so you can see the reflection in the flat mirror. Adjust the position and angle of both mirrors until you can see a clear image. Use a convex lens, such as a hand lens, to examine the reflection that appears in the flat mirror. Try magnifying the image further by using more mirrors. What advantages or disadvantages would your reflecting telescope have compared with a homemade refracting telescope?

Reflecting Telescope
Primary mirror — Eyepiece — Focal point — Incoming light — Secondary mirror

347 | 348

FYI

Twinkle, Twinkle, Little Planet
Why do stars twinkle but planets do not? Planets are closer to Earth than stars are. Stars are so far away that even through a telescope they look like pinpoints of light; through a telescope, planets look like disks of light. The light from the pinpoint of a star seems to shimmer as it comes through the atmosphere, but the different disturbances from across the disk of a planet cancel each other out. So planets seem to shine with a steady light.

HISTORY

Astronomic Photos
Cameras offer a great benefit to astronomy. The use of photography provides a permanent record, and a camera can gather light over long periods so fainter objects can be observed. The moon was photographed as early as 1840, and the sun was photographed in 1845. The stars were first photographed in 1850.

the telescopes. Even with a large telescope, observing celestial objects faces several challenges. Light pollution from cities can brighten the sky, limiting the view of distant light objects. In addition, the atmosphere affects what can be seen. Water vapor, heat, and wind change the atmosphere's composition and change how light passes through it. Light bends whenever it passes through a different substance, and the degree and number of times it bends affects how well an object can be seen. Observatories are often built on mountaintops to minimize these problems.

The idea of having an observatory in space was first suggested in the 1940s. Scientists wanted to place telescopes above Earth's atmosphere in order to get the clearest images possible. In the 1970s, the European Space Agency (ESA) and the National Aeronautics and Space Administration (NASA) began building

W. M. Keck Observatory

349

FYI

Buyer Beware
The more light that a telescope collects, the brighter and more detailed an object will appear. The amount of light a telescope collects depends on its aperture. A telescope's aperture is the diameter of the objective lens or mirror. The larger a telescope's aperture, the more light it can gather and the more detail you can see. During ideal viewing conditions, a refracting telescope's useful magnification is limited to the size of its aperture in millimeters multiplied by 2. However, atmospheric conditions often reduce the useful magnification. In other words, no matter what the advertised magnification, a refracting telescope with a 50 mm aperture is capable of producing clear images only up to 100x magnification.

Hubble Space Telescope above Earth

the Hubble Space Telescope, which today orbits about 600 km above Earth's surface. This telescope features several different instruments that take pictures and record data from light in multiple wavelengths. Hubble has produced images of objects farther away than many telescopes on Earth can detect, and it has done so with greater resolution. Hubble has made many wondrous discoveries possible. Scientists have been able to see stars like the sun at the end of their life cycles, star explosions, and the building blocks of planets forming around young stars in the constellation Orion. Hubble instruments have monitored weather conditions on Mars and enabled greater study of black holes. The Hubble Space Telescope has revealed volcanoes erupting on Io, one of Jupiter's moons, and permitted scientists to witness the Shoemaker-Levy 9 comet breaking apart and crashing into Jupiter as well.

LESSON REVIEW
1. What is the basic difference between refracting and reflecting telescopes?
2. If you were building an optical telescope today, which type of telescope would you build? Why?
3. Why are mountaintops popular locations for observatories?
4. What advantage does the Hubble Space Telescope have over telescopes located on Earth?

350

Name: _____ Date: _____

Lab 7.1.3A Atmospheric Gases

QUESTION: How do the gases in the atmosphere affect people's view of the universe?
HYPOTHESIS: Answers will vary.

EXPERIMENT:

You will need:	• tape	• hot plate
• straight pin	• 2 books	
• aluminum foil	• flashlight	

Steps:
1. Use the straight pin to poke about 20 holes in the aluminum foil.
2. Stand the books upright on a flat surface and tape the aluminum foil between them so light comes through the holes.
3. Darken the room and shine a flashlight through the holes. Record your observations. Answers will vary but should include that the light is clear and steady. 
4. Plug in a hot plate and allow it to heat up. Place the aluminum foil in front of the hot plate.
5. Shine the flashlight over the hot plate through the holes. Record your observations. Answer will vary but should include that the light is wavy.

ANALYZE AND CONCLUDE:
1. What differences did you notice about the light coming through the holes with and without the heat source? Without a heat source, the light is clear and steady. With a heat source, the light is wavy.

Lab 7.1.3A Atmospheric Gases

2. How can you explain the differences between the light coming through the holes with and without the heat source? The warm air from the plate is less dense than the cooler air of the rest of the room. As the light from the holes passes from the denser air into the less dense air, the light's rays are refracted, giving the light a twinkling appearance.

3. Why do you think stars appear to twinkle? Starlight that passes through the layers of Earth's atmosphere is refracted and bent by the different densities of the air layers, making the stars appear to twinkle.

4. Explain whether you think stars would appear to twinkle if you saw them from the moon. Stars would not appear to twinkle if viewed from the moon because the moon has no atmosphere to refract the light.

5. Explain whether you think the twinkling of stars helps or inhibits scientific observation. Possible answers: Twinkling blurs the stars' images; stars don't actually twinkle, so it's not an accurate perception of what stars are really like.

Name: _____ Date: _____

Lab 7.1.3B Refracting Telescope

QUESTION: How does a refracting telescope work?

HYPOTHESIS: Answers will vary.

EXPERIMENT:

You will need:	• convex lens	• aluminum foil
• concave lens	• scissors	• straight pin

Steps:
1. Select a well-lit, distant object to view through the telescope.
2. Hold the concave lens close to one eye. This lens is your eyepiece.
3. Hold the convex lens directly in front of the eyepiece. (The concave lens should be between your eye and the convex lens.) The convex lens is the objective lens.
4. Slowly move the objective lens in a straight line away from the eyepiece until the object you are viewing comes into view through both lenses. Once you are able to see the object, move the objective lens in and out until you can see your object clearly. You may need a partner to move the objective lens out farther than your arm can reach.
5. Compare how the image appears when viewed with and without lenses. Note changes in the image's clarity, brightness, and size, and if it was inverted.

Answers will vary but should include that the image is clearer, larger, and brighter with the lenses and that the image is not inverted with or without the lenses. (Note: The size and quality of the lenses will affect the amount of difference seen.)

6. Cut a square of aluminum foil measuring about 10 cm × 10 cm.
7. Use the straight pin to make a tiny hole approximately in the center of the aluminum foil square.
8. Repeat Steps 1–4 but use the foil square's pinhole as your eyepiece.
9. Compare how the image appears when viewed with and without lenses. Note changes in the image's clarity, brightness, and size, and if it was inverted.

Answers will vary but should include that the image with the lens and foil will be less bright, possibly clearer, smaller, and inverted. (Note: The size and quality of the lens and the definition of the pinhole may affect the amount of difference seen.)

Lab 7.1.3B Refracting Telescope

ANALYZE AND CONCLUDE:

1. Compare your observations using the lens and foil pinhole as eyepieces. Answers will vary.

   Clarity: Possible answer: Using the lens and foil produced a clearer image than using two lenses.

   Brightness: Possible answer: The two lenses produced a brighter image than the lens and foil.

   Size: The image appeared larger with the foil and lens.

   Inversion: Both images were inverted.

2. Using what you observed, what is the eyepiece's purpose? The eyepiece focuses the image produced by the objective lens.

3. Using what you observed, what is the purpose of the objective lens? The objective lens gathers more light than a human eye, so the object seems brighter. The objective lens magnifies the object as well.

4. What do you think accounts for the differences between the way the lens and the foil pinhole function as eyepieces? The concave lens eyepiece bends the light and makes the image appear larger. It captures more light than the foil pinhole, which makes the image appear brighter.

7.1.4 Nonoptical Telescopes

Solar System

OBJECTIVES

Students will be able to
- identify the radiation used and the purposes for nonoptical telescopes.
- explain why the study of the universe is not limited to what people can see.
- summarize the path that data follows from a space telescope to an archive facility.

VOCABULARY

- **electromagnetic radiation** a form of wave energy that has both electrical and magnetic properties
- **electromagnetic spectrum** the entire wavelength range of electromagnetic radiation

MATERIALS

- Flashlight, dry ice, bucket, ultraviolet beads, fluorescent light, slingshot, metal pellets, foam sheet, tape, probes, metric rulers (*Introduction*)

PREPARATION

- Tape a thick foam sheet to a flat wall. (*Introduction*)
- Obtain materials for *Try This: Detect the Infrared*.
- Invite a professor from a local university to share what scientists can learn about the universe from various types of nonoptical telescopes. (*B*)

TRY THIS

Detect the Infrared
- liquid crystal sheet
- cold water
- objects of varying temperatures

(**Possible answers: Infrared detectors show stars' temperatures and hot spots on moons or planets that could be volcanoes. Sunlight decreases the temperature difference between the sheet and the object placed on it, which decreases the sheet's reaction to the object.**)

Introduction

Dim the lights and shine a strong flashlight against a flat surface. Place a piece of dry ice in a bucket and pour warm water over it to create fog. Shine the flashlight through the fog. Have students record their observations of the light with and without the fog.

Have students experiment with ultraviolet beads by holding the beads under a fluorescent light, in the shade, and in direct sunlight. Emphasize that students should cover their beads and allow them to fade back to white after exposing them to each light source. Direct students to record their observations.

Demonstrate the effect of X-ray photons by using a slingshot to shoot metal or wood pellets against the secured sheet of foam. Inform students that X-rays function more like photons than waves. X-ray telescopes direct X-ray photons onto a detecting surface. This surface enables scientists to determine the photons' energy, direction of travel, and the time they struck. Experiment by shooting a few pellets against the foam using different amounts of force and from different directions. Have students use a probe and ruler to measure the depth of each mark on the foam and then label and compare the marks.

Discussion

- Following the fog and ultraviolet beads experiments in *Introduction*, guide a discussion regarding how the experiments are related to using ultraviolet light to study the universe. Ultraviolet light behaves differently as it passes through different mediums, similar to visible light passing through a gas, and ultraviolet light causes some substances to change. Scientists can draw conclusions about the ultraviolet light's source or the substance through which it passes depending on how the light and the substance behave.

- Following the slingshot demonstration in *Introduction*, discuss the relationship between the marks left on the foam and the force and direction used when shooting the pellets. Point out that deeper marks were caused by greater force and that pellets shot from an angle created angular marks. Ask students what scientists might infer from measuring the force of X-ray photons. (**Possible answer: Scientists might infer the distance of the photons' source or the force with which the photons were expelled from the source.**)

Activities

A. Complete *Try This: Detect the Infrared* with students. Discuss students' answers to the questions.

B. Present the invited speaker who will share what scientists learn about the universe by using various types of nonoptical telescopes.

C. Have students find and compare images of a single feature in space, such as a planet, star, or galaxy, produced by different types of telescopes. Students should identify the feature and explain what the different images reveal about it. Alternately, assign student groups to find images of the same space feature from an optical telescope and one type of nonoptical telescope and then compare the images during an in-class presentation.

D. Guide students to research and prepare presentations or reports on a nonoptical telescope, such as one currently in space, one being prepared for launch, or one that is on Earth. Students should include the types of electromagnetic radiation the telescope detects, the objects the telescope examines or looks for, what astronomers have learned or hope to learn from the data the telescope collects, and any new devices or innovations developed to improve some aspect of the telescope's operation.

E. Direct student groups to research and prepare presentations or reports on the impact a very large telescope on Earth has had on its environment and nearby communities. Have students determine the scientific gains the telescope made possible and evaluate whether those gains warrant its socioeconomic cost and any environmental impacts.

Lesson Review

1. What types of electromagnetic radiation can reach Earth? (**visible light, radio waves, infrared light, and some ultraviolet light**)
2. Identify one purpose for each type of nonoptical telescope. (**Possible answers:** *radio*—**measures surface temperatures of planets, identifies chemical components of celestial bodies, identifies distant stars and galaxies;** *infrared*—**detects temperature differences on planet surfaces, determines the gases in a planet's atmosphere;** *ultraviolet*—**observes star formation, determines chemicals in the sun and stars, measures how gases respond to ultraviolet light;** *X-ray*—**detects black holes, measures star and galaxy life cycles, and observes supernovas;** *gamma ray*—**explores black holes, observes supernovas, explores pulsars**)
3. Why are optical telescopes not sufficient tools for thoroughly studying the universe? (**Stars and galaxies emit many types of radiation, most of which is not visible but which still provide important information about its source. Nonoptical telescopes are needed to gather this data.**)
4. Summarize the path data follows from a space telescope to an archive facility. (**Data is stored on a computer chip in the telescope. From the telescope, data is sent to Earth in packets through either a satellite relay system or a ground-based antenna system. The receiving station on Earth sends the data to a processing center. The processing center arranges the data so scientists can use it and stores the usable data in an archive facility.**)

NOTES

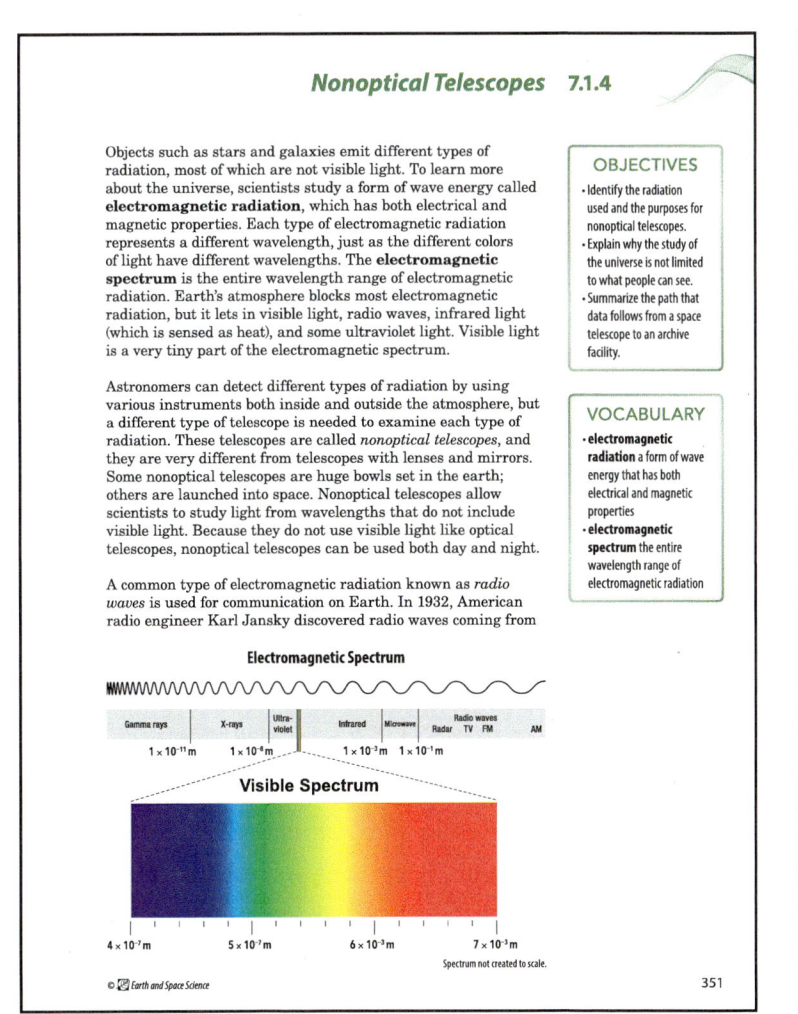

European partner agencies launched the Infrared Astronomical Satellite into space. This infrared telescope detected heat from stars being born. It also found cooling stars that optical telescopes could not find, and it provided evidence that stars other than the sun have planets. Since then, other infrared telescopes have advanced and improved on the observations made by earlier telescopes. Newer infrared telescopes can detect temperature differences on the surface of planets or determine the gases that make up a planet's atmosphere.

The type of electromagnetic radiation that causes suntans is called *ultraviolet radiation*. Earth's ozone layer blocks out most ultraviolet light, so ultraviolet telescopes must be placed at very high elevations or in space. By the 1940s, scientists were launching rockets with ultraviolet detectors on board. Scientists study ultraviolet radiation to learn about star formation and chemicals in the sun and stars. They also study the way different gases respond to ultraviolet light. This research helps them learn about the gases and the objects the gases are associated with.

Earth's atmosphere blocks most X-rays and gamma rays, which are two more types of electromagnetic radiation. Therefore, X-ray and gamma-ray telescopes are usually placed in space. The Chandra X-Ray Observatory, launched in 1999, studies black

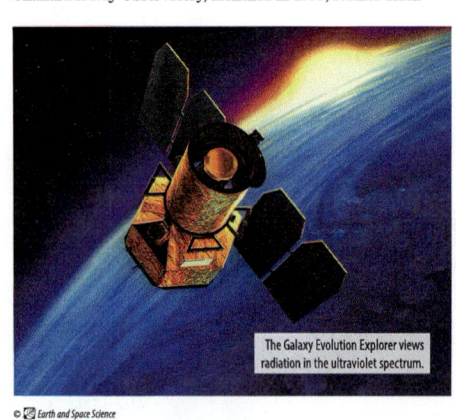

The Galaxy Evolution Explorer views radiation in the ultraviolet spectrum.

FYI

Cosmic Explosions
Gamma-ray bursts are flashes of gamma rays that occur in space. If you could see gamma rays, you would see that one of these bursts would briefly outshine all other sources of gamma rays. Astronomers believe that these gamma-ray bursts indicate the creation of a phenomenon called *a black hole*, which is an extreme gravity field that prevents even light from escaping from it. The length of a gamma-ray burst could indicate how a black hole was formed. A long burst of gamma rays likely comes from a supernova core that collapsed to form a black hole. However, a gamma-ray burst is not emitted by every supernova. Long gamma-ray bursts last between two and several hundred seconds. A short burst of gamma rays is thought to be caused either by two neutron stars colliding and forming a black hole or by a neutron star being sucked into a black hole and creating a bigger black hole. Short gamma-ray bursts last less than two seconds.

FYI

James Webb Space Telescope
In the late 1990s, before the Hubble Space Telescope (HST) was 10 years old, NASA and other international agencies began building the James Webb Space Telescope (JWST) as the HST's successor. The earliest launch date was determined to be sometime in 2018. The JWST's foldable mirror has a 6.4-meter diameter, which is seven times larger than the one on the HST. Eighteen beryllium segments covered in a thin layer of gold comprise the mirror. To protect the telescope from thermal radiation, the agencies designed a five-layer, collapsible sun shield that measures 150 m². Both the mirror and the sun shield's design include the capability to expand once the telescope has traveled its 30-day journey into space. The JWST will reside approximately 1.5 million km away from Earth, on its dark side, and use infrared to record things such as young galaxies and the birth of stars. Future peopled spacecraft will have access to the telescope by way of an external, side-mounted docking ring.

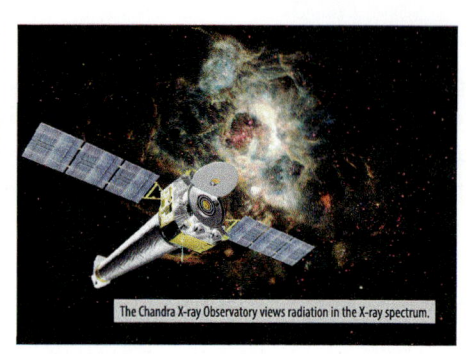

The Chandra X-ray Observatory views radiation in the X-ray spectrum.

holes, star and galaxy life cycles, and exploding stars called *supernovas*. Scientists have discovered X-rays coming from all over the universe. This discovery suggests there may be a great number of radiation sources in the universe that have yet to be identified.

Cosmic gamma rays were discovered accidentally in the 1960s. Artificial satellites designed to detect the gamma rays produced by atomic bomb tests on Earth detected gamma rays beyond Earth's atmosphere. Gamma rays have a very short wavelength, so they can only be detected indirectly. They collide with electrons as they pass through detecting devices, and the collision creates charged particles that can be detected. Gamma-ray telescopes are used to explore black holes, supernovas, and small, dense stars called *pulsars* that give off regular pulses of radio waves, X-rays, or gamma waves.

What happens to all of the data collected by space-based telescopes? Telescopes continuously collect three types of data that are stored on memory chips: system function data, direction, and observation. The data can be sent to Earth in several different ways. In rare cases, the data is transmitted to Earth as soon as it comes in, which is called *real time data*. Most often, however, data is sent to Earth at predetermined times through a satellite relay system or a ground-based antenna system. In the satellite relay system, the telescope sends its data to another satellite, which in turn relays the data to a facility

---

on the ground. With a ground-based antenna system, the data is transmitted directly from the telescope to a receiving facility on the ground. The receiving station then sends the data on to a processing center.

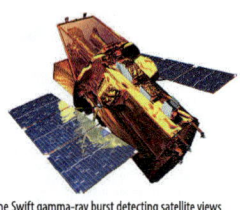

Once the data is at the processing center, it has to be organized. The telescope records and time stamps all its data in groups of bits, called *data packets*. (A bit is a single piece of computer information.) Each of a telescope's systems sends packets of information about how well the telescope was functioning, where the telescope was pointed, or what it saw. A scientist must be able to review all three types of data at the same time to know how good the data is. The job of the data processing center is to take the separate packets and arrange them so a scientist can use them. Data from telescopes in space is eventually made available for anyone to use. To allow multiple systems to understand the data from any telescope without needing a unique program to decode it, a format called *Flexible Image Transport Standard (FITS)* was developed.

Finally, the data is stored in an archive facility. At first, only the scientists who requested the data are permitted to use it. This phase is called *the proprietary phase*, and it lasts about a year. After the proprietary phase ends, the data is made available to anyone in what is called *the public phase*. Typically, the data is then made available through the Internet.

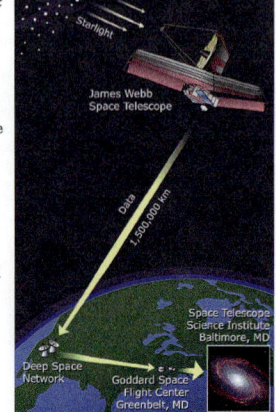

The Swift gamma-ray burst detecting satellite views radiation in the gamma-ray spectrum.

Artist's rendering of the projected data path from the James Webb Space Telescope to an archive facility

LESSON REVIEW
1. What types of electromagnetic radiation can reach Earth?
2. Identify one purpose for each type of nonoptical telescope.
3. Why are optical telescopes not sufficient tools for thoroughly studying the universe?
4. Summarize the path data follows from a space telescope to an archive facility.

7.1.5 Solar System

Solar System

Introduction
Ask students to name the four inner planets in their order from the sun. (**Mercury, Venus, Earth, Mars**) Remind them of the mnemonic device for remembering the inner planets' order: "Monica Visits Every Monday." Challenge students to name the remaining planets in the solar system. (**Jupiter, Saturn, Uranus, Neptune.**) Introduce an addition to the mnemonic sentence to help students remember the order of all the planets: "Monica Visits Every Monday—Just Stays Until Noon."

Discussion
- Convey that in classical terms an argument is not bickering between two people but is a set of reasons given to support an idea in the attempt to persuade others to agree. Ask students, in light of differing views of the age of the solar system and its formation, what gives an argument authority. (**Possible answers: clear evidence using telescopic observations or examination of material from celestial objects; an explanation of a hypothesis using known facts, such as the laws of physics and mathematics or the nature of God as revealed in the Bible**) What might weaken an argument? (**Possible answers: little or no evidence using observations or experimentation, using controversial ideas to support a hypothesis**)

- Have students discuss the creation of the solar system. Invite students to share perspectives they hold or have heard regarding it. Guide students to consider the different views regarding the process God might have used for this part of creation. (Note: A common misconception is that the solar system began with the Big Bang; the Big Bang model addresses the creation of the universe, not the solar system.)

- Lead a discussion regarding the beliefs about God, the Bible, and creation that Christian scientists hold in common, regardless of their views on how God created the solar system. Guide students to consider how these commonly held beliefs can encourage respectful disagreement among scholars.

- Have students consider what the orderliness of different creation theories says about God's nature.

Activities

Lab 7.1.5A Indirect Evidence
- prepared clay balls, 4 per group
- pushpins of different colors, 4 per group
- triple beam balances, 1 per group
- masses for balance
- metric rulers, 1 per group
- bar magnets, 1 per group

After all students have finished the lab chart, direct them to uncover the core of each ball to complete the lab. Encourage students to develop other tests that would give them more or better information. Guide them to consider the implications of indirect evidence when describing the functions, composition, and origins of objects in the solar system.

A. Have students complete **WS 7.1.5A Solar System Diagram**.

B. Assign students to create scale models of the sun and planets' diameters using **BLM 7.1.5A Sun and Planets Measurements** and their own scales.

C. Direct students to use BLM 7.1.5A and the scale of 10^{24} kg = 1 g to determine how many grams the sun and each planet would weigh in a scale model. Inform students that a large, standard paper clip weighs about 1 gram.

OBJECTIVES
Students will be able to
- distinguish among the major celestial bodies in the solar system.
- compare two prominent views of the creation of the solar system.

VOCABULARY
- **asteroid** a rocky object that orbits the sun
- **dwarf planet** a spherical object that orbits the sun but is not large enough to move other objects from its orbit
- **gas giant** a large, gaseous planet of the outer solar system
- **nebula** a vast, moving interstellar cloud of gas and dust
- **planet** a spherical object that orbits the sun and has removed other objects from its orbit
- **terrestrial planet** a small, dense, rocky planet of the inner solar system

MATERIALS
- Compasses, metric rulers, card stock, tape, dowel rods, metric tape measures (*D*)
- Metric tape measures (*E*)
- Metric rulers, city maps (*F*)
- BLM 7.1.5A Sun and Planets Measurements
- WS 7.1.5A Solar System Diagram

PREPARATION
- Obtain materials for *Lab 7.1.5A Indirect Evidence*. For each group, create 4 clay balls of various sizes with one of the following materials in the center of each ball: magnet, rock, table tennis ball, solid clay. The clay around each center should be thick enough that a pushpin cannot touch the object in the center.

NOTES

D. Encourage students to create a scale model of the solar system in which the sun and the planets' diameters and distances from each other are all set to the same scale. (Note: For small classes or limited space, consider having students create only a scale model of the terrestrial planets.) Distribute BLM 7.1.5A. Have each student group use compasses, metric rulers, and card stock to create scale models of the sun and planets. Direct students to cut out the sun and each planet and to tape each to its own dowel rod. Have students in each group hold one of the group's dowel rods. In a very large, open space, guide groups to use metric tape measures to orient each group's planet from the sun.

E. Challenge student groups to use BLM 7.1.5A to calculate the astronomical units (AU) of each planet from the sun. Use the approximate measurement of 150,000,000 km per 1 AU, or challenge students to use Earth's actual mean distance from the sun of 149,597,871 km as 1 AU. Guide students within each group to arrange themselves in a scale model of the planets' distances from the sun using a scale of 1 m per 1 AU and a metric tape measure. Ask students why they will not be making scale models of the planets for this activity. (**Some of the planets would be too small to see without magnification.**)

F. Guide students to make a comparative chart, such as a Venn diagram, comparing terrestrial planets, gas giants, dwarf planets, asteroids, comets, and meteoroids. Challenge students to identify one or two clearly distinguishing characteristics of each category as well as attributes shared between two or more categories.

7.1.5 Solar System

OBJECTIVES
- Distinguish among the major celestial bodies in the solar system.
- Compare two prominent views of the creation of the solar system.

VOCABULARY
- **asteroid** a rocky object that orbits the sun
- **dwarf planet** a spherical object that orbits the sun but is not large enough to move other objects from its orbit
- **gas giant** a large, gaseous planet of the outer solar system
- **nebula** a vast, moving interstellar cloud of gas and dust
- **planet** a spherical object that orbits the sun and has removed other objects from its orbit
- **terrestrial planet** a small, dense, rocky planet of the inner solar system

Exploring the universe begins with Earth's solar system. At the center of the solar system is the sun. Depending on a person's understanding of the solar system, up to five types of celestial bodies orbit the sun. All scientists agree that planets, asteroids, and comets orbit the sun. Many scientists acknowledge a class of planets called *dwarf planets*. Some scientists believe that an additional group of objects, called *the Oort cloud*, exists at the outermost reaches of the solar system.

According to the International Astronomical Union (IAU), a **planet** has three important attributes. First, it orbits the sun; second, its own gravity shaped it into a sphere; and third, its size and gravity are great enough to have cleared its own path around the sun by attracting any smaller objects in its orbit. Eight objects in the solar system qualify as planets. These planets are generally divided into two groups known as *terrestrial planets* and *gas giants*. A **terrestrial planet** is a small, dense, rocky planet with few or no moons that orbits close enough to the sun to receive its warmth. Extending out from the sun, the terrestrial planets are Mercury, Venus, Earth, and Mars. A **gas giant** is a large, gaseous planet with a very low density and many moons that orbits too far from the sun to receive significant warmth. From closest to farthest from the sun, the gas giants are Jupiter, Saturn, Uranus, and Neptune.

In 2006, the IAU created the new category of *dwarf planet* for certain solar system objects. A **dwarf planet** has four attributes.

Earth's solar system: the sun, Mercury, Venus, Earth, Mars, the asteroid belt, Jupiter, Saturn, Uranus, Neptune

Like planets, a dwarf planet orbits the sun and its own gravity shaped it into a sphere. But unlike a true planet, a dwarf planet has not cleared its own path around the sun. And unlike similar celestial bodies, a dwarf planet does not orbit around any celestial body except the sun. Currently known dwarf planets are made almost entirely of rock and ice, but this is not a necessary condition for being classified as a dwarf planet. Pluto was once considered a planet, but it is now classified as a dwarf planet because it has not cleared its own path around the sun. Other dwarf planets in the solar system include Ceres and Eris.

An **asteroid** is a rocky object that can be a variety of sizes and that orbits the sun. Most asteroids are located between the orbits of Mars and Jupiter in what is called *the astroid belt*. Similar to asteroids are large objects composed mostly of ice, which are called *comets*. Each comet has its own orbit around the sun. Smaller objects that are thought to have broken off asteroids or comets are called *meteoroids*. When a meteoroid enters Earth's atmosphere, it is called a *meteor*. If a meteoroid strikes Earth's surface, then it is called a *meteorite*.

Meteor shower

The Oort cloud is a theoretical sphere of icy bodies surrounding the solar system. Some scientists believe that comets may begin in the Oort cloud. However, other scientists do not believe that the Oort cloud exists at all.

How exactly did God create the solar system? According to young-earth theory, which states that God created the universe out of nothing in six literal days, Earth was created first as a featureless planet covered by water and surrounded by darkness. The first day was established with the creation of light and regular intervals of light and dark on Earth. On the second day, God created the atmosphere. On the third day, God created dry land and bodies of water along with all plants, and on the fourth day, God separated the light into the sun, moon, and stars. God finished His creation in two more days by creating all creatures for the water and air on the fifth day and all land creatures and humans on the sixth day. According to this view, God made all things in an instant. Therefore, there is no need to explain how atoms may have bonded to form various things such as the sun

Lesson Review

1. What are the five major types of celestial bodies in the solar system? (**planets, dwarf planets, asteroids, comets, and meteoroids**)
2. Why is Pluto classified as a dwarf planet? (**Its gravity is not sufficient to clear other objects from its orbital path around the sun.**)
3. What do the terrestrial planets have in common? (**The terrestrial planets are small, dense, rocky planets that orbit close enough to the sun to receive its warmth.**)
4. In what order were the sun and Earth created according to both the young-earth theory and the old-earth theory? (**In the young-earth theory, Earth was created before the sun, but in the old-earth theory, the sun was formed before Earth.**)
5. According to both young-earth and old-earth theories, what did God create out of nothing? (**In the young-earth theory, God created all of the celestial bodies out of nothing. In the old-earth theory, God created particles out of nothing.**)
6. Identify one observation that young-earth scientists claim supports their theory of how the solar system was created and one observation that old-earth scientists claim supports their theory of how the solar system was formed. (**Possible answers:** *young earth*—**the relatively slow rotation of the sun, the presence of comets with observable orbits that have not yet disintegrated, and the existence of very large planets in other solar systems that are much closer to their stars than the old-earth theory claims is possible;** *old earth*—**other nebulae in the universe appear to have stars forming within them, the planets' orbits lie nearly in a plane with the sun at the center, the planets move around the sun in the same direction, most planets spin in the same direction with most of their axes pointing nearly perpendicular to their orbital plane**)

NOTES

growing by attracting the dust and debris in their path. As the planetesimals grew larger, their collisions would have became more violent. These more violent collisions might explain phenomena such as Venus's slow and backward rotation and Earth's moon. Small planetesimals that did not crash into planets may have been thrown to the outer edge of the solar system by the gravity of the larger planets. According to this theory, asteroids and comets are larger planetesimals that neither adhered to a planet nor moved beyond the planets. Solar radiation then blew away whatever gas and dust that did not adhere to the planets or condense to form asteroids and comets.

Thor's Helmet nebula
Credit: ESO/B. Bailleul

In contrast, other scientists do not believe the God of the Bible was involved in the creation of the solar system. Some may believe that is was the god of another faith who created the solar system. Others who do not believe in God's existence theorize that the solar system came into being by chance through the random interaction of gases, particles, and gravity following the creation of the universe through what is called *the Big Bang*.

However, God's creation of the solar system displays His craftsmanship. Like a potter with clay, He shaped trillions of tons of matter into a fiery ball of gas and billions of planets, moons, asteroids, comets, and meteors. And into this harmonious handiwork He carefully designed and set into place a dwelling place for His people.

LESSON REVIEW
1. What are the five major types of celestial bodies in the solar system?
2. Why is Pluto classified as a dwarf planet?
3. What do the terrestrial planets have in common?
4. In what order were the sun and Earth created according to both the young-earth theory and the old-earth theory?
5. According to both the young-earth and the old-earth theories, what did God create out of nothing?
6. Identify one observation that young-earth scientists claim supports their theory of how the solar system was created and one observation that old-earth scientists claim supports their theory of how the solar system was formed.

Name: _____ Date: _____

Lab 7.1.5A Indirect Evidence

QUESTION: How helpful is indirect evidence in drawing conclusions about an object?

HYPOTHESIS: Answers will vary.

EXPERIMENT:

You will need:		
• 4 clay balls from your teacher	• triple beam balance	• bar magnet
• 4 pushpins of different colors	• masses for balance	
	• metric ruler	

Steps:
1. Place a different colored pushpin into each planet to identify it.
2. Use a triple beam balance and masses to determine the mass of each planet in grams. Record the data.
3. Use a metric ruler and the formula to the right to determine the volume of each planet in cubic centimeters. Record the data. $V = \dfrac{4}{3}\pi r^3$
4. Use the formula Density = mass/Volume to the right to calculate each planet's density in g/cm³. Record the data. $D = \dfrac{m}{V}$
5. Use the magnet to test each planet for magnetic properties.
6. Using the information you gathered, hypothesize what is at the core of each planet. Explain your reason for each hypothesis.

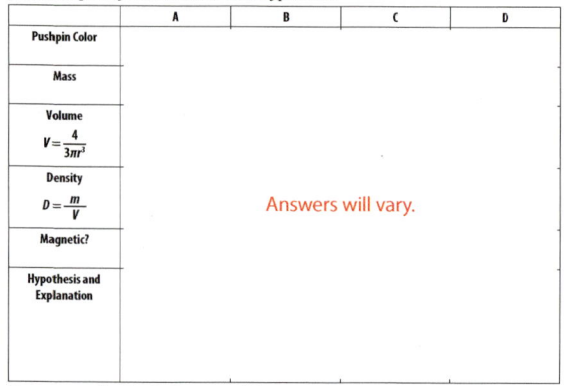

	A	B	C	D
Pushpin Color				
Mass				
Volume $V = \dfrac{4}{3}\pi r^3$		Answers will vary.		
Density $D = \dfrac{m}{V}$				
Magnetic?				
Hypothesis and Explanation				

7. Deconstruct the clay balls to reveal their cores.

Lab 7.1.5A Indirect Evidence

ANALYZE AND CONCLUDE:
1. What difficulties did you encounter when trying to determine the core of each planet? Possible answers: The cores were hidden; several balls seemed to have no unusual characteristics such as magnetism or high density.

2. How accurate were your hypotheses about each planet? Answers will vary.

3. How accurate was your initial hypothesis in response to the question, "How helpful is indirect evidence in drawing conclusions about an object"? Answers will vary.

4. Revise your initial hypothesis if necessary and support it by referring to your experiences in completing this lab. Answers will vary but should include an acknowledgement that indirect evidence will always contain a degree of uncertainty.

Solar System

7.1.6 Sun

Introduction
Use a thermometer to demonstrate the power of infrared radiation by comparing temperatures in full sun to those in the shade. Ask students how Earth's atmosphere is similar to a sunscreen. (**Both the atmosphere and sunscreen block some of the radiation from the sun.**)

Show students a golf ball to represent Earth. A golf ball has a diameter of about 4 cm. Have students use calculators to determine the relative diameter of the sun, which is about 109 times Earth's diameter. (**4.36 m**) Have a student volunteer measure a distance of 4.36 m. Place the golf ball in the center of this measured distance and compare the relative size of Earth to the sun. Inform students that if the line measured the diameter of the sun, about 1.3 million golf balls could fit inside of it.

Use a hair dryer pointed upward to suspend a table tennis ball in its air stream. (Note: Consider practicing this demonstration before class.) The air stream of the dryer balances the ball, keeping it aloft; gravity pulls the ball down, keeping it from floating away. Explain that this demonstration is similar to the balancing forces working inside the sun that keep it stable and prevent it from blowing apart or collapsing in on itself. Fusion inside the sun pushes energy and particles outward, but the massive gravity of the sun counteracts the force of the expanding energy, maintaining a stable sun.

Discussion
- Remind students that Chinese documents record observations of sunspots seen with the unaided eye more than 2,400 years ago. When Galileo rediscovered sunspots with his telescope, his findings were disbelieved. People believed that the sun was without blemish. Lead a discussion regarding how scientific knowledge changes as methods and instruments of observation and measurement improve, such as telescopes.

- Ask the following questions:
  1. Why is the sun so hot? (**Hydrogen propels nuclear fusion in the sun's core. Nuclear fusion produces energy by fusing two light atomic nuclei to form a single heavier nucleus, which is what happens when lightweight hydrogen is converted to helium. The sun is hot because this process produces a lot of energy and it happens continuously.**)
  2. Which of the sun's layers emits visible light? (**photosphere**)
  3. Who demonstrated that energy and matter are interchangeable, and what equation represents this theory? (**Albert Einstein; $E = mc^2$**)
  4. How do astronomers know that the sun's surface is in constant motion? (**The sun's surface has to be in motion because sunspots, prominences, and solar flares occur. These are all caused by magnetic fields on the sun's surface. The sun's rotation and the circulation of gases produces the magnetic fields.**)
  5. What is the solar wind and how is it connected to an aurora? (**The solar wind is the flow of charged particles from the sun's corona. When these charged particles interact with the upper layers of Earth's atmosphere, bands of light called *an aurora* are produced.**)

Activities

Lab 7.1.6A Diameter of the Sun
- white unruled index cards, 2 per group
- pushpins, 1 per group
- metersticks, 1 per group
- tape

A. Complete *Try This: Sunspot Observation* with students. Ask the following questions:
  1. Why are sunspots darker than the rest of the sun's surface? (**Sunspots are cooler than the surrounding region.**)

OBJECTIVES
Students will be able to
- model the structure of the sun.
- describe the sun's physical properties.

VOCABULARY
- **aurora** a band of colored or white light in the atmosphere caused by charged particles from the sun interacting with Earth's upper atmosphere
- **chromosphere** the first layer of the sun's atmosphere
- **convective zone** the zone on the sun where convection currents bring energy to the sun's surface and take gas back into the sun
- **corona** the outer layer of the sun's atmosphere
- **photosphere** the sun's surface, which radiates visible light
- **prominence** a fiery burst of gas from the sun that rises thousands of kilometers into space
- **radiative zone** the zone on the sun where energy from the sun's core moves toward the sun's surface

MATERIALS
- Thermometer, golf ball, meterstick, hair dryer, table tennis ball (*Introduction*)
- TM 7.1.6A Auroras
- WS 7.1.6A Sun Diagram

PREPARATION
- Obtain materials for *Lab 7.1.6A Diameter of the Sun*.
- Gather materials for *Try This: Sunspot Observatory*.

TRY THIS

Sunspot Observation
- refracting telescope with objective lens smaller than 50 mm in diameter
- tripod
- cardboard
- scissors
- bright white paper
- tape
- clipboard or hardcover book

2. What force is especially strong at a sunspot? (**magnetism**)
3. What two other phenomena happen at intense magnetic fields on the sun? (**prominences and solar flares**)

B. Have students complete **WS 7.1.6A Sun Diagram**.

C. Guide students to research how the solar wind interacts with Earth's atmosphere to produce different colors in auroras. Have students watch auroras or examine **TM 7.1.6A Auroras** and use their research to identify elements in the atmosphere and approximately how far into the atmosphere the solar wind penetrated. In general, a green or yellowish light indicates oxygen between 100 and 300 km above the earth. Oxygen glows red around 200 to 300 km. Blue light indicates nitrogen between 100 and 300 km, and nitrogen below 100 km produces a purplish light.

D. Assign students to create cutaway scale models of the sun's layers, sunspots, and prominences. Encourage students to be creative in their choice of materials. Inform students that the approximate radii from the sun's center of each layer are as follows: *core*—138,400 km; *radiative zone*—513,400 km; *convective zone*—695,200 km. The photosphere is approximately 400 km thick from the convective zone to the chromosphere; the chromosphere is about 2,100 km above the photosphere, and the corona extends indefinitely above the chromosphere. Direct students to develop a scale for their models, to label each layer of the sun, and to display their scale and calculations to determine the radius of each layer.

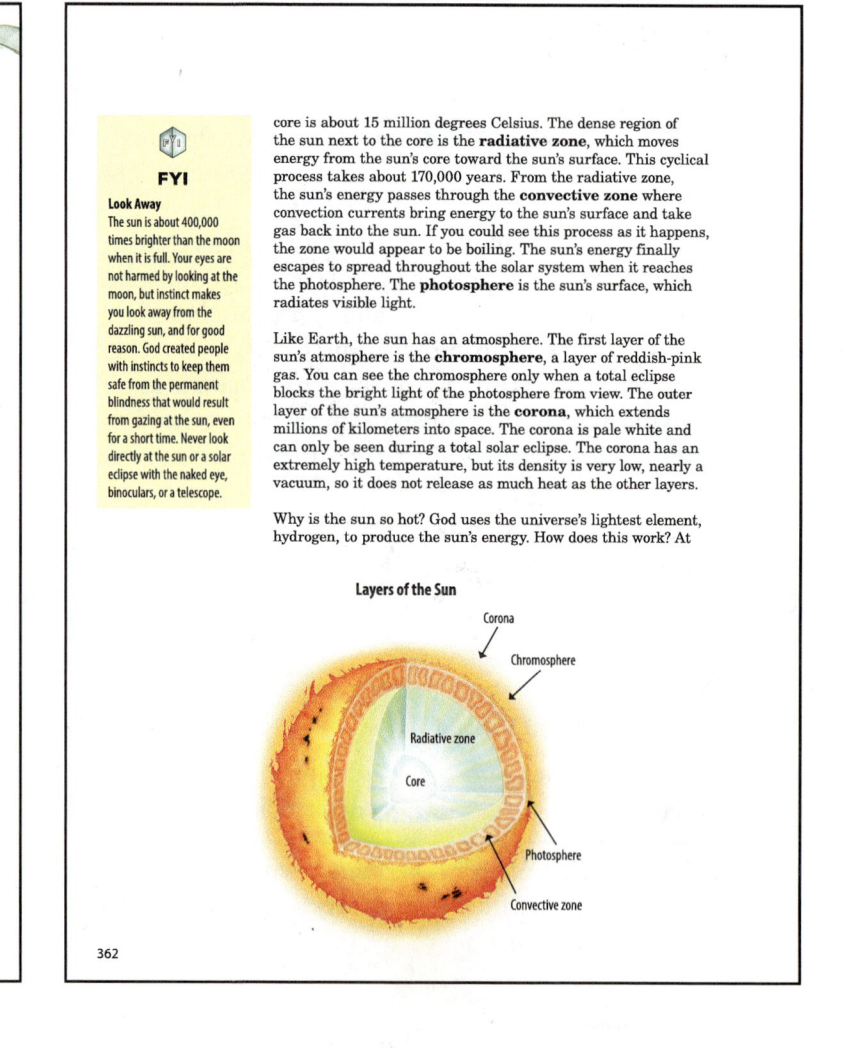

E. Challenge students to research the effects of solar flares on television and radio broadcasts or other forms of wireless communication. Have students report on any technology used to counter the effects of solar flares on communication systems and devices and evaluate whether the technology is scientifically sound.

NOTES

Lesson Review

1. What is the source of the sun's energy? (**The enormous energy of the sun is produced by nuclear fusion reactions, which convert hydrogen into helium.**)
2. Describe the different layers of the sun and of the sun's atmosphere. (**Deep inside is the sun's core where nuclear fusion produces heat and light. The energy from this intensely heated core radiates up through the radiative zone. Streams of superheated gas transport energy up to the convective zone, where convection currents take the gas back into the sun's interior. The photosphere is a thin layer of hot gas that radiates visible light. The chromosphere is a layer of reddish-pink gas above the photosphere. The corona is pale white and extends millions of kilometers into space. The chromosphere and corona are only visible during a solar eclipse.**)
3. List and describe several solar activities. (**Sunspots are cooler regions on the surface of the sun caused by increased magnetism. Prominences are fiery bursts of hot gas rising hundreds of thousands of kilometers into space. The shapes and paths of prominences are determined by the magnetic fields. Solar flares are explosions of magnetic energy that hurl enormous amounts of radiation into space. Solar wind is the flow of charged particles away from the sun.**)
4. How do disturbances on the sun affect Earth? (**They can disrupt communication devices and power grids, and they can cause auroras to appear.**)

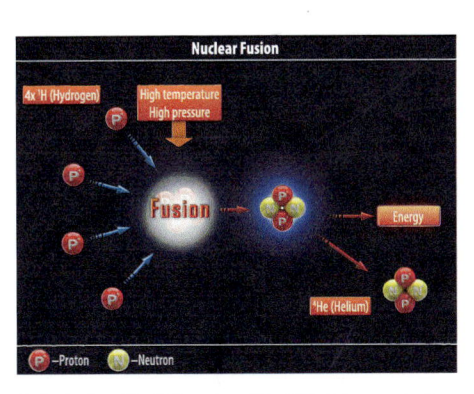

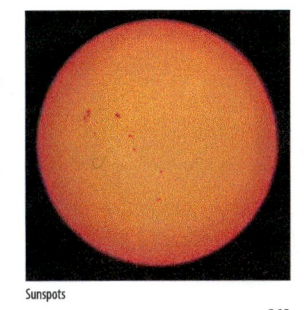

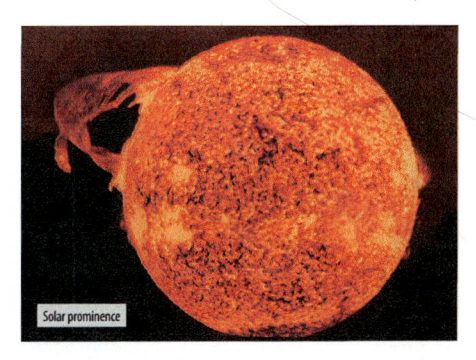

of prominences are determined by the same kinds of magnetic fields that create sunspots.

Violent disturbances on the sun's surface caused by explosions of magnetic energy are called *solar flares*. These explosions send out tremendous bursts of radiation and particles. Solar flares strengthen the flow of charged particles from the sun's corona, which is called *the solar wind*. Gusts of solar particles cause disturbances in Earth's magnetosphere, which works with the other layers of the atmosphere to protect Earth from most of the sun's radiation. An increased disturbance from the solar wind is called *a geomagnetic storm*. These storms can be strong enough to disrupt communication devices and power grids.

Charged particles from the solar wind also create **auroras**, which are bands of colored or white light in the atmosphere caused by charged particles from the sun interacting with Earth's upper atmosphere. In the Northern Hemisphere, auroras are called *the Northern Lights*, and in the Southern Hemisphere they are called *the Southern Lights*.

LESSON REVIEW
1. What is the source of the sun's energy?
2. Describe the different layers of the sun and of the sun's atmosphere.
3. List and describe several solar activities.
4. How do disturbances on the sun affect Earth?

FYI

The Ozone Layer
The sun radiates X-rays, ultraviolet light, and gamma radiation into space. These three forms of radiation are deadly to living things, but God wove a filter system into the atmosphere to prevent harmful amounts from reaching Earth. The ozone layer in Earth's stratosphere keeps most of the sun's ultraviolet radiation from reaching Earth's surface. In recent years, however, the ozone layer has thinned in some areas, increasing the importance of using preventative measures to reduce the risk of skin diseases such as cancer. Such preventative measures include using sunblock, wearing protective clothing, and reducing the amount of time unprotected skin is exposed to the sun.

Name: _____ Date: _____

Lab 7.1.6A Diameter of the Sun

QUESTION: What is the sun's diameter?
HYPOTHESIS: Answers will vary.

EXPERIMENT:

You will need:	• pushpin	• tape
• 2 white unruled index cards	• meterstick	

WARNING: Never look directly at the sun.

Steps:
1. Fold a third of one index card lengthwise.
2. Use the pushpin to make a pinhole in the center of the large section of the folded index card.
3. Tape the smaller section of the card to the end of the meterstick so the fold is lined up with the stick's 0 mark. The folded side with the pinhole in its center should be perpendicular to the stick.
4. On the center of the second index card, draw two parallel lines 0.7 cm apart. This measure will be the width of the sun's image.
5. Position the meterstick so the sun's rays pass through the pinhole.
6. Move the second card on the meterstick so an image of the sun appears.
7. Move the second card along the meterstick until the sun's image just fits between the 0.7 cm marks.
8. Record the distance between the two cards. Answers will vary but should be close to 74.5 cm.
9. Write the proportion of the size of the image to the distance between the cards. 0.7 cm/74.5 cm
10. Write the proportion of the sun's diameter to the distance between Earth and the sun. Let d represent the sun's diameter. The approximate distance between Earth and the sun is 150,000,000 km. d km/150,000,000 km
11. Solve the proportion for d.
Compare proportions: 0.7 cm/74.5 cm = d km/150,000,000 km
Cross multiply: 74.5d cm km = 105,000,000 cm km
Divide both sides by 74.5 cm leaves an answer in km for d, the sun's diameter: 1,409,395 km
Actual diameter of sun is 1,391,400 km.

Lab 7.1.6A Diameter of Sun

ANALYZE AND CONCLUDE:

1. Why do you think your calculations do not exactly match the actual diameter of the sun? Possible answers: The sun's actual diameter is not a perfect constant, so an average is used; the image produced by the pinhole was not perfectly focused; the given image diameter of 0.7 cm may not be perfectly accurate.

2. Do you think it is possible to obtain a perfectly accurate measurement of the sun's diameter? Why? Answers will vary but should acknowledge that because the sun is gaseous it does not maintain a perfectly stable shape, so measurements will likely vary from time to time. Students should also acknowledge that techniques used for measuring may not be perfectly accurate given the distance between them and the sun as well as the fluctuating nature of light.

3. How accurate do you think measurements are of the sun's layers? Why? Answers will vary but should acknowledge that measuring the sun's layers will include some variations due to the variable nature of gas and the impossibility of boring through the layers.

7.2.0 Planets

Chapter 7.2 Summary
The solar system is an intricately designed system of planets and other smaller orbiting bodies. Kepler's laws of planetary motion describe the orbits of these bodies, and Sir Isaac Newton's discovery of the force of gravity explains why these bodies orbit as they do. Although all orbiting bodies follow Kepler's and Newton's laws, these bodies are not all the same. The planets are generally divided into the inner and outer planets. The inner planets have solid surfaces, few or no moons, and are warmed to varying degrees by the sun. The outer planets are gaseous, surrounded by rings of debris, significantly larger than any of the inner planets, and have multiple moons. Despite sharing these general characteristics, each planet is unique. Only Earth is capable of sustaining life. Interspersed among the planets are smaller orbiting bodies. Comets are composed of ice, dust, and rock; many scientists believe comets originated in the Kuiper belt or the Oort cloud. Asteroids are composed of rock and are located primarily in the asteroid belt between Mars and Jupiter. However, they are not limited to that region, and some pass relatively near Earth. Meteoroids are pieces broken off comets or asteroids. Numerous very small meteoroids enter Earth's atmosphere each day where they flare and burn up as meteors. In contrast, meteorites are meteoroids that survive their descent through Earth's atmosphere and land intact on Earth's surface.

Background
Lesson 7.2.1 – Planetary Motion
Each planet rotates and revolves in an orderly fashion. Most planets rotate in a direct rotation and spin counterclockwise; Venus and Uranus rotate in a retrograde rotation and spin clockwise. Planets tilt to varying degrees on their axes. Earth's axial tilt, for example, is about 23.4°. (This tilt helps determine the seasons, which students will study in the next chapter.) The larger, less dense planets rotate more quickly than the smaller, denser planets. Jupiter, for example, is the largest planet, yet it has the shortest period of rotation; Venus is slightly smaller than Earth, yet it has the longest period of rotation

Planets revolve around the sun in ellipses. A planet's eccentricity is the extent to which the ellipse differs from a circle. For example, an eccentricity of 0 would be a circle. The orbits of most planets are close to 0. Venus has the most circular orbit; Mercury has the least circular.

The planets' orbits are explained by Kepler's three laws of planetary motion. The first law states that planets orbit the sun in elliptical paths with the sun at one focus of the ellipse and nothing at the other focus. This law was first published in 1609, shortly before the astronomical telescope was invented. Before this, it was assumed that every heavenly body moved in either a circular orbit or an epicycle, which is a combination of circular orbits.

According to Kepler's second law, a line between a planet or comet and the sun sweeps out equal areas of space in equal amounts or time. Therefore, a planet near the sun moves quickly in its orbit, but a planet farther from the sun moves slowly. The third law of motion relates that the orbital period (P) squared is directly proportional to the average distance of a planet from the sun (a) cubed. The third law can be used to determine the distance of a planet from the Sun if its orbital period is known. The mean distance from Earth to the sun is known as *an astronomical unit* (AU). The third law can be written $P^2 = a^3$, where P is in years and a is in astronomical units.

Isaac Newton also contributed to the knowledge of planetary motion. People had observed that objects released near the Earth's surface fall to the ground; Earth attracts masses near its surface to itself. In 1687, Isaac Newton published the law of universal gravitation, which states that the force of attraction between two masses is directly proportional to the product of their masses and inversely proportional to the square of the distance between their centers. This principle is central to the understanding of planetary motion.

LOOKING AHEAD
- For **Lesson 7.2.1**, obtain a field chalk line marker.

SUPPLEMENTAL MATERIALS
- BLM 7.2.2A Game: Planet Mixer

- Lab 7.2.1A Ellipse
- Lab 7.2.2A Venus

- WS 7.2.1A Kepler's Third Law
- WS 7.2.1B Gravity
- WS 7.2.2A Inner Planets' Properties
- WS 7.2.3A Outer Planets' Properties
- WS 7.2.4A Triple Venn Diagram

- Chapter Test 7.2

BLMs, TMs, and tests are available to download. See Understanding Purposeful Design Earth and Space Science at the front of this book for the web address.

> **WORLDVIEW**
>
> - The planets and other orbiting bodies in our solar system display amazing variety and precision. Although there are similarities among various planets, each one is unique. However, Earth alone is able to sustain life, which is partially because of its specific placement within the solar system. The orderly nature of the solar system and Earth's unique nature within it should cause all people to wonder if there is some grand purpose behind the solar system. Psalm 136 declares that God created the universe as one of a series of acts to demonstrate His enduring love for His people. The psalmist begins by asserting God's supremacy, then moves to review His acts of creation, protection, redemption, and preservation. Inserted after each statement of God's action is the affirmation, "His love endures forever." How marvelous it is that God chose to create one small place in the vast cosmos on which to place beings made especially in His image, whom He particularly loves!

Lesson 7.2.2 – Inner Planets

Although the inner planets are the most visible from Earth, most of what is known about them comes from data sent back from deep space probes, which have flown past, photographed, and even landed on and taken samples from the inner planets. Despite the inner planets' similarities to Earth, the differences are great.

Named after the speedy messenger of the Roman gods, Mercury is the fastest moving planet. Mercury's sky is always black because it has no atmosphere to scatter the light even though it is the planet closest to the sun. Despite its proximity to the sun, Mercury is not the hottest inner planet. Lacking an atmosphere to trap heat, Mercury can drop to −180°C at night, although during the day temperatures can reach 430°C. The surface of Mercury is heavily cratered. These craters are impact craters, not volcanic craters. The craters have remained unchanged since they were created because Mercury does not have wind or precipitation to weather them. Some large craters are separated by broad plains; long lines of cliffs, called *scarps*, cross these plains. Mercury has a magnetic field that is 100 times weaker than Earth's field. This presence of this magnetic field suggests an inner core of iron similar to Earth's core.

Venus was named after the Roman goddess of beauty and love. The planet has a unique retrograde rotation and extremely harsh surface conditions. Spacecraft that have landed there functioned only a few hours, but they sent back information before they failed. Venus's atmosphere is primarily composed of carbon dioxide and yellow clouds of sulfuric acid. Venus experiences winds of up to 360 kph. Venus's surface is covered with basalt lava, which appears to have nearly filled in many older craters. However, there are some impact craters that appear to have been created some time after lava covered much of the planet. Venus also has mountains that may have been formed by ancient volcanoes. Venus is unbearably hot with an average temperature of 470°C. Venus's heavy atmosphere produces a greenhouse effect; sunlight penetrates its thick clouds and heats the planet. The infrared heat cannot easily escape back through the thick clouds of carbon dioxide, so the heat reflects to the surface. As a result, Venus is hotter than Mercury although it is farther from the sun.

Unlike the other planets, which are named for Roman or Greek gods, Earth's name comes from words for "ground" in Old English (*ertha*) and German (*erde*). Earth is a haven for life in the solar system. Earth is unique because its abundance of water, breathable gases, and soil make it capable of sustaining life. Many precise conditions are required to support life. If even one of these conditions was different, Earth would be a barren planet.

Mars, named after the Roman god of war, is a cold, rusty planet whose surface appears to have once been very active. It has several large volcanoes, now seemingly dormant, that once spewed lava across Mars's plains. Running water appears to have cut channels on the surface of Mars. This evidence of water erosion suggests that in the past Mars's atmosphere was dense enough for liquid water to exist. Currently, Mars's atmospheric pressure is about 100 times less than that on Earth. The atmosphere supports high seasonal winds, which are related to the cyclical solar heating of the surface. These winds cause dust storms that erode Mars's surface. The planet's moons, Phobos and Deimos, are named for the attendants of the Greek god of war. They are the only moons aside from Earth's moon in the inner solar system.

Lesson 7.2.3 – Outer Planets

The outer planets are called *the gas giants*, or *Jovian planets*. Scientists believe the gas giants do not have large solid surfaces; their gaseous material simply gets denser with depth. They may, however, have small, hard cores. The visible areas of these planets are the tops of clouds high in their atmospheres.

Jupiter, the largest planet in the solar system, was named for the Roman king of the gods. Jupiter radiates more energy than it receives from the sun. The planet is about 90% hydrogen and 10% helium, but traces of methane, ammonia, ethane, and acetylene are also present. Scientists theorize that great pressure closer to the planet's core strips it of its electrons and condenses hydrogen into a liquid, which causes the hydrogen to behave like a metal. The liquid hydrogen then conducts electricity and is the source of Jupiter's magnetic field. Jupiter has 53 confirmed moons and 4 faint rings.

Named after the early Roman god of agriculture, Saturn was the last planet known to ancient astronomers. Saturn is thought to be made up of approximately 25% helium, 73% hydrogen, and about 2% of other elements, such as methane and ammonia. Saturn also has a liquid metallic hydrogen layer and a molecular hydrogen layer.

Uranus was the first planet discovered in relatively modern times. Named after the early Greek and Roman god of the heavens, it was discovered by William Herschel on March 13, 1781. Although he first thought Uranus was a comet, he came to realize that it was actually a planet. Uranus had been seen by other astronomers, but was misidentified as simply another star. Herschel named it *Georgium Sidus* in honor of his patron, King George III of England. The name *Uranus*, proposed in conformity with the other planetary names from classical mythology, did not come into common use until 1850. Uranus has an unusual magnetic field. The field is not aligned with the planet's center, and it is tilted almost 60° with respect to the axis of rotation. Most of the planets spin on axes nearly perpendicular to the plane of the ecliptic, but Uranus' axis is almost parallel to the ecliptic. Uranus has 27 confirmed moons. Unlike most of the other bodies in the solar system, which have names from Greek and Roman mythology, Uranus's moons are named for characters from the works of Shakespeare and Alexander Pope.

Named for the Roman god of the sea, Neptune was first observed in 1846 by German astronomer Johann Gottfried Galle, who based his observations on the calculations of Urbain-Jean Joseph Le Verrier. However, subsequent observations of Neptune revealed that its calculated orbit diverged from its actual orbit. For example, observations made 75 years earlier or later would not have found Neptune near its predicted location. Neptune's composition is similar to that of Uranus, and like Jupiter and Saturn, it has an internal heat source. Although Uranus is larger, Neptune has more mass. Neptune probably has a small core about the mass of Earth that contains hydrogen, helium, water, and some rocky material, but no solid surface. Its atmosphere is mostly hydrogen and helium but includes a small amount of methane. People can see Neptune with a telescope as a tiny bluish green disk if they know exactly where to look.

In 1930, astronomers added a ninth planet. Pluto, a small icy planet, has an orbit that typically extends beyond Neptune. In 2006, International Astronomical Union scientists decided to add a new distinct class of objects called *dwarf planets*. They put Pluto in this class, along with Ceres, which was previously classified as an asteroid, and Eris. The discovery of Eris and its similarity to Pluto prompted astronomers to develop the dwarf planet class. Since then, other similar celestial bodies have been discovered that meet the requirements to be classified as dwarf planets.

Lesson 7.2.4 – Smaller Orbiting Bodies

Along with the major planets and dwarf planets, several types of smaller bodies orbit the sun. One such category is the comet. From Earth, comets are visible for periods ranging from a few days to several months. In 1705, Edmond Halley concluded that a comet observed in 1682 was the same one that had been described in 1531 and 1607. He predicted that the comet would return again in 1758, and it did on Christmas Day. What became known as *Halley's Comet* returned again in 1835, 1910, and 1986; its next appearance is expected in 2061. A comet can have a highly elongated orbit that takes thousands of years or a much shorter orbit of less than 20 years. When it is positioned far from the sun, a comet consists of a nucleus, which is a dense, solid body that is approximately several kilometers in diameter. Frozen gases interspersed with particles of heavier substances form the nucleus. As the comet nears its perihelion, it forms two tails. The dust tail is composed of small dust particles carried from the nucleus by escaping gases. The dust tail shines with reflected sunlight and is the part of a comet that is the most visible. A long ion tail forms from gas ionized by the sun's ultraviolet radiation. The ion tail is pushed away from the sun by the solar wind, as the brighter dust tail traces the comet's curved orbit. Usually the two tails point in slightly different directions. A comet loses material and brightness when it passes the sun. Some of this material forms a stream of meteoroids in the comet's orbit. When Earth passes through such a path, it causes a meteor shower.

The origin of comets is still uncertain. One modern theory is that long-lived comets originate in the Oort cloud. Dutch astronomer J.H. Oort (1900–1992) proposed that a shell of numerous comets surrounds the solar system at a distance of about one light-year from the sun. He theorized that passing stars may force the orbits of slow-moving comets into the inner part of the solar system. In 1951, Dutch-American astronomer G.P. Kuiper suggested that a disk-shaped region of minor planets outside the orbit of Neptune was the source of short-lived comets. This area is now known as *the Kuiper belt* and it includes dwarf planets, asteroids, and comets. In 1992, the first of these Kuiper belt objects (KBOs) was discovered, and astronomers have now identified about 1,500 KBOs in the Kuiper belt.

An asteroid is a large rock that orbits the sun. The orbits of most asteroids are found in an area called *the asteroid belt*, which lies partially between the orbits of Mars and Jupiter. Astronomers estimate that the asteroid belt contains more than 1.1 million asteroids measuring at least 1 km. New asteroids are discovered almost every week, and not all asteroids are found in the main asteroid belt. Several thousand of the largest asteroids in this belt have been given names. Large asteroids range in size from that of a small mountain to more than one quarter the size of Earth's moon. The smallest asteroids are the size of pebbles. Asteroids that pass near Earth are categorized into several groups. The Amor asteroids lie just outside of Earth's orbit. The Apollo and Atens asteroids cross Earth's orbit. The Atira asteroids remain inside Earth's orbit. The chances of an asteroid colliding with Earth are minuscule, although some do approach Earth. Hermes, for example, comes within 742,000 km of Earth. However, smaller asteroids that are fragments of larger asteroids may collide and be thrown from their stable orbits into the inner solar system where they could threaten Earth. No known asteroids present a threat to Earth.

Meteoroids can originate from either comets or asteroids. A meteor is a meteoroid that burns up when it enters Earth's upper atmosphere. Every day, Earth encounters several tons of meteoroids. Most burn up completely before they hit Earth, although some reach the ground as meteorites. On February 15, 2013, a meteorite hit Chelyabinsk, Russia, causing widespread destruction and injury. Most meteoroids are about the size of pebbles, but they can occasionally be larger. However, scientists estimate that anything 25 m across or smaller will completely burn up in Earth's mesosphere, which extends 50–85 km above Earth's surface. A meteorite is the rare meteor that survives its descent through Earth's atmosphere to strike the surface. However, even these are typically fairly small by the time they reach the ground. The largest meteorite recorded to date is the 66-ton Hoba meteorite, which was found by a farmer near Grootfontein, Namibia, in 1920.

Meteors are most visible after midnight. Larger meteor showers can be seen when Earth crosses through bands of debris left behind by comets. The place from which meteors of a meteor shower all seem to originate is called *the radiant point*. Meteor showers are named for the constellation that appears behind the radiant point. For example, the Perseid meteor shower's radiant point is the Perseus constellation.

Planets

7.2.1 Planetary Motion

Introduction
Divide the class into small groups and go outside for the *Introduction* activities. Direct each group to use chalk powder to mark two foci on the ground about 60 cm apart. Then, have students place a stake in each focus and designate one stake as the sun by tying an inflated yellow balloon to it with string. (Note: It may be necessary to have students hold the stakes in place.) Have each group tie the ends of a 15-meter rope together and loop it around the stakes. Direct one student from each group to pull the rope tight against the stakes, maintain tension on the rope, and walk around the stakes. Have students trace the ellipse with chalk powder as they walk. Next, guide students to select various points on the ellipse. For each point, direct them to measure the distance to each foci and to add the two distances together. Guide students to create a chart for recording their measurements and calculations.

With one hand, grasp the end of the string to which a ball is tied. Hold the center of the string with your other hand and swing the ball in a circle. Demonstrate increasing and decreasing the ball's radius by allowing the string to slip through your fingers and then pulling it back. Invite student volunteers to swing the ball. Challenge student volunteers to swing the ball as slowly as possible as they keep it traveling in a circle. Encourage volunteers to create the shortest radius possible.

Select something nearby to be a target. Challenge students to hit the target with a kickball while spinning. Have students attempt the challenge singly and then in pairs with one person spinning and the other telling the spinner when to throw the ball.

Discussion
- Have students share their experiences with creating ellipses and adding the distances from points to foci from *Introduction*. Ask students why the results might vary from group to group. (**Possible answers: The stakes did not remain perfectly straight; tension was not consistently applied to the rope; the knot in the rope caught on a stake; the chalk powder was not consistently applied directly beneath the rope.**)

- Review the *Introduction* activity in which a ball was swung on a string. Ask students what was necessary in order to maintain a slow orbital rate. (**The string had to be as long as possible.**) How was it possible to create a short period of revolution? (**The ball had to spin faster.**) How does this activity correspond to Kepler's second law of planetary motion? (**Answers will vary but should include that planets and the ball have to move faster when they are close to their foci, but less speed is needed to keep them moving when they are farther from their foci.**)

- Review the *Introduction* activity in which students tried to release a ball when spinning to hit a target. Ask students how they determined when to release the ball to hit the target. (**The best time to release the ball is when we faced the target and the ball was in a straight line with the target.**) If the ball represents a planet and your body represents the sun, what force does your arm represent? (**gravity**) What would happen to the planets if gravity suddenly stopped working? (**They would travel in a straight line away from the sun.**)

Activities

Lab 7.2.1A Ellipse
- pushpins, 2 per group
- cardboard pieces, 1 per group
- metric rulers, 1 per group
- 30-centimeter pieces of string, 1 per group

Note: Answers depend on the string circle's circumference after the string ends are tied together.

OBJECTIVES
Students will be able to
- model the two ways a planet moves as it travels around the sun.
- summarize Kepler's laws of planetary motion.
- explain the role of gravity in planetary orbits.

VOCABULARY
- **aphelion** the point in a planet's orbit when it is farthest from the sun
- **astronomical unit** the average distance between Earth and the sun
- **ellipse** a closed curve along which the sum of the distances between two fixed points is always the same
- **perihelion** the point in a planet's orbit when it is closest to the sun

MATERIALS
- Chalk powder, metric tape measures, stakes, string, yellow balloons, rope, small brightly colored balls, kickballs (*Introduction*)
- Scientific calculators (*B*)
- WS 7.2.1A Kepler's Third Law
- WS 7.2.1B Gravity

PREPARATION
- Obtain a field chalk line marker. Tie a small brightly colored ball to a string approximately 60 cm long. (*Introduction*)
- Gather materials for Lab 7.2.1A Ellipse.
- Obtain materials for *Try This: Foucault Pendulum*.

TRY THIS
Foucault Pendulum
- small weights, 1 per group
- string
- ring stands, 1 per group
- turntables or lazy Susans, 1 per group
(continued)

ⓘ TRY THIS

(continued from previous page)
• *masking tape*
(The pendulum seems to follow the tape; the change in Earth's tilt at any particular location as Earth rotates causes a Foucault pendulum to change direction.)

A. Divide the class into small groups. Complete *Try This: Foucault Pendulum* with students.

B. Direct students to use scientific calculators to complete **WS 7.2.1A Kepler's Third Law**. Challenge students to work the equations backward, starting with the orbital period. Ask students why their answers are not exactly the same as the distances they were given. (**Some accuracy is lost when the answer to each step of the equation is rounded.**)

C. Guide students to complete **WS 7.2.1B Gravity**.

D. Assign students to use creative materials to build a model that demonstrates a planet's period of rotation and the period of revolution as it orbits the sun. Have students identify the following aspects: planet's aphelion, planet's perihelion, orbit's major axis, orbit's semimajor orbit, both foci, direction of rotation, and direction of revolution.

Lesson Review

1. Describe the two ways that planets move. (**A planet spins on its own axis as it travels in an elliptical orbit around the sun.**)
2. Summarize Kepler's laws of planetary motion. (**The planets move around the sun in elliptical orbits, the planets move faster when they are closer to the sun than they do when they are farther away, and the time a planet takes to orbit the sun once is directly related to its distance from the sun.**)
3. What keeps the sun's gravity from pulling Earth into the sun? (**Earth's centrifugal force**)
4. What keeps Earth from traveling in a straight line away from the sun? (**the sun's gravity**)

7.2.1 Planetary Motion

OBJECTIVES
• Model the two ways a planet moves as it travels around the sun.
• Summarize Kepler's laws of planetary motion.
• Explain the role of gravity in planetary orbits.

VOCABULARY
• **aphelion** the point in a planet's orbit when it is farthest from the sun
• **astronomical unit** the average distance between Earth and the sun
• **ellipse** a closed curve along which the sum of the distances between two fixed points is always the same
• **perihelion** the point in a planet's orbit when it is closest to the sun

Around the sun are the planets, which move in orderly ways. All planets rotate, or spin, around an imaginary line called *an axis*. Mercury is the only planet with an axis that is perpendicular to its orbital path. Most of the planets' axes are tilted, and Uranus appears to lie on its side because its axis is tilted so far. As a planet rotates, only half of it faces the sun at any given time. The half that faces the sun experiences daylight while the other half of the planet is in darkness. Because all planets rotate, all have day and night just as Earth does. The time that it takes a planet to rotate on its axis is called its *period of rotation*. Earth's period of rotation is about 24 hours. In contrast, Jupiter's period of rotation is 10 hours, and Venus's period of rotation is 5,832 hours, or 243 Earth days. Most planets rotate counterclockwise. However, Venus and Uranus have a retrograde, or backward, rotation, and they rotate clockwise. On these planets, the sun seems to rise in the west and set in the east.

Planets also travel around the sun in a path called *an orbit*. All planets orbit the sun in the same direction. The time that it takes a planet to orbit the sun is called its *period of revolution*. The period of revolution is a planet's year. For example, Earth's period of revolution is very nearly 365 days, 6 hours, and 9 minutes.

German astronomer Johannes Kepler described three laws that describe the motion of planets. Kepler's first law states that planets move around the sun not in circles but in elongated ovals called *ellipses*. An **ellipse** is a closed curve along which the sum of the distances between two fixed points (each called *a focus*, plural *foci*) is always the same. In a planet's orbit, the sun is

Rotation and Revolution

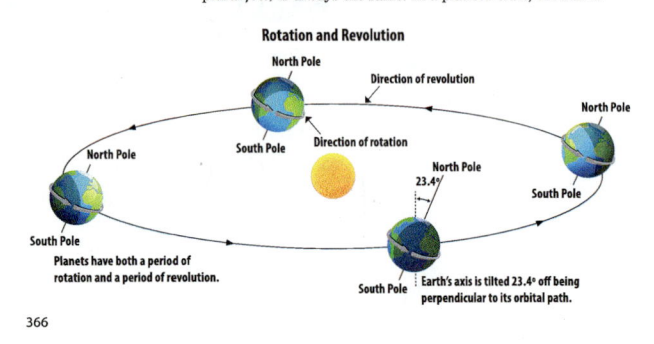

Planets have both a period of rotation and a period of revolution.

Earth's axis is tilted 23.4° off being perpendicular to its orbital path.

one focus, and an invisible point in space is the other focus. The sun, therefore, is off-center in the ellipse, and a planet's distance from the sun varies as it moves through its orbit. The maximum diameter of an ellipse is called *the major axis*, and half of the major axis is known as *the semimajor axis*. The semimajor axis is a planet's average distance from the sun. For example, the semimajor axis of Earth's orbit is about 149,600,000 km. This average distance between Earth and the sun is called an **astronomical unit** (AU). Other planets' distances from the sun are usually measured using this unit. The planet Mercury is 0.387 AU from the sun, so its distance is 0.387 AU × 149,597,871 km = 57,894,376 km.

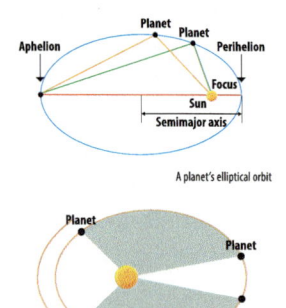

Kepler's second law of planetary motion states that planets move faster when they are closer to the sun than they do when they are farther from the sun. The point in a planet's orbit when it is closest to the sun is its **perihelion**. The point when a planet is farthest from the sun in its orbit is known as its **aphelion**. Because a planet's speed changes in relation to its distance from the sun, the area it covers during any set amount of time will be the same no matter where the planet is in its orbit. Imagine an elastic string connecting a planet to the sun. When the planet is far from the sun, the string will cover a long, narrow space, and when the planet is close to the sun, the string will cover a short, wide space. The calculation of each space's area during the same amount of time will be equal, however.

Kepler's third law states that the time a planet takes to orbit the sun once is directly related to its distance from the sun. If you know the time it takes a planet to orbit the sun, you can determine the distance from the planet to the sun using $P^2 = a^3$. The semimajor axis is measured in astronomical units (*a*), and the period of revolution is expressed in years (*P*). For example, Saturn takes approximately 29 years to orbit the sun.

$$P^2 = a^3, \quad (29 \text{ years})^2 = 841, \quad a^3 = 841$$

$$a = \sqrt[3]{841} = 9.44 \text{ AU}$$

So Saturn is 9.44 AU from the sun, which is more than nine times the distance from Earth to the sun.

A planet's elliptical orbit

Planets cover equal areas in equal amounts of time.

FYI

Top Speed
Distances outside the solar system—between stars, for example—are measured in light-years. Distances within the solar system can be measured in light-minutes and light-hours. Light travels about 300,000 km per second in space. If you could move that fast, you would travel around Earth 7.5 times in one second. Light travels nearly 18,000,000 km in one minute. This distance is called *a light-minute*. It takes light from the sun 8.3 minutes to reach Earth. The distance from Earth to the sun is 8.3 light-minutes.

The third law also explains that a distant planet takes more time to orbit the sun than a planet near the sun. For example, Mercury is the closest planet to the sun, and its orbit takes 88 Earth days. In contrast, Neptune's year is about 165 Earth years.

Kepler described the planets' orbits, but he never figured out why they behave as they do. Nearly 70 years later, Sir Isaac Newton answered that question by describing gravity. Newton deduced that an attraction existed between all objects. Masses of all sizes are attracted to each other. Furthermore, the strength of an object's attractive force depends on its mass. Newton reasoned that small objects fall to Earth because Earth and the objects are attracted to each other by the force of gravity. According to Newton's law of universal gravitation, the force of gravity is described by the product of the objects' masses divided by the square of the distance between them. The equation looks like this:

$$\frac{\text{Object 1 mass} \times \text{Object 2 mass}}{(\text{distance between objects})^2}$$

This principle means that if two objects that are next to each other are moved 3 times as far apart, then the gravitational attraction between them would be 9 times weaker ($3^2 = 9$). If the objects are moved 100 times farther apart, the gravitational attraction would be 10,000 times less ($100^2 = 10,000$).

Gravity pulls the moon toward Earth, and it pulls Earth toward the sun. However, the moon does not crash into Earth, and Earth does not collide with the sun because of centrifugal force. For example, imagine a person twirling a ball on a string. The ball's attempt to move away from the person is centrifugal force. As long as the person holds the string, however, the ball will orbit the person's hand. Earth's centrifugal force keeps the planet from being drawn into the sun by the sun's strong gravity. In this example, the string acts like gravity. If the string was cut, the ball would fly off in a straight line. The planets would behave in the same way without gravity pulling them toward the sun.

LESSON REVIEW
1. Describe the two ways that planets move.
2. Summarize Kepler's laws of planetary motion.
3. What keeps the sun's gravity from pulling Earth into the sun?
4. What keeps Earth from traveling in a straight line away from the sun?

TRY THIS

Foucault Pendulum
In 1851, Jean-Bernard-Léon Foucault suspended a pendulum from the center of the inner dome of a building in Paris. The long wire attached to the pendulum allowed the device an unrestricted swing in any direction. Although the pendulum seemed to change its path during the day, it was actually the earth beneath the pendulum that was moving. The device was called *a Foucault pendulum*.

Tie a small weight to a string. Tie the other end of the string to the clamp on a ring stand. Hang the pendulum over a turntable or a lazy Susan. Put a small piece of masking tape on one side of the turntable. The center of the turntable represents the North Pole. Swing the pendulum and slowly turn the turntable. What direction does the pendulum appear to swing in relation to the tape? Infer what kind of movement causes a real Foucault pendulum to change directions.

Name: _____ Date: _____

Lab 7.2.1A Ellipse

QUESTION: What is an ellipse?

HYPOTHESIS: Answers will vary.

EXPERIMENT:

You will need:	• cardboard	• 30-centimeter piece of string
• 2 pushpins	• metric ruler	

Steps:
1. Stick two pushpins into the cardboard about 10 cm apart.
2. Loop the string around the pushpins and tie the ends together. One person should hold the pushpins in place. Another person should place a pencil inside the string and trace an ellipse; keep the string tight at all times. Label this *Ellipse A*.
3. Move the pushpins until they are 5 cm apart and trace another ellipse. Label this *Ellipse B*.
4. Choose three points on Ellipse A. Measure each point's distance to each focus (pushpin) and add the two distances. Record your measurements in the chart.
5. Choose three points on Ellipse B. Measure each point's distance to each focus (pushpin) and add the two distances. Record your measurements in the chart.

	Point 1 Sum	Point 2 Sum	Point 3 Sum
Ellipse A	Answers will vary. The measurement should be approximately 14.5 cm and all points should have the same measurement.		
Ellipse B	Answers will vary. The measurement should be approximately 19.5 cm and all points should have the same measurement.		

Lab 7.2.1A Ellipse

ANALYZE AND CONCLUDE:
1. How did the distance between the two foci (fixed points) affect the ellipses' shape?
   Placing the pushpins farther apart made the shape more of an oval.

2. What shape would you draw if you removed one focus? a circle

3. What does any point on an ellipse have in common with any other point on an ellipse? Every point on an ellipse is the same total distance from the two foci.

4. Why might some of your sums not be identical for all three points on an ellipse?
   Possible answers: The string was not tight at all times when drawing the ellipse; one or both of the pushpins moved; the string's knot caught on one of the pushpins; the pencil's angle changed slightly; the string slipped on the pencil.

Use the Internet to find the answers to the questions below.

5. Which planet's orbit is the most elliptical? Mercury
6. Which planet's orbit is the most circular? Venus

7.2.2 Inner Planets

Planets

OBJECTIVES

Students will be able to
- distinguish among the solar system's inner planets.
- describe the features of Earth that make life possible.
- evaluate the inner planets' abilities to sustain life.

VOCABULARY

- **crater** a large circular indentation on a planet's surface

MATERIALS

- Planetary artwork by Dr. Mark Garlick (*Introduction*)
- Index cards (*A*)
- BLM 7.2.2A Game: Planet Mixer
- WS 7.2.2A Inner Planets' Properties

PREPARATION

- Search for artwork by Dr. Mark Garlick on the Internet or in scientific publications. (*Introduction*)
- Obtain materials for *Lab 7.2.2A Venus*.
- Write planet facts on index cards. Prepare one card per student. (*A*)

Introduction

Display planetary artwork by Dr. Mark Garlick, British astronomer and artist. Encourage students to write or sketch about the artwork.

Discussion

- Read **Romans 1:20**. Lead a discussion about Earth's unique nature as the only naturally habitable planet. Ask what Earth teaches about God's nature. (**Answers will vary.**)

- Discuss how an inner planet aside from Earth could be made habitable. Ask which inner planet would be the most logical for colonization. (**Answers will vary.**) Query students' knowledge about Elon Musk's efforts through SpaceX to create a human colony on Mars.

- Ask the following questions:
  1. Why are Venus's and Mercury's periods of rotation not conducive to life? (**Organisms, particularly plants, do not receive enough sunlight to photosynthesize during these planets' extensive nights.**)
  2. Why is Venus sometimes called *Earth's twin*? (**Venus has nearly the same diameter, density, mass, and period of revolution as Earth.**)
  3. What effect does Mars's distance from the sun have on its ability to sustain life? (**Mars is far enough from the sun that the sun's rays do not warm it to the same extent as they warm Earth, which causes Mars's temperatures to frequently drop below the freezing point of water.**)
  4. In what ways does Earth's atmosphere make life possible? (**Possible answers: The atmosphere protects Earth from meteoroids; it helps maintain a temperature range in which organisms can live; it deflects or absorbs harmful radiation, it is not dense enough to squash organisms.**)

Activities

Lab 7.2.2A Venus

- books, 2 per group
- balloons, 1 per group
- rubber bands, 1 per group
- thermometers, 2 per group

Carbon dioxide and water vapor are vital ingredients for the atmosphere to sustain life. These substances allow the sun's heat to reach Earth's surface, and they trap this heat to maintain warmth when the sun sets. During this experiment, the thermometer with the balloon around its bulb should record a significantly higher temperature. The balloon, like Earth's atmosphere, allows heat to reach the thermometer but also traps the heat, raising the temperature within the balloon.

Option: Experiment with different colors of balloons to determine which absorbs and holds heat best. Compare the results to the effects of the opaqueness of planets' atmospheres.

A. Play the game on **BLM 7.2.2A Game: Planet Mixer** with students. Limit the game's focus to the inner planets.

B. Have students complete **WS 7.2.2A Inner Planets' Properties**. Remind students that circumference = π × diameter. To determine a student's age on another planet, use the following equation: (365/number of days in period of revolution for planet) × age of student in years. For example, on Mercury, the equation would be (365/88) × 11 = 45.6 Mercury years.

C. Direct student groups to develop an advertisement or infomercial for one of the inner planets using the information they listed on WS 7.2.2A.

D. Assign students to research the meaning and the origins of the inner planets' names. Encourage students to explain the extent to which the planets' names reflect their features.

E. Have students invent a creature that could live on Mercury, Venus, or Mars and explain what kind of features this creature would have to possess in order to survive.

F. Challenge students to conduct further research and to create a presentation that focuses on the unique characteristics God has given Earth in order to sustain life.

G. Assign students to research a spacecraft or probe that visited one of the inner planets. Have students report on what was learned about the planet as a result of the mission and to share pictures of the spacecraft or probe as well as any pictures taken of the planet. Encourage students to include images made using any form of electromagnetic radiation. Direct students to explain what the images reveal about the planet. Provide time for students to present their information to the class.

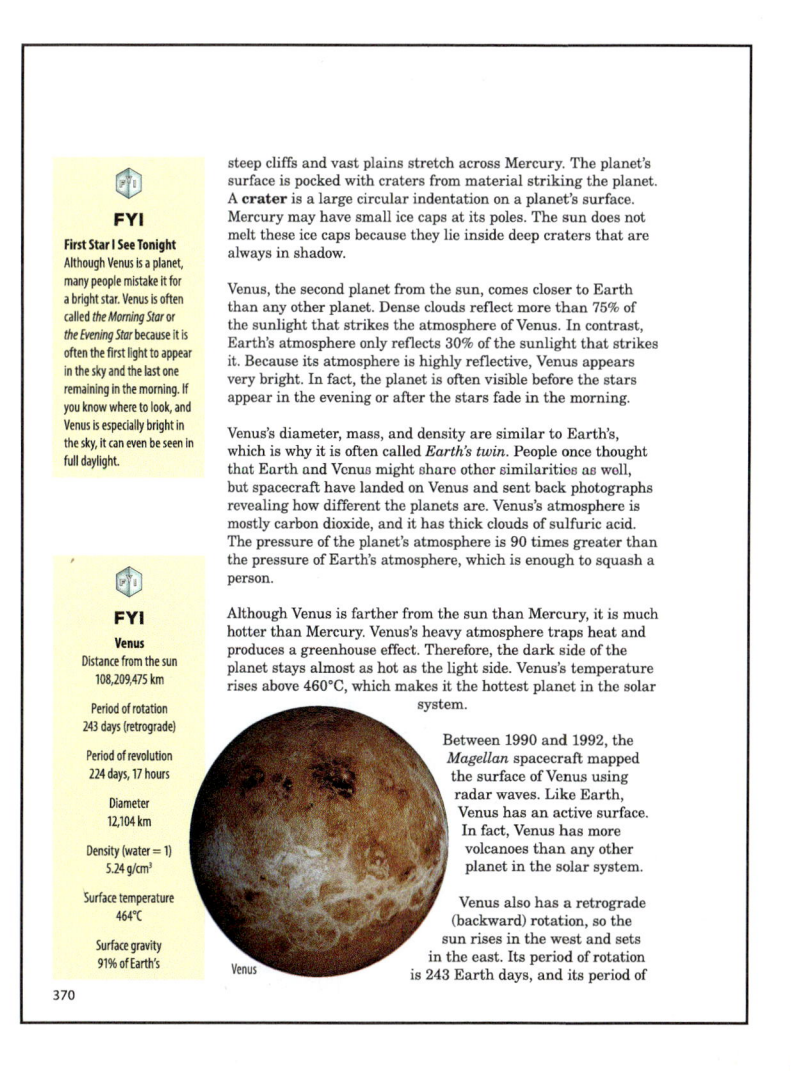

NOTES

Lesson Review
1. Which of the inner planets, other than Earth, is most suitable for life? Why? (**Mars is most suitable for life. It has some atmosphere, unlike Mercury, but the atmosphere is not so dense**

Inner Planets 7.2.2

All the planets follow the same laws of motion, yet the planets themselves are different. Planets in the solar system are divided into inner and outer planets separated by the asteroid belt. The orbits of the inner planets are more closely spaced together than the orbits of the outer planets. The inner planets are also known as *terrestrial planets* because they are all relatively dense, rocky planets. The inner planets are Mercury, Venus, Earth, and Mars.

Mercury is the closest planet to the sun. Because it orbits so close to the sun, it never strays far from the sun in Earth's sky. Visible for only a brief time before sunrise or after sunset, it looks like a bright star shining just above the horizon. In fact, Mercury is so close to the sun that the sun would appear more than three times larger from Mercury than it appears from Earth.

Mercury is smaller than Earth, and the force of gravity is reduced. Because gravity is weaker on less massive planets, people would weigh less on Mercury than on Earth. When combined with the sun's fierce heat, Mercury's low gravity contributes to its lack of an atmosphere.

Mercury is the solar system's fastest planet. Its year, or period of revolution, is 88 Earth days. If it did not move fast, the planet would fall into the sun. Although Mercury revolves very quickly, it rotates very slowly. It rotates only three times for every two revolutions it makes around the sun. Mercury has a sunrise only every 175 Earth days because of its slow rotation and fast revolution.

Mercury's lack of atmosphere and long period of rotation allow the daytime side of the planet to reach a temperature of more than 400°C. On the nighttime side, temperatures can drop below −170°C. Consequently, Mercury is one of the solar system's hottest and coldest planets.

It is hard to observe Mercury from Earth because it is so close to the sun. People did not know very much about it until the spacecraft *Mariner 10* flew by it in 1975 and sent back information. Long,

OBJECTIVES
- Distinguish among the solar system's inner planets.
- Describe the features of Earth that make life possible.
- Evaluate the inner planets' abilities to sustain life.

VOCABULARY
- **crater** a large circular indentation on a planet's surface

FYI
Mercury
Distance from the sun
57,909,227 km

Period of rotation
58 days, 16 hours

Period of revolution
88 days

Diameter
4,879 km

Density (water = 1)
5.43 g/cm³

Surface temperature
−173°C to 427°C

Surface gravity
38% of Earth's

Mercury

369

---

steep cliffs and vast plains stretch across Mercury. The planet's surface is pocked with craters from material striking the planet. A **crater** is a large circular indentation on a planet's surface. Mercury may have small ice caps at its poles. The sun does not melt these ice caps because they lie inside deep craters that are always in shadow.

Venus, the second planet from the sun, comes closer to Earth than any other planet. Dense clouds reflect more than 75% of the sunlight that strikes the atmosphere of Venus. In contrast, Earth's atmosphere only reflects 30% of the sunlight that strikes it. Because its atmosphere is highly reflective, Venus appears very bright. In fact, the planet is often visible before the stars appear in the evening or after the stars fade in the morning.

Venus's diameter, mass, and density are similar to Earth's, which is why it is often called *Earth's twin*. People once thought that Earth and Venus might share other similarities as well, but spacecraft have landed on Venus and sent back photographs revealing how different the planets are. Venus's atmosphere is mostly carbon dioxide, and it has thick clouds of sulfuric acid. The pressure of the planet's atmosphere is 90 times greater than the pressure of Earth's atmosphere, which is enough to squash a person.

Although Venus is farther from the sun than Mercury, it is much hotter than Mercury. Venus's heavy atmosphere traps heat and produces a greenhouse effect. Therefore, the dark side of the planet stays almost as hot as the light side. Venus's temperature rises above 460°C, which makes it the hottest planet in the solar system.

Between 1990 and 1992, the *Magellan* spacecraft mapped the surface of Venus using radar waves. Like Earth, Venus has an active surface. In fact, Venus has more volcanoes than any other planet in the solar system.

Venus also has a retrograde (backward) rotation, so the sun rises in the west and sets in the east. Its period of rotation is 243 Earth days, and its period of

FYI
First Star I See Tonight
Although Venus is a planet, many people mistake it for a bright star. Venus is often called *the Morning Star* or *the Evening Star* because it is often the first light to appear in the sky and the last one remaining in the morning. If you know where to look, and Venus is especially bright in the sky, it can even be seen in full daylight.

FYI
Venus
Distance from the sun
108,209,475 km

Period of rotation
243 days (retrograde)

Period of revolution
224 days, 17 hours

Diameter
12,104 km

Density (water = 1)
5.24 g/cm³

Surface temperature
464°C

Surface gravity
91% of Earth's

Venus

370

NOTES

as to crush organisms, unlike Venus. Mars's temperature is quite cold, but it may be easier to accommodate cold than Venus's extreme heat, and Mars does not have the same extremes of temperature that Mercury has. Mars has water in the form of ice at its poles, and there may be more ice water under its surface. Neither Mercury nor Venus has water available in the same quantity.)

2. Which of the inner planets has the fastest period of revolution? What benefit does this provide? (Mercury. It's speedy period of revolution counteracts the pull of the sun's gravity.)

3. Which of the inner planets has the most dense atmosphere? Is this atmosphere conducive to life? Why? (Venus has the most dense atmosphere. It is not conducive to life because its pressure would squash most organisms and its greenhouse effect is so extreme that temperatures are far too great for life.)

4. Describe two unique features of Earth that make life possible. (Possible answers: Earth's distance from the sun ensures that Earth receives the right amount of light and warmth to sustain a livable biosphere. Earth's atmosphere is thick enough to deflect or burn up most meteorites, it is thin enough to keep everything beneath it from being crushed, and it has sufficient oxygen for life. Gravity on Earth's surface is balanced to allow living organisms to function properly.)

5. Which inner planet, other than Earth, is confirmed to have water? (Mars)

---

revolution is almost 225 Earth days. If people could be born on Venus, some of them could celebrate two birthdays between one sunrise and the next!

The third planet from the sun is Earth. From space, Earth looks like a blue marble. Swirls of clouds blanket the brown and blue of Earth. As far as scientists know, Earth is the only planet that can support life. The idea that God placed Earth at just the right distance from the sun and specifically designed Earth's systems to support life is called *the anthropic principle*. Earth's distance from the sun ensures that Earth receives the right amount of light and warmth to sustain its biosphere. Earth's atmosphere is perfectly composed to support life, and it is thick enough to deflect or burn up most meteorites. At the same time, the atmosphere is thin enough to keep everything beneath it from being crushed. Earth is the only planet in the solar system with an atmosphere containing the oxygen necessary for life. In addition, gravity on Earth is balanced. If it were stronger or weaker, both living organisms and the atmosphere would not function properly. If even one of these or many other important factors were different, then Earth could not support life. God filled Earth with life and uses these qualities to sustain it.

Mars, sometimes known as *the Red Planet*, was studied for years by telescope before spacecraft contributed significantly more information. *Viking 1* and *Viking 2* landed on the surface of Mars and sent back detailed photographs. The *Viking* spacecraft also scooped up and analyzed the soil, which is coated in iron oxide (rust) and accounts for the planet's red color. In 2012, *Curiosity* landed on Mars and conducted more studies of its soil and rocks.

Mars is similar to Earth in several ways. Mars is tilted on its axis slightly more than Earth, which gives it seasons similar to Earth. Mars also has wind and weather patterns. The wind can be strong enough to create dust storms that erode Mars's surface and turn its atmosphere dark pink.

Earth and moon

FYI
It Is Just a Phase
From Earth, Venus appears to have phases, just like the moon. The crescent phases are visible with binoculars. When Venus is nearest to Earth, it appears as a crescent. When it is farthest from Earth, Venus would appear to be round, but it is not visible from Earth because the sun is in the way.

FYI
Earth
Distance from the sun
149,596,262 km

Period of rotation
23 hours, 56 minutes

Period of revolution
365 days, 6 hours

Diameter
12,756 km

Density (water = 1)
5.52 g/cm^3

Surface temperature
−88°C to 58°C

Surface gravity
100% of Earth's

Like Earth, the surface of Mars displays large craters, deep canyons, and inactive volcanoes. Mars has several regions marked by former volcanic activity. The Tharsis Montes region is the largest, and it stretches 4,000 km across Mars. It contains Olympus Mons, the largest mountain in the solar system at 25 km tall and 624 km wide. This shield volcano is similar to Hawaii's Mauna Loa. Scientists believe that because Mars's crust does not move around like Earth's crust, volcanoes there build up in the same spots instead of forming in chains.

Mars even has water, but what instruments can detect is all frozen. Mars has two polar icecaps containing frozen water and frozen carbon dioxide. Scientists believe that additional water may be frozen beneath the soil. Mars has features that look like dry riverbeds, which indicates that liquid water existed on Mars in the past. The past presence of liquid water suggests that Mars was once a warmer place. Liquid water cannot collect in observable amounts on Mars because the atmosphere is mainly carbon dioxide, and it is about 100 times thinner than Earth's atmosphere. The thin atmosphere does not trap heat on the planet's surface. Also, the sun does not warm Mars as much as it does Earth because Mars is farther from the sun. Therefore,

FYI
Mars
Distance from the sun
227,943,824 km

Period of rotation
24 hours, 37 minutes

Period of revolution
1 year, 322 days

Diameter
6,779 km

Density (water = 1)
3.93 g/cm^3

Surface temperature
−153°C to 20°C

Surface gravity
38% of Earth's

FYI
No Pressure
Water boils at 100°C, right? Yes, as long as the water is at sea level on Earth's surface. The boiling point of water depends on atmospheric pressure. Water boils at a lower temperature on top of a high mountain because the atmospheric pressure is less at high altitudes. The atmospheric pressure on Mars is too low for liquid water to exist. If ice does happen to melt, it almost instantly evaporates.

HISTORY
System Error
On December 11, 1998, the multimillion dollar space probe *Mars Climate Orbiter* was launched. About nine months later, on September 23, 1999, it reached Mars and was lost. The orbiter mistakenly steered to within 57 km of the Martian surface, and atmospheric friction burned it up. How could a mistake like this happen? Some of the directional commands to the spacecraft were sent in English units rather than the metric units the spacecraft was programmed for. The spacecraft interpreted the digits as metric and entered the Mars atmosphere incorrectly.

temperatures on Mars remain very cold, ranging between –153°C and 20°C, which would keep water frozen most of the time. Whenever ice does melt on Mars, it boils away quickly because the atmospheric pressure on the planet is so low that water boils at only 10°C.

Mars has two very small moons, Phobos and Deimos. On average, Phobos is only about 22 km wide, and Deimos is only about 14 km wide. The moons' small size suggests that they may have been asteroids that were captured by Mars's gravity.

Deimos

Phobos

LESSON REVIEW
1. Which of the inner planets, other than Earth, is most suitable for life? Why?
2. Which of the inner planets has the fastest period of rotation? What benefit does this provide?
3. Which of the inner planets has the most dense atmosphere? Is this atmosphere conducive to life? Why?
4. Describe two unique features of Earth that make life possible.
5. Which inner planet, other than Earth, is confirmed to have water?

FYI

What is a Day?
Earth's period of rotation is not exactly 24 hours. This statement may be confusing because a day is measured in 24 hours. A day is the time between two successive passages of the sun over a location on Earth, such as from noon to noon in Cairo, Egypt. The length of a solar day varies slightly during the course of a year, but it averages approximately 24 hours. A day is assigned exactly 24 hours for purposes of time measurement. Astronomers measure days as Earth's rotation relative to the fixed stars, which is called *a sidereal day*.

Mars

373

Name: _____ Date: _____

Lab 7.2.2A Venus

QUESTION: Why is the temperature of Venus so hot?
HYPOTHESIS:

EXPERIMENT:

You will need:	• 2 thermometers	• rubber band
• 2 books	• balloon	

Steps:
1. Set two books side by side in direct sunlight next to a window.
2. Place a thermometer on each book with the thermometer's bulb facing the windows and overhanging the books by several centimeters.
3. Inflate a balloon and use a rubber band to secure its end over the bulb of one thermometer.
4. Record both thermometers' readings every minute for 10 minutes.

Time	Thermometer: No Balloon	Thermometer: Balloon
1 min		
2 min		
3 min		
4 min		
5 min	Answers will vary.	
6 min		
7 min		
8 min		
9 min		
10 min		

Lab 7.2.2A Venus

ANALYZE AND CONCLUDE:
1. Explain the differences in temperature. The balloon traps heat.

2. Compared to Earth, the temperature of Venus is very hot. Why might this be? Venus is hotter than Earth partly because its thicker atmosphere traps more heat than Earth's thinner atmosphere.

3. How might the relationship between Venus's atmosphere and temperature help scientists better understand conditions on Earth relating to the atmosphere and changing temperature trends? If Earth's atmosphere is experiencing rising levels of a certain atmospheric ingredient that composes most of Venus's atmosphere, we can expect Earth's temperatures to rise.

7.2.3 Outer Planets

Planets

OBJECTIVES

Students will be able to
- identify the solar system's outer planets.
- describe the distinguishing features of the outer planets' largest moons.
- generalize characteristics common to the gas giants.
- evaluate Pluto's official status as a dwarf planet.

VOCABULARY

- **Kuiper belt** the region of the solar system outside Neptune's orbit

MATERIALS

- BLM 7.2.2A Game: Planet Mixer
- WS 7.2.3A Outer Planets' Properties

PREPARATION

- Obtain materials for *Try This: Jupiter's Spot*.
- Write planet facts on index cards. Prepare one card per student. (*B*)

TRY THIS

Jupiter's Spot
- *plastic spoons, 1 per group*
- *white glue, 250 mL per group*
- *bowls, 1 per group*
- *red food coloring*
- *yellow food coloring*
- *straws, 1 per student*

(**The colors mixed together. When air was blown across the middle of the bowl, the colors pushed out toward the sides and eventually returned to the bowl's center. When air was blown along the side of the bowl, the colors created a swirl pattern.**)

Introduction

Direct students to picture a professional football stadium. Then, have students imagine an ant inside the stadium. Ask students what this might model. (**Answers will vary.**) Explain that this is a model of a hydrogen atom. The stadium is the atom's size compared to the single electron, which is the ant. Most of what makes up an atom is actually space. Ask students what they think makes up most of the universe. (**Answers will vary but should include space.**) Explain that most of what is "out there" is space. According to NASA, less than 5% of the universe is visible matter. Most space charts leave out the largest part of the universe: the apparent empty space. The vast nothingness seems to convey that humans are insignificant in the scope of the universe. Point out that David wrestled with the same thought in Psalm 8:1–4. Convey that compared to the unbelievable scope of the universe, people might seem unimportant, but that is not God's view. Remind students that each individual is precious to the Creator. Although a person may feel insignificant in comparison to the vastness of the universe, the very emptiness of space actually conveys that each person is miraculously important.

Discussion

- Guide students to identify common characteristics among the outer planets and to determine whether Pluto should be included in this category.

- Lead a discussion regarding Pluto's change in status from a full planet to a dwarf planet. Ask what the purpose of a classification system is. (**Answers will vary but should include that a classification system provides consistent standards that are used by scientists around the world.**) Why are these systems modified occasionally? (**Answers will vary but should include that new information leads scientists to modify how objects are classified.**)

- Discuss the planetary rings of the gas giants. Point out that astronomers did not know that all the gas giants had rings until *Voyager 2* confirmed that Neptune had rings in 1989. Ask what the rings are made of. (**icy particles ranging in size from a grain of sand to over a kilometer across**) Who first discovered the rings of Saturn? (**Galileo**) Approximately how many rings does Saturn have? (**between 500 and 1,000**) Which planets' rings are very faint? (**Jupiter and Neptune**)

- Discuss the many moons of the gas giant planets. Ask which two planets have the most moons. (**Jupiter and Saturn**) Which of the outer planets' major moons has an atmosphere? (**Saturn's Titan**) Which is the solar system's largest moon? (**Jupiter's Ganymede**) Encourage discussion about why the outer planets have so many more moons than the inner planets. Ask which moon is about half the size of its planet. (**Pluto's Charon**) Do you think more moons will be discovered around Pluto? (**Answers will vary.**)

- Ask the following questions:
  1. What creates Jupiter's magnetosphere? (**metallic liquid hydrogen**)
  2. In what way is Uranus like Venus? (**They both have retrograde rotations.**)
  3. How was Neptune discovered? (**Astronomers noticed that Uranus's orbit was being altered by another object. Calculations directed astronomers where to look and what to look for, and these led to the discovery of Neptune in 1846.**)
  4. What do astronomers think causes wind on Neptune? (**rising hot gas and falling cool gas**)
  5. In what region of space is Pluto located? (**the Kuiper belt**)

Activities

A. Complete *Try This: Jupiter's Spot* with students. Ask students how wind storms are created on Earth. (**Warm air rises and is replaced with cooler air.**) How is a wind storm on Earth similar to Jupiter's Great Red Spot? (**Just like on Earth, warm gases rise and are replaced with cooler gases to create wind.**)

B. Play the game on **BLM 7.2.2A Game: Planet Mixer** with students. Limit the game to focus on the outer planets or include all the planets as a review.

C. Have students complete **WS 7.2.3A Outer Planets' Properties**. Remind students that circumference = π × diameter. To determine the student's age on another planet, use the following equation: (365/number of days in period of revolution for planet) × age of student in years. For example, on Jupiter, the equation would be (365/4,333) × 11 = 0.93 Jupiter years.

D. Direct student groups to develop an advertisement or infomercial for one of the outer planets using the information they listed on WS 7.2.3A.

E. Assign students to research the meaning and the origins of the outer planets' names. Encourage students to explain the extent to which the planets' names reflect their features.

F. Encourage students to research a spacecraft or probe that visited one of the outer planets. Have students report on what was learned about the planet as a result of the mission and to share pictures of the spacecraft or probe as well as pictures taken of the planet. Guide students to include images made using any form of electromagnetic radiation. Direct students to explain what the images reveal about the planet.

G. Have students design a vehicle that can be used to explore one of the outer planets. Encourage them to consider what unique challenges the vehicle will need to overcome as it explores their chosen planet.

NOTES

7.2.3 Outer Planets

OBJECTIVES
- Identify the solar system's outer planets.
- Describe the distinguishing features of the outer planets' largest moons.
- Generalize characteristics common to the gas giants.
- Evaluate Pluto's official status as a dwarf planet.

VOCABULARY
- **Kuiper belt** the region of the solar system outside Neptune's orbit

Beyond Mars lies the asteroid belt, which will be discussed in the next lesson. The planets beyond the asteroid belt are called *the outer planets*. These planets are very different from the inner planets both in size and in composition. All of the outer planets are also known as *gas giants* because they are significantly larger than the inner planets and have no known solid surfaces. The gas giants include Jupiter, Saturn, Uranus, and Neptune.

Jupiter is the solar system's largest and most colorful planet. It is a true giant! More than 1,300 Earths could fit inside Jupiter. However, Jupiter's mass is only that of about 318 Earths because it has a relatively low density. Jupiter has a very short day of only 9 hours and 55 minutes. Because this gaseous planet rotates so quickly, it bulges at the equator and flattens at the poles.

Like the sun, Jupiter is primarily hydrogen and helium. Jupiter sends more heat into space than it receives from the sun, but how this heat is generated is unknown. The outer part of Jupiter's atmosphere is composed primarily of colored bands of methane and ammonia. The planet's thick cloud cover exerts enough pressure on the planet to change Jupiter's hydrogen into a liquid. As pressure increases closer to Jupiter's core, the liquid hydrogen acts like a metal. This liquid metallic layer creates a magnetic field called *the magnetosphere*, which stretches for millions of kilometers beyond Jupiter.

Jupiter's clouds are very active and stormy. A hurricane-like storm was first observed on Jupiter through a telescope. Over the years, this storm has changed color, grown, shrunk, and even disappeared several times. Some scientists question whether only one storm has been observed or whether different storms are appearing and disappearing. The storm as it can currently be seen is known as *the Great Red Spot*. It has been measured at over three times the size of Earth.

Extending out from Jupiter are four faintly visible rings. Fifty-three confirmed moons orbit Jupiter among its rings, and another 14 objects appear to be moons of Jupiter as well. Galileo discovered Jupiter's four largest moons in 1610. These moons are known as *the Galilean satellites*. Io, the closest to Jupiter of the Galilean satellites, has a high sulfur content and is covered with active volcanoes. Its blotchy orange, red, and yellow appearance reminded one scientist of a pepperoni pizza. In contrast, Europa, an ice-covered moon, has a bright and extremely smooth surface. Ganymede, Jupiter's largest moon, is the largest moon in the entire solar system. It is even larger than Mercury. Half water and half ice, Ganymede seems to have cracks that scientists believe resulted from upheavals similar to earthquakes. Callisto is the solar system's most heavily cratered object.

Saturn is the second largest planet in the solar system. It has about 755 times the volume and 95 times the mass of Earth. Saturn can be seen without a telescope, but it is less dense than water, which means that if Saturn was placed in enough water, it would float. Saturn is like Jupiter because it is mostly hydrogen and helium with lesser amounts of methane and ammonia, it gives off more heat than it receives from the sun, and it spins very rapidly.

Although all of the gas giants have rings, Saturn's rings are the most famous. Galileo first discovered the presence of rings around Saturn. Later astronomers determined that the rings are composed of little bits of ice and ice-covered matter in orbit around the planet. Saturn has at least seven major rings, which are made up of thousands of thin rings. The rings start about 6,700 km from the top of Saturn's clouds and extend approximately 420,000 km from Saturn (more than twice the diameter of Saturn itself). They can be as thin as 100 m. The icy particles that make up the rings range in size from a grain of sand to over a kilometer across. These magnificent rings actually contain very little material. If the material was compressed into a sphere, the sphere's diameter would be only about 400 km across.

Like Jupiter, Saturn has at least 53 moons. The moon Phoebe travels east to west in a retrograde direction, but the others travel west to east in a direct rotation. Titan, Saturn's largest moon, is the only moon in the solar system known

FYI
Jupiter
Distance from the sun
778,340,821 km

Period of rotation
9 hours, 55 minutes

Period of revolution
11 years, 318 days

Diameter
116,464 km

Density (water = 1)
1.33 g/cm³

Temperature
−148°C

Gravity
236% of Earth's

TRY THIS
Jupiter's Spot
Using a spoon, mix 250 mL of white glue and 1 L of water in a bowl. Place two drops of red food coloring and two drops of yellow food coloring on the mixture to represent Jupiter's atmosphere. Take turns blowing on the surface of the mixture through a straw. Experiment with blowing across the middle of the bowl and along the side of the bowl. What happened?

FYI
Saturn
Distance from the sun
1,426,666,422 km

Period of rotation
10 hours, 39 minutes

Period of revolution
29 years, 160 days

Diameter
120,536 km

Density (water = 1)
0.69 g/cm³

Temperature
−178°C

Gravity
92% of Earth's

NOTES

H. Direct students to research arguments for and against Pluto's change in classification from a planet to a dwarf planet, choose one position, and explain why they made their choice.

I. Challenge student groups to imagine that they are planning to travel to a distant planet. Point out that students will need to determine the survival resources they will need for such a voyage and to discuss how they will obtain these resources. Have groups compare their conclusions. Lead a discussion regarding how fragile such a mission would be. Guide students to draw parallels between their imaginary journey and the precision with which God designed planet Earth to sustain life.

Lesson Review

1. What sets Jupiter apart from all the other planets? (**Possible answers: It is the largest planet; it has the fastest period of rotation; it has the largest moon in the solar system.**)
2. Identify one distinguishing feature of each of Jupiter's four Galilean satellites. (***Io***: **covered with active volcanoes;** ***Europa***: **ice-covered, has a bright and extremely smooth surface;** ***Ganymede***: **largest moon in the entire solar system;** ***Callisto***: **the solar system's most heavily cratered object**)
3. What characteristics are common to the gas giants? (**The gas giants are much larger than the terrestrial planets; they have no known solid surfaces; they are composed of gases; they all contain hydrogen, helium, methane, and ammonia; they have many moons; and they all have rings.**)
4. Which outer planets are thought to have internal heat sources? (**Jupiter, Saturn, and Neptune**)
5. What two elements are present in great quantities in all the gas giants? (**hydrogen and helium**)

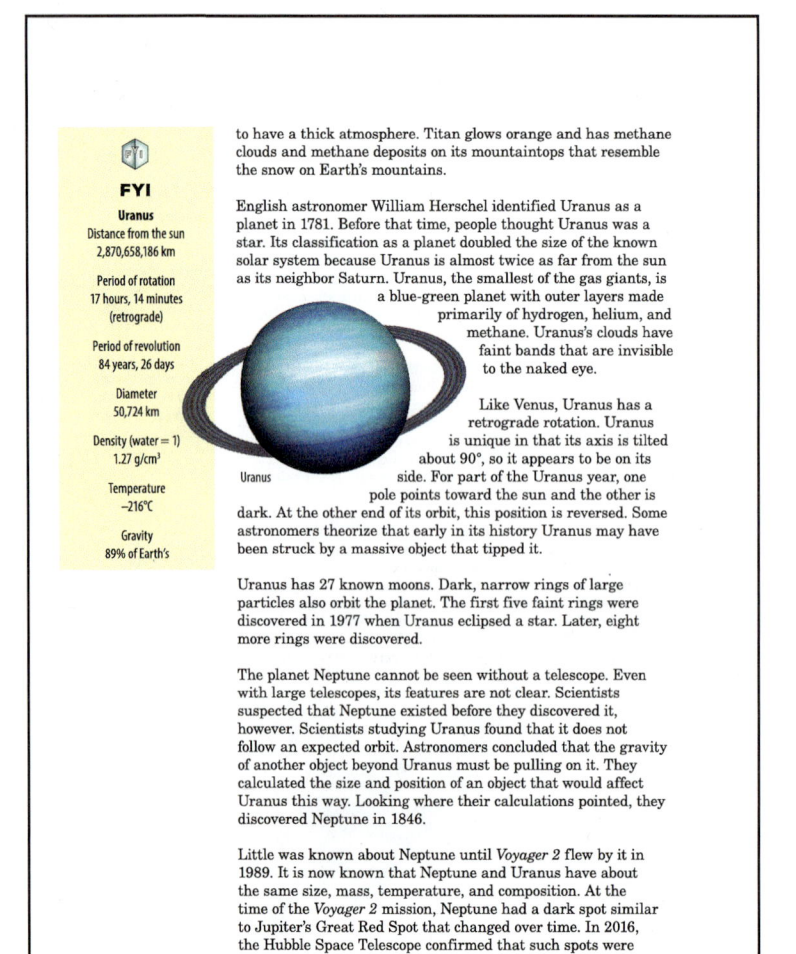

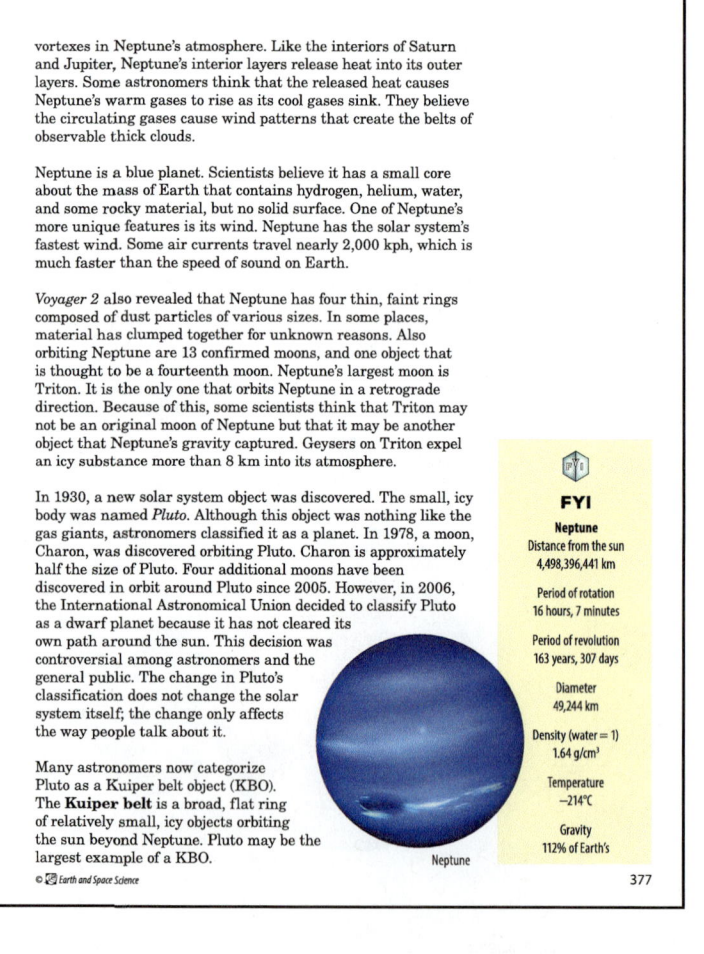

FYI

Uranus

Distance from the sun
2,870,658,186 km

Period of rotation
17 hours, 14 minutes
(retrograde)

Period of revolution
84 years, 26 days

Diameter
50,724 km

Density (water = 1)
1.27 g/cm³

Temperature
−216°C

Gravity
89% of Earth's

to have a thick atmosphere. Titan glows orange and has methane clouds and methane deposits on its mountaintops that resemble the snow on Earth's mountains.

English astronomer William Herschel identified Uranus as a planet in 1781. Before that time, people thought Uranus was a star. Its classification as a planet doubled the size of the known solar system because Uranus is almost twice as far from the sun as its neighbor Saturn. Uranus, the smallest of the gas giants, is a blue-green planet with outer layers made primarily of hydrogen, helium, and methane. Uranus's clouds have faint bands that are invisible to the naked eye.

Like Venus, Uranus has a retrograde rotation. Uranus is unique in that its axis is tilted about 90°, so it appears to be on its side. For part of the Uranus year, one pole points toward the sun and the other is dark. At the other end of its orbit, this position is reversed. Some astronomers theorize that early in its history Uranus may have been struck by a massive object that tipped it.

Uranus has 27 known moons. Dark, narrow rings of large particles also orbit the planet. The first five faint rings were discovered in 1977 when Uranus eclipsed a star. Later, eight more rings were discovered.

The planet Neptune cannot be seen without a telescope. Even with large telescopes, its features are not clear. Scientists suspected that Neptune existed before they discovered it, however. Scientists studying Uranus found that it does not follow an expected orbit. Astronomers concluded that the gravity of another object beyond Uranus must be pulling on it. They calculated the size and position of an object that would affect Uranus this way. Looking where their calculations pointed, they discovered Neptune in 1846.

Little was known about Neptune until *Voyager 2* flew by it in 1989. It is now known that Neptune and Uranus have about the same size, mass, temperature, and composition. At the time of the *Voyager 2* mission, Neptune had a dark spot similar to Jupiter's Great Red Spot that changed over time. In 2016, the Hubble Space Telescope confirmed that such spots were

376

vortexes in Neptune's atmosphere. Like the interiors of Saturn and Jupiter, Neptune's interior layers release heat into its outer layers. Some astronomers think that the released heat causes Neptune's warm gases to rise as its cool gases sink. They believe the circulating gases cause wind patterns that create the belts of observable thick clouds.

Neptune is a blue planet. Scientists believe it has a small core about the mass of Earth that contains hydrogen, helium, water, and some rocky material, but no solid surface. One of Neptune's more unique features is its wind. Neptune has the solar system's fastest wind. Some air currents travel nearly 2,000 kph, which is much faster than the speed of sound on Earth.

Voyager 2 also revealed that Neptune has four thin, faint rings composed of dust particles of various sizes. In some places, material has clumped together for unknown reasons. Also orbiting Neptune are 13 confirmed moons, and one object that is thought to be a fourteenth moon. Neptune's largest moon is Triton. It is the only one that orbits Neptune in a retrograde direction. Because of this, some scientists think that Triton may not be an original moon of Neptune but that it may be another object that Neptune's gravity captured. Geysers on Triton expel an icy substance more than 8 km into its atmosphere.

In 1930, a new solar system object was discovered. The small, icy body was named *Pluto*. Although this object was nothing like the gas giants, astronomers classified it as a planet. In 1978, a moon, Charon, was discovered orbiting Pluto. Charon is approximately half the size of Pluto. Four additional moons have been discovered in orbit around Pluto since 2005. However, in 2006, the International Astronomical Union decided to classify Pluto as a dwarf planet because it has not cleared its own path around the sun. This decision was controversial among astronomers and the general public. The change in Pluto's classification does not change the solar system itself; the change only affects the way people talk about it.

Many astronomers now categorize Pluto as a Kuiper belt object (KBO). The **Kuiper belt** is a broad, flat ring of relatively small, icy objects orbiting the sun beyond Neptune. Pluto may be the largest example of a KBO.

FYI

Neptune

Distance from the sun
4,498,396,441 km

Period of rotation
16 hours, 7 minutes

Period of revolution
163 years, 307 days

Diameter
49,244 km

Density (water = 1)
1.64 g/cm³

Temperature
−214°C

Gravity
112% of Earth's

377

6. How is Pluto different from the gas giants? (**Pluto is classed as a dwarf planet, it is much smaller than the gas giants, it is icy instead of gaseous, and it has only one moon in contrast to the many moons of each gas giant.**)

NOTES

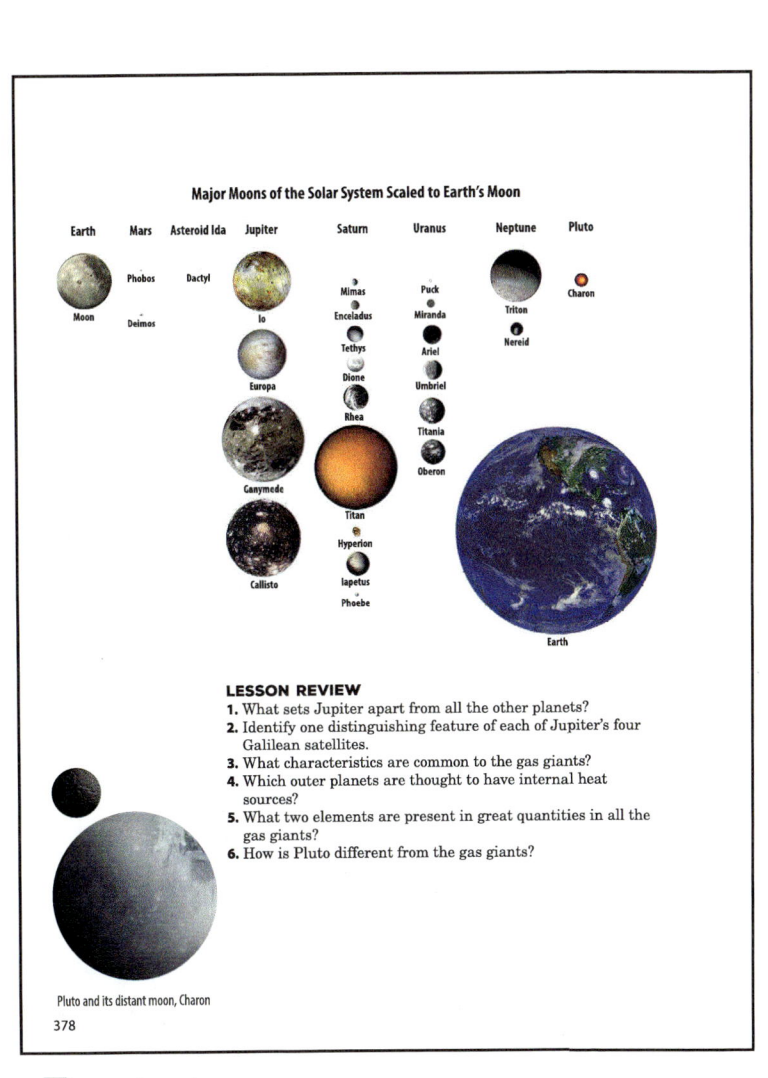

7.2.4 Smaller Orbiting Bodies

Planets

OBJECTIVES

Students will be able to
- explain why a comet changes during its orbit.
- explain the effect of atmospheric friction on a falling body.
- determine when an object should be classified as a meteoroid, meteor, and meteorite.
- distinguish among comets, asteroids, and meteoroids.

VOCABULARY

- **comet** a frozen chunk of ice, dust, and rock that orbits the sun
- **friction** the force that resists motion between two surfaces that are in contact with each other
- **meteor** a meteoroid that enters the earth's atmosphere and burns up
- **meteorite** a meteoroid that enters Earth's atmosphere and strikes the ground
- **meteoroid** a rock fragment from an asteroid or comet

MATERIALS

- Dry ice, cooler, basin, plastic bag, wooden spoon, 500 mL beaker, 100 mL beaker, gravel, soil, dark corn syrup, insulated rubber gloves, newspaper, heat lamp, paper, textbooks (*Introduction*)
- WS 7.2.4A Triple Venn Diagram

PREPARATION

- Crush dry ice and store it in a cooler. Line a basin with a sturdy plastic bag. With a wooden spoon, mix 500 mL of water, 50 mL of fine gravel, 50 mL of soil, and 10 mL of dark corn syrup in the lined basin. Put on insulated rubber gloves. Add 500 mL of crushed dry ice and mix well. (Note: Do not touch dry ice without gloves.) Close the bag around the mixture and form a compact round ball (a "comet") that is frozen enough to hold its shape. (*Introduction*) (*continued*)

Introduction

Remove the prepared simulated comet from the bag and place it on a newspaper. Have students observe the comet. Explain that the dry ice changes directly from a solid to a gas in the same way a comet's ice sublimates as it approaches the sun. Ask students whether a comet would change much when its orbit takes it farther from the sun's heat. (**Answers will vary but should include that the comet's nucleus would not lose as much or any material.**) Point out that eventually the comet will become a crater-filled ice ball as the carbon dioxide sublimates before the ice melts. If possible, invite students to check the comet throughout the day. As an alternative, set the comet near a heat lamp so students can see more immediate changes.

Guide students in a demonstration of atmospheric friction. Have students rub their hands together lightly until they feel heat. Next, have them press their hands more firmly together as they rub until they feel heat. Ask students which method enabled them to feel heat more quickly. (**pressing hands together and rubbing**) What is the definition of friction? (**the force that resists motion between two surfaces in contact with each other**) Explain the relationship among resistance, friction, and heat: more resistance creates greater friction, and greater friction results in more heat.

Direct students to get out two sheets of paper and to hold one with the edge toward the floor and the other with the broad side toward the floor; the edge and side closest to the floor should be at the same height. Have students drop the papers at the same time and observe which landed first. (**the sheet held with the edge toward the floor**) Have students explain the result. (**The sheet with its broad side toward the floor encountered more resistance from the air as it fell, so it fell more slowly.**)

Direct students to hold a textbook in one hand and a sheet of paper in the other hand. Both items should have their broadest sides parallel to the floor and should be held at the same height. Have students drop their books and papers at the same time and observe which landed first. Direct students to place the paper on top of the book and then drop them and observe the result. Ask students which object landed first when they were held separately. (**the book**) Which object landed first when they were stacked? (**They landed simultaneously.**) Explain that the book's greater mass enabled it to force its way through the air's resistance faster than the paper when the book and paper were held separately. The book and paper fell at the same rate when they were stacked because the book broke through the air's resistance for the paper, enabling the paper to fall at the same rate because the force of gravity is a constant.

Discussion

- Discuss how a comet changes during its orbit. Point out that a comet has tails of dust and charged particles as it passes near the sun. The solar wind radiates away from the sun, which causes a comet's ion tail to point away from the sun. Ask what process causes a comet to have tails only when it is near the sun. (**sublimation**) Explain that when a comet is near the sun, the sun's heat causes the comet's ice to sublimate, which creates the coma and tails. A comet is also quite bright when it is near the sun. When a comet is far from the sun, it does not reflect light. A comet has no halo or tails when it is far from the sun because its gas and liquids are frozen.

- Guide a discussion exploring the relationship among resistance, friction, and heat. Have students explain how the activities in *Introduction* demonstrate why a meteoroid begins to burn up as it enters Earth's atmosphere. Ask what force causes an object to produce heat and light as it passes through Earth's atmosphere. (**friction**)

- Discuss the classification of meteoroids, meteors, and meteorites.

- Compare and contrast comets, asteroids, and meteoroids. Encourage students to complete the graphic organizer on **WS 7.2.4A Triple Venn Diagram** during the discussion.

- Ask the following questions:
  1. Why does the moon have more meteorite craters than Earth? (**The moon doesn't have an atmosphere to destroy meteoroids and doesn't have oceans to absorb meteorite impacts.**)
  2. For what accomplishment was Maria Mitchell honored as an astronomer? (**She was the first American to identify a new comet.**)
  3. Prior to Tycho Brahe's demonstration that comets are independent celestial bodies, what did many people believe about comets? (**They were believed to be bad omens.**)
  4. Who demonstrated that comets orbit the sun? (**Edmund Halley**)

Activities

A. Complete *Try This: Turn Up the Heat* with students. Guide students to compare the activity to the friction a meteoroid experiences when it encounters Earth's atmosphere.

B. Assign students to research a comet or an asteroid and to deliver a presentation describing its appearance, orbit, and composition.

C. Challenge students to design an experiment to demonstrate the effect of atmospheric friction on a falling body.

PREPARATION
(continued from previous page)
- Obtain materials for *Try This: Turn Up the Heat*.

TRY THIS

Turn Up the Heat
- buckets, 1 per group
- spatulas, 1 per group

(**Answers will vary but should include that greater resistance equals greater friction. Meteoroids experience greater friction than aircraft because they have a broad surface, but an aircraft's streamlined shape minimizes resistance; a skydiver experiences less friction than a meteor because he or she moves more slowly and creates less resistance.**)

Smaller Orbiting Bodies 7.2.4

The major planets are not the only objects that revolve around the sun. Smaller bodies such as comets, asteroids, and meteoroids also orbit the sun. A **comet** is a frozen chunk of ice, dust, and rock that gives off tails of dust and charged particles as it passes near the sun during its orbit. Although the orbits of most planets are nearly circular, the orbits of comets are very elliptical. Astronomers often describe comets as *dirty snowballs*. The *snow* part is frozen water and gases; the *dirt* part is rocks and dust. The solid center of a comet, called *the nucleus*, is typically about 10 km in diameter, but it can be larger or smaller.

When a comet passes close to the sun, solar radiation heats the ice, changing it directly from a solid to a gas in a process called *sublimation*. This process releases the dust and rock trapped in the ice. The gas and dust create a halo, called *a coma*, around the comet. As it approaches the sun, the comet changes from a cold, dark object into one so bright that people can see it from Earth.

A comet forms two tails as its gas sublimates. One tail is made of dust, and the other is made of charged particles called *ions*. Because the solar wind radiates away from the sun, it causes a comet's ion tail to point away from the sun. A comet's dust tail tends to follow the comet's orbit around the sun and does not always point away from the sun. When a comet is close to the sun, its tails can extend millions of kilometers through space. The dust tail leaves behind pebble-sized debris that other celestial bodies, such as Earth, encounter on a regular basis during their orbits.

OBJECTIVES
- Explain why a comet changes during its orbit.
- Explain the effect of atmospheric friction on a falling body.
- Determine when an object should be classified as a meteoroid, meteor, and meteorite.
- Distinguish among comets, asteroids, and meteoroids.

VOCABULARY
- **comet** a frozen chunk of ice, dust, and rock that orbits the sun
- **friction** the force that resists motion between two surfaces that are in contact with each other
- **meteor** a meteoroid that enters the earth's atmosphere and burns up
- **meteorite** a meteoroid that enters Earth's atmosphere and strikes the ground
- **meteoroid** a rock fragment from an asteroid or comet

Comet

A comet's ion tail always points away from the sun.

FYI

New Comets
Scientists continue to discover new comets, many of which can be seen only with the use of a telescope. As comets first become visible, they look like a blob of light; they brighten and grow a tail as they approach the sun. New comets are named after the people and spacecraft they are discovered by. For example, the Shoemaker-Levy 9 comet, discovered in 1993, was discovered by Carolyn Shoemaker, Eugene Shoemaker, and David Levy, who realized that the comet was going to collide with Jupiter. This unusual comet had been captured by Jupiter's gravity and broken into 21 distinct pieces. Over a period of seven days, these pieces impacted Jupiter one at a time. One of the flares from these explosions shot about 3,000 km above Jupiter's clouds. Another piece of the comet created a temporary dark spot on Jupiter that was larger than Earth.

HISTORY

No Omen
Comets have long been objects of superstition. In fact, the word *disaster* comes from the Latin words *dis*, meaning "negation" and *astrum*, meaning "star." Combined, these words mean "bad star." Cultures around the world believed that comets signaled some form of disaster. Comets were considered evil omens that appeared right before wars, famines, floods, or other disasters. For example, many English citizens believed that Halley's Comet caused the Black Death. Some people today still believe that comets signal disaster.

Although it had been proposed that comets were heavenly bodies like the planets as early as the 1st century, it was not proved until the 16th century by Tycho Brahe. Today people can easily predict the appearance of known comets, and comets are largely understood as part of an orderly creation rather than as bad omens.

Comets appear to originate in the outer solar system. Some astronomers believe they are leftovers from the process of planet formation. If this theory is true, then each comet is a sample of the early solar system, and learning about comets can help scientists piece together the chemical and physical aspects of the solar system's history. Many scientists believe that a sphere of icy bodies called *the Oort cloud* surrounds the solar system and that comets with long periods of revolution come from that region. If the Oort cloud exists, then it probably lies 40,000–50,000 AU from the sun. Comets with relatively short periods of revolution originate in the Kuiper belt. Movies and television programs often show comets zooming across the sky. However, because comets are so far away from Earth, they actually appear to move very slowly in comparison to the stars.

Other small orbiting bodies are known as **asteroids**. Most asteroids are irregularly shaped bodies of various sizes that are pitted with craters from colliding with other space debris. Most asteroids are located in the region between Mars and Jupiter known as *the asteroid belt*. Through a telescope, asteroids look like stars; the word *asteroid* means "starlike."

Some astronomers think that asteroids are leftover material from the solar system's creation. One theory regarding

Composite photo of asteroids Lutetia, Gaspra, and Ida

NOTES

D. Direct students to research predictable meteor showers to determine what left the debris, when the next shower will occur, and the factors that affect the extent of the meteor shower.

E. Have students research what scientists believe they can learn about the composition and origin of celestial bodies from evidence provided by meteorites.

Lesson Review

1. Compare how comets look when they are near the sun to how they look when they are far from the sun. (**When a comet is near the sun, the sun's heat causes the comet's ice to sublimate, which creates a coma and two tails; the comet is also quite bright. When a comet is far from the sun, it does not reflect light, and it has no halo or tails because its gas and liquids are frozen.**)
2. Why does a meteor appear as a bright streak in Earth's atmosphere? (**The friction between a meteoroid and Earth's atmosphere produces heat and light.**)
3. What is the term for a rock that comes from space and lands on the earth? (**meteorite**)
4. What is the difference between a comet and an asteroid? (**Possible answers: A comet is mostly ice, but an asteroid is mostly rock; a comet develops a coma and tails as it approaches the sun, but an asteroid does not change.**)
5. How is a meteoroid related to comets and asteroids? (**Meteoroids are thought to be pieces broken off of either comets or asteroids.**)

FYI

Unlikely Target
Although scientists find all asteroids interesting, some scientists are particularly fascinated with asteroids that could hit Earth. If an asteroid measuring 1 km or more hit Earth, it could potentially cause global environmental disasters. However, it is very unlikely that an asteroid will ever collide with Earth. Genesis 8:22 states that as long as Earth remains, day and night will continue. Scientists have calculated asteroid orbits several hundred years into the future, and their work does not indicate any large asteroids on a collision course with Earth.

of most animal life on Earth, including the dinosaurs. It is likely that the dust from the collision would have blocked the sunlight on Earth for months, changing the climate. Young-earth scientists hold that the changes on Earth after the Flood led to the extinction of the dinosaurs. The moon and Mars are covered with meteorite craters, yet such craters are rare on Earth. On February 15, 2013, a meteorite caused widespread destruction and injury when it hit Chelyabinsk, Russia. Earth's atmosphere prevents most meteoroids from striking the ground as meteorites. In addition, oceans cover more than 70% of Earth's surface, so it is very rare for meteorites to fall on land.

Meteor shower

LESSON REVIEW

1. Compare how comets look when they are near the sun to how they look when they are far from the sun.
2. Why does a meteor appear as a bright streak in Earth's atmosphere?
3. What is the term for a rock that comes from space and lands on the earth?
4. What is the difference between a comet and an asteroid?
5. How is a meteoroid related to comets and asteroids?

Meteor Crater in Arizona, United States

7.3.0 Sun, Earth, and Moon

LOOKING AHEAD

- For **Lesson 7.3.1**, schedule *Lab 7.3.1A Moon Width* when the full moon will be visible in daylight. Obtain a video clip of the first moon landing.
- For **Lesson 7.3.2**, assign *Try This: Moonrise* approximately seven days prior to teaching the lesson.
- For **Lesson 7.3.4**, obtain multiple lamps with bare bulbs and several globes for *Try This: Change the Seasons*.

SUPPLEMENTAL MATERIALS

- BLM 7.3.2A Instructions: Moon Phases Board

- TM 7.1.2A Moon Phases
- TM 7.3.1A Moon Features
- TM 7.3.3A Solar Eclipse Phases
- TM 7.3.3B Lunar Eclipse Phases
- TM 7.3.3C Solar Eclipse
- TM 7.3.3D Lunar Eclipse
- TM 7.3.4A June Solstice
- TM 7.3.4B December Solstice
- TM 7.3.4C Earth's Seasons

- Lab 7.3.1A Moon Width
- Lab 7.3.1B Regolith Formation

- WS 7.3.1A Moon Map
- WS 7.3.2A Going Through a Phase
- WS 7.3.3A Solar Eclipse Diagram
- WS 7.3.3B Lunar Eclipse Diagram
- WS 7.3.4A Changing Seasons

- Chapter 7.3 Test
- Unit 7 Test

BLMs, TMs, and tests are available to download. See Understanding Purposeful Design Earth and Space Science at the front of this book for the web address.

Chapter 7.3 Summary

Earth's moon has evoked mystery and wonder for thousands of years. Why are people so fascinated with the moon? Earth's nearest neighbor is prominent in the night sky, and its phases and eclipses are very conspicuous. Along with the sun, stars, and planets, the moon has served as reference throughout history for measuring the passage of time. God did not create the moon to be worshipped or to be the object of superstition; He made it a key part of His design for Earth.

The sun, Earth, and the moon are not islands in the solar system. They were created to interact with each other in such a way as to create day and night, seasons, perception of time, eclipses, and tides. God's design for the solar system accounts for these phenomena; they are not random or accidental. The orderly nature of eclipses, moon phases, and seasons testifies to God's sovereign hand over His creation.

Background

Lesson 7.3.1 – Structure of the Moon

The moon is a key part of God's plan for Earth. God designed the moon to have distinctive features and characteristics. The moon moves at an average speed of 1.022 km/s. The moon's surface temperature ranges from 127°C to below −173°C; the moon's slow rotation helps account for some of this temperature swing. At 3,476 km, the moon's diameter is roughly one-fourth (27.25%) that of Earth. The moon's mass is about one-eightieth of Earth's, and the moon is three-fifths as dense. Because of the moon's low mass, it has low gravity; the moon's gravitational force is about one-sixth that of Earth. With such weak gravity, the moon cannot maintain a significant atmosphere. Therefore, the moon lacks weather, clouds, rain, wind, and surface water. The lack of atmosphere also means that there is no breathable air on the moon, so the astronauts who landed there wore portable life support systems to provide them with oxygen. They also communicated through radio because the moon has no air to carry sound. Unlike Earth, the moon does not have a significant magnetic field. The weak magnetic field and atmosphere do very little to protect the moon from meteorites or the solar wind, which continuously bombards the lunar regolith. The sun constantly embeds chemical elements such as hydrogen ions into the lunar surface. Therefore, scientists can learn about the sun as well as the moon by studying the moon's surface.

The moon's terrain is a combination of smooth maria, or plains, and heavily cratered highlands. Maria are lowland plains that contain hardened lava. Appearing as dark regions on 16% of the moon's surface, maria are found mainly on the near side of the moon. Many of the maria formed when heat beneath the moon's crust melted rock into lava, which welled up and flowed into impact craters left by meteorites. Some scientists believe that this heat was at least partially caused by radioactive materials.

In contrast to the dark maria, the rocky highlands appear bright and are found mainly on the far side of the moon. The highlands are older than the maria. One theory about their origin is that they were sculpted early in the moon's history when some scientists believe crystals relocated from space to the moon's surface. Thousands of craters scar the highlands, and some craters even overlap or lie within larger craters. Craters are also found on the maria, although maria have far fewer craters than the highlands. Bright streaks called *rays* radiate outward in many directions from certain craters. The rays are believed to be a mixture of broken rocks, which were hurled from the craters, and rock fragments from the secondary craters. The fact that the rays cross over maria, highlands, and other craters suggests that ray craters were formed relatively late in the moon's history.

Rilles are another prominent surface feature of the moon. Rilles are valleys or channels that are typically 5 km or less in width but can extend up to several hundred kilometers long. Most rilles are straight. Scientists are not sure about their origin, but believe straight rilles were probably created when the moon's outer crust cracked because of erupting gases or when sections of the surface dropped down during a moonquake. Sinuous rilles are winding channels that look like dry riverbeds. They probably were formed by the flow of lava on the maria.

Scientists have various theories to explain how the moon was created. The fission theory was proposed in 1878 by George Howard Darwin. This theory suggests that during its early years, Earth became increasingly elongated because of its high rate of spin, and the sun's gravity eventually pulled off a chunk, which became the moon. The capture theory states that the moon was once a wandering planet that was captured by Earth's gravity. The cocreation, or "sister," theory hypothesizes that Earth and the moon formed separately but were side by side and made from the same materials contained in the rest of the solar system's planets. One modern model is the impact theory, which proposes that the moon formed from the debris ejected from Earth when a meteorite at least the size of Mars collided with Earth. It is thought that the energy from the collision vaporized the meteorite on impact and ejected material from Earth's crust into space. The theory speculates that material from the impactor and ejecta consolidated to form the moon, which would explain the low density of the moon. It also explains the moon's tiny core, because Earth's mantle has very little iron. The moon may have gotten its iron core from smaller and later impacts. Finally, it accounts for the angular momentum of the Earth and moon. However, the impact theory does not explain the close similarity of both bodies' chemical compositions. Another theory seeks to explain the similar compositions of Earth and the moon by suggesting that two objects, each about half the size of Earth, collided, rebounded, collided again, and briefly merged. Eventually, this new body broke apart to become Earth and a disk of material orbiting Earth, which later coalesced to become the moon. The Bible records that God created the moon as its own entity on the fourth day of creation. According to this account, God uniquely created each of the moon's features and properties.

Lesson 7.3.2 – Phases of the Moon

The phases of the moon are determined by the relative positions of the sun, Earth, and moon. To understand these phases, it is important to understand how the moon orbits Earth. Because the moon changes its position each day as it orbits from west to east, the moon rises and sets about 50 minutes later each day. The moon and Earth have a geosynchronous orbit, meaning the time the moon takes to rotate once is the same length of time it takes for the moon to orbit Earth. Because of this, people always see the same side of the moon, which is called *the near side*. Although sometimes referred to as *the dark side of the moon*, the other side is more accurately called *the far side of the moon*, because it is lit up half the time just as the near side is. The moon orbits Earth once in approximately 27.3 days. Over time, a little more than 50% of the moon's surface can be seen because the moon orbits Earth in an ellipse. Closer to Earth, the moon orbits a little faster; when farther out, it orbits a little slower.

A lunar phase is the measurement of how much of the illuminated surface of the moon can be seen from Earth. A new moon occurs when the moon moves between Earth and the sun; in this phase none of the illuminated portion is visible from Earth. Half an orbit from the new moon, when the moon is on the opposite side of Earth from the sun, its entire illuminated side is visible as a full moon. A half-moon occurs when the moon has completed one-quarter or three-quarters of its orbit from new moon to new moon and the moon is at quadrature, or at 90°, to the sun. The half-moon that falls between the new moon and the full moon is called *the first quarter*, and the half-moon that falls between the full moon and new moon is called *the last quarter*. More than half of the lighted side is visible in the gibbous phase, which falls between the first quarter and the full moon and between the full moon and the last quarter. A harvest moon is the full moon that occurs closest to the autumnal equinox. Because of the relation of the moon's path to the horizon, the moon rises at nearly the same time for several nights in a row, and if the sky is clear, there is full moonlight from dusk almost to dawn.

WORLDVIEW

- James Irwin was one of twelve U.S. astronauts to walk on the moon. He said, "I felt an overwhelming sense of the presence of God on the moon. I cannot imagine a holier place." God is sovereign everywhere, even in the vast expanse of the solar system. In Deuteronomy 10:14–15, Moses declared, "To the Lord your God belong the heavens, even the highest heavens, the earth and everything in it. Yet the Lord set His affection on your ancestors and loved them, and He chose you, their descendants, above all the nations." Those who trust Jesus for salvation should find it comforting that God is sovereign over all His creation, even the far reaches of space. We are part of that creation, and we not only belong to God, but we are loved by Him as well. For those who do not love God, however, the concept of God's eternal sovereignty should drive them to seek God and repent. Jeremiah 23:23–24 states that no matter where people go, they cannot hide from God. Even in the outer reaches of the universe, God is the supreme authority.

The moon's gravitational pull on Earth causes the oceans on the side of Earth facing the moon to bulge outward. On the opposite side, the oceans also bulge outward because of the weak gravitational attraction of the moon on that side. These bulges form Earth's tides. Earth rotates from west to east, and the tidal bulge appears to flow in a westwardly direction around Earth. Meanwhile, Earth rotates on its axis once every 24 hours, exposing different regions of Earth's oceans to the moon's gravitational pull. During this rotational period, the moon completes 1/27 of its orbit around Earth. The combination of Earth's rotation and the moon's revolution around Earth exposes all areas of the ocean to the moon's direct gravitational pull once every 24 hours and 50 minutes.

Because of Earth's rotation on its axis, the bulges of the tides are slightly ahead of the moon's actual position. The friction between the oceans and Earth's surface slows the response of the oceans to the moon's gravitational pull. The sun also exerts a gravitational pull on the ocean, although its pull is only about 44% of the moon's pull. At times, the gravitational pull of the sun and moon work together to produce larger tides, yet at other times they are in opposition to each other.

Lesson 7.3.3 – Eclipses

During an eclipse, one celestial body casts its shadow on the surface of another. Solar and lunar eclipses are not the only eclipses seen from Earth. Binary stars can eclipse each other as they rotate around one another. Jupiter's moons regularly form eclipses during their orbits, and their eclipses are so regular that the Danish astronomer Ole Rømer used them to calculate the speed of light in 1675.

Both solar and lunar eclipses cast two shadows. The umbra is the dark, central portion of a shadow that completely blocks out the sun's light. The penumbra is an area of partially blocked light surrounding the complete shadow. During a solar eclipse, the moon comes between the sun and Earth and casts a narrow shadow on Earth. If the solar eclipse is total, then Earth passes through both the moon's umbra and penumbra. As a result, only the sun's corona is visible around the moon during a total eclipse. A total solar eclipse lasts only about 7.5 minutes because of the Earth's rotation speed. However, total solar eclipses last longer closer to the equator because of the angle of the sunlight striking Earth. Few people see a total solar eclipse because the range of the eclipse is relatively narrow. Those who do experience a total solar eclipse will notice that stars become visible and the temperature drops when the sun's light is blocked. If a solar eclipse occurs when Earth is at its apogee, which is when the Earth is farthest from the sun, then an annular eclipse is seen. During an annular eclipse, the rim of the sun is visible around the moon because the moon is too far from the sun to cover it completely. A complete solar eclipse in 1919 was used to verify Einstein's theory of relativity, because stars that should have been behind the sun were visible because of the way space and light bend in response to gravity.

A lunar eclipse occurs when Earth passes between the sun and the moon and casts a shadow on the moon. Lunar eclipses are visible to more people because Earth's shadow is large enough to cover the entire moon. In fact, Earth's shadow extends well beyond the moon. Because the size of Earth's shadow is significantly greater than the size of the moon's shadow, a lunar eclipse lasts much longer than a solar eclipse. A total lunar eclipse can last over an hour as the moon passes through Earth's penumbra. Although the sun becomes nearly invisible during a total solar eclipse, the moon may change color rather than disappear. Light refracting through Earth's atmosphere can make the moon's color vary from gray to copper red.

Eclipses do not occur at every new and full moon because the moon's orbit of Earth is tilted about 5° from the plane in which Earth orbits the sun. The difference in the orbital planes prevents all three celestial bodies from entering a perfect alignment more than about seven times per year. The moon and Earth have well-defined orbits. The Babylonians correctly determined that the sun, moon, and Earth will return to the same location every 18 years 11⅓ days. Therefore, eclipses can be predicted well in advance, and the dates for past eclipses can be accurately determined as well. If a historical event was marked using an eclipse, the date of the event can be calculated.

Lesson 7.3.4 – Seasons

Although it may seem logical to expect the warmest days of the year to occur when Earth is closest to the sun, Earth's distance from the sun has very little to do with its temperature or its seasons. At its aphelion, Earth is approximately 152.1 million km from the sun, and at its perihelion it is about 147.1 million km from the sun. With regard to Earth's seasons, the aphelion and perihelion shift over thousands of years. Currently, aphelion occurs in early July, and perihelion occurs in early January. Therefore, at aphelion the Northern Hemisphere experiences summer, but the Southern Hemisphere experiences winter. The seasons are reversed during perihelion.

The true cause of Earth's seasons is its axial tilt of approximately 23.4°. When sunlight strikes Earth at a steep angle, it travels through less atmosphere, so it retains most of its heat. During the June solstice, the sun strikes the Tropic of Cancer at a 90° angle, providing the strongest rays of sunlight to the Northern Hemisphere. During the December solstice, the sun strikes the Tropic of Capricorn at a 90° angle, giving the most warmth to the Southern Hemisphere. Furthermore, the hemispheres warm up or cool down in response to the current length of the day. Earth's tilt causes the sun to appear above the horizon longest for the Northern Hemisphere during the June solstice, and the sun stays above the horizon longest during the December solstice in the Southern Hemisphere.

The equator maintains a fairly constant temperature because the angle at which sunlight strikes it changes very little over the course of a year. The temperature in the hemispheres also evens out during the equinoxes, when the sun's rays strike the equator at a 90° angle. The sun's angle, in addition to the nearly equal periods of day and night during the equinoxes, ensures that temperatures are fairly mild in both the Northern and Southern Hemispheres.

The hottest and coldest days of the year do not typically occur during a solstice. At the time of the summer solstice, the ground has not had sufficient time to store up sufficient heat to raise temperatures to their maximum. Likewise, during the winter solstice, the ground has not lost sufficient heat to allow temperatures to drop to their yearly minimum.

7.3.1 Structure of the Moon

Sun, Earth, and Moon

OBJECTIVES

Students will be able to
- compare properties of the moon and Earth.
- simulate the creation of lunar regolith and craters.
- identify features on the moon's surface.

VOCABULARY

- **lunar regolith** a loose layer of rock and dust on the surface of the moon
- **mare** a flat, lowland plain on the moon's surface filled with hardened lava
- **rille** a channel on the moon's surface

MATERIALS

- Video clip of the first moon landing (C)
- TM 7.3.1A Moon Features
- WS 7.3.1A Moon Map

PREPARATION

- Gather materials for *Try This* activities. (Introduction)
- Obtain materials for *Lab 7.3.1A Moon Width* and *Lab 7.3.1B Regolith Formation*. Schedule *Lab 7.3.1A* when the full moon will be visible in daylight. For *Lab 7.3.1B*, prepare 1 ice cube that contains sand for each group.
- Obtain a video clip of the first moon landing. (C)

TRY THIS

Moon Gravity
- *sticky notes, several per group*
- *meterticks, 1 per group*
- *bathroom scales, 1 per group*

(**Possible answer: I could lift heavier objects and jump higher.**)

Make a List
No additional materials are needed.

Mapping the Moon
No additional materials are needed.
(*continued*)

Introduction

To introduce the difference in the force of gravity between the moon and Earth, have students complete *Try This: Moon Gravity*.

Discussion

- Guide a discussion of some of the differences between the moon and Earth by completing *Try This: Make a List* with students.
    1. Portable radio (**No. Sound waves cannot travel through a vacuum.**)
    2. Flashlight (**Yes. The moon has day and night cycles.**)
    3. Matches (**No. The moon has no oxygen, so matches would not burn.**)
    4. Down jacket (**Yes. The moon is very cold.**)
    5. Umbrella (**No. The moon has no atmosphere; therefore, it has no weather.**)
    6. Kite (**No. The moon has no atmosphere; therefore, it has no wind.**)
    7. Swimming suit (**No. The moon has no liquid water.**)
    8. Suntan lotion (**Yes. The sun shines on the moon.**)
    9. Compass (**No. The moon has no magnetic field, so a compass would not work.**)
    10. Seeds to plant (**No. The moon has no organic material in the soil, so the soil is infertile; it does not rain on the moon; the moon's atmosphere does not contain the components plants need, such as carbon dioxide.**)

- Ask the following questions:
    1. What differences exist between the moon and Earth because of the moon's smaller mass? (**The force of gravity is less on the moon than on Earth; the weaker gravity results in a thinner atmosphere, which keeps oxygen and liquid water from existing in amounts necessary to sustain life.**)
    2. About how much would an object weigh on the moon if its mass on Earth is 300 kg? (**50 kg**)
    3. What kind of erosion is possible on the moon? (**meteorites striking the lunar surface**)
    4. Which lunar surface features likely prompted past astronomers to think that water might exist on the moon? (**Maria look like bodies of water, and rilles appear very similar to stream beds.**)

Activities

Lab 7.3.1A Moon Width

- straight pins, 1 per group
- 10 cm × 7 cm cardboard pieces, 2 per group
- tape
- meterticks, 1 per group

Option: Assign students to complete this lab at home when the full moon will be visible in the night sky.

Lab 7.3.1B Regolith Formation

- slices of toasted wheat bread, 5 per group
- sandpaper pieces, 1 per group
- slices of toasted white bread, 2 per group
- trays, 1 per group
- ice cubes containing sand, 1 per group
- fist-sized rocks, 1 per group
- meterticks, 1 per group

A. Complete *Try This: Impact Craters* and *Try This: Mapping the Moon* with students. For *Try This: Mapping the Moon*, students may use **WS 7.3.1A Moon Map** to create their maps. The answer key for WS 7.3.1A identifies several of the moon's more prominent features. Have students provide verification of features not labeled on the WS 7.3.1A answer key.

B. Display **TM 7.3.1A Moon Features**. Point out major features visible on the near side of the moon. Inform students that maria have Latin names, lunar mountains are often named after mountains on Earth, and craters are frequently named after famous scientists and astronauts.

C. Show students a video clip of the first moon landing.

D. Have students creatively use various materials to make models of Earth and the moon with proportional diameters and distances between them.

E. Assign students to research theories regarding how and when the moon was created. Encourage students to present their findings to the class.

Lesson Review

1. Compare the properties of the moon to those of Earth. (**diameter 27.25% of Earth's; mass 1% of Earth's; smaller core; weaker gravity; crust layered like Earth's; weaker magnetic field; thinner atmosphere; no weather, wind, oxygen, or liquid water; wide daily temperature swings; no life**)
2. Why are oxygen and liquid water not found on the moon? (**The moon's atmosphere is too thin to retain oxygen and liquid water.**)
3. What do scientists believe formed the moon's craters and regolith? (**many meteorite strikes**)
4. Is it likely that the footprints left on the moon by astronauts are still there? (**Yes.**) Why? (**The moon has no wind or rain, and therefore does not experience erosion.**)
5. What are the large, lowland parts of the moon called? (**maria**)
6. What covers most of the maria and some of the moon's highlands? (**solid lava**)

> **TRY THIS**
> (continued from previous page)
> **Impact Craters**
> - boxes, 1 per group
> - large trash bags, 1 per group
> - flour, enough to fill each group's box 8 cm
> - powdered paint
> - marbles, several per group
> - objects of various sizes, several per group
>
> (**If the marbles or other objects broke on impact, then the debris would be scattered around the impact site with larger pieces creating small craters.**)

7.3.1 Structure of the Moon

OBJECTIVES
- Compare properties of the moon and Earth.
- Simulate the creation of lunar regolith and craters.
- Identify features on the moon's surface.

VOCABULARY
- **lunar regolith** a loose layer of rock and dust on the surface of the moon
- **mare** a flat, lowland plain on the moon's surface filled with hardened lava
- **rille** a channel on the moon's surface

A natural or artificial body that orbits a larger astronomical body is called a *satellite*. The moon is Earth's only natural satellite. Occasionally, it is called by its Latin name, *luna*, which is more commonly used to talk about the moon's features, such as its lunar surface.

The moon is a planned part of the heavens, declaring God's glory along with the rest of creation. God created the moon with a purpose in mind. It is a light for the night (Genesis 1:16–18). The moon's phases help mark the passage of time, and the moon generates the ocean's tides. However, the moon is very different from Earth.

What is the moon like? The moon's diameter is 27.25% of Earth's diameter. Its mass is only about 1% of Earth's. Part of the reason the moon has such a low mass is because it has a relatively small core. The moon has a weaker gravitational pull because it has less mass. If you threw a ball on the moon, it would fall six times more slowly than it would fall on Earth. You could lift six times more mass and jump six times higher on the moon than on Earth. You would also weigh one-sixth as much. The spacesuit and life support system that the astronauts used on the moon weighed about 113 kg on Earth. However, these did not weigh down the astronauts, because they weighed only about 18 kg on the moon.

The astronauts who walked on the moon discovered that the weak gravity affected their walking. They bounced a little with each step, causing their stride to be extra long. Such weak gravity can attract only an extremely thin atmosphere to the lunar surface. Because of its lack of atmosphere, the moon has no weather. It has no clouds, no rain, and no wind. Earth's atmosphere traps heat to regulate the temperature, but the moon experiences wide daily temperature swings. It is very hot during the day and very cold at night. Without an atmosphere, the moon also lacks oxygen and liquid water, which means that life cannot exist on the moon. The astronauts who explored the moon did not even find any evidence of bacterial life.

Like Earth, the moon consists of layers. The top layer of the moon is its crust, which is its thinnest layer. Underneath the crust is the mantle. The innermost layer is a small iron core. Earth's large magnetic core gives the planet a magnetic field,

Scale comparison of Earth and the moon

Earthrise from the moon

Clementine, a small spacecraft launched in 1994, found indirect evidence of ice in the Tycho Crater at the moon's southern pole.

> **TRY THIS**
> **Moon Gravity**
> To determine how high you can jump, hold a sticky note and face a blank wall. Then jump as high as you can and place the sticky note on the wall. Use a meterstick to measure the sticky note's height above the floor. Then weigh an object of your choosing. Calculate how high you could jump with it and how much the object would weigh on the moon. How would these differences affect you?

> **HISTORY**
> **First Footprint**
> "The surface is fine and powdery. I can kick it up loosely with my toe.... I can see the footprints of my boots and the treads in the fine sandy particles." —Neil Armstrong
>
> On July 20, 1969, Neil Armstrong left the first footprint on the moon.

FYI

Moon

Period of rotation
27 days, 8 hours

Period of revolution
27 days, 8 hours

Diameter
3,476 km

Density (water = 1)
3.34 g/cm³

Surface temperature
−153°C to 107°C

Surface gravity
16.5% of Earth's

TRY THIS

Make a List
Think about the differences between the moon and Earth. Explain whether each of the following items would be useful on the moon:

1. Portable radio
2. Flashlight
3. Matches
4. Down jacket
5. Umbrella
6. Kite
7. Swimming suit
8. Suntan lotion
9. Compass
10. Seeds to plant

which is what makes compasses point north. The magnetic field on the moon, however, is almost nonexistent; it is 10,000 times weaker than Earth's.

Until *Apollo 11* landed on the moon in 1969, scientists had only theorized what the surface of the moon was like. But as they suspected, the surface was covered with **lunar regolith**, a loose layer of rock and dust. Regolith consists of tiny rock fragments that scientists believe were broken up by the impact of many meteorites. Only the largest meteorites could reach Earth's surface without being completely burned up by atmospheric friction, but on the moon, meteorites can strike the moon's surface at over 48,000 kph. Such an impact pulverizes both the

This thin rock sample from the moon was brought back to Earth by the Apollo 12 mission. The picture was taken using a microscope with a special light to show the different colors and textures of the minerals.

Craters and a rille on the moon

meteorite and the moon's surface, creating dust. Most lunar regolith particles are the size of silt or sand, but some are the size of pebbles.

The Apollo missions to the moon gathered huge amounts of data for scientists on Earth to sift through. Scientists have analyzed 380 kg of moon rock and soil samples, along with photographs and other data sent back to Earth from space instruments. Scientists now know that the moon is made mostly of silicates and oxides that are similar to those found on Earth.

The moon's plains show signs of faulting. Seismic instruments set up by astronauts have registered tremors on the moon's surface called *moonquakes*. Along with rock and soil samples, this data helps scientists form theories about what the inside of the moon is like, what the moon used to be like, and what it might be like in the future.

Without wind or rain to cause weathering on the moon, its surface stays the same. Most changes that do occur on the moon's surface are a result of rock or metal meteorite strikes that form craters of various sizes. Especially large impacts produced features such as the Mare Imbrium, or the Sea of Showers.

TRY THIS

Impact Craters
Line a box with a trash bag and add about 8 cm of flour. Dust the flour surface with powdered paint. Drop marbles one at a time to bombard the surface. Sketch each crater that forms. Drop objects of various sizes from different heights and angles. Sketch the craters. How would the craters change if the marbles or other objects broke on impact?

The moon has dark and light regions. People once thought that the dark regions were seas; these are still called by their Latin name, *mar* meaning "sea." But the moon has no liquid water. A **mare** (pronounced "mah'-ray") is a flat, lowland plain containing hardened lava. Many of the maria (plural for mare) formed when large meteorites created vast impact craters. Lava then welled up and flowed into these craters.

Channels called **rilles** run across the moon. Some rilles wind across the moon's surface. These rilles were formed by the flow of magma on the maria. Other rilles are fairly straight; scientists are not sure how they formed.

The dark areas on the moon are maria, and the light areas are highlands.

The light areas on the moon that are visible from Earth are the rough, pockmarked highlands. Some of the highlands are covered by a type of lava associated with explosive volcanic eruptions. The highlands are all peppered with craters and covered in a layer of broken rock from meteorites.

LESSON REVIEW

1. Compare the properties of the moon to those of Earth.
2. Why are oxygen and liquid water not found on the moon?
3. What do scientists believe formed the moon's craters and regolith?
4. Is it likely that the footprints left on the moon by astronauts are still there? Why?
5. What are the large, lowland parts of the moon called?
6. What covers most of the maria and some of the moon's highlands?

FYI

Many Moons
Even though people talk about Earth's moon as though it were the only one, God created the solar system with more than 173 moons. Some of their diameters can be measured in tens of meters. Others are bigger than the smallest planets. Ganymede, one of Jupiter's moons, is larger than Mercury. The solar system's moons represent the variety in God's creation.

TRY THIS

Mapping the Moon
You can look at the moon on a clear night and identify its features. Impact craters are most visible during or near the first or last quarter phases when the moon looks like a filled semicircle. When the moon is near first or last quarter, draw the visible portion of the moon on a large circle. Include any features that you can see with the unaided eye. Then look at the moon through binoculars or a small telescope and complete the drawing. Consult a labeled map of the moon and identify the more prominent features.

Name: _____ Date: _____

Lab 7.3.1A Moon Width

QUESTION: How wide is the moon?

HYPOTHESIS: Answers will vary.

EXPERIMENT:

You will need:	• two 10 cm × 7 cm cardboard rectangles	• tape
• straight pin		• meterstick

Steps:
1. Poke a pinhole into the center of one piece of cardboard.
2. Tape the cardboard to the 0 mark on a meterstick in such a way that the pinhole is visible when you hold the meterstick in front of you.
3. Create a 0.5-centimeter hole in the center of the second piece of cardboard.
4. Hold the meterstick with the 0 mark away from you. Look at the moon through the pinhole in the cardboard taped to the meterstick. Move the second card back and forth along the meterstick until the moon fits exactly inside the 0.5 cm hole.
5. Record the distance in centimeters between the two cards on the meterstick.

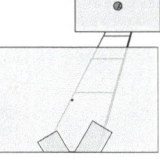

Answers will vary but should be approximately 55 cm.

ANALYZE AND CONCLUDE:
1. The average distance between Earth and the moon is about 382,500 km. Use the following equation to determine the diameter of the moon to the nearest kilometer: moon's diameter × distance between cards = cut circle's diameter × distance from Earth to moon. Answers will vary but should be approximately 3,476 km.

Lab 7.3.1A Moon Width

2. Compare your answer in Question 1 with answers your classmates obtained. Why might the answers be somewhat different? Possible answers: Holes in the second card were not all cut to exactly 0.5 cm in diameter; people made different determinations of when the moon fills the 0.5 cm hole; people made mathematical errors.

3. The diameter of the moon is nearly 27.25% the diameter of Earth, which is about 12,756 km. Calculate the actual diameter of the moon to the nearest kilometer.
3,476 km

4. Compare your answers in Questions 1 and 3. How accurate was your measurement? Why might differences exist? Answers will vary but should include that the answers in Questions 1 and 3 are close but not exact. Differences may exist because the sighting of the moon approximates the actual distance between Earth and the moon on the day of the sighting, but the calculation performed in Question 1 uses the average distance between Earth and the moon; therefore, the calculation of the moon's diameter in Question 1 will likely be somewhat different from its actual diameter found in Question 3.

5. How could you obtain more accurate results if you repeated the experiment? Possible answers: I could obtain a very sharp hole punch that is exactly 0.5 cm in diameter, I could rest the meterstick on a stable surface while sighting on the moon, I could look up the moon's exact distance from Earth on the day of the sighting.

Name: _____ Date: _____

Lab 7.3.1B Regolith Formation

QUESTION: How does moon regolith form?
HYPOTHESIS: Answers will vary.

EXPERIMENT:

You will need:		
• 5 slices of toasted wheat bread	• sandpaper	• fist-sized rock
• tray	• ice cube containing sand	• meterstick
	• 2 slices of toasted white bread	

Steps:

1. Hold a slice of wheat toast above a tray and blow on the toast. Describe what falls into the tray. Answers will vary but should include a few crumbs fell off the toast.

2. Rub the slice of wheat toast from Step 1 with sandpaper over a tray. Describe what falls into the tray. Answers will vary but should include that a great number of crumbs fell off the toast.

3. Hold an ice cube that contains sand underneath running water until it melts. Describe the remaining particles. Answers will vary but should include that the particles settled in the students' hands as the ice cube melted; some particles will likely wash away.

4. Place 2 slices of white toast on top of 4 slices of wheat toast on a tray.

5. Hold a rock 30 cm above the stack of toast and drop it. Describe the bread slices and the crumbs. Answers will vary but should include that parts of each slice of toast were crushed, but the top 2 slices of white toast were crushed the most; some bread crumbs likely scattered across the tray, but most will have stayed near the impact site.

Lab 7.3.1B Regolith Formation

6. Drop the rock onto the layers of toast 25 times. Which crumbs are visible at the surface? Why? Compare the thickness of the crumb layers now to the results noted following Step 5. Possible answers: More white crumbs are visible at the surface because white slices of toast were on top when the rock was dropped; the thickness of the crumb layers surrounding the impact site is deeper after successive drops than after one drop, but the thickness of toast at the impact site itself is less after successive drops because of compression.

ANALYZE AND CONCLUDE:

1. What processes in this activity represent how regolith is formed on Earth?
blowing on the toast, rubbing the toast with sandpaper, holding the ice cube under running water

2. What processes represent how regolith is formed on the moon?
dropping the rock on the toast

3. What process is not a factor of regolith formation on the moon? Why?
Erosion is not part of regolith formation because there is no wind or water to move particles around.

7.3.2 Phases of the Moon

Sun, Earth, and Moon

OBJECTIVES

Students will be able to
- explain why only one side of the moon can be viewed from Earth.
- identify and define the different phases of the moon.
- simulate the moon's orbit of Earth as Earth orbits the sun.

VOCABULARY

- **apogee** the point in the moon's orbit at which it is farthest from Earth
- **full moon** the phase when the entire near side of the moon is illuminated
- **new moon** the phase when the moon is directly between Earth and the sun
- **perigee** the point in the moon's orbit at which it is closest to Earth
- **phase** each different shape of the moon made visible by reflected sunlight
- **waning moon** the phase after the full moon and before the new moon, when its appearance is shrinking
- **waxing moon** the phase after the new moon and before the full moon, when its appearance is growing

MATERIALS

- Stool (*Introduction*)
- BLM 7.3.2A Instructions: Moon Phases Board
- TM 7.1.2A Moon Phases
- WS 7.3.2A Going Through a Phase

PREPARATION

- Assign *Try This: Moonrise* about seven days prior to this lesson.
- Obtain materials for *Try This: Phases*.

TRY THIS

Moonrise
No additional materials are needed.

Phases
- ball
- lamp
(0°, 90°, 180°, 270°)

Introduction

Demonstrate the moon's orbit around Earth with the same hemisphere always toward Earth. First walk around a stool or chair, continually facing it. Ask students how many rotations they observed. (**1**) How many revolutions did you observe? (**1**) Walk around the stool, continually facing the same direction. Ask students how many rotations they observed. (**0**) How many revolutions did you observe? (**1**) (Note: If students have difficulty seeing the difference in rotations, repeat the demonstration without the stool, which will make the rotation and nonrotation more evident.) Ask students how much of the moon could be seen from Earth if the moon did not rotate. (**We could see all of the moon during the month.**)

Discussion

- Lead a discussion on the apparent changing size of the moon. Ask students how the full moon appears when it is just rising above the horizon as compared to when it is higher in the sky. (**Answers will vary but should include the moon appears much larger closer to the horizon.**) Explain that when the moon is lower in the sky, it appears much larger, and it appears much smaller at a higher altitude. The changing size of the moon is an optical illusion; observers compare the moon on the horizon to closer objects and perceive it to be larger, but higher in the sky, observers have no objects for comparison. Challenge students to hold up a dime to the next full moon when it is just rising and again when it is high in the sky and to compare their observations.

- Discuss why moonrise and moonset may not be exactly 50 minutes later on consecutive days. Explain that on average, moonrise or moonset is 50 minutes later than the previous day, but the timing varies depending on a location's latitude. In some places, the moon may rise in less than 50 minutes on consecutive days; in other places, it may take longer. Encourage students to brainstorm about why there may be differences in the timing. (**Answers will vary.**) Point out that the moon's tilt on its axis is different from Earth's tilt. Explain that Earth's plane of revolution around the sun and the moon's plane of revolution around the earth are at different angles; there is a difference of nearly 29°. In addition, Earth's orbit is almost circular and the moon's is elliptical, which also affects the timing. The moon does not travel at a constant speed because of its elliptical orbit; the difference between its rising will vary from the average depending on where the moon is.

- Ask the following questions:
  1. Is it correct to refer to the far side of the moon as *the dark side of the moon*? (**No.**) Why? (**Half of the moon is light at all times, but people cannot always see all the reflected light because of the moon's position in relation to the sun and Earth.**)
  2. Approximately what angle is formed by the sun, Earth, and the moon when the moon is in its first quarter phase? (**90°**)
  3. Approximately how long would one side of the moon experience daylight? (**14 days**) Why? (**The moon takes about 27.3 days to complete one rotation, and any given point is in daylight only about half that time.**)

Activities

A. Complete *Try This* activities with students. For *Try This: Moonrise*, ask them what the difference was between each night's moonrise. (**Answers will vary depending on the season.**) Direct a student to sketch an ellipse on the board with a focus near one end. Ask students what happens when an object travels toward its perigee. (**It speeds up.**) What happens as an object travels away from its perigee? (**It slows down.**) Remind students that the variations in the moon's speed affect the timing of its appearance at night.

B. Use **BLM 7.3.2A Instructions: Moon Phases Board** to construct a moon phases board to demonstrate the view of the moon from Earth. Optionally, have student groups create the board.

C. Assign students to complete **WS 7.3.2A Going Through a Phase**.

D. Display **TM 7.1.2A Moon Phases**. Have students identify images displaying the following phases: *waxing crescent*, *first quarter*, *waxing gibbous*, and *full moon*. Ask students why the new moon is not depicted. (**Possible answers: It cannot be seen; the image would just be black.**) Direct students to sketch similar moon phases but to go from the full moon to the new moon. Ensure that students draw the bright portion on the left and the dark portion on the right.

E. Direct groups of three students to simulate the orbit of the moon around the Earth as Earth orbits the sun. Have every student in each group take turns being the sun, the moon, and Earth.

F. Challenge students to research the laser technology used to make accurate measurements of the distance between Earth and the moon.

Lesson Review

1. Why do people always see the same side of the moon? (**The moon rotates on its axis at the same rate it orbits Earth.**)
2. Why does the moon have phases? (**As the moon orbits Earth, how much of each side people see depends on the relative positions of the sun, moon, and Earth.**)
3. Name the eight phases of the moon. (**new moon, waxing crescent, first quarter, waxing gibbous, full moon, waning gibbous, last quarter, waxing crescent**)
4. How much of the moon can people see when the moon is directly between Earth and the sun? (**none of it**)

NOTES

be seen because all of its light reflects away from Earth. As the moon orbits Earth, it moves away from being in line with the sun and Earth. This movement allows people to see some of the sun's light reflecting off the moon's near side. Eventually, the moon moves until Earth is between it and the sun at which time people see the **full moon**, which is the phase when the entire near side of the moon is illuminated.

After the new moon and before the full moon, when the moon's appearance is growing, is a **waxing moon**. When a sliver of the moon becomes visible, the moon is said to be in its *waxing crescent phase*. As the moon eventually moves through one-quarter of its total orbit around Earth, it looks like a filled semicircle of light. The phase after the new moon when the moon appears as this filled semicircle of light is called *the first-quarter phase*. As the moon continues on its orbit, the visible part becomes larger than a semicircle. This stage is called *the waxing gibbous phase*. This term comes from the Latin word *gibbus*, which means "hunchback."

The full moon marks the halfway point of the moon's orbit around Earth. Once past the full moon, the lighted portion of the moon steadily decreases. The moon after the full moon and before the new moon, when its appearance is shrinking, is called a **waning moon**. The stage in which the moon appears to be shrinking but still appears larger than a semicircle is called *the waning gibbous phase*.

When the moon has completed three-quarters of its orbit, it once again becomes a semicircle of light. The phase after the full moon when the moon appears as a semicircle of light is called *the last-quarter phase*. From here, the lighted portion appears to shrink down to a sliver, called *the waning crescent phase*. After this stage, the moon moves between Earth and the sun to become a new moon and begin another revolution around Earth.

LESSON REVIEW
1. Why do people always see the same side of the moon?
2. Why does the moon have phases?
3. Which phase is the moon in if it is getting bigger and appears more than half-full?
4. How much of the moon can people see when the moon is directly between Earth and the sun?

TRY THIS
Phases
Demonstrate the phases of the moon using a ball to represent the moon and a bright lamp to represent the sun. Since people see the moon from Earth, your head will represent Earth. At about what angles are the sun, Earth, and the moon lined up during the new moon, first-quarter phase, full moon, and last-quarter phase?

FYI
Feeling Blue
The phrase "once in a blue moon" refers to something that rarely happens. A couple of different interpretations for blue moon exist, but both pertain to full moons. The modern interpretation of the phrase is that a blue moon is the second full moon in one month. In contrast, the earlier interpretation is that a blue moon is the third full moon when there are four full moons in a season. Since neither event happens very often, the expression means the same regardless of how *blue moon* is interpreted.

Sun, Earth, and Moon

7.3.3 Eclipses

Introduction
Display a large beach ball on a stool or table near the center of the classroom. Distribute a softball to each student or small group of students. Direct students to hold their balls at arm's length and move around the room until they can block the beach ball from their vision using the softball. (Note: If students experience difficulty seeing the eclipse, direct them to cover one eye.) Have students use metric tape measures to determine the distance between the beach ball and their softballs. Repeat the activity using golf balls. Next, direct students to stand approximately 2 m away from the beach ball and use the softball and the golf ball to eclipse the beach ball by moving the balls closer to or farther away from their faces. Have students use metersticks to measure the approximate distance between each ball and their eyes.

Display **TM 7.3.3A Solar Eclipse Phases** and **TM 7.3.3B Lunar Eclipse Phases** to illustrate the phases of both a solar and a lunar eclipse. Inform students that these images represent the sun and the moon respectively becoming totally eclipsed and then moving out of eclipse.

Discussion
- Ask students how the distance between the beach ball and their softballs compared to the distance between the beach ball and their golf balls during the eclipse activity in *Introduction*. (**The golf balls had to be farther away from the beach ball than the softballs in order to eclipse the beach ball.**) What connection does this activity have to solar and lunar eclipses. (**Possible answer: Earth is larger than the moon, so it eclipses the sun more easily than the moon.**) How did the distance between your face and the softball and golf ball compare when you stood about 2 m from the beach ball to form an eclipse? (**The golf ball had to be closer to my face than the softball.**) How does this activity relate to Earth and the moon during eclipses? (**Possible answer: The moon will eclipse the sun more easily when it is closest to Earth.**)

- Ask students what the dark shape represents in TM 7.3.3A and TM 7.3.3B displayed in *Introduction*. (**In TM 7.3.3A, the dark shape is the moon's shadow, and in TM 7.3.3B the dark shape is the Earth's shadow.**) In what phase is the moon in TM 7.3.3A? (**new moon**) How do you know? (**because solar eclipses can only occur during a new moon**) In what phase is the moon in TM 7.3.3B? (**full moon**) How do you know? (**because lunar eclipses can only happen during a full moon**)

- Guide students to compare the phases of the solar and lunar eclipses displayed in *Introduction*. Have students especially notice colors, shapes, and the degree to which the sun and the moon are visible at the height of their respective eclipses.

Activities
A. Complete *Try This* activities with students. After students complete *Try This: Figure Out Earth's Shape*, have them explain whether a circular shadow is the only evidence a person would need to show that Earth is a sphere. (**A circular shadow is insufficient evidence that Earth is a sphere because a flat disc can also make a circular shadow.**)

B. Direct students to complete **WS 7.3.3A Solar Eclipse Diagram** and **WS 7.3.3B Lunar Eclipse Diagram**. Display **TM 7.3.3C Solar Eclipse** and **TM 7.3.3D Lunar Eclipse** for students to check their answers. Ask students what would happen if the moon and Earth were farther from the sun in the solar eclipse diagram. (**The moon's shadow might not reach Earth.**) Would the Earth's distance to the sun affect a lunar eclipse? (**Yes, but only if Earth was considerably farther from the sun.**)

C. Distribute a softball, golf ball, and flashlight to each group of students and explain that they represent Earth, the moon, and the sun respectively. Direct students to use the objects to model

OBJECTIVES
Students will be able to
- label the parts of a solar and a lunar eclipse.
- distinguish among different types of eclipses.
- model a solar and a lunar eclipse.

VOCABULARY
- **annular eclipse** a solar eclipse during which the outer ring of the sun is visible around the moon
- **eclipse** the casting of one celestial body's shadow on the surface of another celestial body
- **lunar eclipse** an event that occurs when Earth passes directly between the sun and the moon, causing Earth's shadow to block the sun's light from the moon
- **penumbra** an area of partially blocked light surrounding the complete shadow
- **solar eclipse** an event that occurs when the moon passes directly between the sun and Earth, causing the moon's shadow to block the sun's light from a portion of Earth
- **umbra** the dark, central portion of a shadow that completely blocks the sun's light

MATERIALS
- Beach ball, stool, softballs, metric tape measures, golf balls, metersticks, (*Introduction*)
- Softballs, golf balls, flashlights(C)
- TM 7.3.3A Solar Eclipse Phases
- TM 7.3.3B Lunar Eclipse Phases
- TM 7.3.3C Solar Eclipse
- TM 7.3.3D Lunar Eclipse
- WS 7.3.3A Solar Eclipse Diagram
- WS 7.3.3B Lunar Eclipse Diagram

PREPARATION
- Obtain materials for *Try This* activities.

TRY THIS

Figure Out Earth's Shape
- paper circle cutouts, 1 per group
- tennis balls, 1 per group
- various small objects of different shapes, several per group
- lamp

(**Circular objects held with their flat side to the light and spherical objects make circular shadows. The flat side of the cutout makes a circular shadow when held to the light, but its edge makes a straight line.**)

Create an Eclipse
- small bouncy balls, 1 per group
- tennis balls, 1 per group
- flashlights, 1 per group

(**It is a new moon because it must be between the sun and Earth; it is a full moon because the earth must be between it and the sun.**)

---

the following eclipses: *total solar eclipse*, *partial solar eclipse*, *annular solar eclipse*, *total lunar eclipse*, and *partial lunar eclipse*. Direct students to sketch what the sun and moon look like during their respective eclipses and to diagram the positions of the three objects for each eclipse.

D. Encourage students to research when the next solar and lunar eclipses will be visible.

E. Assign students to research how one of the following cultures explained eclipses: Chinese, Persian, Greek, Roman, Maya, Sioux, Aborigines, or Tahitian.

Lesson Review

1. Distinguish between an umbra and a penumbra and identify the types of eclipses each produces. (**An umbra is the dark, central portion of a shadow; a penumbra is an area of partially blocked light surrounding the umbra. The umbra produces a total eclipse, but the penumbra produces a partial eclipse.**)

2. Why do lunar eclipses last longer than solar eclipses? (**Earth's shadow, which produces a lunar eclipse, is much larger than the moon's shadow, which produces a solar eclipse, and Earth's rotation carries it relatively quickly through the moon's shadow.**)

3. What is the difference between an annular solar eclipse and a partial solar eclipse? (**During an annular solar eclipse, the outer ring of the sun is visible around the moon, so the moon does not completely block out the sun. A partial eclipse occurs when a portion of Earth only passes through the moon's penumbra, which causes the sun to appear as if only part of it is covered.**)

4. How are the sun, moon, and Earth arranged in both a solar eclipse and a lunar eclipse? (*solar eclipse*: **sun, moon, Earth**; *lunar eclipse*: **sun, Earth, moon**)

7.3.3 Eclipses

OBJECTIVES
- Label the parts of a solar and a lunar eclipse.
- Distinguish among different types of eclipses.
- Model a solar and a lunar eclipse.

VOCABULARY
- **annular eclipse** a solar eclipse during which the outer ring of the sun is visible around the moon
- **eclipse** the casting of one celestial body's shadow on the surface of another celestial body
- **lunar eclipse** an event that occurs when Earth passes directly between the sun and the moon, causing Earth's shadow to block the sun's light from the moon
- **penumbra** an area of partially blocked light surrounding the complete shadow
- **solar eclipse** an event that occurs when the moon passes directly between the sun and Earth, causing the moon's shadow to block the sun's light from a portion of Earth
- **umbra** the dark, central portion of a shadow that completely blocks the sun's light

The sun sheds light in all directions. Celestial bodies such as moons and planets are lit up on the side that faces the sun. These celestial bodies cast long shadows on their opposite sides. An **eclipse** is the casting of one celestial body's shadow on the surface of another celestial body. When the second body becomes completely covered by the shadow, it seems to disappear. This phenomenon is similar to watching someone on an evening walk step from a well-lit street into a building's shadow.

A **solar eclipse** is the casting of the moon's shadow on the Earth's surface. This event occurs when the moon passes between the sun and Earth and part of the moon's shadow falls on Earth. Shadows cast by Earth and the moon have two parts. The **umbra** is the dark, central portion of a shadow that completely blocks the sun's light. Surrounding the umbra is the **penumbra**, an area of partially blocked light surrounding the complete shadow.

The umbra of the moon's shadow falls on a very small region. People within the umbra experience a total solar eclipse in which the moon blocks out all sunlight. Because the umbra is so small, total solar eclipses for a given area are rare. Earth's rotation causes the moon's shadow to move quickly, so a total solar eclipse rarely lasts more than seven minutes at any one place. During a total solar eclipse, the sky becomes so dark that only stars and planets can be seen, and the temperature may drop 10° or more. Around the dark shadow of the moon, the sun's corona glows with a pearly luster. In contrast, the penumbra of the moon's shadow creates a partial solar eclipse. During a partial solar eclipse, the sky only dims and sunlight may take on a slightly different hue.

Solar Eclipse

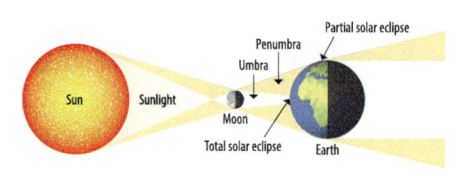

The new moon passes between the sun and Earth during a solar eclipse.

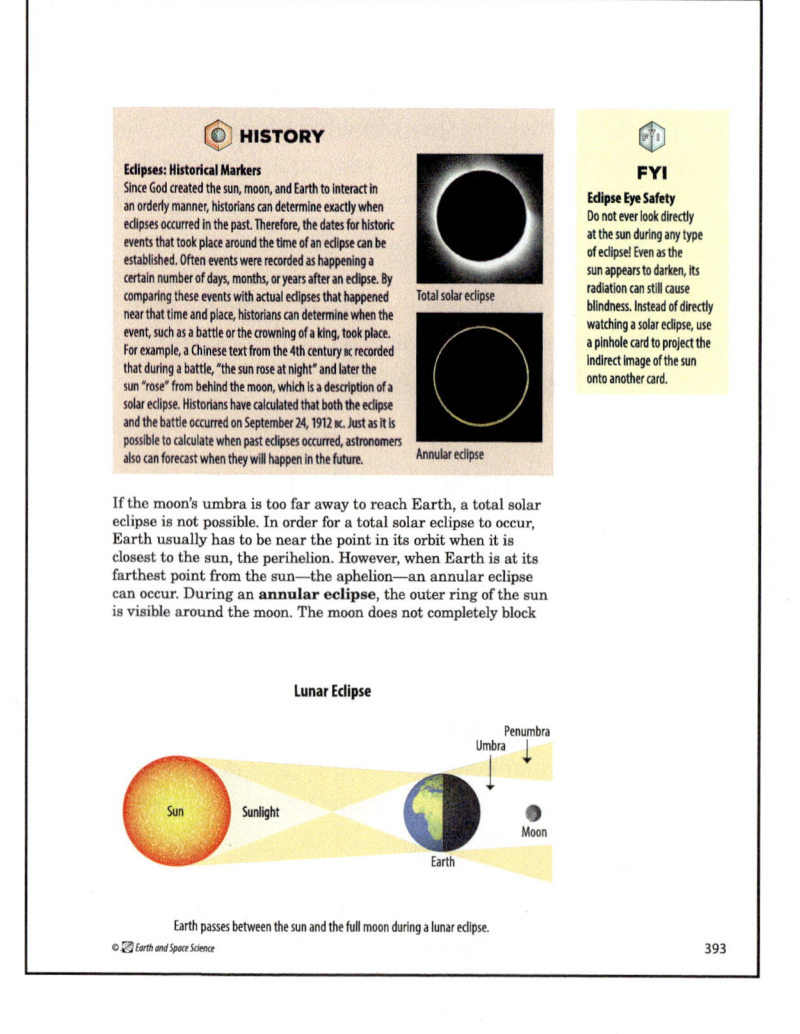

HISTORY

Eclipses: Historical Markers
Since God created the sun, moon, and Earth to interact in an orderly manner, historians can determine exactly when eclipses occurred in the past. Therefore, the dates for historic events that took place around the time of an eclipse can be established. Often events were recorded as happening a certain number of days, months, or years after an eclipse. By comparing these events with actual eclipses that happened near that time and place, historians can determine when the event, such as a battle or the crowning of a king, took place. For example, a Chinese text from the 4th century BC recorded that during a battle, "the sun rose at night" and later the sun "rose" from behind the moon, which is a description of a solar eclipse. Historians have calculated that both the eclipse and the battle occurred on September 24, 1912 BC. Just as it is possible to calculate when past eclipses occurred, astronomers also can forecast when they will happen in the future.

Total solar eclipse

Annular eclipse

FYI

Eclipse Eye Safety
Do not ever look directly at the sun during any type of eclipse! Even as the sun appears to darken, its radiation can still cause blindness. Instead of directly watching a solar eclipse, use a pinhole card to project the indirect image of the sun onto another card.

If the moon's umbra is too far away to reach Earth, a total solar eclipse is not possible. In order for a total solar eclipse to occur, Earth usually has to be near the point in its orbit when it is closest to the sun, the perihelion. However, when Earth is at its farthest point from the sun—the aphelion—an annular eclipse can occur. During an **annular eclipse**, the outer ring of the sun is visible around the moon. The moon does not completely block

Lunar Eclipse

Earth passes between the sun and the full moon during a lunar eclipse.

HISTORY

Stop Fighting
Predicted by the Greek astronomer Thales, one of the most famous eclipses in ancient times (calculated to have occurred on May 28, 585 BC) ended a five-year war between two Middle Eastern armies. The historian Herodotus recorded that the Lydians and the Medes were engaged in battle when "the day turned to night" during a solar eclipse. The eclipse persuaded both armies to make peace.

TRY THIS

Figure Out Earth's Shape
Although various people throughout history believed that Earth was flat, Greek scientist and philosopher Aristotle (384–322 BC) used his observations of eclipses to support the argument that Earth is a sphere.

Obtain a paper circle cutout, a tennis ball, and various small objects of different shapes, such as wooden blocks or pens. Hold each object in front of a lamp so its shadow falls on the circle of white paper. Rotate the object in different directions. Record the shapes cast by the shadows. Which objects always make a circular shadow? What shapes do the shadows of the circular paper cutout make?

the sun during an annular eclipse. Instead, a beautiful, bright ring of light around the edge of the moon's shadow is visible.

A **lunar eclipse** is the casting of Earth's shadow on the moon. During a lunar eclipse, Earth's shadow moves across the lighted moon. A lunar eclipse only happens during a full moon. Although the moon loses its direct light from the sun during an eclipse, it does not get completely dark. Sunlight bends as it passes through Earth's atmosphere, causing the moon's color to range from gray to a copper color. Lunar eclipses can last for several hours because Earth's shadow is much larger than the moon's shadow.

When the moon orbits completely into Earth's umbra, a total lunar eclipse occurs. A partial lunar eclipse happens when only a portion of the moon passes through Earth's umbra. Another form of lunar eclipse, called *a penumbral eclipse*, occurs when the moon is shadowed by only the Earth's penumbra. During a penumbral eclipse, the moon becomes only slightly darker because the sun's light is not completely blocked.

Although there are more solar eclipses than lunar eclipses each year, most people see more lunar eclipses. During solar eclipses, the moon's shadow covers a small portion of Earth, which means that solar eclipses are visible only along a

Sunlight bends as it passes through Earth's atmosphere during a full lunar eclipse.

TRY THIS

Create an Eclipse
Darken the room. Use a small bouncy ball to represent the moon, a tennis ball to represent Earth, and a flashlight to represent the sun. Manipulate these objects to create the conditions necessary for observers on Earth to experience a partial solar eclipse and a total solar eclipse. Describe or sketch these conditions. What phase must the moon be in for a solar eclipse? Why?

Next, manipulate the sun, moon, and Earth to create a lunar eclipse. What phase must the moon be in for a lunar eclipse? Why? Describe or sketch the conditions necessary for a total lunar eclipse and a partial lunar eclipse.

narrow path. In contrast, lunar eclipses are visible to everyone on the dark half of Earth because Earth's shadow can completely cover the moon.

Why are eclipses not more common? The moon passes through the plane of Earth's orbit at least twice a year. However, the sun, Earth, and moon have to be aligned for an eclipse to form. A solar eclipse happens if the moon crosses Earth's orbital plane when the moon is between the sun and Earth. A lunar eclipse occurs if the moon crosses Earth's orbital plane when Earth is between the sun and the moon. Usually the sun, the moon, and Earth are not directly lined up when the moon's orbit crosses Earth's, so no eclipse occurs.

LESSON REVIEW
1. Distinguish between an umbra and a penumbra and identify the types of eclipses each produces.
2. Why do lunar eclipses last longer than solar eclipses?
3. What is the difference between an annular solar eclipse and a partial solar eclipse?
4. How are the sun, moon, and Earth arranged in both a solar eclipse and a lunar eclipse?

Stages of a lunar eclipse

BIBLE CONNECTION

Miraculous Darkness
Solar eclipses occur only during a new moon. Some people believe that the darkness that covered Earth during Jesus's crucifixion was caused by a solar eclipse. However, Jesus was crucified during Passover, which always occurs during a full moon.

7.3.4 Seasons

Sun, Earth, and Moon

OBJECTIVES

Students will be able to
- model the way in which Earth's tilt and revolution around the sun cause different seasons.
- identify Earth's primary lines of latitude.
- differentiate between a solstice and an equinox.

VOCABULARY

- **equinox** a point in Earth's orbit at which the sun crosses Earth's equator, causing the hours of day and night to be nearly equal everywhere on Earth
- **solstice** one of the two days of the year in which the sun's most direct rays reach farthest north or farthest south

MATERIALS

- Thin dowel rods; large Styrofoam balls; purple, blue, red, yellow, and orange paints; paint brushes; lamp with bare bulb; metersticks; protractors (*Introduction*)
- TM 7.3.4A June Solstice
- TM 7.3.4B December Solstice
- TM 7.3.4C Earth's Seasons
- WS 7.3.4A Changing Seasons

PREPARATION

- Obtain materials for *Try This: Change the Seasons*.

TRY THIS

Change the Seasons
- globes, 1 per group
- lights, 1 per group
- thermometers, 1 per group

(The sun's rays at noon feel hotter. The angle at which they strike the earth is steeper close to sunset.)

Introduction

Guide student groups to model and sketch Earth's seasons as the planet revolves around the sun. Direct students to insert a thin dowel rod through the center of a large Styrofoam ball until the dowel's ends protrude from either side of the ball. Have students approximate the Arctic Circle with purple paint, the Tropic of Cancer with blue paint, the equator with red paint, the Tropic of Capricorn with yellow paint, and the Antarctic Circle with orange paint. Turn on the lamp with the bare bulb. Direct students to hold their models with the axes straight up and down near the lamp and at the same time to rotate the models counterclockwise. Then direct students to rotate their models and simultaneously slowly walk around the lamp at a radius of about 1 m. Have students sketch how the light shines on their models.

Guide students to use protractors to tilt their models about 23° on their axes. Have students select a fixed point at which to keep their axes pointed as they carry their models in a revolution around the lamp. Direct students to sketch and label the light shining on their models in the following scenarios:
- When the Arctic Circle is tilted toward the bulb, label the sketch *June solstice*.
- When the Antarctic Circle is tilted toward the bulb, label the sketch *December solstice*.
- When one side of the model is directly opposite the bulb, label the sketch *Spring* or *Autumnal equinox* (depending on the school's geographic location).
- When the other side of the model is directly opposite the bulb, label the sketch *Autumnal* or *Spring equinox* (depending on the school's geographic location).

Discussion

- Direct students to refer to their sketches and the activity in *Introduction* to answer the following questions:
  1. How was the illuminated portion of the models with upright axes different from the illuminated portion of the models with tilted axes? (**Possible answer: The light shone evenly over both hemispheres when the axes were upright, but when the axes were tilted, the light shone more on one hemisphere or another depending on whether the axes were pointed toward or away from the light.**)
  2. How would the difference in illumination affect Earth's temperatures if Earth's axis was upright? (**The temperature in a region would be fairly constant because it would receive the same amount of sunlight every day.**)
  3. When do places north of the equator experience longer days? (**when Earth is moving from the equinox in March toward the equinox in September**)
  4. When do places south of the equator experience longer days? (**when Earth is moving from the equinox in September toward the equinox in March**)
  5. Why does the length of a day at the equator change very little? (**The angle at which the sun strikes the equator changes very little regardless of where Earth is in its orbit.**)

Activities

A. Complete *Try This: Change the Seasons* with students.

B. Display **TM 7.3.4A June Solstice** and **TM 7.3.4B December Solstice**. Direct students' attention to the lines of latitude and the number of daylight hours each latitude experiences during each solstice. Ask students how seasons during each solstice would be affected if Earth was tilted more than 23.4° on its axis. (**Possible answers: Summer and winter in each hemisphere would experience more severe temperature changes; places near the equator would experience seasonal changes.**)

C. Display **TM 7.3.4C Earth's Seasons**. Select student volunteers to identify the solstices, equinoxes, and seasons in both the Northern and Southern Hemispheres relevant to each solstice and equinox. Ask students how seasons during each equinox would be affected if Earth was tilted more than 23.4° on its axis. (**The seasons would be more extreme if Earth's tilt increased.**)

D. Assign students to complete **WS 7.3.4A Changing Seasons**.

E. Challenge students to design an experiment to determine which affects temperature on Earth more: Earth's distance from the sun or the angle at which sunlight strikes Earth.

F. Have students research ancient structures that were built to align with the sun's position during either a solstice or an equinox.

Lesson Review

1. How much does Earth's distance from the sun affect temperatures on Earth? (**Earth's distance from the sun does not affect Earth's temperatures.**)
2. Explain what causes Earth's seasons. (**Earth is tilted on its axis, so as it revolves around the sun, the angle at which sunlight hits the ground changes because of Earth's tilt. The part of Earth tipped toward the sun has hot weather; the part tipped away from the sun has cold weather.**)
3. Which line of latitude does the sun strike at a 90° angle at the beginning of summer in the Northern Hemisphere? (**Tropic of Cancer**) At the beginning of winter? (**Tropic of Capricorn**)
4. What is the difference between an equinox and a solstice? (**An equinox is a point in Earth's orbit where the sun crosses Earth's equator and the number of day and night hours are nearly equal. A solstice is a day of the year in which the sun's most direct rays reach the farthest north or south.**)

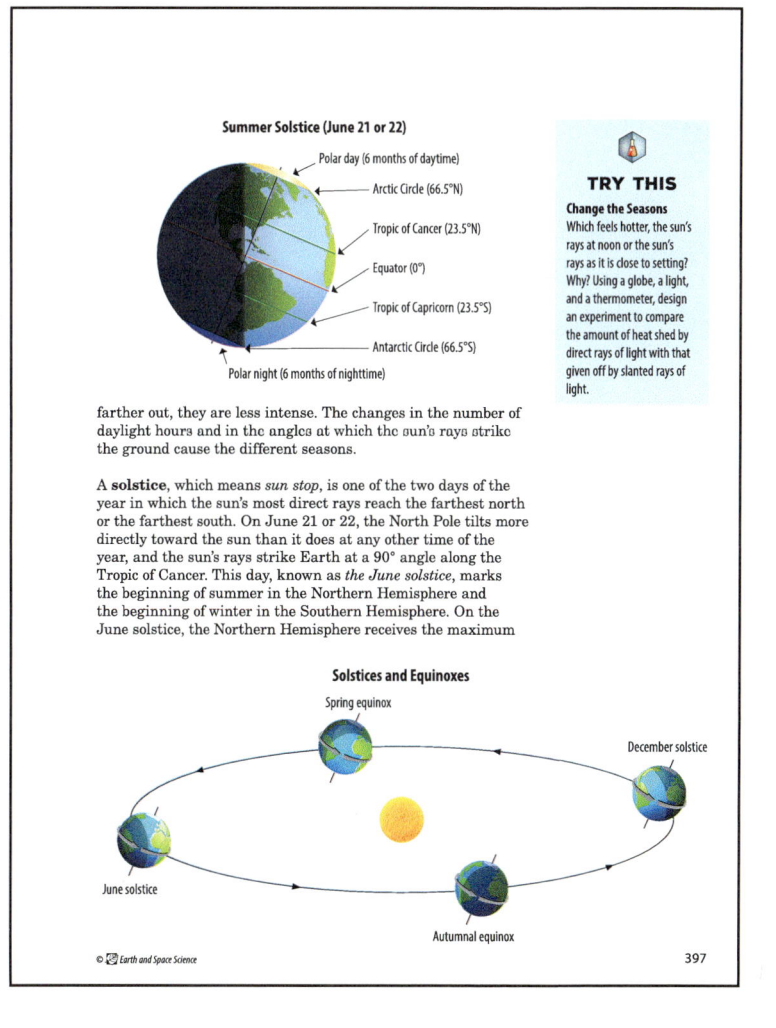

NOTES

7.3.4 Seasons

OBJECTIVES
- Model the way in which Earth's tilt and revolution around the sun cause different seasons.
- Identify Earth's primary lines of latitude.
- Differentiate between a solstice and an equinox.

VOCABULARY
- **equinox** a point in Earth's orbit at which the sun crosses Earth's equator, causing the hours of day and night to be nearly equal everywhere on Earth
- **solstice** one of the two days of the year in which the sun's most direct rays reach farthest north or farthest south

You cannot feel Earth's movement directly, but you can observe its effects. For example, as part of Earth's rotation you can see the sun rising in the east, crossing the sky during daylight hours, and setting in the west. One half of Earth is always bathed in sunlight, and one half is always under the cover of darkness.

You can also see the effects of Earth's revolution around the sun. On January 3, the Earth is at its perihelion. At this point, it is about 147 million km from the sun. Earth's aphelion occurs on July 5, when Earth is more than 152 million km away from the sun. But Earth's distance from the sun is not responsible for seasons. If this were the case, July 5 would be in a cold season not only in the Southern Hemisphere but also in the Northern Hemisphere.

The changing seasons are caused by Earth's tilt as it orbits around the sun. Earth is tilted about 23.4° on its axis. During part of the year, the North Pole is tipped toward the sun. During the other part of the year, the South Pole is tipped toward the sun. When the South Pole is exposed to the sun, the **Southern Hemisphere enjoys longer daylight hours and warmer temperatures.** The angle of the sun's rays hitting the ground in that hemisphere is more direct and intense. At the same time, the Northern Hemisphere receives less daylight because its pole is facing away from the sun. The angle of the sun's rays striking the Northern Hemisphere decreases. When the sun's rays spread farther out, they are less intense. The changes in the number of daylight hours and in the angles at which the sun's rays strike the ground cause the different seasons.

A **solstice**, which means *sun stop*, is one of the two days of the year in which the sun's most direct rays reach the farthest north or the farthest south. On June 21 or 22, the North Pole tilts more directly toward the sun than it does at any other time of the year, and the sun's rays strike Earth at a 90° angle along the Tropic of Cancer. This day, known as *the June solstice*, marks the beginning of summer in the Northern Hemisphere and the beginning of winter in the Southern Hemisphere. On the June solstice, the Northern Hemisphere receives the maximum

Summer Solstice (June 21 or 22)
- Polar day (6 months of daytime)
- Arctic Circle (66.5°N)
- Tropic of Cancer (23.5°N)
- Equator (0°)
- Tropic of Capricorn (23.5°S)
- Antarctic Circle (66.5°S)
- Polar night (6 months of nighttime)

TRY THIS

Change the Seasons
Which feels hotter, the sun's rays at noon or the sun's rays as it is close to setting? Why? Using a globe, a light, and a thermometer, design an experiment to compare the amount of heat shed by direct rays of light with that given off by slanted rays of light.

When I consider Your heavens, the work of Your fingers, the moon and the stars, which You have set in place, what is mankind that You are mindful of them, human beings that You care for them?
—Psalm 8:3–4

Solstices and Equinoxes
- Spring equinox
- December solstice
- June solstice
- Autumnal equinox

> **CHALLENGE**
>
> **Dr. Martin Luther King, Jr.**
> Dr. Martin Luther King, Jr., had the following to say about science and faith: "Science investigates; religion interprets. Science gives man knowledge which is power; religion gives man wisdom which is control. Science deals mainly with facts; religion deals mainly with values. The two are not rivals. They are complementary. Science keeps religion from sinking into the valley of crippling irrationalism and paralyzing obscurantism. Religion prevents science from falling into the marsh of obsolete materialism and moral nihilism." Do you agree? Why?

number of daylight hours, and the Southern Hemisphere receives the minimum number of daylight hours. At this time, the sun stays above the horizon for 24 hours north of the Arctic Circle, but it does not rise south of the Antarctic Circle.

On December 21 or 22, the sun's rays strike at a 90° angle along the Tropic of Capricorn. This day is called *the December solstice*. On this day, the North Pole is tilted away from the sun, but the South Pole is inclined toward the sun. During the December solstice, the daylight hours and intensity in the hemispheres are reversed from the June solstice, so it is winter in the Northern Hemisphere and summer in the Southern Hemisphere.

Earth also experiences two equinoxes. The word *equinox* means "equal night." An **equinox** is a point in Earth's orbit at which the sun crosses Earth's equator, which causes the hours of day and night to be nearly equal everywhere on Earth. The equinoxes happen on March 20 or 21 and again on September 22 or 23. During each equinox, the sun's rays strike the equator at a 90° angle. The March equinox is called *the spring* or *vernal equinox* and the September equinox is known as *the autumnal equinox* in the Northern Hemisphere. Because the seasons in the Southern Hemisphere are the reverse of those in the Northern Hemisphere, the terms are reversed there as well.

LESSON REVIEW
1. How much does Earth's distance from the sun affect temperatures on Earth?
2. Explain what causes Earth's seasons.
3. Which line of latitude does the sun strike at a 90° angle at the beginning of summer in the Northern Hemisphere? At the beginning of winter?
4. What is the difference between an equinox and a solstice?

UNIT 8

Chapter 1: *Stars and the Universe*
Chapter 2: *Space Exploration*

Key Ideas

Unifying Concepts and Processes
- Systems, order, and organization
- Evidence, models, and explanation
- Change, constancy, and measurement
- Evolution and equilibrium
- Form and function

Science as Inquiry
- Abilities necessary to do scientific inquiry
- Understandings about scientific inquiry

Earth and Space Science
- Origin and evolution of the universe

Science and Technology
- Abilities of technological design
- Understandings about science and technology

Science in Personal and Social Perspectives
- Science and technology in society

History and Nature of Science
- Science as a human endeavor
- Nature of science
- History of science
- Nature of scientific knowledge
- Historical perspectives

The Great Expanse

Vocabulary

absolute magnitude	galaxy cluster	propellant	spin-off
apparent magnitude	inertia	pulsar	star cluster
binary star system	light-year	rocket	supernova
black hole	neutron star	satellite	thrust
Doppler effect	parallax	space probe	
galaxy	payload	space station	

SCRIPTURE

Lift up your eyes and look to the heavens: Who created all these? He who brings out the starry host one by one and calls forth each of them by name. Because of His great power and mighty strength, not one of them is missing.

Isaiah 40:26

8.1.0 Stars and the Universe

LOOKING AHEAD

- For **Lesson 8.1.1**, assign students to predict the number of stars they will be able to count on a clear night and then to count the number of visible stars prior to introducing this lesson. Collect cardboard tubes or encourage students to collect tubes for *Try This: Make a Spectroscope*. Gather diffraction grating from a science supply store.
- For **Lesson 8.1.2**, collect chalk dust and a lamp with a 100-watt light bulb for the *Introduction* demonstration.
- For **Lesson 8.1.3**, arrange a field trip to a local planetarium.
- For **Lesson 8.1.4**, obtain a Wiffle ball and a battery-operated buzzer.

SUPPLEMENTAL MATERIALS

- TM 8.1.1A Absorption Spectrum
- TM 8.1.2A H-R Diagram
- TM 8.1.3A The Milky Way
- TM 8.1.3B Constellations in the Northern Hemisphere
- TM 8.1.3C Galaxy Types

- Lab 8.1.1A Star Spectrums

- WS 8.1.1A Parallax
- WS 8.1.2A Life Cycle of a Star

- Chapter 8.1 Test

BLMs, TMs, and tests are available to download. See Understanding Purposeful Design Earth and Space Science at the front of this book for the web address.

Chapter 8.1 Summary

Many years ago, God created the universe. In His time, God crafted stars, planets, and many other celestial bodies. Just as humans are born, grow up, and eventually die, scientific evidence shows that the Creator designed stars with life cycles that may span billions of years. God formed stars of many sizes with varying degrees of energy. He grouped billions of these enormous infernos into galaxies and then gathered billions of galaxies together in enormous clusters that are dominated by a common gravitational attraction.

Nearly 400 years ago on a rooftop in Florence, Italy, Galileo Galilei began to uncover part of God's hidden creation. Pointing a telescope into the night sky, Galileo discovered thousands of stars never before seen and gave humans their first glimpse of God's universe of stars and galaxies. Since that time, curious individuals have continued to explore the universe with a host of powerful telescopes, satellites, and deep-space probes. As astronomers gather more data about distant stars and galaxies, they uncover more mysteries in this vast universe. Christians are able to rejoice in new discoveries and stand in awe of God's unlimited power and creativity through a study of the universe.

Background

Lesson 8.1.1 – Stars

Even when examined by the unaided eye, stars evoke wonder. It can be difficult to grasp the distance, power, and size of stars. Distances between stars are so great that scientists created a new measurement unit—the light-year. A light-year is the distance that light can travel in a vacuum in one year, approximately 9.461×10^{12} km. The basic unit of length for measuring interstellar and intergalactic distances is called *a parsec*. One parsec equals 3.26 light-years (3.08×10^{13} km). A kiloparsec is 1,000 parsecs; a mega parsec is 1 million parsecs.

Nuclear fusion is the powerful source of stars' tremendous energy. During the process of fusion, hydrogen is converted into helium, and enormous amounts of energy are released. Stars are fueled by the nuclei of hydrogen. The extreme temperatures of a star strip away electrons from any atoms, leaving only ions. Under normal circumstances, two protons would repel each other because both are positively charged. However, the extreme heat and pressure inside stars forces two protons together and transforms one of the protons into a neutron. This combination of a proton and a neutron is called *deuterium*. When deuterium forms, large amounts of energy are released. Eventually, the deuterium nucleus, or deuteron, collides with another proton to form a helium isotope. The collision releases more energy. When two of these helium nuclei collide to form a normal helium atom made of two protons and two neutrons, two protons are released in the process. The energy given off during fusion either heats the star or escapes from the star as radiation.

Individual stars vary widely in mass, size, temperature, and luminosity. The sun has a mass of about 1.989×10^{30} kg, a radius of about 6.957×10^{8} m, a surface temperature of about 5,777 K, and a luminosity of about 3.846×10^{26} W. A star's diameter can range from a few kilometers to hundreds of times that of the sun. For example, red giant stars are hundreds of times larger than the sun. If the sun were that large, it would engulf Mars. White dwarfs are about the size of Earth, and neutron stars are several kilometers in diameter.

A star's color is determined by its surface temperature. Light travels from a star in a spectrum of wavelengths. Hotter stars have shorter wavelengths and cooler stars have longer wavelengths. Each wavelength of light produces a specific color. When heated, the different elements contained in stars also emit different wavelengths of light. Each star has a unique set of wavelengths. Astronomers can identify a star's or galaxy's elemental composition by separating the light it produces. Using spectrographs, astronomers apply a set of colors, called *emission lines*, that identify each element.

The most luminous stars expend about 1 million times more energy than the sun. In contrast, the least luminous stars are only one-hundredth as powerful. Stars vary in brightness, but how radiant they appear to people on Earth may not reflect their true luminosity as measured by astronomers. Along with distance, a star's mass, chemical composition, and amount of energy can determine brightness. Most stars are more than 90% hydrogen, so variations in chemical composition have little effect. Hotter stars are brighter than colder stars. Astronomers usually determine a star's absolute magnitude, or intrinsic brightness, by calculating how bright the star would appear if all stars were the same distance away.

Lesson 8.1.2 – Life Cycle of Stars

A star's fusion process may change during the course of its life. It can also produce spectacular changes in the appearance and composition of the star. Stellar evolution relates to the types of changes that stars undergo, which depends on their size.

The theory of stellar evolution states that a star changes as it consumes hydrogen during its power-producing nuclear reactions. When its capability for nuclear reactions is exhausted, the star dies. The heavy atoms produced during supernovas disperse, eventually joining the interstellar matter from which new stars are continuously created. Almost every natural chemical element (with the exceptions of hydrogen and helium) originates from the variety of nuclear reactions that occur in the late phases of a massive star's life.

A star is held in place by the balance of opposing forces: the inward pressure of gravity and the outward pressure of the star's radiation. A star begins in a gas and dust cloud called *a nebula* that is many light-years across. Scientist believe that gravity pulls the gas and dust particles together in a giant sphere. The temperature of the sphere is extremely hot because as the gas and dust collide, they are under intense pressure from the surrounding material. As the temperature nears 15 million °C, the pressure at the center of the nebula is very great. Electrons are stripped from the atoms, which creates plasma—a gas consisting of positively charged ions. Because the nuclei approach each other so fast, they overcome the electrical repulsion of their protons and fuse together. This fusion produces massive energy, which pours from the core. The energy creates an outward pressure that balances gravity. The released energy passes through the outer layers of gas and dust, moves into space in the form of electrical radiation, and causes the star to shine.

However, the hydrogen in a star is eventually depleted. Without the outward pressure generated from nuclear fusion, the star's gravity becomes overwhelming and the star begins to collapse. The star's temperature and pressure increase. This newly generated heat counteracts gravity temporarily, pushing the star's surface layers outward. The star expands exponentially to form a red giant.

Stars' life cycles can be plotted on the Hertzsprung-Russell (H-R) diagram, which was created in the early 20th century to compare magnitude and spectra. On the diagram, a group called *the main sequence* extends in a rough diagonal pattern from the upper left (hot, bright stars) to the lower right (dim, cool stars). The giant sequence of large, bright, cool stars appears in the upper right portion of the diagram. White dwarfs are dim, small, and hot, so they appear in the lower left area of the diagram. The sun is located near the middle of the main sequence.

A star's mass dictates how it progresses through its life cycle. For example, the outer layers of a medium-sized star continue to expand as its core contracts. As the core's helium atoms fuse into carbon, the star sheds its outer layers in a planetary nebula until only about 20% of the star's initial mass remains. The star then cools and shrinks into a white dwarf. Because the inward pull of gravity is balanced by the electrons in the star's core, the white dwarf star is stable. Although no fuel is left, the hot star radiates for many years until it becomes a cold, dark mass.

WORLDVIEW

- God has a plan for everything He created. This principle is evident in every aspect of the universe. For example, the movement and behavior of star clusters, galaxies, white dwarfs, and pulsars in response to mutual gravitational attractions illustrates the orderly and creative nature of God. He created universal systems that work together to establish order and purpose. In Isaiah 40:26, we are reminded that God created the heavens. He caused the stars to shine and knows each one by name. His great power and mighty strength ensure that people can live with purpose. If God did not care about the people on Earth, He would not have built such a vast, complex, sound universe. We can be assured that just as God knows each of the stars by name, He knows us intimately as well. In response to God's outpouring of love, we much cherish the heavens, foster a sense of wonder about the universe in others, and in all things work to fulfill our purpose in Christ.

Larger stars go through a different cycle. After a star becomes a red supergiant, its core yields to gravity and shrinks, becoming hotter and denser. Nuclear reactions temporarily keep the core from collapsing. As the core becomes iron, fusion stops. Shortly after that occurs, the star begins its final collapse. The star's nuclei repel each other, and overcome gravity. The core recoils into an explosive shock wave. This wave heats the material in the outer layers, which fuse to form new elements and radioactive isotopes. These materials are propelled away during the supernova, and the core becomes a dense ball of neutrons, which may remain intact as a neutron star. Electrons near the magnetic poles that are not aligned with the axis may travel outward from the star. When the electrons reach the point where they would have to travel faster than the speed of light in order to rotate with the star, they emit energy in the form of X-rays and gamma rays. The emissions are perceived as pulses of radiation. Such neutron stars that emit pulses are called *pulsars*; they pulse with a fixed, repetitive rate.

Scientists believe black holes form when neutrons of extremely massive stars do not survive the core collapse of a supernova. Black holes cannot be directly observed, but scientists point to the behavior of stars and other nearby materials as evidence of their existence. The density of matter and the curvature of space-time in a black hole become infinite—nothing, not even light, can escape a black hole's gravitational attraction.

Lesson 8.1.3 – Star Systems
Stars systems are bound together by gravity, which means they attract planets and other stars. Binary star systems consist of a pair of stars held together by their mutual gravitational attraction. The more massive star of a binary is called *the primary* (A). The smaller star is called *the secondary* (B). Through observations of their motions and the use of Newton's law of gravitation, scientists can calculate the individual mass of both stars in a binary system. Astronomers then use the information collected from observing binary stars to deduce the masses of similar stars.

A star cluster is a group of stars that are in close proximity and that move at the same rate in the same direction. By observing star clusters, astronomers better understand the distance scale of the universe. The two types of clusters are open clusters and globular clusters. Open clusters, which exist in regions of gas and dust, may include from a few dozen to 1,000 loosely scattered stars. In contrast, globular clusters are densely packed with hundreds of thousands of stars. The Milky Way includes about 150 known globular clusters, including Omega Centauri and 47 Tucanae, which are visible in the southern skies, and M13, which appears in the northern sky. Scientists observe globular clusters to learn about the life span of stars. For example, astronomers have inferred that massive stars change more rapidly over time than smaller stars.

A galaxy is a large aggregation of stars, gas, and dust. Held together by gravitational attraction, a galaxy's rotational motion keeps it from collapsing on itself. Galaxies contain billions of stars. They were not recognized as independent star systems until Edwin Hubble completed his study of the Andromeda Galaxy, which indicated the staggeringly distant locations of other galaxies. Often, smaller satellite galaxies are in close proximity to larger galaxies. For example, the galaxies nearest the Milky Way galaxy form a cluster. This local group includes the Andromeda Galaxy and the Magellanic Clouds, which are both satellite galaxies of the Milky Way.

Gravitation can bind clusters of galaxies together. For example, the cluster containing the Milky Way, which is called *the Local Group*, has a diameter of 10 million light-years and contains about 30 galaxies. Astronomers have discovered over 10,000 galaxy clusters. In addition, galaxy clusters are bound into superclusters. The flattened collection of approximately 100 galaxy clusters that includes the Local Group is called *the Local Supercluster*, which spans about 110 million light-years.

Lesson 8.1.4 – Creation of the Universe

All the calculated ages for the universe rely on assumptions—young-earth as well as old-earth. Young-earth Christian scientists point to the account of Creation in Genesis 1:1–31. Many other astronomers believe the universe is old because of the measurement of celestial objects and events. Old-earth astronomers estimate that the universe is 13.82 billion years old.

Young-earth proponents point to scientific data that indicates that the universe is young, however. For example, particle physicists contend that the very hot conditions of the Big Bang should have created massive magnetic particles called *monopoles*, which would have had single poles. However, these stable particles have not been found, which suggests that the temperature of the universe was never that hot and contradicts the Big Bang model of a cataclysmic, hot beginning. According to physics, the creation of matter in the form of hydrogen and helium gas from the energy of the universe's expansion should have created an equal amount of antimatter. However, the universe contains primarily matter with only tiny amounts of antimatter present. Spiral galaxies also indicate that the universe is younger than many supporters of the Big Bang believe. The centers of these galaxies rotate faster than the outer arms, which causes the spiral structure to tighten rather than unwind. Young-earth astronomers point out that if these most-common galaxies were billions of years old, they would no longer have a spiral shape—they would be compressed into balls or disks. Researchers also point to the existence of few supernova remnants as evidence that the universe is young. For example, the Large Magellanic Cloud would have evidence of about 340 supernovas if it was billions of years old, and it would have remnants of about 24 supernovas if the galaxy was about 7,000 years old. The actual number of supernova remnants found in the Large Magellanic Cloud is 29, which indicates the galaxy is only thousands of years old.

According to the Big Bang model, the universe began with an instantaneous expansion of space that is still occurring and is gradually increasing the distance between the Milky Way galaxy and other galaxies. The universe's expansion "stretches" light rays, so blue light is converted into red light and red light into infrared light. Distant galaxies that are moving away from Earth appear redder than others by comparison. However, gravity slows the expansion of the universe. If the universe is dense enough, the expansion of the universe might eventually reverse and the universe will collapse. If the density is too low, the expansion will continue indefinitely. Therefore, the density of the universe determines its end. Current evidence suggests that the universe is open, which indicates that it will continue to expand and the galaxies will continue to move outward. If this is true, the stars will eventually die out as they use up all their energy. Another piece of supporting evidence for the Big Bang model is that the Milky Way is being bombarded with radiation of equal intensity from every part of the universe. Many scientists believe this radiation has cooled over the past 10–20 billion years to a temperature of about 4 K, which is consistent with radiation temperatures for the Big Bang model.

There are many misconceptions about the Big Bang model. The Big Bang was not an explosion of matter into a previously empty space, but rather an explosion of space itself. The universe has no center. Galaxies all move away from each other, not away from a common point in space. It is theorized that the Big Bang was the very early state of the universe when the distances between all objects were much less than today's. The Big Bang model involves the increase of space, not the movement of objects. Space itself is expanding between galaxies.

Although widely accepted in the scientific community for much of the past century, the Big Bang model may be replaced by a new model. For example, a new model published in 2015 indicates that the universe is without beginning or end. This new model corrects the issues that have arisen with Einstein's theory of general relativity, which is the mathematical foundation for the Big Bang theory. The new model of an eternal universe eliminates the Big Bang model's singularities and may account for dark matter and dark energy. Research is ongoing.

NOTES

8.1.1 Stars

Stars and the Universe

OBJECTIVES

Students will be able to
- relate the color of stars to the stars' temperature and composition.
- compare continuous, emission line, and absorption line spectra.
- explain the different ways scientists classify stars.

VOCABULARY

- **absolute magnitude** the brightness of a star measured by an observer who is a standard 32.6 light-years away
- **apparent magnitude** the brightness of an object as observed from Earth
- **light-year** the distance light travels in a vacuum in one year, approximately 9.46×10^{12} km
- **parallax** the apparent shift in an object's direction when viewed from two geographically distant locations

MATERIALS

- Scissors, yarn, metersticks, protractors (*B*)
- TM 8.1.1A Absorption Spectrum
- WS 8.1.1A Parallax

PREPARATION

- Before this lesson, assign students to predict the number of stars they will be able to count on a clear night and then to count the number of visible stars. (*Introduction*)
- Obtain materials for *Lab 8.1.1A Star Spectrums*.
- Gather materials for *Try This: Make a Spectroscope*.

TRY THIS

Make a Spectroscope
- *black paper pieces, 1 per student*
- *scissors, 1 pair per student*
- *cardboard tubes, 1 per student*
- *tape or rubber bands, 1 per student*
- *box cutters or scalpels, 1 per student*
- *diffraction grating, 1 piece per student*

Introduction

Encourage students to relate how many stars they predicted could be counted. (**Answers will vary.**) Have students indicate how many stars they counted. (**Answers will vary.**) As a class, determine if students' predictions were accurate. Explain that about 3,000 stars are visible to the unaided eye.

Challenge students to remember a time when an object they saw, such as a tall building or a hot air balloon, turned out to be much larger than it initially appeared. Ask students why the objects seemed rather small. (**Possible answer: The objects were far away.**) Point out that the closer they got to the object, the larger it appeared. Relate this idea to stars. Explain that some stars that are actually very bright and large may seem dim and small because they are located a great distance from Earth.

Discussion

- Discuss issues associated with measuring extreme heights and distances. Have students brainstorm creative ways to measure tall objects. (**Answers will vary.**) Challenge students to use one of their suggestions to measure the height of a tall tree nearby.

- Lead a discussion on the difference between apparent magnitude and absolute magnitude. Explain that the absolute magnitude of a star is its brightness measured at a standard distance of 32.6 light-years. In contrast, apparent magnitude is how bright an object appears to an observer on Earth. Ask what condition would have to occur for the absolute brightness and apparent brightness of a star to be the same. (**In order for the for the absolute and apparent brightness to be the same, the star and all the surrounding stars would have to be a standard distance of 32.6 light-years away.**)

- Discuss the relative size of the sun. Point out that compared to other stars, Earth's sun is just an average-sized star. Remind students of the importance of the sun to Earth. Ask why it is significant that God placed the sun at the distance it is from Earth. (**Answers will vary but should include that the sun's distance from Earth makes it possible for life to exist.**) Point out that Venus, the next closest planet to the sun, is scorching hot, and Mars, the fourth planet from the sun, is freezing cold.

- Ask the following questions:
  1. What principle explains the apparent shift of close stars in the sky? (**parallax**) When must astronomers account for parallax? (**when they are attempting to measure the distance of a star from Earth**)
  2. What two basic elements compose all stars? (**hydrogen and helium**)

Activities

Lab 8.1.1A Star Spectrums

- large paper clips, 1 per group
- Bunsen burners, 1 per group
- striker or matches
- spectroscopes, 1 per group
- chalk dust
- baking soda
- ground calcium carbonate
- salt

Use metal paper clips without a coating for this lab.

A. Complete *Try This: Make a Spectroscope* with students. Ask students what type of spectrum is visible on the inside of the tube. (**continuous spectrum**)

B. Assign **WS 8.1.1A Parallax** to groups of three students. Distribute 20 m of yarn to each group.

C. Display **TM 8.1.1A Absorption Spectrum**. Have students examine the spectra produced by the four elements. Ask students what elements produce the spectrum of the sun. (**hydrogen, helium, and sodium**) What element produces the spectrum of the star Sirius? (**hydrogen**) What elemental component do the sun and Sirius share? (**hydrogen**)

Lesson Review

1. Explain how a spectrum helps classify stars. (**The light produced by each element in a star has a unique set of emission lines, which helps scientists identify the elements that form each star.**)
2. What type of spectrum shows all of the visible color? (**continuous spectrum**)
3. Explain how emission lines form. (**When electrons in a star's atoms jump around, the activity releases a photon of light. If split, this light only has a few colors or emission lines.**)
4. Why does starlight show absorption lines? (**The gases in a star's atmosphere are cooler than the gases on the surface or inside of the star. The cooler atmospheric gases absorb some of the star's light, removing certain colors from the continuous spectrum.**)
5. How have stars been classified? (**Scientists have used stars' spectra, temperatures, brightness, and magnitudes to classify stars.**)
6. Explain which is brighter: a star with an apparent magnitude of +5.3 or the sun, which has an apparent magnitude of −26.8. (**A positive apparent magnitude value indicates a relatively dim star. Very bright stars have negative numbers. The sun is brighter than a star with an apparent magnitude of +5.3.**)

NOTES

8.1.1 Stars

OBJECTIVES
- Relate the color of stars to the stars' temperature and composition.
- Compare continuous, emission line, and absorption line spectra.
- Explain the different ways scientists classify stars.

VOCABULARY
- **absolute magnitude** the brightness of a star measured by an observer who is a standard 32.6 light-years away
- **apparent magnitude** the brightness of an object as observed from Earth
- **light-year** the distance light travels in a vacuum in one year, approximately 9.46 x 10^{12} km
- **parallax** the apparent shift in an object's direction when viewed from two geographically distant locations

Stars are enormous balls of hot gases—primarily hydrogen and helium. They also contain varying amounts of heavy elements: carbon, nitrogen, oxygen, and magnesium. Stars are fueled by nuclear fusion, which occurs when two light nuclei join together to form a single heavy nucleus. A large amount of energy is released during such fusion reactions. The energy that heats the star originates primarily in the core before escaping into space as radiation.

Have you ever wondered how scientists know such details about stars that are trillions of kilometers from Earth? Actually, much of the information about stars can be found by studying the light they emit. For example, a star's color can indicate how hot it is. When you look at a fire, you see that flames are a variety of colors. Bluish flames are hotter than yellow flames, and red flames are the coolest. In the same way, the color of a star can provide clues to the star's temperature. When you look at the stars, it is hard to distinguish star colors because the cones in your eyes that help you perceive colors do not work well in low light. However, on a clear night, you may notice that although many stars appear white, some appear blue or red.

A star's color not only indicates its temperature but also its elemental components. Identifying the elements present in a star involves separating the different colors in its starlight. Think about a rainbow. When you see a rainbow, you are seeing white light split into many colors. The color separation happens because raindrops act as prisms. The array of colors produced when white light passes through a prism is called *a spectrum*. The main colors in a spectrum are red, orange, yellow, green, blue, indigo, and violet. A hot solid object, such as the wire inside an incandescent light bulb, gives off all the colors necessary to produce white light. This range of color is called *a continuous spectrum*.

Continuous Spectrum

Emission Line Spectrum

Absorption Line Spectrum

Magellanic Cloud colors indicate varying degrees of heat.

Just as a prism or a raindrop spreads the continuous spectrum into many colors to form a rainbow, an instrument called *a spectrograph* splits starlight into all its colors. However, the light spectrum emitted by a star is not a continuous spectrum. When electrons in the star's atoms are excited during combustion, they quickly "jump" to a higher energy level and then promptly return to their initial lower energy level. This activity releases a photon of light. If the light is split with a spectrograph, only a few colors appear. These colors form what are known as *emission lines*. The light produced by each element in a star has a unique set of emission lines, which helps scientists identify the elements that comprise each star.

Another type of spectrum is produced when white light passes through the cold gas of an element. When that happens, the gas blocks some of the light's colors. If the light is passed through a spectrograph, the spectrum would include a series of black lines where an element's colors were blocked by the gas. The absent colors are the ones emitted when the element is heated. These black lines are called *absorption lines*.

Starlight shows absorption lines because the gases in a star's atmosphere are cooler than the gases on the surface or interior

Spectroscopy

400 | 401

BIOGRAPHY

Annie Jump Cannon
Annie Jump Cannon was born on December 11, 1863, in Dover, Delaware. She attended Wellesley College where she studied physics and astronomy, and she later studied at Harvard Observatory. Annie was the first woman to receive an honorary doctorate from the University of Oxford. Annie was left almost completely deaf after a bout of scarlet fever. She is best known for her invention of the unique star classification system that assigned classes O, B, A, F, G, K, and M to stars according to their spectra, which is still used today. Over the course of Annie's career, she discovered hundreds of variable stars and classified hundreds of thousands of others. Known for being very diligent, enthusiastic, and patient, Annie was the first woman elected to an officer position in the American Astronomical Society. In her honor, the society created the Annie Jump Cannon Award, which is given to female astronomers who are distinguished in their field.

of a star. The cooler atmospheric gases absorb some of the star's light. This absorption removes certain colors of light from the continuous spectrum of the star—the same colors that the star would emit if it were heated. Because stars are created from a mixture of elements, all of the different absorption lines for these elements appear together in a star's spectrum.

In the 1880s, scientists began to classify stars according to their spectra. This original classification system attempted to measure the strength of element absorption lines, but this method proved unreliable. Stars were then classified according to temperature and designated in order of decreasing warmth by one of the following letters: O, B, A, F, G, K, M, R, N, and S. In this system, classifications O and B are the bluest and hottest, and M through S are the reddest and coolest. However, even the coolest stars are fiery infernos with temperatures that reach 2,700°C. The

Spectral Class	Temperature (K)	Size and Color
O	41,000	
B	31,000	
A	9,500	
F	7,240	
G	5,920	
K	5,300	
M	3,850	

sun has a surface temperature of about 6,000°C. Hotter stars have surfaces between 10,000°C and 25,000°C. The hottest stars' temperatures reach between 25,000°C and 50,000°C on their surfaces. The inside of a star is much hotter than its surface. For example, astronomers think that the temperature inside the sun is 15 million °C.

Stars have also been classified by brightness. Some ancient astronomers called the brightest stars *first-magnitude stars* and the faintest stars *sixth-magnitude stars*. When telescopes became available, stars that astronomers could not see before became visible; the sixth-magnitude stars did not look so dim anymore. Later, scientists began classifying stars using a system of magnitudes. They adjusted the temperature classification system that they were already using and assigned positive values to dimmer stars and negative values to brighter stars. Modern astronomers use a combination of temperature and luminosity.

Polaris is the final star in the handle of the Little Dipper.

How bright a star appears depends on how far it is from the observation location. For example, if you are sitting in your front yard at night, your porch light will look brighter than the porch light across the street. In the night sky, some closer small stars look brighter than much larger stars that are farther away. How bright an object appears to an observer on Earth is called **apparent magnitude**. Some stars that look dim from Earth are actually bright stars that are more distant. Astronomers use a star's apparent magnitude and the distance the star is from Earth to calculate absolute magnitude. The **absolute magnitude** of a star is the brightness measured by an observer that is a standard 32.6 light-years away. For both measurements, the dimmer the star, the greater the number. Therefore, very bright stars have negative numbers. For example, from the perspective of Earth, the sun is the brightest star in the sky. It has an apparent magnitude of −26.8 but an absolute magnitude of 4.8. Sirius, the brightest star in the sky other than the sun, has an apparent magnitude of −1.44.

Determining a star's absolute magnitude requires knowing how far away the star is from Earth. Distances in the universe are much too great to be measured in kilometers, so astronomers created a new unit of measurement known as *a light-year*. A **light-year** is the distance that light travels in a vacuum in one

TRY THIS

Make a Spectroscope
Cut a circle of black paper with a diameter larger than the opening of one end of a cardboard tube. Secure the piece of paper to the end of the cardboard tube with tape or a rubber band. Cut a long slit in the middle of the piece of paper. Hold a piece of diffraction grating over the open end of the cardboard tube. Look through the end of the tube with the slit pointing toward a light source. Rotate the grating until a band of color appears on the inside of the tube. Tape the grating in place. Observe the bands of color produced by other light sources.

Parallax

The locations of nearer stars seem to shift more than stars that are farther away. The shift can be measured to determine the distance to the nearer stars.

year, which is approximately 9.46×10^{12} km. Light, the fastest known thing in all of creation, travels about 300,000 km/sec. Light-years help astronomers avoid using large, cumbersome numbers. For example, Proxima Centauri, the closest star to Earth other than the sun, is more than 40 trillion km away, but astronomers refer to its distance as 4.25 light-years.

To determine the distance to a nearby star, astronomers measure its position in relation to other more distant stars. Six months later, when Earth is on the other side of the sun, they measure the star's relative position again. From Earth, it will look as though the star has moved. As the earth revolves around the sun, stars near it seem to move more than stars that are farther away. Astronomers then measure the star's apparent movement against the background of the distant stars. The apparent shift in an object's direction when viewed from two geographically distant locations is called the **parallax**. The different positions of a star as viewed from Earth and from the sun are called *the annual parallax*. Astronomers use the diameter of Earth's

orbit, the angle of the parallax, and geometry to determine the distance from Earth to a star.

Because Earth rotates on its axis, different parts of the planet's surface face the sun. As a result, as Earth revolves, individuals see different stars in the sky at different times during the year. Earth's rotation makes the sun and most of the stars seem to move across the sky. The rest of the stars all seem to rotate around the North Star (Polaris), which is directly above Earth's North Pole. From the perspective of an observer on Earth, the North Star does not appear to move, which is why it has always been an important navigational aid. Each star actually does move in space, and over thousands of years, star patterns change shape. However, because the stars are so far away, this movement is not obvious.

LESSON REVIEW
1. Explain how a spectrum helps classify stars.
2. What type of spectrum shows all of the visible color?
3. Explain how emission lines form.
4. Why does starlight show absorption lines?
5. How have stars been classified?
6. Explain which is brighter: a star with an apparent magnitude of +5.3 or the sun, which has an apparent magnitude of −26.8.

Star trails

Name: _____ Date: _____

Lab 8.1.1A Star Spectrums

QUESTION: How do scientists determine the elemental components of stars?

HYPOTHESIS: Answers will vary.

EXPERIMENT:

You will need:	• striker or matches	• baking soda
• large paper clip	• spectroscope	• ground calcium carbonate
• Bunsen burner	• chalk dust	• salt

Steps:
1. Bend a large paper clip to form a small loop on one end.
2. Place the loop over a lit Bunsen burner for a few seconds to burn off contaminants.
3. Moisten the loop and dip it into the chalk dust.
4. With the overhead lights turned off, place the loop and chalk dust over the flame. Using the spectroscope, observe the spectrum produced by the chalk dust. In the box below, illustrate the spectrum produced.

Chalk Dust

Drawings will vary.

5. Repeat Steps 2 through 4 using baking soda, ground calcium carbonate, and salt. In the boxes below, illustrate the spectra observed.

Baking Soda

Drawings will vary.

Calcium Carbonate

Drawings will vary.

Salt

Drawings will vary.

Lab 8.1.1A Star Spectrums

ANALYZE AND CONCLUDE:

1. Compare and contrast the spectra of each substance that was tested.
Answers will vary.

2. Which spectra were the most similar? The spectra produced by baking soda and salt should be similar. The spectra produced by the calcium carbonate and chalk dust should be similar.

3. What might the spectra indicate about the substances? The spectra indicate what elements comprise each substance.

4. What information could scientists get from studying the spectra of stars? The spectra of stars can indicate the stars' composition.

8.1.2 Life Cycle of Stars

Stars and the Universe

OBJECTIVES

Students will be able to
- interpret H-R diagrams.
- illustrate the life cycle of a star.

VOCABULARY

- **black hole** a massive celestial object with gravity so strong that not even light can escape
- **neutron star** an extremely small, dense star composed primarily of neutrons
- **pulsar** a spinning neutron star that gives off pulses of radiation at regular intervals
- **supernova** the violent explosion of a star

MATERIALS

- Lamp with a 100-watt light bulb, ash or chalk dust (*Introduction*)
- TM 8.1.2A H-R Diagram
- WS 8.1.2A Life Cycle of a Star

⚠ SAFETY

Check school records for students' allergies or health conditions.

⚠ TRY THIS

Find the Main Sequence
Divide the class into groups.

Introduction ⚠

Ask students to describe the life cycles of things such as people, plants, and animals. Share with students that stars have life cycles as well. Relate stars to fire by having students imagine a glowing campfire that warms toes and melts marshmallows. Ask students what keeps the campfire burning. (**fuel such as wood, oxygen, and heat**) When fuel is no longer available, what happens to a campfire? (**The light becomes dim and the fire goes out.**) Share with students that nuclear reactions continue to occur in stars as long as there is sufficient fuel. Explain that stars are fueled by hydrogen. When a star's hydrogen is depleted, it has reached the end of its life cycle.

Convey that scientists believe the sun is a middle-aged yellow dwarf star. Point out that the life cycle of a star is estimated to take billions of years. The sun's next phase will be as a red giant. Inform students that as red giants age, the thin outer layers of gas escape and drift into space. These gases produce a tremendous amount of heat. Simulate the hot core of a red giant and its surrounding gases by placing an unshaded lamp with a 100-watt light bulb in the center of the room. With the overhead lights turned off, sprinkle ash or chalk dust above the bulb. The heat from the bulb will repel the ash or dust and scatter it around the room. Ask students what part of a red giant the floating ash or dust represented. (**the layers of gas that escape from the star**)

Discussion

- Display **TM 8.1.2A H-R Diagram**. Remind students that very bright stars have absolute magnitudes that are negative numbers. Point out that the star temperatures from coolest to hottest are red, orange, yellow, white, and blue. Ask students the following questions:
  1. Where are the hottest, brightest stars located on the diagram? (**top left**)
  2. What color are these stars? (**blue**)
  3. Where are the hottest, most faint stars located on the diagram? (**bottom left**)
  4. What color are these stars? (**white**)
  5. Why are supergiants so bright? (**They have a large mass and emit a large amount of energy.**)
  6. Where is the sun located? (**in the main sequence**)
  7. Describe the stars located in the bottom, right portion of the H-R diagram. (**They are small red stars that are relatively cool and dim.**)

- Ask the following questions:
  1. What force maintains a star's size? (**gravity**)
  2. On the H-R diagram, the stars in the main sequence are different colors. What star property determines its position on the main sequence of the H-R diagram? (**mass**)
  3. Name the last stages of the life cycle of a star. (**giants, supergiants, white dwarfs, and supernovas**)
  4. What can result from the compression of supernovas? (**neutron stars, pulsars, or black holes**)

Activities ⚠ ⚠

A. Complete *Try This: Find the Main Sequence* with students. As a class, compare graph results.

B. Direct students to complete **WS 8.1.2A Life Cycle of a Star**. Lead a class discussion on the processes that must occur for one stage of a star's life cycle to advance to the next stage.

C. Have students plot famous stars on an H-R diagram. Select two stars and assign students to compare their temperatures, absolute magnitudes, and spectral types.

D. Direct students to create a mnemonic device to help them remember the order of the classes of stars during their life cycle.

E. Assign students to write a story or to create a piece of art depicting the life of a star that includes all of the life cycle stages. Schedule time for students to read or display their work.

Lesson Review

1. What two factors are being compared in the H-R diagram? (**a star's spectrum and its absolute magnitude**)
2. In what portion of the H-R diagram will a star appear for the majority of its life? (**the main sequence**)
3. If a star has an absolute magnitude of 0 and a surface temperature that is cooler than the sun's, where would it be located on the H-R diagram? (**above the main sequence on the right side of the diagram**)
4. How is a star born? (**Large clouds of gas and dust, called *nebulae*, condense from their own gravity. As their density increases, the nebulae become much hotter. When enough heat is present in the center, nuclear fusion combines hydrogen atoms to form helium atoms, and the star is born.**)
5. How are white dwarfs created? (**Average-sized stars lose their outer layer of gases over time. Eventually, a small, hot, very faint core is left behind, which is called *a white dwarf*.**)
6. Why are black holes not visible? (**Black holes have such a strong gravitational pull that not even light can escape, so they cannot be seen directly.**)

NOTES

8.1.2 Life Cycle of Stars

OBJECTIVES
- Interpret H-R diagrams.
- Illustrate the life cycle of a star.

VOCABULARY
- **black hole** a massive celestial object with gravity so strong that not even light can escape
- **neutron star** an extremely small, dense star composed primarily of neutrons
- **pulsar** a spinning neutron star that gives off pulses of radiation at regular intervals
- **supernova** the violent explosion of a star

If you look at the stars night after night, individual stars do not seem to be any different from one another. However, stars do change, although it takes an exceptionally long time. Like other parts of God's creation, stars are born, then they mature and die. Individual stars go through this process in different ways. You may be wondering how astronomers discovered that stars have a life cycle.

One tool astronomers use is the Hertzsprung-Russell diagram (H-R diagram). The H-R diagram is a graph that plots the relationship between a star's spectrum and its absolute magnitude. It not only compares the brightness and temperature of stars but is useful for showing how stars change. In 1911, Danish astronomer Ejnar Hertzsprung developed a method of plotting the temperature and brightness of stars. Two years later, Henry Norris Russell developed a similar method. Their findings were combined into the H-R diagram.

The H-R diagram shows temperature along the *x*-axis and absolute magnitude (brightness) along the *y*-axis. Hot, blue stars are located at the left portion of the chart, and cool, red stars are located at the right portion of the chart. Bright stars are at the top of the chart and dim stars are at the bottom. A band of stars seems to stretch from the top left to the bottom right.

TRY THIS
Find the Main Sequence
For a group of students, record each student's height and foot length. Plot the data on a graph. Is there a pattern on the graph? How might the variables be related?

Hertzsprung-Russell Diagram

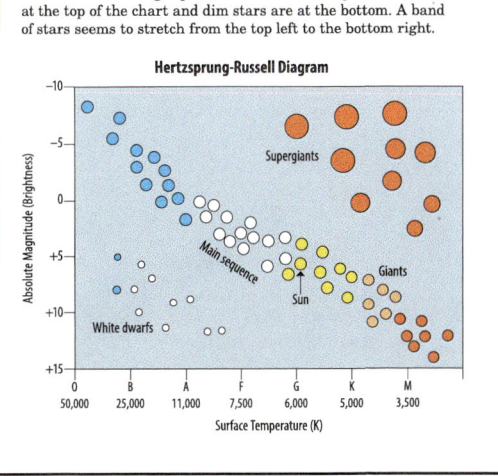

Stellar nursery in a nebula

This diagonal pattern of stars on the H-R diagram is called *the main sequence*. During most of a star's life, it is classified as a main-sequence star. As a star changes over time, it is moved across the H-R diagram. The life cycle of a star is depicted by the graph. Most stars, including the sun, fall into the main sequence. Although it seems bright from Earth, the sun is about 1 million times dimmer than the brightest stars.

Stars are born in nebulae—large clouds of interstellar material known as *stellar nurseries*. These raw materials begin to cluster together. The sphere of particles begins to contract from its own

BIOGRAPHY

Subrahmanyan Chandrasekhar
Chandrasekhar was an Indian-American astrophysicist who is best known for his work on the theoretical structure and evolution of stars and the later life cycle stages of massive stars. He attended Presidency College, the University of Madras, and Trinity College. In 1983, Chandrasekhar shared the Nobel Prize in physics with William Fowler for discovering that a star that exceeds 1.44 times the mass of the sun does not form a white dwarf but instead continues to collapse, explodes in a supernova event, and eventually becomes a neutron star. This discovery established what is known as *the Chandrasekhar limit*. He concluded that more massive stars continue to collapse and form black holes, which he attempted to describe mathematically. As a professor at the University of Chicago, Chandrasekhar conducted research regarding star energy transfer by radiation and convection processes on solar surfaces.

Red supergiant star cluster

Supernova

gravity. It grows hotter as it becomes denser. When it is hot enough in the center (about 15 million °C), a nuclear reaction begins. As nuclear fusion begins to turn hydrogen into helium, a star is born.

Once a star's mass is stabilized, the force of gravity maintains the star's size. At this point, most stars would belong on the main sequence of the H-R diagram. This stage is the longest in a star's life cycle. A star's position on the main sequence during its prime depends almost entirely on its mass. Scientists estimate that a star with the mass of the sun stays at this main-sequence stage for about 10 billion years. Stars with less mass are located at the lower right of the main sequence, and more massive stars are found at the upper left.

As they age, most main-sequence stars move up and to the right on the H-R diagram to become giants or supergiants. A star at the end of its life cycle that expands and cools as it runs out of hydrogen is called *a red giant*. As the outer layers of such a star expand from its helium core, the star widens and becomes larger, cooler, and redder. At this point it begins to burn excess helium. Eventually, average-sized stars like the sun lose their outer layer of gases and eject materials, forming a planetary nebula. What remains is called *a white dwarf*, a small, hot, faint star at the end of its life. White dwarfs fall into the lower left corner of the H-R diagram because they are hot but dim.

Massive blue stars use their hydrogen much faster than cooler stars. They generate a lot more energy, which makes them very hot and causes them to appear blue. Their existence is short compared to the life span of other stars, however. When these massive stars age, they leave the main sequence in a very dramatic way compared to average stars. A blue star may collapse under its own weight and explode in a brilliant flash of light. Such a violent explosion is called a **supernova**. At this point, its absolute magnitude may reach a peak brightness that is a billion times greater than the sun. From the earth, it may look as though a bright new star has appeared in the sky.

What happens to a supernova? The leftover materials in the core of a supernova pack together to form a star with a diameter of about 20 km. This extremely small, dense star is composed primarily of neutrons, so it is called a **neutron star**. Imagine how dense such a small star must be to have a mass comparable to 1.35 suns compressed in such a small space! The material in a neutron star is so dense that on Earth a teaspoon of the star matter would weigh almost one billion metric tons.

A **pulsar** is a spinning neutron star that gives off pulses of radiation at regular intervals—picture something like the flashes of a lighthouse. Radio telescopes can detect these beams. Pulsars emit radio pulses and other radiation at regular intervals with a frequency of around 1,400 pulses per second.

If the leftover materials from a supernova are too massive to become a neutron star, they may collapse to form a **black hole**—a

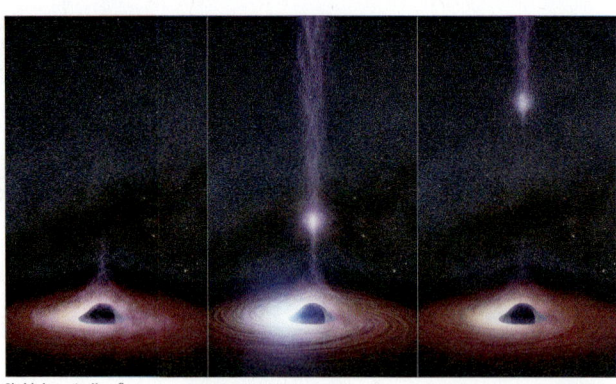

Pulsar

Black hole creating X-ray flares

massive celestial object with gravity so strong that not even light can escape. To produce a black hole, a star needs to be at least three times as massive as the sun. Astronomers theorize that black holes are more than three solar masses squeezed into a ball that is only about 64 km across. Because black holes permanently trap light and other forms of energy and matter, they cannot be directly observed. Black holes are only detected indirectly, such as when they are orbited by a visible star or when dust or gas from a nearby star spirals into the black holes. In this case, the materials may emit X-rays, which astronomers can use to detect the existence of black holes.

LESSON REVIEW
1. What two factors are being compared in the H-R diagram?
2. In what portion of the H-R diagram will a star appear for the majority of its life?
3. If a star has a absolute magnitude of 0 and a surface temperature that is cooler than the sun's, where would it be located on the H-R diagram?
4. How is a star born?
5. How are white dwarfs created?
6. Why are black holes not visible?

Stars and the Universe

8.1.3 Star Systems

Introduction
Divide the class into pairs and provide each student with a toothpick and a small amount of salt. Allow students 10 minutes to count as many grains of salt as they can. Suggest that grains be separated into piles of 10 grains each. After the time has elapsed, tally the total number of salt grains counted by the class. Compare the number of salt grains to the 100 billion stars in the Milky Way galaxy. Share with students that if 10,000 grains of salt could be counted in 10 minutes, it would take 20 students 200 years working 24-hours a day to count the same number of salt grains as there are stars in the Milky Way. Point out that astronomers estimate the number of stars in a galaxy by observing the galaxy's size, brightness, and mass.

Display **TM 8.1.3A The Milky Way**. Share that both images are the Milky Way galaxy, but each reflects a different perspective. Ask students if it would be possible for astronomers to assign the wrong type of identification to a galaxy. (**It is possible to assign the wrong classification if the classification was only determined by a single view.**) How can astronomers avoid misclassifying galaxies? (**by using views from multiple perspectives**)

Discussion
- Display **TM 8.1.3B Constellations of the Northern Hemisphere**. Ask students if they are familiar with any common constellations. (**Answers will vary.**) Do the constellations appear in the same spot of the sky throughout the year? (**No.**) Are the constellations visible all year? Why? (**Some constellations are only visible during certain times of the year because of Earth's changing position in its orbit around the sun.**) Discuss how ancient people used the changing positions of the constellations. For example, they planned when to plant and harvest according to the position of the constellations.

- Discuss the units of measure astronomers use. Remind students that using larger units sometimes makes measuring easier. For example, fabric is purchased by the meter or the yard rather than by the centimeter or the inch. Ask students what unit is generally used to measure distances to objects that are nearby in space, such as planets. (**light-year**) Convey that when measuring galaxies, a light-year is too small; it would be like measuring fabric using millimeters rather than meters. Point out that astronomers prefer to measure galaxies using parsecs and kiloparsecs because galaxies are extremely large.

- Ask the following questions:
  1. What holds binary stars in place? (**mutual gravitational attraction**)
  2. Why does the central bulge of a spiral galaxy appear redder? (**The stars located in the central bulge are older and cooler.**)
  3. What type of stars do elliptical galaxies contain? (**old stars**)
  4. Why are the stars behind dark nebulae not visible? (**Dark nebulae are very dense and block the light of stars behind them.**)
  5. Why do astronomers study quasars? (**Quasars give clues to how galaxies form and change over time. They are the most distant objects in the known universe.**)

Activities
A. Display **TM 8.1.3C Galaxy Types**. Ask students to describe the identifying characteristics for each type of galaxy.

B. Take the class on a trip to a local planetarium. Provide classroom discussion afterward about the star systems seen.

C. Have students share their collected galaxy names and pictures with the class. As a class, classify the galaxies and arrange each type of galaxy on individual poster boards that are labeled *Spiral*,

---

OBJECTIVES
Students will be able to
- identify various types of star systems.
- discuss basic features of different star systems.
- describe how astronomers classify galaxies.

VOCABULARY
- **binary star system** a pair of stars that orbit their common center of mass
- **galaxy** large systems of stars and their solar systems, gas, and dust held together by gravity
- **galaxy cluster** a group of thousands of galaxies under the same gravitational influence
- **star cluster** groups of stars that have a common origin and are held together by mutual gravitational attraction

MATERIALS
- Toothpicks, salt (*Introduction*)
- Poster boards (*C*)
- TM 8.1.3A The Milky Way
- TM 8.1.3B Constellations in the Northern Hemisphere
- TM 8.1.3C Galaxy Types

PREPARATION
- Arrange a field trip to a local planetarium. (*B*)
- Assign students to collect galaxy names and pictures. (*C*)

NOTES

Elliptical, or *Irregular*. Encourage students to explain why they assigned each image to a particular galaxy type. Ask students to indicate what type of galaxy they found most difficult to classify.

D. Assign a different ancient people group from each of the continents. Have students research how these groups interpreted the stars and constellations in times past and how these interpretations affected their daily living.

E. Challenge students to study the night sky by having them observe a particular group of stars in order to create a new constellation. Have students outline their new constellations on a star map.

F. Encourage students to choose a constellation and research the stars that comprise it. Direct students to include the name of each star and when it was discovered. Have students create a model of the constellation to suspend in the classroom. As a class, analyze if each constellation would be easily recognized if viewed from a different perspective.

G. Direct students to create models of each type of galaxy or to model a different star system.

Lesson Review

1. What is the difference between a star system and a constellation? (**A star system is a group of stars that forms close together and influence each other gravitationally. Constellations are groups of stars that form a pattern when observed from Earth, but the stars do not have any physical or gravitational relationship with each other.**)

Star Systems 8.1.3

When many people hear the term *star system*, they probably think it refers to constellations, but constellations are not really star systems. A constellation is a group of stars that forms a recognizable pattern. Individuals named the patterns they observed. The stars in constellations do not have any physical or gravitational relationship with each other. In fact, many of them are separated by great distances. Many of the constellations were recognized by ancient people who grouped stars together to form shapes of objects, animals, and people. They pictured a bear, dog, dragon, lion, and a winged horse. They saw a king, a queen, and a princess. People told stories of Orion (the great hunter) and Ursa Major (the bear). They also viewed objects such as a cup and a scale. Each culture and nation developed its own stories to account for the pictures seen in the sky, and these stories developed into mythology. Astronomers still use the Latin names of constellations.

In contrast, true star systems are groups of stars that influence each other gravitationally. Stars that form close to one another are bound to each other by mutual gravitational attraction. In other words, their individual gravity attracts them to each other, and they move as one body. Many stars are part of a **binary star system**—a pair of stars that orbit their common center of mass. Binary star systems are quite diverse. Some binary stars are extremely close to each other. Other pairs are separated by great distances, such as one-third of a light-year. Binary stars in close proximity may circle each other in a day. Those that are widely separated may orbit one another over perhaps even millions of years. Some binary star systems are orbited by planets.

OBJECTIVES
- Identify various types of star systems.
- Discuss basic features of different star systems.
- Describe how astronomers classify galaxies.

VOCABULARY
- **binary star system** a pair of stars that orbit their common center of mass
- **galaxy** large systems of stars and their solar systems, gas, and dust held together by gravity
- **galaxy cluster** a group of thousands of galaxies under the same gravitational influence
- **star cluster** groups of stars that have a common origin and are held together by mutual gravitational attraction

Binary star system

411

Some stars are part of a triple or quadruple star system, but these systems are not as common as binary systems. Triple systems appear to be binary star systems that have trapped a single star in their orbit. Scientists believe quadruple systems are composed of two binary systems. Like binary star systems, the stars in these systems orbit a common center of mass. These types of multiple star systems are rather unstable. For example, if the three stars in a triple system orbit too closely, the more massive stars will push the smallest star out of its gravity range. The system would then become a binary system.

A large system of stars and its solar systems, gas, and dust that is held together by gravity is called a **galaxy**. God sustains a wide variety of galaxies in all shapes and sizes. Most galaxies are found in groups. Because galaxies are so enormous, astronomers prefer to measure them using large units called *parsecs* and *kiloparsecs*. A parsec (pc) is equal to 3.26 light-years and a kiloparsec (kpc), is equivalent to 3,260 light-years. The Milky Way galaxy is about 30 kiloparsecs across, but there are far larger galaxies. The larger galaxies contain over a trillion stars, whereas smaller galaxies may have only a few million stars. To estimate how many stars might be in a galaxy, astronomers observe its size, brightness, and mass.

Some scientists believe that there could be 100 billion galaxies spread throughout the heavens. These galaxies can be one of

Sombrero Galaxy

Spiral galaxy

412

2. Describe the basic features of three types of star systems. (**Possible answers:** *binary star systems*—have two stars that orbit a common center of mass, can be various distances apart, can be orbited by planets. *triple or quadruple star systems*—have three or four stars that orbit a common center of mass, are less common than binary star systems, are unstable. *galaxies*—have three types; are in constant motion; tend to congregate; can contain a trillion stars; contain large systems of stars, solar systems, gas, and dust. *star clusters*—have stars with a common origin, have two types. *galaxy clusters*—can contain tens of thousands of galaxies, arranged in superclusters that have tens of thousands of galaxies)
3. Explain the basic classifications of galaxies. (*spiral*—have a reddish central bulge of older cooler stars and bluish arms of hot stars, contain nebulae, are the most common type; *elliptical*—have very bright centers, can be elongated, contain very little gas or dust, have only old stars, do not contain nebulae; *irregular*—have no definite shape, contain very little mass, can have lots of gas and young stars, contain nebulae)
4. How many kilometers across is a galaxy that measures 20 kiloparsecs?
   (**20 kpc × 3,260 light-years × 9.46 × 10^{12} km = 6.17 × 10^{17} km**)
5. What is the difference between an open cluster and a globular cluster? (**Open clusters are groups of a few hundred young stars loosely held together by gravity. Globular clusters are dense, ball-shaped groups of older stars.**)

NOTES

HISTORY

Xochicalco
Throughout history, many civilizations have marveled over the patterns of star alignments in the visible universe. The movement of constellations has been used to determine when to plant crops. Mariners also used constellations to navigate. Some groups of people even designed massive ritual structures patterned after star alignments. One such structure in central Mexico is known as *Xochicalco*. Built in the 8th century, this site was an important ceremonial center. It became a cultural link between the earlier Maya civilizations and the Aztecs. Much of the monument is decorated with images of rulers, priests, gods, and astronomers. In addition, calendar signs and a series of 21 calendar altars recorded the months and days of the ceremonial year. Caves located in the hillsides were used for ceremonies and served as underground observatories designed for tracking the sun's movement. Xochicalco was abandoned in the late 9th century.

three basic types: spiral, elliptical, or irregular. Most galaxies appear to be spiral galaxies, which contain lots of gas, dust, and young stars and have a central bulge. Revolving around the bulge are spiral arms that look blue because hot blue stars are found there. The central bulge is redder because the stars there are older and cooler. Although it is hard to tell from Earth's perspective, astronomers believe the Milky Way is a spiral galaxy. They estimate the galaxy's arms are 100,000 light-years across and contain about 100 billion stars. Most of the Milky Way's stars are in the central bulge. Since Earth's solar system is about 30,000 light-years away from the center of the Milky Way, the sun takes about 220,000,000 years to orbit around the galaxy's center.

Elliptical galaxies have very bright centers and very little gas or dust. Some appear elongated like a rugby football, but others look spherical. The apparent shape of an elliptical galaxy depends on a viewer's perspective. In other words, what appears to be a spherical galaxy may actually be the end of an elongated elliptical galaxy. Because these galaxies have little gas, they do not create new stars, so elliptical galaxies contain only old stars.

FYI

Constellations in the Northern Hemisphere
Astronomers divide the northern celestial hemisphere into four quadrants: NQ1, NQ2, NQ3, and NQ4. Thirty-six constellations are found within these quadrants.

NQ4
Aquila
Cepheus
Cygnus
Delphinus

Equuleus
Lacerta
Lyra
Pegasus

Sagitta
Vulpecula

NQ1
Andromeda
Aries

Cassiopeia
Orion
Perseus

Pisces
Taurus
Triangulum

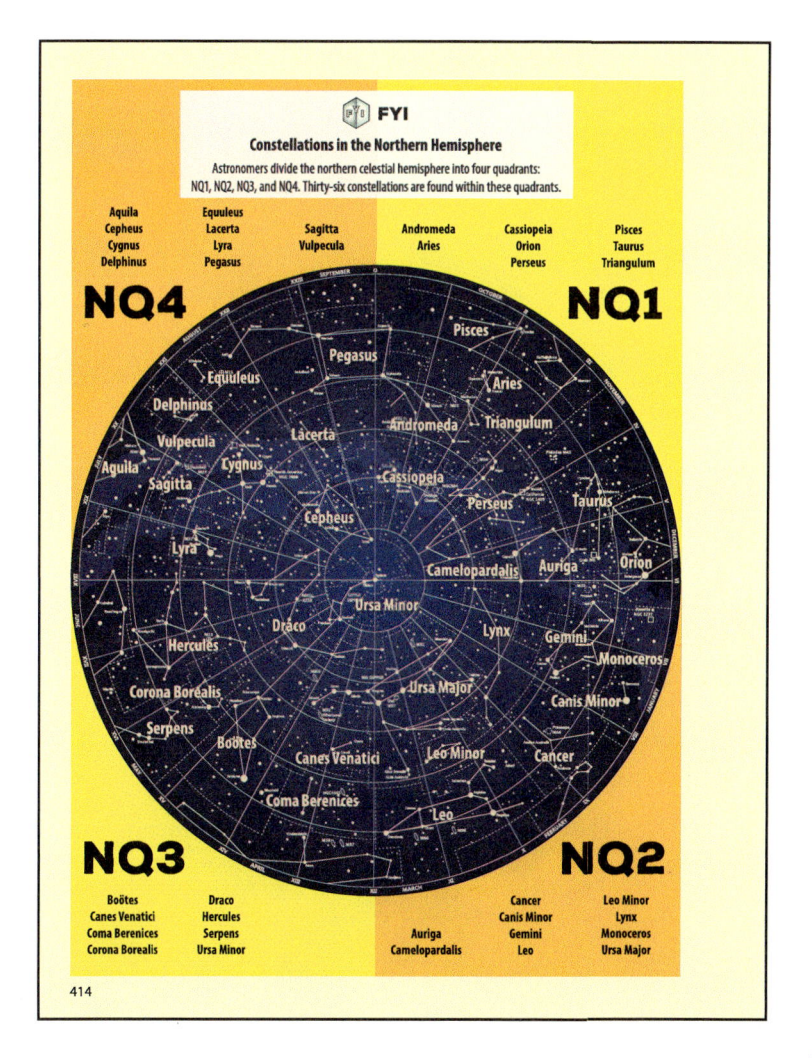

NQ3
Boötes
Canes Venatici
Coma Berenices
Corona Borealis

Draco
Hercules
Serpens
Ursa Minor

NQ2
Auriga
Camelopardalis

Cancer
Canis Minor
Gemini
Leo

Leo Minor
Lynx
Monoceros
Ursa Major

Irregular galaxy

Irregular galaxies have no definite shape and contain very little mass. Like spiral galaxies, some irregular galaxies have a lot of gas and young stars but they do not have arms. Other irregular galaxies have a generally distorted look. Scientists believe that these galaxies may have collided with others in the past.

Regardless of their shape, all galaxies are in constant motion. The Milky Way zooms through space at a speed of 230 kph. Galaxies tend to congregate and can move through space as a unit spanning hundreds of millions of light-years across.

The gas and dust that galaxies contain is called *interstellar matter*. This matter forms the nebulae that give birth to stars. There are two basic types of nebulae: dark and bright. Dark nebulae are very dense and cold. They block the light of the stars behind them. Bright nebulae emit or reflect light from nearby stars. Spiral and irregular galaxies generally contain nebulae, but elliptical galaxies do not.

Star clusters are groups of stars that have a common origin and are held together by mutual gravitational attraction. There are two types of star clusters: open and globular. Open clusters are groups of a few hundred young stars that are loosely held together by gravity. Open clusters are usually located in the bulge of a spiral galaxy. The Hyades in the Taurus constellation

Globular star cluster

Rosette nebula

is the nearest visible open cluster in the Northern Hemisphere; it is 150 light-years away. New open clusters contain young blue stars.

Globular clusters are dense, ball-shaped groups of older stars. The Milky Way galaxy is illuminated by more than 150 globular clusters. Most of these clusters are located at the center of the galaxy. Omega Centauri, visible in the Southern Hemisphere, is one of the largest globular clusters in the galaxy. This globular cluster stretches across a distance of 150 light-years.

Galaxies, although extremely large, are not the largest units in the universe. A **galaxy cluster** is a group of thousands of galaxies under the same gravitational influence. A galaxy cluster may contain tens of thousands of galaxies. The Local Group, the galaxy cluster that includes the Milky Way, contains more than 20 galaxies. Galaxy clusters are arranged into superclusters, which are groups of galaxy clusters. Superclusters contain tens of thousands of galaxies, often millions of light-years across.

An artist's rendering of the most distant quasar

Looking through space is like looking back in time. Astronomers view distant galaxies and other celestial objects to learn how galaxies form and change over time. For example, astronomers study quasars, some of the most distant observable objects. Scientists believe that when the strong gravity of a black hole begins compressing an active galaxy, the galaxy becomes an intense stream of light and radio waves known as *a quasar*. From Earth, quasars look like tiny points of light. They are extremely bright because they drown out the light of surrounding stars. In fact, quasars are so bright they can be observed from billions of light-years away even though they are small.

LESSON REVIEW
1. What is the difference between a star system and a constellation?
2. Describe the basic features of three types of star systems.
3. Explain the basic classifications of galaxies.
4. How many kilometers across is a galaxy that measures 20 kiloparsecs?
5. What is the difference between an open cluster and a globular cluster?

Stars and the Universe

8.1.4 Creation of the Universe

Introduction
Read **1 Chronicles 29:10b–11** to the class. Discuss the enormity of God's power that is evident in His creation of the vast universe. Ask students what the universe can tell people about God. (**Answers will vary but should include that the size, complexity, and design of the universe indicate the genius of God; it indicates that God is orderly, communicative, goal-directed, creative, and resourceful; the balance evidenced in the universe that supports life on Earth provides evidence that God has a purpose for everyone's life.**)

Introduce the subject of the Big Bang model and identify any misconceptions that students may have. Have students share any biases they believe people may have when discussing this subject. Remind students that although they may have differing views on the creation of the universe, they should be respectful of one another. Encourage students to search for common threads rather than to focus on divergent opinions. Challenge students to listen well and act humbly.

Discussion
- Discuss why scientists are interested in discovering information about the early universe. Ask students what the value of such information may be to scientists who are Christians. (**Possible answer: They want to learn more about God's creation.**) What about the value to scientists in general? (**Answer will vary.**) Point out that scientists may wonder about the future of the universe. They may also want to understand current astronomical events.

- Demonstrate the Doppler effect by placing a battery-operated buzzer inside a hollow Wiffle ball. Attach a string to the ball, turn on the buzzer, and swing the ball around your head. Ask students to describe the change in the pitch of the buzzer as the ball swings around. Encourage students to swing the ball above their heads. Point out that the pitch does not really change. Ask students how they could determine the position of the ball without observing its movement. (**Answers will vary but should include that the pitch of the buzzer is lower as it travels away from the observer.**) Challenge students to think of real-life examples that demonstrate the Doppler effect. (**Answers will vary but should include the sound of moving ambulances, trains, or cars that are blowing their sirens, whistles, or horns.**) Explain that the Doppler effect also applies to light waves. When lit objects approach an observer, the emitted light shifts toward the blue end of the electromagnetic spectrum. When lit objects move away from an observer, the light moves toward the red end of the spectrum. Considering that fact, ask students what the Doppler effect tells scientists about stars. (**The Doppler effect indicates the speed and direction of a star's motion.**)

- Ask the following questions:
  1. What is the main focus of the Big Bang model? (**The universe is continually expanding.**)
  2. Describe the movement of matter according to the Big Bang model. (**Matter is moving away from other matter because the space that separates matter is expanding. There is not one central point from which all matter is escaping.**)
  3. Are scientists certain the Big Bang model accurately explains the method of creation of the universe? (**The Big Bang is only a model that people use to explain how the universe was created. It does not explain the exact age of the universe. Also, it does not provide details as to whether the universe will continue to expand or will eventually collapse back on itself.**)

Activities
A. Complete *Try This: Expanding Universe* with students. As a class, discuss the results.

B. Have student groups research the concept of an open, closed, and flat universe according to the Big Bang model. Encourage students to relate their findings to the future of the universe. Direct students to explain their concept's position on whether the universe will continue to expand forever, snap back because of gravitational forces, or eventually stop expanding altogether.

OBJECTIVES
Students will be able to
- summarize the predominant young-earth and old-earth explanations of how the universe was created.
- analyze the evidence used to support the main models of how the universe was created.

VOCABULARY
- **Doppler effect** the apparent shift in frequency of waves emitted by a moving source

MATERIALS
- Battery-operated buzzer, Wiffle ball, string (*Discussion*)

PREPARATION
- Obtain materials for *Try This: Expanding Universe*.

TRY THIS
Expanding Universe
- *balloons, 1 per student*
- *markers, 1 per student*

NOTES

Direct students to create an educational brochure comparing and contrasting the three models. Schedule time for groups to present their brochure and for a class discussion of the findings.

C. Challenge students to list the strengths and weaknesses of the Big Bang model. Ask students how they would address the Big Bang model with individuals who believe the universe is only a few thousand years old. (**Answers will vary but should include conveying to those people that the Big Bang theory is only a model used to explain the evolution of the universe.**) Point out that although the model does not address all aspects of the creation of the universe, it does serve a useful purpose regarding the expansion of space and new star discoveries. Remind students that the Big Bang model does not necessarily deny the existence of God as Creator.

D. Assign students to research the process that led scientists to believe the universe is over 13 billion years old. Ask students to explain how a fast universal expansion rate affected the original ideas about the age of the universe. (**Answers will vary but should include that a faster expansion rate shortens the age of the universe.**) Convey that scientists have had to adjust their original estimate that globular star clusters were at least 15 billion years old to fit the new universe age model.

E. Share something you have learned recently about the universe that inspired a sense of wonder. Challenge students to think about something they have learned in their study of astronomy that caused them to feel wonder at what God created. Assign students to share their personal wonder-producing examples with someone outside of class. Convey that the goal of this activity is not to give the person scientific information, but to express their personal sense of wonder and to practice engaging others in discussions about science. Have students relate their experiences.

Creation of the Universe 8.1.4

God is the Creator and Sustainer of the vast universe. The book of Genesis gives readers a glimpse into the complexity of God's creation, which can be appreciated even more deeply through astronomy. Likewise, many of the psalms marvel at the vastness of the universe. The Bible states that God spoke and brought creation into being. Some Christian scientists believe that when God spoke creation into being, it was formed much like it looks today. However, others believe that God created the universe through a longer process, possibly even over billions of years.

Scientists cannot do experiments on events in the past to perfectly prove the age of the universe. All the calculated ages for the universe rely on assumptions—young earth as well as old earth. Scientists who believe Earth and the universe are young point to the biblical account of Creation. According to the account in Genesis 1:1–19, God created Earth before He created the sun, moon, stars, and other planets, so nothing in the universe is older than planet Earth itself. If Earth is young, every other thing in the universe is younger still. Genesis 1:14–16 explains that God created the sun, moon, and stars individually and then placed them in the heavens. This order of creation directly contradicts the idea that these celestial bodies formed before Earth—or that they formed by a random interaction of dust and gas in the void of space.

Many astronomers and scientists assume the universe is ancient because of their measurements of celestial objects and events. By interpreting their findings, they theorize that the universe is 13.82 billion years old. One scientific model is the Big Bang theory. According to this model, all the contents of the universe were held under tremendous pressure, density, and temperature until finally there was an enormous expansion of space. This event marked the beginning of energy, time, space, and matter. Physical events before the Big Bang are unknown, and scientists do not try to explain them. According to the model, after the Big Bang, gravity began affecting the matter that was expanding outward in all directions. The force of gravity began pulling matter into clumps. These clumps became galaxies, which continued to move outward into the universe and formed galaxy clusters.

In contrast to what you might visualize, objects in the universe are not flying out from one central point. Rather, all celestial objects are moving away from one another because the space between them is expanding. Picture the dough for a loaf of raisin bread: after the dough is mixed, the raisins are a certain distance apart. As the dough rises and expands, each raisin moves away from every other raisin. The universe is like the rising bread dough in that it is expanding in all directions. And like the raisins in the dough, every distant galaxy is moving away from all the other galaxies.

Scientists use the principles they observe in creation to support the Big Bang model. For example, astronomers have determined that most galaxies are moving away from each other and that the universe is expanding at a tremendous rate. How do they know this? To answer that, think of what happens when a fire truck speeds past. As it moves away from you, the pitch of the siren becomes lower because the sound wavelengths are longer. The apparent shift in the frequency of waves emitted by a moving source is called the **Doppler effect**. The Doppler effect applies to light waves as well as sound waves. If a star is moving away from Earth, its light waves will be slightly elongated, or stretched, as they reach Earth. Longer light waves occur on the red end of the light spectrum. So, the light of a star moving away from Earth shifts toward the red end of the spectrum. This color change is called *the redshift*. Because the light from all distant galaxies and quasars has shifted toward red, astronomers have concluded that the universe is still moving outward. It is not known if the universe will continue

OBJECTIVES
- Summarize the predominant young-earth and old-earth explanations of how the universe was created.
- Analyze the evidence used to support the main models of how the universe was created.

VOCABULARY
- **Doppler effect** the apparent shift in frequency of waves emitted by a moving source

TRY THIS

Expanding Universe
With a marker, draw randomly placed dots on a deflated balloon. Slowly inflate the balloon and observe the position of the stars. What happens to the distances between the dots as the balloon expands? Is there a center from which all of the dots are moving or an edge toward which they are all moving?

The universe expansion

The Doppler effect

Lesson Review

1. Summarize the young-earth and old-earth explanations for how the universe was created. (*young earth*: Genesis 1:1–19 states that God spoke and brought creation into being; He created Earth before the sun, moon, and stars; Scripture says that God created the Earth, sun, moon, and stars individually; *old earth*: All the contents of the universe were compressed under tremendous pressure, density, and temperature until there was a great expansion of space; gravity began affecting the expanding matter and pulling it into clumps, which became galaxies; the universe is constantly expanding outward.)
2. What are two pieces of evidence used to support the Big Bang model? (Possible answers: Because very distant galaxies and quasars look red, astronomers believe the universe is expanding. As stars move away from Earth, their wavelengths are elongated by the Doppler effect and shifts toward red on the light spectrum. Cosmic microwave background radiation is believed to be a product of the Big Bang because temperature readings of the radiation are consistent with the estimated temperature of radiation produced by the Big Bang. This type of radiation is found throughout space.)
3. What are two pieces of evidence that suggest the universe is young? (Possible answers: There is no evidence the universe was ever hotter than it is now. There is only a tiny amount of antimatter present in the universe. Spiral galaxies are not compressed into balls or disks. The number of supernova remnants found indicates that the universe is young.)
4. What conclusions can you draw about how the universe was created? Why? (Answers will vary but should include that God is the Creator and Sustainer.)

NOTES

---

to expand or will eventually collapse in on itself, but current evidence suggests that it will continue to expand.

Another piece of evidence used to support the Big Bang model emerged in 1964 when two scientists using a low-noise antenna heard annoying static and tried unsuccessfully to get rid of it. The intensity of the static did not vary, and it came from every direction. The static was identified as cosmic microwave background radiation that was coming from every part of the universe. Many scientists believe this radiation is a product of the Big Bang because its temperature readings match what they estimate those of the Big Bang radiation would be. In addition, this type of radiation is found throughout space.

However, because scientists cannot do experiments on events in the past, the Big Bang model is not testable or repeatable through laboratory science. No scientific method can prove the age of the universe. Young-earth scientists point to evidence that the temperature of the universe is uniform and was never hotter than it is now, which contradicts the Big Bang model of a cataclysmic, hot beginning. According to physics, the creation of matter in the form of hydrogen and helium gas from the energy of the universe's expansion should have created an equal amount of antimatter. However, the universe contains primarily matter with only tiny amounts of antimatter present. Spiral galaxies also indicate that the universe is younger than many supporters of the Big Bang believe. The centers of these galaxies rotate faster than the outer arms, which causes the spiral structure to tighten rather than unwind. Young-earth astronomers point out that if these most-common galaxies were billions of years old, they would no longer have a spiral shape—they would be compressed into balls or disks. Yet the heavens are filled with beautiful spiral galaxies. Other researchers point to the fact that if the universe was very old, there would be evidence of many supernovas. For example, the Large Magellanic Cloud, a nearby galaxy, would have evidence of about 340 supernovas if it was billions of years old, and it would have remnants of about 24 supernovas if the galaxy was about 7,000 years old. The actual number of supernova remnants found in the Large Magellanic Cloud is 29, which indicates the galaxy is only thousands of years old. Even scientists who do not believe in God have argued that something had to have sparked the instant of creation that is represented by the Big Bang. In fact, some atheistic scientists disagree with the theory because they feel the Big Bang model points to the need for a Creator rather than explaining how the universe appeared without one.

The Wide Field Infrared Survey Telescope is scheduled to launch in 2020 to search for evidence of how the universe began.

Spiral galaxy NGC 1232

Supernova remnant NGC 2060 in the Large Magellanic Cloud

Although much of the scientific community has accepted the Big Bang model as a workable explanation for the beginning of the universe, the model cannot answer many questions, such as the universe's exact age or the future expansion or collapse of the universe. Ecclesiastes 3:11 states, "no one can fathom what God has done from beginning to end." It is important to remember that models are tools created by people in an attempt to understand how the eternal God works in His creation.

LESSON REVIEW
1. Summarize the young-earth and old-earth explanations for how the universe was created.
2. What are two pieces of evidence used to support the Big Bang model?
3. What are two pieces of evidence that suggest the universe is young?
4. What conclusions can you draw about how the universe was created? Why?

8.2.0 Space Exploration

Chapter 8.2 Summary

Space has been called *the final frontier*. Since the late 1950s, scientists have advanced from observing the nearby planets through a telescope on Earth to examining soil samples taken from Venus and Mars. Each new accomplishment—rockets, satellites, occupied spacecraft, space probes, visits to the moon, and space stations—has been an amazing feat. Humanity has moved from dreaming about the moon and planets to landing on the moon and sending spacecraft to most of the neighboring worlds. Today, space exploration seems almost commonplace. New technologies are being developed by government space agencies and by private industry as well. The exploration of God's outermost creation will continue as technology advances. Scientists in both the public and private sector are working toward visiting Mars, and in the more distant future, perhaps colonizing it.

Background

Lesson 8.2.1 – Rockets

Rockets are associated with fireworks and military weapons but are most often used for space exploration. The key considerations in rocket design are the propulsion system, the propellant and the exit nozzle, and the number of stages required to lift the payload. Propellants compose about 90% of a rocket's weight. Modern propellants include liquefied gases, which are very powerful, and solid explosives, which are very reliable. In a typical liquid-propellant engine, hydrogen or another liquefied gas serves as the fuel, and oxygen is the oxidizer. A typical solid-propellant engine mixes an explosive, such as nitroglycerin, with a binder or polymer. Liquid-propellant engines require an ignition system only if the propellants do not ignite on contact, whereas solid-propellant engines always require an ignition system. Rocket engines vary in efficiency. All rockets are very inefficient for the first few seconds after liftoff because high power consumption is required to overcome the rocket's inertia and because air resistance is greater in the lower atmosphere. Rockets gain efficiency with altitude.

According to Newton's third law of motion, a rocket's thrust is equal to the mass ejected per second times the velocity of the expelled gases. The action, or the flow of ejected gases, generates the forward motion of the rocket. Rocket acceleration occurs because the gases exert tremendous pressure on all the walls except the exit nozzle. The unbalanced force pushes the rocket ahead.

Early rockets had only one stage of ignition and could not reach orbital velocity or Earth's escape velocity, which is the speed necessary for the rocket to break free of Earth's gravity and continue moving without further propulsion. Therefore, space exploration required the use of multistage rockets. A multistage rocket consists of two or more rockets assembled in tandem that are ignited in turn. Once the lower stage's fuel is exhausted, it detaches and falls back to Earth.

Studies are underway to replace chemical fuels with alternative propulsion systems. Ion propulsion would create plasma using an electric discharge and expel it by an electric field. Such an engine could provide a low thrust efficiently for long periods, allowing for high velocities on a lengthy flight. An ion drive would be a good choice for a journey to the outer planets. Nuclear power propulsion has also been considered. In the 1960s, a nuclear-propelled rocket was developed before scientists understood that the highly radioactive exhaust gases could not be used in Earth's atmosphere. The project was deferred in the 1970s; however, NASA's Nuclear Cryogenic Propulsion Stage team is exploring the future viability of nuclear propulsion.

Lesson 8.2.2 – Space Probes

Space probes easily reach greater distances than humans can travel in space because they do not need as much equipment or safety protection. They also do not have to return to Earth. A space probe is an unmanned space vehicle that explores various aspects of the solar system. Space probes are launched with sufficient energy to escape Earth's gravity and to navigate between the planets. Although spacecraft that orbit the earth usually do not need navigation once they are in orbit, space probes

---

LOOKING AHEAD

- For **Lesson 8.2.1**, obtain water bottle launchers. Plan a rocket launch day. Prior to holding the event, have student groups research designs for model rockets and construct a rocket.
- For **Lesson 8.2.3**, gather 10-centimeter glass tubes for *Lab 8.2.3A Centripetal Force*.
- For **Lesson 8.2.4**, obtain samples of space food or dehydrated food.

SUPPLEMENTAL MATERIALS

- BLM 8.2.2A Space Probe Stress Tests
- BLM 8.2.4A Moon Myths

- Lab 8.2.1A Rockets
- Lab 8.2.3A Centripetal Force

- WS 8.2.2A Astronomical Units
- WS 8.2.3A Orbits

- Chapter 8.2 Test
- Unit 8 Test

BLMs, TMs, and tests are available to download. See Understanding Purposeful Design Earth and Space Science at the front of this book for the web address.

> **WORLDVIEW**
>
> - When King David gazed up at the clear night sky filled with stars, he was inspired to worship the Creator of the heavenly bodies: "When I consider Your heavens, the work of Your fingers, the moon and the stars, which You have set in place, what is mankind that You are mindful of them, human beings that You care for them?" (Psalm 8:3–4). From David's vantage point on the surface of the earth, the stars were magnificent, a testimony of the greatness of God, and a reminder of the "smallness" of man. Today, because we can see more of the celestial bodies than David ever did, the stars and the moon seem so much closer. God has given us the ability to see more of His creation in greater detail. People have walked on the moon, explored deep space, and seen other moons and planets up close. Scientists continue to develop spacecraft and technologies that will take people deeper into space, with the hope of some day colonizing Mars. However, in the process of reaching for the stars, it is tempting to be proud of what human endeavor has done and to shift our focus to ourselves and the created instead of maintaining a sense of awe and wonder for God the Creator. God gave humans the ability to explore the heavens, but He is a jealous god and will not share His glory with anything or anyone. Let us continue to keep the focus where it belongs, on God. Like King David, we should thank the Father above for the beauty of His creation and seek to honor Him as we explore and study the expanse of His universe.

need proper guidance. Many probes require years to reach their destinations, which makes extremely precise prelaunch calculations necessary for a successful mission. Instruments channel data back to Earth via radio contact between a control station and the space probes. Probes are designed to orbit a planet, to deposit instrument packages on a planet, or to fly by one or more planets.

Space probes have greatly increased the scientific knowledge of the solar system and universe. Often these new discoveries have enhanced the understanding of Earth as well. Historically, most space probes have been launched by the United States and the former Soviet Union. Although many of these missions failed or were incomplete, many were also successes. The United States' Mariner missions to the inner planets included visits to Venus, which measured its heavy atmospheric pressure, and observations of Mercury, which revealed that the heavily cratered planet is more massive than previously believed.

The United States' Pioneer missions to the sun, moon, and gas giants were also informative. The program's first missions did not meet their objective of close-up photographs of the moon, but they did gather information on the area between the earth and moon, including data on the Van Allen radiation belts. Launched in 1973, *Pioneer 11* transmitted data for over 20 years and gathered data on Saturn. The Lunar Orbiter probe program greatly enhanced the understanding of the surface features of the moon, especially its far side. NASA used the information these probes gathered to select landing sites for the Apollo program. Later NASA's Surveyor mission landed probes on the moon to determine whether the Apollo lunar module could land on the moon without sinking into the surface, which was thought to be covered in a deep layer of dust.

The United States' Viking missions focused on Mars. The first Viking probe was programmed to search for microscopic life on Mars, but found none. Not all probe missions include landing on a planet's surface. The United States' Voyager missions were outer planet flybys. These two probes greatly multiplied the knowledge of the four gas giants by discovering new moons and Jupiter's rings. They also collected data on the atmospheres of the outer planets and their moons, surface information from the moons, and atmospheric conditions about the gas giants. The two probes are traveling at the outer boundary of the solar system to find the location of the heliopause—the edge of the region where solar wind exists—well beyond the outer planets. If the Voyager probes locate the heliopause, they will be the first spacecraft to reach it.

Space agencies other than NASA have also used probes. The Soviet Luna missions were the first to impact and orbit the moon. The Venera missions to Venus returned data from the atmosphere, surface, and color images of that surface. The *Venera 3*, launched in 1965, crash-landed on Venus, becoming the first spacecraft to impact another planet. The Soviet Union sent 16 Venera probes to Venus from 1961 to 1983. European planetary probes include *Giotto*, which approached Halley's Comet and successfully flew by Comet Grigg-Skjellerup. Japan's planetary missions include *Hiten*, which orbited the moon, and *Sakigate*, which monitors interplanetary magnetic fields, solar wind, and plasma waves.

Lesson 8.2.3 – Satellites

Newton's first law of motion states that a moving body travels along a straight path with constant speed unless it is acted on by an outside force. In order for circular motion to occur, a constant force must act on a body and push it toward the center of the circular path. This force is the centripetal force. For a planet orbiting the sun or a satellite orbiting a celestial body, the centripetal force is gravitational. The force is not limited to satellites, however. When an object is twirled on a string, the force is mechanical; for an electron orbiting an atom, the force is electrical. Centripetal force (F) equals the mass (m) of the body times its velocity squared (v^2) divided by the radius (r) of its path.

$$F = \frac{mv^2}{r}$$

The predictability of both gravity and inertia make the use of satellites possible because scientists depend on these laws to keep artificial satellites in orbit. Initially, a rocket lifts the satellite from Earth's surface, but once it is out of the atmosphere, the satellite maintains its motion without assistance. Artificial satellites remain in orbit indefinitely if they orbit high enough to escape the atmosphere's friction. The atmosphere slows satellites that are orbiting under 320 km; these satellites will inevitably be burned up by the atmosphere. Multistage rockets help satellites achieve sufficient altitude and a velocity of 8 kps, the speed necessary to balance gravitational acceleration.

Satellites' orbits are usually elliptical. An equatorial orbit lies in the plane of Earth's orbit, and a polar orbit lies in the plane that passes through the North and South Poles. A satellite's height above Earth determines the time required to complete one revolution, or period. The higher the satellite, the longer the period. To achieve a 24-hour period, a satellite must orbit approximately 36,000 km above Earth; such an orbit, whose period is the same amount of time of one rotation of Earth, is called *geosynchronous*. If a satellite's orbit is also equatorial, it will remain stationary over one point on Earth's surface.

Precise methods are necessary to track the thousands of satellites that are currently in orbit. Optical and radar tracking are used during the launch, and radio tracking is used once the satellite achieves orbit. Optical tracking uses cameras to follow satellites that are illuminated by the sun or laser beams. In radar tracking, microwaves are directed at the satellite and the reflected echo pinpoints the satellite's direction and distance. Most satellites broadcast their positions to NASA's global network of stations called *STADAN* (*Satellite Tracking and Data Acquisition Network*), which tracks, monitors, and controls artificial satellites.

For decades, the military has used satellites for national security purposes. In the late 1950s, the Soviet Union boasted it had intercontinental missiles capable of crossing the ocean. Satellites were used to check on nations' compliance to weapon agreements. For example, in October 1963, the United States launched six satellites named *Vela* to detect violations of the Nuclear Test Ban Treaty. Since that time, the U.S. military has worked to develop new satellite technologies for military use.

Weather satellites are among the most powerful weather instruments. The first weather satellite, *TIROS* (*Television and Infrared Observation Satellite*), was sent into orbit in 1960. Newer satellites carry high-resolution radiometers that map clouds. Because clouds outline storms, these satellite images can be used to determine weather patterns globally.

The launch of *Sputnik* revealed that the Doppler effect could plot a satellite's exact position. In 1958, scientists began working on the U.S. Navy's Transit system, which used the Doppler effect to position satellites. Since 1967, satellites have been used to determine the positions of offshore oil wells, schools of fish, and vessels at sea. To improve the accuracy of the Transit system, the U.S. Navy and Air Force developed the Navstar/GPS (Navigation Satellite for Time and Ranging/Global Positioning System).

In 1965, the first Intelsat (International Telecommunications Satellite Organization) satellite was launched into orbit. Global communication satellite research is booming; fleets of satellites make it possible to reach almost anyone on Earth by cellphone.

Some satellites are designed for scientific study, such as the *Lageos*, which was put into orbit in 1976 for geophysics research. Other satellites have been placed in orbit to study Earth's atmosphere and magnetic field. GPS satellites help seismologists detect movements in the ground that are too weak for even seismographs to detect. Many aspects of oceanography also rely on satellites like the *TOPEX/Poseidon*, a collaboration between the U.S. and France. This satellite measured ocean topography with near perfect accuracy. The satellite has microwave detectors that measure ice flows during the long polar night and radar to measure wind and waves.

NOTES

Satellites have been instrumental in the fields of astronomy, geophysics, and oceanography. In August 1989, *Hipparcos*, the European Space Agency's High Precision Parallax Collecting Satellite, was launched to measure the precise locations of 120,000 stars. The Hubble Space Telescope (HST), launched by the space shuttle *Discovery* in 1990, has provided improved distance measurements and spectacular photographs of deep space. Currently, NASA, the European Space Agency, and the Canadian Space Agency are conducting tests on the James Webb Space Telescope, the planned successor of the HST. It is scheduled to launch in 2018.

Lesson 8.2.4 – Working in Space

Working in space affords great learning opportunities because the environment is not under the constraints of Earth. For example, gravity influences all experiments on Earth. By conducting such studies on the space station, scientists have the opportunity to conduct long-term experiments away from gravity's influence. For example, space station labs can test and improve medical procedures and equipment. In addition, the loss of bone density from microgravity that astronauts experience offers the opportunity for study on the mechanisms of osteoporosis. Space stations can also house space telescopes for viewing the universe without Earth's atmospheric interference.

People's opinions about the space program vary widely. Some are fascinated by it, many ignore it, and others see it as useless. Space exploration supporters see many reasons to study space. One good reason, of course, is to learn more about creation and the wonders of the Creator. However, because of issues that need study and funding here on Earth, some people have concluded that the realms of space are best left alone. But outer space is not unknown to God; it is part of His creation and through it He reveals himself. "For since the creation of the world God's invisible qualities—His eternal power and divine nature—have been clearly seen, being understood from what has been made, so that people are without excuse" (Romans 1:20). Although people can observe God's power by just looking around the earth, the mysteries of space only widen the awe of God's invisible qualities.

Many people who question the wisdom and practicality of the space program have a limited understanding of it. The space program consists of more than spending billions of dollars on missions to the moon or other planets to take pictures or collect rocks. God has given humanity the intelligence and the curiosity to explore and enjoy His creation, to find new frontiers, and to glean new knowledge. Successfully exploring space or any new frontier requires political will, sufficient funds, popular support, scientific knowledge, and applicable technologies. For decades, the Cold War rivalry between the United States and the Soviet Union justified both countries' aggressive space exploration programs. After the Cold War, funding for many space agencies decreased despite claims that voyages for discovery, moon colonization, and interstellar travel were justification enough for continuing to explore.

The benefits of space exploration do not lie only in the future; past missions have already improved the quality of daily life in ways that are not often realized. Spin-offs such as solar-powered devices, freeze-dried foods, and athletic clothing are direct and indirect products of the space program. Space technology also has applications for archeology—researchers from four organizations used a spin-off from the Hubble Space Telescope called *charged coupled devices* (CCDs) to help decipher previously unreadable portions of the Dead Sea Scrolls. Space archeologists also use satellite images to help locate such things as ancient cities or tombs that are buried or covered with vegetation. The space program has advanced medical technology with voice-controlled wheelchairs, "invisible" braces, implantable heart aids, and many other things. The space program also creates jobs, not only in the aerospace industry but in companies applying space technology spin-offs.

Space Exploration

8.2.1 Rockets

Introduction
Inflate a balloon and release it above students' heads. Ask students why the balloon moves. (**The force of the air moving out of the balloon pushes the balloon forward.**) Have students suggest theories about how rockets work. After students' have shared their ideas, explain that the same principle that moved the balloon in the demonstration applies to rocket propulsion. When fuel burns in the combustion chamber of a rocket, the escaping gas thrusts the rocket forward. The first stage of a multistage rocket lifts the rocket off the ground. The first stage of the rocket is then jettisoned, and the second stage of the rocket places the rocket into orbit. Point out that the force that lifts a rocket off the ground and puts it into orbit is called *thrust*.

Discussion
- Show footage from the Apollo 11 moon landing and the *Challenger* or *Columbia* explosions. Discuss the advantages and disadvantages of exploring space. Ask students why they believe people want to explore space. (**Answers will vary.**) Why do you think people might be opposed to exploring space? (**Answers will vary.**) What part does risk play in scientific endeavor? (**Answers will vary.**)

- Discuss the history of rockets. Point out that until the 1950s, rockets were used as weaponry rather than for space flight. Refer to how World War II advanced the space program in the U.S. Explain that after the war, rocket scientists from Germany, who worked on weaponry during the war, joined the team of rocket scientists from the United States. Ask students if they know which superpower launched *Sputnik*. (**the U.S.S.R.**) Is the Space Race still going on? (**Answers will vary.**) What parts of space would you like to explore? (**Answers will vary.**)

- Ask the following questions:
  1. What is used for rocket fuel? (**Liquefied gases, such as hydrogen and oxygen, are used in liquid-propellant engines. Solid-propellant motors use explosives, such as nitroglycerin, which are combined with a binder.**)
  2. Theoretically, what do you think would be an advantage to using nuclear power to fuel rockets? (**Answers will vary.**) What disadvantage do you think such a fuel may have? (**Answers will vary but should include nuclear power would result in dangerous radiation.**) If nuclear-powered rockets were the only way to reach other planets, do you think they should be used? Why? (**Answers will vary.**)
  3. What is the difference between escape velocity and orbital velocity? (**Escape velocity is the speed and direction necessary for a rocket to break away from the pull of Earth's gravity; orbital velocity is the speed and direction needed for a rocket to orbit Earth.**)
  4. In what cases might a spacecraft be launched with each type of velocity? (**A spacecraft that is going to visit another celestial body would be launched with escape velocity; a satellite designed to orbit Earth would be launched with orbital velocity.**)

Activities

Lab 8.2.1A Rockets
- fishing line
- scissors, 1 pair per group
- tape
- straight drinking straws, 2 per group
- large balloons, 3 per group
- spring clothespins, 2 per group
- metersticks, 1 per group
- marbles or paper clips
- mass scales, 1 per group
- materials for cargo containers (small cardboard boxes or paper cups)

Have extra fishing line, straws, and balloons on hand to use as replacements or for modifications.

OBJECTIVES
Students will be able to
- summarize the history of the rocket.
- explain how a rocket is propelled.

VOCABULARY
- **payload** the cargo or equipment carried by a spacecraft
- **propellant** a fuel used to produce the hot gases that power a rocket
- **rocket** a machine that uses escaping gas to move
- **thrust** the force that causes a rocket to accelerate

MATERIALS
- Balloon (*Introduction*)
- Video footage of Apollo 11 moon landing and of a space shuttle explosion (*Discussion*)
- Rubber balls (*B*)
- Rocket engines for model rockets (*C*)

PREPARATION
- Gather materials for *Lab 8.2.1A Rockets*.
- Obtain materials for *Try This: Bottle Rocket*.
- Schedule a rocket launch day. Prior to holding the event, have student groups research designs for model rockets and construct a rocket. (*C*)

TRY THIS
Bottle Rocket
- water bottle launchers, 1 per group
- bicycle pumps, 1 per group
- 2 L bottles, 1 per group

(**Thrust propelled the rocket, according to Newton's law. This is similar to a rocket because thrust is used to propel both the bottle rocket and a space rocket. A water rocket is different from a space rocket because it uses air pressure and water to build force. A space rocket uses the energy from the reaction of the propellants.**)

NOTES

A. Complete *Try This: Bottle Rocket* with students. Discuss safety procedures and proper use of the water bottle launcher. Extend the activity by discussing how to improve the bottle rockets so they fly straighter and higher.

B. Divide the class into groups and give each group a rubber ball. Direct students to bounce the ball and to identify the action and the opposite and equal reaction. (***action*: the ball hits the ground, *opposite and equal reaction*: the ground pushes back on the ball**) Next, have one student in each group throw a ball to another group member. Ask students to identify the action and the opposite and equal reaction that occur when the student catches the ball. (***action*: the ball impacts the student's hand, *opposite and equal reaction*: the student stops the ball**) Challenge students to create five more scenarios with an action and an opposite and equal reaction, clearly identifying each part of the scenario.

C. Conduct a rocket launch day. Create competition categories including the following: most creative, greatest altitude, longest flight, straightest path, and greatest integrity of rocket.

D. Divide the class into four groups. Assign each group to make a space exploration time line using one of the following:
- History of the rocket from the Chinese to Sputnik
- The United States and Soviet missions
- The Apollo missions
- European, Indian, or Japanese space missions

8.2.1 Rockets

OBJECTIVES
- Summarize the history of the rocket.
- Explain how a rocket is propelled.

VOCABULARY
- **payload** the cargo or equipment carried by a spacecraft
- **propellant** a fuel used to produce the hot gases that power a rocket
- **rocket** a machine that uses escaping gas to move
- **thrust** the force that causes a rocket to accelerate

Have you ever dreamed of going to the moon, landing on Mars, or sailing past Jupiter? Throughout most of history, such achievements were possible only in dreams or science fiction. In the early 20th century, science fiction books that described fantastic trips to the moon and into space sparked the interest of many young people who later made important contributions to the field of space flight. But the key to exploring space was the **rocket**. A rocket is a machine that uses escaping gas to move. The first rockets were developed as weapons by the Chinese in the 13th century. For over 700 years, rockets were used for warfare. The concept of rocketry being used for space travel developed gradually.

Self-taught Russian teacher Konstantin Tsiolkovsky (1857–1935) enjoyed the science fiction writing of Jules Verne, who wrote about space flight in books such as *From the Earth to the Moon*. Tsiolkovsky developed the basic theory of rocket propulsion and suggested the use of liquid propellants. A propellant is a fuel used to produce the hot gases that power a rocket. He even calculated how fast a rocket would need to travel to escape Earth's atmosphere and determined how much fuel would be needed. But even though Tsiolkovsky explained how rockets work, he never built any himself.

Robert Goddard (1882–1945) of the United States is considered the father of modern rocketry and space flight. When he was a teenager, he read H. G. Wells's *War of the Worlds* and dreamed of making a device that could go to Mars. In 1914, while he was watching fireworks, he came up with the idea of launching a rocket. His first rocket went 12.5 m into the air. Goddard believed that rockets could travel in a vacuum even though most scientists at the time did not agree. In 1926, he launched the first rocket powered by a liquid propellant. Between 1930 and 1941, Goddard continued to build and test rockets of increasing complexity. Because at the time, rockets were used for weaponry, not space flight, Goddard assisted the U.S. military during both world wars. Goddard's larger rockets reached altitudes over 2.4 km.

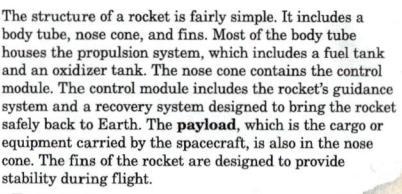

Robert Goddard

During World War II, Germany developed the V-2 rocket, the first long-range guided missile. The V-2, which was used in the bombardment of England, was designed by the rocket engineer Wernher von Braun. During the war, the German government built a center where von Braun and members of his team built and tested rockets as part of the war effort. At the end of the war, von Braun and his team surrendered to the United States, where they hoped to continue their rocket development for space travel rather than military use. The addition of Germany's best rocket scientists was a great benefit to rocket research in the United States in the 1950s.

After World War II, the Cold War began. The Cold War was an arms race between the United States and the Soviet Union (U.S.S.R.), the superpower that included Russia and 15 soviet socialist republics. One outgrowth of the political tension and arms buildup of the Cold War was what is called *the Space Race*. The United States and the U.S.S.R. both wanted to demonstrate their national strength by getting into space. In response to the launch of Sputnik in 1957 by the U.S.S.R., the United States formed the National Aeronautics and Space Administration (NASA) on July 29, 1958. NASA unified the separate teams that were working on rocket development in the U.S. The many accomplishments of NASA include a series of dependable rockets. Until recently, rockets were designed for single launches, although some components of the rockets for the U.S. space shuttle program were retrieved and used on subsequent missions. The entry of private companies into rocket design and production has sparked innovations about how rockets to be landed back on Earth and reused. For example, on April 8, 2016, SpaceX successfully landed the first stage of a Falcon 9 rocket on a ship in the Atlantic Ocean.

The space shuttles used both liquid and solid propellants during launch.

The structure of a rocket is fairly simple. It includes a body tube, nose cone, and fins. Most of the body tube houses the propulsion system, which includes a fuel tank and an oxidizer tank. The nose cone contains the control module. The control module includes the rocket's guidance system and a recovery system designed to bring the rocket safely back to Earth. The **payload**, which is the cargo or equipment carried by the spacecraft, is also in the nose cone. The fins of the rocket are designed to provide stability during flight.

Have each group present its time line to the class. Then as a class, compile all the time lines into one and display it in the classroom.

E. Challenge students to research a specific astronaut or rocket scientist from a country other than your own. Have students write a report and present it to the class.

Lesson Review

1. Who developed the theory of rocket propulsion and liquid propellants? (**Konstantin Tsiolkovsky**)
2. Who is the father of modern rocketry? (**Robert Goddard**)
3. Why was NASA formed? (**The United States and the U.S.S.R. were both involved in the Space Race. Both countries were trying to prove their strength by getting into space. When the U.S.S.R. launched *Sputnik*, the United States started NASA to unify the teams working on the space program.**)
4. Rocket engine propulsion illustrates what law? (**Newton's third law of motion**) What does that law state? (**that every action has an equal and opposite reaction**)
5. What is the force that propels a rocket? (**thrust**)
6. How is thrust produced? (**When the propellant in a rocket ignites, it creates hot gases. These gases are under so much pressure that they exert a tremendous amount of force. As the gases are directed out through the exit nozzle, the rocket is pushed upward.**)

NOTES

Rockets move by propulsion, meaning they are moved forward by a force. Propulsion moves spacecraft or jet planes by pushing something out behind them. For example, when you blow up a balloon and then release it, the air rushing out the open end of the balloon propels it to move. This response occurs according to Sir Isaac Newton's third law of motion, which states that every action has an equal and opposite reaction. In the case of a rocket, the action of the hot gases rushing out through the exit nozzle causes the reaction of the rocket moving in the other direction.

How can any action be large enough to propel a heavy rocket into space? The mass of the rocket, including the fuel that it carries, is much greater than the mass of the hot gases that stream out of the rocket. These hot gases are under so much pressure, however, that they exert a tremendous amount of force on all the interior rocket walls except the exit nozzle. Therefore, as the gases are directed out through the nozzle, the rocket is pushed upward.

Modern **propellants** include liquefied gases, which are very powerful, and solid explosives, which are simpler and very reliable. In a typical liquid engine, hydrogen serves as the fuel and oxygen as the oxidizer; the fuel will not burn without oxygen. The hydrogen and oxygen are stored separately until they are released into a chamber where they ignite. Some liquefied gases burn on contact; others require an ignition system. The reaction of the two propellants produces the hot gas action needed to move the rocket. One advantage to liquid propellant engines is that they can be controlled to allow the engine to stop or restart, which is not possible with solid propellants. Liquefied hydrogen gas was used by NASA, Russia, and the European Space Agency as the fuel in most rockets that launch spacecraft. However, because of its tremendous power, liquid oxygen has become the fuel of choice for SpaceX and other entities that launch satellites.

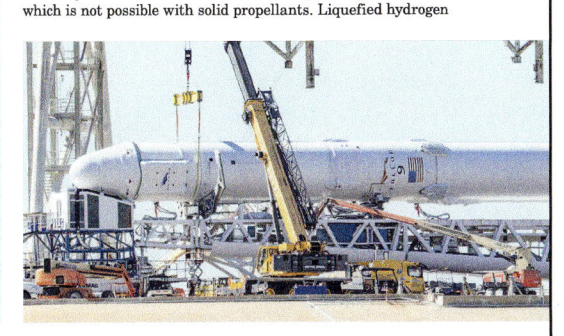

Solid propellants are usually a mixture of an explosive, such as nitroglycerin, with a binder. One common binder is nitrocellulose, which is cellulose that is treated with very flammable nitric acid. The binder solidifies the fuel. A typical solid propellant combination is nitroglycerin and nitrocellulose. Solid propellant mixtures are poured into a mold inside the rocket and allowed to solidify. Then, the mold is removed, which leaves a narrow tunnel through the center of the solid fuel. This long tunnel forms the combustion chamber of the solid rocket. All solid propellants require an ignition system. When the solid fuel in the combustion chamber is ignited, its surface burns at a predictable rate, which provides the thrust necessary for acceleration. Solid propellant motors are used in a variety of ways. For example, solid propellant rockets can be used as the final stage of a spacecraft launch or used to boost payloads such as weather satellites into higher orbits.

The force that causes a rocket to accelerate is called **thrust**. The thrust of a rocket must be sufficient to give the rocket what is called escape velocity, which is the speed and direction necessary for an object to break free of Earth's gravity. Otherwise, a rocket will fall back to Earth. The thrust of a rocket also must be great enough for it to reach what is called orbital velocity, which is the speed and direction needed for a rocket to orbit Earth.

TRY THIS
Bottle Rocket
Rockets are propelled by pressure created in the combustion chamber and directed out the back of the rocket. Create your own rocket using a 2 L bottle, water, and air. Read the package directions before using the water bottle launcher. Wear safety goggles and stand as far from the launcher as possible. Use the bicycle pump to build pressure inside the bottle, but be very careful not to overpressurize the system. Predict how high the rocket will go and how long it will stay aloft. What caused the water bottle rocket to fly into the air? How is the water bottle rocket similar to a space rocket? How is it different from a space rocket?

FYI
Nuclear Propulsion
Researchers at NASA are testing the use of nuclear rocket engines to power rockets. Nuclear rocket engines generate more thrust and are twice as efficient as conventional chemical rocket engines. But, radiation is a real hazard. At the Marshall Space Flight Center in Huntsville, AL, scientists are using a simulator to perform realistic, but nonnuclear, testing. They can test how the nuclear engines will work without risking the hazard of using nuclear materials. NASA's nuclear rocket work may be key to landing human explorers on Mars. The Nuclear Cryogenic Propulsion system could transport people through space more efficiently than conventional spacecraft. It might also reduce the crew's exposure to harmful space radiation and carry heavy payloads.

SpaceX rocket in preparation for launch

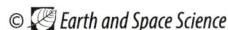

HISTORY

Early Rockets

Basic rocket technology is more than 800 years old. Throughout most of history, rockets were used for war. For example, around 1232 AD, the Chinese stuffed bamboo with a flammable powder to make "fire arrows." The Mongols produced their own rockets and used them in attacks against Japan and Baghdad in the 13th through 15th centuries. The Mongols also may have introduced their rockets to Europe. In the early 1800s, Englishman William Congreve designed rockets that could travel close to 3 km and carry explosive warheads that weighed up to 27 kg. The British used these "Congreve rockets" on the American soldiers at Fort McHenry in 1814. Francis Scott Key, who witnessed the battle, then penned the line "And the rockets' red glare, the bombs bursting in air" for what became the U.S. national anthem.

LESSON REVIEW

1. Who developed the theory of rocket propulsion and liquid propellants?
2. Who is the father of modern rocketry?
3. Why was NASA formed?
4. Rocket engine propulsion illustrates what law? What does that law state?
5. What is the force that propels a rocket?
6. How is thrust produced?

Rocket Parts

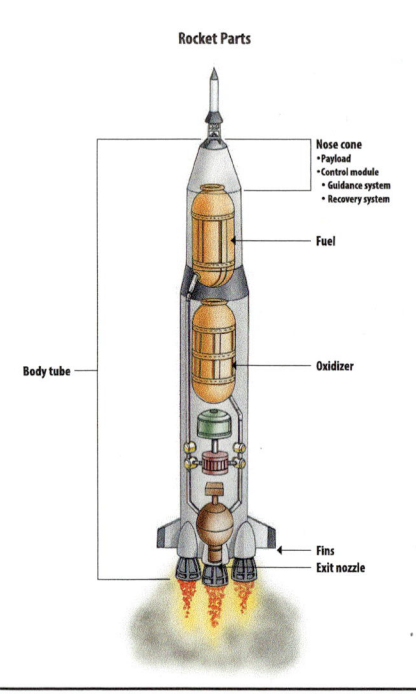

- Nose cone
  - Payload
  - Control module
  - Guidance system
  - Recovery system
- Fuel
- Body tube
- Oxidizer
- Fins
- Exit nozzle

Name: _____ Date: _____

Lab 8.2.1A Rockets

QUESTION: How much cargo can the balloon rocket carry?

HYPOTHESIS: Answers will vary.

EXPERIMENT:

You will need:		
• fishing line	• 2 straight drinking straws	• cargo, such as marbles or paper clips
• scissors	• 3 large balloons	• mass scale
• tape	• 2 spring clothespins	• small cardboard box or paper cup
	• meterstick	

Steps:
1. Cut a length of fishing line that will stretch from the ceiling to the floor. Attach the fishing line to the ceiling with tape. Thread the straw onto the line.
2. Attach the other end of the fishing line to the floor. The line should be vertical and taut.
3. Inflate the balloon to the size of a softball. Do not tie it closed; clip it shut with a clothespin.
4. Tape the balloon to the straw, closed end down, and bring the straw down to touch the floor.
5. Release the clothespin and let go of the balloon. Measure how far it travels. Record the distance. Answers will vary.
6. Remove the balloon from the straw and inflate it so it is twice the size of a softball.
7. Repeat Steps 5 and 6. Record the distance. Answers will vary.
8. Design a rocket to transport cargo. The goal is to transport the heaviest payload possible to the highest height possible. Consider if a multi-stage design is needed. Test and improve your design. You may need to replace your balloons periodically as they lose elasticity. In the Notes section of the table, write your results, any problems you encountered, and what you will change for the next trial.

Trial	Materials Used to Hold Cargo	Cargo and Mass	Distance Traveled	Notes
		Answers will vary.		

Lab 8.2.1A Rockets

Trial	Materials Used to Hold Cargo	Cargo and Mass	Distance Traveled	Notes
		All answers will vary.		

ANALYZE AND CONCLUDE:

1. Explain why the balloon moved. The balloon was propelled by thrust. Newton's third law of motion says that every action has an equal and opposite reaction. As the air was forced out of the balloon, the balloon moved in the opposite direction of the air.
2. What was the maximum payload your rocket could carry? Answers will vary.
3. What problems did you encounter? Answers will vary.
4. How did you correct your problems? Answers will vary.
5. If you could do the experiment again, how would you improve the experiment? Answers will vary.

Space Exploration

8.2.2 Space Probes

Introduction
Have students imagine they are chewing gum and then taking a drink of ice water. Ask what happens to gum in their mouths when cold water hits it. (**Possible answers: The gum gets hard to chew; it loses its elasticity.**) Explain that when *Challenger* exploded, the investigation found that one of the O-rings that sealed the joint between the sections on a solid rocket failed, probably because the temperature on the day of the launch was very cold. Unfortunately, the fact that the silicone-like material of the O-rings became less flexible when it got cold was overlooked. Relate that when spacecraft and probes are sent into space, scientists have to think about the environment they are going to encounter. Brainstorm about what types of circumstances probes going into space may encounter. (**Possible answers: dust or debris, radiation, extreme heat or cold, pressure, acidic gases**) Ask students to consider what a probe destined for Saturn might experience. (**Answers will vary but should include that Saturn's rings are bits of ice and ice-covered matter that are moving very fast.**) How would that be different from what a probe headed to Venus would encounter? (**The atmosphere around Venus is very hot, heavy, and acidic.**) Explain that each probe has to be designed for what it is going to encounter during launch, as it travels, and at its destination.

Have students brainstorm facts about planets. Ask students how they think this information was obtained. (**Answers will vary.**) Explain that space probes have aided in these discoveries. Ask what it takes to get a space probe into space and then to the correct location. (**With the aid of a rocket, a space probe would have to reach escape velocity. The scientists would have to calculate and plot the course of the probe.**)

Discussion
- Place a pan on the floor. Stand on a ladder or step stool and drop a raw egg into the pan. Ask students how a fragile spacecraft might land safely on another planet or how a space probe would safely drop instruments onto a planet. (**Answers will vary.**) Discuss the challenges involved with landing a spacecraft that has been traveling over 30,000 kph.

- Discuss the value of space probes. Ask students how space probes benefit scientific progress. (**Answers will vary but should include probes improve the understanding of other planets and may provide valuable resources.**) Why do you think some people might be opposed to using space probes? (**Answers will vary.**)

- Ask the following questions:
  1. Without a pilot, how do space probes adjust for obstacles in a flight path? (**Scientists on the ground can use radio commands and computers to make adjustments.**)
  2. What instruments are used to gather and send data back to Earth? (**Space probes carry radio transmitters and receivers, magnetometers, and television cameras that are sensitive to infrared, visible, and ultraviolet light. They may also carry devices to detect micrometeorites, gamma rays, and the solar wind.**)
  3. Why does it take years for a space probe to reach a planet? (**because the planets are moving and because they are so far away**)
  4. Scientists believe *Pioneer 10* is on a trajectory headed for Aldebaran. What might prevent it from reaching the red star? (**Answers will vary but should include planets and other space objects are constantly moving and could block or deflect its path.**)

Activities
A. Complete *Try This: Accuracy from a Distance* with students. Discuss how difficult it can be to reach a long-distance target that is in constant motion. Extend the activity by adding orbiting bodies between the launch site and the destination. Increase the distance between the launch site and the destination and have two or three students orbit around the launch site. Direct students to toss the ball into the jar without hitting the orbiting planets. Next, slow everyone down.

OBJECTIVES
Students will be able to
- describe how a space probe reaches its destination.
- evaluate the benefits of using space probes.

VOCABULARY
- **space probe** an unmanned spacecraft designed to gather data

MATERIALS
- Pan, ladder or step stool, egg (*Discussion*)
- BLM 8.2.2A Space Probe Stress Tests
- WS 8.2.2A Astronomical Units

PREPARATION
- Obtain materials for *Try This: Accuracy from a Distance*.

TRY THIS
Accuracy from a Distance
- small balls, 1 per group
- wide-mouthed jars, 1 per group
(Increased distance and movement make the accuracy worse and it is much harder to get the ball into the jar. The ball represents a space probe and the jar represents the probe's destination. Answers will vary but should include variables such as hitting a moving target and passing around and between moving targets.)

NOTES

Choose a student to act as the scientist on Earth who guides the space probe (the ball). Have one student carry the ball toward the destination point by following the scientist's verbal directions.

B. Divide the class into pairs. Distribute two identical sample materials from **BLM 8.2.2A Space Probe Stress Tests** to each pair. Challenge student pairs to conduct a stress test on one of the samples to determine how well the material would withstand the conditions in space. Direct pairs to analyze how the test affected the sample by comparing it to the control and then to report on their findings.

C. Assign **WS 8.2.2A Astronomical Units** for students to complete. For advanced students, direct them to convert the answers into scientific notation.

D. Encourage students to research space probe missions and to make computer presentations explaining the benefits of their selected missions. Provide time for student presentations.

E. Divide the class into groups. Assign a specific space probe for each group to research and model. Assign groups to present their models and to explain the functions of specific components.

F. Challenge students to design a space probe, decide what information it will gather, and plan its mission. Have students consider the following questions:
 1. How does your probe compare to probes that have already been launched?
 2. What are the pros and cons of landing on a planet? Of orbiting it? Of sending pictures? Of gathering samples?

Space Probes 8.2.2

Why are rockets sent into space? One use of rockets is to launch space probes. **Space probes** are unmanned vehicles sent into space to gather information. Space probes have widened the understanding of the solar system. Rockets launch space probes with enough energy to achieve escape velocity and then to navigate among the planets.

Space probes are designed for different missions. Some may fly by or orbit a planet, and others may land instruments on a planetary surface. Because the planets are moving and because they are so far away, it may take years for a probe to reach its destination. Setting a probe on the correct course requires extremely precise calculations. However, by using radio commands and computers, the path of a space probe can be adjusted even after it has been launched. Radio-transmitted commands and computers provide midcourse corrections to a probe's trajectory. Instruments can channel data back to Earth via radio contact between a control station and the space probe. Space probes carry radio transmitters and receivers, magnetometers, and television cameras that are sensitive to infrared, visible, and ultraviolet light. Probes may also carry devices to detect micrometeorites, gamma rays, and solar wind.

Space probes have greatly increased scientists' knowledge of the solar system and the universe. Space probe missions have made discoveries that have enhanced the understanding of Earth as well. The United States and the former Soviet Union have launched the most space probes. Although many of these missions failed or were incomplete, many were also successes.

OBJECTIVES
- Describe how a space probe reaches its destination.
- Evaluate the benefits of using space probes.

VOCABULARY
- **space probe** an unmanned spacecraft designed to gather data

Galileo space probe

427

HISTORY

Astronomical Units
The distances in space are truly astronomical—so large that they are hard to imagine. Over the centuries, many scientists attempted to determine the distance between the sun and the earth. For example, Tycho Brahe estimated the distance at 8 million km, and Johannes Kepler's estimate was 24 million km. In 1672, Giovanni Cassini used the parallax method to determine the distance between the sun and the earth. By comparing the position of Mars in the sky from Paris and from French Guiana in South America, he calculated the distance to be 140 million km. Cassini's figure was very close to the number used today for the average distance to the sun, which is 149,597,871 km. This distance is referred to as *an astronomical unit (AU)*. The AU is handy for measuring distances within the solar system, but going beyond it, the light-year is better.

For example, the United States' early Pioneer missions to the sun, moon, and gas giants did not meet their objective of close-up photographs of the moon, but they did gather data on the area between the earth and moon, including information on the Van Allen radiation belts, which are bands of high-energy radiation in the magnetosphere, where charged particles from the sun are trapped. On later missions, *Pioneer 11* became the first spacecraft to fly by Jupiter and to study Saturn.

Planetary probes enable scientists to learn about the other planets in the solar system. For example, NASA's Mariner missions confirmed that Mars had high temperatures and a carbon dioxide atmosphere. The missions also disclosed that the cratered surface of Mars had huge volcanoes and showed evidence of water. A later Mariner visit to Venus revealed the planet's heavy atmospheric pressure. And Mariner missions gathered data from Mercury that revealed the planet was more massive than previously thought.

Probes are also used to study smaller orbiting bodies and other phenomenas in the solar system. The European planetary probe *Giotto* approached Halley's Comet and successfully flew by the comet Grigg-Skjellerup. Japanese probe missions have included *Sakigate*, which monitors the solar wind, interplanetary magnetic fields, and plasma waves, and *Hiten*, which orbited the moon.

Technicians work on Genesis, which collected solar wind particles.

Scientists use solar probes to monitor the sun. Solar probes gather information about how the

428

3. What are potential hazards the probe might encounter?
4. What kind of calculations will be necessary for the mission?
5. How long will it take the probe to reach its destination?

G. Direct students to write a science fiction story about the discoveries of a space probe on an unknown planet. Encourage students to add illustrations and to present the story to the class.

Lesson Review

1. What is needed for a space probe to get into space? (**Space probes need rockets to launch them with enough energy to achieve escape velocity and then to navigate among the planets.**)
2. Without a pilot, how do scientists direct a space probe to its destination? (**They use radio commands and computers to provide midcourse corrections to a probe's path.**)
3. Why is navigating in space difficult? (**It is difficult because the planets are moving and the destinations are very far away. It can take years for a space probe to reach its destination.**)
4. What did scientists learn about Mars, Venus, and Mercury from the Mariner missions? (*Mars*: **high temperatures, cratered surface with huge volcanoes, water, and carbon dioxide atmosphere;** *Venus*: **heavy atmospheric pressure;** *Mercury*: **heavily cratered and more massive than previously thought**)
5. What were the benefits of the Pioneer missions? (**They gathered information on the area between Earth and the moon, including information on the Van Allen radiation belts.** *Pioneer 11* **was the first spacecraft to fly by Jupiter and study Saturn.**)
6. Which probes gathered information about comets? (**Possible answers:** *Giotto* **and** *Deep Impact*)

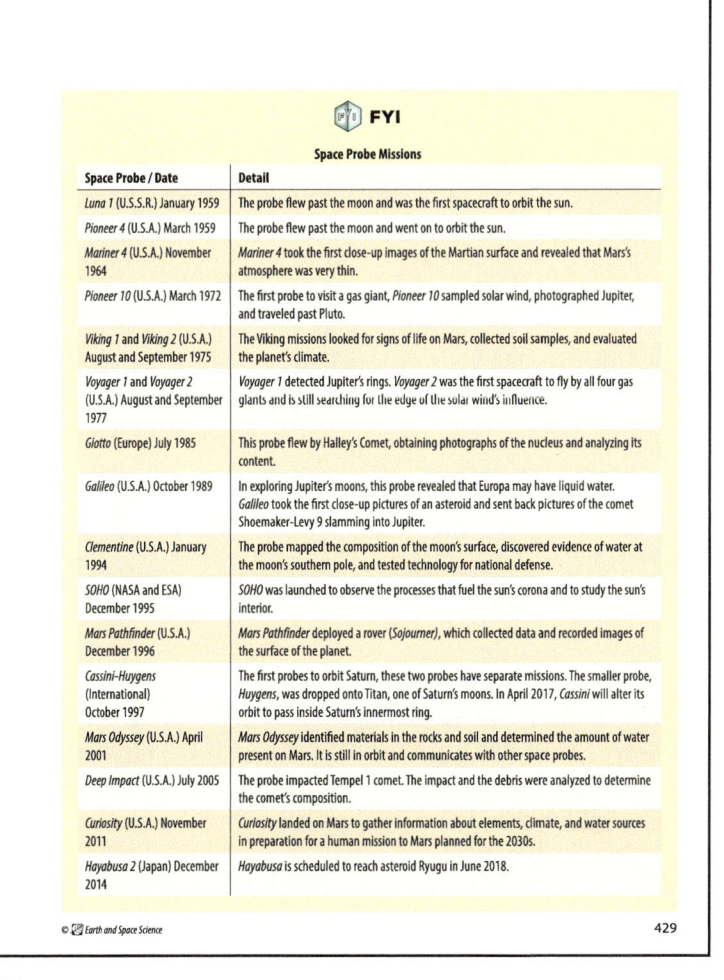

8.2.3 Satellites

Space Exploration

OBJECTIVES

Students will be able to
- explain how a satellite is able to orbit.
- correlate types of satellites with specific orbits.
- compare the types of orbits.
- infer the benefits of satellites.

VOCABULARY

- **inertia** the tendency of a moving object to keep moving in a straight line unless another force acts on it
- **satellite** a natural or artificial object that orbits a larger astronomical body

MATERIALS

- Small bucket (*Introduction*)
- WS 8.2.3A Orbits

PREPARATION

- Fill a small bucket one-quarter full of water. (*Introduction*)
- Obtain materials for *Lab 8.2.3A Centripetal Force*.
- Gather materials for *Try This: Satellites in Orbit*.

TRY THIS

Satellites in Orbit
- tennis balls, 1 per group
- ribbon, 1 m per group
- needles, 1 per group
- thread

(**My hand exerts the force on the ball. The ball would fall down because the only force acting on it would be gravity. The ball is moving in a circular path around my head in the same way a satellite orbits Earth. Unlike a satellite, the tennis ball does not have enough velocity to overcome the pull of gravity or overcome the friction of the air molecules.**)

Introduction

Show students the prepared bucket. Ask students why the water stays in the bucket. (**Answers will vary but should include gravity and the sides of the bucket keep it inside.**) Stand back from students and inform them you are going to hold the bucket upside down over your head. Ask students what will happen. (**Answers will vary but should include the water will fall out and you will get wet.**) Is there any way to hold the bucket in this manner and not get wet? (**Answers will vary.**) Swing the bucket in a circle over your head. Explain that the speed of the orbiting bucket keeps the water at the bottom of the bucket, which prevents the water from falling out. Point out that this is similar to how satellites remain in orbit around Earth. A centripetal force caused by Earth's gravity pulls the satellite in as inertia keeps the satellite moving in a straight line. Gravity and inertia balance each other to keep the satellite in orbit.

Discussion

- Lead a discussion about the launch of *Sputnik 2*, which sent a dog into space. Ask students if they believe sending an animal into space is a good idea. (**Answers will vary.**) Ethically, is it right or wrong? (**Answers will vary.**) Direct students to consider **Genesis 1:26**, "Let Us make mankind in Our image, in Our likeness, so that they may rule over the fish in the sea and the birds in the sky, over the livestock and all the wild animals, and over all the creatures that move along the ground." Ask students how the phrase "rule over" applies to the ethical question of using animals for scientific research. (**Answers will vary.**)

- Discuss the force that holds something in orbit around Earth. Point out that the forces are balanced. Ask students what is necessary to get a satellite into orbit. (**It has to reach orbital velocity.**) How is it able to stay in orbit? (**Inertia keeps the satellite moving, and gravity pulls it into a circular path. These forces combined are centripetal force.**)

- Lead a discussion on how satellite technology benefits human life. Ask students what services they use that are made possible by communications satellites. (**Communications satellites beam television programs, radio messages, telephone conversations, and other kinds of information all over the world.**) What other technologies rely on satellites? (**Answers will vary but should include GPS.**) Explain that navigation satellites do more than tell how to find an address. They send continuous signals to ships and airplanes so navigators can pinpoint their exact position. They also send signals to the global positioning systems, a system of 32 satellites used to pinpoint exact locations on Earth. Canada, the United States, France, and the Soviet Union set up search-and-rescue satellites to determine the positions of ships or aircraft. Point out that weather systems like hurricanes are tracked by weather satellites. These satellites also monitor climate, ocean dynamics, volcanic eruptions, forest fires, search-and-rescue, and global vegetation. Ask students what benefits they think scientific satellites provided. (**Answers will vary.**) Overall, how important to modern life are satellites? (**Answers will vary.**)

- Ask the following questions:
  1. Why do scientific satellites use polar orbits? (**Answers will vary but should include that satellites using the polar orbits give whole earth data every day. Scientists can learn about all parts of the earth.**)
  2. What is a high Earth orbit? (**It is an orbit at 35,780 km above the earth; it is also called a geosynchronous orbit.**)
  3. What is a low Earth orbit? (**It is an orbit 180–2,000 km above the earth. These satellites travel very quickly; they move at about 27,000 kph.**)
  4. Why are high Earth orbits better for satellites than low Earth orbits? (**High Earth orbits have less friction so satellites can stay in orbit longer.**)

Activities

Lab 8.2.3A Centripetal Force
- string
- metric rulers, 1 per group
- scissors, 1 pair per group
- glass tubes, 1 per group
- metal washers, 5 per group
- permanent markers, 1 per group
- stopwatches, 1 per group

Adjust lengths of string as needed. Be sure students follow safety rules when performing the experiment because the last step of the lab can be dangerous.

A. Complete *Try This: Satellites in Orbit* with students. Discuss how the combination of velocity, gravity, and inertia allows satellites to continuously orbit. Clarify that centripetal force is a combination of gravity and velocity that directs an object's circular motion.

B. Have students complete **WS 8.2.3A Orbits**. When finished, direct students to find the name of a current satellite for each type and to include the country responsible for it as well.

C. Show selected clips of the movie *October Sky*, the story of Homer Hickam, a boy from a West Virginia coal-mining town in the 1950s who became a NASA engineer. The launch of *Sputnik 1* inspired Homer to launch his own rocket. Direct students to record the successes and failures of the students in the film. Ask how the movie inspired them. (**Answers will vary.**) How have people benefited from the risks others are willing to take? (**Answers will vary.**)

NOTES

Satellites 8.2.3

In 1955, U.S. President Dwight D. Eisenhower announced that the United States would contribute to international space science by launching a **satellite**, which is a natural or artificial object that orbits a planet. Similar to space probes, satellites are designed to gather data. But unlike space probes, satellites and other spacecraft that orbit the earth usually do not need any navigation once they are in orbit.

The Soviet Union kicked off the Space Race on October 4, 1957, when it launched the first satellite, *Sputnik 1*. This satellite, which weighed only 84 kg and was only 58 cm in diameter, orbited the earth every 96 minutes, and sent information back until January 4, 1958. *Sputnik 1* burned up on reentry into Earth's atmosphere. You might think that the whole world celebrated the triumph of entering the frontier of space, but *Sputnik 1* sparked a fierce competition between the Soviet Union and the United States. The American public saw this Soviet success as a national security threat. If the Soviet Union had the ability to launch a satellite, it might also be able to launch missiles carrying nuclear weapons. As a result, the U.S. began funding its own satellite project and on January 31, 1958, *Explorer 1* was launched. This satellite carried instruments that measured cosmic rays and small dust particles and recorded the temperature of the upper atmosphere. *Explorer 1* also discovered the Van Allen radiation belts.

The idea of an artificial object orbiting Earth was startling in the 1950s. Today, however, satellites are very common. They are launched by governments and by private companies. But what keeps them in orbit? Satellites are placed in orbit high enough so the friction of Earth's atmosphere does not drag them down. At that height, the same laws that keep the moon in orbit also govern artificial satellites. The forward velocity of the satellite plus **inertia**, the tendency of a moving object to keep moving in a straight line, and Earth's gravity keep the satellites from plunging back to Earth or flying off into space. These three forces together cause an object to follow a circular path. This force is called *centripetal force*.

Satellites use four specific types of orbits: low Earth, medium Earth, geosynchronous/high Earth, and polar. Satellites placed into orbit 180–2,000 km above the earth are in low Earth orbit (LEO). These satellites travel very quickly; they move at about 27,000 kph. Their velocity has to be fast enough to overcome the force of Earth's gravity but slow enough to allow the satellite to stay in orbit. In higher orbits, such as medium and high

OBJECTIVES
- Explain how a satellite is able to orbit.
- Correlate types of satellites with specific orbits.
- Compare the types of orbits.
- Infer the benefits of satellites.

VOCABULARY
- **inertia** the tendency of a moving object to keep moving in a straight line unless another force acts on it
- **satellite** a natural or artificial object that orbits a larger astronomical body

FYI

Where No Dog Had Gone Before
In 1957, the Soviets launched *Sputnik 2*, less than a month after *Sputnik 1*. *Sputnik 2* was occupied by a dog named Laika, who was the first living being to orbit the earth. They sent her into orbit in a pressurized space cabin.

Earth orbits, satellites are farther from the earth, so they experience less gravitational pull and move more slowly. They also experience less friction. Medium Earth orbits allow satellites to work in groups to provide global wireless communication. Because the orbit is higher than a low Earth orbit, each satellite in a group can cover a larger area, which means fewer satellites are needed to do the work.

At 35,780 km above the earth, satellites can maintain an orbit speed that matches Earth's rotational speed. This type of orbit is called *a geosynchronous orbit*. In a geosynchronous/high Earth orbit, the satellite is always positioned in the same location above the earth and can observe an entire hemisphere. Satellites with a geosynchronous/high Earth orbit are positioned over the equator because the force of gravity is consistent at this latitude. In contrast, polar orbits allow a satellite to view nearly every part of Earth as it rotates. This type of orbit is useful for monitoring weather patterns or mapping the earth.

Earth's rotation and atmosphere affect satellites. Satellites are usually launched in the direction that Earth rotates (from west to east). By using Earth's rotation to boost the rocket, less rocket fuel is needed to place the satellite in orbit. However, launching satellites into a polar orbit requires more rocket power and more fuel. A nearly polar orbit moves perpendicular to the rotation of the earth; therefore, it cannot use the earth's rotational motion to propel it into orbit. Friction with Earth's atmosphere slows a satellite down, and the lower the orbit, the more friction it

Satellite Orbits
- Polar orbit 830–870 km 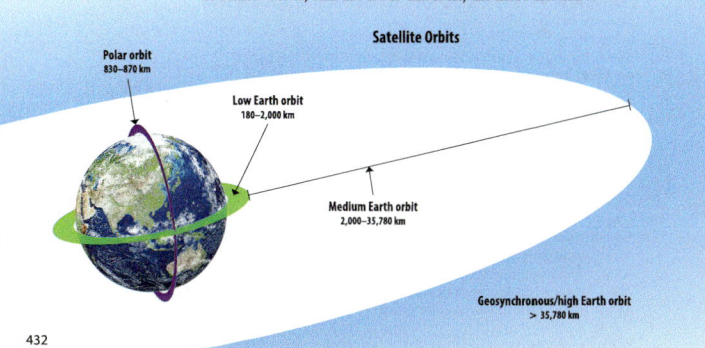
- Low Earth orbit 180–2,000 km
- Medium Earth orbit 2,000–35,780 km
- Geosynchronous/high Earth orbit > 35,780 km

NOTES

D. Divide the class into groups and assign each group a specific type of satellite. Have groups research their satellite type, indicate the satellite's orbit type, make a graphic display, and present their findings to the class.

Lesson Review

1. Explain how a satellite stays in orbit. (**The forward velocity of the satellite, inertia, and Earth's gravity keep the satellite in orbit. These three forces together create centripetal force, which causes an object to follow a circular path.**)
2. What can cause a satellite to fall back to Earth? (**Friction with Earth's atmosphere slows a satellite down with each orbit. Eventually, the satellite slows down so much that gravity pulls it back toward the earth.**)
3. How is a polar orbit different from other orbits? (**A polar orbit is nearly perpendicular to the equator whereas other orbits are more parallel to the equator.**) Why does it take more fuel to place a satellite in a polar orbit than a geosynchronous orbit? (**Launching a satellite into a polar orbit requires more power because it cannot use the earth's motion to help propel it into orbit. The necessary power requires more fuel.**)
4. How is a high Earth orbit different from a low Earth orbit? (**In a high Earth orbit, an object has a higher altitude, travels at a slower velocity, and experiences less friction than in a low Earth orbit.**)
5. What are the benefits of navigation satellites? (**They help ships and airplanes find their way even in a storm, pinpoint exact locations on Earth, and determine the positions of ships or aircraft for search-and-rescue efforts.**)
6. In which orbit are communications satellites found? (**geosynchronous orbit**)

FYI

Nanosatellites
In 2010, NASA started the CubeSat Launch Initiative (CSLI). It provides an opportunity for educational institutions and nonprofit organizations to conduct low-cost research by sending nanosatellites on launches to the *International Space Station*. CubeSats are small spacecraft, averaging about 1,000–3,000 cm³ and weighing approximately 1 kg/1000 cm³. Since the beginning of the program, 46 CubeSats have launched. The *TJ3Sat* was the first built by a high school and the *STMSat-1* was the first built by an elementary school.

experiences. Eventually a satellite slows down so much that gravity pulls it back toward the earth. When this happens, friction with air molecules in the atmosphere causes most of the satellite to burn up in space.

Several different types of satellites orbit the earth, including satellites for communication, weather monitoring, navigation, and scientific discovery.

Communications Satellites. Communications satellites beam television programs, radio messages, telephone conversations, and other kinds of information all over the world. The satellite receives a signal from a transmitting station on Earth, and it beams the information to somewhere else on Earth, perhaps even on the other side of the world. Communications satellites are placed in a geosynchronous/high Earth orbit. NASA's first communications satellite, *Echo 1*, was launched in 1960. By 1962, TV signals could be sent from continent to continent. The first commercial communications satellite, *Intelsat 1*, could handle only 240 voice circuits or one television channel at a time.

TRY THIS

Satellites in Orbit
A center-seeking force that causes an object to follow a circular path is called *centripetal force*. Satellites depend on centripetal force to maintain orbit. Sew one end of a ribbon to a tennis ball. Gently swing the ball to ensure it will stay attached in orbit. Reinforce with more stitches if necessary. Swing the tennis ball over your head, maintaining a constant speed. What exerts the centripetal force on the tennis ball? What would happen to the tennis ball if you continued to hold the ribbon, but stopped exerting force? Why? How is this system like a satellite orbiting the earth? How is it different?

Current satellites can handle tens of thousands of voice circuits or multiple TV channels. About 60% of all global communications pass via satellites. Telecommunications satellites are useful only when they rotate in a certain range. This range is becoming crowded with working satellites and space debris.

Weather Satellites. The United States' first weather satellite, *Tiros 1*, was launched in 1960. This satellite allowed meteorologists to see how Earth and its clouds look from above. Weather satellites, which orbit in a LEO or a polar orbit, contribute to the understanding of how storms develop, how wind patterns are created, and how ocean currents behave. Weather satellites are especially valuable in tracking hurricanes. They give information to improve predictions of hurricane tracks, intensities, and surges, giving people time to prepare. Data about clouds, humidity, and surface properties come from the polar-orbiting Operational Environmental Satellites. These satellites make nearly 14 orbits every day and provide environmental monitoring information on climate, ocean dynamics, volcanic eruptions, forest fires, search and rescue, and global vegetation.

Navigation Satellites. Navigation satellites can be found in medium Earth orbits. They send continuous signals to ships and airplanes so navigators can pinpoint their exact position. This helps them find their way even in a storm when other information may not be accurate. These satellites also send signals to the global positioning systems (GPS), a group of 32

Great Salt Lake, September 2011

satellites used to pinpoint exact locations on Earth. Handheld GPS receivers can identify a person's location within 15 m.

Scientific Satellites. Scientific satellites help scientists make new discoveries about the universe, such as evidence of distant stars and black holes. Satellites have also been helpful for scientific research in meteorology, astronomy, geophysics, and oceanography. From Earth, astronomers have a hard time seeing into space because Earth's atmosphere interferes with light coming from far away. To avoid this problem, satellites are sent outside the atmosphere to take pictures. The Landsat satellite program, started in 1972, has taken millions of images that used to track global and regional changes on Earth. *Landsat 8* was launched in 2013 and is currently providing data on glacier movement to help scientists understand how ice movement is changing globally. Images of the Great Salt Lake in Salt Lake City, Utah, have revealed that it is changing size. Such information would be difficult to obtain at ground level.

LESSON REVIEW
1. Explain how a satellite stays in orbit.
2. What can cause a satellite to fall back to Earth?
3. How is a polar orbit different from other orbits? Why does it take more fuel to place a satellite in a polar orbit than a geosynchronous orbit?
4. How is a high Earth orbit different from a low Earth orbit?
5. What are the benefits of navigation satellites?
6. In which orbit are communications satellites found?

Great Salt Lake, September 2016

Name: _____ Date: _____

Lab 8.2.3A Centripetal Force

QUESTION: How does centripetal force change when the force of gravity changes?

HYPOTHESIS: Answers will vary.

EXPERIMENT:

You will need:	• scissors	• permanent marker
• string	• glass tube, 10 cm in length	• stopwatch
• metric ruler	• 5 metal washers	

Steps:
1. Cut a 38-centimeter length of string. Thread the string through the glass tube. If you cannot get the string through, suck gently on the opposite end of the tube. The suction will pull the string through.
2. Label one end of the tube *Top*.
3. Tie 1 washer to the end of the string coming out of the Top.
4. Pull the string through so 15 cm of string is between the end and the washer.
5. Mark the string at the point it comes out of the tube. Be sure to mark all the way around.

6. Tie 2 washers to the end of the string coming out the other end of the tube.
7. Mark the string at the point where the string comes out. Be sure to mark the string all the way around.
8. Hold the tube horizontally and spin the single washer. The single washer is the satellite and the two washers at the other end are supplying the gravity.
9. Spin the satellite, keeping the mark on the string right at the opening. This will take some practice to maintain a consistent orbit. Hold only the tube, not the string.
10. When you are ready, have a partner use the stopwatch to measure how long it takes for 20 full circles. Do this five times. Record the results and find the average.
11. Tie 2 more washers to the "gravity" end of the string and repeat Step 10.

	Trial 1	Trial 2	Trial 3	Trial 4	Trial 5	Average
Two Washers						
Four Washers						

Answers will vary.

Lab 8.2.3A Centripetal Force

12. For the final step, predict the path the washer will take if the string is cut while the washer is in orbit. Answers will vary.

13. Spin the washer as you did in Step 11 and then cut the string. Illustrate the path the washer took in relation to the original orbit. Use arrows to show the path of the orbit and the path of the cut washer.

Drawings will vary.

ANALYZE AND CONCLUDE:
1. What provided the gravity in the experiment? The end with the two washers and four washers provided the gravity.

What provided the velocity? My hand provided the velocity.

Name: _____ Date: _____

Lab 8.2.3A Centripetal Force continued

2. What is the name of the force that keeps the washer moving in a circular path? centripetal force

3. What was the average speed of 20 orbits with two washer Answers will vary. What was the average speed with four washer Answers will vary.

4. Why do you think the average speed was different? The different number of washers created different amounts of gravity.

5. If you were to compare the two-washer and the four-washer tubes to a high Earth orbit and a low Earth orbit, which one would be which? The two-washer tube would be a high Earth orbit because it exerted less gravitational force. The four-washer tube would be low Earth orbit because it exerted more gravitational force.

6. When you cut the string, did the washer follow the path you predicted? Answers will vary.

7. Explain why the washer did not continue to orbit after the string was cut, but it did continue to move. The washer left orbit because it no longer had gravity pulling on it to keep it moving in a circle. Cutting the string removed the centripetal force. The washer continued to move after the string was cut because it had inertia.

8. What do you think must happen in order for a satellite to leave Earth's orbit and orbit another celestial body? Answers will vary but should include an increase in velocity to overcome the gravitational pull.

8.2.4 Working in Space

Space Exploration

OBJECTIVES

Students will be able to
- infer how science progresses through the work of previous scientists.
- explain the benefits of space research and exploration.
- identify the challenges of working in space.

VOCABULARY

- **space station** a satellite from which vehicles can be launched or scientific research can be conducted
- **spin-off** a by-product or fringe benefit derived from a previous product

MATERIALS

- Dehydrated space food samples (*Introduction*)
- BLM 8.2.4A Moon Myths
- Color-coded slips of paper (*C*)

PREPARATION

- Obtain materials for *Try This: Drinking in Space* and *Try This: Converting Sunlight*.

TRY THIS

Converting Sunlight
- *cans, 2 per group*
- *candles, 1 per group*
- *matches*
- *thermometers, 2 per group*

(**Predictions and final temperatures will vary. The dark can should absorb the most heat. The shiny can should maintain the most heat. Engineers design spacecraft to have black panels to convert sunlight into energy and to have shiny panels to reflect the sun's energy and maintain temperatures. Solar panels are black to absorb energy.**)

Introduction

Distribute dehydrated space food for students to try. Ask students how this food compares to the food they normally eat. (**Answers will vary.**) Inform students that food is one thing that has to be different for people who work in space. Encourage students to brainstorm about other aspects of work and life that would be different in space.

Discussion

- Discuss with students how scientific progress depends on the work of other scientists. The early goals of the United States included putting an explorer into space. This goal required several steps, including launching an astronaut through the atmosphere, sending an astronaut into orbit, and then sending astronauts to the moon. With students' input, create a time line of progress in space exploration. Then, have students brainstorm and infer how previous work spurred on new technology and progress.

- At the beginning of the Space Race, the Soviet Union and the United States were the only countries with space programs. Today, several more countries and private industries are involved in space exploration. Ask students what benefits have come from so many entities being involved in space exploration and research. (**Answers will vary but should include more people means more ideas coming together, progress may move faster, or more people will have jobs.**)

- Ask the following questions:
  1. What conditions exist in space that offer unique opportunities for experimentation? (**weightlessness, temperature extremes, vacuum, and radiation**)
  2. How has the medical field benefited from space research? (**improved understanding of osteoporosis and the muscle weakening that comes with aging or illness, improved cancer treatments, improved treatment for diabetes, and new understanding of nervous system disorders**)
  3. What spin-offs of space exploration are used in the field of medicine? (**blood pressure recorders, "invisible" braces, and pacemakers**)
  4. What do you think would be some of the challenges to living in space for a long period of time? (**Answers will vary but should include having enough fresh water and oxygen, preparing food, growing food, staying healthy, and experiencing loneliness.**)

Activities

A. Complete *Try This* activities with students. For *Try This: Drinking in Space*, discuss the difficulty of doing routine tasks without gravity. For *Try This: Converting Sunlight*, ask students what items they have used that work with solar power. (**Answers will vary.**)

B. Using **BLM 8.2.4A Moon Myths**, direct students to read one of the myths, evaluate the motivation of the author, and analyze the myth's data about space travel and the moon to see if any of the information is correct. Then, challenge students to write a short story about traveling to the moon or to Mars. Direct students to include details about the spacecraft's launch, life onboard, travel time, and what discoveries are made. Extend the activity by directing groups of students to create a skit for one of the myths or one of their stories and present it to the class.

C. Stage a short debate about whether space research and exploration are valuable. Divide the class into two groups, Group 1 and Group 2. Arrange students' desks into two parallel lines facing each other. Provide time for students to list and to develop the arguments supporting their side of the issue and to prepare specific rebuttals to the other group's arguments. Before the debate, place color-coded slips of paper with students' names in a box. As moderator of the debate, draw a name of a student from Group 1's colored paper to present a specific argument for one side of the issue. Set a time limit. Score the student's argument on a scale of one to five.

Allow Group 2 to choose a group member to give a rebuttal, closing the first round of the debate. Continue the debate until each group member has had an opportunity to participate. Arguments should include the following:

Arguments for space research and exploration. They provide technology improvements on Earth in the fields of medicine, archaeology, geology, meteorology, and many others. They create jobs, and now both government and commercial companies are involved. They allow people to explore more of God's creation. In the future, space research and exploration may provide more natural resources for Earth and new places for people to live, alleviating overpopulation.

Arguments against space research and exploration. They are a waste of money and resources. Many missions have been unsuccessful. Space research and exploration are very dangerous; astronauts have died. The rocket launch can damage the ozone, releasing gases, debris like AlO_2, and soot. Space is becoming littered with space debris.

D. Have students design posters titled "Space Exploration—Where Would We Be Without It?" The posters should include five spin-offs, at least two not mentioned in the student text or in class, and at least three additional reasons that the space program is beneficial. Examples may include learning more about the Creator's work, telecommunications, GPS, and the possibility of future discoveries. **Option:** Challenge students to use the categories *Past*, *Present*, and *Future* to organize their posters. For example: *Past*—before the space technology benefits, *Present*—space technology benefits used now, and *Future*—possible space exploration benefits for the future format.

> **⓵ TRY THIS**
>
> **Drinking in Space**
> • glasses of water, 1 per student
> • straws, 1 per student
>
> (Answers will vary but should include my stomach is above my mouth so it is hard to work against gravity to drink.)

8.2.4 Working in Space

OBJECTIVES
- Infer how science progresses through the work of previous scientists.
- Explain the benefits of space research and exploration.
- Identify the challenges of working in space.

VOCABULARY
- **space station** a satellite from which vehicles can be launched or scientific research can be conducted
- **spin-off** a by-product or fringe benefit derived from a previous product

The ambition of the space programs of both the Soviet Union and the United States was to put a human explorer into space. Both countries accomplished that goal in 1961. With each subsequent mission, the scientists from both countries' space programs gained knowledge that was needed for the next mission. Even through failures, information was gained that led to better technology and safer missions. What began in the 1950s Cold War as a competitive space race has become a cooperative endeavor that benefits people worldwide.

The first person to orbit Earth was Soviet cosmonaut Yuri Gagarin on April 12, 1961. On May 5, 1961, U.S. astronaut Alan Shepard was launched into space but not into orbit. This step was a major achievement for the U.S. space program, because many Americans were afraid that they were losing the Cold War to the Soviet Union. When President John F. Kennedy announced the goal of reaching the moon, the announcement inspired the nation. In February 1962, John Glenn was the first American to orbit Earth—an important step toward reaching the moon.

On July 20, 1969, *Apollo 11* took Neil Armstrong, Edwin "Buzz" Aldrin, and Michael Collins to the moon. Neil Armstrong became the first human to set foot on a place other than Earth. The nation and the world watched as he took the first steps and said, "One small step for [a] man. One giant leap for mankind." Although the main objective for *Apollo 11* was to fulfill President Kennedy's goal, it also had a scientific mission; it brought 22 kg of moon rocks back to Earth. Solar devices were also planted on the moon to monitor moonquakes and the solar wind.

Neil Armstrong

📦 FYI

Next Stop, Mars
In the planning stages for the *ISS*, NASA had three goals: establish a human presence in space, foster international cooperation, and conduct research. These goals are still in place, but a new goal has been added—send people to Mars. Before sending people to Mars, research must be conducted on the effects of zero-gravity on humans and potential health risks of long flights. The *ISS* is a great place to test many of the conditions that people will face when traveling into deep space. The *Orion* spacecraft is also being tested in preparation for a mission to Mars. Scheduled to launch in 2018, it will send astronauts beyond the moon and is expected to be the best deep space proving ground yet. On both the *ISS* and the *Orion*, NASA will be testing transportation capabilities, working in space, and staying healthy. Will humans be able to explore deep space? So far only space probes have traveled to such depths.

Apollo Time Line

On April 11, *Apollo 13* was supposed to land on the moon, but an oxygen tank ruptured, leaving the craft with no water, electricity, or heat. *Apollo 13* had to use the moon's gravity to circle the moon without landing and return to Earth earlier than planned, on April 17.

Apollo 17 was the last lunar mission. The crew took three moon walks and left the lunar lander behind.

Apollo 16 landed on the moon, discovered a 11.34 kg rock, and tested the performance of the lunar rover.

1967 — 1968 — 1969 — 1970 — 1971 — 1972

Apollo 7 mission ended tragically when a fire in the command module killed all three astronauts.

In orbit for 11 days, *Apollo 7* did not land on the moon.

Apollo 9 tested orbital docking.

Apollo 14 conducted lunar landing and seismic studies.

Apollo 10 practiced lunar landing within nine miles of the moon's surface.

Apollo 15 landed on the moon and used the first lunar rover to travel on the moon's surface.

Apollo 8 mission was the first trip beyond low Earth orbit. On Christmas Eve, 1968, the crew conducted a live television broadcast while orbiting the moon. The broadcast ended with what they called a message for all people on Earth: a reading of Genesis 1:1–10.

On July 20, *Apollo 11* landed on the moon.

Apollo 12 landed on the moon and perfected precision targeting.

In the early space missions, the spacecraft could only be used once, which was very expensive. Continuing to send people into space required developing a reusable system. In 1972, a space shuttle program was suggested as an economical way to get into space regularly. A space shuttle is a reusable vehicle that is launched like a rocket and lands like an airplane. The first space shuttle, *Columbia*, was launched on April 12, 1981. About two dozen successful missions followed; shuttle launches seemed commonplace until January 28, 1986, when *Challenger* exploded just after takeoff, killing all seven astronauts. In the aftermath of the disaster, no shuttles were launched until *Discovery* in 1988.

After 135 missions, including the loss of *Columbia* and her crew in 2003, the space shuttle program was shut down in 2011. However, NASA and other agencies continue to develop technology to make space travel more economical and practical. In addition, commercial companies are working to create rocket engines that would allow planes to reach orbit, release cargo, return to Earth, and land on an airstrip.

NOTES

E. Assign students to research space missions and space technology developed over the past five years. Have students make a time line and share their research and opinions about the work with the class. Challenge students to relate what direction they think the space program will go or the direction they would like to see the program move.

Lesson Review

1. List in order five space missions that led up to the completion of the *International Space Station*. (**Possible answers: A Soviet cosmonaut orbited the earth; a U.S. astronaut was launched into space; John Glenn orbited the earth; *Apollo 11* landed on the moon; the space shuttle *Columbia* was launched; space shuttle missions continued until 2011; space stations were placed in orbit from 1971 to 1982 by the Soviet Union; the United States placed *Skylab* in orbit; the Russians placed *Mir*; and the *ISS* was finished in 2011.**) How was each mission a step leading to the next mission? (**Answers will vary but should include each mission tested something that could be used in the next mission.**)

2. Why was the space shuttle program started? (**Before the space shuttle, spacecraft could only be used once. Space travel is very costly. To save money, the space shuttle was designed as the first reusable spacecraft.**)

3. Why is it important to conduct research on the *ISS* before sending people to Mars? (**Answers will vary but should include many problems could arise on the trip to Mars or on Mars itself; many conditions can be simulated in safety on the *ISS*; if problems arise in the simulations, people are close enough to Earth that they can be helped; and if the tests were not made ahead of time, unanticipated problems could happen and no one would be able to help the people on the trip.**)

TRY THIS
Drinking in Space
Many common activities become difficult without the presence of gravity. Drinking is one of those activities. Place a glass of water on the floor and lie facedown across a chair. Bend down to take a sip, using your hands to tilt the glass and keeping the glass near the floor. Then bend down to take a drink through a straw. Why is drinking like this so difficult?

The space shuttle astronauts spent up to 19 days in space, but people have lived in space for longer periods on scientific satellites called **space stations**. A space station is a satellite from which vehicles can be launched or scientific research can be conducted. In 1971, the Soviet Union was the first country to place a space station into orbit. By 1982, the Soviets had put up six more space stations. *Skylab* was the United States' first space station, and it orbited Earth from 1973 to 1979. Three successive three-astronaut crews spent a total of 171 days on *Skylab*. They proved that humans could work and live in space for extended periods. Like all satellites in low Earth orbit, *Skylab* eventually spiraled toward Earth and fell into the Indian Ocean in 1979.

In 1986, the Soviet Union launched a new space station, *Mir*. The word *mir* is Russian for "peace." *Mir* was used to conduct scientific and medical experiments. Even astronauts from other countries visited *Mir*. In fact, *Mir* was occupied almost continuously until it was taken out of orbit in 2001.

Perhaps one of the most important accomplishments of *Mir* was that it inspired the construction of the *International Space*

BIOGRAPHY
Katherine Johnson
A brilliant mathematician, Katherine Johnson was born in West Virginia in 1918, a time in history where both her race and her gender made success in the sciences difficult. Because there was no high school for African-American students in her town, Katherine's parents sent her and her siblings to a high school 100 miles away. Katherine was a very advanced student who started high school at age 10. She graduated college at 18 with degrees in mathematics education and French. One of three students chosen to integrate the all-white graduate program at West Virginia University, Katherine left the graduate program when she married and began teaching in the public schools. In 1953, she became a "computer," one of a group of black women with math degrees who worked at the Langley Aeronautical Laboratory. Within weeks, she was assigned to work with the Flight Research Division. This group of engineers eventually was tasked with getting an American into space as part of NASA. Known for her accurate work, Katherine figured out the trajectory for Alan Shepard's flight and double-checked the new electronic computer's calculations for John Glenn's orbital flight as well. She was the first woman ever credited on a report in the Flight Research Division. Katherine worked on the Apollo program as well as the space shuttle. She retired from NASA in 1986. In 2015, she was awarded the Presidential Medal of Freedom, the highest civilian honor, by President Barack Obama.

SpaceX Dragon

Station (*ISS*). The United States, Canada, Russia, Japan, Brazil, and the participating countries of the European Space Agency (ESA) met in 1993 to plan this project. Construction of the *ISS* began in November 1998 with the launch of the Russian module *Zarya*. The building-block assembly was begun in December 1998 when *Zarya* and the American module *Unity* were successfully joined in space, creating a "true" orbiting space station.

The *ISS* was completed in 2011 and represents an unprecedented international cooperation. NASA provided lab modules, solar panels, supporting trusses, and living quarters. Canada's major contribution was the Mobile Servicing System, which keeps the station running. Russia's contributions included a service module, docking modules, life support and research modules, and transportation to and from the station. The ESA and Japan contributed transport vehicles and specialized laboratories. American commercial companies, including SpaceX and Orbital Sciences Corporation, have supplied transportation as well. A planned 2019 commercial addition to the *ISS* by Axiom Space will expand the station's research and make tourism possible.

Even on the *ISS*, living and working in space is challenging. Water and oxygen need to be supplied by the station. Food is generally dehydrated, although some fresh food is grown on the station. The temperature can be too hot or too cold. Astronauts are living in a very small space with people who are not family. It can be very overwhelming. Many astronauts experience loneliness. But the difficulties are not just emotional.

TRY THIS
Converting Sunlight
Spacecraft are fitted with solar panels designed to convert sunlight into energy. Remove the paper from two cans. Blacken the outside of one can with a flame and leave the other can shiny. Fill each can three-quarters full with cold water. Place a thermometer in each can, set the cans on a windowsill, and predict which can you think will absorb the most heat. Monitor the temperature change over the course of 15 minutes. Record the temperatures. Empty the cans and fill them three-quarters full with warm water. Record the starting temperature. Place them in a closet or drawer, make a prediction as to which can will maintain the most heat, and record the final temperature after 15 minutes. Which can absorbed the most heat? Which can maintained the most heat? How could engineers use this method to create energy for a spacecraft? What color do you think solar panels are? Why?

4. How are space research and exploration beneficial? (**Possible answers: Space research and exploration provide new technologies, spin-offs, jobs, possible resources for precious metals, areas for colonization, and opportunities to see more of creation.**)
5. What are three challenges of working in space? (**Possible answers: having small space; not being with family; loneliness; completing daily activities that are challenging because of being in space; working in an environment of weightlessness, which affects the sense of balance, the circulatory system and heart, the sinuses, muscles, and bones**)
6. Is God glorified through space exploration? Why? (**Answers will vary but should include that God is glorified because His wonders are revealed and people learn more about His creation.**)

NOTES

BIBLE CONNECTION

The Heavens Declare
Can you think of a good reason for space exploration? One answer can be found in the Psalm 19. The psalmist wrote, "The heavens declare the glory of God; the skies proclaim the work of His hands" (Psalm 19:1). The wonders uncovered by modern astronomy include a universe of unimaginable size and beauty. Estimates of the size of the universe suggest 200 billion galaxies with new research suggesting this number could be much larger. Each new scientific discovery should remind every person of the Creator's power and creativity. God has given humans the ability to discover the wonders of the universe, and there should be excitement and awe about these discoveries. Loving God includes learning about all parts of His creation.

In space, the lack of gravity makes everyday activities like eating, brushing your teeth, showering, or sleeping much more complicated. The lack of gravity significantly affects the human body, although most of the effects disappear when an astronaut returns to Earth. For example, the body's sense of balance is affected by weightlessness, which causes most space travelers to experience nausea, vomiting, and headaches until they adapt. When they return to Earth, their sense of balance takes a while to recover, so they may lose their balance and fall more. Even the circulatory system is affected by weightlessness. On Earth, blood and other body fluids settle in the lower part of the body and the heart stays strong by pumping against gravity. In space, people's body fluids move up in their bodies, which causes their faces to puff up and their sinuses to block. In fact, astronauts experience what is called *the space sniffles* throughout their time in space. Without gravity to pump against, astronauts' hearts do not work as hard, so they lose muscle mass and shrink. People working in space also lose muscle and bone mass throughout their bodies. The calcium from their bones moves to other parts of their bodies and can cause health issues such as kidney stones. Most astronauts spend several hours a day exercising, but even that amount of activity does not counteract the effects of weightlessness on their bodies. Astronauts are also exposed to radiation from space, which may lead to health problems.

The *ISS* and space exploration are controversial to some people who question the expense and effort required to build and

The first flower grown in ISS's greenhouse facility

Mission Specialist Soichi Noguchi, Japanese astronaut, on ISS

maintain the space station. Some people claim that instead of putting effort into exploring space, solving problems on Earth should be the main priority. But the space program has brought great benefits to Earth. Information gathered by satellites can be used to make people better stewards of Earth or improve people's health. Weather warnings made possible by satellites have saved many lives, and information offered by navigation satellites makes travel easier and safer. Information gathered about Venus's greenhouse effect has helped scientists better understand how large quantities of gases could also affect Earth. The unique conditions of space, including weightlessness, temperature extremes, airlessness, and radiation, give scientists the opportunity to do experiments that would be impossible to conduct on Earth.

Perhaps some of the most valuable knowledge gained has been medical knowledge. Because people in space experience bone and muscle changes from weightlessness, the space program has helped scientists better understand osteoporosis and the muscle weakening that often comes with aging or illness. NASA technology has been applied to cancer treatments and has improved the treatment for diabetes. Space research on the body's balance system has led to a new understanding of nervous system disorders.

Pacemaker

FYI

Spin-Offs

To maximize the benefits of space technology, NASA's Technology Transfer Program makes its technology readily available to the nation. For example, NASA has developed a system to solve vibration issues experienced during a rocket launch. Companies are able to use this technology to test seismic effects on buildings and improve systems to dampen vibrations in bridges, ships, cars, and airplanes.

Some of the technologies created to prepare equipment and people to function in the harsh conditions of space have led to advances in certain materials, electronics, medicine, energy production, manufacturing, transportation, and even food. For example, what do blood pressure recorders, "invisible" braces, cordless power tools, freeze-dried foods, thermal clothing, pacemakers, smoke detectors, and scratch-resistant lenses have in common? They all use technology or materials first developed for the space program.

These products are all **spin-offs**, technologies resulting from research originally intended for use in space. For example, NASA invented scratch-resistant lens material to protect astronaut helmet visors. Heart pacemakers are similar to electronic monitoring systems that were first designed to operate satellites orbiting Earth. You may never travel in space, but you probably benefit from space research—and perhaps even take it for granted. The space program has helped people be better stewards of Earth, save lives, feed people, and enjoy the benefits of better communication. And those are just the benefits that are realized now. Scientists are already working on new possibilities, such as zero-gravity research, space manufacturing, and mining minerals and ores from asteroids.

It is impossible to see into the future and predict how useful space research will be. But it is possible to learn from the past. Each new step in scientific knowledge builds a bridge to exciting possibilities and teaches people more about God's world. Consider all the steps that led to what scientists know today. Many of these steps seemed useless at the time, but they contributed to new medicines, transportation, food production, technology, computers, manufacturing, and much more. As long as God gives people the ability to make new discoveries about His creation, people should be eager to make them!

ISS Expedition 50 crew includes ESA astronaut Thomas Pesquet, NASA astronaut Peggy Whitson, and Russian Cosmonaut Oleg Novitskiy, who left in November 2016.

LESSON REVIEW

1. List in order five space missions that led up to the completion of the *International Space Station*. How was each mission a step leading to the next mission?
2. Why was the space shuttle program started?
3. Why is it important to conduct research in the *ISS* before sending people to Mars?
4. How are space research and exploration beneficial?
5. What are three challenges of working in space?
6. Is God glorified through space exploration? Why?

Glossary

A

aa lava that has a rough surface 3.3.3

abrasion the wearing down of rock surfaces by other rocks or sand particles 2.1.4

absolute magnitude the brightness of a star measured by an observer that is a standard 32.6 light-years away 8.1.1

abyssal plain a large nearly flat region beyond the continental margin 4.2.4

adhesion the force of attraction between different molecules 4.1.1

aeroplankton microscopic organisms that float in the atmosphere 5.1.1

aftershock a tremor that follows a large earthquake 3.2.5

air mass a large body of air with consistent temperature and humidity 5.2.7

algal bloom an explosive growth of algae caused by too many nutrients in the water 4.1.9

amplitude a wave's height or depth measured from the surrounding water level 4.2.2

anemometer an instrument used to measure wind speed 5.2.3

annular eclipse a solar eclipse during which the outer ring of the sun is visible around the moon 7.3.3

aperture an opening through which light passes 7.1.3

aphelion the point in a planet's orbit when it is farthest from the sun 7.2.1

apogee the point in the moon's orbit at which it is farthest from Earth 7.3.2

apparent magnitude the brightness of an object as observed from Earth 8.1.1

aquifer a permeable underground layer of rock 4.1.4

asteroid a rocky object in a variety of sizes that orbits the sun 7.1.5

asthenosphere the layer of the upper mantle composed of low-density rock material that is semiplastic 1.4.2

astronomical unit the average distance between Earth and the sun 7.2.1

astronomy study of physical things beyond the earth's atmosphere 1.1.1

atmosphere a mixture of gases that surrounds the earth 5.1.1

atmospheric pressure the pressure exerted by Earth's atmosphere at any given point 5.1.1

aurora a band of colored or white light in the atmosphere caused by charged particles from the sun interacting with Earth's upper atmosphere 7.1.6

B

barometer an instrument used to measure atmospheric pressure 5.2.2

bedrock the layer of solid rock beneath the soil 2.2.1

binary star system a pair of stars that orbit their common center of mass 8.1.3

biomass derived from organic matter that contains stored energy and is used to produce fuel 6.1.7

black hole a massive celestial object with gravity so strong that not even light can escape 8.1.2

C

caldera a volcanic crater that is greater than 2 km in diameter and is formed by the collapse of surface rock into an empty magma chamber 3.3.4

carbon-14 dating the method used to determine the age of items of organic origin by measuring the radioactivity of their carbon 14 content 1.1.2

carbonation the process in which carbon dioxide from the atmosphere or soil dissolves in water to form carbonic acid 2.1.2

carbonization a process of converting organic material into carbon 2.2.5

CFCs synthetic compounds consisting of carbon, chlorine, and fluorine 5.1.4

channel the path that a stream follows 2.1.5

chemical weathering the breaking down of rocks by chemical processes 2.1.2

chromosphere the first layer of the sun's atmosphere 7.1.6

clastic rock a sedimentary rock made of rock particles and fragments deposited by water, wind, or ice 1.3.2

cleavage a mineral's tendency to split along definite flat surfaces 1.2.3

climate the pattern of weather an area has over a long period of time 5.3.1

climate change any long-term change in Earth's climate 5.3.3

cloud a visible collection of tiny water droplets or ice crystals in the atmosphere 5.2.5

coal a solid fossil fuel formed from decomposed plant remains 6.1.3

coalescence the process of coming together 5.2.6

cohesion the molecular attraction between particles of the same kind 4.1.1

comet a frozen chunk of ice, dust, and rock that orbits the sun 7.2.4

compressional stress the stress produced by two tectonic plates coming together 3.1.3

condensation the change of a substance from a gas to a liquid 4.1.2

conduction the transfer of heat from one substance to another substance through direct contact 5.2.1

contact metamorphism metamorphism that occurs when the heat of magma comes in contact with existing rocks 1.3.3

continental crust the crust on which the continents rest 1.4.3

continental drift the theory that the continents can move apart from each other and have done so in the past 3.1.1

continental glacier a glacier that covers a large area of land in a continuous sheet 4.1.3

continental margin the part of the earth's surface beneath the ocean that is made of continental crust 4.2.4

continental rise the base of the continental slope 4.2.4

continental shelf a broad relatively shallow underwater terrace that slopes outward from the shoreline 4.2.4

continental slope the steepest part of the continental incline located at the edge of the continental shelf 4.2.4

contour farming the plowing of furrows around a hill perpendicular to its slope to reduce erosion 6.1.6

contour line a line on a map that joins points of equal elevation 1.1.3

convection the transfer of heat that occurs in moving fluids, liquids or gases, caused by the circulation of currents from one region to another 5.2.1

convection current the circular movement of heated materials to a cooler area and cooled materials to a warmer area 1.4.2

convective zone the zone on the sun where convection currents bring energy to the sun's surface and take gas back into the sun 7.1.6

core the central portion of the earth 1.4.1

Coriolis effect the curving of moving objects from a straight path because of the Earth's rotation 4.2.2

corona the outer layer of the sun's atmosphere 7.1.6

cover crop fast-growing vegetation planted on bare farmland to prevent erosion 6.1.6

crater a large circular indentation on a planet's surface 7.2.2

crest the highest point of a wave 4.2.2

crop rotation the successive planting of different crops on land to prevent erosion and to improve fertility 6.1.6

crust the thin hard outer layer of the earth 1.4.3

D

dead zone an area that has been depleted of oxygen by eutrophication 4.1.9

deflation hollow a soil depression scooped out by the wind 2.1.4

deformation a change in the shape or volume of rocks 3.1.3

density the mass per unit of volume of a substance 1.2.3

deposition the changing of a gas directly into a solid 5.2.6

desalination the process of removing salt from ocean water to obtain fresh water for drinking, irrigation, or industrial use 4.2.6

desertification the making of new deserts by degrading land that used to be healthy and productive 6.1.5

desiccation a type of fossilization where the organic material becomes dehydrated 2.2.5

divide a ridge or other elevated region that separates watersheds 4.1.5

Doppler effect the apparent shift in frequency of waves emitted by a moving source 8.1.4

drumlin a long tear-shaped mound of till 2.1.6

dwarf planet a spherical object that orbits the sun but is not large enough to move other objects from its orbit 7.1.5

dynamic metamorphism metamorphism that is produced by mechanical force 1.3.3

E

Earth science the study of the earth and the universe around it 1.1.1

earthflow the movement of wet soil down a slope 2.1.3

eclipse the casting of one celestial body's shadow on the surface of another celestial body 7.3.3

electromagnetic radiation a form of wave energy that has both electrical and magnetic properties 7.1.4

electromagnetic spectrum the entire wavelength range of electromagnetic radiation 7.1.4

ellipse a closed curve along which the sum of the distances between two fixed points is always the same 7.2.1

El Niño periodic changes in oceanic and atmospheric conditions in the Pacific Ocean that cause unusually warm surface water 5.3.3

environmental science the study of the relationship between organisms and the environment 1.1.1

epicenter the point on the earth's surface directly above an earthquake's focus 3.2.4

equinox a point in Earth's orbit at which the sun crosses the plane of Earth's equator, causing the hours of day and night to be nearly equal everywhere on Earth 7.3.4

erratic a piece of till that is not native to the place where it was deposited 2.1.6

eutrophication the process by which nitrate or phosphate compounds overenrich a body of water and deplete it of oxygen 4.1.9

evaporation the change of a substance from a liquid to a gas 4.1.2

extrusive rock an igneous rock formed when lava cools on the earth's surface 1.3.1

F

fault a fracture in the earth's crust along which rocks move 2.2.4

faulting the breaking of the earth's crust and the sliding of the blocks of crust along the break 3.1.3

felsic rock a light-colored lightweight igneous rock that is rich in silicon, aluminum, sodium, and potassium 1.3.1

fissure a tear in the crust caused by the friction of a fault 3.2.5

floodplain a flat area along a river formed by sediments deposited when a river overflows 2.1.5

fluorescence the ability of a mineral to glow and change color under ultraviolet light 1.2.4

focus the point inside the earth where an earthquake begins 3.2.4

fog a low-level cloud caused by condensation of warm water vapor as it passes over a cold area 5.2.5

folding the bending of rock layers due to stress in the earth's crust 3.1.3

foliated structure a rock structure with visible layers or bands aligned in planes 1.3.3

footwall the landmass below the fault 3.1.3

fossil the preserved remains or impression of an organism that lived in the past 2.2.5

fossil fuel a source of energy formed from the buried remains of dead plants and animals 6.1.3

fracture a mineral's tendency to break along irregular lines; a break in the Earth's surface 1.2.3

friction the force that resists motion between two surfaces in contact with each other 7.2.4

front the boundary between two air masses 5.2.7

full moon the phase when the entire near side of the moon is illuminated 7.3.2

furrow a ditch in farmland 6.1.6

G

galaxy large systems of stars and their solar systems, gas, and dust held together by gravity 8.1.3

galaxy cluster a group of thousands of galaxies under the same gravitational influence 8.1.3

gas giant a larger gaseous planet of the outer solar system 7.1.5

geocentric centered on or around Earth 7.1.1

geologic column the order of rock layers 2.2.4

geology the study of the solid earth 1.1.1

geothermal energy energy collected from the trapped heat in the earth's crust 6.1.7

geyser a hot spring that periodically erupts 4.1.4

glacial drift the general term for any sediment deposited by a glacier 2.1.6

glacier a large mass of moving ice that forms on land and remains from year to year 2.1.6

greenhouse gas a portion of atmospheric gas molecules that deflect infrared radiation back to Earth's surface that was initially on a path to escape into space 5.3.3

groundwater all the water found underground 4.1.4

gully a narrow ditch cut in the earth by runoff 2.1.5

H

hanging wall the landmass above the fault 3.1.3

hardness a mineral's resistance to being scratched 1.2.3

heat the transfer of energy from one substance to another 5.2.1

heliocentric centered on or around the sun 7.1.1

horizon a layer in a soil profile 2.2.2

hot spot a place on the earth's surface that is directly above a column of rising magma 3.3.1

humidity the amount of water vapor in the air 5.2.6

humus the nutrient-rich organic material in soil 2.2.1

hydroelectric power electricity produced from the power of moving water 4.1.8, 5.1.7, 6.1.7

hydrolysis the breaking down of a substance by a chemical reaction with water 2.1.2

I

ice age a period of time when ice collects in high latitudes and moves toward lower latitudes 5.3.3

ice wedging the mechanical weathering process in which water in the cracks of rocks freezes and expands, widening the cracks 2.1.1

igneous rock rock formed from cooled and hardened magma 1.3.1

incineration the burning of solid waste materials 6.2.1

index fossil a fossil that is useful for dating and correlating the strata in which it is found 2.2.5

inertia the tendency of a moving object to keep moving in a straight line unless another force acts on it 8.2.3

inner core the solid center of the earth 1.4.1

interglacial period warm periods that occur between glacial periods when large ice sheets are absent 5.3.3

intrusion a large mass of igneous rock forced between or through layers of existing rock 2.2.4

intrusive rock an igneous rock formed when magma cools beneath the earth's surface 1.3.1

ion an atom with an electrical charge that has gained or lost one or more electrons 5.1.2

isobar a line that connects points of equal atmospheric pressure 5.2.9

isostasy the equilibrium in the earth's crust maintained by a flow of rock material in the asthenosphere 3.1.1

K

Kuiper belt the region of the solar system outside Neptune's orbit 7.2.3

L

landslide the rapid downhill movement of a large amount of rock and soil 2.1.3

latitude the distance in degrees north or south of the equator 1.1.3

lava the magma that has reached the earth's surface 1.3.1

law of superposition a law that states that layers found lower in the sedimentary rock formation are older than layers found closer to the top of the formation 2.2.4

leachate a solution formed when pollutants from sanitary landfills are dissolved in rainwater and seep into the groundwater 6.2.1

levee a structure built to prevent a river from overflowing 4.1.8

lightning the electric discharge of energy from storm clouds 5.2.8

light-year the distance light travels in a vacuum in one year, approximately 9.46×10^{12} km 8.1.1

liquefaction the process by which soil loses strength and acts as a liquid instead of a solid 3.2.5

lithification the process that transforms layers of rock fragments into sedimentary rock 1.3.2

lithosphere the outermost rigid layer of the earth, composed of the stiff upper layer of the mantle and the crust 1.4.2

lodestone a piece of magnetite that naturally acts as a magnet 1.2.4

longitude the distance in degrees east or west of the prime meridian 1.1.3

Love wave a fast surface wave that moves in a side-to-side pattern as it travels forward 3.2.4

lunar calendar a calendar based on the phases of the moon 7.1.2

lunar eclipse an event that occurs when Earth passes directly between the sun and the moon, causing Earth's shadow to block the sun's light from the moon 7.3.3

lunar regolith a loose layer of rock and dust on the surface of the moon 7.3.1

luster the way a mineral's surface reflects light 1.2.3

M

mafic rock the dark-colored, heavy igneous rock that is rich in iron, magnesium, and calcium 1.3.1

magma the melted rock beneath the earth's surface 1.3.1

magnetosphere the area around the earth that is affected by the earth's magnetic field 1.4.1

magnitude the strength of an earthquake 3.2.1

mantle the portion of the earth's interior extending from the bottom of the crust to the outer core 1.4.2

mare a flat, lowland plain on the moon's surface filled with hardened lava 7.3.1

mass wasting the downhill movement of rocks and soil caused by gravity 2.1.3

mature river a meandering river located at low elevations 4.1.6

mechanical weathering the breaking down of rocks by physical processes 2.1.1

metamorphic rock rock formed when the structure and mineral composition of existing rocks change because of heat, pressure, or chemical reactions 1.3.3

metamorphism the process of change in the structure and mineral composition of a rock 1.3.3

meteor a meteoroid that enters the earth's atmosphere and burns up 7.2.4

meteorite a meteoroid that enters Earth's atmosphere and strikes the ground 7.2.4

meteoroid a rock fragment from an asteroid or comet 7.2.4

meteorology the study of the atmosphere 1.1.1

microclimate unique climate conditions that exist over small areas of land within larger climate regions 5.3.2

mineral a naturally occurring, inorganic solid with a definite chemical composition and a crystalline structure 1.2.1

mineralogy the study of minerals 1.2.1

Moho the boundary between the crust and the mantle 1.4.2

moraine an accumulated deposit of till 2.1.6

mudflow the rapid, downhill movement of a large mass of mud and debris 2.1.3

N

natural gas a mixture of methane and other gases formed from decomposed marine organisms 6.1.3

natural resource any substance, organism, or energy form found in nature that can be used by living things 6.1.1

naturalism the belief that matter and energy are all that exist and that undirected natural processes formed the universe 1.1.2

neap tide a tide that occurs when the sun and the moon are at right angles to each other 4.2.3

nebula a vast, moving interstellar cloud of gas and dust 7.1.5

neutron star an extremely small, dense star composed primarily of neutrons 8.1.2

new moon the phase when the moon is directly between Earth and the sun 7.3.2

nonfoliated structure a rock structure with no visible layers or bands 1.3.3

nonrenewable resource a resource that cannot be replaced once it is used or can be replaced only over an extremely long period of time 6.1.1

nonsilicate mineral a mineral composed of elements or bonded groups of elements other than bonded silicon and oxygen 1.2.2

normal fault a fault in which the hanging wall slides down the footwall 3.2.3

nuclear energy energy that comes from the changes in the nuclei of atoms of radioactive elements 6.1.7

O

oceanic crust the crust beneath the oceans 1.4.3

oceanic ridge the mountain chains that form on the ocean floor where tectonic plates pull apart 3.1.2, 4.2.4

oceanography the study of the earth's oceans 1.1.1

old river a slow-moving, flat river 4.1.6

ore a naturally occurring mineral from which a useful metal or mineral is recovered 6.1.4

orogenesis the process of mountain formation 3.1.4

outer core the liquid layer of Earth's core that surrounds the inner core 1.4.1

oxbow lake a lake formed when a bend of a river is cut off from the main river 4.1.6

oxidation a chemical change in which a substance combines with oxygen 2.1.2

ozone a three-atom form of oxygen gas (O^3) that protects Earth from UV radiation 5.1.4

P

pahoehoe lava that has a smooth or billowy surface 3.3.3

parallax the apparent shift in an object's direction when viewed from two geographically distant locations 8.1.1

payload the cargo or equipment carried by the spacecraft 8.2.1

peat a substance made of partially decayed plant matter 6.1.3

penumbra an area of partially blocked light surrounding the complete shadow 7.3.3

perigee the point in the moon's orbit at which it is closest to Earth 7.3.2

perihelion the point in a planet's orbit when it is closest to the sun 7.2.1

petrifaction a process in which the organic portion of an organism is infiltrated or replaced with minerals 2.2.5

petroleum a liquid fossil fuel formed from microscopic plants, animals, and marine organisms 6.1.3

phase each different shape of the moon made visible by reflected sunlight 7.3.2

phosphorescence the ability of some fluorescent minerals to continue to glow after an ultraviolet light is no longer focused on them 1.2.4

photosphere the sun's surface, which radiates visible light 7.1.6

planet a spherical object that orbits the sun and has removed other objects from its orbit 7.1.5

plasma a super-heated gas composed of electrically charged particles 5.1.3

plate boundary the point at which one tectonic plate meets another 3.1.2

plateau a large area of flat-topped rock high above sea level 3.1.4

plug a structure of hardened magma that forms inside a vent 3.3.4

pluton a body of magma that has hardened underground 3.3.3

pore space the amount of space between soil particles 2.2.1

porosity a measure of the open space in rocks 4.1.4

prime meridian an imaginary line that divides the Earth into the Western Hemisphere and the Eastern Hemisphere 1.1.3

projection a system of lines drawn on a flat surface to represent curves 1.1.3

prominence a fiery burst of gas from the sun that rises thousands of kilometers into space 7.1.6

propellant a fuel used to produce the hot gases that power a rocket 8.2.1

pulsar a spinning neutron star that gives off pulses of radiation at regular intervals 8.1.2

pyroclast a solid volcanic material such as ash and rock that has been ejected during an eruption 3.3.3

P wave the fastest seismic wave, which travels through solids, liquids, and gases 1.4.1

Q

quarry locations where rocks are removed from the ground 6.1.4

R

radiation the transfer of energy through space by electromagnetic waves 5.2.1

radiative zone the zone on the sun where energy from the sun's core moves toward the sun's surface 7.1.6

radioactivity the ability of an element to give off nuclear radiation as a result of a change in the atom's nucleus 1.2.4

Rayleigh wave a slower surface wave that moves in an elliptical pattern while it travels forward 3.2.4

reflecting telescope a telescope that uses a series of mirrors to magnify objects 7.1.3

refracting telescope a telescope that uses a series of lenses to magnify objects 7.1.3

regional metamorphism metamorphism that occurs when large regions of the earth's crust are affected by high temperatures and pressures 1.3.3

regolith a loose layer of rock and soil 2.1.3

rejuvenated river a river with an increased stream gradient and power to erode 4.1.6

renewable resource a resource that is constantly available or that can be replaced in a relatively short period of time through natural processes 6.1.1

reservoir a natural or artificial lake used to store and regulate water 4.1.8

reverse fault a fault in which the hanging wall climbs up the footwall 3.2.3

rift a deep crack that forms between two tectonic plates as they separate 3.3.2

rille a channel on the moon 7.3.1

rock cycle the process by which one rock type changes into another 1.3.4

rocket a machine that uses escaping gas to move 8.2.1

rock pedestal a mushroom-shaped rock formed by the erosion of the rock's base 2.1.4

runoff water from precipitation that flows over the land 2.1.5

S

salinity the amount of dissolved salt in a given quantity of liquid 4.2.1

satellite a natural or artificial object that orbits a larger astronomical body 8.2.3

seafloor spreading the process by which a new oceanic lithosphere is formed at a mid-ocean ridge as older materials are pulled away from the ridge 3.1.2

seamount an underwater volcanic mountain that rises at least 1,000 m above the abyssal plain 4.2.4

sedimentary rock a rock formed from sediments that have been compacted and cemented together 1.3.2

sediments particles of minerals, rock fragments, shells, leaves, and the remains of once-living things 1.3.2

seismic wave a wave of energy that travels through the earth 1.4.1

seismograph an instrument that measures and records seismic waves 3.2.4

shale a clastic rock composed of silt- and clay-sized particles in flat layers 1.3.2

shearing stress the stress produced by two tectonic plates sliding past each other horizontally 3.1.3

shelterbelt a barrier of trees or shrubs designed to protect crops from wind damage 5.2.10

silicate mineral a mineral formed by bonded silicon and oxygen atoms 1.2.2

sinkhole a hole in the ground that forms when an underground cave collapses 4.1.4

sludge the solid waste leftovers from sewage treatment 6.2.1

smog a dense, brownish haze formed when hydrocarbons and nitrogen oxides react in the presence of sunlight 6.2.2

soil creep the extremely slow, downhill slide of soil 2.1.3

soil profile a cross section of soil layers and bedrock in a particular region 2.2.2

solar calendar a calendar uses the amount of time Earth takes to orbit the sun 7.1.2

solar eclipse an event that occurs when the moon passes directly between the sun and Earth, causing the moon's shadow to block the sun's light from a portion of Earth 7.3.3

solar energy radiation from the sun that causes chemical reactions, generates electricity, and produces heat 6.1.7

solar wind the continuous flow of plasma from the sun 5.1.3

solstice one of the two days of the year in which the sun's most direct rays reach farthest north or farthest south 7.3.4

space probe an unmanned spacecraft designed to gather data 8.2.2

space station a satellite from which vehicles can be launched or scientific research can be conducted 8.2.4

spin-off a by-product or fringe benefit derived from a previous product 8.2.4

spring tide a tide that occurs when the sun, moon, and Earth are aligned 4.2.3

star cluster groups of stars that have a common origin and are held together by mutual gravitational attraction 8.1.3

stewardship the attentive management of something entrusted to one's care 6.2.4

streak the color of the powder left by a mineral when it is rubbed against a hard, rough surface 1.2.3

strip cropping the planting of alternating bands of crops and cover vegetation in a planned rotation that are of equal widths 6.1.6

subduction the process of one tectonic plate being pushed under another tectonic plate 3.1.2

subduction zone a place where one tectonic plate is pushed under another tectonic plate 3.3.2

sublimation the change of a substance from a solid to a gas without passing through the liquid state 4.1.2

submersible a small underwater vessel 4.2.5

subsoil soil that is rich in minerals that have drained from the topsoil 2.2.2

supernova a violent explosion of a star 8.1.2

surface tension the force that pulls molecules on the surface of a liquid together to form a layer 4.1.1

S wave the seismic wave that travels only through solids 1.4.1

T

tectonics the study of the movement and changes in the rocks that make up the earth's crust 3.1.1

telescope an instrument used to make distant objects appear closer 7.1.3

temperature the measure of energy in the molecules of a substance 5.2.1

tensional stress the stress produced by two tectonic plates moving apart 3.1.3

terrace farming the construction of steplike ridges that are built into the slope of the land 6.1.6

terrestrial planet a smaller, dense, rocky planet of the inner solar system 7.1.5

theism the belief that the universe was created purposefully by a supernatural being 1.1.2

thrust the force that causes a rocket to accelerate 8.2.1

thunder the sound that results from the rapid heating and expansion of air that accompanies lightning 5.2.8

tidal range the difference in water height between high and low tide 4.2.3

till unsorted rocks and sediments left behind when a glacier melts 2.1.6

tiltmeter an instrument that uses liquid to register changes in the earth 3.3.5

topography the surface features of a place or region 2.1.1, 3.1.5

topsoil rich soil formed from mineral fragments, air, water, and organic materials 2.2.2

trace fossil a fossil of a track, trail, burrow, or other trace of an organism 2.2.5

transpiration the loss of water by plants 4.1.2

trench a deep underwater valley 3.1.2, 4.2.4

tributary a stream or river that flows into a larger stream or river 4.1.5

trough the lowest point of a wave 4.2.2

tsunami a very large ocean wave caused by an underwater earthquake or volcanic eruption 3.2.5

U

umbra the dark, central portion of a shadow that completely blocks the sun's light 7.3.3

unconformity the eroded surface that lies between two groups of strata 2.2.4

UV ultraviolet radiation from the sun 5.1.4

V

valley glacier a long, narrow, U-shaped mass of ice that takes shape as ice moves down a mountain and through a valley area 4.1.3

varves light and dark layers of sediments deposited in a yearly cycle 1.1.2

virga a streak of precipitation that evaporates before reaching the ground 5.2.5

volcanic bomb a fragment of molten rock that is shot into the air by a volcano 3.3.3

volcano a vent in the earth's crust through which lava, steam, ashes, and gases are forced 3.3.1

W

waning moon the phase after the full moon and before the new moon, when its appearance is shrinking 7.3.2

water budget the relationship between the input and the output of all the water on Earth 4.1.2

water table the boundary between unsaturated and saturated ground 4.1.4

watershed an area of land that drains into a particular river system 3.1.4, 4.1.5

wavelength the distance between identical points on two back-to-back waves 4.2.2

waxing moon the phase after the new moon and before the full moon, when its appearance is growing 7.3.2

Y

youthful river a fast-flowing, irregular river with a steep, V-shaped channel 4.1.6

Z

zone of aeration the underground region where pore spaces contain both air and water 4.1.4

zone of saturation the underground region where pore spaces are saturated with groundwater 4.1.4

Index

A

aa 3.3.3
abrasion 2.1.4
absolute magnitude 8.1.1
absorption lines 8.1.1
abyssal plain 4.2.4
acid rain 6.2.2
adhesion 4.1.1
aeroplankton 5.1.1
aftershock 3.2.5
agronomist 2.2.2
air mass 5.2.7
Aldebaran 8.2.2
Aldrin, Edwin 8.2.4
algal bloom 4.1.9
alloy 6.1.4
alluvial fan 2.1.5
amplitude 4.2.2
Anasazi 2.2.4
anemometer 5.2.3
annular eclipse 7.3.3
anthropic principle 7.2.2
anticlines 3.1.3
aperture 7.1.3
aphelion 7.2.1, 7.3.3, 7.3.4
apogee 7.3.1
Apollo program 8.2.4
apparent magnitude 8.1.1
aquifer 4.1.4, 6.1.2
Archimedes 1.4.3
Aristotle 5.1.2, 7.1.1
Armstrong, Neil 7.3.1, 8.2.4
asteroid 7.1.5, 7.2.4
asthenosphere 1.4.2
astrolabe 7.1.1
astronomical unit (AU) 7.2.4, 8.2.2
astronomy 1.1.1
atmosphere 5.1.1
atmospheric pressure 5.1.1, 5.2.2

aurora 5.1.2, 7.1.6
AUV 4.2.0
axis 7.2.1

B

barometer 5.2.2
Barton, D. C. 2.1.1
Bascom, Florence 1.3.1
batholith 3.3.3
bedrock 2.2.1
Big Bang 8.1.4
binary star system 8.1.3
biogas 6.1.7
bioluminescence 4.1.9
biomass 6.1.7
bioremediation 6.2.3
Biosphere II 5.1.1
black hole 7.1.4, 8.1.2
blizzard 5.2.8
blue moon 7.3.2
Brahe, Tycho 7.1.1, 7.2.4, 8.2.2
breaker 4.2.2

C

caldera 3.3.4
calendar 7.1.2
Callisto 7.2.3
Cannon, Annie Jump 8.1.1
capillary action 4.1.1
capture theory 7.3.0
carbon-14 dating 1.1.2
carbonation 2.1.2
carbon cycle 6.1.2
carbonization 2.2.5
centrifugal force 7.2.1
centripetal force 8.2.3
Ceres 7.2.3
CFCs (chlorofluorocarbons) 5.1.4
Chandrasekhar, Subrahmanyan 8.1.2
channel 2.1.5, 4.1.5

Charon 7.2.3
chemical weathering 2.1.2
chromatic aberration 7.1.3
chromosphere 7.1.6
clastic rock 1.3.2
cleavage 1.2.3
Cleopatra's Needle 2.1.2
climate 5.3.1
climate change 5.3.3
clouds 5.2.5
cloud seeding 5.2.10
coal 6.1.3
coalescence 5.2.6
cocreation theory 7.3.0
cohesion 4.1.1
Collins, Michael 8.2.4
comet 7.1.5, 7.2.4
compressional stress 3.1.3
condensation 4.1.2
conduction 5.2.1
conservationist 6.1.1
constellation 8.1.3
contact metamorphism 1.3.3
continental crust 1.4.3, 3.1.1
continental drift 3.1.1
continental glacier 4.1.3
continental margin 4.2.4
continental rise 4.2.4
continental shelf 4.2.4
continental slope 4.2.4
contour farming 6.1.6
contour line 1.1.3
convection 5.2.1
convection current 1.4.2, 3.2.1
convective zone 7.1.6
convergent boundary 1.3.2, 3.2.2
Copernicus, Nicolaus 7.1.1
core 1.4.1, 7.1.6
Coriolis effect 4.2.2, 5.2.4

corona 7.1.6
Cousteau, Jacques-Yves 4.2.5
cover crop 6.1.6
crater 3.3.1, 7.2.2, 7.3.1
crest 4.2.2
crevasse 4.1.3
crop rotation 6.1.6
crust 1.4.3
Curiosity 5.1.3
currents 4.2.2

D

dam 2.1.5, 4.1.8
dead zone 4.1.9
deflation hollow 2.1.4
deformation 3.1.3
Deimos 7.2.2
delta 2.1.5
density 1.2.3
deposition 5.2.6
desalination 4.2.6, 6.1.2
desertification 6.1.5
desiccation 2.2.5
dike 3.3.3, 3.3.4, 4.1.8
distributary 2.1.5
divergent boundary 3.1.2, 3.2.2
divide 4.1.5
Dobson, Gordon M. B. 5.1.4
Doppler effect 8.1.4
downwelling 4.2.0
drumlin 2.1.6
ductility 6.1.4
Dust Bowl 2.1.4
dwarf planet 7.1.5, 7.2.3
dynamic metamorphism 1.3.3
dynamo theory 5.1.3
dysphotic zone 4.2.1

E

Earth 7.2.2

earthflow 2.1.3
earthquake
  cause 3.2.1
  focus 3.2.2, 3.2.4
  preparedness 3.2.5
  San Francisco 3.2.2
  scales 3.2.4
  seismic waves 1.4.1, 3.2.1, 3.2.4
Earth's age 1.1.2, 2.1.6, 2.2.4, 2.2.5, 3.1.1, 3.1.4, 5.1.3, 5.3.3, 7.1.5, 8.1.4
earth science 1.1.1
Easter 7.3.2
Echo 1 8.2.3
eclipse 7.3.3
Einstein, Albert 7.1.6
electromagnetic radiation 7.1.4
electromagnetic spectrum 7.1.4
ellipse 7.2.1
El Niño 5.3.3
environmental science 1.1.1
epicenter 3.2.4
equator 7.3.4
equinox 7.3.4
Eratosthenes 7.1.1
Eris 7.2.0
erosion 2.1.1, 2.1.4, 2.1.5, 2.1.6
erratic 2.1.6
ESA (European Space Agency) 8.2.4
escape velocity 8.2.1
Eudoxus 7.1.1
euphotic zone 4.2.1
Europa 7.2.3
eutrophication 4.1.9
evaporation 4.1.2
exfoliation 2.1.1
exosphere 5.1.2
Explorer 1 5.1.3
extrusive rock 1.3.1
Exxon *Valdez* 6.2.3

F

fault 2.1.4, 3.1.3, 3.2.1, 3.2.3
  creep 3.2.3, 3.2.5
  dip-slip 3.1.0, 3.2.3
  Hayward 3.2.3
  normal 3.2.3
  reverse 3.2.3
  San Andreas 3.2.3
  strike-slip 3.1.0, 3.2.3
  thrust 3.2.3
faulting 3.1.3
felsic rock 1.3.1
Fertile Crescent 6.1.5
fetch 4.2.0
fissure 3.2.5
floodplain 2.1.5
fluorescence 1.2.4
focus 3.2.3, 7.2.1
fog 5.2.5
folding 3.1.3
folds 3.2.2
foliated structure 1.3.3
footwall 3.1.3, 3.2.3
forecasts 5.2.9
fossil 1.3.2, 2.2.5
fossil fuel 6.1.3
Foucault, Jean-Bernard-Léon 7.2.1
fracture 1.2.3, 3.1.0
friction 7.2.4
front 5.2.7
Fujita scale 5.2.8
Fujita, Tetsuya Theodore 5.2.8
furrow 6.1.6

G

Gagarin, Yuri 8.2.4
galaxy 7.1.1, 8.1.3
galaxy cluster 8.1.3
Galilean satellites 7.2.3
Galilei, Galileo 7.1.1, 7.2.3
Galle, Johann Gottfried 7.2.0

Ganymede 7.2.3
Garlick, Mark, Dr. 7.2.2, 7.2.3
gas giant 7.1.5, 7.2.3
Geiger counter 5.1.3
Geikie, Archibald 2.1.1
geocentric 7.1.1
geologic column 2.2.4
geology 1.1.1
geomagnetic storm 5.1.3, 7.1.6
geosynchronous orbit 8.2.3
geothermal energy 6.1.7
geyser 4.1.4
glacial drift 2.1.6
glacier 2.1.5, 2.1.6, 4.1.3
Glenn, John 8.2.4
Goddard, Robert 8.2.1
Gondwanaland 3.1.1
GPS (Global Positioning Satellite) 6.1.6, 8.2.3
graben 3.1.3, 3.2.3
gravity 7.1.1. 7.2.1
Great Divide 4.1.5
Great Lakes 2.1.6
Great Red Spot 7.2.3
greenhouse gas 5.3.3
groundwater 4.1.4
Gulf Stream 4.2.2
gully 2.1.5, 4.1.5

H

Hadrian's Wall 3.3.3
half-life 1.1.2
Halley, Sir Edmund 7.2.4
hanging wall 3.1.3, 3.2.3
hardness 1.2.3
harvest moon 7.3.2
heat 5.2.1
heliocentric 7.1.1
Herschel, Sir William 7.1.1, 7.1.4, 7.2.3
Hertzsprung, Ejnar 8.1.2

Hess, Henry 3.1.2
heterosphere 5.1.2
Hillary, Edmund 3.1.4
homosphere 5.1.2
Hooke, Robert 3.1.3
horizon 2.2.2
horse latitudes 5.2.4
horst 3.1.3, 3.2.3
hot spot 3.3.1
hot spring 4.1.4
Howard, Luke 5.2.5
H-R diagram 8.1.2
Hubble, Edwin 7.1.1
Hubble Space Telescope 7.1.3
humidity 5.2.6
humus 2.2.1
hurricane 5.2.8
Hutton, James 1.3.4, 2.2.0
hydrocarbons 6.2.2
hydroelectric power 4.1.8, 6.1.7
hydrogen bonding 4.1.1
hydrogeologist 4.1.4
hydrolysis 2.1.2
hydrothermal vent 4.2.1

I

ice age 2.1.6, 5.3.3
ice wedging 2.1.1
igneous rock 1.3.1
impact theory 7.3.0
incineration 6.2.1
index fossil 2.2.5
Industrial Revolution 6.2.2
inertia 8.2.3
inner core 1.4.1
Intelsat 1 8.2.3
intensity 3.2.4
interglacial period 5.3.3
International date line 1.1.3
intrusion 2.2.4

intrusive rock 1.3.1
Io 7.2.3
ion 4.2.1, 5.1.2, 7.2.4
ionosphere 5.1.2
isobars 5.2.9
isostasy 3.1.1
ISS (International Space Station) 8.2.4

J

Jansky, Karl 7.1.4
jet stream 5.1.2
Johnson, Katherine 8.2.4
Journey to the Center of the Earth 1.4.1
Julius Caesar 7.1.2
Jupiter 7.2.3

K

karst topography 4.1.4
Kepler, Johannes 7.1.1
Kepler's laws of planetary motion 7.2.1
kiloparsecs 8.1.3
Kola Peninsula 1.4.1
Köppen-Geiger climate classification 5.3.2
Kuiper belt 7.2.3, 7.2.4

L

laccoliths 3.3.3
landfill 6.2.1
Landsat 8.2.3
landslide 2.1.3
laser beams 5.2.10
latitude 1.1.3
Laurasia 3.1.1
lava 1.3.1
law of superposition 2.2.4
law of thermodynamics 6.1.3
leachate 6.2.1
Lehmann, Inge 3.2.2
LEO (low Earth orbit) 8.2.3
levee 2.1.5, 4.1.8

Le Verrier, Urbain Joseph 7.2.3
Levy, David 7.2.4
lidar 3.2.5
lightning 5.2.8
light-year 7.2.1, 8.1.1
liquefaction 3.2.5
lithification 1.3.2
lithosphere 1.4.2
lodestone 1.2.4, 5.1.3
longitude 1.1.3
Love wave 3.2.4
lunar calendar 7.1.2
lunar eclipse 7.3.3
lunar regolith 7.3.1
luster 1.2.3

M

mafic rock 1.3.1
Magellan 7.2.2
magma 1.3.1
magnetosphere 1.4.1, 5.1.3
magnitude 3.2.1
   earthquake 3.2.1, 3.2.2, 3.2.4
   star 8.1.1, 8.1.2
major axis 7.2.1
malleability 6.1.4
mantle 1.4.2
map projections 1.1.3
mare 7.3.1
Mariana Trench 4.2.4
Mariner 10 7.2.2
Mars 7.2.2
mass wasting 2.1.3
mature river 4.1.6
mechanical weathering 2.1.1
Mercalli, Giuseppe 3.2.4
Mercator, Gerardus 1.1.0
Mercury 7.2.2
mesopause 5.1.2
mesosphere 5.1.2

metal 6.1.4
metamorphic rock 1.3.3
metamorphism 1.3.3
meteor 7.1.5, 7.2.4
Meteor Crater 7.2.4
meteorite 7.1.5, 7.2.4
meteoroid 7.1.5, 7.2.4
meteorology 1.1.1
microclimate 5.3.2
Mid-Atlantic Ridge 3.2.2, 3.3.2, 4.2.4
midnight zone 4.2.1
Milankovitch, Milutin 2.1.6, 5.3.3
Milankovitch Theory 2.1.6, 5.3.3
Milky Way 8.1.3
Minamata Convention 4.2.1
mineral 1.2.1, 6.1.4
mineralogy 1.2.1
mining 6.1.4
Mir 8.2.4
Mitchell, Maria 7.2.4
Modified Mercalli Intensity Scale 3.2.4
Moho 1.4.2
Mohorovicic, Andrija 1.4.2
Mohs, Friedrich 1.2.3
Mohs hardness scale 1.2.3
moment magnitude scale 3.2.4
Montreal Protocol 5.1.4
moon 7.3.1, 7.3.2, 7.3.3
   lunar calendar 7.1.2
   lunar eclipse 7.3.3
   phases 7.3.2
moraine 2.1.6
Mouchot, Augustin-Bernard 6.1.0
mountain
   breezes 5.2.3
   ranges 3.1.0, 3.1.4
   Matterhorn 2.1.6
   Mount Saint Helens 3.3.5
mudflow 2.1.3

N

NASA 8.2.1

natural gas 6.1.3
naturalism 1.1.2
natural resource 6.1.1
neap tide 4.2.3
nebula 7.1.1, 7.1.5, 8.1.2
   solar 7.1.5
Neptune 7.2.3
neutron star 8.1.2
new moon 7.3.1
Newton, Sir Isaac 7.1.1, 7.2.1
nitrogen 6.1.2
nitrogen cycle 5.1.1
Noguchi, Soichi 8.2.4
nonfoliated structure 1.3.3
nonrenewable resource 6.1.1
nonsilicate mineral 1.2.2
Norgay, Tenzing 3.1.4
normal fault 3.2.3
Northern Lights 7.1.6
no-till farming 6.1.6
nuclear energy 6.1.7
nuclear fusion 7.1.5, 7.1.6, 8.1.1

O

objective lens 7.1.3
observatory 7.1.3
ocean
   gases 4.2.1
   ridge 3.1.2, 4.2.4
   sea breezes 5.2.3
   zones 4.2.4
oceanic crust 1.4.3
oceanic ridge 3.1.2, 4.2.4
oceanography 1.1.1, 4.2.5
oil spill 6.2.3
old Earth 1.1.2, 2.1.6, 2.2.4, 2.2.5, 3.1.1, 3.1.4, 5.1.3, 5.3.3, 7.1.5, 8.1.4
old river 4.1.6
Olympus Mons 7.2.2
Oort cloud 7.1.5, 7.2.4
orbital velocity 8.2.1

ore 6.1.4
orogenesis 3.1.4
outer core 1.4.1
oxbow lake 4.1.6
oxidation 2.1.2
oxygen 6.1.2
oxygen cycle 5.1.1
ozone 5.1.4
  hole 5.1.4
  layer 7.1.6
  pollution 6.2.2

P

pahoehoe 3.3.3
paleontologist 2.2.5
Pangaea 3.1.1
parallax 8.1.1
Paricutin 3.3.4
particulate matter 6.2.2
Passover 7.3.2, 7.3.3
Pathfinder 7.2.2
payload 8.2.1
peat 6.1.3
penumbra 7.3.3
perigee 7.3.2
perihelion 7.2.1, 7.3.3, 7.3.4
period of revolution 7.2.1
period of rotation 7.2.1
permafrost 2.1.3
petrifaction 2.2.5
petroleum 6.1.3
phase 7.3.2
Phobos 7.2.2
Phoebe 7.2.3
phosphorescence 1.2.4
photosphere 7.1.6
pH Scale 2.2.3
phytoplankton 4.2.1
Pioneer 10 8.2.2
planet 7.1.1, 7.1.5, 7.2.2, 7.2.3
planetesimals 7.1.5

plasma 5.1.3
plateau 3.1.4
plate boundary 1.3.2, 3.1.2, 3.2.2
plates 3.2.1, 3.2.2
  Australian 3.2.2
  Eurasian 3.2.2
  movement 1.3.2, 3.2.1, 3.2.2
  Nazca 3.2.3
  North American 3.2.2
  Pacific 3.2.2
  South American 3.2.3
Plato 5.1.2
Pliny the Elder 3.3.4
Pliny the Younger 3.3.4
plug 3.3.4
Pluto 7.2.3
pluton 3.3.3
polar vortex 5.1.4
pollutant 6.2.1
Pompeii 3.3.4
Pope Gregory XIII 7.1.2
pore space 2.2.1
porosity 4.1.4
precipitation 5.2.6
  acid rain 6.2.2
  fog 5.2.5
  freezing rain 5.2.6
  hail 5.2.6
  sleet 5.2.6
  snow 5.2.6
prime meridian 1.1.3
projection 1.1.3
prominence 7.1.6
propellant 8.2.1
Ptolemy 7.1.1
pulsar 8.1.2
P wave 1.4.1, 3.2.4
pyroclast 3.3.3
Pythagoras 7.1.1
Pytheas of Massilia 4.2.3

Q

quarry 6.1.4

quasar 8.1.3

R

radar 3.2.5, 5.2.9
radiant 7.2.4
radiation 5.2.1
radiative zone 7.1.6
radioactive wastes 6.2.1
radioactivity 1.2.4
radiometric dating 1.1.2
radiosonde 5.2.9
radio waves 7.1.4
radon gas 3.2.5
rapid-decay theory 5.1.3
Rayleigh waves 3.2.4
red shift 8.1.4
red tide 4.1.9
reflecting telescope 7.1.3
refracting telescope 7.1.3
regional metamorphism 1.3.3
regolith 2.1.3
rejuvenated river 4.1.6
relative humidity 5.2.6
renewable resources 6.1.1
reservoir 4.1.8
reverse fault 3.2.3
Richter, Charles F. 3.2.4
Richter scale 3.2.4
rift 3.3.2
rille 7.3.1
Ring of Fire 3.3.2
rip current 4.2.2
river
  Cuyahoga 6.2.3
  Ganges 2.1.5
  mature 4.1.6
  Mississippi 2.1.5
  Nile 2.1.5
  old 4.1.6
  rejuvenated 4.1.6
  salt content 4.2.1
  youthful 4.1.6

rock
  clastic 1.3.2
  cycle 1.3.4
  extrusive 1.3.1
  felsic 1.3.1
  igneous 1.3.1
  intrusive 1.3.1
  mafic 1.3.1
  metamorphic 1.3.3
  pedestal 2.1.4
  sedimentary 1.3.2

rock cycle 1.3.4

rocket 8.2.1

rock pedestal 2.1.4

Røemer, Ole 7.3.0

Roosevelt, Theodore 6.1.5

runoff 2.1.5, 4.1.5

Russell, Henry Norris 8.1.2

S

Sabbath year 6.1.6

salinity 4.2.1

salinization 6.1.5

Salyut 1 8.2.4

San Andreas fault 3.2.3

satellite 7.3.1, 8.2.3

Saturn 7.2.3

scarps 7.2.2

seafloor spreading 3.1.2

seamount 4.2.4

sedimentary rock 1.3.2

sedimentologist 1.3.2

sediments 1.3.2

seismic gap 3.2.5

seismic wave 1.4.1, 3.2.1, 3.2.4

seismogram 3.2.4

seismograph 3.2.4

seismologist 3.2.1, 3.2.4

seismology 3.2.4

semimajor axis 7.2.1

shadow rule 5.1.4

shale 1.3.2

shearing stress 3.1.3

shelterbelt 5.2.10

Shepard, Alan 8.2.4

Shoemaker, Carolyn 7.2.4

Shoemaker, Eugene 7.2.4

silicate mineral 1.2.2

sills 3.3.3

sinkhole 4.1.4

Skylab 8.2.4

sludge 6.2.1

slump 2.1.3

smog 6.2.2

smoking 6.2.2

soil creep 2.1.3

soil profile 2.2.2

solar calendar 7.1.2

solar cell 6.1.7

solar eclipse 7.3.3

solar energy 6.1.7

solar flares 5.1.3, 7.1.6

solar radiation 7.1.5

solar system 7.1.5

solar wind 5.1.3, 7.1.6

solstice 7.3.4

Sosigenes of Alexandria 7.1.2

Southern Lights 7.1.6

space probe 8.2.2

space shuttle 8.2.4

space station 8.2.4

SpaceX 8.2.4

specific heat 4.1.1

spectrograph 8.1.1

spectrum 8.1.1

spin-off 8.2.4

spring tide 4.2.3

Sputnik 8.2.1, 8.2.3

star 8.1.1
  chart 1.1.3
  cluster 8.1.3
  magnitude 8.1.1, 8.1.2

star cluster 8.1.3

stewardship 6.2.4

stratified drift 2.1.6

stratosphere 5.1.2

streak 1.2.3

stress 3.1.3

strip cropping 6.1.6

subduction 3.1.2

subduction zone 3.1.2, 3.3.2

sublimation 4.1.2, 7.2.4

submersible 4.2.5

subsoil 2.2.2

sunspot 7.1.6

supernova 8.1.2

surface tension 4.1.1

surface waves 3.2.4

Surtsey Island 3.3.2

S waves 1.4.1, 3.2.4

synclines 3.1.3

T

tectonics 3.1.1

telescope 7.1.3
  gamma ray 7.1.4
  Hubble Space 7.1.3
  infrared 7.1.4
  radio 7.1.4
  reflecting 7.1.3
  refracting 7.1.3
  ultraviolet 7.1.4
  x-ray 7.1.4

Telkes, Mária 6.1.7

temperature 5.2.1
  inversion 6.2.2

tensional stress 3.1.3

terrace farming 6.1.6

terrestrial planet 7.1.5, 7.2.2

theism 1.1.2

theory
 Big Bang 8.1.4
 capture 7.3.0
 cocreation 7.3.0
 dynamo 5.1.3
 impact 7.3.0
 Milankovitch 2.1.6, 5.3.3
 rapid-decay 5.1.3

thermal pollution 6.2.3

thermal stratification 4.1.7

thermohaline current 4.2.1

thermosphere 5.1.2

thrust 8.2.1

thunder 5.2.8

tidal range 4.2.3

till 2.1.6

tiltmeter 3.3.5

time zones 1.1.3

Tiros 1 8.2.3

Titan 7.2.3

topography 2.1.1, 2.1.5, 4.1.5

topsoil 2.2.2

tornado 5.2.8

toxic waste 6.2.1

trace fossil 2.2.5

transform boundary 3.1.2, 3.2.2

transpiration 4.1.2

trench 3.1.2, 4.2.4

tributary 4.1.5

Triton 7.2.3

Tropic of Cancer 7.3.4

Tropic of Capricorn 7.3.4

tropopause 5.1.2

troposphere 5.1.2

trough 4.2.2

Tsiolkovsky, Konstantin 8.2.1

tsunami 3.2.5, 4.2.2

twilight zone 4.2.1

U

umbra 7.3.3

unconformity 2.2.4

undertow 4.2.2

unglazed solar collector 6.1.7

uniformitarianism 1.1.2

upwelling 4.2.0

uranium 6.1.7

Uranus 7.2.3

UV 5.1.4

V

valley glacier 4.1.3

Van Allen, James 5.1.3

Van Allen radiation belts 5.1.3

varves 1.1.2

Venus 7.2.2

Viking 1 7.2.2

virga 5.2.5

viscosity 3.3.3

volcanic bomb 3.3.3

volcano 3.3.1, 3.3.3, 3.3.4, 3.3.5
 ash 3.3.3
 eruption types 3.3.4
 Mount Saint Helens 3.3.5
 Volcanic Explosivity Index (VEI) 3.3.4

volcanologist 3.3.5

von Braun, Wernher 8.2.1

Voyager 2 7.2.3

W

waning moon 7.3.2

water
 budget 4.1.2
 cycle 4.1.2
 hard 4.1.5
 ocean 4.2.1
 table 3.2.5, 4.1.4

water budget 4.1.2

water cycle 4.1.2

watershed 4.1.4

water table 3.2.5, 4.1.4

wavelength 4.2.2

waves
 body 3.2.4
 capillary 4.2.2
 Love 3.2.4
 P 1.4.1, 3.2.4
 radio 7.1.4
 Rayleigh 3.2.4
 S 1.4.1, 3.2.4
 seismic 1.4.1, 3.2.1, 3.2.4
 surface 3.2.4

waxing moon 7.3.2

weather 5.2.9

weathering 2.1.1

Wegener, Alfred 3.1.1

wetland 6.1.5

Whewell, William 1.1.2

wind
 effect on waves 4.2.2
 global 5.2.4
 jet stream 5.1.2
 land breezes 5.2.3
 local 5.2.3
 mountain breezes 5.2.3
 polar easterlies 5.2.4
 sea breezes 5.2.3
 solar 5.1.3, 7.1.6
 trade 5.2.4
 turbine 6.1.7
 types 3.2.3, 3.2.4
 valley breezes 5.2.3
 westerlies 5.2.4

Wind in the Willows 4.1.6

Winkler, E. M. 2.1.1, 6.1.6

Worldviews
 beauty from ashes 3.3.0
 Earth's uniqueness shows God's love 7.2.0
 foundations 2.2.0
 glimpse of God's beauty 1.2.0
 God is omnipresent 7.3.0
 God is the only constant 2.1.0
 God remains the same 5.3.0
 humility, agree to disagree 3.1.0
 immeasurable gifts 6.1.0
 judgment and love 3.2.0
 knowing Jesus more important than being right 5.1.0
 living water 4.1.0
 made in His image 7.1.0

more powerful than the ocean 4.2.0
purposeful design 1.4.0
science and faith inseparable 1.1.0
stewardship 6.2.0
Sustainer of Earth and His people 1.3.0
universe displays God's nature 8.1.0
wind and waves of false doctrine 5.2.0
worship the creator, not the created 8.2.0

X

Xochicalco 8.1.3

Y

young Earth 1.1.2, 2.1.6, 2.2.4, 2.2.5, 3.1.1, 3.1.4, 5.1.3, 5.3.3, 7.1.5, 8.1.4

youthful river 4.1.6

Z

zone
 ablation 4.1.3
 accumulation 4.1.3
 aeration 4.1.4
 aphotic 4.2.1
 convective 7.1.6
 dead 4.1.9
 dysphotic 4.2.1
 euphotic 4.2.1
 midnight 4.2.1
 ocean 4.2.4
 radiative 7.1.6
 saturation 4.1.4
 subduction 3.1.2, 3.3.2
 time 1.1.3
 twilight 4.2.1

zone of aeration 4.1.4

zone of saturation 4.1.4

Credits

Page numbers refer to those in the Student Edition.

Unit 1
Chapter 1
William Whewell, **page 7**, Wellcome Library, London, CC-BY-4.0

varves, **page 9**, James St. John, CC-BY-2.0

landscape, **pages 10–11**, Sasha Sormann, CC-BY-ND 2.0

Chatham Islands map, **page 15**, Alexrk, CC-BY-SA 2.5

Mercator map, **pages 15–16**, Daniel R. Strebe, CC-BY-SA 3.0, August 15, 2011

tectonic map of Europe, **page 18**, Polyethylen and Woudloper

topographic map, **page 18**, NPS

bathymetric map, **page 19**, NGDC/NOAA/GLOBE/Great Lakes Bathymetry/Darekk2/CC-BY-SA 4.0

weather map, **page 19**, Department of Commerce/NOAA Central Library Data Imaging Project

Hydra star chart, **page 19**, IAU and Sky & Telescope, CC BY 3.0

Crux star chart, **page 19**, IAU and Sky & Telescope, CC BY 3.0

Chapter 2
salt crystals, **page 20**, Mark Schellhase, CC-BY-SA-3.0

oyster pearl, **page 21**, Keith Pomakis, CC-BY-SA-2.5

New Jerusalem fresco, **page 24**, Warburg, CC-BY-SA-3.0

marble wall of Ruskeala, **page 25**, Aleksander Kaasik, CC-BY-SA-4.0

diamond with adamantine luster, **page 29**, Natasha Ptukhina, CC-BY-SA-3.0

turquoise, **page 29**, Mike Beauregard, CC-BY-2.0

malachite, **page 30**, Jonathan Zander, CC-BY-SA-3.0

streak plates, **page 31**, Ra'ike, CC-BY-SA-3.0

Friedrick Mohs, **page 32**, Joseph Kriehuber/Peter Geymayer

chrysotile serpentine, **page 33**, Raimond Spekking/CC-BY-SA-4.0

rutile quartz, **page 33**, unforth/CC-BY-SA-2.0

halite, **page 34**, Didier Descouens/CC-BY-SA-4.0

sylvite, **page 34**, Chris857, CC-BY-SA-3.0

lodestone, **page 35**, Teravolt/Adam Munich/CC-BY-3.0

wernerite in daylight, **page 35**, H. Zell/CC-BY-SA-3.0

wernerite under ultraviolet light, **page 35**, H. Zell/CC-BY-SA-3.0

Iceland Spar calcite, **page 36**, ArniEin/CC-BY-SA-3.0

zircon, **page 37**, Parent Géry, CC-BY-SA-3.0

Chapter 3
rhyolite, **page 39**, Ji-Elle, CC-BY-SA-3.0

gabbro, **page 40**, Mark A. Wilson, Department of Geology, The College of Wooster

mason jar soil, **page 42**, Judith Browning

White Hoodoos near Wahweap Creek, **page 44**, Mark Stacey/NOAA, CC-BY-2.0

West Texas oil pumpjack, **page 47**, Eric Kounce/TexasRaiser

Mount Sodom salt cave, **page 47**, Mark A. Wilson, Department of Geology, The College of Wooster, CC-BY-SA-3.0

gneiss, **page 48**, Huhulenik, CC-BY-SA-3.0

pink quartzite, **page 48**, Amcyrus2012, CC-BY-SA-4.0

schist, **page 51**, Michael C. Rygel/CC-BY-SA 3.0

oil shale, **page 51**, Georgialh/CC-BY-SA 3.0

slate, **page 51**, Jonathan Zander/CC-BY-SA 3.0

igneous intrusion, **page 53**, Arlette1, CC-BY-SA-3.0

Chapter 4
Kola Peninsula drilling site, **page 54**, Andre Belozeroff, CC-BY-SA-3.0

aurora borealis, **page 57**, Nelly Volkovich

Mount Everest, **page 60**, Jone Jones, CC-BY-SA-3.0

Dead Sea, **page 61**, Mark A. Wilson, Department of Geology, The College of Wooster, CC-BY-SA-3.0

Unit 2
Chapter 1
broken rock, **page 65**, Till Niermann

exfoliated rock, **page 66**, Wing-Chi Poon, CC-BY-SA 2.5

plant roots, **page 66**, Diego Delso, delso.photo, CC-BY-SA 4.0

weathered rock erosion, **page 67**, Wilson44691/Mark A. Wilson (Department of Geology, The College of Wooster)

chemical weathering, **page 70**, ThinkGeoEnergy, CC-BY-2.0

landslide, **page 71**, HerbyThyme, CC-BY-SA 4.0

El Salvador landslide, **page 74**, USGS

New Dell Creek channel draining, **page 74**, US Air Force/Master Sgt. Paul Gorman

dust storm, **page 77**, Library of Congress, Prints & Photographs Division, FSA/OWI Collection, LC-USF34-004072-E

water runoff, **page 80**, Forest and Kim Starr, CC BY 3.0

levee breach (cropped), **page 82**, Forest and Kim Starr, CC BY 3.0

Namibia floodplain, **page 82**, NASA Earth Observatory, EO-1 team

river delta, **page 83**, Justin Hall, CC-BY-2.0

alluvial fan, **page 83**, Mikenorton, CC-BY-SA 3.0

Himalaya valley, **page 84**, DanHobley, CC-BY-SA 3.0

tarn, **page 85**, G310ScottS, CC-BY-SA 3.0

drumlin, **page 85**, Boschfoto, CC-BY-SA 3.0

erratic, **page 85**, Coaxial at English Wikipedia, CC-BY-SA 3.0

kettle lake, **page 85**, Algkalv

Chapter 2

silt (cropped), **page 91**, Infrogmation, CC-BY-SA-3.0

agronomist, **page 95**, AusAID/Australian Department of Foreign Affairs and Trade, CC-BY-2.0

global variation in soil map, **page 97**, Ninjatacoshell, CC-BY-SA-3.0

biological soil crusts, **page 98**, Nihonjoe, CC-BY-SA-3.0

intrusive rock, **page 102**, Jonathan, s. kt

angular unconformity, **page 104**, Jimmy Thomas, CC-BY-SA-2.0

fossil jewel beetle, **page 106**, Torsten Wappler, Hessisches Landesmuseum Darmstadt, CC-BY-SA-3.0

ammonite mold and cast, **page 107**, Mike Viney, The Virtual Petrified Wood Museum, www.petrifiedwoodmuseum.org

baby mammoth, **page 109**, Thomas Quine, CC-BY-SA-2.0

paleontologist (modified), **page 110**, NPS

Unit 3
Chapter 1

Wegener world map, **page 114**, Joseph Hutchins Colton/Geographicus Rare Antique Maps

Du Toit's continents, **page 115**, LennyWikidata, CC-BY-SA-3.0

oceanic spreading, **page 116**, Surachit/USGS

etching of James Hutton, **page 116**, CC-BY-4.0, Wellcome Library, London

Abraham Ortelius, **page 116**, Plantin-Moretus Museum

Principles of Geology, **page 116**, by Sir Charles Lyell, displayed at Northeastern Illinois University, courtesy of Nobel International University

Clarence Dutton, **page 116**, NOAA/Department of Commerce

Alfred Wegener, **page 117**, Picture: Alfred Wegener Institute

Oreo cookies, **page 118**, Evan-Amos

Harry Hammond Hess, **page 120**, Orren Jack Turner/Seeley G. Mudd Manuscript Library at Princeton University

Mid-Atlantic Ridge, **page 120**, NGDC/NOAA/NCEI/US Department of Commerce

subduction, **page 120**, Booyabazooka

seafloor spreading, **page 121**, USGS

types of faults, **page 124**, Jesús Gómez Fernández & Gregors, CC-BY-SA-3.0

Hooke's law springs, **page 124**, Svjo, CC-BY-SA-3.0

folding rock, **page 125**, Patrick McGillycuddy

Chapter 2

Nepal earthquake, **page 130**, Hilmi Hacaloğlu

Mark Twain, **page 131**, A.F. Bradley, New York

San Andreas fault, **page 131**, Ian Kluft, CC-BY-SA-4.0

plate boundaries, **page 132**, USGS

San Francisco earthquake, **page 136**, Arnold Genthe/LOC

Hayward fault creep, **page 138**, Kai Schreiber/CC-BY-SA 2.0

seismograph, **page 141**, Daderot, CC0-1.0

Charles Richter, **page 142**, USGS

optical electromagnetic seismograph, **page 142**, Apple2000, CC-BY-SA-4.0

portable seismograph, **page 143**, Dieter Franke/Germany

sand boils, **page 144**, Timothy Musson, CC-BY-SA-2.0

tsunami damage, **page 145**, Lance Cpl. Garry Welch/US Marine Corps

base isolators, **page 146**, Mike Renlund, CC-BY-2.0

Chapter 3

Mount Etna, **page 149**, Mstysalv Chernov, CC-BY-SA-3.0

Teide volcano, **page 151**, NASA/JPL

Mid-Atlantic Ridge, **page 152**, Underlying data source: Müller, R.D., M. Sdrolias, C. Gaina, and W.R. Roest 2008. Age, spreading rates and spreading symmetry of the world's ocean crust, Geochem. Geophys. Geosyst., 9, Q04006, doi:10.1029/2007GC001743. image developer: Mr. Elliot Lim, CIRES & NOAA/NCEI, Boulder CO

Iceland Mid-Atlantic Ridge map, **page 153**, USGS

Ring of Fire, **page 154**, Eric Gaba/Sting/UCLA

pahoehoe, **page 155**, USGS/HVO

pumice, **page 155**, Robert DuHamel, CC-BY-SA-3.0

dike in Makhtesh Ramon, **page 156**, Stéphanie Gromann

volcanic ash, **page 157**, Austin Post/USGS

volcanic bomb, **page 157**, Mark A. Wilson, Department of Geology, The College of Wooster

Roque de Agando, **page 159**, Diego Delso, delso.photo, License CC-BY-SA-4.0

Perícutin, **page 160**, James Allan, 1985, Smithsonian Institution

Aniakchak, **page 161**, USGS

volcanic eruption (cropped), **page 161**, Dr. Wolfgang Beyer, CC-BY-SA-3.0,

David A. Johnson, **page 162**, USGS

ash-covered car, **page 162**, USGS

Spirit Lake, **page 163**, Stephan Schulz, CC-BY-SA-3.0

bulge on Mount Saint Helens **page 164**, USGS

volcanologist, **page 164**, USGS/HVO

Mount Saint Helens monitor installation, **page 165**, USGS

volcanic soil, **page 166**, Paxson Woelber, CC-BY-SA-4.0

Unit 4
Chapter 1

hydrogen bonding, **page 168**, Michal Maňas, Qwerter, CC-BY-SA-3.0

surface tension, **page 169**, Booyabazooka

meniscus, **page 170**, PRHaney, CC-BY-SA-3.0

Greenland ice thickness map (modified), **page 177**, Eric Gaba—Wikimedia Commons user: Sting (http://commons.wikimedia.org/wiki/User:Sting)

Guadalupe River Watershed map, **page 183**, Water Board Staff of the California state government.

Big Horn River tributaries, **page 184**, NASA

Congo-Nile Divide, **page 184**, Imagico and Aymatth2/CC-BY-SA 2.5

oxbow lake, **page 188**, Massimo Tava, CC-BY-3.0

Goosenecks State Park, **page 188**, Michael Rissi, CC-By-SA-3.0

beaver dam, **page 191**, Hugo.arg, CC-BY-SA-4.0

Kelimutu Crater Lake, **page 192**, Rosino, CC-BY-SA-2.0

levee break, **page 193**, US Army Corps of Engineers

Hetch Hetchy, **page 194**, Inklein, CC-BY-SA-3.0

hydroelectric power station, **page 195**, Dr. Bernd Gross

phytoplankton, **page 197**, USGS/NASA/Landsat 7, CC-BY-2.0

Chapter 2

hydrothermal vents, **page 201**, Pacific Ring of Fire 2004 Expedition. NOAA Office of Ocean Exploration; Dr. Bob Embley, NOAA PMEL, Chief Scientist

Portishead docks low tide, **page 208**, David Webb/CC BY 3.0

Portishead docks high tide, **page 209**, David Webb/CC BY 3.0

River Winster tidal bore, **page 211**, ©Copyright Don Burgess and licensed for reuse under this Creative Commons License/CC-BY-SA 2.0

echo sounding, **page 216**, US Navy

ROV *Hercules*, **page 217**, Mountains in the Sea Research Team; the IFE crew; and NOAA/OAR/OER

Delaware beach nourishment, **page 219**, US Army Corps of Engineers

manganese nodule, **page 221**, Koelle aka Walter Kölle/CC-BY-SA 3.0

OTEC, **page 222**, Bluerise BV aka Bkleute/CC-BY-SA 4.0

Unit 5
Chapter 1

Solar Max, **page 230**, NASA and Marshall Space Flight Center

Curiosity, **page 232**, NASA/JPL-Caltech/MSSS

James Van Allen, **page 234**, NASA/JPL

filament eruption, **page 235**, NASA

ozone hole, **page 239**, NASA/Goddard Space Flight Center

Chapter 2

flowstone/dripstone, **page 246**, US National Park Service/NPS photo by Jason Walz

hurricane hunter plane, **page 267**, Michael Black/NOAA

Chapter 3

Hawaii, **page 279**, Jacques Descloitres/NASA/MODIS Land Rapid Response Team/GSFC

convergence zone, **page 281**, NOAA-NASA GOES Project

world map of Köppen-Geiger climate classification, **page 281**, Dr. Markus Kottek and Dr. Franz Rubel, http://koeppen-geiger.vu-wien.ac.at/shifts.htm

volcanoes, **page 289**, USGS

Unit 6
Chapter 1

Mária Telkes, **page 316**, Photo of Dr. Mária Telkes courtesy of New York World-Telegram and the Sun Newspaper Photograph Collection, Prints and Photographs Division, Library of Congress, LC-USZ62-113268

Chapter 2

controlled burn, **page 331**, Chief Petty Officer John Masson, U.S. Coast Guard Atlantic Area

Unit 7
Chapter 1

Edwin Hubble, **page 343**, HUB 1042 (4), Edwin Powell Hubble Papers, The Huntington Library, San Marino, California

Spitzer space telescope, **page 352**, NASA/JPL-Caltech/R. Hurt (SSC)

GALEX, **page 353**, NASA/JPL-Caltech

Chandra X-Ray Observatory, **page 354**, NASA

Swift satellite, **page 355**, NASA E/PO, Sonoma State University/Aurore Simmonnet

JWST data path, **page 355**, NASA/STScI

Thor's helmet nebula, **page 360**, ESO/B. Bailleul

Chapter 2

solar system moons, **page 378**, NASA

Chapter 3

Tycho crater, **page 385**, NASA/GSFC/Arizona State University

Edwin E. "Buzz" Aldrin, Jr., **page 390**, NASA

lunar eclipse, **page 395**, CC0 1.0 Universal/Jake Hills

Unit 8
Chapter 1

supernova, **page 408**, NASA/JPL-Caltech/UCLA

black hole flare, **page 410**, NASA/JPL-Caltech

Sombrero Galaxy, **page 412**, NASA/JPL-Caltech

quasar rendering, **page 416**, ESO/M. Kornmesser

WFIRST telescope, **page 419**, NASA/GSFC/Conceptual Image Lab

spiral galaxy NGC 1232, **page 420**, ESO

supernova remnant NGC 2060, **page 421**, ESO

Chapter 2

Robert H. Goddard, **page 422**, NASA Goddard Space Flight Center's Flickr Photostream, CC-BY-2.0

SpaceX rocket, **pages 424–425**, Michael Seeley/We Report Space,CC-BY-2.0

Genesis solar arrays, **page 428**, NASA, Kennedy Space Center

Galileo, **page 430**, NASA

Great Salt Lake, **pages 434–435**, NASA/Joshua Stevens/US Geological Survey

Neil Armstrong, **page 436**, NASA

Katherine Johnson, **page 438**, NASA

SpaceX *Dragon*, **page 439**, NASA

space station flower, **page 440**, NASA

Soichi Noguchi, **page 441**, NASA

Expedition 50 crew, **page 442**, NASA/Bill Ingalls

Transparency Masters
Unit 1
Chapter 1

Earth rising from moon, **TM 1.1.1A**, NASA and Kulandru mor

orthographic projection, **TM 1.1.3A**, Daniel R. Strebe/August 15, 2011/CC-BY-SA-3.0

conical projection, **TM 1.1.3A**, Daniel R. Strebe/August 15, 2011/CC-BY-SA-3.0

cylindrical equal-area projection, **TM 1.1.3A**, Daniel R. Strebe/August 15, 2011/CC-BY-SA-3.0

Chapter 2

hornfels, **TM 1.2.1A**, Piotr Sosnowski, CC-BY-SA-4.0

nepheline, **TM 1.2.3A**, Eurico Zimbres, CC-BY-SA-2.5

tremolite, **TM 1.2.3A**, Didier Descouens, CC-BY-SA-4.0

angelstite, **TM 1.2.3A**, Didier Descouens, CC-BY-SA-3.0

Chapter 3

basalt, **TM 1.3.1A**, USGS

felsite, **TM 1.3.1A**, Aram Dulyan

volcanic tuff, **TM 1.3.1B**, Qfl247, CC-BY-SA-3.0

peridotite, **TM 1.3.1B**, Woudloper, CC-BY-SA-1.0

gabbro, **TM 1.3.1B**, Mark A. Wilson, Department of Geology, The College of Wooster

Chapter 4

Moho map, **TM 1.4.2A**, AllenMcC., CC-BY-SA-3.0

Unit 2
Chapter 2

dinosaur nest, **TM 2.2.5A**, Daderot/CC0 1.0 Universal

Sandstone slab with *Chirotherium storetonense* trackway, **TM 2.2.5A**, FunkMonk as uploaded by Ballista at en.wikipedia/CC-BY-SA 3.0

Unit 3
Chapter 1

Wegener world map, **TM 3.1.1A**, Joseph Hutchins Colton/Geographicus Rare Antique Maps

Ortelius's World Map 1570, **TM 3.1.1A**, Abraham Ortelius

world map 2011, **TM 3.1.1A**, Colomet, CC-BY-SA-3.0

compression stress, **TM 3.1.3A**, Neuda4nik, CC-BY-SA-3.0

tensional stress, **TM 3.1.3A**, Florian Grossir, CC-BY-SA-3.0

shearing stress, **TM3.1.3A**, NPS/Robb Hannawacker

Mount Imam, **TM3.1.4A**, ئاسۆ/Yas, CC-BY-SA-4.0

plateau, **TM 3.1.4A**, Mark Iverson, CC-BY-SA-2.0

Chapter 2

car crash 1, **TM 3.2.2A**, Thue

car crash 2, **TM 3.2.2A**, Biswarup Ganguly, CC-BY-SA-3.0

car crash 3, **TM 3.2.2A**, Brady Holt, CC-BY-SA-3.0

car crash 4, **TM 3.2.2A**, ChengH, CC-BY-SA-3.0

Chapter 3

Ring of fire, **TM 3.3.2A**, Eric Gaba/Sting/UCLA

Sierra Grande shield volcano, **TM 3.3.4A**, Phillip M. Stewart aka Pmsyyz/CC-BY-SA 3.0

composite volcano, **TM 3.3.4B**, USGS

buried house, **TM 3.3.5A**, Hajotthu, CC-BY-3.0

dead trees, **TM 3.3.5A**, Bas Wallet, CC-BY-2.0

damaged town near Soufriere Hills volcano, **TM 3.3.5A**, Wailunip, CC-BY-SA-2.5

Unit 4
Chapter 1

Muir Glacier 2010, **TM 4.1.3A**, Eric E. Castro, CC-BY-2.0

Muir Glacier 1915, **TM 4.1.3A**, John Muir and Samuel Hall Young

hydroelectric dam, **TM 4.1.8A**, Tennessee Valley Authority

gravity dam, **TM 4.1.8C**, David Brodbeck, CC-BY-2.0

Summersville Dam, **TM 4.1.8C**, Ken Thomas

buttress dam, **TM 4.1.8C**, versgui, CC-BY-SA-3.0

earth dam, **TM 4.1.8C**, WillMcMinn, CC-BY-SA-3.0

arch dam, **TM 4.1.8C**, Licko

embankment, **TM 4.1.8C**, ProfessorX, CC-BY-SA-3.0

Chapter 2

bathymetric world map, **TM 4.2.5A**, NOAA/NCEI

Turtle (modified), **TM 4.2.5B**, Zenit/CC-BY-SA 3.0

Deep Flight Super Falcon (modified), **TM 4.2.5B**, Steve Jurvetson/CC-BY-2.0

Mir 1 submersible, **TM 4.2.5B**, NOAA/L. Murphy/http://oceanexplorer.noaa.gov/technology/subs/mir/mir-front.html

Johnson-Sea Link, **TM 4.2.5B**, US Geological Survey/photo by Cheryl Morrison

Alvin, **TM 4.2.5B**, Photo courtesy of the Woods Hole Oceanographic Institution Archives

bathyscaph *Trieste*, **TM 4.2.5B**, US Navy Electronics Laboratory/ San Diego, California

Unit 5
Chapter 3
world map of Köppen-Geiger climate classification, **TM 5.3.2A**, Dr. Markus Kottek and Dr. Franz Rubel, http://koeppen-geiger.vu-wien.ac.at/shifts.htm

ice core segment, **TM 5.3.3B**, NASA

coral growth rings, **TM 5.3.3B**, Eric Matson/AIMS

air sample archive, **TM 5.3.3B**, Mark Fergus/CC-BY-3.0

Unit 6
Chapter 1
water resources map (modified), **TM 6.1.2B**, CC-BY-4.0/MDPI and World Bank Group

Unit 8
Chapter 1
elliptical galaxy, **TM 8.1.3C**, NASA/JPL-Caltech

Blackline Masters
Unit 7
faith and science, **BLM 7.1.1A**, "Do Faith and Science Conflict?" used with permission of www.teachFASTly.com.

moon phases board, **BLM 7.3.2A**, Shayna A., Science Teaching Junkie, Inc. (https://www.teacherspayteachers.com/Store/Science-Teaching-Junkie-Inc) and (http://www.scienceteachingjunkie.com/2013/03/clearest-way-to-teach-moon-phasesever.html)

Worksheets
Unit 1
Chapter 1
Astana, Kazakhstan, **WS 1.1.3A**, Askar9992, CC-BY-SA-4.0

Unit 2
Chapter 1
mechanical weathering, **WS 2.1.2A**, Jessica Reid and Kren Leidecker

Unit 3
Chapter 1
mountain description activity, **WS 3.1.4B**, adapted and used with permission of www.teachFASTly.com

Unit 7
Chapter 1
Galileo and the sun, **WS 7.1.1A**, "Galileo and the Sun" used with permission of www.teachFASTly.com.

Galileo Goes to Jail article, **WS 7.1.1B**, "Galileo Goes to Jail" used with permission of www.teachFASTly.com.

Creative Commons licensing does not necessarily sponsor, authorize, or endorse this textbook.

Mondelēz International or its affiliates does not necessarily sponsor, authorize, or endorse this textbook.

NASA does not necessarily sponsor, authorize, or endorse this textbook.

The appearance of U.S. Department of Defense (DoD) visual information does not imply or constitute DoD endorsement.

Permission to reproduce copyrighted items must be secured from the copyright owner(s).